Student's Solutions Guide

to accompany

Discrete Mathematics and Its Applications

Sixth Edition

Kenneth H. Rosen
AT&T Laboratories

Prepared by
Jerrold Grossman
Oakland University

Boston Burr Ridge, IL Dubuque, IA New York San Francisco St. Louis
Bangkok Bogotá Caracas Kuala Lumpur Lisbon London Madrid Mexico City
Milan Montreal New Delhi Santiago Seoul Singapore Sydney Taipei Toronto

The McGraw·Hill Companies

Student's Solutions Guide to accompany
DISCRETE MATHEMATICS AND ITS APPLICATIONS, SIXTH EDITION
KENNETH H. ROSEN

Published by McGraw-Hill Higher Education, an imprint of The McGraw-Hill Companies, Inc., 1221 Avenue of the Americas, New York, NY 10020.

6 7 8 9 0 WDQ/WDQ 10

ISBN 978-0-07-310779-0
MHID 0-07-310779-4

www.mhhe.com

Preface

This *Student's Solutions Guide for Discrete Mathematics and Its Applications*, sixth edition, contains several useful and important study aids.

- **SOLUTIONS TO ODD-NUMBERED EXERCISES**

  The bulk of this work consists of **solutions to all the odd-numbered exercises in the text**. These are not just answers, but complete, worked-out solutions, showing how the principles discussed in the text and examples are applied to the problems. I have also added bits of wisdom, insights, warnings about errors to avoid, and extra comments that go beyond the question as posed. Furthermore, at the beginning of each section you will find some general words of advice and hints on approaching the exercises in that section.

- **REFERENCES FOR CHAPTER REVIEWS**

  Exact page references, theorem and example references, and answers are provided as a **guide for all the chapter review questions in the text**. This will make reviewing for tests and quizzes particularly easy.

- **A GUIDE TO WRITING PROOFS**

  Near the end of this book is a **section on writing proofs**, a skill that most students find difficult to master. Proofs are introduced formally in Chapter 1 (and proofs by mathematical induction are studied in Chapter 4), but exercises throughout the text ask for proofs of propositions. Reading this section when studying Sections 1.5–1.7, and then periodically thereafter, will be rewarded, as your proof-writing ability grows.

- **REFERENCES AND ADVICE ON THE WRITING PROJECTS**

  Near the end of this book you will find some general **advice on the Writing Projects** given at the end of each chapter. There is a discussion of various resources available in the mathematical sciences (such as *Mathematical Reviews* and the Internet), tips on doing library research, and some suggestions on how to write well. In addition, there is a rather **extensive bibliography of books and articles** that will be useful when researching these projects. We also provide specific hints and suggestions for each project, with pointers to the references; these can be found in the solutions section of this manual, at the end of each chapter.

- **SAMPLE CHAPTER TESTS**

  Near the end of this book you will find a sequence of **12 chapter tests**, comparable to what might be given in a course. You can take these sample tests in a simulated test setting as practice for the real thing. Complete solutions are provided, of course.

- **PROBLEM-SOLVING TIPS AND LIST OF COMMON MISTAKES**

  People beginning any endeavor tend to make the same kinds of mistakes. This is especially true in the study of mathematics. I have included **a detailed list of common misconceptions** that students of discrete mathematics often have and the kinds of errors they tend to make. Specific examples are given. It will be useful for you to review this list from time to time, to make sure that you are not falling into these common traps. Also included in this section is general advice on solving problems in mathematics, which you will find helpful as you tackle the exercises.

- **CRIB SHEETS**

  Finally, I have prepared a set of **12 single-page "crib sheets,"** one for each chapter of the book. They provide a quick summary of all the important concepts, definitions, and theorems in the chapter. There are at least three ways to use these. First, they can be used as a reference source by someone who wants to brush up on the material quickly or reveal gaps in old knowledge. Second, they provide an excellent review sheet for studying for tests and quizzes, especially useful for glancing over in the last few minutes. And third, a copy of this page (augmented by your own notes in the margins) is ideal in those cases where an instructor allows the students to come to a test with notes.

Several comments about the solutions in this volume are in order. In many cases, more than one solution to an exercise is presented, and sometimes the solutions presented here are not the same as the answers given in the back of the text. Indeed, there is rarely only one way to solve a problem in mathematics. You may well come upon still other valid ways to arrive at the correct answers. Even if you have solved a problem completely, you will find that reviewing the solutions presented here will be valuable, since there is insight to be gained from seeing how someone else handles a problem that you have just solved.

Exercises often ask that answers be justified or verified, or they ask you to show or prove a particular statement. In all these cases your solution should be a proof, i.e., a mathematical argument based on the rules of logic. Such a proof needs to be complete, convincing, and correct. Read your proof after finishing it. Ask yourself whether you would understand and believe it if it were presented to you by your instructor.

Although I cannot personally discuss with you my philosophy on learning discrete mathematics by solving exercises, let me include a few general words of advice. The best way to learn mathematics is by solving problems, and it is crucial that you first try to work these exercises independently. Consequently, do not use this *Guide* as a crutch. Do not look at the solution (or even the answer) to a problem before you have worked on it yourself. Resist the temptation to consult the solution as soon as the going gets rough. Make a real effort to work the problem completely on your own—preferably to the point of *writing down* a complete solution—before checking your work with the solutions presented here. If you have not been able to solve a problem and have reached the point where you feel it necessary to look at the answer or solution, try reading it only casually, looking for a hint as to how you might proceed; then try working on the exercise again, armed with this added information. As a last resort, study the solution in detail and make sure you could explain it to a fellow student.

I want to thank Jerry Grossman for his extensive advice and assistance in the preparation of this entire *Guide*, Paul Lorczak, Georgia Mederer, and Suzanne Zeitman for double-checking the solutions, Ron Morash for preparing the advice on writing proofs, and students at Monmouth College and Oakland University for their input on preliminary versions of these solutions.

A tremendous amount of effort has been devoted to ensuring the accuracy of these solutions, but it is possible that a few scattered errors remain. I would appreciate hearing about all that you find, be they typographical or mathematical. You can reach me using the Reporting of Errata link on the companion website's Information Center at `http://www.mhhe.com/rosen`.

One final note: In addition to this *Guide*, you will find the companion website created for *Discrete Mathematics and Its Applications* an invaluable resource. Included here are a Web Resources Guide with links to external websites keyed to the textbook, numerous Extra Examples to reinforce important topics, Interactive Demonstration Applets for exploring key algorithms, Self Assessment question banks to gauge your understanding of core concepts, and many other helpful resources. See the section titled "The MathZone Companion Website" on page xviii of the textbook for more details. The address is `http://www.mhhe.com/rosen`.

Kenneth H. Rosen

Contents

CHAPTER 1
The Foundations: Logic and Proofs

SECTION 1.1 Propositional Logic

Manipulating propositions and constructing truth tables are straightforward. A truth table is constructed by finding the truth values of compound propositions from the inside out; see the solution to Exercise 27, for instance. This exercise set also introduces **fuzzy logic**.

1. Propositions must have clearly defined truth values, so a proposition must be a declarative sentence with no free variables.

a) This is a true proposition.

b) This is a false proposition (Tallahassee is the capital).

c) This is a true proposition.

d) This is a false proposition.

e) This is not a proposition (it contains a variable; the truth value depends on the value assigned to x).

f) This is not a proposition, since it does not assert anything.

3. a) Today is not Thursday.

b) There is pollution in New Jersey.

c) $2 + 1 \neq 3$

d) It is not the case that the summer in Maine is hot and sunny. In other words, the summer in Maine is not hot and sunny, which means that it is not hot *or* it is not sunny. It is not correct to negate this by saying "The summer in Maine is not hot and not sunny." [For this part (and in a similar vein for part **(b)**) we need to assume that there are well-defined notions of hot and sunny; otherwise this would not be a proposition because of not having a definite truth value.]

5. This is pretty straightforward, using the normal words for the logical operators.

a) Sharks have not been spotted near the shore.

b) Swimming at the New Jersey shore is allowed, and sharks have been spotted near the shore.

c) Swimming at the New Jersey shore is not allowed, or sharks have been spotted near the shore.

d) If swimming at the New Jersey shore is allowed, then sharks have not been spotted near the shore.

e) If sharks have not been spotted near the shore, then swimming at the New Jersey shore is allowed.

f) If swimming at the New Jersey shore is not allowed, then sharks have not been spotted near the shore.

g) Swimming at the New Jersey shore is allowed if and only if sharks have not been spotted near the shore.

h) Swimming at the New Jersey shore is not allowed, and either swimming at the New Jersey shore is allowed or sharks have not been spotted near the shore. Note that we were able to incorporate the parentheses by using the word "either" in the second half of the sentence.

7. a) Here we have the conjunction $p \wedge q$.

b) Here we have a conjunction of p with the negation of q, namely $p \wedge \neg q$. Note that "but" logically means the same thing as "and."

c) Again this is a conjunction: $\neg p \land \neg q$.

d) Here we have a disjunction, $p \lor q$. Note that $\lor$ is the inclusive *or*, so the "(or both)" part of the English sentence is automatically included.

e) This sentence is a conditional statement, $p \to q$.

f) This is a conjunction of propositions, both of which are compound: $(p \lor q) \land (p \to \neg q)$.

g) This is the biconditional $p \leftrightarrow q$.

9. a) This is just the negation of p, so we write $\neg p$.

b) This is a conjunction ("but" means "and"): $p \land \neg q$.

c) The position of the word "if" tells us which is the antecedent and which is the consequence: $p \to q$.

d) $\neg p \to \neg q$

e) The sufficient condition is the antecedent: $p \to q$.

f) $q \land \neg p$

g) "Whenever" means "if": $q \to p$.

11. a) "But" is a logical synonym for "and" (although it often suggests that the second part of the sentence is likely to be unexpected). So this is $r \land \neg p$.

b) Because of the agreement about precedence, we do not need parentheses in this expression: $\neg p \land q \land r$.

c) The outermost structure here is the conditional statement, and the conclusion part of the conditional statement is itself a biconditional: $r \to (q \leftrightarrow \neg p)$.

d) This is similar to part **(b)**: $\neg q \land \neg p \land r$.

e) This one is a little tricky. The statement that the condition is necessary is a conditional statement in one direction, and the statement that this condition is not sufficient is the negation of the conditional statement in the other direction. Thus we have the structure (safe $\to$ conditions) $\land \neg$(conditions $\to$ safe). Fleshing this out gives our answer: $(q \to (\neg r \land \neg p)) \land \neg((\neg r \land \neg p) \to q)$. There are some logically equivalent correct answers as well.

f) We just need to remember that "whenever" means "if" in logic: $(p \land r) \to \neg q$.

13. In each case, we simply need to determine the truth value of the hypothesis and the conclusion, and then use the definition of the truth value of the conditional statement. The conditional statement is true in every case except when the hypothesis (the "if" part) is true and the conclusion (the "then" part) is false.

a) Since the hypothesis is true and the conclusion is false, this conditional statement is false.

b) Since the hypothesis is false and the conclusion is true, this conditional statement is true.

c) Since the hypothesis is false and the conclusion is false, this conditional statement is true. Note that the conditional statement is false in both part **(b)** and part **(c)**; as long as the hypothesis is false, we need look no further to conclude that the conditional statement is true.

d) Since the hypothesis is false, this conditional statement is true.

15. a) Presumably the diner gets to choose only one of these beverages, so this is an exclusive *or*.

b) This is probably meant to be inclusive, so that long passwords with many digits are acceptable.

c) This is surely meant to be inclusive. If a student has had both of the prerequisites, so much the better.

d) At first glance one might argue that no one would pay with both currencies simultaneously, so it would seem reasonable to call this an exclusive *or*. There certainly could be cases, however, in which the patron would pay a portion of the bill in dollars and the remainder in euros. Therefore, an inclusive *or* seems better.

17. a) If this is an inclusive *or*, then it is allowable to take discrete mathematics if you have had calculus or computer science or both. If this is an exclusive *or*, then a person who had both courses would not be allowed to take discrete mathematics—only someone who had taken exactly one of the prerequisites would be allowed in. Clearly the former interpretation is intended; if anything, the person who has had both calculus and computer science is even better prepared for discrete mathematics.

b) If this is an inclusive *or*, then you can take the rebate, or you can sign up for the low-interest loan, or you can demand both of these incentives. If this is an exclusive *or*, then you will receive one of the incentives but not both. Since both of these deals are expensive for the dealer or manufacturer, surely the exclusive *or* was intended.

c) If this is an inclusive *or*, you can order two items from column A (and none from B), or three items from column B (and none from A), or five items (two from A and three from B). If this is an exclusive *or*, which it surely is here, then you get your choice of the two A items or the three B items, but not both.

d) If this is an inclusive *or*, then lots of snow, or extreme cold, or a combination of the two will close school. If this is an exclusive *or*, then one form of bad weather would close school but if both of them happened then school would meet. This latter interpretation is clearly absurd, so the inclusive *or* is intended.

19. **a)** If the wind blows from the northeast, then it snows. ["Whenever" means "if."]

b) If it stays warm for a week, then the apple trees will bloom. [Sometimes word order is flexible in English, as it is here. Other times it is not—"The man bit the dog" does not have the same meaning as "The dog bit the man."]

c) If the Pistons win the championship, then they beat the Lakers.

d) If you get to the top of Long's Peak, then you must have walked eight miles. [The necessary condition is the conclusion.]

e) If you are world famous, then you will get tenure as a professor. [The sufficient condition is the antecedent.]

f) If you drive more than 400 miles, then you will need to buy gasoline. [The word "then" is sometimes omitted in English sentences, but it is still understood.]

g) If your guarantee is good, then you must have bought your CD player less than 90 days ago. [Note that "only if" does not mean "if"; the clause following the "only if" is the conclusion, not the antecedent.]

h) If the water is not too cold, then Jan will go swimming. [Note that "unless" really means "if not." It also can be taken to mean "or."]

21. In each case there will be two statements. It is being asserted that the first one holds true if and only if the second one does. The order doesn't matter, but often one order is more colloquial English.

a) You buy an ice cream cone if and only if it is hot outside.

b) You win the contest if and only if you hold the only winning ticket.

c) You get promoted if and only if you have connections.

d) Your mind will decay if and only if you watch television.

e) The train runs late if and only if it is a day I take the train.

23. Many forms of the answers for this exercise are possible.

a) One form of the converse that reads well in English is "I will ski tomorrow only if it snows today." We could state the contrapositive as "If I don't ski tomorrow, then it will not have snowed today." The inverse is "If it does not snow today, then I will not ski tomorrow."

b) The proposition as stated can be rendered "If there is going to be a quiz, then I will come to class." The converse is "If I come to class, then there will be a quiz." (Or, perhaps even better, "I come to class only if there's going to be a quiz.") The contrapositive is "If I don't come to class, then there won't be a quiz." The inverse is "If there is not going to be a quiz, then I don't come to class."

c) There is a variable ("a positive integer") in this sentence, so technically it is not a proposition. Nevertheless, we can treat sentences such as this in the same way we treat propositions. Its converse is "A positive integer is a prime if it has no divisors other than 1 and itself." (Note that this can be false, since the number 1 satisfies the hypothesis but not the conclusion.) The contrapositive of the original proposition is "If a positive integer has a divisor other than 1 and itself, then it is not prime." (We are simplifying a bit here, replacing "does not have no divisors" by "has a divisor." Note that this is always true, assuming that we are talking about positive divisors.) The inverse is "If a positive integer is not prime, then it has a divisor other than 1 and itself."

25. A truth table will need 2^n rows if there are n variables.
a) $2^1 = 2$ **b)** $2^4 = 16$ **c)** $2^6 = 64$ **d)** $2^4 = 16$

27. To construct the truth table for a compound proposition, we work from the inside out. In each case, we will show the intermediate steps. In part **(d)**, for example, we first construct the truth table for $p \vee q$, then the truth table for $p \wedge q$, and finally combine them to get the truth table for $(p \vee q) \to (p \wedge q)$. For parts **(a)** and **(b)** we have the following table (column three for part **(a)**, column four for part **(b)**).

p	$\neg p$	$p \wedge \neg p$	$p \vee \neg p$
T	F	F	T
F	T	F	T

For part **(c)** we have the following table.

p	q	$\neg q$	$p \vee \neg q$	$(p \vee \neg q) \to q$
T	T	F	T	T
T	F	T	T	F
F	T	F	F	T
F	F	T	T	F

For part **(d)** we have the following table.

p	q	$p \vee q$	$p \wedge q$	$(p \vee q) \to (p \wedge q)$
T	T	T	T	T
T	F	T	F	F
F	T	T	F	F
F	F	F	F	T

For part **(e)** we have the following table. This time we have omitted the column explicitly showing the negations of p and q. Note that this true proposition is telling us that a conditional statement and its contrapositive always have the same truth value.

p	q	$p \to q$	$\neg q \to \neg p$	$(p \to q) \leftrightarrow (\neg q \to \neg p)$
T	T	T	T	T
T	F	F	F	T
F	T	T	T	T
F	F	T	T	T

For part **(f)** we have the following table. The fact that this proposition is not always true tells us that knowing a conditional statement in one direction does not tell us that the conditional statement is true in the other direction.

p	q	$p \to q$	$q \to p$	$(p \to q) \to (q \to p)$
T	T	T	T	T
T	F	F	T	T
F	T	T	F	F
F	F	T	T	T

29. To construct the truth table for a compound proposition, we work from the inside out. In each case, we will show the intermediate steps. In part **(a)**, for example, we first construct the truth table for $p \vee q$, then the truth table for $p \oplus q$, and finally combine them to get the truth table for $(p \vee q) \rightarrow (p \oplus q)$. For parts **(a)**, **(b)**, and **(c)** we have the following table (column five for part **(a)**, column seven for part **(b)**, column eight for part **(c)**).

p	q	$p \vee q$	$p \oplus q$	$(p \vee q) \rightarrow (p \oplus q)$	$p \wedge q$	$(p \oplus q) \rightarrow (p \wedge q)$	$(p \vee q) \oplus (p \wedge q)$
T	T	T	F	F	T	T	F
T	F	T	T	T	F	F	T
F	T	T	T	T	F	F	T
F	F	F	F	T	F	T	F

For part **(d)** we have the following table.

p	q	$\neg p$	$p \leftrightarrow q$	$\neg p \leftrightarrow q$	$(p \leftrightarrow q) \oplus (\neg p \leftrightarrow q)$
T	T	F	T	F	T
T	F	F	F	T	T
F	T	T	F	T	T
F	F	T	T	F	T

For part **(e)** we need eight rows in our truth table, because we have three variables.

p	q	r	$\neg p$	$\neg r$	$p \leftrightarrow q$	$\neg p \leftrightarrow \neg r$	$(p \leftrightarrow q) \oplus (\neg p \leftrightarrow \neg r)$
T	T	T	F	F	T	T	F
T	T	F	F	T	T	F	T
T	F	T	F	F	F	T	T
T	F	F	F	T	F	F	F
F	T	T	T	F	F	F	F
F	T	F	T	T	F	T	T
F	F	T	T	F	T	F	T
F	F	F	T	T	T	T	F

For part **(f)** we have the following table.

p	q	$\neg q$	$p \oplus q$	$p \oplus \neg q$	$(p \oplus q) \rightarrow (p \oplus \neg q)$
T	T	F	F	T	T
T	F	T	T	F	F
F	T	F	T	F	F
F	F	T	F	T	T

31. The techniques are the same as in Exercises 27–30. For parts **(a)** and **(b)** we have the following table (column four for part **(a)**, column six for part **(b)**).

p	q	$\neg q$	$p \rightarrow \neg q$	$\neg p$	$\neg p \leftrightarrow q$
T	T	F	F	F	F
T	F	T	T	F	T
F	T	F	T	T	T
F	F	T	T	T	F

For parts **(c)** and **(d)** we have the following table (columns six and seven, respectively).

p	q	$p \rightarrow q$	$\neg p$	$\neg p \rightarrow q$	$(p \rightarrow q) \vee (\neg p \rightarrow q)$	$(p \rightarrow q) \wedge (\neg p \rightarrow q)$
T	T	T	F	T	T	T
T	F	F	F	T	T	F
F	T	T	T	T	T	T
F	F	T	T	F	T	F

For parts **(e)** and **(f)** we have the following table (this time we have not explicitly shown the columns for negation). Column five shows the answer for part **(e)**, and column seven shows the answer for part **(f)**.

p	q	$p \leftrightarrow q$	$\neg p \leftrightarrow q$	$(p \leftrightarrow q) \vee (\neg p \leftrightarrow q)$	$\neg p \leftrightarrow \neg q$	$(\neg p \leftrightarrow \neg q) \leftrightarrow (p \leftrightarrow q)$
T	T	T	F	T	T	T
T	F	F	T	T	F	T
F	T	F	T	T	F	T
F	F	T	F	T	T	T

33. The techniques are the same as in Exercises 27–32, except that there are now three variables and therefore eight rows. For part **(a)**, we have

p	q	r	$\neg q$	$\neg q \vee r$	$p \to (\neg q \vee r)$
T	T	T	F	T	T
T	T	F	F	F	F
T	F	T	T	T	T
T	F	F	T	T	T
F	T	T	F	T	T
F	T	F	F	F	T
F	F	T	T	T	T
F	F	F	T	T	T

For part **(b)**, we have

p	q	r	$\neg p$	$q \to r$	$\neg p \to (q \to r)$
T	T	T	F	T	T
T	T	F	F	F	T
T	F	T	F	T	T
T	F	F	F	T	T
F	T	T	T	T	T
F	T	F	T	F	F
F	F	T	T	T	T
F	F	F	T	T	T

Parts **(c)** and **(d)** we can combine into a single table.

p	q	r	$p \to q$	$\neg p$	$\neg p \to r$	$(p \to q) \vee (\neg p \to r)$	$(p \to q) \wedge (\neg p \to r)$
T	T	T	T	F	T	T	T
T	T	F	T	F	T	T	T
T	F	T	F	F	T	T	F
T	F	F	F	F	T	T	F
F	T	T	T	T	T	T	T
F	T	F	T	T	F	T	F
F	F	T	T	T	T	T	T
F	F	F	T	T	F	T	F

For part **(e)** we have

p	q	r	$p \leftrightarrow q$	$\neg q$	$\neg q \leftrightarrow r$	$(p \leftrightarrow q) \vee (\neg q \leftrightarrow r)$
T	T	T	T	F	F	T
T	T	F	T	F	T	T
T	F	T	F	T	T	T
T	F	F	F	T	F	F
F	T	T	F	F	F	F
F	T	F	F	F	T	T
F	F	T	T	T	T	T
F	F	F	T	T	F	T

Finally, for part **(f)** we have

p	q	r	$\neg p$	$\neg q$	$\neg p \leftrightarrow \neg q$	$q \leftrightarrow r$	$(\neg p \leftrightarrow \neg q) \leftrightarrow (q \leftrightarrow r)$
T	T	T	F	F	T	T	T
T	T	F	F	F	T	F	F
T	F	T	F	T	F	F	T
T	F	F	F	T	F	T	F
F	T	T	T	F	F	T	F
F	T	F	T	F	F	F	T
F	F	T	T	T	T	F	F
F	F	F	T	T	T	T	T

35. This time the truth table needs $2^4 = 16$ rows. Note the systematic order in which we list the possibilities.

p	q	r	s	$p \leftrightarrow q$	$r \leftrightarrow s$	$(p \leftrightarrow q) \leftrightarrow (r \leftrightarrow s)$
T	T	T	T	T	T	T
T	T	T	F	T	F	F
T	T	F	T	T	F	F
T	T	F	F	T	T	T
T	F	T	T	F	T	F
T	F	T	F	F	F	T
T	F	F	T	F	F	T
T	F	F	F	F	T	F
F	T	T	T	F	T	F
F	T	T	F	F	F	T
F	T	F	T	F	F	T
F	T	F	F	F	T	F
F	F	T	T	T	T	T
F	F	T	F	T	F	F
F	F	F	T	T	F	F
F	F	F	F	T	T	T

37. a) bitwise OR = 111 1111; bitwise AND = 000 0000; bitwise XOR = 111 1111
b) bitwise OR = 1111 1010; bitwise AND = 1010 0000; bitwise XOR = 0101 1010
c) bitwise OR = 10 0111 1001; bitwise AND = 00 0100 0000; bitwise XOR = 10 0011 1001
d) bitwise OR = 11 1111 1111; bitwise AND = 00 0000 0000; bitwise XOR = 11 1111 1111

39. For "Fred is not happy," the truth value is $1 - 0.8 = 0.2$.
For "John is not happy," the truth value is $1 - 0.4 = 0.6$.

41. For "Fred is happy, or John is happy," the truth value is $\max(0.8, 0.4) = 0.8$.
For "Fred is not happy, or John is not happy," the truth value is $\max(0.2, 0.6) = 0.6$ (using the result of Exercise 39).

43. One great problem-solving strategy to try with problems like this, when the parameter is large (100 statements here) is to lower the parameter. Look at a simpler problem, with just two or three statements, and see if you can figure out what's going on. That was the approach used to discover the solution presented here.
a) Some number of these statements are true, so in fact exactly one of the statements must be true and the other 99 of them must be false. That is what the 99th statement is saying, so it is true and the rest are false.
b) The 100th statement cannot be true, since it is asserting that all the statements are false. Therefore it must be false. That makes the first statement true. Now if the 99th statement were true, then we would conclude that statements 2 through 100 were false, which contradicts the truth of statement 99. So statement 99 must be false. That means that statement 2 is true. We continue in this way and conclude that statements 1 through 50 are all true and statements 51 through 100 are all false.

c) If there are an odd number of statements, then we'd run into a contradiction when we got to the middle. If there were just three statements, for example, then statement 3 would have to be false, making statement 1 true, and now the truth of statement 2 would imply its falsity and its falsity would imply its truth. Therefore this situation cannot occur with three (or any odd number of) statements. It is a logical paradox, showing that in fact these are not statements after all.

45. There are many correct answers to this problem, but all involve some sort of double layering, or combining a question about the kind of person being addressed with a question about the information being sought. One solution is to ask this question: "If I were to ask you whether the right branch leads to the ruins, would you say 'yes'?" If the villager is a truth-teller, then of course he will reply "yes" if and only if the right branch leads to the ruins. Now let us see what the liar says. If the right branch leads to the ruins, then he would say "no" if asked whether the right branch leads to the ruins. Therefore, the truthful answer to your convoluted question is "no." Since he always lies, he will reply "yes." On the other hand, if the right branch does not lead to the ruins, then he would say "yes" if asked whether the right branch leads to the ruins; and so the truthful answer to your question is "yes"; therefore he will reply "no." Note that in both cases, he gives the same answer to your question as the truth-teller; namely, he says "yes" if and only if the right branch leads to the ruins. A more detailed discussion can be found in Martin Gardner's *Scientific American Book of Mathematical Puzzles and Diversions* (Simon and Schuster, 1959), p. 25; reprinted as *Hexaflexagons and Other Mathematical Diversions: The First Scientific American Book of Puzzles and Games* (University of Chicago Press, 1988).

47. **a)** Since "whenever" means "if," we have $q \to p$.
b) Since "but" means "and," we have $q \land \neg p$.
c) This sentence is saying the same thing as the sentence in part **(a)**, so the answer is the same: $q \to p$.
d) Again, we recall that "when" means "if" in logic: $\neg q \to \neg p$.

49. Let m, n, k, and i represent the propositions "The system is in multiuser state," "The system is operating normally," "The kernel is functioning," and "The system is in interrupt mode," respectively. Then we want to make the following expressions simultaneously true by our choice of truth values for m, n, k, and i:

$$m \leftrightarrow n, \quad n \to k, \quad \neg k \lor i, \quad \neg m \to i, \quad \neg i$$

In order for this to happen, clearly i must be false. In order for $\neg m \to i$ to be true when i is false, the hypothesis $\neg m$ must be false, so m must be true. Since we want $m \leftrightarrow n$ to be true, this implies that n must also be true. Since we want $n \to k$ to be true, we must therefore have k true. But now if k is true and i is false, then the third specification, $\neg k \lor i$ is false. Therefore we conclude that this system is not consistent.

51. Let s be "The router can send packets to the edge system"; let a be "The router supports the new address space"; let r be "The latest software release is installed." Then we are told $s \to a$, $a \to r$, $r \to s$, and $\neg a$. Since a is false, the first conditional statement tells us that s must be false. From that we deduce from the third conditional statement that r must be false. If indeed all three propositions are false, then all four specifications are true, so they are consistent.

53. This is similar to Example 17, about universities in New Mexico. To search for beaches in New Jersey, we could enter NEW **AND** JERSEY **AND** BEACHES. If we enter (JERSEY **AND** BEACHES) **NOT** NEW, then we'll get websites about beaches on the isle of Jersey, except for sites that happen to use the word "new" in a different context (e.g., a recently opened beach there). If we were sure that the word "isle" was in the name of the location, then of course we could enter ISLE **AND** JERSEY **AND** BEACHES.

55. If A is a knight, then he is telling the truth, in which case B must be a knave. Since B said nothing, that is certainly possible. If A is a knave, then he is lying, which means that his statement that at least one of them is a knave is false; hence they are both knights. That is a contradiction. So we can conclude that A is a knight and B is a knave.

57. If A is a knight, then he is telling the truth, in which case B must be a knight as well, since A is not a knave. (If $p \lor q$ and $\neg p$ are both true, then q must be true.) Since B said nothing, that is certainly possible. If A is a knave, then his statement is patently true, but that is a contradiction to the behavior of knaves. So we can conclude that A is a knight and B is a knight.

59. If A is a knight, then he should be telling the truth, but he is asserting that he is a knave. So that cannot be. If A is a knave, then in order for his statement to be false, B must be a knight. So we can conclude that A is a knave and B is a knight.

61. Because of the first piece of information that Steve has, let's assume first that Fred is not the highest paid. Then Janice is. Therefore Janice is not the lowest paid, so by the second piece of information that Steve has, Maggie is the highest paid. But that is a contradiction. Therefore we know that Fred *is* the highest paid. Next let's assume that Janice is not the lowest paid. Then our second fact implies that Maggie is the highest paid. But that contradicts the fact that Fred is the highest paid. Therefore we know that Janice *is* the lowest paid. So it appears that the only hope of a consistent set of facts is to have Fred paid the most, Maggie next, and Janice the least. (We have just seen that any other assumption leads to a contradiction.) This assumption does not contradict either of our two facts, since in both cases, the hypothesis is false.

63. Let's use the letters B, C, G, and H for the statements that the butler, cook, gardner, and handyman are telling the truth, respectively. We can then write each fact as a true proposition: $B \to C$; $\neg(C \land G)$, which is equivalent to $\neg C \lor \neg G$ (see the discussion of De Morgan's law in Section 1.2); $\neg(\neg G \land \neg H)$, which is equivalent to $G \lor H$; and $H \to \neg C$. Suppose that B is true. Then it follows from the first of our propositions that C must also be true. This tells us (using the second proposition) that G must be false, whence the third proposition makes H true. But now the fourth proposition is violated. Therefore we conclude that B cannot be true. If fact, the argument we have just given also proves that C cannot be true. Therefore we know that the butler and the cook are lying. This much already makes the first, second, and fourth propositions true, regardless of the truth of G or H. Thus either the gardner or the handyman could be lying or telling the truth; all we know (from the third proposition) is that at least one of them is telling the truth.

65. This is often given as an exercise in constraint programming, and it is difficult to solve by hand. The following table shows a solution consistent with all the clues, with the houses listed from left to right. Reportedly the solution is unique.

NATIONALITY	Norwegian	Italian	Englishman	Spaniard	Japanese
COLOR	Yellow	Blue	Red	White	Green
PET	Fox	Horse	Snail	Dog	Zebra
JOB	Diplomat	Physician	Photographer	Violinist	Painter
DRINK	Water	Tea	Milk	Juice	Coffee

In this solution the Japanese man owns the zebra, and the Norwegian drinks water. The logical reasoning needed to solve the problem is rather extensive, and the reader is referred to the following website containing the solution to a similar problem: `http://mathforum.org/dr.math/problems/joseph8.5.97.html`.

SECTION 1.2 Propositional Equivalences

The solutions to Exercises 1–10 are routine; we use truth tables to show that a proposition is a tautology or that two propositions are equivalent. The reader should do more than this, however; think about what the equivalence is saying. See Exercise 11 for this approach. Some important topics not covered in the text are introduced in this exercise set, including the notion of the **dual** *of a proposition,* **disjunctive normal form** *for propositions,* **functional completeness**, **satisfiability**, *and two other logical connectives,* NAND *and* NOR. *Much of this material foreshadows the study of Boolean algebra in Chapter 11.*

1. First we construct the following truth tables, for the propositions we are asked to deal with.

p	$p \wedge \mathbf{T}$	$p \vee \mathbf{F}$	$p \wedge \mathbf{F}$	$p \vee \mathbf{T}$	$p \vee p$	$p \wedge p$
T	T	T	F	T	T	T
F	F	F	F	T	F	F

The first equivalence, $p \wedge \mathbf{T} \equiv p$, is valid because the second column $p \wedge \mathbf{T}$ is identical to the first column p. Similarly, part **(b)** comes from looking at columns three and one. Since column four is a column of F's, and column five is a column of T's, part **(c)** and part **(d)** hold. Finally, the last two parts follow from the fact that the last two columns are identical to the first column.

3. We construct the following truth tables.

p	q	$p \vee q$	$q \vee p$	$p \wedge q$	$q \wedge p$
T	T	T	T	T	T
T	F	T	T	F	F
F	T	T	T	F	F
F	F	F	F	F	F

Part **(a)** follows from the fact that the third and fourth columns are identical; part **(b)** follows from the fact that the fifth and sixth columns are identical.

5. We construct the following truth table and note that the fifth and eighth columns are identical.

p	q	r	$q \vee r$	$p \wedge (q \vee r)$	$p \wedge q$	$p \wedge r$	$(p \wedge q) \vee (p \wedge r)$
T	T	T	T	T	T	T	T
T	T	F	T	T	T	F	T
T	F	T	T	T	F	T	T
T	F	F	F	F	F	F	F
F	T	T	T	F	F	F	F
F	T	F	T	F	F	F	F
F	F	T	T	F	F	F	F
F	F	F	F	F	F	F	F

7. De Morgan's laws tell us that to negate a conjunction we form the disjunction of the negations, and to negate a disjunction we form the conjunction of the negations.

a) This is the conjunction "Jan is rich, and Jan is happy." So the negation is "Jan is not rich, or Jan is not happy."

b) This is the disjunction "Carlos will bicycle tomorrow, or Carlos will run tomorrow." So the negation is "Carlos will not bicycle tomorrow, and Carlos will not run tomorrow." We could also render this as "Carlos will neither bicycle nor run tomorrow."

c) This is the disjunction "Mei walks to class, or Mei takes the bus to class." So the negation is "Mei does not walk to class, and Mei does not take the bus to class." (Maybe she gets a ride with a friend.) We could also render this as "Mei neither walks nor takes the bus to class."

d) This is the conjunction "Ibrahim is smart, and Ibrahim is hard working." So the negation is "Ibrahim is not smart, or Ibrahim is not hard working."

9. We construct a truth table for each conditional statement and note that the relevant column contains only T's. For parts **(a)** and **(b)** we have the following table (column four for part **(a)**, column six for part **(b)**).

p	q	$p \wedge q$	$(p \wedge q) \to p$	$p \vee q$	$p \to (p \vee q)$
T	T	T	T	T	T
T	F	F	T	T	T
F	T	F	T	T	T
F	F	F	T	F	T

For parts **(c)** and **(d)** we have the following table (columns five and seven, respectively).

p	q	$\neg p$	$p \to q$	$\neg p \to (p \to q)$	$p \wedge q$	$(p \wedge q) \to (p \to q)$
T	T	F	T	T	T	T
T	F	F	F	T	F	T
F	T	T	T	T	F	T
F	F	T	T	T	F	T

For parts **(e)** and **(f)** we have the following table (this time we have not explicitly shown the columns for negation). Column five shows the answer for part **(e)**, and column seven shows the answer for part **(f)**.

p	q	$p \to q$	$\neg(p \to q)$	$\neg(p \to q) \to p$	$\neg q$	$\neg(p \to q) \to \neg q$
T	T	T	F	T	F	T
T	F	F	T	T	T	T
F	T	T	F	T	F	T
F	F	T	F	T	T	T

11. Here is one approach: Recall that the only way a conditional statement can be false is for the hypothesis to be true and the conclusion to be false; hence it is sufficient to show that the conclusion must be true whenever the hypothesis is true. An alternative approach that works for some of these tautologies is to use the equivalences given in this section and prove these "algebraically." We will demonstrate this second method in some of the solutions.

a) If the hypothesis is true, then by the definition of $\wedge$ we know that p is true. Hence the conclusion is also true. For an algebraic proof, we exhibit the following string of equivalences, each one following from one of the laws in this section: $(p \wedge q) \to p \equiv \neg(p \wedge q) \vee p \equiv (\neg p \vee \neg q) \vee p \equiv (\neg q \vee \neg p) \vee p \equiv \neg q \vee (\neg p \vee p) \equiv \neg q \vee \mathbf{T} \equiv \mathbf{T}$. The first logical equivalence is the first equivalence in Table 7 (with $p \wedge q$ playing the role of p, and p playing the role of q); the second is De Morgan's law; the third is the commutative law; the fourth is the associative law; the fifth is the negation law (with the commutative law); and the sixth is the domination law.

b) If the hypothesis p is true, then by the definition of $\vee$, the conclusion $p \vee q$ must also be true.

c) If the hypothesis is true, then p must be false; hence the conclusion $p \to q$ is true, since *its* hypothesis is false. Symbolically we have $\neg p \to (p \to q) \equiv \neg\neg p \vee (\neg p \vee q) \equiv p \vee (\neg p \vee q) \equiv (p \vee \neg p) \vee q \equiv \mathbf{T} \vee q \equiv \mathbf{T}$.

d) If the hypothesis is true, then by the definition of $\wedge$ we know that q must be true. This makes the conclusion $p \to q$ true, since *its* conclusion is true.

e) If the hypothesis is true, then $p \to q$ must be false. But this can happen only if p is true, which is precisely what we wanted to show.

f) If the hypothesis is true, then $p \to q$ must be false. But this can happen only if q is false, which is precisely what we wanted to show.

13. We first construct truth tables and verify that in each case the two propositions give identical columns. The fact that the fourth column is identical to the first column proves part **(a)**, and the fact that the sixth column is identical to the first column proves part **(b)**.

p	q	$p \wedge q$	$p \vee (p \wedge q)$	$p \vee q$	$p \wedge (p \vee q)$
T	T	T	T	T	T
T	F	F	T	T	T
F	T	F	F	T	F
F	F	F	F	F	F

Alternately, we can argue as follows.

a) If p is true, then $p \vee (p \wedge q)$ is true, since the first proposition in the disjunction is true. On the other hand, if p is false, then both parts of the disjunction are false. Hence $p \vee (p \wedge q)$ always has the same truth value as p does, so the two propositions are logically equivalent.

b) If p is false, then $p \wedge (p \vee q)$ is false, since the first proposition in the conjunction is false. On the other hand, if p is true, then both parts of the conjunction are true. Hence $p \wedge (p \vee q)$ always has the same truth value as p does, so the two propositions are logically equivalent.

15. We need to determine whether we can find an assignment of truth values to p and q to make this proposition false. Let us try to find one. The only way that a conditional statement can be false is for the hypothesis to be true and the conclusion to be false. Hence we must make $\neg p$ false, which means we must make p true. Furthermore, in order for the hypothesis to be true, we will need to make q false, so that the first part of the conjunction will be true. But now with p true and q false, the second part of the conjunction is false. Therefore the entire hypothesis is false, so this assignment will not yield a false conditional statement. Since we have argued that no assignment of truth values can make this proposition false, we have proved that this proposition is a tautology. (An alternative approach would be to construct a truth table and see that its final column had only T's in it.) This tautology is telling us that if we know that a conditional statement is true, and that its conclusion is false, then we can conclude that its antecedent is also false.

17. The proposition $\neg(p \leftrightarrow q)$ is true when p and q do not have the same truth values, which means that p and q have different truth values (either p is true and q is false, or vice versa). These are exactly the cases in which $p \leftrightarrow \neg q$ is true. Therefore these two expressions are true in exactly the same instances, and therefore are logically equivalent.

19. The proposition $\neg p \leftrightarrow q$ is true when $\neg p$ and q have the same truth values, which means that p and q have different truth values (either p is true and q is false, or vice versa). By the same reasoning, these are exactly the cases in which $p \leftrightarrow \neg q$ is true. Therefore these two expressions are true in exactly the same instances, and therefore are logically equivalent.

21. This is essentially the same as Exercise 17. The proposition $\neg(p \leftrightarrow q)$ is true when $p \leftrightarrow q$ is false. Since $p \leftrightarrow q$ is true when p and q have the same truth value, it is false when p and q have different truth values (either p is true and q is false, or vice versa). These are precisely the cases in which $\neg p \leftrightarrow q$ is true.

23. We'll determine exactly which rows of the truth table will have F as their entries. In order for $(p \rightarrow r) \wedge (q \rightarrow r)$ to be false, we must have at least one of the two conditional statements false, which happens exactly when r is false and at least one of p and q is true. But these are precisely the cases in which $p \vee q$ is true and r is false, which is precisely when $(p \vee q) \rightarrow r$ is false. Since the two propositions are false in exactly the same situations, they are logically equivalent.

25. We'll determine exactly which rows of the truth table will have F as their entries. In order for $(p \rightarrow r) \vee (q \rightarrow r)$ to be false, we must have both of the two conditional statements false, which happens exactly when r is false and both p and q are true. But this is precisely the case in which $p \wedge q$ is true and r is false, which is precisely when $(p \wedge q) \rightarrow r$ is false. Since the two propositions are false in exactly the same situations, they are logically equivalent.

27. This fact was observed in Section 1.1 when the biconditional was first defined. Each of these is true precisely when p and q have the same truth values.

29. We will show that if $p \to q$ and $q \to r$ are both true, then $p \to r$ is true. Thus we want to show that if p is true, then so is r. Given that p and $p \to q$ are both true, we conclude that q is true; from that and $q \to r$ we conclude that r is true, as desired. This can also be done with a truth table.

31. To show that these are *not* logically equivalent, we need only find one assignment of truth values to p, q, and r for which the truth values of $(p \to q) \to r$ and $p \to (q \to r)$ differ. One such assignment is F for all three. Then $(p \to q) \to r$ is false and $p \to (q \to r)$ is true.

33. To show that these are *not* logically equivalent, we need only find one assignment of truth values to p, q, r, and s for which the truth values of $(p \to q) \to (r \to s)$ and $(p \to r) \to (q \to s)$ differ. Let us try to make the first one false. That means we have to make $r \to s$ false, so we want r to be true and s to be false. If we let p and q be false, then each of the other three simple conditional statements ($p \to q$, $p \to r$, and $q \to s$) will be true. Then $(p \to q) \to (r \to s)$ will be $\mathrm{T} \to \mathrm{F}$, which is false; but $(p \to r) \to (q \to s)$ will be $\mathrm{T} \to \mathrm{T}$, which is true.

35. We apply the rules stated in the preamble.

a) $p \vee \neg q \vee \neg r$ **b)** $(p \vee q \vee r) \wedge s$ **c)** $(p \wedge \mathbf{T}) \vee (q \wedge \mathbf{F})$

37. If we apply the operation for forming the dual twice to a proposition, then every symbol returns to what it originally was. The $\wedge$ changes to the $\vee$, then changes back to the $\wedge$. Similarly the $\vee$ changes to the $\wedge$, then back to the $\vee$. The same thing happens with the **T** and the **F**. Thus the dual of the dual of a proposition s, namely $(s^*)^*$, is equal to the original proposition s.

39. Let p and q be two compound propositions involving only the operators $\wedge$, $\vee$, and $\neg$; we can also allow them to involve the constants **T** and **F**. We want to show that if p and q are logically equivalent, then p^* and q^* are logically equivalent. The trick is to look at $\neg p$ and $\neg q$. They are certainly logically equivalent if p and q are. Now if p is a conjunction, say $r \wedge s$, then $\neg p$ is logically equivalent, by De Morgan's law, to $\neg r \vee \neg s$; a similar statement applies if p is a disjunction. If r and/or s are themselves compound propositions, then we apply De Morgan's laws again to "push" the negation symbol $\neg$ deeper inside the formula, changing $\wedge$ to $\vee$ and $\vee$ to $\wedge$. We repeat this process until all the negation signs have been "pushed in" as far as possible and are now attached to the atomic (i.e., not compound) propositions in the compound propositions p and q. Call these atomic propositions p_1, p_2, etc. Now in this process De Morgan's laws have forced us to change each $\wedge$ to $\vee$ and each $\vee$ to $\wedge$. Furthermore, if there are any constants **T** or **F** in the propositions, then they will be changed to their opposite when the negation operation is applied: $\neg\mathbf{T}$ is the same as **F**, and $\neg\mathbf{F}$ is the same as **T**. In summary, $\neg p$ and $\neg q$ look just like p^* and q^*, except that each atomic proposition p_i within them is replaced by its negation. Now we agreed that $\neg p \equiv \neg q$; this means that for *every* possible assignment of truth values to the atomic propositions p_1, p_2, etc., the truth values of $\neg p$ and $\neg q$ are the same. But assigning T to p_i is the same as assigning F to $\neg p_i$, and assigning F to p_i is the same as assigning T to $\neg p_i$. Thus, for every possible assignment of truth values to the atomic propositions, the truth values of p^* and q^* are the same. This is precisely what we wanted to prove.

41. There are three ways in which exactly two of p, q, and r can be true. We write down these three possibilities as conjunctions and join them by $\vee$ to obtain the answer: $(p \wedge q \wedge \neg r) \vee (p \wedge \neg q \wedge r) \vee (\neg p \wedge q \wedge r)$. See Exercise 42 for a more general result.

43. Given a compound proposition p, we can construct its truth table and then, by Exercise 42, write down a proposition q in disjunctive normal form that is logically equivalent to p. Since q involves only $\neg$, $\wedge$, and $\vee$, this shows that $\neg$, $\wedge$, and $\vee$ form a functionally complete collection of logical operators.

45. Given a compound proposition p, we can, by Exercise 43, write down a proposition q that is logically equivalent to p and uses only $\neg$, $\wedge$, and $\vee$. Now by De Morgan's law we can get rid of all the $\wedge$'s by replacing each occurrence of $p_1 \wedge p_2 \wedge \cdots \wedge p_n$ with the equivalent proposition $\neg(\neg p_1 \vee \neg p_2 \vee \cdots \vee \neg p_n)$.

47. The proposition $\neg(p \wedge q)$ is true when either p or q, or both, are false, and is false when both p and q are true; since this was the definition of $p \mid q$, the two are logically equivalent.

49. The proposition $\neg(p \vee q)$ is true when both p and q are false, and is false otherwise; since this was the definition of $p \downarrow q$, the two are logically equivalent.

51. A straightforward approach, using the results of Exercise 50, parts **(a)** and **(b)**, is as follows: $(p \to q) \equiv (\neg p \vee q) \equiv ((p \downarrow p) \vee q) \equiv (((p \downarrow p) \downarrow q) \downarrow ((p \downarrow p) \downarrow q))$. If we allow the constant **F** in our expression, then a simpler answer is $\mathbf{F} \downarrow ((\mathbf{F} \downarrow p) \downarrow q)$.

53. This is clear from the definition, in which p and q play a symmetric role.

55. A truth table for a compound proposition involving p and q has four lines, one for each of the following combinations of truth values for p and q: TT, TF, FT, and FF. Now each line of the truth table for the compound proposition can be either T or F. Thus there are two possibilities for the first line; for each of those there are two possibilities for the second line, giving $2 \cdot 2 = 4$ possibilities for the first two lines; for each of those there are two possibilities for the third line, giving $4 \cdot 2 = 8$ possibilities for the first three lines; and finally for each of those, there are two possibilities for the fourth line, giving $8 \cdot 2 = 16$ possibilities altogether. This sort of counting will be studied extensively in Chapter 5.

57. Let do, mc, and in stand for the propositions "The directory database is opened," "The monitor is put in a closed state," and "The system is in its initial state," respectively. Then the given statement reads $\neg in \to (do \to mc)$. By the third line of Table 7 (twice), this is equivalent to $in \vee (\neg do \vee mc)$. In words, this says that it must always be true that either the system is in its initial state, or the data base is not opened, or the monitor is put in a closed state. Another way to render this would be to say that if the database is open, then either the system is in its initial state or the monitor is put in a closed state.

59. Disjunctions are easy to make true, since we just have to make sure that at least one of the things being "or-ed" is true. In this problem, we notice that $\neg p$ occurs in four of the disjunctions, so we can satisfy all of them by making p false. Three of the remaining disjunctions contain r, so if we let r be true, those will be taken care of. That leaves only $p \vee \neg q \vee s$ and $q \vee \neg r \vee \neg s$, and we can satisfy both of those by making q and s both true. This assignment, then, makes all nine of the disjunctions true.

61. A compound proposition c is a tautology if every assignment of truth values to its variables makes c true. That means that every assignment of truth values to its variables makes $\neg c$ false, in other words, that $\neg c$ is not satisfiable. If we had an algorithm to determine whether or not a compound proposition is satisfiable, then we could apply that algorithm to $\neg c$ to determine whether c is a tautology. If the algorithm says that $\neg c$ is satisfiable, then we report that c is not a tautology, and if the algorithm says that $\neg c$ is not satisfiable, then we report that c is a tautology.

SECTION 1.3 Predicates and Quantifiers

The reader may find quantifiers hard to understand at first. Predicate logic (the study of propositions with quantifiers) is one level of abstraction higher than propositional logic (the study of propositions without quantifiers). Careful attention to this material will aid you in thinking more clearly, not only in mathematics but in other areas as well, from computer science to politics. Keep in mind exactly what the quantifiers mean: $\forall x$ *means "for all* x*" or "for every* x*," and* $\exists x$ *means "there exists an* x *such that" or "for some* x*." It is good practice to read every such sentence aloud, paying attention to English grammar as well as meaning. It is very important to understand how the negations of quantified statements are formed, and why this method is correct; it is just common sense, really.*

The word "any" in mathematical statements can be ambiguous, so it is best to avoid using it. In negative contexts it almost always means "some" (existential quantifier), as in the statement "You will be suspended from school if you are found guilty of violating any of the plagiarism rules" (you don't have to violate all the rules to get into trouble—breaking one is sufficient). In positive contexts, however, it can mean either "some" (existential quantifier) or "every" (universal quantifier), depending on context. For example, in the sentence "The fraternity will be put on probation if any of its members is found intoxicated," the use is existential (one drunk brother is enough to cause the sanction); but in the sentence "Any member of the sorority will be happy to lead you on a tour of the house," the use is universal (every member is able to be the guide). Another interesting example is an exercise in a mathematics textbook that asks you to show that "the sum of any two odd numbers is even." The author clearly intends the universal interpretation here—you need to show that the sum of two odd numbers is always even. If you interpreted the question existentially, you might say, "Look, $3+5=8$*, so I've shown it is true—you said I could do it for any numbers, and those are the ones I chose."*

1. a) T, since $0 \le 4$ **b)** T, since $4 \le 4$ **c)** F, since $6 \not\le 4$

3. a) This is true.
b) This is false, since Lansing, not Detroit, is the capital.
c) This is false (but $Q(\text{Boston}, \text{Massachusetts})$ is true).
d) This is false, since Albany, not New York, is the capital.

5. a) There is a student who spends more than five hours every weekday in class.
b) Every student spends more than five hours every weekday in class.
c) There is a student who does not spend more than five hours every weekday in class.
d) No student spends more than five hours every weekday in class. (Or, equivalently, every student spends less than or equal to five hours every weekday in class.)

7. a) This statement is that for every x, if x is a comedian, then x is funny. In English, this is most simply stated, "Every comedian is funny."
b) This statement is that for every x in the domain (universe of discourse), x is a comedian *and* x is funny. In English, this is most simply stated, "Every person is a funny comedian." Note that this is not the sort of thing one wants to say. It really makes no sense and doesn't say anything about the existence of boring comedians; it's surely false, because there exist lots of x for which $C(x)$ is false. This illustrates the fact that you rarely want to use conjunctions with universal quantifiers.
c) This statement is that there exists an x in the domain such that if x is a comedian then x is funny. In English, this might be rendered, "There exists a person such that if s/he is a comedian, then s/he is funny." Note that this is not the sort of thing one wants to say. It really makes no sense and doesn't say anything about the existence of funny comedians; it's surely true, because there exist lots of x for which $C(x)$ is false (recall

the definition of the truth value of $p \rightarrow q$). This illustrates the fact that you rarely want to use conditional statements with existential quantifiers.

d) This statement is that there exists an x in the domain such that x is a comedian and x is funny. In English, this might be rendered, "There exists a funny comedian" or "Some comedians are funny" or "Some funny people are comedians."

9. **a)** We assume that this sentence is asserting that the same person has both talents. Therefore we can write $\exists x(P(x) \wedge Q(x))$.

b) Since "but" really means the same thing as "and" logically, this is $\exists x(P(x) \wedge \neg Q(x))$

c) This time we are making a universal statement: $\forall x(P(x) \vee Q(x))$

d) This sentence is asserting the nonexistence of anyone with either talent, so we could write it as $\neg\exists x(P(x) \vee Q(x))$. Alternatively, we can think of this as asserting that everyone fails to have either of these talents, and we obtain the logically equivalent answer $\forall x \neg(P(x) \vee Q(x))$. Failing to have either talent is equivalent to having neither talent (by De Morgan's law), so we can also write this as $\forall x((\neg P(x)) \wedge (\neg Q(x))$. Note that it would *not* be correct to write $\forall x((\neg P(x)) \vee (\neg Q(x))$ nor to write $\forall x \neg(P(x) \wedge Q(x))$.

11. **a)** T, since $0 = 0^2$ **b)** T, since $1 = 1^2$ **c)** F, since $2 \neq 2^2$
d) F, since $-1 \neq (-1)^2$ **e)** T (let $x = 1$) **f)** F (let $x = 2$)

13. **a)** Since adding 1 to a number makes it larger, this is true.

b) Since $2 \cdot 0 = 3 \cdot 0$, this is true.

c) This statement is true, since $0 = -0$.

d) As was explained in Example 13, this is true for the integers.

15. Recall that the integers include the positive and negative integers and 0.

a) This is the well-known true fact that the square of a real number cannot be negative.

b) There are two *real* numbers that satisfy $n^2 = 2$, namely $\pm\sqrt{2}$, but there do not exist any *integers* with this property, so the statement is false.

c) If n is a positive integer, then $n^2 \geq n$ is certainly true; it's also true for $n = 0$; and it's trivially true if n is negative. Therefore the universally quantified statement is true.

d) Squares can never be negative; therefore this statement is false.

17. Existential quantifiers are like disjunctions, and universal quantifiers are like conjunctions. See Examples 11 and 16.

a) We want to assert that $P(x)$ is true for some x in the universe, so either $P(0)$ is true or $P(1)$ is true or $P(2)$ is true or $P(3)$ is true or $P(4)$ is true. Thus the answer is $P(0) \vee P(1) \vee P(2) \vee P(3) \vee P(4)$. The other parts of this exercise are similar. Note that by De Morgan's laws, the expression in part **(c)** is logically equivalent to the expression in part **(f)**, and the expression in part **(d)** is logically equivalent to the expression in part **(e)**.

b) $P(0) \wedge P(1) \wedge P(2) \wedge P(3) \wedge P(4)$

c) $\neg P(0) \vee \neg P(1) \vee \neg P(2) \vee \neg P(3) \vee \neg P(4)$

d) $\neg P(0) \wedge \neg P(1) \wedge \neg P(2) \wedge \neg P(3) \wedge \neg P(4)$

e) This is just the negation of part **(a)**: $\neg(P(0) \vee P(1) \vee P(2) \vee P(3) \vee P(4))$

f) This is just the negation of part **(b)**: $\neg(P(0) \wedge P(1) \wedge P(2) \wedge P(3) \wedge P(4))$

19. Existential quantifiers are like disjunctions, and universal quantifiers are like conjunctions. See Examples 11 and 16.

a) We want to assert that $P(x)$ is true for some x in the universe, so either $P(1)$ is true or $P(2)$ is true or $P(3)$ is true or $P(4)$ is true or $P(5)$ is true. Thus the answer is $P(1) \vee P(2) \vee P(3) \vee P(4) \vee P(5)$.

b) $P(1) \wedge P(2) \wedge P(3) \wedge P(4) \wedge P(5)$

c) This is just the negation of part **(a)**: $\neg(P(1) \vee P(2) \vee P(3) \vee P(4) \vee P(5))$

d) This is just the negation of part **(b)**: $\neg(P(1) \wedge P(2) \wedge P(3) \wedge P(4) \wedge P(5))$

e) The formal translation is as follows: $(((1 \neq 3) \rightarrow P(1)) \wedge ((2 \neq 3) \rightarrow P(2)) \wedge ((3 \neq 3) \rightarrow P(3)) \wedge ((4 \neq 3) \rightarrow P(4)) \wedge ((5 \neq 3) \rightarrow P(5))) \vee (\neg P(1) \vee \neg P(2) \vee \neg P(3) \vee \neg P(4) \vee \neg P(5))$. However, since the hypothesis $x \neq 3$ is false when x is 3 and true when x is anything other than 3, we have more simply $(P(1) \wedge P(2) \wedge P(4) \wedge P(5)) \vee (\neg P(1) \vee \neg P(2) \vee \neg P(3) \vee \neg P(4) \vee \neg P(5))$. Thinking about it a little more, we note that this statement is always true, since if the first part is not true, then the second part must be true.

21. a) One would hope that if we take the domain to be the students in your class, then the statement is true. If we take the domain to be all students in the world, then the statement is clearly false, because some of them are studying only other subjects.

b) If we take the domain to be United States Senators, then the statement is true. If we take the domain to be college football players, then the statement is false, because some of them are younger than 21.

c) If the domain consists of just Princes William and Harry of Great Britain (sons of the late Princess Diana), then the statement is true. It is also true if the domain consists of just one person (everyone has the same mother as him- or herself). If the domain consists of all the grandchildren of Queen Elizabeth II of Great Britain (of whom William and Harry are just two), then the statement is false.

d) If the domain consists of Bill Clinton and George W. Bush, then this statement is true because they do not have the same grandmother. If the domain consists of all residents of the United States, then the statement is false, because there are many instances of siblings and first cousins, who have at least one grandmother in common.

23. In order to do the translation the second way, we let $C(x)$ be the propositional function "x is in your class." Note that for the second way, we always want to use conditional statements with universal quantifiers and conjunctions with existential quantifiers.

a) Let $H(x)$ be "x can speak Hindi." Then we have $\exists x\, H(x)$ the first way, or $\exists x(C(x) \wedge H(x))$ the second way.

b) Let $F(x)$ be "x is friendly." Then we have $\forall x\, F(x)$ the first way, or $\forall x(C(x) \rightarrow F(x))$ the second way.

c) Let $B(x)$ be "x was born in California." Then we have $\exists x\, \neg B(x)$ the first way, or $\exists x(C(x) \wedge \neg B(x))$ the second way.

d) Let $M(x)$ be "x has been in a movie." Then we have $\exists x\, M(x)$ the first way, or $\exists x(C(x) \wedge M(x))$ the second way.

e) This is saying that everyone has failed to take the course. So the answer here is $\forall x\, \neg L(x)$ the first way, or $\forall x(C(x) \rightarrow \neg L(x))$ the second way, where $L(x)$ is "x has taken a course in logic programming."

25. Let $P(x)$ be "x is perfect"; let $F(x)$ be "x is your friend"; and let the domain (universe of discourse) be all people.

a) This means that everyone has the property of being not perfect: $\forall x\, \neg P(x)$. Alternatively, we can write this as $\neg \exists x\, P(x)$, which says that there does not exist a perfect person.

b) This is just the negation of "Everyone is perfect": $\neg \forall x\, P(x)$.

c) If someone is your friend, then that person is perfect: $\forall x(F(x) \rightarrow P(x))$. Note the use of conditional statement with universal quantifiers.

d) We do not have to rule out your having more than one perfect friend. Thus we have simply $\exists x(F(x) \wedge P(x))$. Note the use of conjunction with existential quantifiers.

e) The expression is $\forall x(F(x) \wedge P(x))$. Note that here we did use a conjunction with the universal quantifier, but the sentence is not natural (who could claim this?). We could also have split this up into two quantified statements and written $(\forall x\, F(x)) \wedge (\forall x\, P(x))$.

f) This is a disjunction. The expression is $(\neg\forall x\, F(x)) \vee (\exists x\, \neg P(x))$.

27. In all of these, we will let $Y(x)$ be the propositional function that x is in your school or class, as appropriate.

a) If we let $V(x)$ be "x has lived in Vietnam," then we have $\exists x\, V(x)$ if the universe is just your schoolmates, or $\exists x(Y(x) \wedge V(x))$ if the universe is all people. If we let $D(x,y)$ mean that person x has lived in country y, then we can rewrite this last one as $\exists x(Y(x) \wedge D(x, \text{Vietnam}))$.

b) If we let $H(x)$ be "x can speak Hindi," then we have $\exists x\, \neg H(x)$ if the universe is just your schoolmates, or $\exists x(Y(x) \wedge \neg H(x))$ if the universe is all people. If we let $S(x,y)$ mean that person x can speak language y, then we can rewrite this last one as $\exists x(Y(x) \wedge \neg S(x, \text{Hindi}))$.

c) If we let $J(x)$, $P(x)$, and $C(x)$ be the propositional functions asserting x's knowledge of Java, Prolog, and C++, respectively, then we have $\exists x(J(x) \wedge P(x) \wedge C(x))$ if the universe is just your schoolmates, or $\exists x(Y(x) \wedge J(x) \wedge P(x) \wedge C(x))$ if the universe is all people. If we let $K(x,y)$ mean that person x knows programming language y, then we can rewrite this last one as $\exists x(Y(x) \wedge K(x, \text{Java}) \wedge K(x, \text{Prolog}) \wedge K(x, \text{C++}))$.

d) If we let $T(x)$ be "x enjoys Thai food," then we have $\forall x\, T(x)$ if the universe is just your classmates, or $\forall x(Y(x) \rightarrow T(x))$ if the universe is all people. If we let $E(x,y)$ mean that person x enjoys food of type y, then we can rewrite this last one as $\forall x(Y(x) \rightarrow E(x, \text{Thai}))$.

e) If we let $H(x)$ be "x plays hockey," then we have $\exists x\, \neg H(x)$ if the universe is just your classmates, or $\exists x(Y(x) \wedge \neg H(x))$ if the universe is all people. If we let $P(x,y)$ mean that person x plays game y, then we can rewrite this last one as $\exists x(Y(x) \wedge \neg P(x, \text{hockey}))$.

29. Our domain (universe of discourse) here is all propositions. Let $T(x)$ mean that x is a tautology and $C(x)$ mean that x is a contradiction. Since a contingency is just a proposition that is neither a tautology nor a contradiction, we do not need a separate predicate for being a contingency.

a) This one is just the assertion that tautologies exist: $\exists x\, T(x)$.

b) Although the word "all" or "every" does not appear here, this sentence is really expressing a universal meaning, that the negation of a contradiction is always a tautology. So we want to say that if x is a contradiction, then $\neg x$ is a tautology. Thus we have $\forall x(C(x) \rightarrow T(\neg x))$. Note the rare use of a logical symbol (negation) applied to a variable (x); this is purely a coincidence in this exercise because the universe happens itself to be propositions.

c) The words "can be" are expressing an existential idea—that there exist two contingencies whose disjunction is a tautology. Thus we have $\exists x \exists y(\neg T(x) \wedge \neg C(x) \wedge \neg T(y) \wedge \neg C(y) \wedge T(x \vee y))$. The same final comment as in part **(b)** applies here. Also note the explanation about contingencies in the preamble.

d) As in part **(b)**, this is the universal assertion that whenever x and y are tautologies, then so is $x \wedge y$; thus we have $\forall x \forall y((T(x) \wedge T(y)) \rightarrow T(x \wedge y))$.

31. In each case we just have to list all the possibilities, joining them with $\vee$ if the quantifier is $\exists$, and joining them with $\wedge$ if the quantifier is $\forall$.

a) $Q(0,0,0) \wedge Q(0,1,0)$ **b)** $Q(0,1,1) \vee Q(1,1,1) \vee Q(2,1,1)$

c) $\neg Q(0,0,0) \vee \neg Q(0,0,1)$ **d)** $\neg Q(0,0,1) \vee \neg Q(1,0,1) \vee \neg Q(2,0,1)$

33. In each case we need to specify some predicates and identify the domain (universe of discourse).

a) Let $T(x)$ be the predicate that x can learn new tricks, and let the domain be old dogs. Our original statement is $\exists x\, T(x)$. Its negation is $\neg\exists x\, T(x)$, which we must to rewrite in the required manner as $\forall x\, \neg T(x)$. In English this reads "Every old dog is unable to learn new tricks" or "All old dogs can't learn new tricks."

(Note that this does *not* say that not all old dogs can learn new tricks—it is saying something stronger than that.) More colloquially, we can say "No old dogs can learn new tricks."

b) Let $C(x)$ be the predicate that x knows calculus, and let the domain be rabbits. Our original statement is $\neg\exists x\, C(x)$. Its negation is, of course, simply $\exists x\, C(x)$. In English this reads "There is a rabbit that knows calculus."

c) Let $F(x)$ be the predicate that x can fly, and let the domain be birds. Our original statement is $\forall x\, F(x)$. Its negation is $\neg\forall x\, F(x)$ (i.e., not all birds can fly), which we must to rewrite in the required manner as $\exists x\, \neg F(x)$. In English this reads "There is a bird who cannot fly."

d) Let $T(x)$ be the predicate that x can talk, and let the domain be dogs. Our original statement is $\neg\exists x\, T(x)$. Its negation is, of course, simply $\exists x\, T(x)$. In English this reads "There is a dog that talks."

e) Let $F(x)$ and $R(x)$ be the predicates that x knows French and knows Russian, respectively, and let the domain be people in this class. Our original statement is $\neg\exists x(F(x) \wedge R(x))$. Its negation is, of course, simply $\exists x(F(x) \wedge R(x))$. In English this reads "There is someone in this class who knows French and Russian."

35. a) As we saw in Example 13, this is true, so there is no counterexample.

b) Since 0 is neither greater than nor less than 0, this is a counterexample.

c) This proposition says that 1 is the only integer—that every integer equals 1. If is obviously false, and any other integer, such as -111749, provides a counterexample.

37. In each case we need to make up predicates. The answers are certainly not unique and depend on the choice of predicate, among other things.

a) $\forall x((F(x, 25000) \vee S(x, 25)) \to E(x))$, where $E(x)$ is "Person x qualifies as an elite flyer in a given year," $F(x, y)$ is "Person x flies more than y miles in a given year," and $S(x, y)$ is "Person x takes more than y flights in a given year"

b) $\forall x(((M(x) \wedge T(x, 3)) \vee (\neg M(x) \wedge T(x, 3.5))) \to Q(x))$, where $Q(x)$ is "Person x qualifies for the marathon," $M(x)$ is "Person x is a man," and $T(x, y)$ is "Person x has run the marathon in less than y hours"

c) $M \to ((H(60) \vee (H(45) \wedge T)) \wedge \forall y\, G(\mathrm{B}, y))$, where M is the proposition "The student received a masters degree," $H(x)$ is "The student took at least x course hours," T is the proposition "The student wrote a thesis," and $G(x, y)$ is "The person got grade x or higher in his course y"

d) $\exists x\, ((T(x, 21) \wedge G(x, 4.0))$, where $T(x, y)$ is "Person x took more than y credit hours" and $G(x, p)$ is "Person x earned grade point average p" (we assume that we are talking about one given semester)

39. In each case we pretty much just write what we see.

a) If there is a printer that is both out of service and busy, then some job has been lost.

b) If every printer is busy, then there is a job in the queue.

c) If there is a job that is both queued and lost, then some printer is out of service.

d) If every printer is busy and every job is queued, then some job is lost.

41. In each case we need to make up predicates. The answers are certainly not unique and depend on the choice of predicate, among other things.

a) $(\exists x\, F(x, 10)) \to \exists x\, S(x)$, where $F(x, y)$ is "Disk x has more than y kilobytes of free space," and $S(x)$ is "Mail message x can be saved"

b) $(\exists x\, A(x)) \to \forall x(Q(x) \to T(x))$, where $A(x)$ is "Alert x is active," $Q(x)$ is "Message x is queued," and $T(x)$ is "Message x is transmitted"

c) $\forall x((x \neq \text{main console}) \to T(x))$, where $T(x)$ is "The diagnostic monitor tracks the status of system x"

d) $\forall x(\neg L(x) \to B(x))$, where $L(x)$ is "The host of the conference call put participant x on a special list" and $B(x)$ is "Participant x was billed"

43. A conditional statement is true if the hypothesis is false. Thus it is very easy for the second of these propositions to be true—just have $P(x)$ be something that is not always true, such as "The integer x is a multiple of 2." On the other hand, it is certainly not always true that if a number is a multiple of 2, then it is also a multiple of 4, so if we let $Q(x)$ be "The integer x is a multiple of 4," then $\forall x(P(x) \rightarrow Q(x))$ will be false. Thus these two propositions can have different truth values. Of course, for some choices of P and Q, they will have the same truth values, such as when P and Q are true all the time.

45. Both are true precisely when at least one of $P(x)$ and $Q(x)$ is true for at least one value of x in the domain (universe of discourse).

47. We can establish these equivalences by arguing that one side is true if and only if the other side is true. For both parts, we will look at the two cases: either A is true or A is false.

a) Suppose that A is true. Then the left-hand side is logically equivalent to $\forall x P(x)$, since the conjunction of any proposition with a true proposition has the same truth value as that proposition. By similar reasoning the right-hand side is equivalent to $\forall x P(x)$. Therefore the two propositions are logically equivalent in this case; each one is true precisely when $P(x)$ is true for every x. On the other hand, suppose that A is false. Then the left-hand side is certainly false. Furthermore, for every x, $P(x) \wedge A$ is false, so the right-hand side is false as well. Thus in all cases, the two propositions have the same truth value.

b) This problem is similar to part **(a)**. If A is true, then both sides are logically equivalent to $\exists x P(x)$. If A is false, then both sides are false.

49. We can establish these equivalences by arguing that one side is true if and only if the other side is true. For both parts, we will look at the two cases: either A is true or A is false.

a) Suppose that A is true. Then for each x, $P(x) \rightarrow A$ is true, because a conditional statement with a true conclusion is always true; therefore the left-hand side is always true in this case. By similar reasoning the right-hand side is always true in this case (here we used the fact that the domain is nonempty). Therefore the two propositions are logically equivalent when A is true. On the other hand, suppose that A is false. There are two subcases. If $P(x)$ is false for every x, then $P(x) \rightarrow A$ is vacuously true (a conditional statement with a false hypothesis is true), so the left-hand side is vacuously true. The same reasoning shows that the right-hand side is also true, because in this subcase $\exists x P(x)$ is false. For the second subcase, suppose that $P(x)$ is true for some x. Then for that x, $P(x) \rightarrow A$ is false (a conditional statement with a true hypothesis and false conclusion is false), so the left-hand side is false. The right-hand side is also false, because in this subcase $\exists x P(x)$ is true but A is false. Thus in all cases, the two propositions have the same truth value.

b) This problem is similar to part **(a)**. If A is true, then both sides are trivially true, because the conditional statements have true conclusions. If A is false, then there are two subcases. If $P(x)$ is false for some x, then $P(x) \rightarrow A$ is vacuously true for that x (a conditional statement with a false hypothesis is true), so the left-hand side is true. The same reasoning shows that the right-hand side is true, because in this subcase $\forall x P(x)$ is false. For the second subcase, suppose that $P(x)$ is true for every x. Then for every x, $P(x) \rightarrow A$ is false (a conditional statement with a true hypothesis and false conclusion is false), so the left-hand side is false (there is no x making the conditional statement true). The right-hand side is also false, because it is a conditional statement with a true hypothesis and a false conclusion. Thus in all cases, the two propositions have the same truth value.

51. We can show that these are not logically equivalent by giving an example in which one is true and the other is false. Let $P(x)$ be the statement "x is odd" applied to positive integers. Similarly let $Q(x)$ be "x is even." Then since there exist odd numbers and there exist even numbers, the statement $\exists x P(x) \wedge \exists x Q(x)$ is true. On the other hand, no number is both odd and even, so $\exists x(P(x) \wedge Q(x))$ is false.

53. a) This is certainly true: if there is a unique x satisfying $P(x)$, then there certainly is *an* x satisfying $P(x)$.

b) Unless the domain (universe of discourse) has fewer than two items in it, the truth of the hypothesis implies that there is more than one x such that $P(x)$ holds. Therefore this proposition need not be true. (For example, let $P(x)$ be the proposition $x^2 \geq 0$ in the context of the real numbers. The hypothesis is true, but there is not a unique x for which $x^2 \geq 0$.)

c) This is true: if there is an x (unique or not) such that $P(x)$ is false, then we can conclude that it is not the case that $P(x)$ holds for all x.

55. A Prolog query returns a yes/no answer if there are no variables in the query, and it returns all values that make the query true if there are.

a) One of the facts was that Chan was the instructor of Math 273, so the response is `yes`.

b) None of the facts was that Patel was the instructor of CS 301, so the response is `no`.

c) Prolog returns the names of the people enrolled in CS 301, namely `juana` and `kiko`.

d) Prolog returns the names of the courses Kiko is enrolled in, namely `math273` and `cs301`.

e) Prolog returns the names of the students enrolled in courses which Grossman is the instructor for (which is just CS 301), namely `juana` and `kiko`.

57. Following the idea and syntax of Example 28, we have the following rule: `sibling(X,Y) :- mother(M,X), mother(M,Y), father(F,X), father(F,Y)`. Note that we used the comma to mean "and"; `X` and `Y` must have the same mother and the same father in order to be (full) siblings.

59. a) This is the statement that every person who is a professor is not ignorant. In other words, for every person, if that person is a professor, then that person is not ignorant. In symbols: $\forall x(P(x) \to \neg Q(x))$. This is not the only possible answer. We could equivalently think of the statement as asserting that there does not exist an ignorant professor: $\neg\exists x(P(x) \land Q(x))$.

b) Every person who is ignorant is vain: $\forall x(Q(x) \to R(x))$.

c) This is similar to part **(a)**: $\forall x(P(x) \to \neg R(x))$.

d) The conclusion (part **(c)**) does not follow. There may well be vain professors, since the premises do not rule out the possibility that there are vain people besides the ignorant ones.

61. a) This is asserting that every person who is a baby is necessarily not logical: $\forall x(P(x) \to \neg Q(x))$.

b) If a person can manage a crocodile, then that person is not despised: $\forall x(R(x) \to \neg S(x))$.

c) Every person who is not logical is necessarily despised: $\forall x(\neg Q(x) \to S(x))$.

d) Every person who is a baby cannot manage a crocodile: $\forall x(P(x) \to \neg R(x))$.

e) The conclusion follows. Suppose that x is a baby. Then by the first premise, x is illogical, and hence, by the third premise, x is despised. But the second premise says that if x could manage a crocodile, then x would not be despised. Therefore x cannot manage a crocodile. Thus we have proved that babies cannot manage crocodiles.

SECTION 1.4 Nested Quantifiers

Nested quantifiers are one of the most difficult things for students to understand. The theoretical definition of limit in calculus, for example, is hard to comprehend because it has three levels of nested quantifiers. Study the examples in this section carefully before attempting the exercises, and make sure that you understand the solutions to the exercises you have difficulty with. Practice enough of these until you feel comfortable. The effort will be rewarded in such areas as computer programming and advanced mathematics courses.

1. a) For every real number x there exists a real number y such that x is less than y. Basically, this is asserting that there is no largest real number—for any real number you care to name, there is a larger one.

b) For every real number x and real number y, if x and y are both nonnegative, then their product is nonnegative. Or, more simply, the product of nonnegative real numbers is nonnegative.

c) For every real number x and real number y, there exists a real number z such that $xy = z$. Or, more simply, the real numbers are closed under multiplication. (Some authors would include the uniqueness of z as part of the meaning of the word *closed*.)

3. It is useful to keep in mind that x and y can be the same person, so sending messages to oneself counts in this problem.

a) Formally, this says that there exist students x and y such that x has sent a message to y. In other words, there is some student in your class who has sent a message to some student in your class.

b) This is similar to part **(a)** except that x has sent a message to everyone, not just to at least one person. So this says there is some student in your class who has sent a message to every student in your class.

c) Note that this is not the same as part **(b)**. Here we have that for every x there exists a y such that x has sent a message to y. In other words, every student in your class has sent a message to at least one student in your class.

d) Note that this is not the same as part **(c)**, since the order of quantifiers has changed. In part **(c)**, y could depend on x; in other words, the recipient of the messages could vary from sender to sender. Here the existential quantification on y comes first, so it's the same recipient for all the messages. The meaning is that there is a student in your class who has been sent a message by every student in your class.

e) This is similar to part **(c)**, with the role of sender and recipient reversed: every student in your class has been sent a message from at least one student in your class. Again, note that the sender can depend on the recipient.

f) Every student in the class has sent a message to every student in the class.

5. a) This simply says that Sarah Smith has visited `www.att.com`.

b) To say that an x exists such that x has visited `www.imdb.org` is just to say that someone (i.e., at least one person) has visited `www.imdb.org`.

c) This is similar to part **(b)**. Jose Orez has visited some website.

d) This is asserting that a y exists that both of these students has visited. In other words, Ashok Puri and Cindy Yoon have both visited the same website.

e) When there are two quantifiers of opposite types, the sentence gets more complicated. This is saying that there is a person (y) other than David Belcher who has visited all the websites that David has visited (i.e., for every website z, if David has visited z, then so has this person). Note that it is not saying that this person has visited only websites that David has visited (that would be the converse conditional statement)—this person may have visited other sites as well.

f) Here the existence of two people is being asserted; they are said to be unequal, and for every website z, one of these people has visited z if and only if the other one has. In plain English, there are two different people who have visited exactly the same websites.

7. a) Abdallah Hussein does not like Japanese cuisine.

b) Note that this is the conjunction of two separate quantified statements. Some student at your school likes Korean cuisine, and everyone at your school likes Mexican cuisine.

c) There is some cuisine that either Monique Arsenault or Jay Johnson likes.

d) Formally this says that for every x and z, there exists a y such that if x and z are not equal, then it is not the case that both x and z like y. In simple English, this says that for every pair of distinct students at your school, there is some cuisine that at least one them does not like.

e) There are two students at your school who have exactly the same tastes (i.e., they like exactly the same cuisines).

f) For every pair of students at your school, there is some cuisine about which they have the same opinion (either they both like it or they both do not like it).

9. We need to be careful to put the lover first and the lovee second as arguments in the propositional function L.

a) $\forall x L(x, \text{Jerry})$

b) Note that the "somebody" being loved depends on the person doing the loving, so we have to put the universal quantifier first: $\forall x \exists y L(x, y)$.

c) In this case, one lovee works for all lovers, so we have to put the existential quantifier first: $\exists y \forall x L(x, y)$.

d) We could think of this as saying that there does not exist anyone who loves everybody ($\neg\exists x \forall y L(x, y)$), or we could think of it as saying that for each person, we can find a person whom he or she does not love ($\forall x \exists y \neg L(x, y)$. These two expressions are logically equivalent.

e) $\exists x \neg L(\text{Lydia}, x)$

f) We are asserting the existence of an individual such that everybody fails to love that person: $\exists x \forall y \neg L(y, x)$.

g) In Exercise 52 of Section 1.3, we saw that there is a notation for the existence of a unique object satisfying a certain condition. Employing that device, we could write this as $\exists! x \forall y L(y, x)$. In Exercise 52 of the present section we will discover a way to avoid this notation in general. What we have to say is that the x asserted here exists, and that every z satisfying this condition (of being loved by everybody) must equal x. Thus we obtain $\exists x\big(\forall y L(y, x) \wedge \forall z((\forall w L(w, z)) \rightarrow z = x)\big)$. Note that we could have used y as the bound variable where we used w; since the scope of the first use of y had ended before we came to this point in the formula, reusing y as the bound variable would cause no ambiguity.

h) We want to assert the existence of two distinct people, whom we will call x and y, whom Lynn loves, as well as make the statement that everyone whom Lynn loves must be either x or y: $\exists x \exists y\big(x \neq y \wedge L(\text{Lynn}, x) \wedge L(\text{Lynn}, y) \wedge \forall z(L(\text{Lynn}, z) \rightarrow (z = x \vee z = y))\big)$.

i) $\forall x L(x, x)$ (Note that nothing in our earlier answers ruled out the possibility that variables or constants with different names might be equal to each other. For example, in part **(a)**, x could equal Jerry, so that statement includes as a special case the assertion that Jerry loves himself. Similarly, in part **(h)**, the two people whom Lynn loves either could be two people other than Lynn (in which case we know that Lynn does not love herself), or could be Lynn herself and one other person.)

j) This is asserting that the one and only one person who is loved by the person being discussed is in fact that person: $\exists x \forall y(L(x, y) \leftrightarrow x = y)$.

11. a) We might want to assert that Lois is a student and Michaels is a faculty member, but the sentence doesn't really say that, so the simple answer is just $A(\text{Lois}, \text{Professor Michaels})$.

b) To say that every student (as opposed to every person) has done this, we need to restrict our universally quantified variable to being a student. The easiest way to do this is to make the assertion being quantified a conditional statement. *As a general rule of thumb, use conditional statements with universal quantifiers and conjunctions with existential quantifiers (see part* **(d)**, *for example).* Thus our answer is $\forall x(S(x) \rightarrow A(x, \text{Professor Gross}))$.

c) This is similar to part **(b)**: $\forall x(F(x) \rightarrow (A(x, \text{Professor Miller}) \vee A(\text{Professor Miller}, x)))$. Note the need for parentheses in these answers.

d) There is a student such that for every faculty member, that student has not asked that faculty member a question. Note how we need to include the S and F predicates: $\exists x(S(x) \wedge \forall y(F(y) \rightarrow \neg A(x, y)))$. We could also write this as $\exists x(S(x) \wedge \neg\exists y(F(y) \wedge A(x, y)))$.

e) This is very similar to part **(d)**, with the role of the players reversed: $\exists x(F(x) \wedge \forall y(S(y) \rightarrow \neg A(y, x)))$.

f) This is a little ambiguous in English. If the statement is that there is a very inquisitive student, one who has gone around and asked a question of every professor, then this is similar to part **(d)**, without the negation: $\exists x(S(x) \land \forall y(F(y) \to A(x,y)))$. On the other hand, the statement might be intended as asserting simply that for every professor, there exists some student who has asked that professor a question. In other words, the questioner might depend on the questionee. Note how the meaning changes with the change in order of quantifiers. Under the second interpretation the answer is $\forall y(F(y) \to \exists x(S(x) \land A(x,y)))$. The first interpretation is probably the intended one.

g) This is pretty straightforward, except that we have to rule out the possibility that the askee is the same as the asker. Our sentence needs to say that there exists a faculty member such that for every other faculty member, the first has asked the second a question: $\exists x(F(x) \land \forall y((F(y) \land y \neq x) \to A(x,y)))$.

h) There is a student such that every faculty member has failed to ask him a question: $\exists x(S(x) \land \forall y(F(y) \to \neg A(y,x)))$.

13. Be careful to put in parentheses where needed; otherwise your answer can be either ambiguous or wrong.

a) Clearly this is simply $\neg M(\text{Chou}, \text{Koko})$.

b) We can give two answers, which are equivalent by De Morgan's law: $\neg(M(\text{Arlene}, \text{Sarah}) \lor T(\text{Arlene}, \text{Sarah}))$ or $\neg M(\text{Arlene}, \text{Sarah}) \land \neg T(\text{Arlene}, \text{Sarah})$.

c) Clearly this is simply $\neg M(\text{Deborah}, \text{Jose})$.

d) Note that this statement includes the assertion that Ken has sent himself a message: $\forall x\, M(x, \text{Ken})$.

e) We can write this in two equivalent ways, depending on whether we want to say that everyone has failed to phone Nina or to say that there does not exist someone who has phoned her: $\forall x \neg T(x, \text{Nina})$ or $\neg \exists x\, T(x, \text{Nina})$.

f) This is almost identical to part **(d)**: $\forall x(T(x, \text{Avi}) \lor M(x, \text{Avi}))$.

g) To get the "else" in there, we have to make sure that y is different from x in our answer: $\exists x \forall y(y \neq x \to M(x,y))$.

h) This is almost identical to part **(g)**: $\exists x \forall y(y \neq x \to (M(x,y) \lor T(x,y)))$.

i) We need to assert the existence of two distinct people who have sent e-mail both ways: $\exists x \exists y(x \neq y \land M(x,y) \land M(y,x))$.

j) Only one variable is needed: $\exists x\, M(x,x)$.

k) This poor soul (x in our expression) has a certain thing happen for every person y other than himself: $\exists x \forall y(x \neq y \to (\neg M(y,x) \land \neg T(y,x)))$.

l) Here y is "another student": $\forall x \exists y(x \neq y \land (M(y,x) \lor T(y,x)))$.

m) This is almost identical to part **(i)**: $\exists x \exists y(x \neq y \land M(x,y) \land T(y,x))$.

n) Note how the "everyone else" means someone different from both x and y in our expression (and note that there are four possibilities for how each such person z might be contacted): $\exists x \exists y(x \neq y \land \forall z((z \neq x \land z \neq y) \to (M(x,z) \lor M(y,z) \lor T(x,z) \lor T(y,z))))$.

15. The answers presented here are not the only ones possible; other answers can be obtained using different predicates and different variables, or by varying the domain (universe of discourse).

a) $\forall x N(x, \text{discrete mathematics})$, where $N(x,y)$ is "computer science x needs a course in subject y"

b) $\exists x O(x, \text{personal computer})$, where $O(x,y)$ is "x owns y," and the domain for x is students in this class

c) $\forall x \exists y P(x,y)$, where $P(x,y)$ is "x has taken y"; x ranges over students in this class, and y ranges over computer science courses

d) $\exists x \exists y P(x,y)$, with the environment of part **(c)** (i.e., the same definition of P and the same domain)

e) $\forall x \forall y P(x,y)$, where $P(x,y)$ is "x has been in y"; x ranges over students in this class, and y ranges over buildings on campus

f) $\exists x \exists y \forall z (P(z,y) \rightarrow Q(x,z))$, where $P(z,y)$ is "z is in y" and $Q(x,z)$ is "x has been in z"; x ranges over students in this class, y ranges over buildings on campus, and z ranges over rooms

g) $\forall x \forall y \exists z (P(z,y) \wedge Q(x,z))$, with the environment of part **(f)**

17. a) We need to rule out the possibility that the user has access to another mailbox different from the one that is guaranteed: $\forall u \exists m (A(u,m) \wedge \forall n (n \neq m \rightarrow \neg A(u,n)))$, where $A(u,m)$ means that user u has access to mailbox m.

b) $\exists p \forall e (H(e) \rightarrow S(p, \text{running})) \rightarrow S(\text{kernel}, \text{working correctly})$, where $H(e)$ means that error condition e is in effect and $S(x,y)$ means that the status of x is y. Obviously there are other ways to express this with different choices of predicates. Note that "only if" is the converse of "if," so the kernel's working properly is the conclusion, not the hypothesis.

c) $\forall u \forall s (E(s, \texttt{.edu}) \rightarrow A(u,s))$, where $E(s,x)$ means that website s has extension x, and $A(u,s)$ means that user u can access website s

d) This is tricky, because we have to interpret the English sentence first, and different interpretations would lead to different answers. We will assume that the specification is that there exist two distinct systems such that they monitor every remote server, and no other system has the property of monitoring every remote system. Thus our answer is $\exists x \exists y (x \neq y \wedge \forall z ((\forall s\, M(z,s)) \leftrightarrow (z = x \vee z = y)))$, where $M(a,b)$ means that system a monitors remote server b. Note that the last part of our expression serves two purposes—it says that x and y do monitor all servers, and it says that no other system does. There are at least two other interpretations of this sentence, which would lead to different legitimate answers.

19. a) $\forall x \forall y ((x < 0) \wedge (y < 0) \rightarrow (x + y < 0))$

b) What does "necessarily" mean in this context? The best explanation is to assert that a certain universal conditional statement is not true. So we have $\neg \forall x \forall y ((x > 0) \wedge (y > 0) \rightarrow (x - y > 0))$. Note that we do not want to put the negation symbol inside (it is not true that the difference of two positive integers is never positive), nor do we want to negate just the conclusion (it is not true that the sum is always nonpositive). We could rewrite our solution by passing the negation inside, obtaining $\exists x \exists y ((x > 0) \wedge (y > 0) \wedge (x - y \leq 0))$.

c) $\forall x \forall y\, (x^2 + y^2 \geq (x+y)^2)$

d) $\forall x \forall y\, (|xy| = |x||y|)$

21. $\forall x \exists a \exists b \exists c \exists d\, ((x > 0) \rightarrow x = a^2 + b^2 + c^2 + d^2)$, where the domain (universe of discourse) consists of all integers

23. a) $\forall x \forall y ((x < 0) \wedge (y < 0) \rightarrow (xy > 0))$ **b)** $\forall x\, (x - x = 0)$

c) To say that there are exactly two objects that meet some condition, we must have two existentially quantified variables to represent the two objects, we must say that they are different, and then we must say that an object meets the conditions if and only if it is one of those two. In this case we have $\forall x \exists a \exists b\, (a \neq b \wedge \forall c (c^2 = x \leftrightarrow (c = a \vee c = b)))$.

d) $\forall x\, ((x < 0) \rightarrow \neg \exists y\, (x = y^2))$

25. a) This says that there exists a real number x such that for every real number y, the product xy equals y. That is, there is a multiplicative identity for the real numbers. This is a true statement, since $x = 1$ is the identity.

b) The product of two negative real numbers is always a positive real number.

c) There exist real numbers x and y such that x^2 exceeds y but x is less than y. This is true, since we can take $x = 2$ and $y = 3$, for instance.

d) This says that for every pair of real numbers x and y, there exists a real number z that is their sum. In other words, the real numbers are closed under the operation of addition, another true fact. (Some authors would include the uniqueness of z as part of the meaning of the word *closed*.)

27. Recall that the integers include the positive and negative integers and 0.

a) The import of this statement is that no matter how large n might be, we can always find an integer m bigger than n^2. This is certainly true; for example, we could always take $m = n^2 + 1$.

b) This statement is asserting that there is an n that is smaller than the square of *every* integer; note that n is not allowed to depend on m, since the existential quantifier comes first. This statement is true, since we could take, for instance, $n = -3$, and then n would be less than every square, since squares are always greater than or equal to 0.

c) Note the order of quantifiers: m here is allowed to depend on n. Since we can take $m = -n$, this statement is true (additive inverses exist for the integers).

d) Here one n must work for all m. Clearly $n = 1$ does the trick, so the statement is true.

e) The statement is that the equation $n^2 + m^2 = 5$ has a solution over the integers. This is true; in fact there are eight solutions, namely $n = \pm 1$, $m = \pm 2$, and vice versa.

f) The statement is that the equation $n^2 + m^2 = 6$ has a solution over the integers. There are only a small finite number of cases to try, since if $|m|$ or $|n|$ were bigger than 2 then the left-hand side would be bigger than 6. A few minutes reflection shows that in fact there is no solution, so the existential statement is false.

g) The statement is that the system of equations $\{n + m = 4,\ n - m = 1\}$ has a solution over the integers. By algebra we see that there is a unique solution to this system, namely $n = 2\frac{1}{2}$, $m = 1\frac{1}{2}$. Since there do not exist *integers* that make the equations true, the statement is false.

h) The statement is that the system of equations $\{n + m = 4,\ n - m = 2\}$ has a solution over the integers. By algebra we see that there is indeed an integral solution to this system, namely $n = 3$, $m = 1$. Therefore the statement is true.

i) This statement says that the average of two integers is always an integer. If we take $m = 1$ and $n = 2$, for example, then the only p for which $p = (m + n)/2$ is $p = 1\frac{1}{2}$, which is not an integer. Therefore the statement is false.

29. **a)** $P(1,1) \wedge P(1,2) \wedge P(1,3) \wedge P(2,1) \wedge P(2,2) \wedge P(2,3) \wedge P(3,1) \wedge P(3,2) \wedge P(3,3)$

b) $P(1,1) \vee P(1,2) \vee P(1,3) \vee P(2,1) \vee P(2,2) \vee P(2,3) \vee P(3,1) \vee P(3,2) \vee P(3,3)$

c) $(P(1,1) \wedge P(1,2) \wedge P(1,3)) \vee (P(2,1) \wedge P(2,2) \wedge P(2,3)) \vee (P(3,1) \wedge P(3,2) \wedge P(3,3))$

d) $(P(1,1) \vee P(2,1) \vee P(3,1)) \wedge (P(1,2) \vee P(2,2) \vee P(3,2)) \wedge (P(1,3) \vee P(2,3) \vee P(3,3))$

Note the crucial difference between parts **(c)** and **(d)**.

31. As we push the negation symbol toward the inside, each quantifier it passes must change its type. For logical connectives we either use De Morgan's laws or recall that $\neg(p \to q) \equiv p \wedge \neg q$.

a)
$$\begin{aligned}\neg\forall x\exists y\forall z\, T(x,y,z) &\equiv \exists x\neg\exists y\forall z\, T(x,y,z)\\ &\equiv \exists x\forall y\neg\forall z\, T(x,y,z)\\ &\equiv \exists x\forall y\exists z\, \neg T(x,y,z)\end{aligned}$$

b)
$$\begin{aligned}\neg(\forall x\exists y\, P(x,y) \vee \forall x\exists y\, Q(x,y)) &\equiv \neg\forall x\exists y\, P(x,y) \wedge \neg\forall x\exists y\, Q(x,y)\\ &\equiv \exists x\neg\exists y\, P(x,y) \wedge \exists x\neg\exists y\, Q(x,y)\\ &\equiv \exists x\forall y\, \neg P(x,y) \wedge \exists x\forall y\, \neg Q(x,y)\end{aligned}$$

c)
$$\begin{aligned}\neg\forall x\exists y(P(x,y) \wedge \exists z\, R(x,y,z)) &\equiv \exists x\neg\exists y(P(x,y) \wedge \exists z\, R(x,y,z))\\ &\equiv \exists x\forall y\, \neg(P(x,y) \wedge \exists z\, R(x,y,z))\\ &\equiv \exists x\forall y(\neg P(x,y) \vee \neg\exists z\, R(x,y,z))\\ &\equiv \exists x\forall y(\neg P(x,y) \vee \forall z\, \neg R(x,y,z))\end{aligned}$$

d)

$$\begin{aligned}\neg\forall x\exists y(P(x,y)\to Q(x,y)) &\equiv \exists x\,\neg\exists y(P(x,y)\to Q(x,y))\\ &\equiv \exists x\forall y\,\neg(P(x,y)\to Q(x,y))\\ &\equiv \exists x\forall y(P(x,y)\wedge\neg Q(x,y))\end{aligned}$$

33. We need to use the transformations shown in Table 2 of Section 1.3, replacing $\neg\forall$ by $\exists\neg$, and replacing $\neg\exists$ by $\forall\neg$. In other words, we push all the negation symbols inside the quantifiers, changing the sense of the quantifiers as we do so, because of the equivalences in Table 2 of Section 1.3. In addition, we need to use De Morgan's laws (Section 1.2) to change the negation of a conjunction to the disjunction of the negations and to change the negation of a disjunction to the conjunction of the negations. We also use the double negation law.

a) $\exists x\exists y\,\neg P(x,y)$ **b)** $\exists y\forall x\,\neg P(x,y)$

c) We can think of this in two steps. First we transform the expression into the equivalent expression $\exists y\exists x\,\neg(P(x,y)\vee Q(x,y))$, and then we use De Morgan's law to rewrite this as $\exists y\exists x(\neg P(x,y)\wedge\neg Q(x,y))$.

d) First we apply De Morgan's law to write this as a disjunction: $(\neg\exists x\exists y\,\neg P(x,y))\vee(\neg\forall x\forall y\,Q(x,y))$. Then we push the negation inside the quantifiers, and note that the two negations in front of P then cancel out $(\neg\neg P(x,y)\equiv P(x,y))$. So our final answer is $(\forall x\forall y\,P(x,y))\vee(\exists x\exists y\,\neg Q(x,y))$.

e) First we push the negation inside the outer universal quantifier, then apply De Morgan's law, and finally push it inside the inner quantifiers: $\exists x\,\neg(\exists y\forall z\,P(x,y,z)\wedge\exists z\forall y\,P(x,y,z))\equiv\exists x(\neg\exists y\forall z\,P(x,y,z)\vee\neg\exists z\forall y\,P(x,y,z))\equiv\exists x(\forall y\exists z\,\neg P(x,y,z)\vee\forall z\exists y\,\neg P(x,y,z))$.

35. If the domain (universe of discourse) has at least four members, then no matter what values are assigned to x, y, and z, there will always be another member of the domain, different from those three, that we can assign to w to make the statement true. Thus we can use a domain such as United States Senators. On the other hand, for any domain with three or fewer members, if we assign all the members to x, y, and z (repeating some if necessary), then there will be nothing left to assign to w to make the statement true. For this we can use a domain such as your biological parents.

37. In each case we need to specify some predicates and identify the domain (universe of discourse).

a) To get into the spirit of the problem, we should let $T(x,y)$ be the predicate that x has taken y, where x ranges over students in this class and y ranges over mathematics classes at this school. Then our original statement is $\forall x\exists y\exists z(y\neq z\wedge T(x,y)\wedge T(x,z)\wedge\forall w(T(x,w)\to(w=y\vee w=z)))$. Here y and z are the two math classes that x has taken, and our statement says that these are different and that if x has taken any math class w, then w is one of these two. We form the negation by using Table 2 of Section 1.3 and De Morgan's law to push the negation symbol that we place before the entire expression inwards, to achieve $\exists x\forall y\forall z(y=z\vee\neg T(x,y)\vee\neg T(x,z)\vee\exists w(T(x,w)\wedge w\neq y\wedge w\neq z))$. This can also be expressed as $\exists x\forall y\forall z(y\neq z\to(\neg T(x,y)\vee\neg T(x,z)\vee\exists w(T(x,w)\wedge w\neq y\wedge w\neq z)))$. Note that we formed the negation of a conditional statement by asserting that the hypothesis was true and the conclusion was false. In simple English, this last statement reads "There is someone in this class for whom no matter which two distinct math courses you consider, these are not the two and only two math courses this person has taken."

b) Let $V(x,y)$ be the predicate that x has visited y, where x ranges over people and y ranges over countries. The statement seems to be asserting that the person identified here has visited country y if and only if y is not Libya. So we can write this symbolically as $\exists x\forall y(V(x,y)\leftrightarrow y\neq \text{Libya})$. One way to form the negation of $P\leftrightarrow Q$ is to write $P\leftrightarrow\neg Q$; this can be seen by looking at truth tables. Thus the negation is $\forall x\exists y(V(x,y)\leftrightarrow y=\text{Libya})$. Note that there are two ways for a biconditional to be true; therefore in English this reads "For every person there is a country such that either that country is Libya and the person has visited it, or else that country is not Libya and the person has not visited it." More simply, "For every person, either that person has visited Libya or else that person has failed to visit some country other than Libya." If we are willing to keep the negation in front of the quantifier in English, then of course we could just say "There is nobody who has visited every country except Libya," but that would not be in the spirit of the exercise.

c) Let $C(x,y)$ be the predicate that x has climbed y, where x ranges over people and y ranges over mountains in the Himalayas. Our statement is $\neg\exists x\forall y\, C(x,y)$. Its negation is, of course, simply $\exists x\forall y\, C(x,y)$. In English this reads "Someone has climbed every mountain in the Himalayas."

d) There are different ways to approach this, depending on how many variables we want to introduce. Let $M(x,y,z)$ be the predicate that x has been in movie z with y, where the domains for x and y are movie actors, and for z is movies. The statement then reads: $\forall x((\exists z\, M(x, \text{Kevin Bacon}, z)) \vee (\exists y \exists z_1 \exists z_2 (M(x,y,z_1) \wedge M(y, \text{Kevin Bacon}, z_2))))$. The negation is formed in the usual manner: $\exists x((\forall z\, \neg M(x, \text{Kevin Bacon}, z)) \wedge (\forall y \forall z_1 \forall z_2 (\neg M(x,y,z_1) \vee \neg M(y, \text{Kevin Bacon}, z_2))))$. In simple English this means that there is someone who has neither been in a movie with Kevin Bacon nor been in a movie with someone who has been in a movie with Kevin Bacon.

39. **a)** Since the square of a number and its additive inverse are the same, we have many counterexamples, such as $x=2$ and $y=-2$.

b) This statement is saying that every number has a square root. If x is negative (like $x=-4$), or, since we are working in the domain of the integers, x is not a perfect square (like $x=6$), then the equation $y^2=x$ has no solution.

c) Since negative numbers are not larger than positive numbers, we can take something like $x=17$ and $y=-1$ for our counterexample.

41. We simply want to say that a certain equation holds for all real numbers: $\forall x\forall y\forall z((x\cdot y)\cdot z = x\cdot(y\cdot z))$.

43. We want to say that for each pair of coefficients (the m and the b in the expression $mx+b$), as long as m is not 0, there is a unique x making that expression equal to 0. So we write $\forall m\forall b(m\neq 0 \rightarrow \exists x(mx+b=0 \wedge \forall w(mw+b=0 \rightarrow w=x)))$. Notice that the uniqueness is expressed by the last part of our proposition.

45. This statement says that every number has a multiplicative inverse.

a) In the universe of nonzero real numbers, this is certainly true. In each case we let $y=1/x$.

b) Integers usually don't have inverses that are integers. If we let $x=3$, then no integer y satisfies $xy=1$. Thus in this setting, the statement is false.

c) As in part **(a)** this is true, since $1/x$ is positive when x is positive.

47. We use the equivalences explained in Table 2 of Section 1.3, twice:
$\neg\exists x\forall y P(x,y) \equiv \forall x\neg\forall y P(x,y) \equiv \forall x\exists y\neg P(x,y)$

49. **a)** We prove this by arguing that whenever the first proposition is true, so is the second; and that whenever the second proposition is true, so is the first. So suppose that $\forall x P(x) \wedge \exists x Q(x)$ is true. In particular, P always holds, and there is some object, call it y, in the domain (universe of discourse) that makes Q true. Now to show that the second proposition is true, suppose that x is any object in the domain. By our assumptions, $P(x)$ is true. Furthermore, $Q(y)$ is true for the particular y we mentioned above. Therefore $P(x)\wedge Q(y)$ is true for this x and y. Since x was arbitrary, we have showed that $\forall x\exists y(P(x)\wedge Q(y))$ is true, as desired. Conversely, suppose that the second proposition is true. Letting x be any member of the domain allows us to assert that there exists a y such that $P(x)\wedge Q(y)$ is true, and therefore $Q(y)$ is true. Thus by the definition of existential quantifiers, $\exists x Q(x)$ is true. Furthermore, our hypothesis tells us in particular that $\forall x P(x)$ is true. Therefore the first proposition, $\forall x P(x) \wedge \exists x Q(x)$ is true.

b) This is similar to part **(a)**. Suppose that $\forall x P(x) \vee \exists x Q(x)$ is true. Thus either P always holds, or there is some object, call it y, in the domain that makes Q true. In the first case it follows that $P(x) \vee Q(y)$ is true for all x, and so we can conclude that $\forall x \exists y (P(x) \vee Q(y))$ is true (it does not matter in this case whether $Q(y)$ is true or not). In the second case, $Q(y)$ is true for this particular y, and so $P(x) \vee Q(y)$ is true regardless of what x is. Again, it follows that $\forall x \exists y (P(x) \vee Q(y))$ is true. Conversely, suppose that the second proposition is true. If $P(x)$ is true for all x, then the first proposition must be true. If not, then $P(x)$ fails for some x, but for this x there must be a y such that $P(x) \vee Q(y)$ is true; hence $Q(y)$ must be true. Therefore $\exists y Q(y)$ holds, and thus the first proposition is true.

51. This will essentially be a proof by (structural) mathematical induction (see Sections 4.1–4.3), where we show how a long expression can be put into prenex normal form if the subexpressions in it can be put into prenex normal form. First we invoke the result of Exercise 45 from Section 1.2 to assume without loss of generality that our given proposition uses only the logical connectives $\vee$ and $\neg$. Then every proposition must either be a single propositional variable (like P), the disjunction of two propositions, the negation of a proposition, or the universal or existential quantification of a predicate. (There is a small technical point that we are sliding over here; disjunction and negation need to be defined for predicates as well as for propositions, since otherwise we would not be able to write down such things as $\forall x (P(x) \wedge Q(x))$. We assume that all that we have done for propositions applies to predicates as well.)

Certainly every proposition that involves no quantifiers is already in prenex normal form; this is the base case of our induction. Next suppose that our proposition is of the form $\mathrm{Q} x P(x)$, where Q is a quantifier. Then $P(x)$ is a shorter expression than the given proposition, so (by the inductive hypothesis) we can put it into prenex form, with all of its quantifiers coming at the beginning. Then $\mathrm{Q}x$ followed by this prenex form is again in prenex form and is equivalent to the original proposition. Next suppose that our proposition is of the form $\neg P$. Again, we can invoke the inductive hypothesis and assume that P is already in prenex form, with all of its quantifiers coming at its front. We now slide the negation symbol past all the quantifiers, using the equivalences in Table 2 of Section 1.3. For example, $\neg \forall x \exists y R(x,y)$ becomes $\exists x \forall y \neg R(x,y)$, which is in prenex form.

Finally, suppose that our given proposition is a disjunction of two propositions, $P \vee Q$, each of which can (again by the inductive hypothesis) be assumed to be in prenex normal form, with their quantifiers at the front. There are several cases. If only one of P and Q has quantifiers, then we invoke the result of Exercise 46 of Section 1.3 to bring the quantifier in front of both. We then apply our process to what remains. For example, $P \vee \forall x Q(x)$ is equivalent to $\forall x (P \vee Q(x))$, and then $P \vee Q(x)$ is put into prenex form. Another case is that the proposition might look like $\exists x R(x) \vee \exists x S(x)$. In this case, by Exercise 45 of Section 1.3, the proposition is equivalent to $\exists x (R(x) \vee S(x))$. Once again, by the inductive hypothesis we can then put $R(x) \vee S(x)$ into prenex form, and so the entire proposition can be put into prenex form. Similarly, using Exercise 48 of the present section we can transform $\forall x R(x) \vee \forall x S(x)$ into the equivalent $\forall x \forall y (R(x) \vee S(y))$; putting $R(x) \vee S(y)$ into prenex form then brings the entire proposition into prenex form. Finally, if the proposition is of the form $\forall x R(x) \vee \exists x Q(x)$, then we invoke Exercise 49b of the present section and apply the same construction.

Note that this proof actually gives us the process for finding the proposition in prenex form equivalent to the given proposition—we just work from the inside out, dealing with one logical operation or quantifier at a time. Here is an example:

$$\begin{aligned}
\forall x P(x) \vee \neg \exists x (Q(x) \vee \forall y R(x,y)) &\equiv \forall x P(x) \vee \neg \exists x \forall y (Q(x) \vee R(x,y)) \\
&\equiv \forall x P(x) \vee \forall x \exists y \neg (Q(x) \vee R(x,y)) \\
&\equiv \forall x \forall z (P(x) \vee \exists y \neg (Q(z) \vee R(z,y))) \\
&\equiv \forall x \forall z \exists y (P(x) \vee \neg (Q(z) \vee R(z,y)))
\end{aligned}$$

SECTION 1.5 Rules of Inference

This section lays the groundwork for understanding proofs. You are asked to understand the logical rules of inference behind valid arguments, and you are asked to construct some highly stylized proofs using these rules. The proofs will become more informal in the next section and throughout the remainder of this book (and your mathematical studies).

1. This is modus ponens. The first statement is $p \to q$, where p is "Socrates is human" and q is "Socrates is mortal." The second statement is p. The third is q. Modus ponens is valid. We can therefore conclude that the conclusion of the argument (third statement) is true, because the hypotheses (the first two statements) are true.

3. a) This is the addition rule. We are concluding from p that $p \vee q$ must be true, where p is "Alice is a mathematics major" and q is "Alice is a computer science major."
b) This is the simplification rule. We are concluding from $p \wedge q$ that p must be true, where p is "Jerry is a mathematics major" and q is "Jerry is a computer science major.".
c) This is modus ponens. We are concluding from $p \to q$ and p that q must be true, where p is "it is rainy" and q is "the pool will be closed."
d) This is modus tollens. We are concluding from $p \to q$ and $\neg q$ that $\neg p$ must be true, where p is "it will snow today" and q is "the university will close today."
e) This is hypothetical syllogism. We are concluding from $p \to q$ and $q \to r$ that $p \to r$ must be true, where p is "I will go swimming," q is "I will stay in the sun too long," and r is "I will sunburn."

5. Let w be the proposition "Randy works hard," let d be the proposition "Randy is a dull boy," and let j be the proposition "Randy will get the job." We are given premises w, $w \to d$, and $d \to \neg j$. We want to conclude $\neg j$. We set up the proof in two columns, with reasons, as in Example 6.

Step	Reason
1. w	Hypothesis
2. $w \to d$	Hypothesis
3. d	Modus ponens using (2) and (3)
4. $d \to \neg j$	Hypothesis
5. $\neg j$	Modus ponens using (3) and (4)

7. First we use universal instantiation to conclude from "For all x, if x is a man, then x is mortal" the special case of interest, "If Socrates is a man, then Socrates is mortal." Then we use modus ponens to conclude that Socrates is mortal.

9. a) Because it was sunny on Tuesday, we assume that it did not rain or snow on Tuesday (otherwise we cannot do anything with this problem). If we use modus tollens on the universal instantiation of the given conditional statement applied to Tuesday, then we conclude that I did not take Tuesday off. If we now apply disjunctive syllogism to the disjunction in light of this conclusion, we see that I took Thursday off. Now use modus ponens on the universal instantiation of the given conditional statement applied to Thursday; we conclude that it rained or snowed on Thursday. One more application of disjunctive syllogism tells us that it rained on Thursday.
b) Using modus tollens we conclude two things—that I did not eat spicy food and that it did not thunder. Therefore by the conjunction rule of inference (Table 1), we conclude "I did not eat spicy food and it did not thunder."
c) By disjunctive syllogism from the first two hypotheses we conclude that I am clever. The third hypothesis gives us no useful information.

d) We can apply universal instantiation to the conditional statement and conclude that if Ralph (respectively, Ann) is a CS major, then he (she) has a PC. Now modus tollens tells us that Ralph is not a CS major. There are no conclusions to be drawn about Ann.

e) The first two conditional statements can be phrased as "If x is good for corporations, then x is good for the U.S." and "If x is good for the U.S., then x is good for you." If we now apply universal instantiation with x being "for you to buy lots of stuff," then we can conclude using modus ponens twice that for you to buy lots of stuff is good for the U.S. and is good for you.

f) The given conditional statement is "For all x, if x is a rodent, then x gnaws its food." We can form the universal instantiation of this with x being a mouse, a rabbit, and a bat. Then modus ponens allows us to conclude that mice gnaw their food; and modus tollens allows us to conclude that rabbits are not rodents. We can conclude nothing about bats.

11. We are asked to show that whenever $p_1, p_2, \ldots, p_n$ are true, then $q \to r$ must be true, given that we know that whenever $p_1, p_2, \ldots, p_n$ and q are true, then r must be true. So suppose that $p_1, p_2, \ldots, p_n$ are true. We want to establish that $q \to r$ is true. If q is false, then we are done, vacuously. Otherwise, q is true, so by the validity of the given argument form, we know that r is true.

13. In each case we set up the proof in two columns, with reasons, as in Example 6.

a) Let $c(x)$ be "x is in this class," let $j(x)$ be "x knows how to write programs in JAVA," and let $h(x)$ be "x can get a high paying job." We are given premises $c(\text{Doug})$, $j(\text{Doug})$, and $\forall x(j(x) \to h(x))$, and we want to conclude $\exists x(c(x) \wedge h(x))$.

Step	Reason
1. $\forall x(j(x) \to h(x))$	Hypothesis
2. $j(\text{Doug}) \to h(\text{Doug})$	Universal instantiation using (1)
3. $j(\text{Doug})$	Hypothesis
4. $h(\text{Doug})$	Modus ponens using (2) and (3)
5. $c(\text{Doug})$	Hypothesis
6. $c(\text{Doug}) \wedge h(\text{Doug})$	Conjunction using (4) and (5)
7. $\exists x(c(x) \wedge h(x))$	Existential generalization using (6)

b) Let $c(x)$ be "x is in this class," let $w(x)$ be "x enjoys whale watching," and let $p(x)$ be "x cares about ocean pollution." We are given premises $\exists x(c(x) \wedge w(x))$ and $\forall x(w(x) \to p(x))$, and we want to conclude $\exists x(c(x) \wedge p(x))$. In our proof, y represents an unspecified particular person.

Step	Reason
1. $\exists x(c(x) \wedge w(x))$	Hypothesis
2. $c(y) \wedge w(y)$	Existential instantiation using (1)
3. $w(y)$	Simplification using (2)
4. $c(y)$	Simplification using (2)
5. $\forall x(w(x) \to p(x))$	Hypothesis
6. $w(y) \to p(y)$	Universal instantiation using (5)
7. $p(y)$	Modus ponens using (3) and (6)
8. $c(y) \wedge p(y)$	Conjunction using (4) and (7)
9. $\exists x(c(x) \wedge p(x))$	Existential generalization using (8)

c) Let $c(x)$ be "x is in this class," let $p(x)$ be "x owns a PC," and let $w(x)$ be "x can use a word processing program." We are given premises $c(\text{Zeke})$, $\forall x(c(x) \to p(x))$, and $\forall x(p(x) \to w(x))$, and we want to conclude $w(\text{Zeke})$.

Step	**Reason**
1. $\forall x(c(x) \to p(x))$	Hypothesis
2. $c(\text{Zeke}) \to p(\text{Zeke})$	Universal instantiation using (1)
3. $c(\text{Zeke})$	Hypothesis
4. $p(\text{Zeke})$	Modus ponens using (2) and (3)
5. $\forall x(p(x) \to w(x))$	Hypothesis
6. $p(\text{Zeke}) \to w(\text{Zeke})$	Universal instantiation using (5)
7. $w(\text{Zeke})$	Modus ponens using (4) and (6)

d) Let $j(x)$ be "x is in New Jersey," let $f(x)$ be "x lives within fifty miles of the ocean," and let $s(x)$ be "x has seen the ocean." We are given premises $\forall x(j(x) \to f(x))$ and $\exists x(j(x) \wedge \neg s(x))$, and we want to conclude $\exists x(f(x) \wedge \neg s(x))$. In our proof, y represents an unspecified particular person.

Step	**Reason**
1. $\exists x(j(x) \wedge \neg s(x))$	Hypothesis
2. $j(y) \wedge \neg s(y)$	Existential instantiation using (1)
3. $j(y)$	Simplification using (2)
4. $\forall x(j(x) \to f(x))$	Hypothesis
5. $j(y) \to f(y)$	Universal instantiation using (4)
6. $f(y)$	Modus ponens using (3) and (5)
7. $\neg s(y)$	Simplification using (2)
8. $f(y) \wedge \neg s(y)$	Conjunction using (6) and (7)
9. $\exists x(f(x) \wedge \neg s(x))$	Existential generalization using (8)

15. **a)** This is correct, using universal instantiation and modus ponens.
b) This is invalid. After applying universal instantiation, it contains the fallacy of affirming the conclusion.
c) This is invalid. After applying universal instantiation, it contains the fallacy of denying the hypothesis.
d) This is valid by universal instantiation and modus tollens.

17. We know that *some* x exists that makes $H(x)$ true, but we cannot conclude that Lola is one such x. Maybe only Suzanne is happy and everyone else is not happy. Then $\exists x\, H(x)$ is true, but $H(\text{Lola})$ is false.

19. **a)** This is the fallacy of affirming the conclusion, since it has the form "$p \to q$ and q implies p."
b) This reasoning is valid; it is modus tollens.
c) This is the fallacy of denying the hypothesis, since it has the form "$p \to q$ and $\neg p$ implies $\neg q$."

21. Let us give an argument justifying the conclusion. By the second premise, there is some lion that does not drink coffee. Let us call him Leo. Thus we know that Leo is a lion and that Leo does not drink coffee. By simplification this allows us to assert each of these statements separately. The first premise says that all lions are fierce; in particular, if Leo is a lion, then Leo is fierce. By modus ponens, we can conclude that Leo is fierce. Thus we conclude that Leo is fierce and Leo does not drink coffee. By the definition of the existential quantifier, this tells us that there exist fierce creatures that do not drink coffee; in other words, that some fierce creatures do not drink coffee.

23. The error occurs in step (5), because we cannot assume, as is being done here, that the c that makes P true is the same as the c that makes Q true.

25. We are given the premises $\forall x(P(x) \to Q(x))$ and $\neg Q(a)$. We want to show $\neg P(a)$. Suppose, to the contrary, that $\neg P(a)$ is not true. Then $P(a)$ is true. Therefore by universal modus ponens, we have $Q(a)$. But this contradicts the given premise $\neg Q(a)$. Therefore our supposition must have been wrong, and so $\neg P(a)$ is true, as desired.

27. We can set this up in two-column format.

Step	**Reason**
1. $\forall x(P(x) \wedge R(x))$	Premise
2. $P(a) \wedge R(a)$	Universal instantiation using (1)
3. $P(a)$	Simplification using (2)
4. $\forall x(P(x) \rightarrow (Q(x) \wedge S(x)))$	Premise
5. $Q(a) \wedge S(a)$	Universal modus ponens using (3) and (4)
6. $S(a)$	Simplification using (5)
7. $R(a)$	Simplification using (2)
8. $R(a) \wedge S(a)$	Conjunction using (7) and (6)
9. $\forall x(R(x) \wedge S(x))$	Universal generalization using (5)

29. We can set this up in two-column format. The proof is rather long but straightforward if we go one step at a time.

Step	**Reason**
1. $\exists x \neg P(x)$	Premise
2. $\neg P(c)$	Existential instantiation using (1)
3. $\forall x(P(x) \vee Q(x))$	Premise
4. $P(c) \vee Q(c)$	Universal instantiation using (3)
5. $Q(c)$	Disjunctive syllogism using (4) and (2)
6. $\forall x(\neg Q(x) \vee S(x))$	Premise
7. $\neg Q(c) \vee S(c)$	Universal instantiation using (6)
8. $S(c)$	Disjunctive syllogism using (5) and (7), since $\neg\neg Q(c) \equiv Q(c)$
9. $\forall x(R(x) \rightarrow \neg S(x))$	Premise
10. $R(c) \rightarrow \neg S(c)$	Universal instantiation using (9)
11. $\neg R(c)$	Modus tollens using (8) and (10), since $\neg\neg S(c) \equiv S(c)$
12. $\exists x \neg R(x)$	Existential generalization using (11)

31. Let p be "It is raining"; let q be "Yvette has her umbrella"; let r be "Yvette gets wet." Then our assumptions are $\neg p \vee q$, $\neg q \vee \neg r$, and $p \vee \neg r$. Using resolution on the first two assumptions gives us $\neg p \vee \neg r$. Using resolution on this and the third assumption gives us $\neg r$, so Yvette does not get wet.

33. Assume that this proposition is satisfiable. Using resolution on the first two clauses allows us to conclude $q \vee q$; in other words, we know that q has to be true. Using resolution on the last two clauses allows us to conclude $\neg q \vee \neg q$; in other words, we know that $\neg q$ has to be true. This is a contradiction. So this proposition is not satisfiable.

35. This argument is valid. We argue by contradiction. Assume that Superman does exist. Then he is not impotent, and he is not malevolent (this follows from the fourth sentence). Therefore by (the contrapositives of) the two parts of the second sentence, we conclude that he is able to prevent evil, and he is willing to prevent evil. By the first sentence, we therefore know that Superman does prevent evil. This contradicts the third sentence. Since we have arrived at a contradiction, our original assumption must have been false, so we conclude finally that Superman does not exist.

SECTION 1.6 Introduction to Proofs

This introduction applies jointly to this section and the next (1.7).

Learning to construct good mathematical proofs takes years. There is no algorithm for constructing the proof of a true proposition (there is actually a deep theorem in mathematical logic that says this). Instead, the construction of a valid proof is an art, honed after much practice. There are two problems for the beginning student—figuring out the key ideas in a problem (what is it that really makes the proposition true?) and writing down the proof in acceptable mathematical language.

Here are some general things to keep in mind in constructing proofs. First, of course, you need to find out exactly what is going on—why the proposition is true. This can take anywhere from ten seconds (for a really simple proposition) to a lifetime (some mathematicians have spent their entire careers trying to prove certain conjectures). For a typical student at this level, tackling a typical problem, the median might be somewhere around 15 minutes. This time should be spent looking at examples, making tentative assumptions, breaking the problem down into cases, perhaps looking at analogous but simpler problems, and in general bringing all of your mathematical intuition and training to bear.

It is often easiest to give a proof by contradiction, since you get to assume the most (all the hypotheses as well as the negation of the conclusion), and all you have to do is to derive a contradiction. Another thing to try early in attacking a problem is to separate the proposition into several cases; proof by cases is a valid technique, if you make sure to include all the possibilities. In proving propositions, all the rules of inference are at your disposal, as well as axioms and previously proved results. Ask yourself what definitions, axioms, or other theorems might be relevant to the problem at hand. The importance of constantly returning to the definitions cannot be overstated!

Once you think you see what is involved, you need to write down the proof. In doing so, pay attention both to content (does each statement follow logically? are you making any fallacious arguments? are you leaving out any cases or using hidden assumptions?) and to style. There are certain conventions in mathematical proofs, and you need to follow them. For example, you must use complete sentences and say exactly what you mean. (An equation is a complete sentence, with "equals" as the verb; however, a good proof will usually have more English words than mathematical symbols in it.) The point of a proof is to convince the reader that your line of argument is sound, and that therefore the proposition under discussion is true; put yourself in the reader's shoes, and ask yourself whether you are convinced.

Most of the proofs called for in this exercise set are not extremely difficult. Nevertheless, expect to have a fairly rough time constructing proofs that look like those presented in this solutions manual, the textbook, or other mathematics textbooks. The more proofs you write, utilizing the different methods discussed in this section, the better you will become at it. As a bonus, your ability to construct and respond to nonmathematical arguments (politics, religion, or whatever) will be enhanced. Good luck!

1. We must show that whenever we have two odd integers, their sum is even. Suppose that a and b are two odd integers. Then there exist integers s and t such that $a = 2s + 1$ and $b = 2t + 1$. Adding, we obtain $a + b = (2s + 1) + (2t + 1) = 2(s + t + 1)$. Since this represents $a + b$ as 2 times the integer $s + t + 1$, we conclude that $a + b$ is even, as desired.

3. We need to prove the following assertion for an arbitrary integer n: "If n is even, then n^2 is even." Suppose that n is even. Then $n = 2k$ for some integer k. Therefore $n^2 = (2k)^2 = 4k^2 = 2(2k^2)$. Since we have written n^2 as 2 times an integer, we conclude that n^2 is even.

5. We can give a direct proof. Suppose that $m + n$ is even. Then $m + n = 2s$ for some integer s. Suppose that $n + p$ is even. Then $n + p = 2t$ for some integer t. If we add these [this step is inspired by the fact that we want to look at $m + p$], we get $m + p + 2n = 2s + 2t$. Subtracting $2n$ from both sides and factoring, we have

$m+p = 2s+2t-2n = 2(s+t-n)$. Since we have written $m+p$ as 2 times an integer, we conclude that $m+p$ is even, as desired.

7. The difference of two squares can be factored: $a^2-b^2 = (a+b)(a-b)$. If we can arrange for our given odd integer to equal $a+b$ and for $a-b$ to equal 1, then we will be done. But we can do this by letting a and b be the integers that straddle $n/2$. For example, if $n=11$, then we take $a=6$ and $b=5$. Specifically, if $n=2k+1$, then we let $a=k+1$ and $b=k$. Here, then, is our proof. Since n is odd, we can write $n=2k+1$ for some integer k. Then $(k+1)^2-k^2 = k^2+2k+1-k^2 = 2k+1 = n$. This expresses n as the difference of two squares.

9. The proposition to be proved here is as follows: If r is a rational number and i is an irrational number, then $s=r+i$ is an irrational number. So suppose that r is rational, i is irrational, and s is rational. Then by Example 7 the sum of the rational numbers s and $-r$ must be rational. (Indeed, if $s=a/b$ and $r=c/d$, where a, b, c, and d are integers, with $b \neq 0$ and $d \neq 0$, then by algebra we see that $s+(-r) = (ad-bc)/(bd)$, so that patently $s+(-r)$ is a rational number.) But $s+(-r) = r+i-r = i$, forcing us to the conclusion that i is rational. This contradicts our hypothesis that i is irrational. Therefore the assumption that s was rational was incorrect, and we conclude, as desired, that s is irrational.

11. To disprove this proposition it is enough to find a counterexample, since the proposition has an implied universal quantification. We know from Example 10 that $\sqrt{2}$ is irrational. If we take the product of the irrational number $\sqrt{2}$ and the irrational number $\sqrt{2}$, then we obtain the rational number 2. This counterexample refutes the proposition.

13. We give an proof by contraposition. The contrapositive of this statement is "If $1/x$ is rational, then x is rational" so we give a direct proof of this contrapositive. Note that since $1/x$ exists, we know that $x \neq 0$. If $1/x$ is rational, then by definition $1/x = p/q$ for some integers p and q with $q \neq 0$. Since $1/x$ cannot be 0 (if it were, then we'd have the contradiction $1 = x \cdot 0$ by multiplying both sides by x), we know that $p \neq 0$. Now $x = 1/(1/x) = 1/(p/q) = q/p$ by the usual rules of algebra and arithmetic. Hence x can be written as the quotient of two integers with the denominator nonzero. Thus by definition, x is rational.

15. We will prove the contrapositive (that if it is not true that $x \geq 1$ or $y \geq 1$, then it is not true that $x+y \geq 2$), using a direct argument. Assume that it is not true that $x \geq 1$ or $y \geq 1$. Then (by De Morgan's law) $x < 1$ and $y < 1$. Adding these two inequalities, we obtain $x+y < 2$. This is the negation of $x+y \geq 2$, and our proof is complete.

17. **a)** We must prove the contrapositive: If n is odd, then n^3+5 is even. Assume that n is odd. Then we can write $n = 2k+1$ for some integer k. Then $n^3+5 = (2k+1)^3+5 = 8k^3+12k^2+6k+6 = 2(4k^3+6k^2+3k+3)$. Thus n^3+5 is two times some integer, so it is even.

b) Suppose that n^3+5 is odd and that n is odd. Since n is odd, and the product of odd numbers is odd, in two steps we see that n^3 is odd. But then subtracting we conclude that 5, being the difference of the two odd numbers n^3+5 and n^3, is even. This is not true. Therefore our supposition was wrong, and the proof by contradiction is complete.

19. The proposition we are trying to prove is "If 0 is a positive integer greater than 1, then $0^2 > 0$." Our proof is a vacuous one, exactly as in Example 5. Since the hypothesis is false, the conditional statement is automatically true.

21. The proposition we are trying to prove is "If a and b are positive real numbers, then $(a+b)^1 \geq a^1 + b^1$." Our proof is a direct one. By the definition of exponentiation, any real number to the power 1 is itself. Hence $(a+b)^1 = a+b = a^1 + b^1$. Finally, by the addition rule, we can conclude from $(a+b)^1 = a^1 + b^1$ that $(a+b)^1 \geq a^1 + b^1$ (the latter being the disjunction of $(a+b)^1 = a^1 + b^1$ and $(a+b)^1 > a^1 + b^1$). One might also say that this is a trivial proof, since we did not use the hypothesis that a and b are positive (although of course we used the hypothesis that they are numbers).

23. We give a proof by contradiction. If there were nine or fewer days on each day of the week, this would account for at most $9 \cdot 7 = 63$ days. But we chose 64 days. This contradiction shows that at least ten of the days must be on the same day of the week.

25. One way to prove this is to use the rational root test from high school algebra: Every rational number that satisfies a polynomial with integer coefficients is of the form p/q, where p is a factor of the constant term of the polynomial, and q is a factor of the coefficient of the leading term. In this case, both the constant and the leading coefficient are 1, so the only possible values for p and q are ± 1. Therefore the only possible rational roots are $\pm 1/(\pm 1)$, which means that 1 and -1 are the only possible rational roots. Clearly neither of them is a root, so there are no rational roots.

Alternatively, we can follow the hint. Suppose by way of contradiction that a/b is a rational root, where a and b are integers and this fraction is in lowest terms (that is, a and b have no common divisor greater than 1). Plug this proposed root into the equation to obtain $a^3/b^3 + a/b + 1 = 0$. Multiply through by b^3 to obtain $a^3 + ab^2 + b^3 = 0$. If a and b are both odd, then the left-hand side is the sum of three odd numbers and therefore must be odd. If a is odd and b is even, then the left-hand side is odd + even + even, which is again odd. Similarly, if a is even and b is odd, then the left-hand side is even + even + odd, which is again odd. Because the fraction a/b is in simplest terms, it cannot happen that both a and b are even. Thus in all cases, the left-hand side is odd, and therefore cannot equal 0. This contradiction shows that no such root exists.

27. We must prove two conditional statements. First, we assume that n is odd and show that $5n+6$ is odd (this is a direct proof). By assumption, $n = 2k+1$ for some integer k. Then $5n+6 = 5(2k+1)+6 = 10k+11 = 2(5k+5)+1$. Since we have written $5n+6$ as 2 times an integer plus 1, we have showed that $5n+6$ is odd, as desired. Now we give an proof by contraposition of the converse. Suppose that n is not odd—in other words, that n is even. Then $n = 2k$ for some integer k. Then $5n+6 = 10k+6 = 2(5k+3)$. Since we have written $5n+6$ as 2 times an integer, we have showed that $5n+6$ is even. This completes the proof by contraposition of this conditional statement.

29. This proposition is true. We give a proof by contradiction. Suppose that m is neither 1 nor -1. Then mn has a factor (namely $|m|$) larger than 1. On the other hand, $mn = 1$, and 1 clearly has no such factor. Therefore we conclude that $m = 1$ or $m = -1$. It is then immediate that $n = 1$ in the first case and $n = -1$ in the second case, since $mn = 1$ implies that $n = 1/m$.

31. Perhaps the best way to do this is to prove that all of them are equivalent to x being even, which one can discover easily enough by trying a few small values of x. If x is even, then $x = 2k$ for some integer k. Therefore $3x+2 = 3 \cdot 2k + 2 = 6k+2 = 2(3k+1)$, which is even, since it has been written in the form $2t$, where $t = 3k+1$. Similarly, $x+5 = 2k+5 = 2k+4+1 = 2(k+2)+1$, so $x+5$ is odd; and $x^2 = (2k)^2 = 2(2k^2)$, so x^2 is even. For the converses, we will use a proof by contraposition. So assume that x is not even; thus x is odd and we can write $x = 2k+1$ for some integer k. Then $3x+2 = 3(2k+1)+2 = 6k+5 = 2(3k+2)+1$, which is odd (i.e., not even), since it has been written in the form $2t+1$, where $t = 3k+2$. Similarly, $x+5 = 2k+1+5 = 2(k+3)$, so $x+5$ is even (i.e., not odd). That x^2 is odd was already proved in Example 1. This completes the proof.

33. It is easiest to give proofs by contraposition of $(i) \rightarrow (ii)$, $(ii) \rightarrow (i)$, $(i) \rightarrow (iii)$, and $(iii) \rightarrow (i)$. For the first of these, suppose that $3x+2$ is rational, namely equal to p/q for some integers p and q with $q \neq 0$. Then we can write $x = ((p/q)-2)/3 = (p-2q)/(3q)$, where $3q \neq 0$. This shows that x is rational. For the second conditional statement, suppose that x is rational, namely equal to p/q for some integers p and q with $q \neq 0$. Then we can write $3x+2 = (3p+2q)/q$, where $q \neq 0$. This shows that $3x+2$ is rational. For the third conditional statement, suppose that $x/2$ is rational, namely equal to p/q for some integers p and q with $q \neq 0$. Then we can write $x = 2p/q$, where $q \neq 0$. This shows that x is rational. And for the fourth conditional statement, suppose that x is rational, namely equal to p/q for some integers p and q with $q \neq 0$. Then we can write $x/2 = p/(2q)$, where $2q \neq 0$. This shows that $x/2$ is rational.

35. The steps are valid for obtaining possible solutions to the equations. *If* the given equation is true, then we can conclude that $x = 1$ or $x = 6$, since the truth of each equation implies the truth of the next equation. However, the steps are not all reversible; in particular, the squaring step is not reversible. Therefore the possible answers must be checked in the original equation. We know that no other solutions are possible, but we do not know that these two numbers are in fact solutions. If we plug in $x = 1$ we get the true statement $2 = 2$; but if we plug in $x = 6$ we get the false statement $3 = -3$. Therefore $x = 1$ is the one and only solution of $\sqrt{x+3} = 3 - x$.

37. Suppose that we have proved $p_1 \rightarrow p_4 \rightarrow p_2 \rightarrow p_5 \rightarrow p_3 \rightarrow p_1$. Imagine these conditional statements arranged around a circle. Then to prove that each one of these propositions (say p_i) implies each of the others (say p_j), we just have to follow the circle, starting at p_i, until we come to p_j, using hypothetical syllogism repeatedly.

39. We can give a very satisfying proof by contradiction here. Suppose instead that all of the numbers a_1, a_2, $\ldots$, a_n are less than their average, which we can denote by A. In symbols, we have $a_i < A$ for all i. If we add these n inequalities, we see that

$$a_1 + a_2 + \cdots + a_n < nA\,.$$

By definition,

$$A = \frac{a_1 + a_2 + \cdots + a_n}{n}\,.$$

The two displayed formulae clearly contradict each other, however: they imply that $nA < nA$. Thus our assumption must have been incorrect, and we conclude that at least one of the numbers a_i is greater than or equal to their average.

41. We can prove that these four statements are equivalent in a circular way: $(i) \rightarrow (ii) \rightarrow (iii) \rightarrow (iv) \rightarrow (i)$. For the first, we want to show that if n is even, then $n+1$ is odd. Assume that n is even. Then $n = 2k$ for some integer k. Thus $n+1 = 2k+1$, so by definition $n+1$ is odd. This completes the first proof. Next we give a direct proof of $(ii) \rightarrow (iii)$. Suppose that $n+1$ is odd, say $n+1 = 2k+1$. Then $3n+1 = 2n+(n+1) = 2n+2k+1 = 2(n+k)+1$. Since this shows that $3n+1$ is 2 times an integer plus 1, we conclude that $3n+1$ is odd, as desired. For the next proof, suppose that $3n+1$ is odd, say $3n+1 = 2k+1$. Then $3n = (3n+1)-1 = (2k+1)-1 = 2k$. Therefore by definition $3n$ is even. Finally, we must prove that if $3n$ is even, then n is even. We will do this using a proof by contraposition. Suppose that n is not even. Then n is odd, so we can write $n = 2k+1$ for some integer k. Then $3n = 3(2k+1) = 6k+3 = 2(3k+1)+1$. This exhibits $3n$ as 2 times an integer plus 1, so $3n$ is odd, completing the proof by contraposition.

SECTION 1.7 Proof Methods and Strategy

The preamble to the solutions for Section 1.6 applies here as well, so you might want to reread it at this time. In addition, the section near the back of this Guide*, entitled "A Guide to Proof-Writing," provides an excellent tutorial, with many additional examples. Don't forget to take advantage of the many additional resources on the website for this text, as well.*

If you are majoring in mathematics, then proofs are the bread and butter of your field. Most likely you will take a course devoted entirely to learning how to read and write proofs, using one of the many textbooks available on this subject. For a review of many of them (as well as reviews of hundreds of mathematics books), see this site provided by the Mathematical Association of America: `http://www.maa.org/reviews/`.

1. We give an exhaustive proof—just check the entire domain. For $n = 1$ we have $1^2 + 1 = 2 \geq 2 = 2^1$. For $n = 2$ we have $2^2 + 1 = 5 \geq 4 = 2^2$. For $n = 3$ we have $3^2 + 1 = 10 \geq 8 = 2^3$. For $n = 4$ we have $4^2 + 1 = 17 \geq 16 = 2^4$. Notice that for $n \geq 5$, the inequality is no longer true.

3. Following the hint, we consider the two cases determined by the relative sizes of x and y. First suppose that $x \geq y$. Then by definition $\max(x, y) = x$ and $\min(x, y) = y$. Therefore in this case $\max(x, y) + \min(x, y) = x + y$, exactly as desired. For the second (and final) case, suppose that $x < y$. Then $\max(x, y) = y$ and $\min(x, y) = x$. Therefore in this case $\max(x, y) + \min(x, y) = y + x = x + y$, again the desired conclusion. Hence in all cases, the equality holds.

5. There are several cases to consider. If x and y are both nonnegative, then $|x| + |y| = x + y = |x + y|$. Similarly, if both are negative, then $|x| + |y| = (-x) + (-y) = -(x + y) = |x + y|$, since $x + y$ is negative in this case. The complication (and strict inequality) comes if one of the variables is nonnegative and the other is negative. By the symmetry of the roles of x and y here (strictly speaking, by the commutativity of addition), we can assume without loss of generality that it is x that is nonnegative and y that is negative. So we have $x \geq 0$ and $y < 0$.

Now there are two subcases to consider within this case, depending on the relative sizes of the nonnegative numbers x and $-y$. First suppose that $x \geq -y$. Then $x + y \geq 0$. Therefore $|x + y| = x + y$, and this quantity is a nonnegative number smaller than x (since y is negative). On the other hand $|x| + |y| = x + |y|$ is a positive number bigger than x. Therefore we have $|x + y| < x < |x| + |y|$, as desired.

Finally, consider the possibility that $x < -y$. Then $|x + y| = -(x + y) = (-x) + (-y)$ is a positive number less than or equal to $-y$ (since $-x$ is nonpositive). On the other hand $|x| + |y| = |x| + (-y)$ is a positive number greater than or equal to $-y$. Therefore we have $|x + y| \leq -y \leq |x| + |y|$, as desired.

7. We want to find consecutive squares that are far apart. If n is large enough, then $(n + 1)^2$ will be much bigger than n^2, and that will do it. Let's take $n = 100$. Then $100^2 = 10000$ and $101^2 = 10201$, so the 201 consecutive numbers 10001, 10002, ..., 10200 are not perfect squares. The first 100 of these will satisfy the requirements of this exercise. Our proof was constructive, since we actually exhibited the numbers.

9. We try some small numbers and discover that $8 = 2^3$ and $9 = 3^2$. In fact, this is the only solution, but the proof of this fact is not trivial.

11. One way to solve this is the following nonconstructive proof. Let $x = 2$ (rational) and $y = \sqrt{2}$ (irrational). If $x^y = 2^{\sqrt{2}}$ is irrational, we are done. If not, then let $x = 2^{\sqrt{2}}$ and $y = \sqrt{2}/4$; x is rational by assumption, and y is irrational (if it were rational, then $\sqrt{2}$ would be rational). But now $x^y = (2^{\sqrt{2}})^{\sqrt{2}/4} = 2^{\sqrt{2}\cdot(\sqrt{2})/4} = 2^{1/2} = \sqrt{2}$, which is irrational, as desired.

13. **a)** This statement asserts the existence of x with a certain property. If we let $y = x$, then we see that $P(x)$ is true. If y is anything other than x, then $P(x)$ is not true. Thus x is the unique element that makes P true.

b) The first clause here says that there is an element that makes P true. The second clause says that whenever two elements both make P true, they are in fact the same element. Together this says that P is satisfied by exactly one element.

c) This statement asserts the existence of an x that makes P true and has the further property that whenever we find an element that makes P true, that element is x. In other words, x is the unique element that makes P true. Note that this is essentially the same as the definition given in the text, except that the final conditional statement has been replaced by its contrapositive.

15. The equation $|a-c| = |b-c|$ is equivalent to the disjunction of two equations: $a-c = b-c$ or $a-c = -b+c$. The first of these is equivalent to $a = b$, which contradicts the assumptions made in this problem, so the original equation is equivalent to $a - c = -b + c$. By adding $b + c$ to both sides and dividing by 2, we see that this equation is equivalent to $c = (a+b)/2$. Thus there is a unique solution. Furthermore, this c is an integer, because the sum of the odd integers a and b is even.

17. We are being asked to solve $n = (k-2) + (k+3)$ for k. Using the usual, reversible, rules of algebra, we see that this equation is equivalent to $k = (n-1)/2$. In other words, this is the one and only value of k that makes our equation true. Since n is odd, $n-1$ is even, so k is an integer.

19. If x is itself an integer, then we can take $n = x$ and $\epsilon = 0$. No other solution is possible in this case, since if the integer n is greater than x, then n is at least $x+1$, which would make $\epsilon \geq 1$. If x is not an integer, then round it up to the next integer, and call that integer n. We let $\epsilon = n - x$. Clearly $0 \leq \epsilon < 1$, this is the only ϵ that will work with this n, and n cannot be any larger, since ϵ is constrained to be less than 1.

21. If $x = 5$ and $y = 8$, then the harmonic mean is $2 \cdot 5 \cdot 8/(5+8) \approx 6.15$, and the geometric mean is $\sqrt{5 \cdot 8} \approx 6.32$. If $x = 10$ and $y = 100$, then the harmonic mean is $2 \cdot 10 \cdot 100/(10+100) \approx 18.18$, and the geometric mean is $\sqrt{10 \cdot 100} \approx 31.62$. We conjecture that the harmonic mean of x and y is always less than their geometric mean if x and y are distinct positive real numbers (clearly if $x = y$ then both means are this common value). So we want to verify the inequality $2xy/(x+y) < \sqrt{xy}$. Multiplying both sides by $(x+y)/(2\sqrt{xy})$ gives us the equivalent inequality $\sqrt{xy} < (x+y)/2$, which is proved in Example 14.

23. The key point here is that *the parity (oddness or evenness) of the sum of the numbers written on the board never changes*. If j and k are both even or both odd, then their sum and their difference are both even, and we are replacing the even sum $j+k$ by the even difference $|j-k|$, leaving the parity of the total unchanged. If j and k have different parities, then erasing them changes the parity of the total, but their difference $|j-k|$ is odd, so adding this difference restores the parity of the total. Therefore the integer we end up with at the end of the process must have the same parity as $1 + 2 + \cdots + (2n)$. It is easy to compute this sum. If we add the first and last terms we get $2n+1$; if we add the second and next-to-last terms we get $2 + (2n-1) = 2n+1$; and so on. In all we get n sums of $2n+1$, so the total sum is $n(2n+1)$. If n is odd, this is the product of two odd numbers and therefore is odd, as desired.

25. Without loss of generality we can assume that n is nonnegative, since the fourth power of an integer and the fourth power of its negative are the same. To get a handle on the last digit of n, we can divide n by 10, obtaining a quotient k and remainder l, whence $n = 10k + l$, and l is an integer between 0 and 9, inclusive. Then we compute n^4 in each of these ten cases. We get the following values, where ?? is some integer that is a multiple of 10, whose exact value we do not care about.

$$\begin{aligned}
(10k+0)^4 &= 10000k^4 = 10000k^4 + 0 \\
(10k+1)^4 &= 10000k^4 + ??\cdot k^3 + ??\cdot k^2 + ??\cdot k + 1 \\
(10k+2)^4 &= 10000k^4 + ??\cdot k^3 + ??\cdot k^2 + ??\cdot k + 16 \\
(10k+3)^4 &= 10000k^4 + ??\cdot k^3 + ??\cdot k^2 + ??\cdot k + 81 \\
(10k+4)^4 &= 10000k^4 + ??\cdot k^3 + ??\cdot k^2 + ??\cdot k + 256 \\
(10k+5)^4 &= 10000k^4 + ??\cdot k^3 + ??\cdot k^2 + ??\cdot k + 625 \\
(10k+6)^4 &= 10000k^4 + ??\cdot k^3 + ??\cdot k^2 + ??\cdot k + 1296 \\
(10k+7)^4 &= 10000k^4 + ??\cdot k^3 + ??\cdot k^2 + ??\cdot k + 2401 \\
(10k+8)^4 &= 10000k^4 + ??\cdot k^3 + ??\cdot k^2 + ??\cdot k + 4096 \\
(10k+9)^4 &= 10000k^4 + ??\cdot k^3 + ??\cdot k^2 + ??\cdot k + 6561
\end{aligned}$$

Since each coefficient indicated by ?? is a multiple of 10, the corresponding term has no effect on the ones digit of the answer. Therefore the ones digits are 0, 1, 6, 1, 6, 5, 6, 1, 6, 1, respectively, so it is always a 0, 1, 5, or 6.

27. Because $n^3 > 100$ for all $n > 4$, we need only note that $n = 1$, $n = 2$, $n = 3$, and $n = 4$ do not satisfy $n^2 + n^3 = 100$.

29. Since $5^4 = 625$, for there to be positive integer solutions to this equation both x and y must be less than 5. This means that each of x^4 and y^4 is at most $4^4 = 256$, so their sum is at most 512 and cannot be 625.

31. We give a proof by contraposition. Assume that it is not the case that $a \le \sqrt[3]{n}$ or $b \le \sqrt[3]{n}$ or $c \le \sqrt[3]{n}$. Then it must be true that $a > \sqrt[3]{n}$ and $b > \sqrt[3]{n}$ and $c > \sqrt[3]{n}$. Multiplying these inequalities of positive numbers together we obtain $abc < (\sqrt[3]{n})^3 = n$, which implies the negation of our hypothesis that $n = abc$.

33. The idea is to find a small irrational number to add to the smaller of the two given rational numbers. Because we know that $\sqrt{2}$ is irrational, we can use a small multiple of $\sqrt{2}$. Here is our proof: By finding a common denominator, we can assume that the given rational numbers are a/b and c/b, where b is a positive integer and a and c are integers with $a < c$. In particular, $(a+1)/b \le c/b$. Thus $x = (a + \frac{1}{2}\sqrt{2})/b$ is between the two given rational numbers, because $0 < \sqrt{2} < 2$. Furthermore, x is irrational, because if x were rational, then $2(bx - a) = \sqrt{2}$ would be as well, in violation of Example 10 in Section 1.6.

35. **a)** Without loss of generality, we may assume that the x sequence is already sorted into nondecreasing order, since we can relabel the indices. There are only a finite number of possible orderings for the y sequence, so if we can show that we can increase the sum (or at least keep it the same) whenever we find y_i and y_j that are out of order (i.e., $i < j$ but $y_i > y_j$) by switching them, then we will have shown that the sum is largest when the y sequence is in nondecreasing order. Indeed, if we perform the swap, then we have added $x_iy_j + x_jy_i$ to the sum and subtracted $x_iy_i + x_jy_j$. The net effect, then, is to have added $x_iy_j + x_jy_i - x_iy_i - x_jy_j = (x_j - x_i)(y_i - y_j)$, which is nonnegative by our ordering assumptions.

b) This is similar to part **(a)**. Again we assume that the x sequence is already sorted into nondecreasing order. If the y sequence is not in nonincreasing order, then $y_i < y_j$ for some $i < j$. By swapping y_i and y_j we increase the sum by $x_iy_j + x_jy_i - x_iy_i - x_jy_j = (x_j - x_i)(y_i - y_j)$, which is nonpositive by our ordering assumptions.

37. In each case we just have to keep applying the function f until we reach 1, where $f(x) = 3x + 1$ if x is odd and $f(x) = x/2$ if x is even.

a) $f(6) = 3$, $f(3) = 10$, $f(10) = 5$, $f(5) = 16$, $f(16) = 8$, $f(8) = 4$, $f(4) = 2$, $f(2) = 1$. We abbreviate this to $6 \to 3 \to 10 \to 5 \to 16 \to 8 \to 4 \to 2 \to 1$.

b) $7 \to 22 \to 11 \to 34 \to 17 \to 52 \to 26 \to 13 \to 40 \to 20 \to 10 \to 5 \to 16 \to 8 \to 4 \to 2 \to 1$

c) $17 \to 52 \to 26 \to 13 \to 40 \to 20 \to 10 \to 5 \to 16 \to 8 \to 4 \to 2 \to 1$

d) $21 \to 64 \to 32 \to 16 \to 8 \to 4 \to 2 \to 1$

39. We give a constructive proof. Without loss of generality, we can assume that the upper left and upper right corners of the board are removed. We can place three dominoes horizontally to fill the remaining portion of the first row, and we can place four dominoes horizontally in each of the other seven rows to fill them.

41. The number of squares in a rectangular board is the product of the number of squares in each row and the number of squares in each column. We are given that this number is even, so there is either an even number of squares in each row or an even number of squares in each column. In the former case, we can tile the board in the obvious way by placing the dominoes horizontally, and in the latter case, we can tile the board in the obvious way by placing the dominoes vertically.

43. We follow the suggested labeling scheme. Clearly we can rotate the board if necessary to make the removed squares be 1 and 16. Square 2 must be covered by a domino. If that domino is placed to cover squares 2 and 6, then the following domino placements are forced in succession: 5-9, 13-14, and 10-11, at which point there is no way to cover square 15. Otherwise, square 2 must be covered by a domino placed at 2-3. Then the following domino placements are forced: 4-8, 11-12, 6-7, 5-9, and 10-14, and again there is no way to cover square 15.

45. Remove the two black squares adjacent to one of the white corners, and remove two white squares other than that corner. Then no domino can cover that white corner, because neither of the squares adjacent to it remains.

47. **a)** It is not hard to find the five patterns:

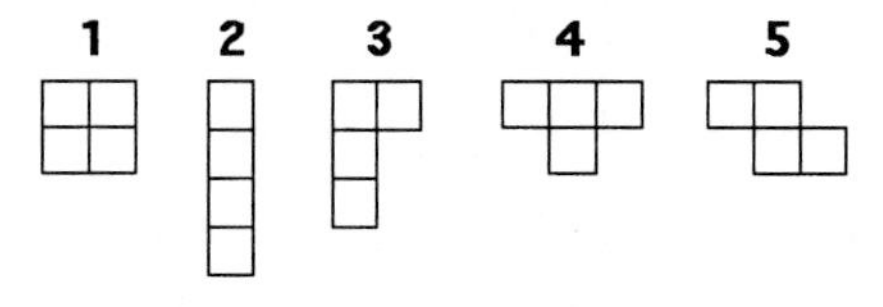

b) It is clear that the pattern labeled 1 and the pattern labeled 2 will tile the checkerboard. It is harder to find the tiling for patterns 3 and 4, but a little experimentation shows that it is possible.

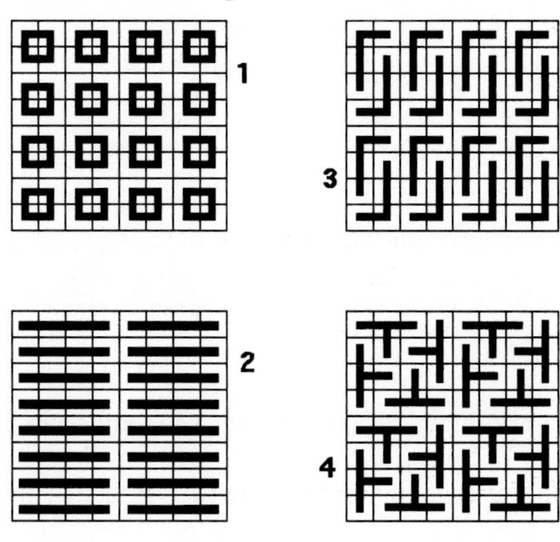

It remains to argue that pattern 5 cannot tile the checkerboard. Label the squares from 1 to 64, one row at a time from the top, from left to right in each row. Thus square 1 is the upper left corner, and square 64 is the lower right. Suppose we did have a tiling. By symmetry and without loss of generality, we may suppose that the tile is positioned in the upper left corner, covering squares 1, 2, 10, and 11. This forces a tile to be adjacent to it on the right, covering squares 3, 4, 12, and 13. Continue in this manner and we are forced to have a tile covering squares 6, 7, 15, and 16. This makes it impossible to cover square 8. Thus no tiling is possible.

GUIDE TO REVIEW QUESTIONS FOR CHAPTER 1

1. a) See p. 3. **b)** This is not a boring course.

2. a) See pp. 4, 5, 6, and 9.
b) Disjunction: "I'll go to the movies tonight or I'll finish my discrete mathematics homework." Conjunction: "I'll go to the movies tonight and I'll finish my discrete mathematics homework." Exclusive or: "I'll go to the movies tonight or I'll finish my discrete mathematics homework, but not both." Conditional statement: "If I'll go to the movies tonight, then I'll finish my discrete mathematics homework." Biconditional: "I'll go to the movies tonight if and only if I'll finish my discrete mathematics homework."

3. a) See p. 6. **b)** See p. 8.
c) Converse: "If I go for a walk in the woods tomorrow, then it will be sunny." Contrapositive: "If I don't go for a walk in the woods tomorrow, then it will not be sunny."

4. a) See p. 22.
b) using truth tables; symbolically, using identities in Tables 6–8 in Section 1.2; by giving a valid argument about the possible truth values of the propositional variables involved
c) Use the fact that $r \to \neg q \equiv \neg r \vee \neg q$, or use truth tables.

5. a) Each line of the truth table corresponds to exactly one combination of truth values for the n atomic propositions involved. We can write down a conjunction that is true precisely in this case, namely the conjunction of all the atomic propositions that are true and the negations of all the atomic propositions that are false. If we do this for *each* line of the truth table for which the value of the compound proposition is to be true, and take the disjunction of the resulting propositions, then we have the desired proposition in its disjunctive normal form. See Exercise 42 in Section 1.2.
b) See Exercise 43 in Section 1.2.
c) See Exercises 50 and 52 in Section 1.2.

6. The negation of $\forall x P(x)$ is $\exists x \neg P(x)$, and the negation of $\exists x P(x)$ is $\forall x \neg P(x)$.

7. a) In the second, x can depend on y. In the first, the same x must "work" for every y.
b) See Example 4 in Section 1.4.

8. See pp. 63–64. This is a valid argument because it uses the valid rule of inference called modus tollens.

9. This is a valid argument because it uses the universal modus ponens rule of inference. Therefore if the premises are true, the conclusion must be true.

10. **a)** See pp. 76, 77, and 80.

b) For a direct proof, the hypothesis implies that $n = 2k$ for some k, whence $n+4 = 2(k+2)$, so $n+4$ is even. For a proof by contraposition, suppose that $n+4$ is odd; hence $n+4 = 2k+1$ for some k. Then $n = 2(k-2)+1$, so n is odd, hence not even. For a proof by contradiction, assume that $n = 2k$ and $n+4 = 2l+1$ for some k and l. Subtracting gives $4 = 2(l-k)+1$, which means that 4 is odd, a contradiction.

11. **a)** See p. 82.

b) Suppose that $3n+2$ is odd, so that $3n+2 = 2k+1$ for some k. Multiply both sides by 3 and subtract 1, obtaining $9n+5 = 6k+2 = 2(3k+1)$. This shows that $9n+5$ is even. We prove the converse by contraposition. Suppose that $3n+2$ is not odd, i.e., that it is even. Then $3n+2 = 2k$ for some k. Multiply both sides by 3 and subtract 1, obtaining $9n+5 = 6k-1 = 2(3k-1)+1$. This shows that $9n+5$ is odd.

12. No—we could add to these $p_2 \to p_3$ and $p_1 \to p_4$, for example.

13. **a)** Find a counterexample, i.e., an object c such that $P(c)$ is false. **b)** $n = 1$ is a counterexample.

14. See p. 91.

15. See p. 92.

16. See Example 4 in Section 1.7.

SUPPLEMENTARY EXERCISES FOR CHAPTER 1

1. **a)** $q \to p$ (note that "only if" does not mean "if")

b) $q \wedge p$ **c)** $\neg q \vee \neg p$ (assuming inclusive use of the English word "or" is intended by the speaker)

d) $q \leftrightarrow p$ (this is another way to say "if and only if" in English words)

3. We could use truth tables, but we can also argue as follows.

a) Since q is false but the conditional statement $p \to q$ is true, we must conclude that p is also false.

b) The disjunction says that either p or q is true. Since p is given to be false, it follows that q must be true.

5. The inverse of $p \to q$ is $\neg p \to \neg q$. Therefore the converse of the inverse is $\neg q \to \neg p$. Note that this is the same as the contrapositive of the original statement. The converse of $p \to q$ is $q \to p$. Therefore the converse of the converse is $p \to q$, which was the original statement. The contrapositive of $p \to q$ is $\neg q \to \neg p$. Therefore the converse of the contrapositive is $\neg p \to \neg q$, which is the same as the inverse of the original statement.

7. The straightforward approach is to use disjunctive normal form. There are four cases in which exactly three of the variables are true. The desired proposition is $(p\wedge q\wedge r\wedge \neg s) \vee (p\wedge q\wedge \neg r\wedge s) \vee (p\wedge \neg q\wedge r\wedge s) \vee (\neg p\wedge q\wedge r\wedge s)$.

9. Translating these statements into symbols, using the obvious letters, we have $\neg t \to \neg g$, $\neg g \to \neg q$, $r \to q$, and $\neg t \wedge r$. Assume the statements are consistent. The fourth statement tells us that $\neg t$ must be true. Therefore by modus ponens with the first statement, we know that $\neg g$ is true, hence (from the second statement) that $\neg q$ is true. Also, the fourth statement tells us that r must be true, and so again modus ponens (third statement) makes q true. This is a contraction: $q \wedge \neg q$. Thus the statements are inconsistent.

11. We are told that exactly one of these people committed the crime, and exactly one (the guilty party) is a knight. We look at the three cases to determine who the knight is. If Amy were the knight, then her protestations of innocence would be true, but that cannot be, since we know that the knight is guilty. If Claire were the knight, then her statement that Brenda is not a normal is true; and since Brenda cannot be the knight in this situation, Brenda must be a knave. That means that Brenda is lying when she says that Amy was telling the truth; therefore Amy is lying. This means that Amy is guilty, but that cannot be, since Amy isn't the knight. So Brenda must be the knight. Amy is an innocent normal who is telling the truth when she says she is innocent; Brenda is telling the truth when she says that Amy is telling the truth; and Claire is a normal who is telling the truth when she says that Brenda is not a normal. So Brenda committed the crime.

13. The definition of valid argument is an argument in which the truth of all the premises forces the truth of conclusion. In this example, the two premises can never be true simultaneously, because they are contradictory, irrespective of the true status of the tooth fairy. Therefore it is (vacuously) true that whenever both of the premises are true, the conclusion is also true (irrespective of your luck at finding gold at the end of the rainbow). Because the premises are not both true, we cannot conclude that the conclusion is true.

15. **a)** F, since 4 does not divide 5 **b)** T, since 2 divides 4
c) F, by the counterexample in part **(a)** **d)** T, since 1 divides every positive integer
e) F, since no number is a multiple of all positive integers (No matter what positive integer n one chooses, if we take $m = n+1$, then $P(m,n)$ is false, since $n+1$ does not divide n.)
f) T, since 1 divides every positive integer

17. The given statement tells us that there are exactly two elements in the domain. Therefore if we let the domain be anything with size other than 2 the statement will be false.

19. For each person we want to assert the existence of two different people who are that person's parents. The most elegant way to do so is $\forall x \exists y \exists z (y \neq z \wedge \forall w (P(w,x) \leftrightarrow (w = y \vee w = z)))$. Note that we are saying that w is a parent of x if and only if w is one of the two people whose existence we asserted.

21. To express the statement that exactly n members of the domain satisfy P, we need to use n existential quantifiers, express the fact that these n variables all satisfy P and are all different, and express the fact that every other member of the domain that satisfies P must be one of these.
a) This is a special case, however. To say that there are no values of x that make P true we can simply write $\neg \exists x P(x)$ or $\forall x \neg P(x)$.
b) This is the same as Exercise 53 in Section 1.4, because $\exists_1 x P(x)$ is the same as $\exists! x P(x)$. Thus we can write $\exists x (P(x) \wedge \forall y (P(y) \rightarrow y = x))$.
c) Following the discussion above, we write $\exists x_1 \exists x_2 (P(x_1) \wedge P(x_2) \wedge x_1 \neq x_2 \wedge \forall y (P(y) \rightarrow (y = x_1 \vee y = x_2)))$.
d) We expand the previous answer to one more variable: $\exists x_1 \exists x_2 \exists x_3 (P(x_1) \wedge P(x_2) \wedge P(x_3) \wedge x_1 \neq x_2 \wedge x_1 \neq x_3 \wedge x_2 \neq x_3 \wedge \forall y (P(y) \rightarrow (y = x_1 \vee y = x_2 \vee y = x_3)))$.

23. Suppose that $\exists x (P(x) \rightarrow Q(x))$ is true. Then for some x, either $Q(x)$ is true or $P(x)$ is false. If $Q(x)$ is true for some x, then the conditional statement $\forall x P(x) \rightarrow \exists x Q(x)$ is true (having true conclusion). If $P(x)$ is false for some x, then again the conditional statement $\forall x P(x) \rightarrow \exists x Q(x)$ is true (having false hypothesis). Conversely, suppose that $\exists x (P(x) \rightarrow Q(x))$ is false. That means that for every x, the conditional statement $P(x) \rightarrow Q(x)$ is false, or, in other words, $P(x)$ is true and $Q(x)$ is false. The latter statement implies that $\exists x Q(x)$ is false. Thus $\forall x P(x) \rightarrow \exists x Q(x)$ has a true hypothesis and a false conclusion and is therefore false.

25. No. For each x there may be just one y making $P(x,y)$ true, so that the second proposition will not be true. For example, let $P(x,y)$ be $x+y=0$, where the domain (universe of discourse) is the integers. Then the first proposition is true, since for each x there exists a y, namely $-x$, such that $P(x,y)$ holds. On the other hand, there is no one x such that $x+y=0$ for *every* y.

27. Let $T(s,c,d)$ be the statement that student s has taken class c in department d. Then, with the domains (universes of discourse) being the students in this class, the courses at this university, and the departments in the school of mathematical sciences, the given statement is $\forall s \forall d \exists c\, T(s,c,d)$.

29. Let $T(x,y)$ mean that student x has taken class y, where the domain is all students in this class. We want to say that there exists exactly one student for whom there exists exactly one class that this student has taken. So we can write simply $\exists! x \exists! y\, T(x,y)$. To do this without quantifiers, we need to expand the uniqueness quantifier using Exercise 52 in Section 1.4. Doing so, we have $\exists x \forall z((\exists y \forall w(T(z,w) \leftrightarrow w=y)) \leftrightarrow z=x)$.

31. By universal instantiation we have $P(a) \to Q(a)$ and $Q(a) \to R(a)$. By modus tollens we then conclude $\neg Q(a)$, and again by modus tollens we conclude $\neg P(a)$.

33. We give a proof by contraposition that if $\sqrt{x}$ is rational, then x is rational, assuming throughout that $x \geq 0$. Suppose that $\sqrt{x} = p/q$ is rational, $q \neq 0$. Then $x = (\sqrt{x})^2 = p^2/q^2$ is also rational (q^2 is again nonzero).

35. We can give a constructive proof by letting $m = 10^{500} + 1$. Then $m^2 = (10^{500}+1)^2 > (10^{500})^2 = 10^{1000}$.

37. The first three positive cubes are 1, 8, and 27. If we want to find a number that cannot be written as the sum of eight cubes, we would look for a number that is 7 more than a small multiple of 8. Indeed, 23 will do. We can use two 8's but then would have to use seven 1's to reach 23, a total of nine numbers. Clearly no smaller collection will do. This counterexample disproves the statement.

39. The first three positive fifth powers are 1, 32, and 243. If we want to find a number that cannot be written as the sum of 36 fifth powers, we would look for a number that is 31 more than a small multiple of 32. Indeed, $7 \cdot 32 - 1 = 223$ will do. We can use six 32's but then would have to use 31 1's to reach 223, a total of 37 numbers. Clearly no smaller collection will do. This counterexample disproves the statement.

WRITING PROJECTS FOR CHAPTER 1

Books and articles indicated by bracketed symbols below are listed near the end of this manual. You should also read the general comments and advice you will find there about researching and writing these essays.

1. An excellent website for this is `http://www.wordsmith.demon.co.uk/paradoxes`. It includes a bibliography.

2. Search your library's on-line catalog for a book with the word *fuzzy* in the title. You might find [BaGo], [DuPr], [Ka], [Ko3], or [McFr], for example.

3. Even if you can't find a set, you may find some articles about it in materials for high school students and teachers, such as old issues of *Mathematics Teacher*, published by the National Council of Teachers of Mathematics. This journal, and possibly even copies of the game, may exist in the education library at your school (if there is one). The company that currently produces it has a website: `http://www.wff-n-proof.com`. See also `http://thinkers.law.umich.edu/files/WPGames/WFFNPRUF.htm`, which includes the rules.

4. Martin Gardner and others have written some books that annotate Carroll's writings quite extensively. Lewis Carroll has become a cult figure in certain circles. See also [Ca1], [Ca2], and [Ca3], for original material.

5. A textbook on logic programming and/or the language PROLOG, such as [Ho2] or [Sa1], would be a logical place to start. Many bookstores have huge computer science sections these days, so that source should not be ignored.

6. A course on computational logic at Stanford in 2005–2006 had a Web page with class notes: `http://logic.stanford.edu/classes/cs157/2005fall/cs157.html`. Enderton's book on logic [En] would be a possible choice for background information.

7. There are books on this subject, such as [Du].

8. A place to start might be a recent article on this topic in *Science* [Re]. As always, a Web search will also turn up more information.

9. The Web has an encyclopedia made up of articles by contributors. Remarkably, it is usually quite good, with accurate information and useful links and cross-references. See their article on Chomp: `http://en.wikipedia.org/wiki/Chomp`.

10. The references given in the text are the obvious place to start. The mathematics education field has bought into Pólya's ideas, especially as they relate to problem-solving. See what the National Council of Teachers of Mathematics (`http://www.nctm.org`) has to say about it.

11. The classic work in this field is [GrSh].

CHAPTER 2
Basic Structures: Sets, Functions, Sequences, and Sums

SECTION 2.1 Sets

This exercise set (note that this is a "set" in the mathematical sense!) reinforces the concepts introduced in this section—set description, subset and containment, cardinality, power set, and Cartesian product. A few of the exercises (mostly some of the even-numbered ones) are a bit subtle. Keep in mind the distinction between "is an element of" and "is a subset of." Similarly, there is a big difference between $\emptyset$ and $\{\emptyset\}$. In dealing with sets, as in most of mathematics, it is extremely important to say exactly what you mean.

1. a) $\{1, -1\}$ **b)** $\{1, 2, 3, 4, 5, 6, 7, 8, 9, 10, 11\}$
c) $\{0, 1, 4, 9, 16, 25, 36, 49, 64, 81\}$ **d)** $\emptyset$ ($\sqrt{2}$ is not an integer)

3. a) Yes; order and repetition do not matter.
b) No; the first set has one element, and the second has two elements.
c) No; the first set has no elements, and the second has one element (namely the empty set).

5. a) Since 2 is an integer greater than 1, 2 is an element of this set.
b) Since 2 is not a perfect square ($1^2 < 2$, but $n^2 > 2$ for $n > 1$), 2 is not an element of this set.
c) This set has two elements, and as we can clearly see, one of those elements is 2.
d) This set has two elements, and as we can clearly see, neither of those elements is 2. Both of the elements of this set are sets; 2 is a number, not a set.
e) This set has two elements, and as we can clearly see, neither of those elements is 2. Both of the elements of this set are sets; 2 is a number, not a set.
f) This set has just one element, namely the set $\{\{2\}\}$. So 2 is not an element of this set. Note that $\{2\}$ is not an element either, since $\{2\} \neq \{\{2\}\}$.

7. a) This is false, since the empty set has no elements.
b) This is false. The set on the right has only one element, namely the number 0, not the empty set.
c) This is false. In fact, the empty set has *no* proper subsets.
d) This is true. Every element of the set on the left is, vacuously, an element of the set on the right; and the set on the right contains an element, namely 0, that is not in the set on the left.
e) This is false. The set on the right has only one element, namely the number 0, not the set containing the number 0.
f) This is false. For one set to be a proper subset of another, the two sets cannot be equal.
g) This is true. Every set is a subset of itself.

9. a) T (in fact x is the only element) **b)** T (every set is a subset of itself)
c) F (the only element of $\{x\}$ is a letter, not a set) **d)** T (in fact, $\{x\}$ is the only element)
e) T (the empty set is a subset of every set) **f)** F (the only element of $\{x\}$ is a letter, not a set)

11. The four months whose names don't contain the letter R form a subset of the set of twelve months, as shown here.

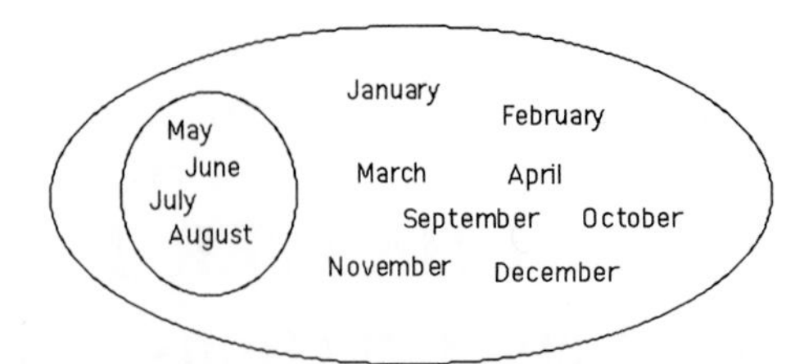

13. We put the subsets inside the supersets. We also put dots in certain regions to indicate that those regions are not empty (required by the fact that these are proper subset relations). Thus the answer is as shown.

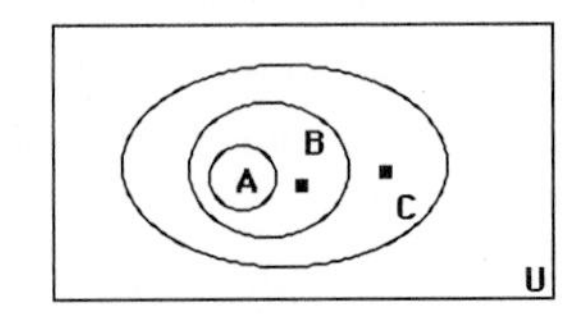

15. We need to show that every element of A is also an element of C. Let $x \in A$. Then since $A \subseteq B$, we can conclude that $x \in B$. Furthermore, since $B \subseteq C$, the fact that $x \in B$ implies that $x \in C$, as we wished to show.

17. The cardinality of a set is the number of elements it has. The number of elements in its elements is irrelevant.
a) 1 **b)** 1 **c)** 2 **d)** 3

19. **a)** $\{\emptyset, \{a\}\}$ **b)** $\{\emptyset, \{a\}, \{b\}, \{a, b\}\}$ **c)** $\{\emptyset, \{\emptyset\}, \{\{\emptyset\}\}, \{\emptyset, \{\emptyset\}\}\}$

21. **a)** Since the set we are working with has 3 elements, the power set has $2^3 = 8$ elements.
b) Since the set we are working with has 4 elements, the power set has $2^4 = 16$ elements.
c) The power set of the empty set has $2^0 = 1$ element. The power set of this set therefore has $2^1 = 2$ elements. In particular, it is $\{\emptyset, \{\emptyset\}\}$. (See Example 14.)

23. In each case we need to list all the ordered pairs, and there are $4 \times 2 = 8$ of them.
a) $\{(a, y), (a, z), (b, y), (b, z), (c, y), (c, z), (d, y), (d, z)\}$
b) $\{(y, a), (y, b), (y, c), (y, d), (z, a), (z, b), (z, c), (z, d)\}$

25. This is the set of triples (a, b, c), where a is an airline and b and c are cities. For example, (TWA, Rochester Hills Michigan, Middletown New Jersey) is an element of this Cartesian product. A useful subset of this set is the set of triples (a, b, c) for which a flies between b and c. For example, (Northwest, Detroit, New York) is in this subset, but the triple mentioned earlier is not.

27. By definition, $\emptyset \times A$ consists of all pairs (x, a) such that $x \in \emptyset$ and $a \in A$. Since there are no elements $x \in \emptyset$, there are no such pairs, so $\emptyset \times A = \emptyset$. Similar reasoning shows that $A \times \emptyset = \emptyset$.

29. The Cartesian product $A \times B$ has mn elements. (This problem foreshadows the general discussion of counting in Chapter 5.) To see that this answer is correct, note that for each $a \in A$ there are n different elements $b \in B$ with which to form the pair (a, b). Since there are m different elements of A, each leading to n different pairs, there must be mn pairs altogether.

31. The only difference between $A \times B \times C$ and $(A \times B) \times C$ is parentheses, so for all practical purposes one can think of them as essentially the same thing. By Definition 10, the elements of $A \times B \times C$ consist of 3-tuples (a, b, c), where $a \in A$, $b \in B$, and $c \in C$. By Definition 9, the elements of $(A \times B) \times C$ consist of ordered pairs (p, c), where $p \in A \times B$ and $c \in C$, so the typical element of $(A \times B) \times C$ looks like $((a, b), c)$. A 3-tuple is a different creature from a 2-tuple, even if the 3-tuple and the 2-tuple in this case convey exactly the same information. To be more precise, there is a natural one-to-one correspondence between $A \times B \times C$ and $(A \times B) \times C$ given by $(a, b, c) \leftrightarrow ((a, b), c)$.

33. **a)** Every real number has its square not equal to -1. Alternatively, the square of a real number is never -1. This is true, since squares of real numbers are always nonnegative.

b) There exists an integer whose square is 2. This is false, since the only two numbers whose squares are 2, namely $\sqrt{2}$ and $-\sqrt{2}$, are not integers.

c) The square of every integer is positive. This is almost true, but not quite, since $0^2 \not> 0$.

d) There is a real number equal to its own square. This is true, since $x = 1$ (as well as $x = 0$) fits the bill.

35. In each case we want the set of all values of x in the domain (the set of integers) that satisfy the given equation or inequality.

a) The only integers whose squares are less than 3 are the integers whose absolute values are less than 2. So the truth set is $\{x \in \mathbf{Z} \mid x^2 < 3\} = \{-1, 0, 1\}$.

b) All negative integers satisfy this inequality, as do all nonnegative integers other than 0 and 1. So the truth set is $\{x \in \mathbf{Z} \mid x^2 > x\} = \mathbf{Z} - \{0, 1\} = \{\ldots, -2, -1, 2, 3, 4, \ldots\}$.

c) The only real number satisfying this equation is $x = -1/2$. Because this value is not in our domain, the truth set is empty: $\{x \in \mathbf{Z} \mid 2x + 1 = 0\} = \emptyset$.

37. First we prove the statement mentioned in the hint. The "if" part is immediate from the definition of equality. The "only if" part is rather subtle. We want to show that if $\{\{a\}, \{a, b\}\} = \{\{c\}, \{c, d\}\}$, then $a = c$ and $b = d$. First consider the case in which $a \neq b$. Then $\{\{a\}, \{a, b\}\}$ has exactly two elements, both of which are sets; exactly one of them contains one element, and exactly one of them contains two elements. Thus $\{\{c\}, \{c, d\}\}$ must have the same property; hence c cannot equal d, and so $\{c\}$ is the element containing one element. Hence $\{a\} = \{c\}$, and so $a = c$. Also in this case the two-element elements $\{a, b\}$ and $\{c, d\}$ must be equal, and since $b \neq a = c$, we must have $b = d$. The other possibility is that $a = b$. Then $\{\{a\}, \{a, b\}\} = \{\{a\}\}$, a set with one element. Hence $\{\{c\}, \{c, d\}\}$ must also have only one element, which can only happen when $c = d$ and the set is $\{\{c\}\}$. It then follows that $a = c$, and hence $b = d$, as well.

Now there is really nothing else to prove. The property that we want ordered pairs to have is precisely the one that we just proved is satisfied by this definition. Furthermore, if we look at the proof, then it is clear how to "recover" both a and b from $\{\{a\}, \{a, b\}\}$. If this set has two elements, then a is the unique element in the one-element element of this set, and b is the unique member of the two-element element of this set other than a. If this set has only one element, then a and b are both equal to the unique element of the unique element of this set.

39. We can do this recursively, using the idea from Section 4.4 of reducing a problem to a smaller instance of the same problem. Suppose that the elements of the set in question are listed: $A = \{a_1, a_2, a_3, \ldots, a_n\}$. First we will write down all the subsets that do not involve a_n. This is just the same problem we are talking about all over again, but with a smaller set—one with just $n - 1$ elements. We do this by the process we are currently describing. Then we write these same subsets down again, but this time adjoin a_n to each one. Each subset of A will have been written down, then—first all those that do not include a_n, and then all those that do.

For example, using this procedure the subsets of $\{p, d, q\}$ would be listed in the order $\emptyset$, $\{p\}$, $\{d\}$, $\{p, d\}$, $\{q\}$, $\{p, q\}$, $\{d, q\}$, $\{p, d, q\}$.

An alternative solution is given in the answer key in the back of the textbook.

SECTION 2.2 Set Operations

Most of the exercises involving operations on sets can be done fairly routinely by following the definitions. It is important to understand what it means for two sets to be equal and how to prove that two given sets are equal—using membership tables, using the definition to reduce the problem to logic, or showing that each is a subset of the other; see, for example, Exercises 5–24. It is often helpful when looking at operations on sets to draw the Venn diagram, even if you are not asked to do so. The **symmetric difference** *is a fairly important set operation not discussed in the section; it is developed in Exercises 32–43. Two other new concepts,* **multisets** *and* **fuzzy sets**, *are also introduced in this set of exercises.*

1. **a)** the set of students who live within one mile of school and walk to class (only students who do both of these things are in the intersection)

b) the set of students who either live within one mile of school or walk to class (or, it goes without saying, both)

c) the set of students who live within one mile of school but do not walk to class

d) the set of students who live more than a mile from school but nevertheless walk to class

3. **a)** We include all numbers that are in one or both of the sets, obtaining $\{0,1,2,3,4,5,6\}$.

b) There is only one number in both of these sets, so the answer is $\{3\}$.

c) The set of numbers in A but not in B is $\{1,2,4,5\}$.

d) The set of numbers in B but not in A is $\{0,6\}$.

5. By definition $\overline{\overline{A}}$ is the set of elements of the universal set that are not in $\overline{A}$. Not being in $\overline{A}$ means being in A. Thus $\overline{\overline{A}}$ is the same set as A. We can give this proof in symbols as follows: $\overline{\overline{A}} = \{x \mid \neg x \in \overline{A}\} = \{x \mid \neg\neg x \in A\} = \{x \mid x \in A\} = A$.

7. These identities are straightforward applications of the definitions and are most easily stated using set-builder notation. Recall that **T** means the proposition that is always true, and **F** means the proposition that is always false.

a) $A \cup U = \{x \mid x \in A \vee x \in U\} = \{x \mid x \in A \vee \mathbf{T}\} = \{x \mid \mathbf{T}\} = U$

b) $A \cap \emptyset = \{x \mid x \in A \wedge x \in \emptyset\} = \{x \mid x \in A \wedge \mathbf{F}\} = \{x \mid \mathbf{F}\} = \emptyset$

9. **a)** We must show that every element (of the universal set) is in $A \cup \overline{A}$. This is clear, since every element is either in A (and hence in that union) or else not in A (and hence in that union).

b) We must show that no element is in $A \cap \overline{A}$. This is clear, since $A \cap \overline{A}$ consists of elements that are in A and not in A at the same time, obviously an impossibility.

11. These follow directly from the corresponding properties for the logical operations *or* and *and*.

a) $A \cup B = \{x \mid x \in A \vee x \in B\} = \{x \mid x \in B \vee x \in A\} = B \cup A$

b) $A \cap B = \{x \mid x \in A \wedge x \in B\} = \{x \mid x \in B \wedge x \in A\} = B \cap A$

13. We will show that these two sets are equal by showing that each is a subset of the other. Suppose $x \in A \cap (A \cup B)$. Then $x \in A$ and $x \in A \cup B$ by the definition of intersection. Since $x \in A$, we have proved that the left-hand side is a subset of the right-hand side. Conversely, let $x \in A$. Then by the definition of union, $x \in A \cup B$ as well. Since both of these are true, $x \in A \cap (A \cup B)$ by the definition of intersection, and we have shown that the right-hand side is a subset of the left-hand side.

15. This exercise asks for a proof of one of De Morgan's laws for sets. The primary way to show that two sets are equal is to show that each is a subset of the other. In other words, to show that $X = Y$, we must show that whenever $x \in X$, it follows that $x \in Y$, and that whenever $x \in Y$, it follows that $x \in X$. Exercises 5–7 could also have been done this way, but it was easier in those cases to reduce the problems to the corresponding problems of logic. Here, too, we can reduce the problem to logic and invoke De Morgan's law for logic, but this problem requests specific proof techniques.

a) This proof is similar to the proof of the dual property, given in Example 10. Suppose $x \in \overline{A \cup B}$. Then $x \notin A \cup B$, which means that x is in neither A nor B. In other words, $x \notin A$ and $x \notin B$. This is equivalent to saying that $x \in \overline{A}$ and $x \in \overline{B}$. Therefore $x \in \overline{A} \cap \overline{B}$, as desired. Conversely, if $x \in \overline{A} \cap \overline{B}$, then $x \in \overline{A}$ and $x \in \overline{B}$. This means $x \notin A$ and $x \notin B$, so x cannot be in the union of A and B. Since $x \notin A \cup B$, we conclude that $x \in \overline{A \cup B}$, as desired.

b) The following membership table gives the desired equality, since columns four and seven are identical.

A	B	$A \cup B$	$\overline{A \cup B}$	$\overline{A}$	$\overline{B}$	$\overline{A} \cap \overline{B}$
1	1	1	0	0	0	0
1	0	1	0	0	1	0
0	1	1	0	1	0	0
0	0	0	1	1	1	1

17. This exercise asks for a proof of a generalization of one of De Morgan's laws for sets from two sets to three.

a) This proof is similar to the proof of the two-set property, given in Example 10. Suppose $x \in \overline{A \cap B \cap C}$. Then $x \notin A \cap B \cap C$, which means that x fails to be in at least one of these three sets. In other words, $x \notin A$ or $x \notin B$ or $x \notin C$. This is equivalent to saying that $x \in \overline{A}$ or $x \in \overline{B}$ or $x \in \overline{C}$. Therefore $x \in \overline{A} \cup \overline{B} \cup \overline{C}$, as desired. Conversely, if $x \in \overline{A} \cup \overline{B} \cup \overline{C}$, then $x \in \overline{A}$ or $x \in \overline{B}$ or $x \in \overline{C}$. This means $x \notin A$ or $x \notin B$ or $x \notin C$, so x cannot be in the intersection of A, B and C. Since $x \notin A \cap B \cap C$, we conclude that $x \in \overline{A \cap B \cap C}$, as desired.

b) The following membership table gives the desired equality, since columns five and nine are identical.

A	B	C	$A \cap B \cap C$	$\overline{A \cap B \cap C}$	$\overline{A}$	$\overline{B}$	$\overline{C}$	$\overline{A} \cup \overline{B} \cup \overline{C}$
1	1	1	1	0	0	0	0	0
1	1	0	0	1	0	0	1	1
1	0	1	0	1	0	1	0	1
1	0	0	0	1	0	1	1	1
0	1	1	0	1	1	0	0	1
0	1	0	0	1	1	0	1	1
0	0	1	0	1	1	1	0	1
0	0	0	0	1	1	1	1	1

19. This is clear, since both of these sets are precisely $\{x \mid x \in A \wedge x \notin B\}$.

21. There are many ways to prove identities such as the one given here. One way is to reduce them to logical identities (some of the associative and distributive laws for $\vee$ and $\wedge$). Alternately, we could argue in each case that the left-hand side is a subset of the right-hand side and vice versa. Another method would be to construct membership tables (they will have eight rows in order to cover all the possibilities). Here we choose the second method. First we show that every element of the left-hand side must be in the right-hand side as well. If $x \in A \cup (B \cup C)$, then x must be either in A or in $B \cup C$ (or both). In the former case, we can conclude that $x \in A \cup B$ and thus $x \in (A \cup B) \cup C$, by the definition of union. In the latter case, x must be either in B or in C (or both). In the former subcase, we can conclude that $x \in A \cup B$ and thus $x \in (A \cup B) \cup C$, by the definition of union; in the latter subcase, we can conclude that $x \in (A \cup B) \cup C$, again using the definition of union.

23. There are many ways to prove identities such as the one given here. One way is to reduce them to logical identities (some of the associative and distributive laws for $\vee$ and $\wedge$). Alternately, we could argue in each case that the left-hand side is a subset of the right-hand side and vice versa. Another method would be to construct membership tables (they will have eight rows in order to cover all the possibilities). Here we choose the third method. We construct the following membership table and note that the fifth and eighth columns are identical.

A	B	C	$B\cap C$	$A\cup(B\cap C)$	$A\cup B$	$A\cup C$	$(A\cup B)\cap(A\cup C)$
1	1	1	1	1	1	1	1
1	1	0	0	1	1	1	1
1	0	1	0	1	1	1	1
1	0	0	0	1	1	1	1
0	1	1	1	1	1	1	1
0	1	0	0	0	1	0	0
0	0	1	0	0	0	1	0
0	0	0	0	0	0	0	0

25. These are straightforward applications of the definitions.

a) The set of elements common to all three sets is $\{4,6\}$.

b) The set of elements in at least one of the three sets is $\{0,1,2,3,4,5,6,7,8,9,10\}$.

c) The set of elements in C and at the same time in at least one of A and B is $\{4,5,6,8,10\}$.

d) The set of elements either in C or in both A and B (or in both of these) is $\{0,2,4,5,6,7,8,9,10\}$.

27. a) In the figure we have shaded the A set with horizontal bars (including the double-shaded portion, which includes both horizontal and vertical bars), and we have shaded the set $B-C$ with vertical bars (that portion inside B but outside C. The intersection is where these overlap—the double-shaded portion (shaped like an arrowhead).

b) In the figure we have shaded the set $A\cap B$ with horizontal bars (including the double-shaded portion, which includes both horizontal and vertical bars), and we have shaded the set $A\cap C$ with vertical bars. The union is the entire region that has any shading at all (shaped like a tilted mustache).

c) In the figure we have shaded the set $A\cap\overline{B}$ with horizontal bars (including the double-shaded portion, which includes both horizontal and vertical bars), and we have shaded the set $A\cap\overline{C}$ with vertical bars. The union is the entire region that has any shading at all (everything inside A except the triangular middle portion where all three sets overlap) portion (shaped like an arrowhead).

(a)

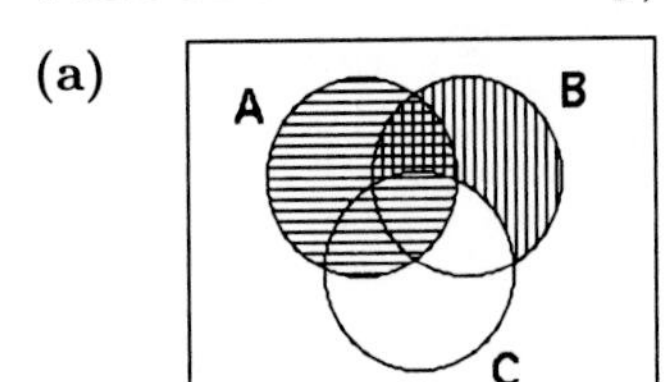

(b)

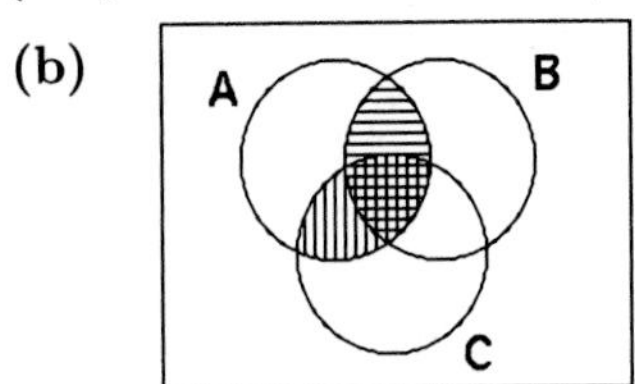

(c)

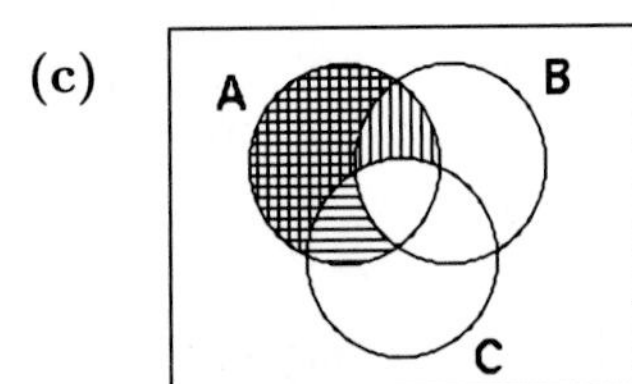

29. a) If B adds nothing new to A, then we can conclude that all the elements of B were already in A. In other words, $B\subseteq A$.

b) In this case, all the elements of A are forced to be in B as well, so we conclude that $A\subseteq B$.

c) This equality holds precisely when none of the elements of A are in B (if there were any such elements, then $A-B$ would not contain all the elements of A). Thus we conclude that A and B are disjoint ($A\cap B=\emptyset$).

d) We can conclude nothing about A and B in this case, since this equality always holds.

e) Every element in $A-B$ must be in A, and every element in $B-A$ must not be in A. Since no item can be in A and not be in A at the same time, there are no elements in both $A-B$ and $B-A$. Thus the only way for these two sets to be equal is if both of them are the empty set. This means that every element of A must be in B, and every element of B must be in A. Thus we conclude that $A=B$.

31. This is the set-theoretic version of the contrapositive law for logic, which says that $p \to q$ is logically equivalent to $\neg q \to \neg p$. We argue as follows.
$A \subseteq B \equiv \forall x(x \in A \to x \in B) \equiv \forall x(x \notin B \to x \notin A) \equiv \forall x(x \in \overline{B} \to x \in \overline{A}) \equiv \overline{B} \subseteq \overline{A}$

33. Clearly this will be the set of students majoring in computer science or mathematics but not both.

35. This is just a restatement of the definition. An element is in $(A \cup B) - (A \cap B)$ if it is in the union (i.e., in either A or B), but not in the intersection (i.e., not in both A and B).

37. We will use the result of Exercise 36 as well as some obvious identities (some of which are in Exercises 6–10).
a) $A \oplus A = (A - A) \cup (A - A) = \emptyset \cup \emptyset = \emptyset$ **b)** $A \oplus \emptyset = (A - \emptyset) \cup (\emptyset - A) = A \cup \emptyset = A$
c) $A \oplus U = (A - U) \cup (U - A) = \emptyset \cup \overline{A} = \overline{A}$ **d)** $A \oplus \overline{A} = (A - \overline{A}) \cup (\overline{A} - A) = A \cup \overline{A} = U$

39. We can conclude that $B = \emptyset$. To see this, suppose that B contains some element b. If $b \in A$, then b is excluded from $A \oplus B$, so $A \oplus B$ cannot equal A. On the other hand, if $b \notin A$, then b must be in $A \oplus B$, so again $A \oplus B$ cannot equal A. Thus in either case, $A \oplus B \neq A$. We conclude that B cannot have any elements.

41. Yes. To show that $A = B$, we need to show that $x \in A$ implies $x \in B$ and conversely. By symmetry, it will be enough to show one direction of this. So assume that $A \oplus C = B \oplus C$, and let $x \in A$ be given. There are two cases to consider, depending on whether $x \in C$. If $x \in C$, then by definition we can conclude that $x \notin A \oplus C$. Therefore $x \notin B \oplus C$. Now if x were *not* in B, then x *would* be in $B \oplus C$ (since $x \in C$ by assumption). Since this is not true, we conclude that $x \in B$, as desired. For the other case, assume that $x \notin C$. Then $x \in A \oplus C$. Therefore $x \in B \oplus C$ as well. Again, if x were *not* in B, then it could not be in $B \oplus C$ (since $x \notin C$ by assumption). Once again we conclude that $x \in B$, and the proof is complete.

43. Yes. Both sides equal the set of elements that occur in an odd number of the sets A, B, C, and D.

45. a) The union of these sets is the set of elements that appear in at least one of them. In this case the sets are "increasing": $A_1 \subseteq A_2 \subseteq \cdots \subseteq A_n$. Therefore every element in any of the sets is in A_n, so the union is $A_n = \{1, 2, \ldots, n\}$.

b) The intersection of these sets is the set of elements that appear in all of them. Since $A_1 = \{1\}$, only the number 1 has a chance to be in the intersection. In fact 1 is in the intersection, since it is in all of the sets. Therefore the intersection is $A_1 = \{1\}$.

47. a) Here the sets are increasing. A bit string of length not exceeding 1 is also a bit string of length not exceeding 2, so $A_1 \subseteq A_2$. Similarly, $A_2 \subseteq A_3 \subseteq A_4 \subseteq \cdots \subseteq A_n$. Therefore the union of the sets $A_1, A_2, \ldots, A_n$ is just A_n itself.

b) Since A_1 is a subset of all the A_i's, the intersection is A_1, the set of all nonempty bit strings of length not exceeding 1, namely $\{0, 1\}$.

49. a) As i increases, the sets get larger: $A_1 \subset A_2 \subset A_3 \cdots$. All the sets are subsets of the set of integers, and every integer is included eventually, so $\bigcup_{i=1}^{\infty} A_i = \mathbf{Z}$. Because A_1 is a subset of each of the others, $\bigcap_{i=1}^{\infty} A_i = A_1 = \{-1, 0, 1\}$.

b) All the sets are subsets of the set of integers, and every nonzero integer is in exactly one of the sets, so $\bigcup_{i=1}^{\infty} A_i = \mathbf{Z} - \{0\}$. Each pair of these sets are disjoint, so no element is common to all of the sets. Therefore $\bigcap_{i=1}^{\infty} A_i = \emptyset$.

c) This is similar to part **(a)**, the only difference being that here we are working with real numbers. Therefore $\bigcup_{i=1}^{\infty} A_i = \mathbf{R}$ (the set of all real numbers), and $\bigcap_{i=1}^{\infty} A_i = A_1 = [-1,1]$ (the interval of all real numbers between -1 and 1, inclusive).

d) This time the sets are getting smaller as i increases: $\cdots \subset A_3 \subset A_2 \subset A_1$. Because A_1 includes all the others, $\bigcup_{i=1}^{\infty} A_i = A_1 = [1,\infty)$ (all real numbers greater than or equal to 1). Every number eventually gets excluded as i increases, so $\bigcap_{i=1}^{\infty} A_i = \emptyset$. Notice that ∞ is not a real number, so we cannot write $\bigcap_{i=1}^{\infty} A_i = \{\infty\}$.

51. The ith digit in the string indicates whether the ith number in the universal set (in this case the number i) is in the set in question.

a) $\{1,2,3,4,7,8,9,10\}$ **b)** $\{2,4,5,6,7\}$ **c)** $\{1,10\}$

53. We are given two bit strings, representing two sets. We want to represent the set of elements that are in the first set but not the second. Thus the bit in the ith position of the bit string for the difference is 1 if the ith bit of the first string is 1 and the ith bit of the second string is 0, and is 0 otherwise.

55. We represent the sets by bit strings of length 26, using alphabetical order. Thus

A is represented by 11 1110 0000 0000 0000 0000 0000,

B is represented by 01 1100 1000 0000 0100 0101 0000,

C is represented by 00 1010 0010 0000 1000 0010 0111, and

D is represented by 00 0110 0110 0001 1000 0110 0110.

To find the desired sets, we apply the indicated bitwise operations to these strings.

a) 11 1110 0000 0000 0000 0000 0000 $\vee$ 01 1100 1000 0000 0100 0101 0000 $=$

11 1110 1000 0000 0100 0101 0000, which represents the set $\{a,b,c,d,e,g,p,t,v\}$

b) 11 1110 0000 0000 0000 0000 0000 $\wedge$ 01 1100 1000 0000 0100 0101 0000 $=$

01 1100 0000 0000 0000 0000 0000, which represents the set $\{b,c,d\}$

c) (11 1110 0000 0000 0000 0000 0000 $\vee$ 00 0110 0110 0001 1000 0110 0110) $\wedge$

(01 1100 1000 0000 0100 0101 0000 $\vee$ 00 1010 0010 0000 1000 0010 0111) $=$

11 1110 0110 0001 1000 0110 0110 $\wedge$ 01 1110 1010 0000 1100 0111 0111 $=$

01 1110 0010 0000 1000 0110 0110, which represents the set $\{b,c,d,e,i,o,t,u,x,y\}$

d) 11 1110 0000 0000 0000 0000 0000 $\vee$ 01 1100 1000 0000 0100 0101 0000 $\vee$

00 1010 0010 0000 1000 0010 0111 $\vee$ 00 0110 0110 0001 1000 0110 0110 $=$

11 1110 1110 0001 1100 0111 0111, which represents the set

$\{a,b,c,d,e,g,h,i,n,o,p,t,u,v,x,y,z\}$

57. We simply adjoin the set itself to the list of its elements.

a) $\{1,2,3,\{1,2,3\}\}$ **b)** $\{\emptyset\}$ **c)** $\{\emptyset,\{\emptyset\}\}$ **d)** $\{\emptyset,\{\emptyset\},\{\emptyset,\{\emptyset\}\}\}$

59. a) The multiplicity of a in the union is the maximum of 3 and 2, the multiplicities of a in A and B. Since the maximum is 3, we find that a occurs with multiplicity 3 in the union. Working similarly with b, c (which appears with multiplicity 0 in B), and d (which appears with multiplicity 0 in A), we find that $A \cup B = \{3 \cdot a, 3 \cdot b, 1 \cdot c, 4 \cdot d\}$.

b) This is similar to part **(a)**, with "maximum" replaced by "minimum." Thus $A \cap B = \{2 \cdot a, 2 \cdot b\}$. (In particular, c and d appear with multiplicity 0—i.e., do not appear—in the intersection.)

c) In this case we subtract multiplicities, but never go below 0. Thus the answer is $\{1 \cdot a, 1 \cdot c\}$.

d) Similar to part **(c)** (subtraction in the opposite order); the answer is $\{1 \cdot b, 4 \cdot d\}$.

e) We add multiplicities here, to get $\{5 \cdot a, 5 \cdot b, 1 \cdot c, 4 \cdot d\}$.

61. Assume that the universal set contains just Alice, Brian, Fred, Oscar, and Rita. We subtract the degrees of membership from 1 to obtain the complement. Thus $\overline{F}$ is $\{0.4 \text{ Alice}, 0.1 \text{ Brian}, 0.6 \text{ Fred}, 0.9 \text{ Oscar}, 0.5 \text{ Rita}\}$, and $\overline{R}$ is $\{0.6 \text{ Alice}, 0.2 \text{ Brian}, 0.8 \text{ Fred}, 0.1 \text{ Oscar}, 0.3 \text{ Rita}\}$.

63. Taking the minimums, we obtain $\{0.4 \text{ Alice}, 0.8 \text{ Brian}, 0.2 \text{ Fred}, 0.1 \text{ Oscar}, 0.5 \text{ Rita}\}$ for $F \cap R$.

SECTION 2.3 Functions

The importance of understanding what a function is cannot be overemphasized—functions permeate all of mathematics and computer science. This exercise set enables you to make sure you understand functions and their properties. Exercise 29 is a particularly good benchmark to test your full comprehension of the abstractions involved. The definitions play a crucial role in doing proofs about functions. To prove that a function $f : A \to B$ is one-to-one, you need to show that $x_1 \neq x_2 \to f(x_1) \neq f(x_2)$ for all $x_1, x_2 \in A$. To prove that such a function is onto, you need to show that $\forall y \in B\, \exists x \in A\, (f(x) = y)$.

1. a) The expression $1/x$ is meaningless for $x = 0$, which is one of the elements in the domain; thus the "rule" is no rule at all. In other words, $f(0)$ is not defined.

b) Things like $\sqrt{-3}$ are undefined (or, at best, are complex numbers).

c) The "rule" for f is ambiguous. We must have $f(x)$ defined uniquely, but here there are two values associated with every x, the positive square root and the negative square root of $x^2 + 1$.

3. a) This is not a function, because there may be no zero bit in S, or there may be more than one zero bit in S. Thus there may be no value for $f(S)$ or more than one. In either case this violates the definition of a function, since $f(S)$ must have a unique value.

b) This is a function from the set of bit strings to the set of integers, since the number of 1 bits is always a clearly defined nonnegative integer.

c) This definition does not tell what to do with a nonempty string consisting of all 0's. Thus, for example, $f(000)$ is undefined. Therefore this is not a function.

5. In each case we want to find the domain (the set on which the function operates, which is implicitly stated in the problem) and the range (the set of possible output values).

a) Clearly the domain is the set of all bit strings. The range is $\mathbf{Z}$; the function evaluated at a string with n 1's and no 0's is n, and the function evaluated at a string with n 0's and no 1's is $-n$.

b) Again the domain is clearly the set of all bit strings. Since there can be any natural number of 0's in a bit string, the value of the function can be $0, 2, 4, \ldots$. Therefore the range is the set of even natural numbers.

c) Again the domain is the set of all bit strings. Since the number of leftover bits can be any whole number from 0 to 7 (if it were more, then we could form another byte), the range is $\{0, 1, 2, 3, 4, 5, 6, 7\}$.

d) As the problem states, the domain is the set of positive integers. Only perfect squares can be function values, and clearly every positive perfect square is possible. Therefore the range is $\{1, 4, 9, 16, \ldots\}$.

7. In each case, the domain is the set of possible inputs for which the function is defined, and the range is the set of all possible outputs on these inputs.

a) The domain is $\mathbf{Z}^+ \times \mathbf{Z}^+$, since we are told that the function operates on pairs of positive integers (the word "pair" in mathematics is usually understood to mean ordered pair). Since the maximum is again a positive integer, and all positive integers are possible maximums (by letting the two elements of the pair be the same), the range is $\mathbf{Z}^+$.

b) We are told that the domain is $\mathbf{Z}^+$. Since the decimal representation of an integer has to have at least one digit, at most nine digits do not appear, and of course the number of missing digits could be any number less than 9. Thus the range is $\{0,1,2,3,4,5,6,7,8,9\}$.

c) We are told that the domain is the set of bit strings. The block 11 could appear no times, or it could appear any positive number of times, so the range is $\mathbf{N}$.

d) We are told that the domain is the set of bit strings. Since the first 1 can be anywhere in the string, its position can be $1,2,3,\ldots$. If the bit string contains no 1's, the value is 0 by definition. Therefore the range is $\mathbf{N}$.

9. The floor function rounds down and the ceiling function rounds up.

a) 1 **b)** 0 **c)** 0 **d)** -1 **e)** 3 **f)** -1 **g)** $\lfloor \frac{1}{2} + \lceil \frac{3}{2} \rceil \rfloor = \lfloor \frac{1}{2} + 2 \rfloor = \lfloor 2\frac{1}{2} \rfloor = 2$
h) $\lfloor \frac{1}{2} \lfloor \frac{5}{2} \rfloor \rfloor = \lfloor \frac{1}{2} \cdot 2 \rfloor = \lfloor 1 \rfloor = 1$

11. We need to determine whether the range is all of $\{a,b,c,d\}$. It is for the function in part **(a)**, but not for the other two functions.

13. a) This function is onto, since every integer is 1 less than some integer. In particular, $f(x+1) = x$.

b) This function is not onto. Since n^2+1 is always positive, the range cannot include any negative integers.

c) This function is not onto, since the integer 2, for example, is not in the range. In other words, 2 is not the cube of any integer.

d) This function is onto. If we want to obtain the value x, then we simply need to start with $2x$, since $f(2x) = \lceil 2x/2 \rceil = \lceil x \rceil = x$ for all $x \in \mathbf{Z}$.

15. An onto function is one whose range is the entire codomain. Thus we must determine whether we can write every integer in the form given by the rule for f in each case.

a) Given any integer n, we have $f(0,n) = n$, so the function is onto.

b) Clearly the range contains no negative integers, so the function is not onto.

c) Given any integer m, we have $f(m,25) = m$, so the function is onto. (We could have used any constant in place of 25 in this argument.)

d) Clearly the range contains no negative integers, so the function is not onto.

e) Given any integer m, we have $f(m,0) = m$, so the function is onto.

17. Obviously there are an infinite number of correct answers to each part. The problem asked for a "formula." Parts **(a)** and **(c)** seem harder here, since we somehow have to fold the negative integers into the positive ones without overlap. Therefore we probably want to treat the negative integers differently from the positive integers. One way to do this with a formula is to make it a two-part formula. If one objects that this is not "a formula," we can counter as follows. Consider the function $g(x) = \lfloor 2^x \rfloor / 2^x$. Clearly if $x \geq 0$, then 2^x is a positive integer, so $g(x) = 2^x/2^x = 1$. If $x < 0$, then 2^x is a number between 0 and 1, so $g(x) = 0/2^x = 0$. If we want to define a function that has the value $f_1(x)$ when $x \geq 0$ and $f_2(x)$ when $x < 0$, then we can use the formula $g(x) \cdot f_1(x) + (1 - g(x)) \cdot f_2(x)$.

a) We could map the positive integers (and 0) into the positive multiples of 3, say, and the negative integers into numbers that are 1 greater than a multiple of 3, in a one-to-one manner. This will give us a function that leaves some elements out of the range. So let us define our function as follows:

$$f(x) = \begin{cases} 3x+3 & \text{if } x \geq 0 \\ 3|x|+1 & \text{if } x < 0 \end{cases}$$

The values of f on the inputs 0 through 4 are then $3,6,9,12,15$; and the values on the inputs -1 to -4 are $4,7,10,13$. Clearly this function is one-to-one, but it is not onto since, for example, 2 is not in the range.

b) This is easier. We can just take $f(x) = |x| + 1$. It is clearly onto, but $f(n)$ and $f(-n)$ have the same value for every positive integer n, so f is not one-to-one.

c) This is similar to part **(a)**, except that we have to be careful to hit all values. Mapping the nonnegative integers to the odds and the negative integers to the evens will do the trick:

$$f(x) = \begin{cases} 2x+1 & \text{if } x \geq 0 \\ 2|x| & \text{if } x < 0 \end{cases}$$

d) Here we can use a trivial example like $f(x) = 17$ or a simple nontrivial one like $f(x) = x^2 + 1$. Clearly these are neither one-to-one nor onto.

19. **a)** One way to determine whether a function is a bijection is to try to construct its inverse. This function is a bijection, since its inverse (obtained by solving $y = 2x + 1$ for x) is the function $g(y) = (y-1)/2$. Alternatively, we can argue directly. To show that the function is one-to-one, note that if $2x + 1 = 2x' + 1$, then $x = x'$. To show that the function is onto, note that $2((y-1)/2) + 1 = y$, so every number is in the range.

b) This function is not a bijection, since its range is the set of real numbers greater than or equal to 1 (which is sometimes written $[1, \infty)$), not all of $\mathbf{R}$. (It is not injective either.)

c) This function is a bijection, since it has an inverse function, namely the function $f(y) = y^{1/3}$ (obtained by solving $y = x^3$ for x).

d) This function is not a bijection. It is easy to see that it is not injective, since x and $-x$ have the same image, for all real numbers x. A little work shows that the range is only $\{\, y \mid 0.5 \leq y < 1 \,\} = [0.5, 1)$.

21. The key here is that larger denominators make smaller fractions, and smaller denominators make larger fractions. We have two things to prove, since this is an "if and only if" statement. First, suppose that f is strictly decreasing. This means that $f(x) > f(y)$ whenever $x < y$. To show that g is strictly increasing, suppose that $x < y$. Then $g(x) = 1/f(x) < 1/f(y) = g(y)$. Conversely, suppose that g is strictly increasing. This means that $g(x) < g(y)$ whenever $x < y$. To show that f is strictly decreasing, suppose that $x < y$. Then $f(x) = 1/g(x) > 1/g(y) = f(y)$.

23. We need to make the function decreasing, but not *strictly* decreasing, so, for example, we could take the trivial function $f(x) = 17$. If we want the range to be all of $\mathbf{R}$, we could define f in parts this way: $f(x) = -x - 1$ for $x < -1$; $f(x) = 0$ for $-1 \leq x \leq 1$; and $f(x) = -x + 1$ for $x > 1$.

25. The function is not one-to-one (for example, $f(2) = 2 = f(-2)$), so it is not invertible. On the restricted domain, the function is the identity function from the set of nonnegative real numbers to itself, $f(x) = x$, so it is one-to-one and onto and therefore invertible; in fact, it is its own inverse.

27. In each case, we need to compute the values of $f(x)$ for each $x \in S$.

a) Note that $f(\pm 2) = \lfloor (\pm 2)^2/3 \rfloor = \lfloor 4/3 \rfloor = 1$, $f(\pm 1) = \lfloor (\pm 1)^2/3 \rfloor = \lfloor 1/3 \rfloor = 0$, $f(0) = \lfloor 0^2/3 \rfloor = \lfloor 0 \rfloor = 0$, and $f(3) = \lfloor 3^2/3 \rfloor = \lfloor 3 \rfloor = 3$. Therefore $f(S) = \{0, 1, 3\}$.

b) In addition to the values we computed above, we note that $f(4) = 5$ and $f(5) = 8$. Therefore $f(S) = \{0, 1, 3, 5, 8\}$.

c) Note this time also that $f(7) = 16$ and $f(11) = 40$, so $f(S) = \{0, 8, 16, 40\}$.

d) $\{f(2), f(6), f(10), f(14)\} = \{1, 12, 33, 65\}$

29. In both cases, we can argue directly from the definitions.

a) Assume that both f and g are one-to-one. We need to show that $f \circ g$ is one-to-one. This means that we need to show that if x and y are two distinct elements of A, then $f(g(x)) \neq f(g(y))$. First, since g is one-to-one, the definition tells us that $g(x) \neq g(y)$. Second, since now $g(x)$ and $g(y)$ are distinct elements of B, and since f is one-to-one, we conclude that $f(g(x)) \neq f(g(y))$, as desired.

b) Assume that both f and g are onto. We need to show that $f \circ g$ is onto. This means that we need to show that if z is any element of C, then there is some element $x \in A$ such that $f(g(x)) = z$. First, since f is onto, we can conclude that there is an element $y \in B$ such that $f(y) = z$. Second, since g is onto and $y \in B$, we can conclude that there is an element $x \in A$ such that $g(x) = y$. Putting these together, we have $z = f(y) = f(g(x))$, as desired.

31. To establish the setting here, let us suppose that $g : A \to B$ and $f : B \to C$. Then $f \circ g : A \to C$. We are told that f and $f \circ g$ are onto. Thus all of C gets "hit" by the images of elements of B; in fact, each element in C gets hit by an element from A under the composition $f \circ g$. But this does not seem to tell us anything about the elements of B getting hit by the images of elements of A. Indeed, there is no reason that they must. For a simple counterexample, suppose that $A = \{a\}$, $B = \{b_1, b_2\}$, and $C = \{c\}$. Let $g(a) = b_1$, and let $f(b_1) = c$ and $f(b_2) = c$. Then clearly f and $f \circ g$ are onto, but g is not, since b_2 is not in its range.

33. We just perform the indicated operations on the defining expressions. Thus $f + g$ is the function whose value at x is $(x^2+1)+(x+2)$, or, more simply, $(f+g)(x) = x^2 + x + 3$. Similarly fg is the function whose value at x is $(x^2+1)(x+2)$; in other words, $(fg)(x) = x^3 + 2x^2 + x + 2$.

35. We simply solve the equation $y = ax + b$ for x. This gives $x = (y-b)/a$, which is well-defined since $a \neq 0$. Thus the inverse is $f^{-1}(y) = (y-b)/a$. To check that our work is correct, we must show that $f \circ f^{-1}(y) = y$ for all $y \in \mathbf{R}$ and that $f^{-1} \circ f(x) = x$ for all $x \in \mathbf{R}$. Both of these are straightforward algebraic manipulations. For the first, we have $f \circ f^{-1}(y) = f(f^{-1}(y)) = f((y-b)/a) = a\big((y-b)/a\big) + b = y$. The second is similar.

37. Let us arrange for S and T to be nonempty sets that have empty intersection. Then the left-hand side will be $f(\emptyset)$, which is the empty set. If we can make the right-hand side nonempty, then we will be done. We can make the right-hand side nonempty by making the codomain consist of just one element, so that $f(S)$ and $f(T)$ will both be the set consisting of that one element. The simplest example is as follows. Let $A = \{1, 2\}$ and $B = \{3\}$. Let f be the unique function from A to B (namely $f(1) = f(2) = 3$). Let $S = \{1\}$ and $T = \{2\}$. Then $f(S \cap T) = f(\emptyset) = \emptyset$, which is a proper subset of $f(S) \cap f(T) = \{3\} \cap \{3\} = \{3\}$.

39. a) We want to find the set of all numbers whose floor is 0. Since all numbers from 0 to 1 (including 0 but not 1) round down to 0, we conclude that $g^{-1}(\{0\}) = \{\, x \mid 0 \leq x < 1 \,\} = [0, 1)$.

b) This is similar to part **(a)**. All numbers from -1 to 2 (including -1 but not 2) round down to -1, 0, or 1; we conclude that $g^{-1}(\{-1, 0, 1\}) = \{\, x \mid -1 \leq x < 2 \,\} = [-1, 2)$.

c) Since $g(x)$ is always an integer, there are no values of x such that $g(x)$ is strictly between 0 and 1. Thus the inverse image in this case is the empty set.

41. Note that the complementation here is with respect to the relevant universal set. Thus $\overline{S} = B - S$ and $\overline{f^{-1}(S)} = A - f^{-1}(S)$. There are two things to prove in order to show that these two sets are equal: that the left-hand side of the equation is a subset of the right-hand side, and that the right-hand side is a subset of the left-hand side. First let $x \in f^{-1}(\overline{S})$. This means that $f(x) \in \overline{S}$, or equivalently that $f(x) \notin S$. Therefore by definition of inverse image, $x \notin f^{-1}(S)$, so $x \in \overline{f^{-1}(S)}$. For the other direction, assume that $x \in \overline{f^{-1}(S)}$. Then $x \notin f^{-1}(S)$. By definition this means that $f(x) \notin S$, which means that $f(x) \in \overline{S}$. Therefore by definition, $x \in f^{-1}(\overline{S})$.

43. There are three cases. Define the "fractional part" of x to be $f(x) = x - \lfloor x \rfloor$. Clearly $f(x)$ is always between 0 and 1 (inclusive at 0, exclusive at 1), and $x = \lfloor x \rfloor + f(x)$. If $f(x)$ is less than $\frac{1}{2}$, then $x - \frac{1}{2}$ will have a value slightly less than $\lfloor x \rfloor$, so when we round up, we get $\lfloor x \rfloor$. In other words, in this case $\lceil x - \frac{1}{2} \rceil = \lfloor x \rfloor$, and indeed that is the integer closest to x. If $f(x)$ is greater than $\frac{1}{2}$, then $x - \frac{1}{2}$ will have a value slightly greater than $\lfloor x \rfloor$, so when we round up, we get $\lfloor x \rfloor + 1$. In other words, in this case $\lceil x - \frac{1}{2} \rceil = \lfloor x \rfloor + 1$, and indeed that is the integer closest to x in this case. Finally, if the fractional part is exactly $\frac{1}{2}$, then x is midway between two integers, and $\lceil x - \frac{1}{2} \rceil = \lfloor x \rfloor$, which is the smaller of these two integers.

45. We can write the real number x as $\lfloor x \rfloor + \epsilon$, where ϵ is a real number satisfying $0 \le \epsilon < 1$. Since $\epsilon = x - \lfloor x \rfloor$, we have $0 \le x - \lfloor x \rfloor < 1$. The first two inequalities, $x - 1 < \lfloor x \rfloor$ and $\lfloor x \rfloor \le x$, follow algebraically. For the other two inequalities, we can write $x = \lceil x \rceil - \epsilon$, where again $0 \le \epsilon < 1$. Then $0 \le \lceil x \rceil - x < 1$, and again the desired inequalities follow by easy algebra.

47. **a)** One direction (the "only if" part) is obvious: If $x < n$, then since $\lfloor x \rfloor \le x$ it follows that $\lfloor x \rfloor < n$. We will prove the other direction (the "if" part) indirectly (we will prove its contrapositive). Suppose that $x \ge n$. Then "the greatest integer not exceeding x" must be at least n, since n is an integer not exceeding x. That is, $\lfloor x \rfloor \ge n$.

b) One direction (the "only if" part) is obvious: If $n < x$, then since $x \le \lceil x \rceil$ it follows that $n < \lceil x \rceil$. We will prove the other direction (the "if" part) indirectly (we will prove its contrapositive). Suppose that $n \ge x$. Then "the smallest integer not less than x" must be no greater than n, since n is an integer not less than x. That is, $\lceil x \rceil \le n$.

49. If n is even, then $n = 2k$ for some integer k. Thus $\lfloor n/2 \rfloor = \lfloor k \rfloor = k = n/2$. If n is odd, then $n = 2k + 1$ for some integer k. Thus $\lfloor n/2 \rfloor = \lfloor k + \frac{1}{2} \rfloor = k = (n-1)/2$.

51. Without loss of generality we can assume that $x \ge 0$, since the equation to be proved is equivalent to the same equation with $-x$ substituted for x. Then the left-hand side is $\lceil -x \rceil$ by definition, and the right-hand side is $-\lfloor x \rfloor$. Thus this problem reduces to Exercise 50. Its proof is straightforward. Write x as $n + \epsilon$, where n is a natural number and ϵ is a real number satisfying $0 \le \epsilon < 1$. Then clearly $\lceil -x \rceil = \lceil -n - \epsilon \rceil = -n$ and $-\lfloor x \rfloor = -\lfloor n + \epsilon \rfloor = -n$ as well.

53. In some sense this question is its own answer—the number of integers strictly between a and b is the number of integers strictly between a and b. Presumably we seek an expression involving a, b, and the floor and/or ceiling function to answer this question. If we round a down and round b up to integers, then we will be looking at the smallest and largest integers just outside the range of integers we want to count, respectively. These values are of course $\lfloor a \rfloor$ and $\lceil b \rceil$, respectively. Then the answer is $\lceil b \rceil - \lfloor a \rfloor - 1$ (just think of counting all the integers between these two values, excluding both ends—if a row of fenceposts one foot apart extends for k feet, then there are $k - 1$ fenceposts not counting the end posts). Note that this even works when, for example, $a = 0.3$ and $b = 0.7$.

55. Since a byte is eight bits, all we are asking for in each case is $\lceil n/8 \rceil$, where n is the number of bits.
a) $\lceil 7/8 \rceil = 1$ **b)** $\lceil 17/8 \rceil = 3$ **c)** $\lceil 1001/8 \rceil = 126$ **d)** $\lceil 28800/8 \rceil = 3600$

57. In each case we need to divide the number of bytes (octets) by 1500 and round up. In other words, the answer is $\lceil n/1500 \rceil$, where n is the number of bytes.
a) $\lceil 150{,}000/1500 \rceil = 100$ **b)** $\lceil 384{,}000/1500 \rceil = 256$ **c)** $\lceil 1{,}544{,}000/1500 \rceil = 1030$
d) $\lceil 45{,}300{,}000/1500 \rceil = 30{,}200$

59. The graph will look exactly like the graph of the function $f(x) = \lfloor x \rfloor$, shown in Figure 10a, except that the picture will be compressed by a factor of 2 in the horizontal direction, since x has been replaced by $2x$.

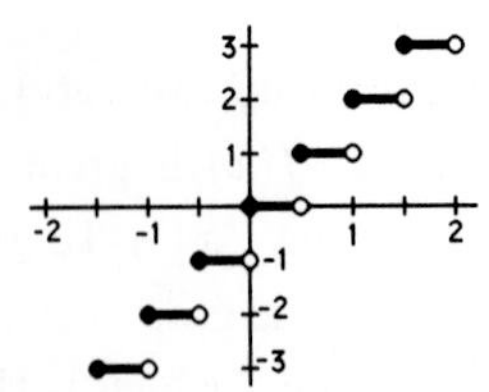

61. This is a step function, with values changing only at the integers. We note the pattern that $f(x)$ jumps by 1 when x passes through an odd integer (because of the $\lfloor x \rfloor$ term), and by 2 when x passes through an even integer (an additional jump caused by the $\lfloor x/2 \rfloor$ term). The result is as shown.

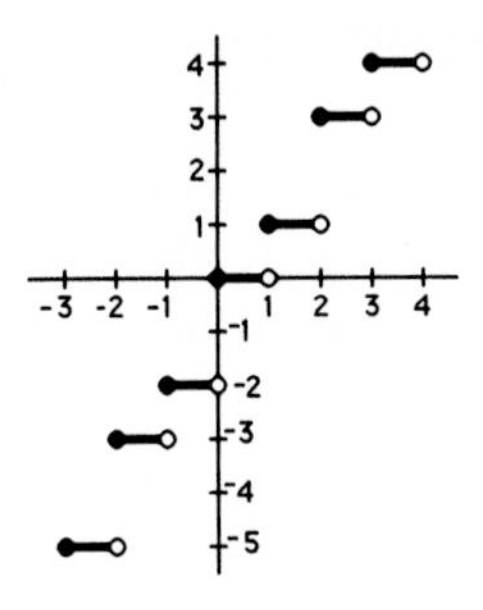

63. **a)** The graph will look exactly like the graph of the function $f(x) = \lfloor x \rfloor$, shown in Figure 10a, except that the picture will be shifted to the left by $\frac{1}{2}$ unit, since x has been replaced by $x + \frac{1}{2} = x - (-\frac{1}{2})$.

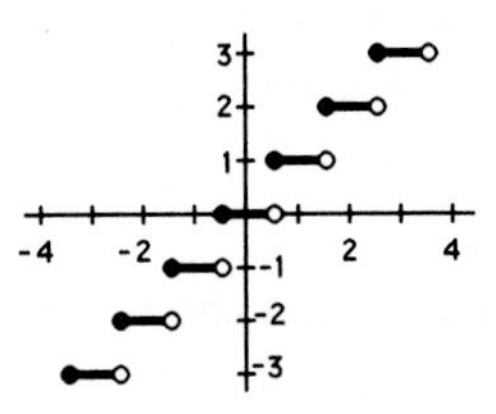

b) The graph will look exactly like the graph of the function $f(x) = \lfloor 2x \rfloor$, shown in the solution to Exercise 59, except that the picture will be shifted to the left by $\frac{1}{2}$ unit, since x has been replaced by $x + \frac{1}{2}$. Alternatively, we can note that $f(x)$ can be rewritten as $\lfloor 2x \rfloor + 1$, so the graph is the graph shown in the solution to Exercise 59 shifted up one unit.

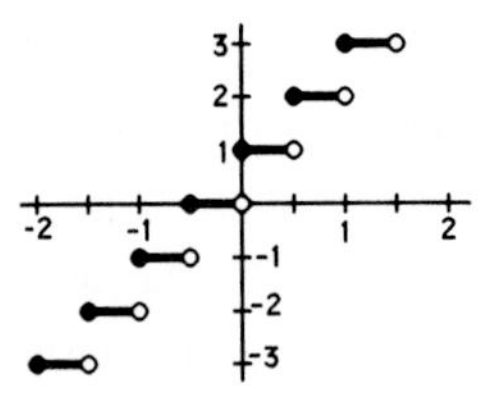

c) The graph will look exactly like the graph of the function $f(x) = \lceil x \rceil$, shown in Figure 10b, except that the x-axis is stretched by a factor of 3. Thus we can use the same picture and just relabel the x-axis.

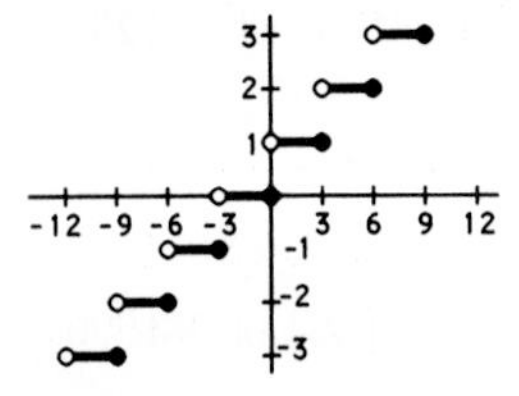

d) The graph is a step version of the usual hyperbola $y = 1/x$. Note that $x = 0$ is not in the domain. The graph can be drawn by first plotting the points at which $1/x$ is an integer ($x = 1, \pm\frac{1}{2}, \pm\frac{1}{3}, \ldots$) and then filling in the horizontal segments, making sure to note that they go to the right (for example, if x is a little bigger than $\frac{1}{2}$, then $1/x$ is a little less than 2, so $f(x) = 2$, since we are rounding up here). Note that $f(x) = 1$ for $x \geq 1$, and $f(x) = 0$ for $x < -1$.

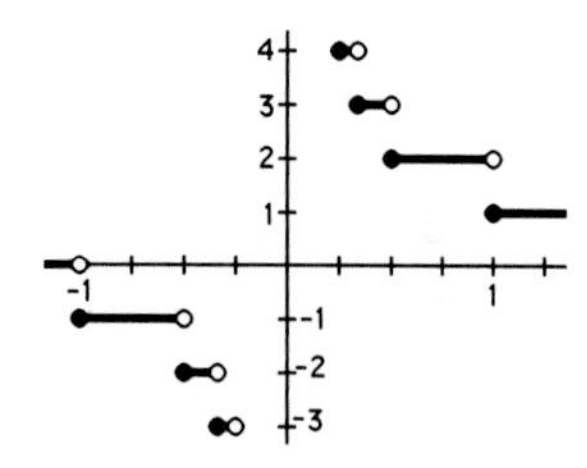

e) The key thing to note is that since we can pull integers outside the floor and ceiling function (identity (4) in Table 1), we can write $f(x)$ more simply as $\lfloor x \rfloor + \lceil x \rceil$. When x is an integer, this is just $2x$. When x is between two integers, however, this has the value of the integer between the two integers $2\lfloor x \rfloor$ and $2\lceil x \rceil$. The graph is therefore as shown here.

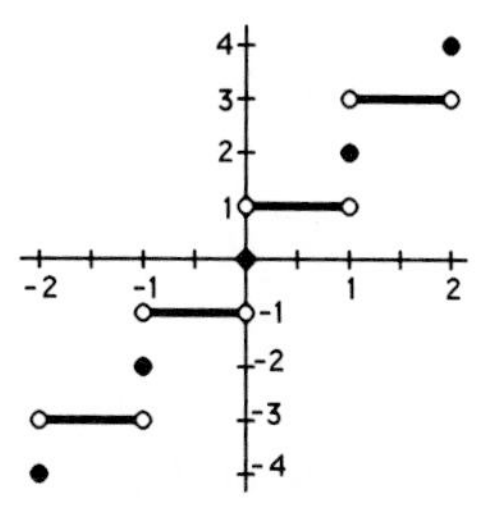

f) The basic shape is the parabola, $y = x^2$. In particular, for x an even integer, $f(x) = x^2$, since the terms inside the floor and ceiling function symbols are integers. However, because of these step functions, the curve is broken into steps. At even integers other than $x = 0$ there are isolated points in the graph. Also, the graph takes jumps at all the integer and half-integer values outside the range $-2 < x < \frac{1}{2}$ (where in fact $f(x) = 0$). The portion of the graph near the origin is shown here.

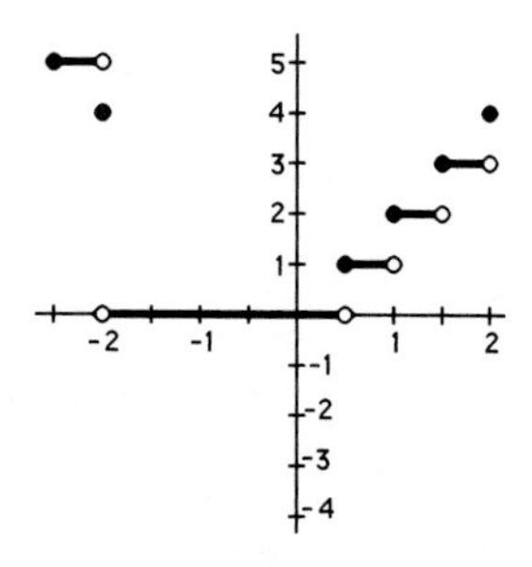

g) Despite the complicated-looking formula, this is really quite similar to part **(a)**; in fact, we'll see that it's identical! First note that the expression inside the outer ceiling function symbols is always going to be a half-integer; therefore we can tell exactly what its rounded-up value will be, namely $\lfloor x - \frac{1}{2} \rfloor + 1$. Furthermore, since identity (4) of Table 1 allows us to move the 1 inside the floor function symbols, we have $f(x) = \lfloor x + \frac{1}{2} \rfloor$. Therefore this is the same function as in part **(a)**.

65. We simply need to solve the equation $y = x^3 + 1$ for x. This is easily done by algebra: $x = (y-1)^{1/3}$. Therefore the inverse function is given by the rule $f^{-1}(y) = (y-1)^{1/3}$ (or, equivalently, by the rule $f^{-1}(x) = (x-1)^{1/3}$, since the variable in the definition is just a dummy variable).

67. We can prove all of these identities by showing that the left-hand side is equal to the right-hand side for all possible values of x. In each instance (except part **(c)**, in which there are only two cases), there are four cases to consider, depending on whether x is in A and/or B.

a) If x is in both A and B, then $f_{A\cap B}(x) = 1$; and the right-hand side is $1\cdot 1 = 1$ as well. Otherwise $x \notin A\cap B$, so the left-hand side is 0, and the right-hand side is either $0\cdot 1$ or $1\cdot 0$ or $0\cdot 0$, all of which are also 0.

b) If x is in both A and B, then $f_{A\cup B}(x) = 1$; and the right-hand side is $1+1-1\cdot 1 = 1$ as well. If x is in A but not B, then $x \in A\cup B$, so the left-hand side is still 1, and the right-hand side is $1+0-1\cdot 0 = 1$, as desired. The case in which x is in B but not A is similar. Finally, if x is in neither A nor B, then the left-hand side is 0, and the right-hand side is $0+0-0\cdot 0 = 0$ as well.

c) If $x \in A$, then $x \notin \overline{A}$, so $f_{\overline{A}}(x) = 0$. The right-hand side equals $1-1=0$ in this case, as well. On the other hand, if $x \notin A$, then $x \in \overline{A}$, so the left-hand side is 1, and the right-hand side is $1-0=1$ as well.

d) If x is in both A and B, then $x \notin A\oplus B$, so $f_{A\oplus B}(x) = 0$. The right-hand side is $1+1-2\cdot 1\cdot 1 = 0$ as well. Next, if $x\in A$ but $x \notin B$, then $x \in A\oplus B$, so the left-hand side is 1. The right-hand side is $1+0-2\cdot 1\cdot 0 = 1$ as well. The case $x\in B \wedge x\notin A$ is similar. Finally, if x is in neither A nor B, then $x\notin A\oplus B$, so the left-hand side is 0; and the right-hand side is also $0+0-2\cdot 0\cdot 0 = 0$.

69. a) This is true. Since $\lfloor x\rfloor$ is already an integer, $\lceil \lfloor x\rfloor \rceil = \lfloor x \rfloor$.

b) A little experimentation shows that this is not always true. To disprove it we need only produce a counterexample, such as $x = \frac{1}{2}$. In that case the left-hand side is $\lfloor 1\rfloor = 1$, while the right-hand side is $2\cdot 0 = 0$.

c) This is true. We prove it by cases. If x is an integer, then by identity (4b) in Table 1, we know that $\lceil x+y\rceil = x + \lceil y\rceil$, and it follows that the difference is 0. Similarly, if y is an integer. The remaining case is that $x = n+\epsilon$ and $y = m+\delta$, where n and m are integers and ϵ and δ are positive real numbers less than 1. Then $x+y$ will be greater than $m+n$ but less than $m+n+2$, so $\lceil x+y\rceil$ will be either $m+n+1$ or $m+n+2$. Therefore the given expression will be either $(n+1)+(m+1)-(m+n+1) = 1$ or $(n+1)+(m+1)-(m+n+2) = 0$, as desired.

d) This is clearly false, as we can find with a little experimentation. Take, for example, $x = 1/10$ and $y = 3$. Then the left-hand side is $\lceil 3/10\rceil = 1$, but the right-hand side is $1\cdot 3 = 3$.

e) Again a little trial and error will produce a counterexample. Take $x = 1/2$. Then the left-hand side is 1 while the right-hand side is 0.

71. a) If x is a positive integer, then the two sides are identical. So suppose that $x = n^2+m+\epsilon$, where n is the largest perfect square integer less than x, m is a nonnegative integer, and $0<\epsilon<1$. For example, if $x = 13.2$, then $n=3$, $m=4$, and $\epsilon = 0.2$. Then both $\sqrt{x}$ and $\sqrt{\lfloor x\rfloor} = \sqrt{n^2+m}$ are between n and $n+1$. Therefore both sides of the equation equal n.

b) If x is a positive integer, then the two sides are identical. So suppose that $x = n^2-m-\epsilon$, where n is the smallest perfect square integer greater than x, m is a nonnegative integer, and ϵ is a real number with $0<\epsilon<1$. For example, if $x = 13.2$, then $n=4$, $m=2$, and $\epsilon = 0.8$. Then both $\sqrt{x}$ and $\sqrt{\lceil x\rceil} = \sqrt{n^2-m}$ are between $n-1$ and n. Therefore both sides of the equation equal n.

73. In each case we easily read the domain and codomain from the notation. The domain of definition is obtained by determining for which values the definition makes sense. The function is total if the domain of definition is the entire domain (so that there are no values for which the partial function is undefined).

a) The domain is $\mathbf{Z}$ and the codomain is $\mathbf{R}$. Since division is possible by every nonzero number, the domain of definition is all the nonzero integers; $\{0\}$ is the set of values for which f is undefined. (It is not total.)

b) The domain and codomain are both given to be $\mathbf{Z}$. Since the definition makes sense for all integers, this is a total function, whose domain of definition is also $\mathbf{Z}$; the set of values for which f is undefined is $\emptyset$.

c) By inspection, the domain is the Cartesian product $\mathbf{Z}\times\mathbf{Z}$, and the codomain is $\mathbf{Q}$. Since fractions cannot have a 0 in the denominator, we must exclude the "slice" of $\mathbf{Z}\times\mathbf{Z}$ in which the second coordinate is 0. Thus the domain of definition is $\mathbf{Z} \times (\mathbf{Z} - \{0\})$, and the function is undefined for all values in $\mathbf{Z} \times \{0\}$. It is not a total function.

d) The domain is given to be $\mathbf{Z} \times \mathbf{Z}$ and the codomain is given to be $\mathbf{Z}$. Since the definition makes sense for all pairs of integers, this is a total function, whose domain of definition is also $\mathbf{Z} \times \mathbf{Z}$; the set of values for which f is undefined is $\emptyset$.

e) Again the domain and codomain are $\mathbf{Z} \times \mathbf{Z}$ and $\mathbf{Z}$, respectively. Since the definition is only stated for those pairs in which the first coordinate exceeds the second, the domain of definition is $\{\,(m,n) \mid m > n\,\}$, and therefore the set of values for which the function is undefined is $\{\,(m,n) \mid m \le n\,\}$. It is not a total function.

75. a) By definition, to say that S has cardinality m is to say that S has exactly m distinct elements. Therefore we can imagine enumerating them, as a child counts objects: assign the first object to 1, the second to 2, and so on. This provides the one-to-one correspondence.

b) By part **(a)**, there is a bijection f from S to $\{1,2,\ldots,m\}$ and a bijection g from T to $\{1,2,\ldots,m\}$. This tells us that g^{-1} is a bijection from $\{1,2,\ldots,m\}$ to T. Then the composition $g^{-1} \circ f$ is the desired bijection from S to T.

77. A little experimentation with this function shows the pattern:

$f(1,1)=1$	$f(2,1)=3$	$f(3,1)=6$	$f(4,1)=10$	$f(5,1)=15$	$f(6,1)=21$
$f(1,2)=2$	$f(2,2)=5$	$f(3,2)=9$	$f(4,2)=14$	$f(5,2)=20$	$f(6,2)=27$
$f(1,3)=4$	$f(2,3)=8$	$f(3,3)=13$	$f(4,3)=19$	$f(5,3)=26$	
$f(1,4)=7$	$f(2,4)=12$	$f(3,4)=18$	$f(4,4)=25$		
$f(1,5)=11$	$f(2,5)=17$	$f(3,5)=24$			
$f(1,6)=16$	$f(2,6)=23$				
$f(1,7)=22$					

We see by looking at the diagonals of this table that the function takes on successive values as $m+n$ increases. When $m+n=2$, $f(m,n)=1$. When $m+n=3$, $f(m,n)$ takes on the values 2 and 3. When $m+n=4$, $f(m,n)$ takes on the values 4, 5, and 6. And so on. It is clear from the formula that the range of values the function takes on for a fixed value of $m+n$, say $m+n=x$, is $\frac{(x-2)(x-1)}{2}+1$ through $\frac{(x-2)(x-1)}{2}+(x-1)$, since m can assume the values $1,2,3,\ldots,(x-1)$ under these conditions, and the first term in the formula is a fixed positive integer when $m+n$ is fixed. To show that this function is one-to-one and onto, we merely need to show that the range of values for $x+1$ picks up precisely where the range of values for x left off, i.e., that $f(x-1,1)+1=f(1,x)$. We compute:

$$f(x-1,1)+1 = \frac{(x-2)(x-1)}{2}+(x-1)+1 = \frac{x^2-x+2}{2} = \frac{(x-1)x}{2}+1 = f(1,x)$$

SECTION 2.4 Sequences and Summations

The first half of this exercise set contains a lot of routine practice with the concept of and notation for sequences. It also discusses **telescoping sums**; *the* **product notation**, *corresponding to the summation notation discussed in the section; the* **factorial function**, *which occurs repeatedly in subsequent chapters; and a few challenging exercises on more complicated sequences. The last part of the exercise set has some fairly challenging exercises involving infinite sets. Do not be surprised if you find this last material very strange and hard to comprehend at first going; mathematicians did not understand it at all until the late nineteenth century, three hundred years after calculus was well understood. To show that an infinite set is countable, you need to find a one-to-one correspondence between the set and the set of positive integers. One way to do this directly is to provide a listing of the elements of the set. (There is no listing, for instance, of the set of real numbers.) Various indirect means are also available, such as showing that the set is a subset of a countable set, or showing that it is the union of a countable collection of countable sets.*

1. **a)** $a_0 = 2 \cdot (-3)^0 + 5^0 = 2 \cdot 1 + 1 = 3$ **b)** $a_1 = 2 \cdot (-3)^1 + 5^1 = 2 \cdot (-3) + 5 = -1$
c) $a_4 = 2 \cdot (-3)^4 + 5^4 = 2 \cdot 81 + 625 = 787$ **d)** $a_5 = 2 \cdot (-3)^5 + 5^5 = 2 \cdot (-243) + 3125 = 2639$

3. In each case we simply evaluate the given function at $n = 0, 1, 2, 3$.
a) $a_0 = 2^0 + 1 = 2$, $a_1 = 2^1 + 1 = 3$, $a_2 = 2^2 + 1 = 5$, $a_3 = 2^3 + 1 = 9$
b) $a_0 = 1^1 = 1$, $a_1 = 2^2 = 4$, $a_2 = 3^3 = 27$, $a_3 = 4^4 = 256$
c) $a_0 = \lfloor 0/2 \rfloor = 0$, $a_1 = \lfloor 1/2 \rfloor = 0$, $a_2 = \lfloor 2/2 \rfloor = 1$, $a_3 = \lfloor 3/2 \rfloor = 1$
d) $a_0 = \lfloor 0/2 \rfloor + \lceil 0/2 \rceil = 0 + 0 = 0$, $a_1 = \lfloor 1/2 \rfloor + \lceil 1/2 \rceil = 0 + 1 = 1$, $a_2 = \lfloor 2/2 \rfloor + \lceil 2/2 \rceil = 1 + 1 = 2$, $a_3 = \lfloor 3/2 \rfloor + \lceil 3/2 \rceil = 1 + 2 = 3$. Note that $\lfloor n/2 \rfloor + \lceil n/2 \rceil$ always equals n.

5. In each case we just follow the instructions.
a) $2, 5, 8, 11, 14, 17, 20, 23, 26, 29$ **b)** $1, 1, 1, 2, 2, 2, 3, 3, 3, 4$ **c)** $1, 1, 3, 3, 5, 5, 7, 7, 9, 9$
d) This requires a bit of routine calculation. For example, the fifth term is $5! - 2^5 = 120 - 32 = 88$. The first ten terms are $-1, -2, -2, 8, 88, 656, 4912, 40064, 362368, 3627776$.
e) $3, 6, 12, 24, 48, 96, 192, 384, 768, 1536$ **f)** $1, 1, 2, 3, 5, 8, 13, 21, 34, 55$
g) For $n = 1$, the binary expansion is 1, which has one bit, so the first term of the sequence is 1. For $n = 2$, the binary expansion is 10, which has two bits, so the second term of the sequence is 2. Continuing in this way we see that the first ten terms are $1, 2, 2, 3, 3, 3, 3, 4, 4, 4$. Note that the sequence has one 1, two 2's, four 3's, eight 4's, as so on, with 2^{k-1} copies of k.
h) The English word for 1 is "one" which has three letters, so the first term is 3. This makes a good brain-teaser; give someone the sequence and ask her or him to find the pattern. The first ten terms are $3, 3, 5, 4, 4, 3, 5, 5, 4, 3$.

7. One pattern is that each term is twice the preceding term. A formula for this would be that the nth term is 2^{n-1}. Another pattern is that we obtain the next term by adding increasing values to the previous term. Thus to move from the first term to the second we add 1; to move from the second to the third we add 2; then add 3, and so on. So the sequence would start out $1, 2, 4, 7, 11, 16, 22, \ldots$. We could also have trivial answers such as the rule that the first three terms are $1, 2, 4$ and all the rest are 17 (so the sequence is $1, 2, 4, 17, 17, 17, \ldots$), or that the terms simply repeat $1, 2, 4, 1, 2, 4, 1, 2, 4, \ldots$. Here is another pattern: Take n points on the unit circle, and connect each of them to all the others by line segments. The inside of the circle will be divided into a number of regions. What is the largest this number can be? Call that value a_n. If there is one point, then there are no lines and therefore just the one original region inside the circle; thus $a_1 = 1$. If $n = 2$, then the one chord divides the interior into two parts, so $a_2 = 2$. Three points give us a triangle, and that makes four regions (the inside of the triangle and the three pieces outside the triangle), so $a_3 = 4$. Careful drawing shows that the sequence starts out $1, 2, 4, 8, 16, 31$. That's right: 31, not 32.

Creative students may well find other rules or patterns with various degrees of appeal.

9. In some sense there are no right answers here. The solutions stated are the most appealing patterns that the author has found.
a) It looks as if we have one 1 and one 0, then two of each, then three of each, and so on, increasing the number of repetitions by one each time. Thus we need three more 1's (and then four 0's) to continue the sequence.
b) A pattern here is that the positive integers are listed in increasing order, with each even number repeated. Thus the next three terms are $9, 10, 10$.
c) The terms in the odd locations (first, third, fifth, etc.) are just the successive terms in the geometric sequence that starts with 1 and has ratio 2, and the terms in the even locations are all 0. The nth term is 0 if n is even and is $2^{(n-1)/2}$ if n is odd. Thus the next three terms are $32, 0, 64$.

d) The first term is 3 and each successive term is twice its predecessor. This is a geometric sequence. The nth term is $3 \cdot 2^{n-1}$. Thus the next three terms are $384, 768, 1536$.

e) The first term is 15 and each successive term is 7 less than its predecessor. This is an arithmetic sequence. The nth term is $22 - 7n$. Thus the next three terms are $-34, -41, -48$.

f) The rule is that the first term is 3 and the nth term is obtained by adding n to the $(n-1)^{\text{th}}$ term. One can actually find a quadratic expression for a sequence in which the successive differences form an arithmetic sequence; here it is $(n^2+n+4)/2$. The easiest way to see this is to note that the nth term is $3+2+3+4+5+6+\cdots+n$. Except for the initial 3 instead of a 1, the nth term is the sum of the first n positive integers, which is $n(n+1)/2$ by a formula in Table 2. Therefore the nth term is $(n(n+1)/2)+2$, as claimed. We see that the next three terms are $57, 68, 80$.

g) One should play around with the sequence if nothing is apparent at first. Here we note that all the terms are even, so if we divide by 2 we obtain the sequence $1, 8, 27, 64, 125, 216, 343, \ldots$. This sequence appears in Table 1; it is the cubes. So the nth term is $2n^3$. Thus the next three terms are $1024, 1458, 2000$.

h) These terms look close to the terms of the sequence whose n^{th} term is $n!$ (see Table 1). In fact, we see that the nth term here is $n!+1$. Thus the next three terms are $362881, 3628801, 39916801$.

11. It is pretty clear that a_n should be approximately equal to $n+\sqrt{n}$, since the sequence is just the sequence of positive integers with perfect squares left out. There are about $\sqrt{n}$ perfect squares up to n, so the count needs to get ahead by about this amount. Proving that this plausibility argument gives the correct formula involves some careful counting.

The sequence begins $2, 3, 5, 6, 7, 8, 10, 11, \ldots$. We can write it as the sequence $a_n = n$ plus a sequence b_n that jumps every time a perfect square is encountered. Thus $\{b_n\}$ begins $1, 1, 2, 2, 2, 2, 3, 3, 3, 3, 3, 3, 4, \ldots$; there are two 1's, four 2's, six 3's, eight 4's, and so on. So we must show that $b_n = \{\sqrt{n}\}$, where $\{\sqrt{n}\}$ means the integer closest to $\sqrt{n}$ (note that there is never an ambiguity here, since this will never be a half-integer). Because of the way the sequence is formed, $b_n \le k$ if and only if $2+4+6+\cdots+2k \ge n$. This is equivalent to $k(k+1) \ge n$. Applying the quadratic formula and recalling that k is an integer, we obtain $b_n = \lceil(-1+\sqrt{1+4n})/2\rceil$. Simplifying, we have $b_n = \left\lceil -\frac{1}{2}+\sqrt{n+\frac{1}{4}}\right\rceil$. Subtracting $\frac{1}{2}$ and then rounding up is the same as rounding to the nearest integer (the smaller one if $\sqrt{n+\frac{1}{4}}$ is a half-integer—see Exercise 43 in Section 2.3), so (with this understanding) $b_n = \left\{\sqrt{n+\frac{1}{4}}\right\}$. But it can never happen that $\sqrt{n} \le m+\frac{1}{2}$ while $\sqrt{n+\frac{1}{4}} > m+\frac{1}{2}$ for some positive integer m—this would imply that $n \le m^2+m+\frac{1}{4}$ and $n > m^2+m$, an impossibility. Therefore $\{\sqrt{n}\} = \left\{\sqrt{n+\frac{1}{4}}\right\}$, and we are done.

An alternative solution is provided in the answer section of the text.

13. **a)** $2+3+4+5+6=20$ **b)** $1-2+4-8+16=11$ **c)** $3+3+\cdots+3=10\cdot 3=30$

d) This series "telescopes": each term cancels part of the term before it (see also Exercise 19). The sum is $(2-1)+(4-2)+(8-4)+\cdots+(512-256) = -1+512 = 511$.

15. We use the formula for the sum of a geometric progression: $\sum_{j=0}^{n} ar^j = a(r^{n+1}-1)/(r-1)$.

a) Here $a=3$, $r=2$, and $n=8$, so the sum is $3(2^9-1)/(2-1) = 1533$.

b) Here $a=1$, $r=2$, and $n=8$. The sum taken over all the values of j from 0 to n is, by the formula, $(2^9-1)/(2-1) = 511$. However, our sum starts at $j=1$, so we must subtract out the term that isn't there, namely 2^0. Hence the answer is $511-1=510$.

c) Again we have to subtract the missing terms, so the sum is $((-3)^9-1)/((-3)-1)-(-3)^0-(-3)^1 = 4921-1-(-3) = 4923$.

d) $2((-3)^9-1)/((-3)-1) = 9842$

17. The easiest way to do these sums, since the number of terms is reasonably small, is just to write out the summands explicitly. Note that the inside index (j) runs through all of its values for each value of the outside index (i).

a) $(1+1)+(1+2)+(1+3)+(2+1)+(2+2)+(2+3)=21$

b) $(0+3+6+9)+(2+5+8+11)+(4+7+10+13)=78$

c) $(1+1+1)+(2+2+2)+(3+3+3)=18$

d) $(0+0+0)+(1+2+3)+(2+4+6)=18$

19. If we just write out what the sum means, we see that parts of successive terms cancel, leaving only two terms:

$$\sum_{j=1}^{n}(a_j-a_{j-1})=a_1-a_0+a_2-a_1+a_3-a_2+\cdots+a_{n-1}-a_{n-2}+a_n-a_{n-1}=a_n-a_0$$

21. a) We use the hint, where $a_k=k^2$:

$$\sum_{k=1}^{n}(2k-1)=\sum_{k=1}^{n}(k^2-(k-1)^2)=n^2-0^2=n^2$$

b) We can use the distributive law to rewrite $\sum_{k=1}^{n}(2k-1)$ (which we know from part **(a)** equals n^2) in terms of the sum we want, $S=\sum_{k=1}^{n}k$:

$$n^2=\sum_{k=1}^{n}(2k-1)=2\sum_{k=1}^{n}k-\sum_{k=1}^{n}1=2S-n\,.$$

Now we solve for S, obtaining $S=(n^2+n)/2$, which is usually expressed as $n(n+1)/2$.

23. This exercise is like Example 15. From Table 2 we know that $\sum_{k=1}^{200}k=200\cdot 201/2=20100$, and $\sum_{k=1}^{99}k=99\cdot 100/2=4950$. Therefore the desired sum is $20100-4950=15150$.

25. If we write down the first few terms of this sum we notice a pattern. It starts $(1+1+1)+(2+2+2+2+2)+(3+3+3+3+3+3+3)+\cdots$. There are three 1's, then five 2'ss, then seven 3'ss, and so on; in general there are $(i+1)^2-i^2=2i+1$ copies of i. So we need to sum $i(2i+1)$ for an appropriate range of values for i. We must find this range. It gets a little messy at the end if m is such that the sequence stops before a complete range of the last value is present. Let $n=\lfloor\sqrt{m}\rfloor-1$. Then there are $n+1$ blocks, and $(n+1)^2-1$ is where the next-to-last block ends. The sum of those complete blocks is $\sum_{i=1}^{n}i(2i+1)=\sum_{i=1}^{n}2i^2+i=n(n+1)(2n+1)/3+n(n+1)/2$. The remaining terms in our summation all have the value $n+1$ and the number of them present is $m-((n+1)^2-1)$. Our final answer is therefore $n(n+1)(2n+1)/3+n(n+1)/2+(n+1)(m-(n+1)^2+1)$.

27. a) 0 (anything times 0 is 0) **b)** $5\cdot 6\cdot 7\cdot 8=1680$

c) Each factor is either 1 or -1, so the product is either 1 or -1. To see which it is, we need to determine how many of the factors are -1. Clearly there are 50 such factors, namely when $i=1, 3, 5, \ldots, 99$. Since $(-1)^{50}=1$, the product is 1.

d) $2\cdot 2\cdots 2=2^{10}=1024$

29. $0!+1!+2!+3!+4!=1+1+1\cdot 2+1\cdot 2\cdot 3+1\cdot 2\cdot 3\cdot 4=1+1+2+6+24=34$

31. a) The negative integers are countable. Each negative integer can be paired with its absolute value to give the desired one-to-one correspondence: $1\leftrightarrow -1$, $2\leftrightarrow -2$, $3\leftrightarrow -3$, etc.

b) The even integers are countable. We can list the set of even integers in the order $0, 2, -2, 4, -4, 6, -6, \ldots$, and pair them with the positive integers listed in their natural order. Thus $1 \leftrightarrow 0$, $2 \leftrightarrow 2$, $3 \leftrightarrow -2$, $4 \leftrightarrow 4$, etc. There is no need to give a formula for this correspondence—the discussion given is quite sufficient; but it is not hard to see that we are pairing the positive integer n with the even integer $f(n)$, where $f(n) = n$ if n is even and $f(n) = 1 - n$ if n is odd.

c) The proof that the set of real numbers between 0 and 1 is not countable (Example 21) can easily be modified to show that the set of real numbers between 0 and $1/2$ is not countable. We need to let the digit d_i be something like 2 if $d_{ii} \neq 2$ and 3 otherwise. The number thus constructed will be a real number between 0 and $1/2$ that is not in the list.

d) This set is countable, exactly as in part (**b**); the only difference is that there we are looking at the multiples of 2 and here we are looking at the multiples of 7. The correspondence is given by pairing the positive integer n with $7n/2$ if n is even and $-7(n-1)/2$ if n is odd.

33. **a)** The bit strings not containing 0 are just the bit strings consisting of all 1's, so this set is $\{\lambda, 1, 11, 111, 1111, \ldots\}$, where λ denotes the empty string (the string of length 0). Thus this set is countable, where the correspondence matches the natural number n with the string of n 1's.

b) This is a subset of the set of rational numbers, so it is countable (see Exercise 36). To find a correspondence, we can just follow the path in Example 20, but omit fractions in the top three rows (as well as continuing to omit those fractions not in lowest terms).

c) This set is uncountable, as can be shown by applying the diagonal argument of Example 21.

d) This set is uncountable, as can be shown by applying the diagonal argument of Example 21.

35. Yes. We need to look at this from the other direction, by noting that $A = B \cup (A - B)$. We are given that B is countable. If $A - B$ were also countable, then, since the union of two countable sets is countable (which we are asked to prove as Exercise 40), we would conclude that A is countable. But we are given that A is not countable. Therefore our assumption that $A - B$ is countable is wrong, and we conclude that $A - B$ is uncountable. (This is an example of a proof by contraposition.)

37. This is just the contrapositive of Exercise 36 and so follows directly from it. In more detail, suppose that B were countable, say with elements $b_1, b_2, \ldots$. Then since $A \subseteq B$, we can list the elements of A using the order in which they appear in this listing of B. Therefore A is countable, contradicting the hypothesis. Thus B is not countable.

39. By what we are given, we know that there are bijections f from A to B and g from C to D. Then we can define a bijection from $A \times C$ to $B \times D$ by sending (a, c) to $(f(a), g(c))$. This is clearly one-to-one and onto, so we have shown that $A \times C$ and $B \times D$ have the same cardinality.

41. Since empty sets do not contribute any elements to unions, we can assume that none of the sets in our given countable collection of countable sets is the empty set. If there are no sets in the collection, then the union is empty and therefore countable. Otherwise let the countable sets be A_1, A_2, ... (If there are only a finite number k of them, then we can still assume that they form an infinite sequence by taking $A_{k+1} = A_{k+2} = \cdots = A_1$.) Since each set A_i is countable and nonempty, we can list its elements in a sequence as a_{i1}, a_{i2}, ...; again, if the set is finite we can list its elements and then list a_{i1} repeatedly to assure an infinite sequence. Now we just need a systematic way to put all the elements a_{ij} into a sequence. We do this by listing first all the elements a_{ij} in which $i + j = 2$ (there is only one such pair, $(1, 1)$), then all the elements in which $i + j = 3$ (there are only two such pairs, $(1, 2)$ and $(2, 1)$), and so on; except that we do not list any element that we have already listed. So, assuming that these elements are distinct, our list starts a_{11}, a_{12}, a_{21}, a_{13}, a_{22}, a_{31}, a_{14}, ... (If any of these terms duplicates a previous term, then it is simply omitted.) The result of this process will be either an infinite sequence or a finite sequence containing all the elements of the union of the sets A_i. Thus that union is countable.

43. There are only a finite number of bit strings of each finite length, so we can list all the bit strings by listing first those of length 0, then those of length 1, etc. The listing might be $\lambda, 0, 1, 00, 01, 10, 11, 000, 001, \ldots$. (Recall that λ denotes the empty string.) Actually this is a special case of Exercise 41: the set of all bit strings is the union of a countable number of countable (actually finite) sets, namely the sets of bit strings of length n for $n = 0, 1, 2, \ldots$.

45. We argued in the solution to Exercise 43 that the set of all strings of symbols from the alphabet $\{0, 1\}$ is countable, since there are only a finite number of bit strings of each length. There was nothing special about the alphabet $\{0, 1\}$ in that argument. For any finite alphabet (for example, the alphabet consisting of all upper and lower case letters, numerals, and punctuation and other mathematical marks typically used in a programming language), there are only a finite number of strings of length 1 (namely the number of symbols in the alphabet), only a finite number of strings of length 2 (namely, the square of this number), and so on. Therefore, using the result of Exercise 41 again, we conclude that there are only countably many strings from any given finite alphabet. Now the set of all computer programs in a particular language is just a subset of the set of all strings over that alphabet (some strings are meaningless jumbles of symbols that are not valid programs), so by Exercise 36, this set, too, is countable.

47. In Exercise 45 we saw that there are only a countable number of computer programs, so there are only a countable number of computable functions. In Exercise 46 we saw that there are an uncountable number of functions. Hence not all functions are computable. Indeed, in some sense, since uncountable sets are so much bigger than countable sets, *almost all* functions are not computable! This is not really so surprising; in real life we deal with only a small handful of useful functions, and these are computable. Note that this is a nonconstructive proof—we have not exhibited even one noncomputable function, merely argued that they have to exist. Actually finding one is much harder, but it can be done. For example, the following function is not computable. Let T be the function from the set of positive integers to $\{0, 1\}$ defined by letting $T(n)$ be 0 if the number 0 is in the range of the function computed by the nth computer program (where we list them in alphabetical order by length) and letting $T(n) = 1$ otherwise.

GUIDE TO REVIEW QUESTIONS FOR CHAPTER 2

1. See p. 114. To prove that A is a subset of B we need to show that an arbitrarily chosen element x of A must also be an element of B.

2. The empty set is the set with no elements. It satisfies the definition of being a subset of every set vacuously.

3. **a)** See p. 116. **b)** See p. 122.

4. **a)** See p. 116. **b)** always **c)** 2^n

5. **a)** See pp. 121 and 123 and the preamble to Exercise 32 in Section 2.2.
b) union: integers that are odd or positive; intersection: odd positive integers; difference: even positive integers; symmetric difference: even positive integers together with odd negative integers

6. **a)** $A = B \equiv (A \subseteq B \wedge B \subseteq A) \equiv \forall x(x \in A \leftrightarrow x \in B)$ **b)** See pp. 124–126.
c) $A \cap \overline{B \cap C} = A \cap (\overline{B} \cup \overline{C}) = (A \cap \overline{B}) \cup (A \cap \overline{C}) = (A - B) \cup (A - C)$; use Venn diagrams

7. Underlying each set identity is a logical equivalence. See, for instance, Example 11 in Section 2.2.

8. **a)** See p. 134. **b)** $\mathbf{Z}$, $\mathbf{Z}$, $\mathbf{Z}^+ = \mathbf{N} - \{0\}$

9. **a)** See p. 136. **b)** See p. 137. **c)** $f(n) = n$ **d)** $f(n) = 2n$ **e)** $f(n) = \lceil n/2 \rceil$
f) $f(n) = 42548$

10. **a)** See p. 139: $f^{-1}(b) = a \equiv f(a) = b$
b) when it is one-to-one and onto
c) yes—itself

11. **a)** See p. 143. **b)** integers

12. Hint: subtract 5 from each term and look at the resulting sequence.

13. See p. 155.

14. Set up a one-to-one correspondence between the set of positive integers and the set of all odd integers, such as $1 \leftrightarrow 1$, $2 \leftrightarrow -1$, $3 \leftrightarrow 3$, $4 \leftrightarrow -3$, $5 \leftrightarrow 5$, $6 \leftrightarrow -5$, and so on.

15. See Example 21 in Section 2.4.

SUPPLEMENTARY EXERCISES FOR CHAPTER 2

1. **a)** $\overline{A} =$ the set of words that are not in A
b) $A \cap B =$ the set of words that are in both A and B
c) $A - B =$ the set of words that are in A but not B
d) $\overline{A} \cap \overline{B} = \overline{(A \cup B)} =$ the set of words that are in neither A nor B
e) $A \oplus B =$ the set of words that are in A or B but not both (can also be written as $(A - B) \cup (B - A)$ or as $(A \cup B) - (A \cap B)$)

3. Yes. We must show that every element of A is also an element of B. So suppose a is an arbitrary element of A. Then $\{a\}$ is a subset of A, so it is an element of the power set of A. Since the power set of A is a subset of the power set of B, it follows that $\{a\}$ is an element of the power set of B, which means that $\{a\}$ is a subset of B. But this means that the element of $\{a\}$, namely a, is an element of B, as desired.

5. We will show that each side is a subset of the other. First suppose $x \in A - (A - B)$. Then $x \in A$ and $x \notin A - B$. Now the only way for x not to be in $A - B$, given that it is in A, is for it to be in B. Thus we have that x is in both A and B, so $x \in A \cap B$. For the other direction, let $x \in A \cap B$. Then $x \in A$ and $x \in B$. It follows that $x \notin A - B$, and so x is in $A - (A - B)$.

7. We need only provide a counterexample to show that $(A - B) - C$ is not necessarily equal to $A - (B - C)$. Let $A = C = \{1\}$, and let $B = \emptyset$. Then $(A - B) - C = (\{1\} - \emptyset) - \{1\} = \{1\} - \{1\} = \emptyset$, whereas $A - (B - C) = \{1\} - (\emptyset - \{1\}) = \{1\} - \emptyset = \{1\}$.

9. This is not necessarily true. For a counterexample, let $A = B = \{1, 2\}$, let $C = \emptyset$, and let $D = \{1\}$. Then $(A - B) - (C - D) = \emptyset - \emptyset = \emptyset$, but $(A - C) - (B - D) = \{1, 2\} - \{2\} = \{1\}$.

11. **a)** Since $\emptyset \subseteq A \cap B \subseteq A \subseteq A \cup B \subseteq U$, we have the order $|\emptyset| \leq |A \cap B| \leq |A| \leq |A \cup B| \leq |U|$.
b) Note that $A - B \subseteq A \oplus B \subseteq A \cup B$. Also recall that $|A \cup B| = |A| + |B| - |A \cap B|$, so that $|A \cup B|$ is always less than or equal to $|A| + |B|$. Putting this all together, we have $|\emptyset| \leq |A - B| \leq |A \oplus B| \leq |A \cup B| \leq |A| + |B|$.

13. a) Yes, f is one-to-one, since each element of the domain $\{1,2,3,4\}$ is sent by f to a different element of the codomain. No, g is not one-to-one, since g sends the two different elements a and d of the domain to the same element, 2.

b) Yes, f is onto, since every element in the codomain $\{a,b,c,d\}$ is the image under f of some element in the domain $\{1,2,3,4\}$. In other words, the range of f is the entire codomain. No, g is not onto, since the element 4 in the codomain is not in the range of g (is not the image under g of any element of the domain $\{a,b,c,d\}$).

c) Certainly f has an inverse, since it is one-to-one and onto. Its inverse is the function from $\{a,b,c,d\}$ to $\{1,2,3,4\}$ that sends a to 3, sends b to 4, sends c to 2, and sends d to 1. (Each element in $\{a,b,c,d\}$ gets sent by f^{-1} to the element in $\{1,2,3,4\}$ that gets sent to it by f.) Since g is not one-to-one and onto, it has no inverse.

15. We need to look at an example in which f is not one-to-one. Suppose we let A be a set with two elements, say 1 and 2, and let B be a set with just one element, say 3. Of course f will be the unique function from A to B. If we let $S=\{1\}$ and $T=\{2\}$, then $f(S\cap T)=f(\emptyset)=\emptyset$, but $f(S)\cap f(T)=\{3\}\cap\{3\}=\{3\}$.

17. The key is to look at sets with just one element. On these sets, the induced functions act just like the original functions. So let x be an arbitrary element of A. Then $\{x\}\in P(A)$, and $S_f(\{x\})=\{\,f(y)\mid y\in\{x\}\,\}=\{f(x)\}$. By the same reasoning, $S_g(\{x\})=\{g(x)\}$. Since $S_f=S_g$, we can conclude that $\{f(x)\}=\{g(x)\}$, and so necessarily $f(x)=g(x)$.

19. This is certainly true if either x or y is an integer, since then this equation is equivalent to the identity (4a) in Table 1 of Section 2.3. Otherwise, write x and y in terms of their integer and fractional parts: $x=n+\epsilon$ and $y=m+\delta$, where $n=\lfloor x\rfloor$, $0<\epsilon<1$, $m=\lfloor y\rfloor$, and $0<\delta<1$. If $\delta+\epsilon<1$, then the equation is true, since both sides equal $m+n$; if $\delta+\epsilon\ge 1$, then the equation is false, since the left-hand side equals $m+n$, but the right-hand side equals $m+n+1$. In summary, the equation is true if and only if either at least one of x and y is an integer or the sum of the fractional parts of x and y is less than 1. (Note that the second condition in the disjunction subsumes the first.)

21. Write x and y in terms of their integer and fractional parts: $x=n+\epsilon$ and $y=m+\delta$, where $n=\lfloor x\rfloor$, $0\le\epsilon<1$, $m=\lfloor y\rfloor$, and $0\le\delta<1$. If $\delta=\epsilon=0$, then both sides equal $n+m$. If $\epsilon=0$ but $\delta>0$, then the left-hand side equals $n+m+1$, but the right-hand side equals $n+m$. If $\epsilon>0$, then the right-hand side equals $n+m+1$, so the two sides will be equal if and only if $\epsilon+\delta\le 1$ (otherwise the left-hand side would be $n+m+2$). In summary, the equation is true if and only if either both x and y are integers, or x is not an integer but the sum of the fractional parts of x and y is less than or equal to 1.

23. If x is an integer, then clearly $\lfloor x\rfloor+\lfloor m-x\rfloor=x+m-x=m$. Otherwise, write x in terms of its integer and fractional parts: $x=n+\epsilon$, where $n=\lfloor x\rfloor$ and $0<\epsilon<1$. In this case $\lfloor x\rfloor+\lfloor m-x\rfloor=\lfloor n+\epsilon\rfloor+\lfloor m-n-\epsilon\rfloor=n+m-n-1=m-1$, because we had to round $m-n-\epsilon$ down to the next smaller integer.

25. Write $n=2k+1$ for some integer k. Then $n^2=4k^2+4k+1$, so $n^2/4=k^2+k+\frac{1}{4}$. Therefore $\lceil n^2/4\rceil=k^2+k+1$. But we also have $(n^2+3)/4=(4k^2+4k+1+3)/4=(4k^2+4k+4)/4=k^2+k+1$.

27. Let us write $x=n+(r/m)+\epsilon$, where n is an integer, r is an nonnegative integer less than m, and ϵ is a real number with $0\le\epsilon<1/m$. In other words, we are peeling off the integer part of x (i.e., $n=\lfloor x\rfloor$) and the whole multiples of $1/m$ beyond that. Then the left-hand side is $\lfloor nm+r+m\epsilon\rfloor=nm+r$. On the right-hand side, the terms $\lfloor x\rfloor$ through $\lfloor x+(m-r-1)/m\rfloor$ are all just n, and the remaining terms, if any, from $\lfloor x+(m-r)/m\rfloor$ through $\lfloor x+(m-1)/m\rfloor$, are all $n+1$. Therefore the right-hand side is $(m-r)n+r(n+1)=nm+r$ as well.

29. This product telescopes. The numerator in the fraction for k cancels the denominator in the fraction for $k+1$. So all that remains of the product is the numerator for $k = 100$ and the denominator for $k = 1$, namely $101/1 = 101$.

31. There is no good way to determine a nice rule for this kind of problem. One just has to look at the sequence and see what seems to be happening. In this sequence, we notice that $10 = 2 \cdot 5$, $39 = 3 \cdot 13$, $172 = 4 \cdot 43$, and $885 = 5 \cdot 177$. We then also notice that $3 = 1 \cdot 3$ for the second and third terms. So each odd-indexed term (assuming that we call the first term a_1) comes from the term before it, by multiplying by successively larger integers. In symbols, this says that $a_{2n+1} = n \cdot a_{2n}$ for all $n > 0$. Then we notice that the even-indexed terms are obtained in a similar way by adding: $a_{2n} = n + a_{2n-1}$ for all $n > 0$. So the next four terms are $a_{13} = 6 \cdot 891 = 5346$, $a_{14} = 7 + 5346 = 5353$, $a_{15} = 7 \cdot 5353 = 37471$, and $a_{14} = 8 + 37471 = 37479$.

WRITING PROJECTS FOR CHAPTER 2

Books and articles indicated by bracketed symbols below are listed near the end of this manual. You should also read the general comments and advice you will find there about researching and writing these essays.

1. A classic source here is [Wil]. It gives a very readable account of many philosophical issues in the foundations of mathematics, including the topic for this essay.

2. Our list of references mentions several history of mathematics books, such as [Bo4] and [Ev3]. You should also browse the shelves in your library, around QA 21.

3. Go to the Encyclopedia's website, `http://www.research.att.com/~njas/sequences/`.

4. A Web search should turn up some useful references here, including an article in *Science News Online.* It gets its name from the fact that a graph describing it looks like the output of an electrocardiogram.

5. A Web search for this phrase will turn up much information.

6. A classic source here is [Wil]. It gives a very readable account of many philosophical issues in the foundations of mathematics, including the topic for this essay. Of course a Web search will turn up lots of useful material, as well.

CHAPTER 3
The Fundamentals: Algorithms, the Integers, and Matrices

SECTION 3.1 Algorithms

Many of the exercises here are actually miniature programming assignments. Since this is not a book on programming, we have glossed over some of the finer points. For example, there are (at least) two ways to pass variables to procedures—by value and by reference. In the former case the original values of the arguments are not changed. In the latter case they are. In most cases we will assume that arguments are passed by reference. None of these exercises are tricky; they just give the reader a chance to become familiar with algorithms written in pseudocode. The reader should refer to Appendix 3 for more details of the pseudocode being used here.

1. Initially max is set equal to the first element of the list, namely 1. The **for** loop then begins, with i set equal to 2. Immediately i (namely 2) is compared to n, which equals 10 for this sequence (the entire input is known to the computer, including the value of n). Since $2 < 10$, the statement in the loop is executed. This is an **if**...**then** statement, so first the comparison in the **if** part is made: max (which equals 1) is compared to $a_i = a_2 = 8$. Since the condition is true, namely $1 < 8$, the **then** part of the statement is executed, so max is assigned the value 8.

The only statement in the **for** loop has now been executed, so the loop variable i is incremented (from 2 to 3), and we repeat the process. First we check again to verify that i is still less than n (namely $3 < 10$), and then we execute the **if**...**then** statement in the body of the loop. This time, too, the condition is satisfied, since $max = 8$ is less than $a_3 = 12$. Therefore the assignment statement $max := a_i$ is executed, and max receives the value 12.

Next the loop variable is incremented again, so that now $i = 4$. After a comparison to determine that $4 < 10$, the **if**...**then** statement is executed. This time the condition fails, since $max = 12$ is not less than $a_4 = 9$. Therefore the **then** part of the statement is not executed. Having finished with this pass through the loop, we increment i again, to 5. This pass through the loop, as well as the next pass through, behave exactly as the previous pass, since the condition $max < a_i$ continues to fail. On the sixth pass through the loop, however, with $i = 7$, we find again that $max < a_i$, namely $12 < 14$. Therefore max is assigned the value 14.

After three more uneventful passes through the loop (with $i = 8$, 9, and 10), we finally increment i to 11. At this point, when the comparison of i with n is made, we find that i is no longer less than or equal to n, so no further passes through the loop are made. Instead, control passes beyond the loop. In this case there are no statements beyond the loop, so execution halts. Note that when execution halts, max has the value 14 (which is the correct maximum of the list), and i has the value 11. (Actually in many programming languages, the value of i after the loop has terminated in this way is undefined.)

3. We will call the procedure *sum*. Its input is a list of integers, just as was the case for Algorithm 1. Indeed, we can just mimic the structure of Algorithm 1. We assume that the list is not empty (an assumption made in Algorithm 1 as well).

```
procedure sum(a_1, a_2, ..., a_n : integers)
sum := a_1
for i := 2 to n
        sum := sum + a_i
{ sum is the sum of all the elements in the list }
```

5. We need to go through the list and find cases when one element is equal to the following element. However, in order to avoid listing the values that occur more than once more than once, we need to skip over repeated duplicates after we have found one duplicate. The following algorithm will do it.

```
procedure duplicates(a_1, a_2, ..., a_n : integers in nondecreasing order)
k := 0 {this counts the duplicates}
j := 2
while j ≤ n
begin
        if a_j = a_{j-1} then
        begin
                k := k + 1
                c_k := a_j
                while (j ≤ n and a_j = c_k)
                        j := j + 1
        end
j := j + 1
end {c_1, c_2, ..., c_k is the desired list}
```

7. We need to go through the list and record the index of the last even integer seen.

```
procedure last even location(a_1, a_2, ..., a_n : integers)
k := 0
for i := 1 to n
        if a_i is even then k := i
end {k is the desired location (or 0 if there are no evens)}
```

9. We just need to look at the list forward and backward simultaneously, going at least half-way through it.

```
procedure palindrome check(a_1 a_2 ... a_n : string)
answer := true
for i := 1 to ⌊n/2⌋
        if a_i ≠ a_{n+1-i} then answer := false
end { answer is true if and only if string is a palindrome}
```

11. We cannot simply write $x := y$ followed by $y := x$, because then the two variables will have the same value, and the original value of x will be lost. Thus there is no way to accomplish this task with just two assignment statements. Three are necessary, and sufficient, as the following code shows. The idea is that we need to save temporarily the original value of x.

```
temp := x
x := y
y := temp
```

13. We will not give these answers in quite the detail we used in Exercise 1.

a) Note that $n = 8$ and $x = 9$. Initially i is set equal to 1. The **while** loop is executed as long as $i \leq 8$ and the ith element of the list is not equal to 9. Thus on the first pass we check that $1 \leq 8$ and that $9 \neq 1$ (since $a_1 = 1$), and therefore perform the statement $i := i + 1$. At this point $i = 2$. We check that $2 \leq 8$ and $9 \neq 3$, and therefore again increment i, this time to 3. This process continues until $i = 7$. At that point the condition "$i \leq 8$ and $9 \neq a_i$" is false, since $a_7 = 9$. Therefore the body of the loop is not executed (so i is still equal to 7), and control passes beyond the loop.

The next statement is the **if**... **then** statement. The condition is satisfied, since $7 \leq 8$, so the statement *location* $:= i$ is executed, and *location* receives the value 7. The **else** clause is not executed. This completes the procedure, so *location* has the correct value, namely 7, which indicates the location of the element x (namely 9) in the list: 9 is the seventh element.

b) Initially i is set equal to 1 and j is set equal to 8. Since $i < j$ at this point, the steps of the **while** loop are executed. First m is set equal to $\lfloor (1+8)/2 \rfloor = 4$. Then since x (which equals 9) is greater than a_4 (which equals 5), the statement $i := m + 1$ is executed, so i now has the value 5. At this point the first iteration through the loop is finished, and the search has been narrowed to the sequence $a_5, \ldots, a_8$.

In the next pass through the loop (there is another pass since $i < j$ is still true), m becomes $\lfloor (5+8)/2 \rfloor = 6$. Since again $x > a_m$, we reset i to be $m+1$, which is 7. The loop is now repeated with $i = 7$ and $j = 8$. This time m becomes 7, so the test $x > a_m$ (i.e., $9 > 9$) fails; thus $j := m$ is executed, so now $j = 7$.

At this point $i \not< j$, so there are no more iterations of the loop. Instead control passes to the statement beyond the loop. Since the condition $x = a_i$ is true, *location* is set to 7, as it should be, and the algorithm is finished.

15. We need to find where x goes, then slide the rest of the list down to make room for x, then put x into the space created. In the procedure that follows, we employ the trick of temporarily tacking $x+1$ onto the end of the list, so that the **while** loop will always terminate. Also note that the indexing in the **for** loop is slightly tricky since we need to work from the end of the list toward the front.

```
procedure insert(x, a1, a2, ..., an : integers)
{the list is in order: a1 ≤ a2 ≤ ··· ≤ an}
a(n+1) := x + 1
i := 1
while x > ai
        i := i + 1 {the loop ends when i is the index for x}
for j := 0 to n − i {shove the rest of the list to the right}
        a(n−j+1) := a(n−j)
ai := x
{x has been inserted into the correct spot in the list, now of length n + 1}
```

17. This algorithm is similar to *max*, except that we need to keep track of the location of the maximum value, as well as the maximum value itself. Note that we need a strict inequality in the test $max < a_i$, since we do not want to change *location* if we find another occurrence of the maximum value. As usual we assume that the list is not empty.

```
procedure first largest(a1, a2, ..., an : integers)
max := a1
location := 1
for i := 2 to n
        if max < ai then
        begin
                max := ai
                location := i
        end
{location is the location of the first occurrence of the largest element in the list}
```

19. We need to handle the six possible orderings in which the three integers might occur. (Actually there are more than six possibilities, because some of the numbers might be equal to each other—we get around this problem by using $\leq$ rather than $<$ for our comparisons.) We will use the **if**... **then**... **else if**... **then**... **else if**... construction. A condition such as $a \leq b \leq c$ is really the conjunction of two conditions: $a \leq b$ and

$b \le c$. (Alternately, we could have handled the cases in a nested fashion.) Note that the mean is computed first, independent of the ordering.

procedure *mean median max min*$(a, b, c$: integers)
$mean := (a + b + c)/3$
if $a \le b \le c$ **then**
begin
  $min := a$
  $median := b$
  $max := c$
end
else if $a \le c \le b$ **then**
begin
  $min := a$
  $median := c$
  $max := b$
end
else if $b \le a \le c$ **then**
begin
  $min := b$
  $median := a$
  $max := c$
end
else if $b \le c \le a$ **then**
begin
  $min := b$
  $median := c$
  $max := a$
end
else if $c \le a \le b$ **then**
begin
  $min := c$
  $median := a$
  $max := b$
end
else if $c \le b \le a$ **then**
begin
  $min := c$
  $median := b$
  $max := a$
end
{the correct values of *mean*, *median*, *max*, and *min* have been assigned}

21. We must assume that the sequence has at least three terms. This is a special case of a sorting algorithm. Our approach is to interchange numbers in the list when they are out of order. It is not hard to see that this needs to be done only three times in order to guarantee that the elements are finally in correct order: test and interchange (if necessary) the first two elements, then test and interchange (if necessary) the second and third elements (insuring that the largest of the three is now third), then test and interchange (if necessary) the first and second elements again (insuring that the smallest is now first).

procedure *first three*$(a_1, a_2, \ldots, a_n$: integers)
if $a_1 > a_2$ **then** interchange a_1 and a_2
if $a_2 > a_3$ **then** interchange a_2 and a_3
if $a_1 > a_2$ **then** interchange a_1 and a_2
{the first three elements are now in nondecreasing order}

23. For notation, assume that $f : A \to B$, where A is the set consisting of the distinct integers $a_1, a_2, \ldots, a_n$, and B is the set consisting of the distinct integers $b_1, b_2, \ldots, b_m$. All $n+m+1$ of these entities (the elements of A, the elements of B, and the function f) are the input to the algorithm. We set up an array called *hit* (indexed by the integers) to keep track of which elements of B are the images of elements of A; thus $hit(b_i)$ equals 0 until we find an a_j such that $f(a_j) = b_i$, at which time we set $hit(b_i)$ equal to 1. Simultaneously we keep track of how many hits we have made (i.e., how many times we changed some $hit(b_i)$ from 0 to 1). If at the end we have made m hits, then f is onto; otherwise it is not. Note that we record the output as a logical value assigned to the variable that has the name of the procedure. This is a common practice in some programming languages.

```
procedure onto(f : function, a_1, a_2, ..., a_n, b_1, b_2, ..., b_m : integers)
for i := 1 to m
        hit(b_i) := 0 {no one has been hit yet}
count := 0 {there have been no hits yet}
for j := 1 to n
        if hit(f(a_j)) = 0 then {a new hit!}
        begin
                hit(f(a_j)) := 1
                count := count + 1
        end
if count = m then onto := true
else onto := false
{f is onto if and only if there have been m hits}
```

25. This algorithm is straightforward.

```
procedure count ones(a_1 a_2 ... a_n : bit string)
count := 0 {no 1's yet}
for i := 1 to n
        if a_i = 1 then count := count + 1
{count contains the number of 1's}
```

27. We start with the pseudocode for binary search given in the text and modify it. In particular, we need to compute two middle subscripts (one third of the way through the list and two thirds of the way through) and compare x with two elements in the list. Furthermore, we need special handling of the case when there are two elements left to be considered. The following pseudocode is reasonably straightforward.

```
procedure ternary search(x : integer, a_1, a_2, ..., a_n : increasing integers)
i := 1
j := n
while i < j - 1
begin
        l := ⌊(i + j)/3⌋
        u := ⌊2(i + j)/3⌋
        if x > a_u then i := u + 1
        else if x > a_l then
        begin
                i := l + 1
                j := u
        end
        else j := l
end
if x = a_i then location := i
else if x = a_j then location := j
else location := 0
{location is the subscript of the term equal to x (0 if not found)}
```

29. The following algorithm will find the first mode in the sequence. At each point in the execution of this algorithm, *modecount* is the number of occurrences of the element found to occur most often so far (which is called *mode*). Whenever a more frequently occurring element is found (the main inner loop), *modecount* and *mode* are updated.

```
procedure find a mode(a_1, a_2, ..., a_n : nondecreasing integers)
modecount := 0
i := 1
while i ≤ n
begin
        value := a_i
        count := 1
        while i ≤ n and a_i = value
        begin
                count := count + 1
                i := i + 1
        end
        if count > modecount then
        begin
                modecount := count
                mode := value
        end
end
{mode is the first value occurring most often, namely modecount times}
```

31. The following algorithm goes through the terms of the sequence one by one, and, for each term, compares it to all previous terms. If it finds a match, then it stores the subscript of that term in *location* and terminates the search. If no match is ever found, then *location* is set to 0.

```
procedure find duplicate(a_1, a_2, ..., a_n : integers)
location := 0 {no match found yet}
i := 2
while i ≤ n and location = 0
begin
        j := 1
        while j < i and location = 0
                if a_i = a_j then location := i
                else j := j + 1
        i := i + 1
end
{location is the subscript of the first value that repeats a previous value in the sequence
  and is 0 if there is no such value}
```

33. The following algorithm goes through the terms of the sequence one by one, and, for each term, checks whether it is less than the immediately preceding term. If it finds such a term, then it stores the subscript of that term in *location* and terminates the search. If no term satisfies this condition, then *location* is set to 0.

```
procedure find decrease(a_1, a_2, ..., a_n : positive integers)
location := 0 {no match found yet}
i := 2
while i ≤ n and location = 0
        if a_i < a_{i-1} then location := i
        else i := i + 1
{location is the subscript of the first value that is less than the immediately preceding one
  and is 0 if there is no such value}
```

35. There are four passes through the list. On the first pass, the 3 and the 1 are interchanged first, then the next two comparisons produce no interchanges, and finally the last comparison results in the interchange of the 7 and the 4. Thus after one pass the list reads $1,3,5,4,7$. During the next pass, the 5 and the 4 are interchanged, yielding $1,3,4,5,7$. There are two more passes, but no further interchanges are made, since the list is now in order.

37. We need to add a Boolean variable to indicate whether any interchanges were made during a pass. Initially this variable, which we will call *still_interchanging*, is set to **true**. If no interchanges were made, then we can quit. To do this neatly, we turn the outermost loop into a **while** loop that is executed as long as $i < n$ and *still_interchanging* is true. Thus our pseudocode is as follows.

```
procedure betterbubblesort(a_1, ..., a_n)
i := 1
still_interchanging := true
while i < n and still_interchanging
begin
    still_interchanging := false
    for j := 1 to n - i
        if a_j > a_{j+1} then
        begin
            still_interchanging := true
            interchange a_j and a_{j+1}
        end
    i := i + 1
end {a_1, ..., a_n is in nondecreasing order}
```

39. We start with $3,1,5,7,4$. The first step inserts 1 correctly into the sorted list 3, producing $1,3,5,7,4$. Next 5 is inserted into $1,3$, and the list still reads $1,3,5,7,4$, as it does after the 7 is inserted into $1,3,5$. Finally, the 4 is inserted, and we obtain the sorted list $1,3,4,5,7$. At each insertion, the element to be inserted is compared with the elements already sorted, starting from the beginning, until its correct spot is found, and then the previously sorted elements beyond that spot are each moved one position toward the back of the list.

41. We assume that when the least element is found at each stage, it is interchanged with the element in the position it wants to occupy.

a) The smallest element is 1, so it is interchanged with the 3 at the beginning of the list, yielding $1,5,4,3,2$. Next, the smallest element among the remaining elements in the list (the second through fifth positions) is 2, so it is interchanged with the 5 in position 2, yielding $1,2,4,3,5$. One more pass gives us $1,2,3,4,5$. At this point we find the fourth smallest element among the fourth and fifth positions, namely 4, and interchange it with itself, again yielding $1,2,3,4,5$. This completes the sort.

b) The process is similar to part **(a)**. We just show the status at the end of each of the four passes: $1,4,3,2,5$; $1,2,3,4,5$; $1,2,3,4,5$; $1,2,3,4,5$.

c) Again there are four passes, but all interchanges result in the list remaining as it is.

43. We carry out the linear search algorithm given as Algorithm 2 in this section, except that we replace $x \neq a_i$ by $x < a_i$, and we replace the **else** clause with **else** *location* $:= n+1$. The cursor skips past all elements in the list less than x, the new element we are trying to insert, and ends up in the correct position for the new element.

45. We are counting just the comparisons of the numbers in the list, not any comparisons needed for the book-keeping in the **for** loop. The second element in the list must be compared with the first and compared with itself (in other words, when $j = 2$ in Algorithm 5, i takes the values 1 and 2 before we drop out of the

while loop). The third element must be compared with the first two and itself, since it exceeds them both. We continue in this way, until finally the nth element must be compared with all the elements. So the total number of comparisons is $2+3+4+\cdots+n$, which can be written as $(n^2+n-2)/2$. This is the worst case for insertion sort in terms of number of comparisons; see Example 6 in Section 3.3. On the other hand, no movements of elements are required, since each new element is already in its correct position.

47. There are two kinds of steps—the searching and the inserting. We assume the answer to Exercise 44, which is to use Algorithm 3 but replace the final check with **if** $x < a_i$ **then** $location := i$ **else** $location := i+1$. So the first step is to find the location for 2 in the list 3, and we insert it in front of the 3, so the list now reads $2, 3, 4, 5, 1, 6$. This took one comparison. Next we use binary search to find the location for the 4, and we see, after comparing it to the 2 and then the 3, that it comes after the 3, so we insert it there, leaving still $2, 3, 4, 5, 1, 6$. Next we use binary search to find the location for the 5, and we see, after comparing it to the 3 and then the 4, that it comes after the 4, so we insert it there, leaving still $2, 3, 4, 5, 1, 6$. Next we use binary search to find the location for the 1, and we see, after comparing it to the 3 and then the 2 and then the 2 again, that it comes before the 2, so we insert it there, leaving $1, 2, 3, 4, 5, 6$. Finally we use binary search to find the location for the 6, and we see, after comparing it to the 3 and then the 4 and then the 5, that it comes after the 5, so we insert it there, giving the final answer $1, 2, 3, 4, 5, 6$. Note that this took 11 comparisons in all.

49. We combine the search technique of Algorithm 3, as modified in Exercises 44 and 47, with the insertion part of Algorithm 5.

```
procedure binary insertion sort(a_1, a_2, ..., a_n : real numbers with n ≥ 2)
for j := 2 to n
begin
        {binary search for insertion location i}
        left := 1
        right := j − 1
        while left < right
        begin
                middle := ⌊(left + right)/2⌋
                if a_j > a_middle then left := middle + 1
                else right := middle
        end
        if a_j < a_left then i := left else i := left + 1
        {insert a_j in location i by moving a_i through a_{j−1} toward back of list}
        m := a_j
        for k := 0 to j − i − 1
                a_{j−k} := a_{j−k−1}
        a_i := m
end {a_1, a_2, ..., a_n are sorted}
```

51. If the elements are in close to the correct order, then we would usually find the correct spot for the next item to be inserted near the upper end of the list of already-sorted elements. Hence the variation from Exercise 50, which starts comparing at that end, would be best.

53. In each case we use as many quarters as we can, then as many dimes to achieve the remaining amount, then as many nickels, then as many pennies.

a) The algorithm uses the maximum number of quarters, two, leaving 1 cent. It then uses the maximum number of dimes (none) and nickels (none), before using one penny.

b) two quarters, leaving 19 cents, then one dime, leaving 9 cents, then one nickel, leaving 4 cents, then four pennies

c) three quarters, leaving 1 cent, then one penny

d) two quarters, leaving 10 cents, then one dime

55. In each case we uses as many quarters as we can, then as many dimes to achieve the remaining amount, then as many pennies.

a) The algorithm uses the maximum number of quarters, two, leaving 1 cent. It then uses the maximum number of dimes (none), before using one penny. Since the answer to Exercise 53a used no nickels anyway, the greedy algorithm here certainly used the fewest coins possible.

b) The algorithm uses two quarters, leaving 19 cents, then one dime, leaving 9 cents, then nine pennies. The greedy algorithm thus uses 12 coins. Since there are no nickels available, we must either use nine pennies or else use only one quarter and four pennies, along with four dimes to reach the needed total of 69 cents. This uses only nine coins, so the greedy algorithm here did not achieve the optimum.

c) The algorithm uses three quarters, leaving 1 cent, then one penny. Since the answer to Exercise 53c used no nickels anyway, the greedy algorithm here certainly used the fewest coins possible.

d) The algorithm uses two quarters, leaving 10 cents, then one dime. Since the answer to Exercise 53c used no nickels anyway, the greedy algorithm here certainly used the fewest coins possible.

57. a) We keep track of a variable f giving the finishing time of the talk last selected, starting out with f equal to the time the hall becomes available. (We ignore any time that might be needed to clear the hall between talks.) We order the talks in increasing order of the ending times, and start at the top of the list. At each stage of the algorithm, we go down the list of talks from where we left off, and find the first one whose starting time is not less than f. We schedule that talk and update f to record its finishing time. (See also Example 11 in Section 4.1.)

b) We schedule the 9:00 talk and set f to 9:45. The talk with the earliest finishing time among those that do not start before 9:45 is the one that starts at 9:50, so we schedule it and set f to 10:15. The talk with the earliest finishing time among those that do not start before 10:15 is the one that starts at 10:15, so we schedule it and set f to 10:45. The talk with the earliest finishing time among those that do not start before 10:45 is the one that starts at 11:00, so we schedule it and set f to 11:15. There are no more talks that start after this, so we are done, having scheduled four talks.

59. a) In the algorithm presented here, the input consists, for each man, of a list of all women in his preference order, and for each woman, of a list of all men in her preference order. At the risk of being sexist, we will let the men be the suitors and the women the suitees (although obviously we could reverse these roles). The procedure needs to have data structures (lists) to keep track, for each man, of his status (rejected or not) and the list of women who have rejected him, and, for each woman, of the men currently on her proposal list.

```
procedure stable(M_1, M_2, ..., M_s, W_1, W_2, ..., W_s,: preference lists)
for i := 1 to s
        mark man i as rejected
for i := 1 to s
        set man i's rejection list to be empty
for j := 1 to s
        set woman j's proposal list to be empty
while rejected men remain
begin
        for i := 1 to s
                if man i is marked rejected then add i to the proposal list
                for the woman j who ranks highest on his preference list
                but does not appear on his rejection list, and mark i as not rejected
        for j := 1 to s
```

if woman j's proposal list is nonempty **then** remove from j's proposal list all men i except the man i_0 who ranks highest on her preference list, and for each such man i mark him as rejected and add j to his rejection list
end
for $j := 1$ **to** s
match j with the one man on j's proposal list
{This matching is stable.}

b) The algorithm will terminate if at some point at the conclusion of the **while** loop, no man is rejected. If this happens, then that must mean that each man has one and only one proposal pending with some woman, because he proposed to only one in that round, and since he was not rejected, his proposal is the only one pending with that woman. It follows that at that point there are s pending proposals, one from each man, so each woman will be matched with a unique man. Finally, we argue that there are at most s^2 iterations of the **while** loop, so the algorithm must terminate. Indeed, if at the conclusion of the **while** loop rejected men remain, then some man must have been rejected, because no man is marked as rejected at the conclusion of the proposal phase (first **for** loop inside the **while** loop). If a man is rejected, then his rejection list grows. Thus each pass through the **while** loop, at least one more of the s^2 possible rejections will have been recorded, unless the loop is about to terminate. (Actually there will be fewer than s^2 iterations, because no man is rejected by the woman with whom he is eventually matched.) There is one more subtlety we need to address. Is it possible that at the end of some round, some man has been rejected by *every* woman and therefore the algorithm cannot continue? We claim not. If at the end of some round some man has been rejected by every woman, then every woman has one pending proposal at the completion of that round (from someone she likes better—otherwise she never would have rejected that poor man), and of course these proposals are all from different men because a man proposes only once in each round. That means s men have pending proposals, so in fact our poor universally-rejected man does not exist.

c) Suppose the assignment is not stable. Then there is a man m and a woman w such that m prefers w to the woman (call her w') with whom he is matched, and w prefers m to the man with whom she is matched. But m must have proposed to w before he proposed to w', since he prefers the former. And since m did not end up matched with w, she must have rejected him. Since women reject a suitor only when they get a better proposal, and they eventually get matched with a pending suitor, the woman with whom w is matched must be better in her eyes than m, contradicting our original assumption. Therefore the matching is stable.

61. The algorithm is simply to run the two programs on their inputs concurrently and wait for one to halt. This algorithm will terminate by the conditions of the problem, and we'll have the desired answer.

SECTION 3.2 The Growth of Functions

The big-O notation is used extensively in computer science and other areas. Think of it as a crude ruler for measuring functions in terms of how fast they grow. The idea is to treat all functions that are more or less the same as one function—one mark on this ruler. Thus, for example, all linear functions are simply thought of as $O(n)$. Although technically the big-O notation gives an upper bound on the growth of a function, in practice we choose the smallest big-O estimate that applies. (This is made more rigorous with the big-Theta notation, also discussed in this section.) In essence, one finds best big-O estimates by discarding lower order terms and multiplicative constants. Furthermore, one usually chooses the simplest possible representative of the big-O (or big-Theta) class (for example, writing $O(n^2)$ rather than $O(3n^2+5)$). A related concept, used in combinatorics and applied mathematics, is the **little-*o* notation**, *dealt with in Exercises 51–59.*

1. Note that the choices of witnesses C and k are not unique.

a) Yes, since $|10| \le |x|$ for all $x > 10$. The witnesses are $C = 1$ and $k = 10$.

b) Yes, since $|3x + 7| \le |4x| = 4|x|$ for all $x > 7$. The witnesses are $C = 4$ and $k = 7$.

c) No. There is no *constant* C such that $|x^2 + x + 1| \le C|x|$ for all sufficiently large x. To see this, suppose this inequality held for all sufficiently large positive values of x. Then we would have $x^2 \le Cx$, which would imply that $x \le C$ for *all* sufficiently large x, an obvious impossibility.

d) Yes. This follows from the fact that $\log x < x$ for all $x > 1$ (which in turn follows from the fact that $x < 2^x$, which can be formally proved by mathematical induction—see Section 4.1). Therefore $|5 \log x| \le 5|x|$ for all $x > 1$. The witnesses are $C = 5$ and $k = 1$.

e) Yes. This follows from the fact that $\lfloor x \rfloor \le x$. Thus $\big|\lfloor x \rfloor\big| \le |x|$ for all $x > 0$. The witnesses are $C = 1$ and $k = 0$.

f) Yes. This follows from the fact that $\lceil x/2 \rceil \le (x/2) + 1$. Thus $\big|\lceil x/2 \rceil\big| \le |(x/2) + 1| \le |x|$ for all $x > 2$. The witnesses are $C = 1$ and $k = 2$.

3. We need to put some bounds on the lower order terms. If $x > 9$ then we have $x^4 + 9x^3 + 4x + 7 \le x^4 + x^4 + x^4 + x^4 = 4x^4$. Therefore $x^4 + 9x^3 + 4x + 7$ is $O(x^4)$, taking witnesses $C = 4$ and $k = 9$.

5. We use long division to rewrite this function:

$$\frac{x^2+1}{x+1} = \frac{x^2-1+2}{x+1} = \frac{x^2-1}{x+1} + \frac{2}{x+1} = x - 1 + \frac{2}{x+1}$$

Now this is certainly less than x as long as $x > 1$, so our function is $O(x)$. The witnesses are $C = 1$ and $k = 1$.

7. a) Since $\log x$ grows more slowly than x, $x^2 \log x$ grows more slowly than x^3, so the first term dominates. Therefore this function is $O(x^3)$ but not $O(x^n)$ for any $n < 3$. More precisely, $2x^3 + x^2 \log x \le 2x^3 + x^3 = 3x^3$ for all x, so we have witnesses $C = 3$ and $k = 0$.

b) We know that $\log x$ grows so much more slowly than x that *every* power of $\log x$ grows more slowly than x. Thus the first term dominates, and the best estimate is $O(x^3)$. More precisely, $(\log x)^4 < x^3$ for all $x > 1$, so $3x^3 + (\log x)^4 \le 3x^3 + x^3 = 4x^3$ for all x, so we have witnesses $C = 4$ and $k = 1$.

c) By long division, we see that $f(x) = x +$ lower order terms. Therefore this function is $O(x)$, so $n = 1$. In fact, $f(x) = x + \frac{1}{x+1} \le 2x$ for all $x > 1$, so the witnesses can be taken to be $C = 2$ and $k = 1$.

d) Again by long division, this quotient has the form $f(x) = 1 +$ lower order terms. Therefore this function is $O(1)$. In other words, $n = 0$. Since $5 \log x < x^4$ for $x > 1$, we have $f(x) \le 2x^4/x^4 = 2$, so we can take as witnesses $C = 2$ and $k = 1$.

9. On the one hand we have $x^2 + 4x + 17 \le x^2 + x^2 + x^2 = 3x^2 \le 3x^3$ for all $x > 17$, so $x^2 + 4x + 17$ is $O(x^3)$, with witnesses $C = 3$ and $k = 17$. On the other hand, if x^3 were $O(x^2 + 4x + 17)$, then we would have $x^3 \le C(x^2 + 4x + 17) \le 3Cx^2$ for all sufficiently large x. But this says that $x \le 3C$, clearly impossible for the constant C to satisfy for all large x. Therefore x^3 is not $O(x^2 + 4x + 17)$.

11. For the first part we have $3x^4 + 1 \le 4x^4 = 8|x^4/2|$ for all $x > 1$; we have witnesses $C = 8$, $k = 1$. For the second part we have $x^4/2 \le 3x^4 \le 1 \cdot |3x^4 + 1|$ for all x; witnesses are $C = 1$, $k = 0$.

13. To show that 2^n is $O(3^n)$ it is enough to note that $2^n \le 3^n$ for all $n > 0$. In terms of witnesses we have $C = 1$ and $k = 0$. On the other hand, if 3^n were $O(2^n)$, then we would have $3^n \le C \cdot 2^n$ for all sufficiently large n. This is equivalent to $C \ge \left(\frac{3}{2}\right)^n$, which is clearly impossible, since $\left(\frac{3}{2}\right)^n$ grows without bound as n increases.

15. A function f is $O(1)$ if $|f(x)| \leq C$ for all sufficiently large x. In other words, f is $O(1)$ if its absolute value is **bounded** for all $x > k$ (where k is some constant).

17. Let C_1, C_2, k_1, and k_2 be numbers such that $|f(x)| \leq C_1|g(x)|$ for all $x > k_1$ and $|g(x)| \leq C_2|h(x)|$ for all $x > k_2$. Let $C = C_1C_2$ and let k be the larger of k_1 and k_2. Then for all $x > k$ we have $|f(x)| \leq C_1|g(x)| \leq C_1C_2|h(x)| = C|h(x)|$, which is precisely what we needed to show.

19. a) The significant terms here are the n^2 being multiplied by the n; thus this function is $O(n^3)$.

b) Since $\log n$ is smaller than n, the significant term in the first factor is n^2. Therefore the entire function is $O(n^5)$.

c) For the first factor we note that $2^n < n!$ for $n \geq 4$, so the significant term is $n!$. For the second factor, the significant term is n^3. Therefore this function is $O(n^3 n!)$.

21. a) First we note that $\log(n^2+1)$ and $\log n$ are in the same big-O class, since $\log n^2 = 2\log n$. Therefore the second term here dominates the first, and the simplest good answer would be just $O(n^2 \log n)$.

b) The first term is in the same big-O class as $O(n^2(\log n)^2)$, while the second is in a slightly smaller class, $O(n^2 \log n)$. (In each case, we can throw away the smaller order terms, since they are dominated by the terms we are keeping—this is the essence of doing big-O estimates.) Therefore the answer is $O(n^2(\log n)^2)$.

c) The only issue here is whether 2^n or n^2 is the faster-growing, and clearly it is the former. Therefore the best big-O estimate we can give is $O(n^{2^n})$.

23. We can use the following rule of thumb to determine what simple big-Theta function to use: throw away all the lower order terms (those that don't grow as fast as other terms) and all constant coefficients.

a) This function is $\Theta(x)$, so it is not $\Theta(x^2)$, since x^2 grows faster than x. To be precise, x^2 is not $O(17x+11)$. For the same reason, this function is not $\Omega(x^2)$.

b) This function is $\Theta(x^2)$; we can ignore the " $+1000$" since it is a lower order term. Of course, since $f(x)$ is $\Theta(x^2)$, it is also $\Omega(x^2)$.

c) This function grows more slowly than x^2, since $\log x$ grows more slowly than x. Therefore $f(x)$ is not $\Theta(x^2)$ or $\Omega(x^2)$.

d) This function grows faster than x^2. Therefore $f(x)$ is not $\Theta(x^2)$, but it is $\Omega(x^2)$.

e) Exponential functions (with base larger than 1) grow faster than all polynomials, so this function is not $O(x^2)$ and therefore not $\Theta(x^2)$. But it is $\Omega(x^2)$.

f) For large values of x, this is quite close to x^2, since both factors are quite close to x. Certainly $\lfloor x \rfloor \cdot \lceil x \rceil$ is always between $x^2/2$ and $2x^2$, for $x > 2$. Therefore this function is $\Theta(x^2)$ and hence also $\Omega(x^2)$.

25. If $f(x)$ is $\Theta(g(x))$, then $|f(x)| \leq C_2|g(x)|$ and $|g(x)| \leq C_1^{-1}|f(x)|$ for all $x > k$. Thus $f(x)$ is $O(g(x))$ and $g(x)$ is $O(f(x))$. Conversely, suppose that $f(x)$ is $O(g(x))$ and $g(x)$ is $O(f(x))$. Then (with appropriate choice of variable names) we may assume that $|f(x)| \leq C_2|g(x)|$ and $|g(x)| \leq C|f(x)|$ for all $x > k$. (The k here will be the larger of the two k's involved in the hypotheses.) If $C > 0$ then we can take $C_1 = C^{-1}$ to obtain the desired inequalities in "$f(x)$ is $\Theta(g(x))$." If $C \leq 0$, then $g(x) = 0$ for all $x > k$, and hence by the first inequality $f(x) = 0$ for all $x > k$; thus we have $f(x) = g(x)$ for all $x > k$, and we can take $C_1 = C_2 = 1$.

27. The definition of "$f(x)$ is $\Theta(g(x))$" is that $f(x)$ is both $O(g(x))$ and $\Omega(g(x))$. That means that there are positive constants C_1, k_1, C_2, and k_2 such that $|f(x)| \leq C_2|g(x)|$ for all $x > k_2$ and $|f(x)| \geq C_1|g(x)|$ for all $x > k_1$. That is practically the same as the statement in this exercise. We need only note that we can take k to be the larger of k_1 and k_2 if we want to prove the "only if" direction, and we can take $k_1 = k_2 = k$ if we want to prove the "if" direction.

29. In the following picture, the wavy line is the graph of the function f. For simplicity we assume that the graph of g is a straight line through the origin. Then the graphs of $C_1 g$ and $C_2 g$ are also straight lines through the origin, as drawn here. The fact that $f(x)$ is $\Theta(g(x))$ is shown by the fact that for $x > k$ the graph of f is confined to the shaded wedge-shaped space between these latter two lines (see Exercise 27). (We assume that $g(x)$ is positive for positive x, so that $|g(x)|$ is the same as $g(x)$.)

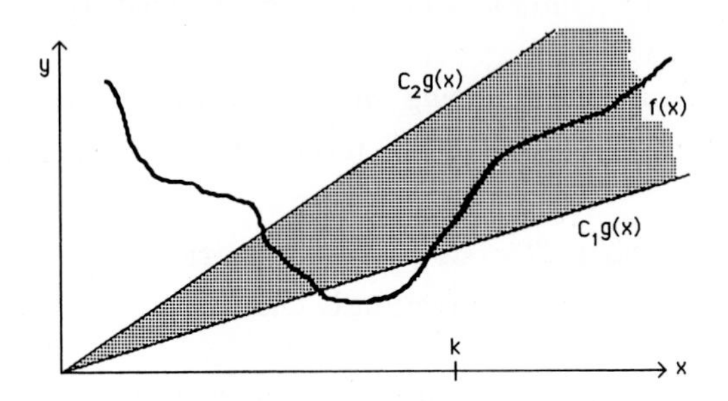

31. Looking at the definition tells us that if $f(x)$ is $\Theta(1)$ then $|f(x)|$ has to be bounded between two positive constants. In other words, $f(x)$ can't get too large (either positive or negative), and it can't get too close to 0.

33. We are given that $|f(x)| \leq C|g(x)|$ for all $x > b$ (we cannot use the variable name k here since we need it later). Hence $|f^k(x)| = |f(x)|^k \leq C^k|g^k(x)|$ for all $x > b$, so $f^k(x)$ is $O(g^k(x))$ (take the constants in the definition to be C^k and b).

35. Since the functions are given to be increasing and unbounded, we may assume that they both take on values greater than 1 for all sufficiently large x. The hypothesis can then be written as $f(x) \leq Cg(x)$ for all $x > k$. If we take the logarithm of both sides, then we obtain $\log f(x) \leq \log C + \log g(x)$. Finally, this latter expression is less than $2\log g(x)$ for large enough x, since $\log g(x)$ is growing without bound. Note that the converse to this problem is not true.

37. By definition there are positive constants C_1, C_1', C_2, C_2', k_1, k_1', k_2, and k_2' such that $f_1(x) \geq C_1|g(x)|$ for all $x > k_1$, $f_1(x) \leq C_1'|g(x)|$ for all $x > k_1'$, $f_2(x) \geq C_2|g(x)|$ for all $x > k_2$, and $f_2(x) \leq C_2'|g(x)|$ for all $x > k_2'$. We are able to omit the absolute value signs on the $f(x)$'s since we are told that they are positive; we are also told here that the $g(x)$'s are positive, but we do not need that. Adding the first and third inequalities we obtain $f_1(x) + f_2(x) \geq (C_1 + C_2)|g(x)|$ for all $x > \max(k_1, k_2)$; and similarly with the second and fourth inequalities we know $f_1(x) + f_2(x) \leq (C_1' + C_2')|g(x)|$ for all $x > \max(k_1', k_2')$. Thus $f_1(x) + f_2(x)$ meets the definition of being $\Theta(g(x))$.

If the f's can take on negative values, then this is no longer true. For example, let $f_1(x) = x^2 + x$, let $f_2(x) = -x^2 + x$, and let $g(x) = x^2$. Then each $f_i(x)$ is $\Theta(g(x))$, but the sum is $2x$, which is not $\Theta(g(x))$.

39. This is not true. It is similar to Exercise 37, and essentially the same counterexample suffices. Let $f_1(x) = x^2 + 2x$, $f_2(x) = x^2 + x$, and $g(x) = x^2$. Then clearly $f_1(x)$ and $f_2(x)$ are both $\Theta(g(x))$, but $(f_1 - f_2)(x) = x$ is not.

41. The key here is that if a function is to be big-O of another, then the appropriate inequality has to hold for *all* large inputs. Suppose we let $f(x) = x^2$ for even x and x for odd x. Similarly, we let $g(x) = x^2$ for odd x and x for even x. Then clearly neither inequality $|f(x)| \leq C|g(x)|$ nor $|g(x)| \leq C|f(x)|$ holds for all x, since for even x the first function is much bigger than the second, while for odd x the second is much bigger than the first.

43. We are given that there are positive constants C_1, C_1', C_2, C_2', k_1, k_1', k_2, and k_2' such that $|f_1(x)| \geq C_1|g_1(x)|$ for all $x > k_1$, $|f_1(x)| \leq C_1'|g_1(x)|$ for all $x > k_1'$, $|f_2(x)| \geq C_2|g_2(x)|$ for all $x > k_2$, and $|f_2(x)| \leq C_2'|g_2(x)|$ for all $x > k_2'$. Since f_2 and g_2 never take on the value 0, we can rewrite the last two of these inequalities as $|1/f_2(x)| \leq (1/C_2)|1/g_2(x)|$ and $|1/f_2(x)| \geq (1/C_2')|1/g_2(x)|$. Now we multiply the first inequality and the rewritten fourth inequality to obtain $|f_1(x)/f_2(x)| \geq (C_1/C_2')|g_1(x)/g_2(x)|$ for all $x > \max(k_1, k_2')$. Working with the other two inequalities gives us $|f_1(x)/f_2(x)| \leq (C_1'/C_2)|g_1(x)/g_2(x)|$ for all $x > \max(k_1', k_2)$. Together these tell us that f_1/f_2 is big-Theta of g_1/g_2.

45. We just make the analogous change in the definition of big-Theta that was made in the definition of big-O: there exist positive constants C_1, C_2, k_1, k_2, k_1', k_2' such that $|f(x,y)| \leq C_1|g(x,y)|$ for all $x > k_1$ and $y > k_2$, and $|f(x,y)| \geq C_2|g(x,y)|$ for all $x > k_1'$ and $y > k_2'$.

47. For all values of x and y greater than 1, each term of the expression inside parentheses is less than x^2y, so the entire expression inside parentheses is less than $3x^2y$. Therefore our function is less than $27x^6y^3$ for all $x > 1$ and $y > 1$. By definition this shows that it is big-O of x^6y^3. Specifically, we take $C = 27$ and $k_1 = k_2 = 1$ in the definition.

49. For all positive values of x and y, we know that $\lfloor xy \rfloor \leq xy$ by definition (since the floor function value cannot exceed the argument). Thus $\lfloor xy \rfloor$ is $O(xy)$ from the definition, taking $C = 1$ and $k_1 = k_2 = 0$. In fact, $\lfloor xy \rfloor$ is also $\Omega(xy)$ (and therefore $\Theta(xy)$); this is easy to see since $\lfloor xy \rfloor \geq (x-1)(y-1) \geq (\frac{1}{2}x)(\frac{1}{2}y) = \frac{1}{4}xy$ for all x and y greater than 2.

51. All that we need to do is determine whether the ratio of the two functions approaches 0 as x approaches infinity.

a) $\lim\limits_{x\to\infty} \dfrac{x^2}{x^3} = \lim\limits_{x\to\infty} \dfrac{1}{x} = 0$

b) $\lim\limits_{x\to\infty} \dfrac{x \log x}{x^2} = \lim\limits_{x\to\infty} \dfrac{\log x}{x} = \lim\limits_{x\to\infty} \dfrac{1}{x \ln 2} = 0$ (using L'Hôpital's rule for the second equality)

c) $\lim\limits_{x\to\infty} \dfrac{x^2}{2^x} = \lim\limits_{x\to\infty} \dfrac{2x}{2^x \ln 2} = \lim\limits_{x\to\infty} \dfrac{2}{2^x (\ln 2)^2} = 0$ (with two applications of L'Hôpital's rule)

d) $\lim\limits_{x\to\infty} \dfrac{x^2+x+1}{x^2} = \lim\limits_{x\to\infty} (1 + \dfrac{1}{x} + \dfrac{1}{x^2}) = 1 \neq 0$

53. The picture shows the graph of $y = x^2$ increasing quite rapidly and $y = x \log x$ increasing less rapidly. The ratio is hard to see on the picture; it rises to about $y = 0.53$ at about $x = 2.7$ and then slowly decreases toward 0. The limit as $x \to \infty$ of $(x \log x)/x^2$ is in fact 0.

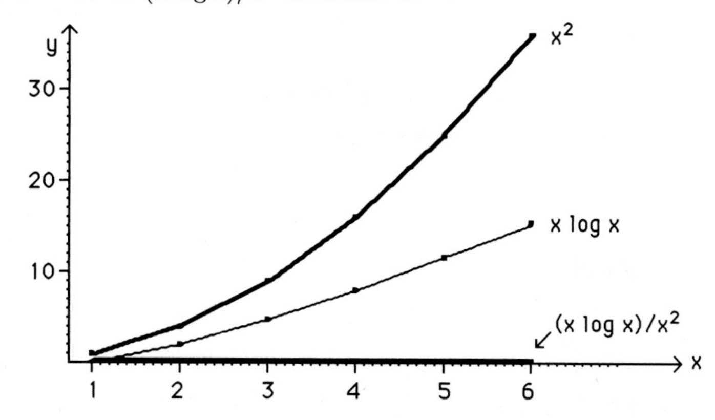

55. No. As one example, take $f(x) = x^{-2}$ and $g(x) = x^{-1}$. Then $f(x)$ is $o(g(x))$, since $\lim\limits_{x\to\infty} x^{-2}/x^{-1} = \lim\limits_{x\to\infty} 1/x = 0$. On the other hand $\lim\limits_{x\to\infty} (2^{x^{-2}}/2^{x^{-1}}) = \lim\limits_{x\to\infty} 2^{x^{-2}-x^{-1}} = 2^0 = 1 \neq 0$.

57. **a)** Since the limit of $f(x)/g(x)$ is 0 (as $x \to \infty$), so too is the limit of $|f(x)|/|g(x)|$. In particular, for x large enough, this ratio is certainly less than 1. In other words $|f(x)| \le |g(x)|$ for sufficiently large x, which meets the definition of "$f(x)$ is $O(g(x))$."

b) We can simply let $f(x) = g(x)$ be any function with positive values. Then the limit of their ratio is 1, not 0, so $f(x)$ is not $o(g(x))$, but certainly $f(x)$ is $O(g(x))$.

59. This follows immediately from Exercise 57a (whereby we can conclude that $f_2(x)$ is $O(g(x))$) and Corollary 1 to Theorem 2.

61. What we want to show is equivalent to the statement that $\log(n^n)$ is at most a constant times $\log(n!)$, which in turn is equivalent to the statement that n^n is at most a constant power of $n!$ (because of the fact that $C \log A = \log(A^C)$—see Appendix 2). We will show that in fact $n^n \le (n!)^2$ for all $n > 1$. To do this, let us write $(n!)^2$ as $(n \cdot 1) \cdot ((n-1) \cdot 2) \cdot ((n-2) \cdot 3) \cdots (2 \cdot (n-1)) \cdot (1 \cdot n)$. Now clearly each product pair $(i+1) \cdot (n-i)$ is at least as big as n (indeed, the ones near the middle are significantly bigger than n). Therefore the entire product is at least as big as n^n, as desired.

63. For $n = 5$ we compute that $\log 5! \approx 6.9$ and $(5 \log 5)/4 \approx 2.9$, so the inequality holds (it actually holds for all $n > 1$). Therefore we can assume that $n \ge 6$. Since $n!$ is the product of all the integers from n down to 1, we certainly have $n! > n(n-1)(n-2) \cdots \lceil n/2 \rceil$ (since at least the term 2 is missing). Note that there are more than $n/2$ terms in this product, and each term is at least as big as $n/2$. Therefore the product is greater than $(n/2)^{(n/2)}$. Taking the log of both sides of the inequality, we have

$$\log n! > \log \left(\frac{n}{2}\right)^{n/2} = \frac{n}{2} \log \frac{n}{2} = \frac{n}{2}(\log n - 1) > (n \log n)/4\,,$$

since $n > 4$ implies $\log n - 1 > (\log n)/2$.

65. In each case we need to evaluate the limit of $f(x)/g(x)$ as $x \to \infty$. If it equals 1, then f and g are asymptotic; otherwise (including the case in which the limit does not exist) they are not. Most of these are straightforward applications of algebra, elementary notions about limits, or L'Hôpital's rule.

a) $f(x) = \log(x^2+1) \ge \log(x^2) = 2 \log x$. Therefore $f(x)/g(x) \ge 2$ for all x. Thus the limit is not 1 (in fact, of course, it's 2), so f and g are not asymptotic.

b) By the algebraic rules for exponents, $f(x)/g(x) = 2^{-4} = 1/16$. Therefore the limit of the ratio is 1/16, not 1, so f and g are not asymptotic.

c) $\displaystyle\lim_{x\to\infty} \frac{2^{2^x}}{2^{x^2}} = \lim_{x\to\infty} 2^{2^x - x^2}$. As x gets large, the exponent grows without bound, so this limit is ∞. Thus f and g are not asymptotic.

d) $\displaystyle\lim_{x\to\infty} \frac{2^{x^2+x+1}}{2^{x^2+2x}} = \lim_{x\to\infty} 2^{1-x}$. As x gets large, the exponent grows in the negative direction without bound, so this limit is 0. Thus f and g are not asymptotic.

SECTION 3.3 Complexity of Algorithms

Some of these exercises involve analysis of algorithms, as was done in the examples in this section. These are a matter of carefully counting the operations of interest, usually in the worst case. Some of the others are algebra exercises that display the results of the analysis in real terms—the number of years of computer time, for example, required to solve a large problem. **Horner's method** *for evaluating a polynomial, given in Exercise 8, is a nice trick to know. It is extremely handy for polynomial evaluation on a pocket calculator (especially if the calculator is so cheap that it does not use the usual precedence rules).*

1. Assuming that the algorithm given to find the smallest element of a list is identical to Algorithm 1 in Section 3.1, except that the inequality is reversed (and the name *max* replaced by the name *min*), the analysis will be identical to the analysis given in Example 1 in the current section. In particular, there will be $2n - 1$ comparisons needed, counting the bookkeeping for the loop.

3. The linear search would find this element after at most 9 comparisons (4 to determine that we have not yet finished with the **while** loop, 4 more to determine if we have located the desired element yet, and 1 to set the value of *location*). Binary search, according to Example 3, will take $2\log 32 + 2 = 2\cdot 5 + 2 = 12$ comparisons. Since $9 < 12$, the linear search will be faster, in terms of comparisons.

5. The algorithm simply scans the bits one at a time. Thus clearly $O(n)$ comparisons are required (perhaps one for bookkeeping and one for looking at the ith bit, for each i from 1 to n).

7. a) Here we have $n = 2$, $a_0 = 1$, $a_1 = 1$, $a_2 = 3$, and $c = 2$. Initially, we set *power* equal to 1 and y equal to 1. The first time through the **for** loop (with $i = 1$), *power* becomes 2 and so y becomes $1 + 1\cdot 2 = 3$. The second and final time through the loop, *power* becomes $2\cdot 2 = 4$ and y becomes $3 + 3\cdot 4 = 15$. Thus the value of the polynomial at $x = 2$ is 15.

b) Each pass through the loop requires two multiplications and one addition. Therefore there are a total of $2n$ multiplications and n additions in all.

9. This is an exercise in algebra, numerical analysis (for some of the parts), and using a calculator. Since each bit operation requires 10^{-9} seconds, we want to know for what value of n there will be at most 10^9 bit operations required. Thus we need to set the expression equal to 10^9, solve for n, and round down if necessary.

a) Solving $\log n = 10^9$, we get (recalling that "log" means logarithm base 2) $n = 2^{10^9}$. By taking $\log_{10}$ of both sides, we find that this number is approximately equal to $10^{300{,}000{,}000}$. Obviously we do not want to write out the answer explicitly!

b) Clearly $n = 10^9$.

c) Solving $n\log n = 10^9$ is not trivial. There is no good formula for solving such **transcendental** equations. An algorithm that works well with a calculator is to rewrite the equation as $n = 10^9/\log n$, enter a random starting value, say $n = 2$, and repeatedly calculate a new value of n. Thus we would obtain, in succession, $n = 10^9/\log 2 = 10^9$, $n = 10^9/\log(10^9) \approx 33{,}447{,}777.3$, $n = 10^9/\log(33{,}447{,}777.3) \approx 40{,}007{,}350.14$, $n = 10^9/\log(40{,}007{,}350.14) \approx 39{,}598{,}061.08$, and so on. After a few more iterations, the numbers stabilize at approximately $39{,}620{,}077.73$, so the answer is $39{,}620{,}077$.

d) Solving $n^2 = 10^9$ gives $n = 10^{4.5}$, which is 31,622 when rounded down.

e) Solving $2^n = 10^9$ gives $n = \log(10^9) \approx 29.9$. Rounding down gives the answer, 29.

f) The quickest way to find the largest value of n such that $n! \le 10^9$ is simply to try a few values of n. We find that $12! \approx 4.8\times 10^8$ while $13! \approx 6.2\times 10^9$, so the answer is 12.

11. In each case, we just multiply the number of seconds per operation by the number of operations (namely 2^{50}). To convert seconds to minutes, we divide by 60; to convert minutes to hours, we divide by 60 again. To convert hours to days, we divide by 24; to convert days to years, we divide by $365\frac{1}{4}$.

a) $2^{50}\times 10^{-6} = 1{,}125{,}899{,}907$ seconds ≈ 36 years

b) $2^{50}\times 10^{-9} = 1{,}125{,}899.907$ seconds ≈ 13 days

c) $2^{50}\times 10^{-12} = 1{,}125.899907$ seconds ≈ 19 minutes

13. If the element is not in the list, then $2n + 2$ comparisons are needed: two for each pass through the loop, one more to get out of the loop, and one more for the statement just after the loop. If the element is in the list, say as the ith element, then we need to enter the loop i times, each time costing two comparisons, and use one comparison for the final assignment of *location*. Thus $2i+1$ comparisons are needed. We need to average the numbers $2i+1$ (for i from 1 to n), to find the average number of comparisons needed for a successful search. This is $(3+5+\cdots+(2n+1))/n = (n+(2+4+\cdots+2n))/n = (n+2(1+2+\cdots+n))/n = (n+2(n(n+1)/2))/n = n+2$. Finally, we average the $2n+2$ comparisons for the unsuccessful search with this average $n+2$ comparisons for a successful search to obtain a grand average of $(2n+2+n+2)/2 = (3n+4)/2$ comparisons.

15. We will count comparisons of elements in the list to x. (This ignores comparisons of subscripts, but since we are only interested in a big-O analysis, no harm is done.) Furthermore, we will assume that the number of elements in the list is a power of 3, say $n = 3^k$. Just as in the case of binary search, we need to determine the maximum number of times the **while** loop is iterated. Each pass through the loop cuts the number of elements still being considered (those whose subscripts are from i to j) by a factor of 3. Therefore after k iterations, the active portion of the list will have length 1; that is, we will have $i = j$. The loop terminates at this point. Now each iteration of the loop requires two comparisons in the worst case (one with a_u and one with a_l). Two more comparisons are needed at the end. Therefore the number of comparisons is $2k+2$, which is $O(k)$. But $k = \log_3 n$, which is $O(\log n)$ since logarithms to different bases differ only by multiplicative constants, so the time complexity of this algorithm (in all cases, not just the worst case) is $O(\log n)$.

17. The algorithm we gave for finding a mode essentially just goes through the list once, doing a little bookkeeping at each step. In particular, between any two successive executions of the statement $i := i+1$ there are at most about six operations (such as comparing *count* with *modecount*, or reinitializing *value*). Therefore at most about $6n$ steps are done in all, so the time complexity in all cases is $O(n)$.

19. The worst case is that in which we do not find any term equal to some previous term. In that case, we need to go through all the terms a_2 through a_n, and for each of those, we need to go through all the terms occurring prior to that term. Thus the inner loop of our algorithm is executed once for $i = 2$ (namely for $j = 1$), twice for $i = 3$ (namely for $j = 1$ and $j = 2$), three times for $i = 4$, and so on, up to $n-1$ times for $i = n$. Thus the number of comparisons that need to be made in the inner loop is $1+2+3+\cdots+(n-1)$. As was mentioned in this section (and will be shown in Section 4.1), that sum is $(n-1)(n-1+1)/2$, which is clearly $O(n^2)$ but no better. Bookkeeping details do not increase this estimate.

21. We needed to go through the sequence only once, making one comparison of terms (and two bookkeeping comparisons) until we found the desired term (or had exhausted the list). Thus this algorithm's time complexity is clearly $O(n)$.

23. We had to read the string (or at least half of it) simultaneously from the front and the back and compare characters to make sure they were equal. Thus we will need at least $\lfloor n/2 \rfloor$ comparisons in the worst case, which is $O(n)$.

25. Since we are doing binary search to find the correct location of the jth element among the first $j-1$ elements, which are already sorted, we need $O(\log n)$ comparisons for each element. Therefore we use $O(n \log n)$ comparisons in all. The cost of swapping of items to make room for the insertions is $O(n^2)$, however (the originally first item may need to move $n-1$ places, the second item $n-2$ places, and so on).

27. **a)** The linear search algorithm uses about n comparisons for a list of length n, and $2n$ comparisons for a list of length $2n$. Therefore the number of comparisons, like the size of the list, doubles.

b) The binary search algorithm uses about $\log n$ comparisons for a list of length n, and $\log(2n) = \log 2 + \log n = 1 + \log n$ comparisons for a list of length $2n$. Therefore the number of comparisons increases by about 1.

SECTION 3.4 The Integers and Division

Number theory is playing an increasingly important role in computer science. This section and these exercises just scratch the surface of what is relevant. Many of these exercises are simply a matter of applying definitions. It is sometimes hard for a beginning student to remember that in order to prove something about a concept (such as modular arithmetic), it is usually necessary to invoke the definition! Exercises 20–25 hint at the rich structure that modular arithmetic has (sometimes resembling real number arithmetic more than integer arithmetic). In many contexts in mathematics and computer science, modular arithmetic is more relevant and convenient than ordinary integer arithmetic.

1. a) yes, since $68 = 17 \cdot 4$ **b)** no, remainder $= 16$
c) yes, since $357 = 17 \cdot 21$ **d)** no, remainder $= 15$

3. If $a \mid b$, then we know that $b = at$ for some integer t. Therefore $bc = a(tc)$, so by definition $a \mid bc$.

5. The given conditions imply that there are integers s and t such that $a = bs$ and $b = at$. Combining these, we obtain $a = ats$; since $a \neq 0$, we conclude that $st = 1$. Now the only way for this to happen is for $s = t = 1$ or $s = t = -1$. Therefore either $a = b$ or $a = -b$.

7. The given condition means that $bc = (ac)t$ for some integer t. Since $c \neq 0$, we can divide both sides by c to obtain $b = at$. This is the definition of $a \mid b$, as desired.

9. In each case we need to find (the unique integers) q and r such that $a = dq + r$ and $0 \leq r < d$, where a and d are the given integers. In each case $q = \lfloor a/d \rfloor$.
a) $19 = 7 \cdot 2 + 5$, so $q = 2$ and $r = 5$ **b)** $-111 = 11 \cdot (-11) + 10$, so $q = -11$ and $r = 10$
c) $789 = 23 \cdot 34 + 7$, so $q = 34$ and $r = 7$ **d)** $1001 = 13 \cdot 77 + 0$, so $q = 77$ and $r = 0$
e) $0 = 19 \cdot 0 + 0$, so $q = 0$ and $r = 0$ **f)** $3 = 5 \cdot 0 + 3$, so $q = 0$ and $r = 3$
g) $-1 = 3 \cdot (-1) + 2$, so $q = -1$ and $r = 2$ **h)** $4 = 1 \cdot 4 + 0$, so $q = 4$ and $r = 0$

11. The given condition, that $a \textbf{ mod } m = b \textbf{ mod } m$, means that a and b have the same remainder when divided by m. In symbols, $a = q_1 m + r$ and $b = q_2 m + r$ for some integers q_1, q_2, and r. Subtracting these two equations gives us $a - b = (q_1 - q_2)m$, which says that m divides (is a factor of) $a - b$. This is precisely the definition of $a \equiv b \pmod{m}$.

13. The quotient n/k lies between two consecutive integers, say $b-1$ and b, possibly equal to b. In symbols, there exists a positive integer b such that $b - 1 < n/k \leq b$. In particular, $\lceil n/k \rceil = b$. Also, since $n/k > b - 1$, we have $n > k(b-1)$, and so (since everything is an integer) $n - 1 \geq k(b-1)$. This means that $(n-1)/k \geq b-1$, so $\lfloor (n-1)/k \rfloor \geq b-1$. On the other hand, $\lfloor (n-1)/k \rfloor \leq (n-1)/k < n/k \leq b$, so $\lfloor (n-1)/k \rfloor < b$. Therefore $\lfloor (n-1)/k \rfloor = b - 1$. The desired conclusion follows.

15. Let's first look at an example or two. If $m = 7$, then the usual set of values we use for the congruence classes modulo m is $\{0, 1, 2, 3, 4, 5, 6\}$. However, we can replace 6 by -1, 5 by -2, and 4 by -3 to get the collection $\{-3, -2, -1, 0, 1, 2, 3\}$. These will be the values with smallest absolute values. Similarly, if $m = 8$, then the collection we want is $\{-3, -2, -1, 0, 1, 2, 3, 4\}$ ($\{-4, -3, -2, -1, 0, 1, 2, 3\}$ would do just as well). In general, in place of $\{0, 1, 2, \ldots, m-1\}$ we can use $\{\lceil -m/2 \rceil, \lceil -m/2 \rceil + 1, \ldots, -1, 0, 1, 2, \ldots, \lceil m/2 \rceil\}$, omitting either $\lceil -m/2 \rceil$ or $\lceil m/2 \rceil$ if m is even. Note that the values in $\{0, 1, 2, \ldots, m-1\}$ greater than $\lceil m/2 \rceil$ have had m subtracted from them to produce the negative values in our answer. As for a formula to produce these values, we can use a two-part formula:

$$f(x) = \begin{cases} x \textbf{ mod } m & \text{if } x \textbf{ mod } m \leq \lceil m/2 \rceil \\ (x \textbf{ mod } m) - m & \text{if } x \textbf{ mod } m > \lceil m/2 \rceil. \end{cases}$$

Note that if m is even, then we can, alternatively, take $f(m/2) = -m/2$.

17. For these problems, we need to perform the division (as in Exercise 9) and report the remainder.
a) $13 = 3 \cdot 4 + 1$, so $13 \textbf{ mod } 3 = 1$ **b)** $-97 = 11 \cdot (-9) + 2$, so $-97 \textbf{ mod } 11 = 2$
c) $155 = 19 \cdot 8 + 3$, so $155 \textbf{ mod } 19 = 3$ **d)** $-221 = 23 \cdot (-10) + 9$, so $-221 \textbf{ mod } 23 = 9$

19. For these problems, we need to divide by 17 and see whether the remainder equals 5. Remember that the quotient can be negative, but the remainder r must satisfy $0 \le r < 17$.
a) $80 = 17 \cdot 4 + 12$, so $80 \not\equiv 5 \pmod{17}$ **b)** $103 = 17 \cdot 6 + 1$, so $103 \not\equiv 5 \pmod{17}$
c) $-29 = 17 \cdot (-2) + 5$, so $80 \equiv 5 \pmod{17}$ **d)** $-122 = 17 \cdot (-8) + 14$, so $80 \not\equiv 5 \pmod{17}$

21. The hypothesis $a \equiv b \pmod{m}$ means that $m \mid (a - b)$. Since we are given that $n \mid m$, Theorem 1(*iii*) implies that $n \mid (a - b)$. Therefore $a \equiv b \pmod{n}$, as desired.

23. a) To show that this conditional statement does not necessarily hold, we need to find an example in which $ac \equiv bc \pmod{m}$, but $a \not\equiv b \pmod{m}$. Let $m = 4$ and $c = 2$ (what is important in constructing this example is that m and c have a nontrivial common factor). Let $a = 0$ and $b = 2$. Then $ac = 0$ and $bc = 4$, so $ac \equiv bc \pmod{4}$, but $0 \not\equiv 2 \pmod{4}$.

b) To show that this conditional statement does not necessarily hold, we need to find an example in which $a \equiv b \pmod{m}$ and $c \equiv d \pmod{m}$, but $a^c \not\equiv b^d \pmod{m}$. If we try a few randomly chosen positive integers, we will soon find one. Let $m = 5$, $a = 3$, $b = 3$, $c = 1$, and $d = 6$. Then $a^c = 3$ and $b^d = 729 \equiv 4 \pmod{5}$, so $3^1 \not\equiv 3^6 \pmod{5}$, even though $3 \equiv 3 \pmod{5}$ and $1 \equiv 6 \pmod{5}$.

25. There are at least two ways to prove this. One way is to invoke Theorem 5 repeatedly. Since $a \equiv b \pmod{m}$, Theorem 5 implies that $a \cdot a \equiv b \cdot b \pmod{m}$, i.e., $a^2 \equiv b^2 \pmod{m}$. Invoking Theorem 5 again, since $a \equiv b \pmod{m}$ and $a^2 \equiv b^2 \pmod{m}$, we obtain $a^3 \equiv b^3 \pmod{m}$. After $k - 1$ applications of this process, we obtain $a^k \equiv b^k \pmod{m}$, as desired. (This is really a proof by mathematical induction, a topic to be considered formally in Chapter 4.)

Alternately, we can argue directly, using the algebraic identity $a^k - b^k = (a-b)(a^{k-1} + a^{k-2}b + \cdots + ab^{k-2} + b^{k-1})$. Specifically, the hypothesis that $a \equiv b \pmod{m}$ means that $m \mid (a - b)$. Therefore by Theorem 1(*ii*), m divides the right-hand side of this identity, so $m \mid (a^k - b^k)$. This means precisely that $a^k \equiv b^k \pmod{m}$.

27. a) We need to compute k **mod** 31 in each case. A good way to do this on a calculator is as follows. Enter k and divide by 31. The result will be a number with an integer part and a decimal fractional part. Subtract off the integer part, leaving a decimal fraction between 0 and 1. This is the remainder expressed as a decimal. To find out what whole number remainder that really represents, multiply by 31. The answer will be a whole number (or nearly so—it may require rounding, say from 4.9999 or 5.0001 to 5), and that number is k **mod** 31.
(i) $317 \textbf{ mod } 31 = 7$ (ii) $918 \textbf{ mod } 31 = 19$ (iii) $007 \textbf{ mod } 31 = 7$
(iv) $100 \textbf{ mod } 31 = 7$ (v) $111 \textbf{ mod } 31 = 18$ (vi) $310 \textbf{ mod } 31 = 0$

b) Take the next available space, where the next space is computed by adding 1 to the space number and pretending that $30 + 1 = 0$.

29. We compute until the sequence begins to repeat:

$$x_1 = 3 \cdot 2 \textbf{ mod } 11 = 6$$
$$x_2 = 3 \cdot 6 \textbf{ mod } 11 = 7$$
$$x_3 = 3 \cdot 7 \textbf{ mod } 11 = 10$$
$$x_4 = 3 \cdot 10 \textbf{ mod } 11 = 8$$
$$x_5 = 3 \cdot 8 \textbf{ mod } 11 = 2$$

Since $x_5 = x_0$, the sequence repeats forever: $2, 6, 7, 10, 8, 2, 6, 7, 10, 8, \ldots$.

31. **a)** The message in numbers is 3–14 13–14–19 15–0–18–18 6–14. Adding 3 to each number (no sum is greater than 25, so there is no need to reduce modulo 26), we obtain the numerical message 6–17 16–17–22 18–3–21–21 9–17. Then translating this back into letters (A= 0, ..., Z= 25), we obtain "GR QRW SDVV JR." Note that this really is easier to do without translating to numbers first—just recite the next three letters in the alphabet after the given letter (e.g., "D goes to E-F-G").

b) We follow the same procedure, except that we add 13 rather than 3, and we need to subtract 26 from some of the sums to reduce modulo 26. The encoded numerical message is 16–1 0–1–6 2–13–5–5 19–1, so the literal encoded message is "QB ABG CNFF TB."

c) This time we need to multiply by 3, add 7, and reduce modulo 26 to obtain the encoded numerical message: 16–23 20–23–12 0–7–9–9 25–23, which gives us "QX UXM AHJJ ZX."

33. Let d be the check digit. Then we know that $1{\cdot}0+2{\cdot}0+3{\cdot}7+4{\cdot}1+5{\cdot}1+6{\cdot}9+7{\cdot}8+8{\cdot}8+9{\cdot}1+10{\cdot}d \equiv 0 \pmod{11}$. This simplifies to $213+10{\cdot}d \equiv 0 \pmod{11}$. But $213 \equiv 4 \pmod{11}$, and $10 \equiv -1 \pmod{11}$, so this is equivalent to $4-d \equiv 0 \pmod{11}$, or $d=4$.

35. The 10-digit ISBN number of this book is 0-07-288008-2. Therefore $1\cdot 0+2\cdot 0+3\cdot 7+4\cdot 2+5\cdot 8+6\cdot 8+7\cdot 0+8\cdot 0+9\cdot 8+10\cdot 2$ should be congruent to $0 \pmod{11}$. The sum is $209 = 11\cdot 19$, so it checks.

SECTION 3.5 Primes and Greatest Common Divisors

The prime numbers are the building blocks for the natural numbers in terms of multiplication, just as the elements (like carbon, oxygen, or uranium) are the building blocks of all matter. Just as we can put two hydrogen atoms and one oxygen atom together to form water, every composite natural number is uniquely constructed by multiplying together prime numbers. Analyzing numbers in terms of their prime factorizations allows us to solve many problems, such as finding greatest common divisors. Prime numbers have fascinated people for millennia, and many easy-to-state questions about them remain unanswered. Students interested in pursuing these topics more should definitely consider taking a course in number theory.

1. In each case we can just use trial division up to the square root of the number being tested.

a) Since $21 = 3\cdot 7$, we know that 21 is not prime.

b) Since $2 \nmid 29$, $3 \nmid 29$, and $5 \nmid 29$, we know that 29 is prime. We needed to check for prime divisors only up to $\sqrt{29}$, which is less than 6.

c) Since $2 \nmid 71$, $3 \nmid 71$, $5 \nmid 71$, and $7 \nmid 71$, we know that 71 is prime.

d) Since $2 \nmid 97$, $3 \nmid 97$, $5 \nmid 97$, and $7 \nmid 97$, we know that 97 is prime.

e) Since $111 = 3\cdot 37$, we know that 111 is not prime.

f) Since $143 = 11\cdot 13$, we know that 143 is not prime.

3. In each case we can use trial division, starting with the smallest prime and increasing to the next prime once we find that a given prime no longer is a divisor of what is left. A calculator comes in handy. Alternatively, one could use a factor tree.

a) We note that 2 is a factor of 88, and the quotient upon division by 2 is 44. We divide by 2 again, and then again, leaving a quotient of 11. Since 11 is prime, we are done, and we have found the prime factorization: $88 = 2^3\cdot 11$.

b) $126 = 2\cdot 63 = 2\cdot 3\cdot 21 = 2\cdot 3\cdot 3\cdot 7 = 2\cdot 3^2\cdot 7$

c) $729 = 3\cdot 243 = 3\cdot 3\cdot 81 = 3\cdot 3\cdot 3\cdot 27 = 3\cdot 3\cdot 3\cdot 3\cdot 9 = 3\cdot 3\cdot 3\cdot 3\cdot 3\cdot 3 = 3^6$

d) $1001 = 7\cdot 143 = 7\cdot 11\cdot 13$

e) $1111 = 11\cdot 101$ (we know that 101 is prime because we have already tried all prime factors less than $\sqrt{101}$)

f) $909090 = 2\cdot 454545 = 2\cdot 3\cdot 151515 = 2\cdot 3\cdot 3\cdot 50505 = 2\cdot 3\cdot 3\cdot 3\cdot 16835 = 2\cdot 3\cdot 3\cdot 3\cdot 5\cdot 3367 = 2\cdot 3\cdot 3\cdot 3\cdot 5\cdot 7\cdot 481 = 2\cdot 3\cdot 3\cdot 3\cdot 5\cdot 7\cdot 13\cdot 37 = 2\cdot 3^3\cdot 5\cdot 7\cdot 13\cdot 37$

5. $10! = 2\cdot 3\cdot 4\cdot 5\cdot 6\cdot 7\cdot 8\cdot 9\cdot 10 = 2\cdot 3\cdot 2^2\cdot 5\cdot (2\cdot 3)\cdot 7\cdot 2^3\cdot 3^2\cdot (2\cdot 5) = 2^8\cdot 3^4\cdot 5^2\cdot 7$

7. We give a proof by contradiction. Suppose that in fact $\log_2 3$ is the rational number p/q, where p and q are integers. Since $\log_2 3 > 0$, we can assume that p and q are positive. Translating the equation $\log_2 3 = p/q$ into its exponential equivalent, we obtain $3 = 2^{p/q}$. Raising both sides to the qth power yields $3^q = 2^p$. Now this is a violation of the Fundamental Theorem of Arithmetic, since it gives two different prime factorizations of the same number. Hence our assumption (that $\log_2 3$ is rational) must be wrong, and we conclude that $\log_2 3$ is irrational.

9. This is simply an existence statement. To prove that it is true, we need only exhibit the primes. Indeed, 3, 5, and 7 satisfy the conditions. (Actually, this is the only example, and a harder problem is to prove that there are no others.)

11. The prime factors of 30 are 2, 3, and 5. Thus we are looking for positive integers less than 30 that have none of these as prime factors. Since the smallest prime number other than these is 7, and 7^2 is already greater than 30, in fact only primes (and the number 1) will satisfy this condition. Therefore the answer is 1, 7, 11, 13, 17, 19, 23, and 29.

13. **a)** Since $\gcd(11,15) = 1$, $\gcd(11,19) = 1$, and $\gcd(15,19) = 1$, these three numbers are pairwise relatively prime.

b) Since $\gcd(15,21) = 3 > 1$, these three numbers are not pairwise relatively prime.

c) Since $\gcd(12,17) = 1$, $\gcd(12,31) = 1$, $\gcd(12,37) = 1$, $\gcd(17,31) = 1$, $\gcd(17,37) = 1$, and $\gcd(31,37) = 1$, these four numbers are pairwise relatively prime. (Indeed, the last three are primes, and the prime factors of the first are 2 and 3.)

d) Again, since no two of 7, 8, 9, and 11 have a common factor greater than 1, this set is pairwise relatively prime.

15. The identity shown in the hint is valid, as can be readily seen by multiplying out the right-hand side (all the terms cancel—telescope—except for 2^{ab} and -1). We will prove the assertion by proving its contrapositive. Suppose that n is *not* prime. Then by definition $n = ab$ for some integers a and b each greater than 1. Since $a > 1$, $2^a - 1$, the first factor in the suggested identity, is greater than 1. Clearly the second factor is greater than 1. Thus $2^n - 1 = 2^{ab} - 1$ is the product of two integers each greater than 1, so it is not prime.

17. We compute $\phi(n)$ here by enumerating the set of positive integers less than n that are relatively prime to n.

a) $\phi(4) = |\{1,3\}| = 2$ **b)** $\phi(10) = |\{1,3,7,9\}| = 4$

c) $\phi(13) = |\{1,2,3,4,5,6,7,8,9,10,11,12\}| = 12$

19. All the positive integers less than or equal to p^k (and there are clearly p^k of them) are less than p^k and relatively prime to p^k unless they are a multiple of p. Since the fraction $1/p$ of them are multiples of p, we have $\phi(p^k) = p^k(1 - 1/p) = p^k - p^{k-1}$.

21. To find the greatest common divisor of two numbers whose prime factorizations are given, we just need to take the smaller exponent for each prime.

a) The first number has no prime factors of 2, so the gcd has no 2's. Since the first number has seven factors of 3, but the second number has only five, the gcd has five factors of 3. Similarly the gcd has a factor of 5^3. So the gcd is $3^5\cdot 5^3$.

b) These numbers have no common prime factors, so the gcd is 1. **c)** 23^{17} **d)** $41 \cdot 43 \cdot 53$

e) These numbers have no common prime factors, so the gcd is 1.

f) The gcd of any positive integer and 0 is that integer, so the answer is 1111.

23. To find the least common multiple of two numbers whose prime factorizations are given, we just need to take the larger exponent for each prime.

a) The first number has no prime factors of 2 but the second number has 11 of them, so the lcm has 11 factors of 2. Since the first number has seven factors of 3 and the second number has five, the lcm has seven factors of 3. Similarly the lcm has a factor of 5^9 and a factor of 7^3. So the lcm is $2^{11} \cdot 3^7 \cdot 5^9 \cdot 7^3$.

b) These numbers have no common prime factors, so the lcm is their product, $2^9 \cdot 3^7 \cdot 5^5 \cdot 7^3 \cdot 11 \cdot 13 \cdot 17$.

c) 23^{31} **d)** $41 \cdot 43 \cdot 53$ **e)** $2^{12} \cdot 3^{13} \cdot 5^{17} \cdot 7^{21}$, as in part **(b)**

f) It makes no sense to ask for a positive multiple of 0, so this question has no answer. Least common multiples are defined only for positive integers.

25. First we find the prime factorizations: $92928 = 2^8 \cdot 3 \cdot 11^2$ and $123552 = 2^5 \cdot 3^3 \cdot 11 \cdot 13$. Therefore $\gcd(92928, 123552) = 2^5 \cdot 3 \cdot 11 = 1056$ and $\text{lcm}(92928, 123552) = 2^8 \cdot 3^3 \cdot 11^2 \cdot 13 = 10872576$. The requested products are $(2^5 \cdot 3 \cdot 11) \cdot (2^8 \cdot 3^3 \cdot 11^2 \cdot 13)$ and $(2^8 \cdot 3 \cdot 11^2) \cdot (2^5 \cdot 3^3 \cdot 11 \cdot 13)$, both of which are $2^{13} \cdot 3^4 \cdot 11^3 \cdot 13 = 11{,}481{,}440{,}256$.

27. The important observation to make here is that the smaller of any two numbers plus the larger of the two numbers is always equal to the sum of the two numbers. Since the exponent of the prime p in $\gcd(a, b)$ is the smaller of the exponents of p in a and in b, and since the exponent of the prime p in $\text{lcm}(a, b)$ is the larger of the exponents of p in a and in b, the exponent of p in $\gcd(a, b)\text{lcm}(a, b)$ is the sum of the smaller and the larger of these two values. Therefore by the observation, it equals the sum of the two values themselves, which is clearly equal to the exponent of p in ab. Since this is true for every prime p, we conclude that $\gcd(a, b)\text{lcm}(a, b)$ and ab have the same prime factorizations and are therefore equal.

29. Obviously there are no definitive answers to these problems, but we present below a reasonable and satisfying rule for forming the sequence in each case.

a) There are 1's in the prime locations and 0's elsewhere. In other words, the nth term of the sequence is 1 if n is a prime number and 0 otherwise.

b) The suspicious 2's occurring every other term and the appearance of the 11 and 13 lead us to discover that the nth term is the smallest prime factor of n (and is 1 when $n = 1$).

c) The nth term is the number of positive divisors of n. For example, the twelfth term is 6, since 12 has the positive divisors 1, 2, 3, 4, 6, and 12. A tip-off to get us going in the right direction is that there are 2's in the prime locations.

d) Perhaps the composer of the problem had something else in mind, but one rule here is that the nth term is 0 if and only if n has a repeated prime factor; the 1's occur at locations for which n is "square-free" (has no factor, other than 1, that is a perfect square). For example, 12 has the square 2^2, so the twelfth term is 0.

e) We note that all the terms (after the first one) are primes. This leads us to guess that the nth term is the largest prime less than or equal to n (and is 1 when $n = 1$).

f) Each term comes from the one before it by multiplying by a certain number. The multipliers are 2, 3, 5, 7, 11, 13, 17, 19, and 23—the primes. So the rule seems to be that we obtain the next term from the nth term by multiplying by the nth prime number (and we start at 1). In other words, the nth term is the product of the smallest $n - 1$ prime numbers.

31. Consider the product $n(n+1)(n+2)$ for some integer n. Since every second integer is even (divisible by 2), this product is divisible by 2. Since every third integer is divisible by 3, this product is divisible by 3. Therefore this product has both 2 and 3 in its prime factorization and is therefore divisible by $2 \cdot 3 = 6$.

33. It is hard to know how to get started on this problem. To some extent, mathematics is an experimental science, so it would probably be a good idea to compute $n^2 - 79n + 1601$ for several positive integer values of n to get a feel for what is happening. Using a computer, or at least a calculator, would be helpful. If we plug in $n = 1, 2, 3, 4,$ and 5, then we get the values 1523, 1447, 1373, 1301, and 1231, all of which are prime. This may lead us to believe that the proposition is true, but it gives us no clue as to how to prove it. Indeed, it seems as if it would be very hard to prove that this expression *always* produces a prime number, since being prime means the absence of nontrivial factors, and nothing in the expression seems to be very helpful in proving such a negative assertion. (The fact that we cannot factor it algebraically is irrelevant—in fact, if it factored algebraically, then it would essentially *never* be prime.) Perhaps we should try some more integers. If we do so, we find a lot more prime numbers, but we are still skeptical. Well, perhaps there is some way to arrange that this expression will have a factor. How about 1601? Well, yes! If we let $n = 1601$, then all three terms will have 1601 as a common factor, so that 1601 is a factor of the entire expression. In fact, $1601^2 - 79 \cdot 1601 + 1601 = 1601 \cdot 1523$. So we have found a counterexample after all, and the proposition is false. Note that this was not a problem in which we could proceed in a calm, calculated way from problem to solution. Mathematics is often like that—lots of false leads and approaches that get us nowhere, and then suddenly a burst of insight that solves the problem. (The smallest n for which this expression is not prime is $n = 80$; this gives the value $1681 = 41 \cdot 41$.)

35. Recall that the proof that there are infinitely many primes starts by assuming that there are only finitely many primes p_1, p_2, $\ldots$, p_n, and forming the number $p_1 p_2 \cdots p_n + 1$. This number is either prime or has a prime factor different from each of the primes p_1, p_2, $\ldots$, p_n; this shows that there are infinitely many primes. So, let us suppose that there are only finitely many primes of the form $4k+3$, namely q_1, q_2, $\ldots$, q_n, where $q_1 = 3$, $q_2 = 7$, and so on.

What number can we form that is not divisible by any of these primes, but that must be divisible by a prime of the form $4k+3$? We might consider the number $4q_1q_2 \cdots q_n + 3$. Unfortunately, this number is not prime, as it is is divisible by 3 (because $q_1 = 3$). Instead we consider the number $Q = 4q_1q_2 \cdots q_n - 1$. Note that Q is of the form $4k+3$ (where $k = q_1q_2 \cdots q_n - 1$). If Q is prime, then we have found a prime of the desired form different from all those listed. If Q is not prime, then Q has at least one prime factor not in the list q_1, q_2, $\ldots$, q_n, because the remainder when Q is divided by q_j is $q_j - 1$, and $q_j - 1 \neq 0$. Therefore $q_j \nmid Q$ for $j = 1, 2, \ldots, n$. Because all odd primes are either of the form $4k+1$ or of the form $4k+3$, and the product of primes of the form $4k+1$ is also of this form (because $(4k+1)(4m+1) = 4(4km+k+m)+1$), there must be a factor of Q of the form $4k+3$ different from the primes we listed. This complete the proof.

37. We need to show that this function is one-to-one and onto. In other words, if we are given a positive integer x, we must show that there is exactly one positive rational number m/n (written in lowest terms) such that $K(m/n) = x$. To do this, we factor x into its prime factorization and then read off the m and n such that $K(m/n) = x$. The primes that occur to even powers are the primes that occur in the prime factorization of m, with the exponents being half the corresponding exponents in x; and the primes that occur to odd powers are the primes that occur in the prime factorization of n, with the exponents being half of one more than the exponents in x. Since this uniquely determines m and n, there is one and only one fraction, in lowest terms, that maps to x under K.

SECTION 3.6 Integers and Algorithms

In addition to calculation exercises on the Euclidean algorithm, the base conversion algorithm, and algorithms for the basic arithmetic operations, this exercise set introduces other forms of representing integers. These are **balanced ternary expansion, Cantor expansion, binary coded decimal** *(or* **BCD***) representation, and* **one's** *and* **two's complement** *representations. Each has practical and/or theoretical importance in mathematics or computer science. If all else fails, one can carry out an algorithm by "playing computer" and mechanically following the pseudocode step by step.*

1. We divide repeatedly by 2, noting the remainders. The remainders are then arranged from right to left to obtain the binary representation of the given number.

a) We begin by dividing 231 by 2, obtaining a quotient of 115 and a remainder of 1. Therefore $a_0 = 1$. Next $115/2 = 57$, remainder 1. Therefore $a_1 = 1$. Similarly $57/2 = 28$, remainder 1. Therefore $a_2 = 1$. Then $28/2 = 14$, remainder 0, so $a_3 = 0$. Similarly $a_4 = 0$, after we divide 14 by 2, obtaining 7 with remainder 0. Three more divisions yield quotients of 3, 1, and 0, with remainders of 1, 1, and 1, respectively, so $a_5 = a_6 = a_7 = 1$. Putting all this together, we see that the binary representation is $(a_7a_6a_5a_4a_3a_2a_1a_0)_2 = (1110\,0111)_2$. As a check we can compute that $2^0 + 2^1 + 2^2 + 2^5 + 2^6 + 2^7 = 231$.

b) Following the same procedure as in part **(a)**, we obtain successive remainders 0, 0, 1, 0, 1, 1, 0, 1, 1, 0, 0, 0, 1. Therefore $4532 = (1\ 0001\ 1011\ 0100)_2$.

c) By the same method we obtain $97644 = (1\ 0111\ 1101\ 0110\ 1100)_2$.

3. a) $(1\ 1111)_2 = 2^4 + 2^3 + 2^2 + 2^1 + 2^0 = 16 + 8 + 4 + 2 + 1 = 31$. An easier way to get the answer is to note that $(1\ 1111)_2 = (10\ 0000)_2 - 1 = 2^5 - 1 = 31$.

b) $(10\ 0000\ 0001)_2 = 2^9 + 2^0 = 513$

c) $(1\ 0101\ 0101)_2 = 2^8 + 2^6 + 2^4 + 2^2 + 2^0 = 256 + 64 + 16 + 4 + 1 = 341$

d) $(110\ 1001\ 0001\ 0000)_2 = 2^{14} + 2^{13} + 2^{11} + 2^8 + 2^4 = 16384 + 8192 + 2048 + 256 + 16 = 26896$

5. Following Example 6, we simply write the binary equivalents of each digit: $(\text{A})_{16} = (1010)_2$, $(\text{B})_{16} = (1011)_2$, $(\text{C})_{16} = (1100)_2$, $(\text{D})_{16} = (1101)_2$, $(\text{E})_{16} = (1110)_2$, and $(\text{F})_{16} = (1111)_2$. Note that the blocking by groups of four binary digits is just for readability by humans.

a) $(\text{80E})_{16} = (1000\ 0000\ 1110)_2$

b) $(\text{135AB})_{16} = (0001\ 0011\ 0101\ 1010\ 1011)_2$

c) $(\text{ABBA})_{16} = (1010\ 1011\ 1011\ 1010)_2$

d) $(\text{DEFACED})_{16} = (1101\ 1110\ 1111\ 1010\ 1100\ 1110\ 1101)_2$

7. Following Example 6, we simply write the binary equivalents of each digit. Since $(\text{A})_{16} = (1010)_2$, $(\text{B})_{16} = (1011)_2$, $(\text{C})_{16} = (1100)_2$, $(\text{D})_{16} = (1101)_2$, $(\text{E})_{16} = (1110)_2$, and $(\text{F})_{16} = (1111)_2$, we see that $(\text{ABCDEF})_{16} = (101010111100110111101111)_2$. Following the convention shown in Exercise 3 of grouping binary digits by fours, we can write this in a more readable form as 1010 1011 1100 1101 1110 1111.

9. Following Example 6, we simply write the hexadecimal equivalents of each group of four binary digits. Thus we have $(1011\ 0111\ 1011)_2 = (\text{B7B})_{16}$.

11. We adopt a notation that will help with the explanation. Adding up to three leading 0's if necessary, write the binary expansion as $(\ldots b_{23}b_{22}b_{21}b_{20}b_{13}b_{12}b_{11}b_{10}b_{03}b_{02}b_{01}b_{00})_2$. The value of this numeral is $b_{00} + 2b_{01} + 4b_{02} + 8b_{03} + 2^4b_{10} + 2^5b_{11} + 2^6b_{12} + 2^7b_{13} + 2^8b_{20} + 2^9b_{21} + 2^{10}b_{22} + 2^{11}b_{23} + \cdots$, which we can rewrite as $b_{00} + 2b_{01} + 4b_{02} + 8b_{03} + (b_{10} + 2b_{11} + 4b_{12} + 8b_{13}) \cdot 2^4 + (b_{20} + 2b_{21} + 4b_{22} + 8b_{23}) \cdot 2^8 + \cdots$. Now $(b_{i3}b_{i2}b_{i1}b_{i0})_2$ translates into the hexadecimal digit h_i. So our number is $h_0 + h_1 \cdot 2^4 + h_2 \cdot 2^8 + \cdots = h_0 + h_1 \cdot 16 + h_2 \cdot 16^2 + \cdots$, which is the hexadecimal expansion $(\ldots h_1h_1h_0)_{16}$.

13. This is exactly the same as what we can do with hexadecimal expansion, replacing groups of four with groups of three. Specifically, group together blocks of three binary digits, adding up to two initial 0's if necessary, and translate each block of three binary digits into a single octal digit. For example, $(011\ 000\ 110)_2 = (306)_8$

15. In each case we follow the method of Example 6, blocking by threes instead of fours. We replace each octal digit of the given numeral by its 3-digit binary equivalent and string the digits together. The first digit is $(7)_8 = (111)_2$, the next is $(3)_8 = (011)_2$, and so on, so we obtain $(111011100101011010001)_2$. For the other direction, we split the given binary numeral into blocks of three digits, adding initial 0's to fill it out: 001 010 111 011. Then we replace each block by its octal equivalent, obtaining the answer $(1273)_8$.

17. Since we have procedures for converting both octal and hexadecimal to and from binary (Example 6 and Exercises 13–15), to convert from octal to hexadecimal, we first convert from octal to binary and then convert from binary to hexadecimal.

19. In effect, this algorithm computes $7 \textbf{ mod } 645$, $7^2 \textbf{ mod } 645$, $7^4 \textbf{ mod } 645$, $7^8 \textbf{ mod } 645$, $7^{16} \textbf{ mod } 645$, ..., and then multiplies (modulo 645) the required values. Since $644 = (1010000100)_2$, we need to multiply together $7^4 \textbf{ mod } 645$, $7^{128} \textbf{ mod } 645$, and $7^{512} \textbf{ mod } 645$, reducing modulo 645 at each step. We compute by repeatedly squaring: $7^2 \textbf{ mod } 645 = 49$, $7^4 \textbf{ mod } 645 = 49^2 \textbf{ mod } 645 = 2401 \textbf{ mod } 645 = 466$, $7^8 \textbf{ mod } 645 = 466^2 \textbf{ mod } 645 = 217156 \textbf{ mod } 645 = 436$, $7^{16} \textbf{ mod } 645 = 436^2 \textbf{ mod } 645 = 190096 \textbf{ mod } 645 = 466$. At this point we see a pattern with period 2, so we have $7^{32} \textbf{ mod } 645 = 436$, $7^{64} \textbf{ mod } 645 = 466$, $7^{128} \textbf{ mod } 645 = 436$, $7^{256} \textbf{ mod } 645 = 466$, and $7^{512} \textbf{ mod } 645 = 436$. Thus our final answer will be the product of 466, 436, and 436, reduced modulo 645. We compute these one at a time: $466 \cdot 436 \textbf{ mod } 645 = 203176 \textbf{ mod } 645 = 1$, and $1 \cdot 436 \textbf{ mod } 645 = 436$. So $7^{644} \textbf{ mod } 645 = 436$. A computer algebra system will verify this; use the command "`7 &^ 644 mod 645;`" in *Maple*, for example. The ampersand here tells *Maple* to use modular exponentiation, rather than first computing the integer 7^{644}, which has over 500 digits, although it could certainly handle this if asked. The point is that modular exponentiation is much faster and avoids having to deal with such large numbers.

21. In effect, this algorithm computes $3 \textbf{ mod } 99$, $3^2 \textbf{ mod } 99$, $3^4 \textbf{ mod } 99$, $3^8 \textbf{ mod } 99$, $3^{16} \textbf{ mod } 99$, ..., and then multiplies (modulo 99) the required values. Since $2003 = (11111010011)_2$, we need to multiply together $3 \textbf{ mod } 99$, $3^2 \textbf{ mod } 99$, $3^{16} \textbf{ mod } 99$, $3^{64} \textbf{ mod } 99$, $3^{128} \textbf{ mod } 99$, $3^{256} \textbf{ mod } 99$, $3^{512} \textbf{ mod } 99$, and $3^{1024} \textbf{ mod } 99$, reducing modulo 99 at each step. We compute by repeatedly squaring: $3^2 \textbf{ mod } 99 = 9$, $3^4 \textbf{ mod } 99 = 81$, $3^8 \textbf{ mod } 99 = 81^2 \textbf{ mod } 99 = 6561 \textbf{ mod } 99 = 27$, $3^{16} \textbf{ mod } 99 = 27^2 \textbf{ mod } 99 = 729 \textbf{ mod } 99 = 36$, $3^{32} \textbf{ mod } 99 = 36^2 \textbf{ mod } 99 = 1296 \textbf{ mod } 99 = 9$, and then the pattern repeats, so $3^{64} \textbf{ mod } 99 = 81$, $3^{128} \textbf{ mod } 99 = 27$, $3^{256} \textbf{ mod } 99 = 36$, $3^{512} \textbf{ mod } 99 = 9$, and $3^{1024} \textbf{ mod } 99 = 81$. Thus our final answer will be the product of 3, 9, 36, 81, 27, 36, 9, and 81. We compute these one at a time modulo 99: $3 \cdot 9$ is 27, $27 \cdot 36$ is 81, $81 \cdot 81$ is 27, $27 \cdot 27$ is 36, $36 \cdot 36$ is 9, $9 \cdot 9$ is 81, and finally $81 \cdot 81$ is 27. So $3^{2003} \textbf{ mod } 99 = 27$.

23. **a)** By Lemma 1, $\gcd(12,18)$ is the same as the gcd of the smaller of these two numbers (12) and the remainder when the larger (18) is divided by the smaller. In this case the remainder is 6, so $\gcd(12,18) = \gcd(12,6)$. Now $\gcd(12,6)$ is the same as the gcd of the smaller of *these* two numbers (6) and the remainder when the larger (12) is divided by the smaller, namely 0. This gives $\gcd(12,6) = \gcd(6,0)$. But $\gcd(x,0) = x$ for all positive integers, so $\gcd(6,0) = 6$. Thus the answer is 6. In brief (the form we will use for the remaining parts), $\gcd(12,18) = \gcd(12,6) = \gcd(6,0) = 6$.

b) $\gcd(111,201) = \gcd(111,90) = \gcd(90,21) = \gcd(21,6) = \gcd(6,3) = \gcd(3,0) = 3$

c) $\gcd(1001,1331) = \gcd(1001,330) = \gcd(330,11) = \gcd(11,0) = 11$

d) $\gcd(12345,54321) = \gcd(12345,4941) = \gcd(4941,2463) = \gcd(2463,15) = \gcd(15,3) = \gcd(3,0) = 3$

e) $\gcd(1000, 5040) = \gcd(1000, 40) = \gcd(40, 0) = 40$

f) $\gcd(9888, 6060) = \gcd(6060, 3828) = \gcd(3828, 2232) = \gcd(2232, 1596) = \gcd(1596, 636) = \gcd(636, 324)$ $= \gcd(324, 312) = \gcd(312, 12) = \gcd(12, 0) = 12$

25. In carrying out the Euclidean algorithm on this data, we divide successively by 34, 21, 13, 8, 5, 3, 2, and 1, so eight divisions are required.

27. The binary expansion of an integer represents the integer as a sum of distinct powers of 2. For example, since $21 = (1\ 0101)_2$, we have $21 = 2^4 + 2^2 + 2^0$. Since binary expansions are unique, each integer can be so represented uniquely.

29. Let the decimal expansion of the integer a be given by $a = (a_{n-1}a_{n-2}\ldots a_1a_0)_{10}$. Thus $a = 10^{n-1}a_{n-1} + 10^{n-2}a_{n-2} + \cdots + 10a_1 + a_0$. Since $10 \equiv 1 \pmod 3$, we have $a \equiv a_{n-1} + a_{n-2} + \cdots + a_1 + a_0 \pmod 3$. Therefore $a \equiv 0 \pmod 3$ if and only if the sum of the digits is congruent to $0 \pmod 3$. Since being divisible by 3 is the same as being congruent to $0 \pmod 3$, we have proved that a positive integer is divisible by 3 if and only if the sum of its decimal digits is divisible by 3.

31. Let the binary expansion of the positive integer a be given by $a = (a_{n-1}a_{n-2}\ldots a_1a_0)_2$. Thus $a = a_0 + 2a_1 + 2^2a_2 + \cdots + 2^{n-1}a_{n-1}$. Since $2^2 \equiv 1 \pmod 3$, we see that $2^k \equiv 1 \pmod 3$ when k is even, and $2^k \equiv 2 \equiv -1 \pmod 3$ when k is odd. Therefore we have $a \equiv a_0 - a_1 + a_2 - a_3 + \cdots \pm a_{n-1} \pmod 3$. Thus $a \equiv 0 \pmod 3$ if and only if the sum of the binary digits in the even-numbered positions minus the sum of the binary digits in the odd-numbered positions is congruent to 0 modulo 3. Since being divisible by 3 is the same as being congruent to $0 \pmod 3$, our proof is complete.

33. a) Since the leading bit is a 1, this represents a negative number. The binary expansion of the absolute value of this number is the complement of the rest of the expansion, namely the complement of 1001, or 0110. Since $(0110)_2 = 6$, the answer is -6.

b) Since the leading bit is a 0, this represents a positive number, namely the number whose binary expansion is the rest of this string, 1101. Since $(1101)_2 = 13$, the answer is 13.

c) The answer is the negative of the complement of 0001, namely $-(1110)_2 = -14$.

d) $-(0000)_2 = 0$; note that 0 has two different representations, 0000 and 1111

35. We must assume that the sum actually represents a number in the appropriate range. Assume that n bits are being used, so that numbers strictly between -2^{n-1} and 2^{n-1} can be represented. The answer is almost, but not quite, that to obtain the one's complement representation of the sum of two numbers, we simply add the two strings representing these numbers using Algorithm 3. Instead, after performing this operation, there may be a carry out of the left-most column; in such a case, we then add 1 more to the answer. For example, suppose that $n = 4$; then numbers from -7 to 7 can be represented. To add -5 and 3, we add 1010 and 0011, obtaining 1101; there was no carry out of the left-most column. Since 1101 is the one's complement representation of -2, we have the correct answer. On the other hand, to add -4 and -3, we add 1011 and 1100, obtaining 1 0111. The 1 that was carried out of the left-most column is instead added to 0111, yielding 1000, which is the one's complement representation of -7. A proof that this method works entails considering the various cases determined by the signs and magnitudes of the addends.

37. If m is positive (or 0), then the leading bit (a_{n-1}) is 0, so the formula reads simply $m = \sum_{i=0}^{n-2} a_i 2^i$, which is clearly correct, since this is the binary expansion of m. (See Section 2.4 for the meaning of summation notation. This symbolism is a shorthand way of writing $a_0 + 2a_1 + 4a_2 + \cdots + 2^{n-2}a_{n-2}$.) Now suppose that m is negative. The one's complement expansion for m has its leading bit equal to 1. By the definition of one's complement, we can think of obtaining the remaining $n-1$ bits by subtracting $-m$, written in binary, from $111\ldots1$ (with $n-1$ 1's), since subtracting a bit from 1 is the same thing as complementing it. Equivalently, if we view the bit string $(a_{n-2}a_{n-1}\ldots a_0)$ as a binary number, then it represents $(2^{n-1}-1)-(-m)$. In symbols, this says that $(2^{n-1}-1)-(-m)=\sum_{i=0}^{n-2} a_i 2^i$. Solving for m gives us the equation we are trying to prove (since $a_{n-1}=1$).

39. Following the definition, if the first bit is a 0, then we just evaluate the binary expansion. If the first bit is a 1, then we find what number x is represented by the remaining four bits in binary; the answer is then $-(2^4-x)$.

a) Since the first bit is a 1, and the remaining bits represent the number 9, this string represents the number $-(2^4-9)=-7$.

b) Since the first bit is a 0 and this is just the binary expansion of 13, the answer is 13.

c) Since the first bit is a 1, and the remaining bits represent the number 1, this string represents the number $-(2^4-1)=-15$.

d) Since the first bit is a 1, and the remaining bits represent the number 15, this string represents the number $-(2^4-15)=-1$. Note that 10000 would represent $-(2^4-0)=-16$, so in fact we can represent one extra negative number than positive number with this notation.

41. The nice thing about two's complement arithmetic is that we can just work as if it were all in base 2, since $-x$ (where x is positive) is represented by 2^n-x; in other words, modulo 2^n, negative numbers represent themselves. However, if overflow occurs, then we must recognize an error. Let us look at some examples, where $n=5$ (i.e., we use five bits to represent numbers between -15 and 15). To add $5+7$, we write $00101+00111=01100$ in base 2, which gives us the correct answer, 12. However, if we try to add $13+7$ we obtain $01101+00111=10100$, which represents -12, rather than 20, so we report an overflow error. (Of course these two numbers are congruent modulo 32.) Similarly, for $5+(-7)$, we write $00101+11001=11110$ in base 2, and 11110 is the two's complement representation of -2, the right answer. For $(-5)+(-7)$, we write $11011+11001=110100$ in base 2; if we ignore the extra 1 in the left-most column (which doesn't exist), then this is the two's complement representation of -12, again the right answer. To summarize, to obtain the two's complement representation of the sum of two integers given in two's complement representation, add them as if they were binary integers, and ignore any carry out of the left-most column. However, if the left-most digits of the two addends agree and the left-most digit of the answer is different from their common value, then an overflow has occurred, and the answer is not valid.

43. If m is positive (or 0), then the leading bit (a_{n-1}) is 0, so the formula reads simply $m=\sum_{i=0}^{n-2} a_i 2^i$, which is clearly correct, since this is the binary expansion of m. (See Section 2.4 for the meaning of summation notation. This symbolism is a shorthand way of writing $a_0+2a_1+4a_2+\cdots+2^{n-2}a_{n-2}$.) Now suppose that m is negative. The two's complement expansion for m has its leading bit equal to 1. By the definition of two's complement, the remaining $n-1$ bits are the binary expansion of $2^{n-1}-(-m)$. In symbols, this says that $2^{n-1}-(-m)=\sum_{i=0}^{n-2} a_i 2^i$. Solving for m gives us the equation we are trying to prove (since $a_{n-1}=1$).

45. Clearly we need $4n$ digits, four for each digit of the decimal representation.

47. To find the Cantor expansion, we will work from left to right. Thus the first step will be to find the largest number n whose factorial is still less than or equal to the given positive integer x. Then we determine the digits in the expansion, starting with a_n and ending with a_1.

procedure *Cantor*(x : positive integer)
$n := 1$; *factorial* $:= 1$
while $(n+1)\cdot \textit{factorial} \le x$
begin

```
        n := n + 1
        factorial := factorial · n
end {at this point we know that there are n digits in the expansion}
y := x {this is just so we do not destroy the original input}
while n > 0
begin
        a_n := ⌊y/factorial⌋
        y := y − a_n · factorial
        factorial := factorial/n
        n := n − 1
end
{we are done: x = a_n n! + a_{n−1}(n − 1)! + ··· + a_2 2! + a_1 1! }
```

49. Note that $n = 5$. Initially the carry is $c = 0$, and we start the **for** loop with $j = 0$. Since $a_0 = 1$ and $b_0 = 0$, we set d to be $\lfloor(1 + 0 + 0)/2\rfloor = 0$; then $s_0 = 1+0+0-2\cdot 0$, which equals 1, and finally $c = 0$. At the end of the first pass, then, the right-most digit of the answer has been determined (it's a 1), and there is a carry of 0 into the next column.

Now $j = 1$, and we compute d to be $\lfloor(a_1 + b_1 + c)/2\rfloor = \lfloor(1 + 1 + 0)/2\rfloor = 1$; whereupon s_1 becomes $1+1+0-2\cdot 1 = 0$, and c is set to 1. Thus far we have determined that the last two bits of the answer are 01 (from left to right), and there is a carry of 1 into the next column.

The next three passes through the loop are similar. As a result of the pass when $j = 2$ we set $d = 1$, $s_2 = 0$, and then $c = 1$. When $j = 3$, we obtain $d = 1$, $s_3 = 0$, and then $c = 1$. Finally, when $j = 4$, we obtain $d = 1$, $s_4 = 1$, and then $c = 1$. At this point the loop is terminated, and when we execute the final step, $s_5 = 1$. Thus the answer is 11 0001.

51. We will assume that the answer is not negative, since otherwise we would need something like the one's complement representation. The algorithm is similar to the algorithm for addition, except that we need to borrow instead of carry. Rather than trying to incorporate the two cases (borrow or no borrow) into one, as was done in the algorithm for addition, we will use an **if**...**then** statement to treat the cases separately. The notation is the usual one: $a = (a_{n-1} \ldots a_1 a_0)_2$ and $b = (b_{n-1} \ldots b_1 b_0)_2$

```
procedure subtract(a, b : nonnegative integers)
borrow := 0
for j := 0 to n − 1
        if a_j − borrow ≥ b_j then
        begin
                s_j := a_j − borrow − b_j
                borrow := 0
        end
        else
        begin
                s_j := a_j + 2 − borrow − b_j
                borrow := 1
        end
{assuming a ≥ b, we have a − b = (s_{n−1} s_{n−2} ... s_1 s_0)_2 }
```

53. To determine which of two integers (we assume they are nonnegative), given in binary as $a = (a_{n-1} \ldots a_1 a_0)_2$ and $b = (b_{n-1} \ldots b_1 b_0)_2$, is larger, we need to compare digits from the most significant end ($i = n - 1$) to the least ($i = 0$), stopping if and when we find a difference. For variety here we record the answer as a character string; in most applications it would probably be better to set *compare* to one of three code values (such as -1, 1, and 0) to indicate which of the three possibilities held.

procedure *compare*$(a, b:$ nonnegative integers)
$i := n - 1$
while $i > 0$ and $a_i = b_i$
    $i := i - 1$
if $a_i > b_i$ **then** *answer* := "$a > b$"
else if $a_i < b_i$ **then** *answer* := "$a < b$"
else *answer* := "$a = b$"
{the answer is recorded in *answer*}

55. There is one division for each pass through the **while** loop. Also, each pass generates one digit in the base b expansion. Thus the number of divisions equals the number of digits in the base b expansion of n. This is just $\lfloor \log_b n \rfloor + 1$ (for example, numbers from 10 to 99, inclusive, have common logarithms in the interval $[1, 2)$). Therefore exactly $\lfloor \log_b n \rfloor + 1$ divisions are required, and this is $O(\log n)$. (We are counting only the actual division operation in the statement $q := \lfloor q/b \rfloor$. If we also count the implied division in the statement $a_k := q \textbf{ mod } b$, then there are twice as many as we computed here. The big-O estimate is the same, of course.)

57. The only time-consuming part of the algorithm is the **while** loop, which is iterated q times. The work done inside is a subtraction of integers no bigger than a, which has $\log a$ bits. The results now follows from Example 8.

SECTION 3.7 Applications of Number Theory

Many of these exercises are reasonably straightforward calculations, but the amount of arithmetic involved in some of them can be formidable. Look at the worked out examples in the text if you need help getting the hang of it. The theoretical exercises, such as #16 and #17 give you a good taste of the kinds of proofs in an elementary number theory course. Look at Exercise 45: it will give you a better understanding of some of the issues underlying this section, by forcing you to confront the difference between calculations that can be done without trial and error and those that seem to require it.

1. a) This first one is easy to do by inspection. Clearly 10 and 11 are relatively prime, so their greatest common divisor is 1, and $1 = 11 - 10 = (-1) \cdot 10 + 1 \cdot 11$.

b) In order to find the coefficients s and t such that $21s + 44t = \gcd(21, 44)$, we carry out the steps of the Euclidean algorithm.

$$44 = 2 \cdot 21 + 2$$

$$21 = 10 \cdot 2 + 1$$

Then we work up from the bottom, expressing the greatest common divisor (which we have just seen to be 1) in terms of the numbers involved in the algorithm, namely 44, 21, and 2. In particular, the last equation tells us that $1 = 21 - 10 \cdot 2$, so that we have expressed the gcd as a linear combination of 21 and 2. But now the first equation tells us that $2 = 44 - 2 \cdot 21$; we plug this into our previous equation and obtain

$$1 = 21 - 10 \cdot (44 - 2 \cdot 21) = 21 \cdot 21 - 10 \cdot 44 .$$

Thus we have expressed 1 as a linear combination (with integer coefficients) of 21 and 44, namely $\gcd(21, 44) = 21 \cdot 21 + (-10) \cdot 44$.

c) Again, we carry out the Euclidean algorithm. Since $48 = 1 \cdot 36 + 12$, and $12 \mid 36$, we know that $\gcd(36, 48) = 12$. From the equation shown here, we can immediately write $12 = (-1) \cdot 36 + 48$.

d) The calculation of the greatest common divisor takes several steps:

$$55 = 1 \cdot 34 + 21$$
$$34 = 1 \cdot 21 + 13$$
$$21 = 1 \cdot 13 + 8$$
$$13 = 1 \cdot 8 + 5$$
$$8 = 1 \cdot 5 + 3$$
$$5 = 1 \cdot 3 + 2$$
$$3 = 1 \cdot 2 + 1$$

Then we need to work our way back up, successively plugging in for the remainders determined in this calculation:

$$\begin{aligned} 1 &= 3 - 2 \\ &= 3 - (5 - 3) = 2 \cdot 3 - 5 \\ &= 2 \cdot (8 - 5) - 5 = 2 \cdot 8 - 3 \cdot 5 \\ &= 2 \cdot 8 - 3 \cdot (13 - 8) = 5 \cdot 8 - 3 \cdot 13 \\ &= 5 \cdot (21 - 13) - 3 \cdot 13 = 5 \cdot 21 - 8 \cdot 13 \\ &= 5 \cdot 21 - 8 \cdot (34 - 21) = 13 \cdot 21 - 8 \cdot 34 \\ &= 13 \cdot (55 - 34) - 8 \cdot 34 = 13 \cdot 55 - 21 \cdot 34 \end{aligned}$$

e) Here are the two calculations—down to the gcd using the Euclidean algorithm, and then back up by substitution until we have expressed the gcd as the desired linear combination of the original numbers.

$$213 = 1 \cdot 117 + 96$$
$$117 = 1 \cdot 96 + 21$$
$$96 = 4 \cdot 21 + 12$$
$$21 = 1 \cdot 12 + 9$$
$$12 = 1 \cdot 9 + 3$$

Since $3 \mid 9$, we have $\gcd(117, 213) = 3$.

$$\begin{aligned} 3 &= 12 - 9 \\ &= 12 - (21 - 12) = 2 \cdot 12 - 21 \\ &= 2 \cdot (96 - 4 \cdot 21) - 21 = 2 \cdot 96 - 9 \cdot 21 \\ &= 2 \cdot 96 - 9 \cdot (117 - 96) = 11 \cdot 96 - 9 \cdot 117 \\ &= 11 \cdot (213 - 117) - 9 \cdot 117 = 11 \cdot 213 - 20 \cdot 117 \end{aligned}$$

f) Clearly $\gcd(0, 223) = 223$, so we can write $223 = s \cdot 0 + 1 \cdot 223$ for any integer s.

g) Here are the two calculations—down to the gcd using the Euclidean algorithm, and then back up by substitution until we have expressed the gcd as the desired linear combination of the original numbers.

$$2347 = 19 \cdot 123 + 10$$
$$123 = 12 \cdot 10 + 3$$
$$10 = 3 \cdot 3 + 1$$

Thus the greatest common divisor is 1.

$$\begin{aligned} 1 &= 10 - 3 \cdot 3 \\ &= 10 - 3 \cdot (123 - 12 \cdot 10) = 37 \cdot 10 - 3 \cdot 123 \\ &= 37 \cdot (2347 - 19 \cdot 123) - 3 \cdot 123 = 37 \cdot 2347 - 706 \cdot 123 \end{aligned}$$

h) Here are the two calculations—down to the gcd using the Euclidean algorithm, and then back up by substitution until we have expressed the gcd as the desired linear combination of the original numbers.

$$4666 = 3454 + 1212$$
$$3454 = 2 \cdot 1212 + 1030$$
$$1212 = 1030 + 182$$
$$1030 = 5 \cdot 182 + 120$$
$$182 = 120 + 62$$
$$120 = 62 + 58$$
$$62 = 58 + 4$$
$$58 = 14 \cdot 4 + 2$$

Since $2 \mid 4$, the greatest common divisor is 2.

$$\begin{aligned}
2 &= 58 - 14 \cdot 4 \\
&= 58 - 14 \cdot (62 - 58) = 15 \cdot 58 - 14 \cdot 62 \\
&= 15 \cdot (120 - 62) - 14 \cdot 62 = 15 \cdot 120 - 29 \cdot 62 \\
&= 15 \cdot 120 - 29 \cdot (182 - 120) = 44 \cdot 120 - 29 \cdot 182 \\
&= 44 \cdot (1030 - 5 \cdot 182) - 29 \cdot 182 = 44 \cdot 1030 - 249 \cdot 182 \\
&= 44 \cdot 1030 - 249 \cdot (1212 - 1030) = 293 \cdot 1030 - 249 \cdot 1212 \\
&= 293 \cdot (3454 - 2 \cdot 1212) - 249 \cdot 1212 = 293 \cdot 3454 - 835 \cdot 1212 \\
&= 293 \cdot 3454 - 835 \cdot (4666 - 3454) = 1128 \cdot 3454 - 835 \cdot 4666
\end{aligned}$$

i) Here are the two calculations—down to the gcd using the Euclidean algorithm, and then back up by substitution until we have expressed the gcd as the desired linear combination of the original numbers.

$$11111 = 9999 + 1112$$
$$9999 = 8 \cdot 1112 + 1103$$
$$1112 = 1103 + 9$$
$$1103 = 122 \cdot 9 + 5$$
$$9 = 5 + 4$$
$$5 = 4 + 1$$

Thus 1 is the greatest common divisor.

$$\begin{aligned}
1 &= 5 - 4 \\
&= 5 - (9 - 5) = 2 \cdot 5 - 9 \\
&= 2 \cdot (1103 - 122 \cdot 9) - 9 = 2 \cdot 1103 - 245 \cdot 9 \\
&= 2 \cdot 1103 - 245 \cdot (1112 - 1103) = 247 \cdot 1103 - 245 \cdot 1112 \\
&= 247 \cdot (9999 - 8 \cdot 1112) - 245 \cdot 1112 = 247 \cdot 9999 - 2221 \cdot 1112 \\
&= 247 \cdot 9999 - 2221 \cdot (11111 - 9999) = 2468 \cdot 9999 - 2221 \cdot 11111
\end{aligned}$$

3. We simply need to show that $15 \cdot 7 \equiv 1 \pmod{26}$, or in other words, that $15 \cdot 7 - 1$ is divisible by 26. But this quantity is 104, which is $26 \cdot 4$.

5. We could look for the inverse by trial and error, since there are only nine possibilities. Alternately, we could write 1 (the greatest common divisor of 4 and 9) as a linear combination of 4 and 9. Indeed, using the techniques shown in Exercise 1, we have $1 = 7 \cdot 4 - 3 \cdot 9$. This tells us that $7 \cdot 4 - 1$ is divisible by 9, which means that $7 \cdot 4 \equiv 1 \pmod{9}$. In other words, 7 is the desired inverse of 4 modulo 9.

7. Using the techniques of Exercise 1, we can determine that $1 = 52 \cdot 19 - 7 \cdot 141$. This immediately tells us that $52 \cdot 19 \equiv 1 \pmod{141}$, so 52 is the desired inverse.

9. We follow the hint. Suppose that we had two inverses of a modulo m, say b and c. In symbols, we would have $ba \equiv 1 \pmod m$ and $ca \equiv 1 \pmod m$. The first congruence says that m divides $ba - 1$, and the second says that m divides $ca - 1$. Therefore m divides the difference $(ba - 1) - (ca - 1) = ba - ca$. (The difference of two multiples of m is a multiple of m.) Thus $ba \equiv ca \pmod m$. It follows immediately from Theorem 2 (the roles of a, b, and c need to be permuted) that $b \equiv c \pmod m$, which is what we wanted to prove.

11. By Exercise 5, we know that 7 is an inverse of 4 modulo 9. Therefore we can multiply both sides of the given congruence by 7 and obtain $x \equiv 7 \cdot 5 = 35 \equiv 8 \pmod 9$. Therefore the solution set consists of all numbers congruent to 8 modulo 9, namely $\ldots, -10, -1, 8, 17, \ldots$. We can check, for example, that $4 \cdot 8 = 32 \equiv 5 \pmod 9$.

13. The hypothesis tells us that m divides $ac - bc$, which is the product $(a - b)c$. Let m' be $m/\gcd(c, m)$. Then m' is a factor of m, so certainly $m' \mid (a - b)c$. Now since all the common factors of m and c were divided out of m to get m', we know that m' is relatively prime to c. It follows from Lemma 1 that $m' \mid a - b$. But this means that $a \equiv b \pmod{m'}$, exactly what we were trying to prove.

15. We want to find numbers x such that $x^2 \equiv 1 \pmod p$, in other words, such that p divides $x^2 - 1$. Factoring this expression, we see that we are seeking numbers x such that $p \mid (x + 1)(x - 1)$. By Lemma 2, this can only happen if $p \mid x + 1$ or $p \mid x - 1$. But these two congruences are equivalent to the statements $x \equiv -1 \pmod p$ and $x \equiv 1 \pmod p$.

17. **a)** If two of these integers were congruent modulo p, say ia and ja, where $1 \le i < j < p$, then we would have $p \mid ja - ia$, or $p \mid (j - i)a$. By Lemma 1, since a is not divisible by p, p must divide $j - i$. But this is impossible, since $j - i$ is a positive integer less than p. Therefore no two of these integers are congruent modulo p.

b) By part **(a)**, since no two of a, $2a$, $\ldots$, $(p - 1)a$ are congruent modulo p, each must be congruent to a different number from 1 to $p - 1$. Therefore if we multiply them all together, we will obtain the same product, modulo p, as if we had multiplied all the numbers from 1 to $p - 1$. In symbols,

$$a \cdot 2a \cdot 3a \cdots (p - 1)a \equiv 1 \cdot 2 \cdot 3 \cdots (p - 1) \pmod p.$$

The left-hand side of this congruence is clearly $(p - 1)! \cdot a^{p-1}$, and the right-hand side is just $(p - 1)!$, as desired.

c) Wilson's Theorem says that $(p - 1)!$ is congruent to -1 modulo p. Therefore the congruence in part **(b)** says that $(-1) \cdot a^{p-1} \equiv -1 \pmod p$. Multiplying both sides by -1, we see that $a^{p-1} \equiv 1 \pmod p$, as desired. Note that we already assumed the hypothesis that $p \not| a$ in part **(a)**.

d) If $p \mid a$, then both sides of $a^p \equiv a \pmod p$ are 0 modulo p, so the congruence holds. If not, then we just multiply the result obtained in part **(c)** by a.

19. Since 2, 3, 5, and 11 are pairwise relatively prime, we can use the Chinese Remainder Theorem. The answer will be unique modulo $2 \cdot 3 \cdot 5 \cdot 11 = 330$. Using the notation in the text, we have $a_1 = 1$, $m_1 = 2$, $a_2 = 2$, $m_2 = 3$, $a_3 = 3$, $m_3 = 5$, $a_4 = 4$, $m_4 = 11$, $m = 330$, $M_1 = 330/2 = 165$, $M_2 = 330/3 = 110$, $M_3 = 330/5 = 66$, $M_4 = 330/11 = 30$. Then we need to find inverses y_i of M_i modulo m_i for $i = 1, 2, 3, 4$. This can be done by inspection (trial and error), since the moduli here are so small, or systematically using the Euclidean algorithm, as in Example 3; we find that $y_1 = 1$, $y_2 = 2$, $y_3 = 1$, and $y_4 = 7$ (for this last one, $30 \equiv 8 \pmod{11}$, so we want to solve $8y_4 = 1 \pmod{11}$, and we observe that $8 \cdot 7 = 56 \equiv 1 \pmod{11}$). Thus our solution is $x = 1 \cdot 165 \cdot 1 + 2 \cdot 110 \cdot 2 + 3 \cdot 66 \cdot 1 + 4 \cdot 30 \cdot 7 = 1643 \equiv 323 \pmod{330}$. So the solutions are all integers of the form $323 + 330k$, where k is an integer.

21. We cannot apply the Chinese Remainder Theorem directly, since the moduli are not pairwise relatively prime. However, we can, using the Chinese Remainder Theorem, translate these congruences into a set of congruences that together are equivalent to the given congruence. Since we want $x \equiv 4 \pmod{12}$, we must have $x \equiv 4 \equiv 1 \pmod 3$ and $x \equiv 4 \equiv 0 \pmod 4$. Similarly, from the third congruence we must have $x \equiv 1 \pmod 3$ and $x \equiv 2 \pmod 7$. Since the first congruence is consistent with the requirement that $x \equiv 1 \pmod 3$, we see that our system is equivalent to the system $x \equiv 7 \pmod 9$, $x \equiv 0 \pmod 4$, $x \equiv 2 \pmod 7$. These can be solved using the Chinese Remainder Theorem (see Exercise 19 or Example 6) to yield $x \equiv 16 \pmod{252}$. Therefore the solutions are all integers of the form $16 + 252k$, where k is an integer.

23. We will argue for the truth of this statement using the Fundamental Theorem of Arithmetic. What we must show is that $m_1m_2\cdots m_n \mid a-b$. Look at the prime factorization of both sides of this proposition. Suppose that p is a prime appearing in the prime factorization of the left-hand side. Then $p \mid m_j$ for some j. Since the m_i's are relatively prime, p does not appear as a factor in any of the other m_i's. Now we know from the hypothesis that $m_j \mid a-b$. Therefore $a-b$ contains the factor p in its prime factorization, and p must appear to a power at least as large as the power to which it appears in m_j. But what we have just shown is that each prime power p^r in the prime factorization of the left-hand side also appears in the prime factorization of the right-hand side. Therefore the left-hand side does, indeed, divide the right-hand side.

25. We are asked to solve the simultaneous congruences $x \equiv 1 \pmod 2$ and $x \equiv 1 \pmod 3$. The solution will be unique modulo $2 \cdot 3 = 6$. By inspection we see that the answer is simply that $x \equiv 1 \pmod 6$. The solution set is $\{\ldots, -11, -5, 1, 7, 13, \ldots\}$.

27. **a)** We calculate $2^{340} = (2^{10})^{34} \equiv 1^{34} = 1 \pmod{11}$, since Fermat's Little Theorem says that $2^{10} \equiv 1 \pmod{11}$.
b) We calculate $2^{340} = (2^5)^{68} = 32^{68} \equiv 1^{68} = 1 \pmod{31}$, since $32 \equiv 1 \pmod{31}$.
c) Since 11 and 31 are relatively prime, and $11 \cdot 31 = 341$, it follows from the first two parts and Exercise 23 that $2^{340} \equiv 1 \pmod{341}$.

29. **a)** By Fermat's Little Theorem we know that $5^6 \equiv 1 \pmod 7$; therefore $5^{1998} = (5^6)^{333} \equiv 1^{333} \equiv 1 \pmod 7$, and so $5^{2003} = 5^5 \cdot 5^{1998} \equiv 3125 \cdot 1 \equiv 3 \pmod 7$. So 5^{2003} **mod** $7 = 3$. Similarly, $5^{10} \equiv 1 \pmod{11}$; therefore $5^{2000} = (5^{10})^{200} \equiv 1^{200} \equiv 1 \pmod{11}$, and so $5^{2003} = 5^3 \cdot 5^{2000} \equiv 125 \cdot 1 \equiv 4 \pmod{11}$. So 5^{2003} **mod** $11 = 4$. Finally, $5^{12} \equiv 1 \pmod{13}$; therefore $5^{1992} = (5^{12})^{166} \equiv 1^{166} \equiv 1 \pmod{13}$, and so $5^{2003} = 5^{11} \cdot 5^{1992} \equiv 48{,}828{,}125 \cdot 1 \equiv 8 \pmod{13}$. So 5^{2003} **mod** $13 = 8$.
b) We now apply the Chinese Remainder Theorem to the results of part **(a)**, as in Example 6. Let $m = 7 \cdot 11 \cdot 13 = 1001$, $M_1 = m/7 = 143$, $M_2 = m/11 = 91$, and $M_3 = m/13 = 77$. We see that 5 is an inverse of 143 modulo 7, since $143 \equiv 3 \pmod 7$, and $3 \cdot 5 = 15 \equiv 1 \pmod 7$. Similarly, 4 is an inverse of 91 modulo 11, and 12 is an inverse of 77 modulo 13. (An algorithm to compute inverses—if we don't want to find them by inspection as we've done here—is given in Theorem 3. See Example 3.) Therefore the answer is $(3 \cdot 143 \cdot 5 + 4 \cdot 91 \cdot 4 + 8 \cdot 77 \cdot 12)$ **mod** $1001 = 10993$ **mod** $1001 = 983$.

31. First note that $2047 = 23 \cdot 89$, so 2047 is composite. To apply Miller's test, we write $2047 - 1 = 2046 = 2 \cdot 1023$, so $s = 1$ and $t = 1023$. We must show that either $2^{1023} \equiv 1 \pmod{2047}$ or $2^{1023} \equiv -1 \pmod{2047}$. To compute, we write $2^{1023} = (2^{11})^{93} = 2048^{93} \equiv 1^{93} = 1 \pmod{2047}$, as desired. (We could also compute this using the modular exponentiation algorithm given in Section 3.6—see Example 11 in that section.)

33. We factor $2821 = 7 \cdot 13 \cdot 31$. We must show that this number meets the definition of Carmichael number, namely that $b^{2820} \equiv 1 \pmod{2821}$ for all b relatively prime to 2821. Note that if $\gcd(b, 2821) = 1$, then $\gcd(b,7) = \gcd(b,13) = \gcd(b,31) = 1$. Using Fermat's Little Theorem we find that $b^6 \equiv 1 \pmod 7$, $b^{12} \equiv 1 \pmod{13}$, and $b^{30} \equiv 1 \pmod{31}$. It follows that $b^{2820} = (b^6)^{470} \equiv 1 \pmod 7$, $b^{2820} = (b^{12})^{235} \equiv$

1 (mod 13), and $b^{2820} = (b^{30})^{94} \equiv 1 \pmod{31}$. By Exercise 23 (or the Chinese Remainder Theorem) it follows that $b^{2820} \equiv 1 \pmod{2821}$, as desired.

35. **a)** If we multiply out this expression, we get $n = 1296m^3 + 396m^2 + 36m + 1$. Clearly $6m \mid n-1$, $12m \mid n-1$, and $18m \mid n-1$. Therefore, the conditions of Exercise 34 are met, and we conclude that n is a Carmichael number.

b) Letting $m = 51$ gives $n = 172{,}947{,}529$. We note that $6m+1 = 307$, $12m+1 = 613$, and $18m+1 = 919$ are all prime.

37. It is straightforward to calculate the remainders when the integers from 0 to 14 are divided by 3 and by 5. For example, the remainders when 10 is divided by 3 and 5 are 1 and 0, respectively, so we represent 10 by the pair $(1,0)$. The exercise is simply asking us to tabulate these remainders, as in Example 7.

$0 = (0,0)$	$3 = (0,3)$	$6 = (0,1)$	$9 = (0,4)$	$12 = (0,2)$
$1 = (1,1)$	$4 = (1,4)$	$7 = (1,2)$	$10 = (1,0)$	$13 = (1,3)$
$2 = (2,2)$	$5 = (2,0)$	$8 = (2,3)$	$11 = (2,1)$	$14 = (2,4)$

39. The method of solving a system of congruences such as this is given in the proof of Theorem 4. Here we have $m_1 = 99$, $m_2 = 98$, $m_3 = 97$, and $m_4 = 95$, so that $m = 99 \cdot 98 \cdot 97 \cdot 95 = 89403930$. We compute the values $M_k = m/m_k$ and obtain $M_1 = 903070$, $M_2 = 912285$, $M_3 = 921690$, and $M_4 = 941094$. Next we need to find the inverses y_k of M_k modulo m_k. To do this we first replace each M_k by its remainder modulo m_k (to make the arithmetic easier), and then apply the technique shown in the solution to Exercise 7. For $k = 1$ we want to find the inverse of 903070 modulo 99, which is the same as the inverse of 903070 **mod** 99, namely 91. To do this we apply the Euclidean algorithm to express 1 as a linear combination of 91 and 99.

$$\begin{aligned}
99 &= 91 + 8 \\
91 &= 11 \cdot 8 + 3 \\
8 &= 2 \cdot 3 + 2 \\
3 &= 2 + 1 \\
\therefore\ 1 &= 3 - 2 \\
&= 3 - (8 - 2 \cdot 3) = 3 \cdot 3 - 8 \\
&= 3 \cdot (91 - 11 \cdot 8) - 8 = 3 \cdot 91 - 34 \cdot 8 \\
&= 3 \cdot 91 - 34 \cdot (99 - 91) = 37 \cdot 91 - 34 \cdot 99
\end{aligned}$$

We therefore conclude that the inverse of 91 modulo 99 is 37, so we have $y_1 = 37$. Similar calculations show that $y_2 = 33$, $y_3 = 24$, and $y_4 = 4$. Continuing with the procedure outlined in the proof of Theorem 4, we now form the sum of the products $a_k M_k y_k$, and this will be our solution. We have

$$65 \cdot 903070 \cdot 37 + 2 \cdot 912285 \cdot 33 + 51 \cdot 921690 \cdot 24 + 10 \cdot 941094 \cdot 4 = 3397886480\,.$$

We want our answer reduced modulo m, so we divide by 89403930 and take the remainder, obtaining 537140. (All of these calculations are not difficult using a scientific calculator.) Finally, let us check our answer: 537140 **mod** 99 = 65, 537140 **mod** 98 = 2, 537140 **mod** 97 = 51, 537140 **mod** 95 = 10.

41. One can compute $\gcd(2^a - 1, 2^b - 1)$ using the Euclidean algorithm. Let us look at what happens when we do so. If $b = 1$, then the answer is just $2^a - 1$, which is the same as $2^{\gcd(a,b)} - 1$ in this case. Otherwise, we reduce the problem to computing $\gcd(2^b - 1, (2^a - 1) \mathbf{~mod~} (2^b - 1))$. Now from Exercise 40 we know that this second argument equals $2^{a \mathbf{~mod~} b} - 1$. Therefore the exponents involved in the continuing calculation are b and a **mod** b—exactly the same quantities that are involved in computing $\gcd(a,b)$! It follows that when the process terminates, the answer must be $2^{\gcd(a,b)} - 1$, as desired.

43. Let q be a (necessarily odd) prime dividing $2^p - 1$. By Fermat's Little Theorem, we know that $q \mid 2^{q-1} - 1$. Then from Exercise 41 we know that $\gcd(2^p - 1, 2^{q-1} - 1) = 2^{\gcd(p,q-1)} - 1$. Since q is a common divisor of $2^p - 1$ and $2^{q-1} - 1$, we know that $\gcd(2^p - 1, 2^{q-1} - 1) > 1$. Hence $\gcd(p, q-1) = p$, since the only other possibility, namely $\gcd(p, q-1) = 1$, would give us $\gcd(2^p - 1, 2^{q-1} - 1) = 1$. Hence $p \mid q - 1$, and therefore there is a positive integer m such that $q - 1 = mp$. Since q is odd, m must be even, say $m = 2k$, and so every prime divisor of $2^p - 1$ is of the form $2kp + 1$. Furthermore, products of numbers of this form are also of this form, since $(2k_1p + 1)(2k_2p + 1) = 4k_1k_2p^2 + 2k_1p + 2k_2p + 1 = 2(2k_1k_2p + k_1 + k_2)p + 1$. Therefore all divisors of $2^p - 1$ are of this form.

45. Suppose that we know $n = pq$ and $(p-1)(q-1)$, and we wish to find p and q. Here is how we do so. Expanding $(p-1)(q-1)$ algebraically we obtain $pq - p - q + 1 = n - p - q + 1$. Thus we know the value of $n - p - q + 1$, and so we can easily calculate the value of $p + q$ (since we know n). But we also know the value of pq, namely n. This gives us two simultaneous equations in two unknowns, and we can solve them using the quadratic formula. Here is an example. Suppose that we want to factor $n = 341$, and we are told that $(p-1)(q-1) = 300$. We want to find p and q. Following the argument just outlined, we know that $p + q = 341 + 1 - 300 = 42$. Plugging $q = 42 - p$ into $pq = 341$ we obtain $p(42 - p) = 341$, or $p^2 - 42p + 341 = 0$. The quadratic formula then tells us that $p = (42 + \sqrt{42^2 - 4 \cdot 341})/2 = 31$, and so the factors are 31 and $42 - 31 = 11$. Note that absolutely no trial divisions were involved here—it was just straight calculation.

47. This problem requires a great amount of calculation. Ideally, one should do it using a computer algebra package, such as *Mathematica* or *Maple*. Let us follow the procedure outlined in Example 12. We need to compute 0667^{937} **mod** $2537 = 1808$, 1947^{937} **mod** $2537 = 1121$, and 0671^{937} **mod** $2537 = 0417$. (These calculations can in principle be done with a calculator, using the fast modular exponentiation algorithm, but it would probably take the better part of an hour and be prone to transcription errors.) Thus the original message is 1808 1121 0417, which is easily translated into letters as SILVER.

49. When we apply the Euclidean algorithm we obtain the following quotients and remainders: $q_1 = 1$, $r_2 = 55$, $q_2 = 1$, $r_3 = 34$, $q_3 = 1$, $r_4 = 21$, $q_4 = 1$, $r_5 = 13$, $q_5 = 1$, $r_6 = 8$, $q_6 = 1$, $r_7 = 5$, $q_7 = 1$, $r_8 = 3$, $q_8 = 1$, $r_9 = 2$, $q_9 = 1$, $r_{10} = 1$, $q_{10} = 2$. Note that $n = 10$. Thus we compute the successive s's and t's as follows, using the given recurrences:

$s_2 = s_0 - q_1s_1 = 1 - 1 \cdot 0 = 1,$ $\quad t_2 = t_0 - q_1t_1 = 0 - 1 \cdot 1 = -1$

$s_3 = s_1 - q_2s_2 = 0 - 1 \cdot 1 = -1,$ $\quad t_3 = t_1 - q_2t_2 = 1 - 1 \cdot (-1) = 2$

$s_4 = s_2 - q_3s_3 = 1 - 1 \cdot (-1) = 2,$ $\quad t_4 = t_2 - q_3t_3 = -1 - 1 \cdot 2 = -3$

$s_5 = s_3 - q_4s_4 = -1 - 1 \cdot 2 = -3,$ $\quad t_5 = t_3 - q_4t_4 = 2 - 1 \cdot (-3) = 5$

$s_6 = s_4 - q_5s_5 = 2 - 1 \cdot (-3) = 5,$ $\quad t_6 = t_4 - q_5t_5 = -3 - 1 \cdot 5 = -8$

$s_7 = s_5 - q_6s_6 = -3 - 1 \cdot 5 = -8,$ $\quad t_7 = t_5 - q_6t_6 = 5 - 1 \cdot (-8) = 13$

$s_8 = s_6 - q_7s_7 = 5 - 1 \cdot (-8) = 13,$ $\quad t_8 = t_6 - q_7t_7 = -8 - 1 \cdot 13 = -21$

$s_9 = s_7 - q_8s_8 = -8 - 1 \cdot 13 = -21,$ $\quad t_9 = t_7 - q_8t_8 = 13 - 1 \cdot (-21) = 34$

$s_{10} = s_8 - q_9s_9 = 13 - 1 \cdot (-21) = 34,$ $\quad t_{10} = t_8 - q_9t_9 = -21 - 1 \cdot 34 = -55$

Thus we have $s_{10}a + t_{10}b = 34 \cdot 144 + (-55) \cdot 89 = 1$, which is $\gcd(144, 89)$.

51. We start with the pseudocode for the Euclidean algorithm (Algorithm 6 in Section 3.6) and add variables to keep track of the s and t values. We need three of them, since the new s depends on the previous two s's, and similarly for t. We also need to keep track of q.

procedure *extended Euclidean*$(a, b$: positive integers)
$x := a$
$y := b$
$oldolds := 1$
$olds := 0$
$oldoldt := 0$
$oldt := 1$
while $y \neq 0$
begin
  $q := x$ **div** y
  $r := x$ **mod** y
  $x := y$
  $y := r$
  $s := oldolds - q \cdot olds$
  $t := oldoldt - q \cdot oldt$
  $oldolds := olds$
  $oldoldt := oldt$
  $olds := s$
  $oldt := t$
end $\{\gcd(a, b)$ is x, and $(oldolds)a + (oldoldt)b = x\}$

53. We need to prove that if the congruence $x^2 \equiv a \pmod{p}$ has any solutions at all, then it has exactly two solutions. So let us assume that s is a solution. Clearly $-s$ is a solution as well, since $(-s)^2 = s^2$. Furthermore, $-s \not\equiv s \pmod{p}$, since if it were, we would have $2s \equiv 0 \pmod{p}$, which means that $p \mid 2s$. Since p is an odd prime, that means that $p \mid s$, so that $s \equiv 0 \pmod{p}$. Therefore $a \equiv 0 \pmod{p}$, contradicting the conditions of the problem.

It remains to prove that there cannot be more than two incongruent solutions. Suppose that s is one solution and that t is a second solution. We have $s^2 \equiv t^2 \pmod{p}$. This means that $p \mid s^2 - t^2$, that is, $p \mid (s+t)(s-t)$. Since p is prime, Lemma 2 guarantees that $p \mid s - t$ or $p \mid s + t$. This means that $t \equiv s \pmod{p}$ or $t \equiv -s \pmod{p}$. Therefore any solution t must be either the first solution or its negative. In other words, there are at most two solutions.

55. There is really almost nothing to prove here. The value $\left(\frac{a}{p}\right)$ depends only on whether or not a is a quadratic residue modulo p, i.e., whether or not the equivalence $x^2 \equiv a \pmod{p}$ has a solution. Obviously, this depends only on the equivalence class of a modulo p.

57. By Exercise 56 we know that $\left(\frac{a}{p}\right)\left(\frac{b}{p}\right) \equiv a^{(p-1)/2}b^{(p-1)/2} = (ab)^{(p-1)/2} \equiv \left(\frac{ab}{p}\right) \pmod{p}$. Since the only values either side of this equivalence can take on are ± 1, being congruent modulo p is the same as being equal.

59. We follow the hint. Working modulo 5, we want to solve $x^2 \equiv 4$. It is easy to see that there are exactly two solutions modulo 5, namely $x = 2$ and $x = 3$. Similarly there are only the solutions $x = 1$ and $x = 6$ modulo 7. Therefore we want to find values of x modulo $5 \cdot 7 = 35$ such that $x \equiv 2$ or $3 \pmod{5}$ and $x \equiv 1$ or $6 \pmod{7}$. We can do this by applying the Chinese Remainder Theorem (as in Example 6) four times, for the four combinations of these values. For example, to solve $x \equiv 2 \pmod{5}$ and $x \equiv 1 \pmod{7}$, we find that $m = 35$, $M_1 = 7$, $M_2 = 5$, $y_1 = 3$, $y_2 = 3$, so $x \equiv 2 \cdot 7 \cdot 3 + 1 \cdot 5 \cdot 3 = 57 \equiv 22 \pmod{35}$. Doing the similar calculation with the other three possibilities yields the other three solutions modulo 35: $x = 8$, $x = 13$, and $x = 27$.

61. Suppose that we use a prime for n. To find a private decryption key from the corresponding public encryption key e, one would need to find a number d that is an inverse for e modulo $n - 1$ so that the calculation shown before Example 12 can go through. But finding such a d is easy using the Euclidean algorithm, because the person doing this would already know $n-1$. (In particular, to find d, one can work backward through the steps of the Euclidean algorithm to express 1 as a linear combination of e and $n-1$; then d is the coefficient of e in this linear combination.) The important point in the actual RSA system is that the person trying to find this inverse will not know $(p-1)(q-1)$ and therefore cannot simply use the Euclidean algorithm.

SECTION 3.8 Matrices

In addition to routine exercises with matrix calculations, there are several exercises here asking for proofs of various properties of matrix operations. In most cases the proofs follow immediately from the definitions of the matrix operations and properties of operations on the set from which the entries in the matrices are drawn. Also, the important notion of the (multiplicative) **inverse** *of a matrix is examined in Exercises 18–21. Keep in mind that some matrix operations are performed "entrywise," whereas others operate on whole rows or columns at a time. The general problem of efficient calculation of multiple matrix products, suggested by Exercises 23–25, is interesting and nontrivial. Exercise 31 foreshadows material in Section 8.4.*

1. a) Since $\mathbf{A}$ has 3 rows and 4 columns, its size is 3×4.

b) The third column of $\mathbf{A}$ is the 3×1 matrix $\begin{bmatrix}1\\4\\3\end{bmatrix}$.

c) The second row of $\mathbf{A}$ is the 1×4 matrix $[2\ 0\ 4\ 6]$.

d) This is the element in the third row, second column, namely 1.

e) The transpose of $\mathbf{A}$ is the 4×3 matrix $\begin{bmatrix}1&2&1\\1&0&1\\1&4&3\\3&6&7\end{bmatrix}$.

3. a) We use the definition of matrix multiplication to obtain the four entries in the product $\mathbf{AB}$. The $(1,1)^{\text{th}}$ entry is the sum $a_{11}b_{11}+a_{12}b_{21}=2\cdot 0+1\cdot 1=1$. Similarly, the $(1,2)^{\text{th}}$ entry is the sum $a_{11}b_{12}+a_{12}b_{22}=2\cdot 4+1\cdot 3=11$; $(2,1)^{\text{th}}$ entry is the sum $a_{21}b_{11}+a_{22}b_{21}=3\cdot 0+2\cdot 1=2$; and $(2,2)^{\text{th}}$ entry is the sum $a_{21}b_{12}+a_{22}b_{22}=3\cdot 4+2\cdot 3=18$. Therefore the answer is $\begin{bmatrix}1&11\\2&18\end{bmatrix}$.

b) The calculation is similar. Again, to get the $(i,j)^{\text{th}}$ entry of the product, we need to add up all the products $a_{ik}b_{kj}$. You can visualize "lifting" the ith row from the first factor ($\mathbf{A}$) and placing it on top of the jth column from the second factor ($\mathbf{B}$), multiplying the pairs of numbers that lie on top of each other, and taking the sum. Here we have

$$\begin{bmatrix}1&-1\\0&1\\2&3\end{bmatrix}\begin{bmatrix}3&-2&-1\\1&0&2\end{bmatrix}=\begin{bmatrix}1\cdot 3+(-1)\cdot 1 & 1\cdot(-2)+(-1)\cdot 0 & 1\cdot(-1)+(-1)\cdot 2\\ 0\cdot 3+1\cdot 1 & 0\cdot(-2)+1\cdot 0 & 0\cdot(-1)+1\cdot 2\\ 2\cdot 3+3\cdot 1 & 2\cdot(-2)+3\cdot 0 & 2\cdot(-1)+3\cdot 2\end{bmatrix}$$

$$=\begin{bmatrix}2&-2&-3\\1&0&2\\9&-4&4\end{bmatrix}.$$

c) The calculation is similar to the previous parts:

$$\begin{bmatrix} 4 & -3 \\ 3 & -1 \\ 0 & -2 \\ -1 & 5 \end{bmatrix} \begin{bmatrix} -1 & 3 & 2 & -2 \\ 0 & -1 & 4 & -3 \end{bmatrix}$$

$$= \begin{bmatrix} 4\cdot(-1)+(-3)\cdot 0 & 4\cdot 3+(-3)\cdot(-1) & 4\cdot 2+(-3)\cdot 4 & 4\cdot(-2)+(-3)\cdot(-3) \\ 3\cdot(-1)+(-1)\cdot 0 & 3\cdot 3+(-1)\cdot(-1) & 3\cdot 2+(-1)\cdot 4 & 3\cdot(-2)+(-1)\cdot(-3) \\ 0\cdot(-1)+(-2)\cdot 0 & 0\cdot 3+(-2)\cdot(-1) & 0\cdot 2+(-2)\cdot 4 & 0\cdot(-2)+(-2)\cdot(-3) \\ (-1)\cdot(-1)+5\cdot 0 & (-1)\cdot 3+5\cdot(-1) & (-1)\cdot 2+5\cdot 4 & (-1)\cdot(-2)+5\cdot(-3) \end{bmatrix}$$

$$= \begin{bmatrix} -4 & 15 & -4 & 1 \\ -3 & 10 & 2 & -3 \\ 0 & 2 & -8 & 6 \\ 1 & -8 & 18 & -13 \end{bmatrix}$$

5. First we need to observe that $\mathbf{A} = [a_{ij}]$ must be a 2×2 matrix; it must have two rows since the matrix it is being multiplied by on the left has two columns, and it must have two columns since the answer obtained has two columns. If we write out what the matrix multiplication means, then we obtain the following system of linear equations:

$$2a_{11} + 3a_{21} = 3$$

$$2a_{12} + 3a_{22} = 0$$

$$1a_{11} + 4a_{21} = 1$$

$$1a_{12} + 4a_{22} = 2$$

Solving these equations by elimination of variables (or other means—it's really two systems of two equations each in two unknowns), we obtain $a_{11} = 9/5$, $a_{12} = -6/5$, $a_{21} = -1/5$, $a_{22} = 4/5$. As a check we compute that, indeed,

$$\begin{bmatrix} 2 & 3 \\ 1 & 4 \end{bmatrix} \begin{bmatrix} 9/5 & -6/5 \\ -1/5 & 4/5 \end{bmatrix} = \begin{bmatrix} 3 & 0 \\ 1 & 2 \end{bmatrix}.$$

7. Since the $(i,j)^{\text{th}}$ entry of $\mathbf{0} + \mathbf{A}$ is the sum of the $(i,j)^{\text{th}}$ entry of $\mathbf{0}$ (namely 0) and the $(i,j)^{\text{th}}$ entry of $\mathbf{A}$, this entry is the same as the $(i,j)^{\text{th}}$ entry of $\mathbf{A}$. Therefore by the definition of matrix equality, $\mathbf{0} + \mathbf{A} = \mathbf{A}$. A similar argument shows that $\mathbf{A} + \mathbf{0} = \mathbf{A}$.

9. We simply look at the $(i,j)^{\text{th}}$ entries of each side. The $(i,j)^{\text{th}}$ entry of the left-hand side is $a_{ij} + (b_{ij} + c_{ij})$. The $(i,j)^{\text{th}}$ entry of the right-hand side is $(a_{ij} + b_{ij}) + c_{ij}$. By the associativity law for real number addition, these are equal. The conclusion follows.

11. In order for **AB** to be defined, the number of columns of **A** must equal the number of rows of **B**. In order for **BA** to be defined, the number of columns of **B** must equal the number of rows of **A**. Thus for some positive integers m and n, it must be the case that **A** is an $m \times n$ matrix and **B** is an $n \times m$ matrix. Another way to say this is to say that **A** must have the same size as $\mathbf{B}^t$ (and/or vice versa).

13. Let us begin with the left-hand side and find its $(i,j)^{\text{th}}$ entry. First we need to find the entries of **BC**. By definition, the $(q,j)^{\text{th}}$ entry of **BC** is $\sum_{r=1}^{k} b_{qr}c_{rj}$. (See Section 2.4 for the meaning of summation notation. This symbolism is a shorthand way of writing $b_{q1}c_{1j} + b_{q2}c_{2j} + \cdots + b_{qk}c_{kj}$.) Therefore the $(i,j)^{\text{th}}$ entry of $\mathbf{A}(\mathbf{BC})$ is $\sum_{q=1}^{p} a_{iq}(\sum_{r=1}^{k} b_{qr}c_{rj})$. By distributing multiplication over addition (for real numbers), we can move the term a_{iq} inside the inner summation, to obtain $\sum_{q=1}^{p}\sum_{r=1}^{k} a_{iq}b_{qr}c_{rj}$. (We are also implicitly using associativity of multiplication of real numbers here, to avoid putting parentheses in the product $a_{iq}b_{qr}c_{rj}$.)

A similar analysis with the right-hand side shows that the $(i,j)^{\text{th}}$ entry there is equal to $\sum_{r=1}^{k}(\sum_{q=1}^{p} a_{iq}b_{qr})c_{rj}$ $= \sum_{r=1}^{k}\sum_{q=1}^{p} a_{iq}b_{qr}c_{rj}$. Now by the commutativity of addition, the order of summation (whether we sum over r first and then q, or over q first and then r) does not matter, so these two expressions are equal, and the proof is complete.

15. Let us begin by computing $\mathbf{A}^n$ for the first few values of n.

$$\mathbf{A}^1 = \begin{bmatrix} 1 & 1 \\ 0 & 1 \end{bmatrix}, \quad \mathbf{A}^2 = \begin{bmatrix} 1 & 2 \\ 0 & 1 \end{bmatrix}, \quad \mathbf{A}^3 = \begin{bmatrix} 1 & 3 \\ 0 & 1 \end{bmatrix}, \quad \mathbf{A}^4 = \begin{bmatrix} 1 & 4 \\ 0 & 1 \end{bmatrix}, \quad \mathbf{A}^5 = \begin{bmatrix} 1 & 5 \\ 0 & 1 \end{bmatrix}.$$

It seems clear from this pattern, then, that $\mathbf{A}^n = \begin{bmatrix} 1 & n \\ 0 & 1 \end{bmatrix}$. (A proof of this fact could be given using mathematical induction, discussed in Section 4.1.)

17. a) The $(i,j)^{\text{th}}$ entry of $(\mathbf{A}+\mathbf{B})^t$ is the $(j,i)^{\text{th}}$ entry of $\mathbf{A}+\mathbf{B}$, namely $a_{ji}+b_{ji}$. On the other hand, the $(i,j)^{\text{th}}$ entry of $\mathbf{A}^t+\mathbf{B}^t$ is the sum of the $(i,j)^{\text{th}}$ entries of $\mathbf{A}^t$ and $\mathbf{B}^t$, which are the $(j,i)^{\text{th}}$ entries of $\mathbf{A}$ and $\mathbf{B}$, again $a_{ji}+b_{ji}$. Hence $(\mathbf{A}+\mathbf{B})^t = \mathbf{A}^t + \mathbf{B}^t$.

b) The $(i,j)^{\text{th}}$ entry of $(\mathbf{AB})^t$ is the $(j,i)^{\text{th}}$ entry of $\mathbf{AB}$, namely $\sum_{k=1}^{n} a_{jk}b_{ki}$. (See Section 2.4 for the meaning of summation notation. This symbolism is a shorthand way of writing $ba_{j1}b_{1i} + a_{j2}b_{2i} + \cdots + a_{jn}b_{ni}$.) On the other hand, the $(i,j)^{\text{th}}$ entry of $\mathbf{B}^t\mathbf{A}^t$ is $\sum_{k=1}^{n} b_{ki}a_{jk}$ (since the $(i,k)^{\text{th}}$ entry of $\mathbf{B}^t$ is b_{ki} and the $(k,j)^{\text{th}}$ entry of $\mathbf{A}^t$ is a_{jk}). By the commutativity of multiplication of real numbers, these two values are the same, so the matrices are equal.

19. All we have to do is form the products $\mathbf{AA}^{-1}$ and $\mathbf{A}^{-1}\mathbf{A}$, using the purported $\mathbf{A}^{-1}$, and see that both of them are the 2×2 identity matrix. It is easy to see that the upper left and lower right entries in each case are $(ad-bc)/(ad-bc)=1$, and the upper right and lower left entries are all 0.

21. We must show that $\mathbf{A}^n(\mathbf{A}^{-1})^n = \mathbf{I}$, where $\mathbf{I}$ is the $n\times n$ identity matrix. Since matrix multiplication is associative, we can write this product as

$$\mathbf{A}^n\left((\mathbf{A}^{-1})^n\right) = \mathbf{A}(\mathbf{A}\ldots(\mathbf{A}(\mathbf{AA}^{-1})\mathbf{A}^{-1})\ldots\mathbf{A}^{-1})\mathbf{A}^{-1}.$$

By dropping each $\mathbf{AA}^{-1} = \mathbf{I}$ from the center as it is obtained, this product reduces to $\mathbf{I}$. Similarly $\left((\mathbf{A}^{-1})^n\right)\mathbf{A}^n = \mathbf{I}$. Therefore by definition $(\mathbf{A}^n)^{-1} = (\mathbf{A}^{-1})^n$. (A more formal proof requires mathematical induction; see Section 4.1.)

23. In order to compute the $(i,j)^{\text{th}}$ entry of the product $\mathbf{AB}$, we need to compute the product $a_{ik}b_{kj}$ for each k from 1 to m_2, requiring m_2 multiplications. Since there are m_1m_3 such pairs (i,j), we need a total of $m_1m_2m_3$ multiplications.

25. There are five different ways to perform this multiplication:

$$(\mathbf{A}_1\mathbf{A}_2)(\mathbf{A}_3\mathbf{A}_4), \quad ((\mathbf{A}_1\mathbf{A}_2)\mathbf{A}_3)\mathbf{A}_4, \quad \mathbf{A}_1(\mathbf{A}_2(\mathbf{A}_3\mathbf{A}_4)), \quad (\mathbf{A}_1(\mathbf{A}_2\mathbf{A}_3))\mathbf{A}_4, \quad \mathbf{A}_1((\mathbf{A}_2\mathbf{A}_3)\mathbf{A}_4).$$

We can use the result of Exercise 23 to find the numbers of multiplications needed in these five cases. For example, in the first case we need $10\cdot 2\cdot 5 = 100$ multiplications to compute the 10×5 matrix $\mathbf{A}_1\mathbf{A}_2$, $5\cdot 20\cdot 3 = 300$ multiplications to compute the 5×3 matrix $\mathbf{A}_3\mathbf{A}_4$, and then $10\cdot 5\cdot 3 = 150$ multiplications to multiply these two matrices together to obtain the final answer. This gives a total of $100+300+150 = 550$ multiplications. Similar calculations for the other four cases yield $10\cdot 2\cdot 5 + 10\cdot 5\cdot 20 + 10\cdot 20\cdot 3 = 1700$, $5\cdot 20\cdot 3 + 2\cdot 5\cdot 3 + 10\cdot 2\cdot 3 = 390$, $2\cdot 5\cdot 20 + 10\cdot 2\cdot 20 + 10\cdot 20\cdot 3 = 1200$, and $2\cdot 5\cdot 20 + 2\cdot 20\cdot 3 + 10\cdot 2\cdot 3 = 380$, respectively. The winner is therefore $\mathbf{A}_1((\mathbf{A}_2\mathbf{A}_3)\mathbf{A}_4)$, requiring 380 multiplications. Note that the worst arrangement requires 1700 multiplications; it will take over four times as long.

27. Using the idea in Exercise 26, we see that the given system can be expressed as $\mathbf{AX} = \mathbf{B}$, where $\mathbf{A}$ is the coefficient matrix, $\mathbf{X}$ is an $n \times 1$ matrix with x_i the entry in its ith row, and $\mathbf{B}$ is the $n \times 1$ matrix of right-hand sides. Specifically we have

$$\begin{bmatrix} 7 & -8 & 5 \\ -4 & 5 & -3 \\ 1 & -1 & 1 \end{bmatrix} \begin{bmatrix} x_1 \\ x_2 \\ x_3 \end{bmatrix} = \begin{bmatrix} 5 \\ -3 \\ 0 \end{bmatrix}.$$

If we can find the inverse $\mathbf{A}^{-1}$, then we can find $\mathbf{X}$ simply by computing $\mathbf{A}^{-1}\mathbf{B}$. But Exercise 18 tells us that $\mathbf{A}^{-1} = \begin{bmatrix} 2 & 3 & -1 \\ 1 & 2 & 1 \\ -1 & -1 & 3 \end{bmatrix}$. Therefore

$$\mathbf{X} = \begin{bmatrix} 2 & 3 & -1 \\ 1 & 2 & 1 \\ -1 & -1 & 3 \end{bmatrix} \begin{bmatrix} 5 \\ -3 \\ 0 \end{bmatrix} = \begin{bmatrix} 1 \\ -1 \\ -2 \end{bmatrix}.$$

We should plug in $x_1 = 1$, $x_2 = -1$, and $x_3 = -2$ to see that these do indeed form the solution.

29. These routine exercises simply require application of the appropriate definitions. Parts **(a)** and **(b)** are entry-wise operations, whereas the operation $\odot$ in part **(c)** is similar to matrix multiplication (the $(i,j)^{\text{th}}$ entry of $\mathbf{A} \odot \mathbf{B}$ depends on the ith row of $\mathbf{A}$ and the jth column of $\mathbf{B}$).

a) $\mathbf{A} \vee \mathbf{B} = \begin{bmatrix} 1 & 1 & 1 \\ 1 & 1 & 1 \\ 1 & 0 & 1 \end{bmatrix}$ **b)** $\mathbf{A} \wedge \mathbf{B} = \begin{bmatrix} 0 & 0 & 1 \\ 1 & 0 & 0 \\ 0 & 0 & 1 \end{bmatrix}$ **c)** $\mathbf{A} \odot \mathbf{B} = \begin{bmatrix} 1 & 1 & 1 \\ 1 & 1 & 1 \\ 1 & 0 & 1 \end{bmatrix}$

31. Note that $\mathbf{A}^{[2]}$ means $\mathbf{A} \odot \mathbf{A}$, and $\mathbf{A}^{[3]}$ means $\mathbf{A} \odot \mathbf{A} \odot \mathbf{A}$. We just apply the definition.

a) $\mathbf{A}^{[2]} = \begin{bmatrix} 1 & 0 & 0 \\ 1 & 1 & 0 \\ 1 & 0 & 1 \end{bmatrix}$ **b)** $\mathbf{A}^{[3]} = \begin{bmatrix} 1 & 0 & 0 \\ 1 & 0 & 1 \\ 1 & 1 & 0 \end{bmatrix}$ **c)** $\mathbf{A} \vee \mathbf{A}^{[2]} \vee \mathbf{A}^{[3]} = \begin{bmatrix} 1 & 0 & 0 \\ 1 & 1 & 1 \\ 1 & 1 & 1 \end{bmatrix}$

33. These are immediate from the commutativity of the corresponding logical operations on variables.

a) $\mathbf{A} \vee \mathbf{B} = [a_{ij} \vee b_{ij}] = [b_{ij} \vee a_{ij}] = \mathbf{B} \vee \mathbf{A}$

b) $\mathbf{B} \wedge \mathbf{A} = [b_{ij} \wedge a_{ij}] = [a_{ij} \wedge b_{ij}] = \mathbf{A} \wedge \mathbf{B}$

35. These are immediate from the distributivity of the corresponding logical operations on variables.

a) $\mathbf{A} \vee (\mathbf{B} \wedge \mathbf{C}) = [a_{ij} \vee (b_{ij} \wedge c_{ij})] = [(a_{ij} \vee b_{ij}) \wedge (a_{ij} \vee c_{ij})] = (\mathbf{A} \vee \mathbf{B}) \wedge (\mathbf{A} \vee \mathbf{C})$

b) $\mathbf{A} \wedge (\mathbf{B} \vee \mathbf{C}) = [a_{ij} \wedge (b_{ij} \vee c_{ij})] = [(a_{ij} \wedge b_{ij}) \vee (a_{ij} \wedge c_{ij})] = (\mathbf{A} \wedge \mathbf{B}) \vee (\mathbf{A} \wedge \mathbf{C})$

37. The proof is identical to the proof in Exercise 13, except that real number multiplication is replaced by $\wedge$, and real number addition is replaced by $\vee$. Briefly, in symbols, $\mathbf{A} \odot (\mathbf{B} \odot \mathbf{C}) = \big[\bigvee_{q=1}^{p} a_{iq} \wedge \big(\bigvee_{r=1}^{k} b_{qr} \wedge c_{rj}\big)\big] = \big[\bigvee_{q=1}^{p} \bigvee_{r=1}^{k} a_{iq} \wedge b_{qr} \wedge c_{rj}\big] = \big[\bigvee_{r=1}^{k} \bigvee_{q=1}^{p} a_{iq} \wedge b_{qr} \wedge c_{rj}\big] = \big[\bigvee_{r=1}^{k} \big(\bigvee_{q=1}^{p} a_{iq} \wedge b_{qr}\big) \wedge c_{rj}\big] = (\mathbf{A} \odot \mathbf{B}) \odot \mathbf{C}$.

GUIDE TO REVIEW QUESTIONS FOR CHAPTER 3

1. a) See p. 168.

b) in English, in a computer language, in pseudocode

c) An algorithm is more of an abstract idea—a method in theory that will solve a problem. A computer program is the implementation of that idea into a specific syntactically correct set of instructions that a real computer can use to solve the problem. It is rather like the difference between a dollar (a certain amount of money, capable in theory of purchasing a certain quantity of goods and services) and a dollar bill.

2. a) See first three lines of text following Algorithm 1 on p. 169. **b)** See Algorithm 1 in Section 3.1.

c) See Example 1 in Section 3.3.

3. a) See p. 180. **b)** $n^2 + 18n + 107 \le n^3 + n^3 + n^3 = 3n^3$ for all $n > 107$.

c) n^3 is not less than a constant times $n^2 + 18n + 107$, since their ratio exceeds $n^3/(3n^2) = n/3$ for all $n > 107$.

4. a) For the sum, take the largest term; for the product multiply the factors together.

b) $g(n) = n^{n-2}2^n = (2n)^n/n^2$

5. a) the largest, average, and smallest number of comparisons used by the algorithm before it stops, among all lists of n integers

b) all are $n - 1$

6. a) See pp. 170–172. **b)** See pp. 194–195.

c) No—it depends on the lists involved. (However, the worst case complexity for binary search is always better than that for linear search for lists of any given size except for very short lists.)

7. a) See p. 173.

b) On the first pass, the 5 bubbles down to the end, producing 24135. On the next pass, the 4 bubbles down to the end, producing 21345. On the next pass, the 1 and the 2 are swapped. No further changes are made on the fourth pass.

c) $O(n^2)$; see Example 5 in Section 3.3

8. a) See p. 174.

b) On the first pass, the 5 is inserted into its correct position relative to the 2, producing 25143. On the next pass, the 1 is inserted into its correct position relative to 25, producing 12543. On the next pass, the 4 is inserted into its correct position relative to 125, producing 12453. On the final pass, the 3 is inserted, producing the sorted list.

c) $O(n^2)$; see Example 6 in Section 3.3

9. a) See pp. 174–175. **b)** See Example 6 and Theorem 1 in Section 3.1.

c) See Exercise 55b in Section 3.1.

10. See p. 211.

11. a) See the bottom of p. 211. **b)** $11^2 \cdot 23 \cdot 29$

12. a) See p. 215.

b) find all the common factors (not a good algorithm unless the numbers are really small); find the prime factorization of each integer (works well if the numbers aren't too big and therefore can be easily factored); use the Euclidean algorithm (really the best method)

c) 1 (use the Euclidean algorithm)

d) $2^3 3^5 5^5 7^3$

13. a) $7 \mid a-b$ **b)** $0 \equiv -7$; $-1 \equiv -8$; $3 \equiv 17 \equiv -11$
c) $(10a+13)-(-4b+20) = 3(a-b)+7(a+b-1)$; note that 7 divides both terms

14. See Algorithm 1 in Section 3.6 with $b = 16$. Use "A" for a remainder of 10, "B" for a remainder of 11, and so on. See also Example 4 in that section.

15. a) Use the Euclidean algorithm; see Example 1 in Section 3.7.
b) $7 = 5 \cdot 119 - 7 \cdot 84$

16. a) $a\bar{a} \equiv 1 \pmod{m}$
b) Express 1 as $sa + tm$ (see Review Question 15). Then s is the inverse of a modulo m.
c) 11

17. a) Multiply each side by the inverse of a modulo m. **b)** $\{\, 10 + 19k \mid k \in \mathbf{Z} \,\}$

18. a) See p. 236. **b)** $\{\, 17 + 140k \mid k \in \mathbf{Z} \,\}$

19. No—n could be a pseudoprime such as 341.

20. a) See p. 241.
b) The amount of shift, k, is kept secret. It is needed both to encode and to decode messages.
c) Although the key for decoding, d, is kept secret, the keys for encoding, n and e, are published.

21. See p. 248. The number of columns of $\mathbf{A}$ must equal the number of rows of $\mathbf{B}$.

22. a) 5 **b)** $(\mathbf{A}_1\mathbf{A}_2)(\mathbf{A}_3\mathbf{A}_4)$; see Exercise 25 in Section 3.8

SUPPLEMENTARY EXERCISES FOR CHAPTER 3

1. a) This algorithm will be identical to the algorithm *first largest* for Exercise 17 of Section 3.1, except that we want to change the value of *location* each time we find another element in the list that is equal to the current value of *max*. Therefore we simply change the strict less than ($<$) in the comparison $max < a_i$ to a less than or equal to ($\leq$), rendering the fifth line of that procedure "**if** $max \leq a_i$ **then**."

b) The number of comparisons used by this algorithm can be computed as follows. There are $n-1$ passes through the **for** loop, each one requiring a comparison of max with a_i. In addition, n comparisons are needed for bookkeeping for the loop (comparison of i with n, as i assumes the values $2, 3, \ldots, n+1$). Therefore $2n-1$ comparisons are needed altogether, which is $O(n)$.

3. a) We will try to write an algorithm sophisticated enough to avoid unnecessary checking. The answer—**true** or **false**—will be placed in a variable called *zeros*, the name of the procedure. (This approach is used by function subprograms in some computer languages, although it is not always permitted to use the procedure name in an expression as we do here.)

procedure *zeros*($a_1a_2\ldots a_n$: bit string)
$i := 1$
zeros := **false** {no pair of zeros found yet}
while $i < n$ and $\neg$*zeros*
    if $a_i = 1$ **then** $i := i + 1$
    else if $a_{i+1} = 1$ **then** $i := i + 2$
    else *zeros* := **true**
{*zeros* was set to **true** if and only if there were a pair of consecutive zeros}

b) The number of comparisons depends on whether a pair of 0's is found and also depends on the pattern of increments of the looping variable i. Without getting into the intricate details of exactly which is the worst case, we note that at worst there are approximately n passes through the loop, each requiring one comparison of a_i with 1 (there may be two comparisons on some passes, but then there will be fewer passes). In addition, n bookkeeping comparisons of i with n are needed (we are ignoring the testing of the logical variable *zeros*). Thus a total of approximately $2n$ comparisons are used, which is $O(n)$.

5. a) and **b)**. We have a variable *min* to keep track of the minimum as well as a variable *max* to keep track of the maximum.

procedure *smallest and largest*($a_1, a_2, \ldots, a_n$: integers)
min := a_1
max := a_1
for $i := 2$ **to** n
begin
    if $a_i <$ *min* **then** *min* := a_i
    if $a_i >$ *max* **then** *max* := a_i
end {*min* is the smallest integer among the input, and *max* is the largest}

c) There are two comparisons for each iteration of the loop, and there are $n-1$ iterations, so there are $2n-2$ comparisons in all.

7. We think of ourselves as observers as some algorithm for solving this problem is executed. We do not care what the algorithm's strategy is, but we view it along the following lines, in effect taking notes as to what is happening and what *we* know as it proceeds. Before any comparisons are done, there is a possibility that each element could be the maximum and a possibility that it could be the minimum. This means that there are $2n$ different possibilities, and $2n-2$ of them have to be eliminated through comparisons of elements, since we need to find the unique maximum and the unique minimum. We classify comparisons of two elements as "nonvirgin" or "virgin," depending on whether or not both elements being compared have been in any previous comparison. A virgin comparison, between two elements that have not yet been involved in any comparison, eliminates the possibility that the larger one is the minimum and that the smaller one is the maximum; thus each virgin comparison eliminates two possibilities, but it clearly cannot do more. A nonvirgin comparison, one involving at least one element that has been compared before, must be between two elements that are still in the running to be the maximum or two elements that are still in the running to be the minimum, and at least one of these elements must *not* be in the running for the other category. For example, we might be comparing x and y, where all we know is that x has been eliminated as the minimum. If we find that $x > y$ in this case, then only one possibility has been ruled out—we now know that y is not the maximum. Thus in the worst case, a nonvirgin comparison eliminates only one possibility. (The cases of other nonvirgin comparisons are similar.) Now there are at most $\lfloor n/2 \rfloor$ comparisons of elements that have never been compared before, each removing two possibilities; they remove $2\lfloor n/2 \rfloor$ possibilities altogether. Therefore we need $2n-2-2\lfloor n/2 \rfloor$ more comparisons that, as we have argued, can remove only one possibility each, in order to find the answers in the worst case, since $2n-2$ possibilities have to be eliminated. This gives us a total of $2n-2-2\lfloor n/2 \rfloor + \lfloor n/2 \rfloor$ comparisons in all. But $2n-2-2\lfloor n/2 \rfloor + \lfloor n/2 \rfloor = 2n-2-\lfloor n/2 \rfloor = 2n-2+\lceil -n/2 \rceil = \lceil 2n-n/2 \rceil -2 = \lceil 3n/2 \rceil -2$, as desired.

Note that this gives us a lower bound on the number of comparisons used in an algorithm to find the minimum and the maximum. On the other hand, Exercise 6 gave us an upper bound of the same size. Thus the algorithm in Exercise 6 is the most efficient algorithm possible for solving this problem.

9. After the comparison and possible exchange of adjacent elements on the first pass, from front to back, the list is $3, 1, 4, 5, 2, 6$, where the 6 is known to be in its correct position. After the comparison and possible exchange of adjacent elements on the second pass, from back to front, the list is $1, 3, 2, 4, 5, 6$, where the 6 and the 1 are known to be in their correct positions. After the next pass, the result is $1, 2, 3, 4, 5, 6$. One more pass finds nothing to exchange, and the algorithm terminates.

11. There are possibly as many as $n-1$ passes through the list (or parts of it—it depends on the particular way it is implemented), and each pass uses $O(n)$ comparisons. Thus there are $O(n^2)$ comparisons in all.

13. Since $\log n < n$, we have $(n \log n + n^2)^3 \leq (n^2 + n^2)^3 \leq (2n^2)^3 = 8n^6$ for all $n > 0$. This proves that $(n \log n + n^2)^3$ is $O(n^6)$, with witnesses $C = 8$ and $k = 0$.

15. In the first factor the x^2 term dominates the other term, since $(\log x)^3$ is $O(x)$. Therefore by Theorem 2 in Section 3.2, this term is $O(x^2)$. Similarly, in the second factor, the 2^x term dominates. Thus by Theorem 3 of Section 3.2, the product is $O(x^2 2^x)$.

17. Let us look at the ratio

$$\frac{n!}{2^n} = \frac{n \cdot (n-1) \cdot (n-2) \cdots 3 \cdot 2 \cdot 1}{2 \cdot 2 \cdot 2 \cdots 2 \cdot 2 \cdot 2} = \frac{n}{2} \cdot \frac{n-1}{2} \cdot \frac{n-2}{2} \cdots \frac{3}{2} \cdot \frac{2}{2} \cdot \frac{1}{2}.$$

Each of the fractions in the final expression is greater than 1 except the last one, so the entire expression is at least $(n/2)/2 = n/4$. Since $n!/2^n$ increases without bound as n increases, $n!$ cannot be bounded by a constant times 2^n, which tells us that $n!$ is not $O(2^n)$.

19. Obviously there are an infinite number of possible answers. The numbers congruent to 5 modulo 17 include $5, 22, 39, 56, \ldots$, as well as $-12, -29, -46, \ldots$.

21. From the hypothesis $ac \equiv bc \pmod{m}$ we know that $ac - bc = km$ for some integer k. Divide both sides by c to obtain the equation $a - b = (km)/c$. Now the left-hand side is an integer, and so the right-hand side must be an integer as well. In other words, $c \mid km$. Letting $d = \gcd(m, c)$, we write $c = de$. Then the way that c divides km is that $d \mid m$ and $e \mid k$ (since no factor of e divides m/d). Thus our equation reduces to $a - b = (k/e)(m/d)$, where both factors on the right are integers. By definition, this means that $a \equiv b \pmod{m/d}$.

23. $\gcd(10223, 33341) = \gcd(10223, 2672) = \gcd(2672, 2207) = \gcd(2207, 465) = \gcd(465, 347) = \gcd(347, 118) = \gcd(118, 111) = \gcd(111, 7) = \gcd(7, 6) = \gcd(6, 1) = \gcd(1, 0) = 1$

25. By Lemma 1 in Section 3.6, $\gcd(2n+1, 3n+2) = \gcd(2n+1, n+1)$, since $2n+1$ goes once into $3n+2$ with a remainder of $n+1$. Now if we divide $n+1$ into $2n+1$, we get a remainder of n, so the answer must equal $\gcd(n+1, n)$. At this point, the remainder when dividing n into $n+1$ is 1, so the answer must equal $\gcd(n, 1)$, which is clearly 1. Thus the answer is 1.

27. This problem is similar to Exercise 7 in Section 3.5. Without loss of generality, we may assume that the given integer is positive (since $n \mid a$ if and only if $n \mid (-a)$, and the case $a = 0$ is trivial). Let the decimal expansion of the integer a be given by $a = (a_{n-1}a_{n-2} \ldots a_1 a_0)_{10}$. Thus $a = 10^{n-1}a_{n-1} + 10^{n-2}a_{n-2} + \cdots + 10a_1 + a_0$. Since $10 \equiv 1 \pmod{9}$, we have $a \equiv a_{n-1} + a_{n-2} + \cdots + a_1 + a_0 \pmod{9}$. Therefore $a \equiv 0 \pmod{9}$ if and only if the sum of the digits is congruent to 0 (mod 9). Since being divisible by 9 is the same as being congruent to 0 (mod 9), we have proved that an integer is divisible by 9 if and only if the sum of its decimal digits is divisible by 9.

29. It might be helpful to read the solution to Exercise 35 in Section 3.5 to see the philosophy behind this approach. Suppose by way of contradiction that $q_1, q_2, \ldots, q_n$ are all the primes of the form $6k+5$. Thus $q_1 = 5$, $q_2 = 11$, and so on. Let $Q = 6q_1q_2\cdots q_n - 1$. We note that Q is of the form $6k+5$, where $k = q_1q_2\cdots q_n - 1$. Now Q has a prime factorization $Q = p_1p_2\cdots p_t$. Clearly no p_i is 2, 3, or any q_j, because the remainder when Q is divided by 2 is 1, by 3 is 2, and by q_j is $q_j - 1$. All odd primes other than 3 are of the form $6k+1$ or $6k+5$, and the product of primes of the form $6k+1$ is also of this form. Therefore at least one of the p_i's must be of the form $6k+5$, a contradiction.

31. a) Since 2 is a factor of all three of these integers, this set is not mutually relatively prime.
b) Since 12 and 25 share no common factors, this set has greatest common divisor 1, so it is mutually relatively prime. (It is possible for every pair of integers in a set of mutually relatively prime integers to have a nontrivial common factor (see Exercise 32), but certainly if two of the integers in a set are relatively prime, then the set is automatically mutually relatively prime.)
c) Since 15 and 28 share no common factors, this set has greatest common divisor 1, so it is mutually relatively prime.
d) Since 21 and 32 share no common factors, this set has greatest common divisor 1, so it is mutually relatively prime.

33. a) We need to find the inverse function. In other words, given $(ap+b) \textbf{ mod } 26$, how does one recover p? Working modulo 26, if we subtract b, then we will have ap. If we then multiply by an inverse of a (which must exist since we are assuming that $\gcd(a, 26) = 1$), we will have p back. Therefore the decryption function is $g(q) = \overline{a}(q-b) \textbf{ mod } 26$, where $\overline{a}$ is an inverse of a modulo 26.
b) Translating the letters into numbers, we see that the message is 11–9–12–10–6 12–6–12–23–5 16–4–23–12–22. By part **(a)** the decryption function is $g(q) = 15(q-10) \textbf{ mod } 26$. (This required computing the inverse of 7 modulo 26, using the technique shown in the solution to Exercise 7 in Section 3.7; it is easy to see that $7 \cdot 15 = 105 \equiv 1 \pmod{26}$.) Applying this function, we obtain the decrypted message 15–11–4–0–18 4–18–4–13–3 12–14–13–4–24. This translates into PLEAS ESEND MONEY, which, after correcting the spacing, is "please send money."

35. The least common multiple of 6 and 15 is 30, so the set of solutions will be given modulo 30 (see Exercise 36). Since the numbers involved here are so small, it is probably best simply to write down the solutions of $x \equiv 4 \pmod 6$ and then see which, if any, of them are also solutions of $x \equiv 13 \pmod{15}$. The solutions of the first congruence, up to 30, are 4, 10, 16, 22, and 28. Only 28 is congruent to 13 modulo 15. Therefore the general solution is all numbers congruent to 28 modulo 30, i.e., $\ldots, -32, -2, 28, 58, \ldots$.

37. We give a proof by contradiction. For this proof we need a fact about polynomials, namely that a nonconstant polynomial can take on the same value only a finite number of times. (Think about its graph—a polynomial of degree n has at most n "wiggles" and so a horizontal line can intersect it at most n times. Alternatively, this statement follows from the Fundamental Theorem of Algebra, which guarantees that a polynomial of degree n has at most n 0's; if $f(x) = a$, then x is a zero of the polynomial $f(x) - a$.) Our given polynomial f can take on the values 0 and ± 1 only finitely many times, so if there is not some y such that $f(y)$ is composite, then there must be some x_0 such that $\pm f(x_0)$ is prime, say p. Now look at $f(x_0 + kp)$. When we plug $x_0 + kp$ in for x in the polynomial and multiply it out, every term will contain a factor of p except for the terms that form $f(x_0)$. Therefore $f(x_0 + kp) = f(x_0) + mp = (m \pm 1)p$ for some integer m. As k varies, this value can be 0 or p or $-p$ only finitely many times; therefore it must be a different multiple of p and therefore a composite number for some values of k, and our proof is complete.

39. Assume that every even integer greater than 2 is the sum of two primes, and let n be an integer greater than 5. If n is odd, then we can write $n = 3 + (n-3)$, decompose $n - 3 = p + q$ into the sum of two primes (since $n-3$ is an even integer greater than 2), and therefore have written $n = 3 + p + q$ as the sum of three primes. If n is even, then we can write $n = 2 + (n-2)$, decompose $n - 2 = p + q$ into the sum of two primes (since $n-2$ is an even integer greater than 2), and therefore have written $n = 2 + p + q$ as the sum of three primes. For the converse, assume that every integer greater than 5 is the sum of three primes, and let n be an even integer greater than 2. By our assumption we can write $n+2$ as the sum of three primes. Since $n+2$ is even, these three primes cannot all be odd, so we have $n+2 = 2+p+q$, where p and q are primes, whence $n = p + q$, as desired.

41. Let us begin by computing $\mathbf{A}^n$ for the first few values of n.

$$\mathbf{A}^1 = \begin{bmatrix} 0 & 1 \\ -1 & 0 \end{bmatrix}, \quad \mathbf{A}^2 = \begin{bmatrix} -1 & 0 \\ 0 & -1 \end{bmatrix}, \quad \mathbf{A}^3 = \begin{bmatrix} 0 & -1 \\ 1 & 0 \end{bmatrix}, \quad \mathbf{A}^4 = \begin{bmatrix} 1 & 0 \\ 0 & 1 \end{bmatrix}.$$

Since $\mathbf{A}^4 = \mathbf{I}$, the pattern will repeat from here: $\mathbf{A}^5 = \mathbf{A}^4\mathbf{A} = \mathbf{IA} = \mathbf{A}$, $\mathbf{A}^6 = \mathbf{A}^2$, $\mathbf{A}^7 = \mathbf{A}^3$, and so on. Thus for all $n \ge 0$ we have

$$\mathbf{A}^{4n+1} = \begin{bmatrix} 0 & 1 \\ -1 & 0 \end{bmatrix}, \quad \mathbf{A}^{4n+2} = \begin{bmatrix} -1 & 0 \\ 0 & -1 \end{bmatrix}, \quad \mathbf{A}^{4n+3} = \begin{bmatrix} 0 & -1 \\ 1 & 0 \end{bmatrix}, \quad \mathbf{A}^{4n+4} = \begin{bmatrix} 1 & 0 \\ 0 & 1 \end{bmatrix}.$$

43. (The notation $c\mathbf{I}$ means the identity matrix $\mathbf{I}$ with each entry multiplied by the real number c; thus this matrix consists of c's along the main diagonal and 0's elsewhere.) Let $\mathbf{A} = \begin{bmatrix} u & v \\ w & x \end{bmatrix}$. We will determine what these entries have to be by using the fact that $\mathbf{AB} = \mathbf{BA}$ for a few judiciously chosen matrices $\mathbf{B}$. First let $\mathbf{B} = \begin{bmatrix} 0 & 1 \\ 0 & 0 \end{bmatrix}$. Then $\mathbf{AB} = \begin{bmatrix} 0 & u \\ 0 & w \end{bmatrix}$, and $\mathbf{BA} = \begin{bmatrix} w & x \\ 0 & 0 \end{bmatrix}$. Since these two must be equal, we know that $0 = w$ and $u = x$. Next choose $\mathbf{B} = \begin{bmatrix} 0 & 0 \\ 1 & 0 \end{bmatrix}$. Then we get $\begin{bmatrix} v & 0 \\ x & 0 \end{bmatrix} = \begin{bmatrix} 0 & 0 \\ u & v \end{bmatrix}$, whence $v = 0$. Therefore the matrix $\mathbf{A}$ must be in the form $\begin{bmatrix} u & 0 \\ 0 & u \end{bmatrix}$, which is just u times the identity matrix, as desired.

45. In an $n \times n$ upper-triangular matrix, all entries a_{ij} are zero unless $i \le j$. Therefore we can store such matrices in about half the space that would be required to store an ordinary $n \times n$ matrix. In implementing something like Algorithm 1 in Section 3.8, then, we need only do the computations for those values of the indices that can produce nonzero entries. The following algorithm does this. We follow the usual notation: $\mathbf{A} = [a_{ij}]$ and $\mathbf{B} = [b_{ij}]$.

```
procedure triangular matrix multiplication(A, B : upper-triangular matrices)
for i := 1 to n
    for j := i to n {since we want j ≥ i}
    begin
        c_ij := 0
        for k := i to j {the only relevant part}
            c_ij := c_ij + a_ik b_kj
    end
{the upper-triangular matrix C = [c_ij] is the product of A and B}
```

47. We simply need to show that the alleged inverse of $\mathbf{AB}$ has the correct defining property—that its product with $\mathbf{AB}$ (on either side) is the identity. Thus we compute

$$(\mathbf{AB})(\mathbf{B}^{-1}\mathbf{A}^{-1}) = \mathbf{A}(\mathbf{BB}^{-1})\mathbf{A}^{-1} = \mathbf{AIA}^{-1} = \mathbf{AA}^{-1} = \mathbf{I},$$

and similarly $(\mathbf{B}^{-1}\mathbf{A}^{-1})(\mathbf{AB}) = \mathbf{I}$. Therefore $(\mathbf{AB})^{-1} = \mathbf{B}^{-1}\mathbf{A}^{-1}$. (Note that the indicated matrix multiplications were all defined, since the hypotheses implied that both $\mathbf{A}$ and $\mathbf{B}$ were $n \times n$ matrices for some (and the same) n.)

49. **a)** The $(i,j)^{\text{th}}$ entry of $\mathbf{A} \odot \mathbf{0}$ is by definition the Boolean sum ($\vee$) of some Boolean products ($\wedge$) of the form $a_{ik} \wedge 0$. Since the latter always equals 0, every entry is 0, so $\mathbf{A} \odot \mathbf{0} = \mathbf{0}$. Similarly $\mathbf{0} \odot \mathbf{A}$ consists of entries that are all 0, so it, too, equals $\mathbf{0}$.

b) Since $\vee$ operates entrywise, the statements that $\mathbf{A} \vee \mathbf{0} = \mathbf{A}$ and $\mathbf{0} \vee \mathbf{A} = \mathbf{A}$ follow from the facts that $a_{ij} \vee 0 = a_{ij}$ and $0 \vee a_{ij} = a_{ij}$.

c) Since $\wedge$ operates entrywise, the statements that $\mathbf{A} \wedge \mathbf{0} = \mathbf{0}$ and $\mathbf{0} \wedge \mathbf{A} = \mathbf{0}$ follow from the facts that $a_{ij} \wedge 0 = 0$ and $0 \wedge a_{ij} = 0$.

51. We assume that someone has chosen a positive integer less than 2^n, which we are to guess. We ask the person to write the number in binary, using leading 0's if necessary to make it n bits long. We then ask "Is the first bit a 1?", "Is the second bit a 1?", "Is the third bit a 1?", and so on. After we know the answers to these n questions, we will know the number, because we will know its binary expansion.

WRITING PROJECTS FOR CHAPTER 3

Books and articles indicated by bracketed symbols below are listed near the end of this manual. You should also read the general comments and advice you will find there about researching and writing these essays.

1. Algorithms are as much a subject of computer science as they are a subject of mathematics. Thus sources on the history of computer science (or perhaps even verbose introductory programming or comprehensive computer science textbooks) might be a good place to look. See also [Ha2].

2. The original is [Ba2]. Good history of mathematics books would be a place to follow up.

3. See [Me1], or do a Web search.

4. Knuth's volume 3 [Kn] is the bible for sorting algorithms.

5. This is a vast and extremely important subject in computer science today. One basic book on the subject is [Gi]; see also an essay in [De2].

6. See the comments for Writing Project 5, above. One basic algorithm to think about doing in parallel is sorting. Imagine a group of school-children asked to arrange themselves in a row by height, and any child can determine whether another child is shorter or taller than him/herself.

7. As usual, the Web is an excellent resource here. Check out the GIMPS page and then follow its links: `http://www.mersenne.org/prime.htm`.

8. These primes are sometimes jokingly referred to as "industrial strength primes." Number theory textbooks, such as [Ro3], would be a place to start. See also the article [Le3].

9. One can often find mathematical news reported in *The New York Times* and other nontechnical media. Search an index to find a story about this topic. (While you're looking at back issues of the *Times*, read the January 31, 1995, article on the solution to Fermat's Last Theorem.)

10. Your essay should mention the RSA-129 project. See *The New York Times*, around the spring of 1994 (use its index). For an expository article, try [Po].

11. There are dozens of books on computer hardware and circuit design that discuss the algorithms and circuits used in performing these operations. If you are a computer science or computer engineering major, you probably have taken (or will take) a course that deals with these topics. See [Ko2] and similar books.

12. A traditional history of mathematics book should be helpful here; try [Bo4] or [Ev3].

13. This topic has taken on a lot of significance recently as randomized algorithms become more and more important. The August/September 1994 issue of *SIAM News* (the newsletter of the Society for Industrial and Applied Mathematics) has a provocative article on the subject. See also [La1], or for older material in a textbook try volume 2 of [Kn].

14. This topic gets into the news on a regular basis. Try *The New York Times* index. See also Writing Project 10, above, and 15, below.

15. If I encrypt my signature with my private key, then I will produce something that will decrypt (using my public key) as my signature. Furthermore, no one else can do this, since no one else knows my private key. Good sources for cryptography include [Be], [MeOo], and [St2].

16. The author's number theory text ([Ro3]) has material on this topic. The amazing mathematician John H. Conway (inventor of the Game of Life, among other things) has devised what he calls the Doomsday Algorithm, and it works quite fast with practice. See [BeCo]. (Conway can determine any day of the week mentally in a second or two.)

17. An advanced algorithms text, such as [Ma2] or [BrBr], is the place to look. One algorithm has the names Schönhage and Strassen associated with it. A related topic is the fast Fourier transform. See also volume 2 of [Kn].

18. This is related to Writing Project 17; see the suggestions there. Strassen also invented (or discovered, depending on your philosophy) a fast matrix multiplication algorithm.

19. Several excellent books have appeared in the past decade on cryptography, such as [MeOo]. Many of them, including that one, will treat this topic.

CHAPTER 4
Induction and Recursion

SECTION 4.1 Mathematical Induction

Understanding and constructing proofs by mathematical induction are extremely difficult tasks for most students. Do not be discouraged, and do not give up, because, without doubt, this proof technique is the most important one there is in mathematics and computer science. Pay careful attention to the conventions to be observed in writing down a proof by induction. As with all proofs, remember that a proof by mathematical induction is like an essay—it must have a beginning, a middle, and an end; it must consist of complete sentences, logically and aesthetically arranged; and it must convince the reader. Be sure that your basis step (also called the "base case") is correct (that you have verified the proposition in question for the smallest value or values of n*), and be sure that your inductive step is correct and complete (that you have derived the proposition for* $k+1$*, assuming the inductive hypothesis that the proposition is true for* k*).*

Some, but not all, proofs by mathematical induction are like Exercises 3–17. In each of these, you are asked to prove that a certain summation has a "closed form" representation given by a certain expression. Here the proofs are usually straightforward algebra. For the inductive step you start with the summation for $P(k+1)$*, find the summation for* $P(k)$ *as its first* k *terms, replace that much by the closed form given by the inductive hypothesis, and do the algebra to get the resulting expression into the desired form. When doing proofs like this, however, remember to include all the words surrounding your algebra—the algebra alone is not the proof. Also keep in mind that* $P(n)$ *is the proposition that the sum equals the closed-form expression, not just the sum and not just the expression.*

Many inequalities can be proved by mathematical induction; see Exercises 18–24, for example. The method also extends to such things as set operations, divisibility, and a host of other applications; a sampling of them is given in other exercises in this set. Some are quite complicated.

One final point about notation. In performing the inductive step, it really does not matter what letter we use. We see in the text the proof of $P(k) \to P(k+1)$*; but it would be just as valid to prove* $P(n) \to P(n+1)$*, since the* k *in the first case and the* n *in the second case are just dummy variables. We will use both notations in this Guide; in particular, we will use* k *for the first few exercises but often use* n *afterwards.*

1. We can prove this by mathematical induction. Let $P(n)$ be the statement that the train stops at station n. We want to prove that $P(n)$ is true for all positive integers n. For the basis step, we are told that $P(1)$ is true. For the inductive step, we are told that $P(k)$ implies $P(k+1)$ for each $k \ge 1$. Therefore by the principle of mathematical induction, $P(n)$ is true for all positive integers n.

3. a) Plugging in $n = 1$ we have that $P(1)$ is the statement $1^2 = 1 \cdot 2 \cdot 3/6$.

b) Both sides of $P(1)$ shown in part **(a)** equal 1.

c) The inductive hypothesis is the statement that

$$1^2 + 2^2 + \cdots + k^2 = \frac{k(k+1)(2k+1)}{6}.$$

d) For the inductive step, we want to show for each $k \ge 1$ that $P(k)$ implies $P(k+1)$. In other words, we want to show that assuming the inductive hypothesis (see part **(c)**) we can show

$$1^2 + 2^2 + \cdots + k^2 + (k+1)^2 = \frac{(k+1)(k+2)(2k+3)}{6}.$$

e) The left-hand side of the equation in part **(d)** equals, by the inductive hypothesis, $k(k+1)(2k+1)/6 + (k+1)^2$. We need only do a bit of algebraic manipulation to get this expression into the desired form: factor out $(k+1)/6$ and then factor the rest. In detail,

$$\begin{aligned}(1^2 + 2^2 + \cdots + k^2) + (k+1)^2 &= \frac{k(k+1)(2k+1)}{6} + (k+1)^2 \quad \text{(by the inductive hypothesis)}\\ &= \frac{k+1}{6}\big(k(2k+1) + 6(k+1)\big) = \frac{k+1}{6}(2k^2 + 7k + 6)\\ &= \frac{k+1}{6}(k+2)(2k+3) = \frac{(k+1)(k+2)(2k+3)}{6}.\end{aligned}$$

f) We have completed both the basis step and the inductive step, so by the principle of mathematical induction, the statement is true for every positive integer n.

5. We proceed by induction. The basis step, $n = 0$, is true, since $1^2 = 1 \cdot 1 \cdot 3/3$. For the inductive step assume the inductive hypothesis that

$$1^2 + 3^2 + 5^2 + \cdots + (2k+1)^2 = \frac{(k+1)(2k+1)(2k+3)}{3}.$$

We want to show that

$$1^2 + 3^2 + 5^2 + \cdots + (2k+1)^2 + (2k+3)^2 = \frac{(k+2)(2k+3)(2k+5)}{3}$$

(the right-hand side is the same formula with $k+1$ plugged in for n). Now the left-hand side equals, by the inductive hypothesis, $(k+1)(2k+1)(2k+3)/3 + (2k+3)^2$. We need only do a bit of algebraic manipulation to get this expression into the desired form: factor out $(2k+3)/3$ and then factor the rest. In detail,

$$\begin{aligned}&\big(1^2 + 3^2 + 5^2 + \cdots + (2k+1)^2\big) + (2k+3)^2\\ &\qquad = \frac{(k+1)(2k+1)(2k+3)}{3} + (2k+3)^2 \quad \text{(by the inductive hypothesis)}\\ &\qquad = \frac{2k+3}{3}\big((k+1)(2k+1) + 3(2k+3)\big) = \frac{2k+3}{3}(2k^2 + 9k + 10)\\ &\qquad = \frac{2k+3}{3}\big((k+2)(2k+5)\big) = \frac{(k+2)(2k+3)(2k+5)}{3}.\end{aligned}$$

7. Let $P(n)$ be the proposition $3 + 3\cdot 5 + 3\cdot 5^2 + \cdots + 3\cdot 5^n = 3(5^{n+1} - 1)/4$. To prove that this is true for all nonnegative integers n, we proceed by mathematical induction. First we verify $P(0)$, namely that $3 = 3(5-1)/4$, which is certainly true. Next we assume that $P(k)$ is true and try to derive $P(k+1)$. Now $P(k+1)$ is the formula

$$3 + 3\cdot 5 + 3\cdot 5^2 + \cdots + 3\cdot 5^k + 3\cdot 5^{k+1} = \frac{3(5^{k+2} - 1)}{4}.$$

All but the last term of the left-hand side of this equation is exactly the left-hand side of $P(k)$, so by the inductive hypothesis, it equals $3(5^{k+1} - 1)/4$. Thus we have

$$\begin{aligned}3 + 3\cdot 5 + 3\cdot 5^2 + \cdots + 3\cdot 5^k + 3\cdot 5^{k+1} &= \frac{3(5^{k+1} - 1)}{4} + 3\cdot 5^{k+1}\\ &= 5^{k+1}\left(\frac{3}{4} + 3\right) - \frac{3}{4} = 5^{k+1}\cdot\frac{15}{4} - \frac{3}{4}\\ &= 5^{k+2}\cdot\frac{3}{4} - \frac{3}{4} = \frac{3(5^{k+2} - 1)}{4}.\end{aligned}$$

9. a) We can obtain a formula for the sum of the first n even positive integers from the formula for the sum of the first n positive integers, since $2+4+6+\cdots+2n = 2(1+2+3+\cdots+n)$. Therefore, using the result of Example 1, the sum of the first n even positive integers is $2(n(n+1)/2) = n(n+1)$.

b) We want to prove the proposition $P(n): 2+4+6+\cdots+2n = n(n+1)$. The basis step, $n=1$, says that $2 = 1\cdot(1+1)$, which is certainly true. For the inductive step, we assume that $P(k)$ is true, namely that

$$2+4+6+\cdots+2k = k(k+1)\,,$$

and try to prove from this assumption that $P(k+1)$ is true, namely that

$$2+4+6+\cdots+2k+2(k+1) = (k+1)(k+2)\,.$$

(Note that the left-hand side consists of the sum of the first $k+1$ even positive integers.) We have

$$\begin{aligned}2+4+6+\cdots+2k+2(k+1) &= (2+4+6+\cdots+2k)+2(k+1)\\ &= k(k+1)+2(k+1) \quad \text{(by the inductive hypothesis)}\\ &= (k+1)(k+2)\,,\end{aligned}$$

as desired, and our proof by mathematical induction is complete.

11. a) Let us compute the values of this sum for $n \le 4$ to see whether we can discover a pattern. For $n=1$ the sum is $\frac{1}{2}$. For $n=2$ the sum is $\frac{1}{2}+\frac{1}{4}=\frac{3}{4}$. For $n=3$ the sum is $\frac{1}{2}+\frac{1}{4}+\frac{1}{8}=\frac{7}{8}$. And for $n=4$ the sum is $15/16$. The pattern seems pretty clear, so we conjecture that the sum is always $(2^n-1)/2^n$.

b) We have already verified that this is true in the base case (in fact, in four base cases). So let us assume it for k and try to prove it for $k+1$. More formally, we are letting $P(n)$ be the *statement* that

$$\frac{1}{2}+\frac{1}{4}+\frac{1}{8}+\cdots+\frac{1}{2^n} = \frac{2^n-1}{2^n},$$

and trying to prove that $P(n)$ is true for all n. We have already verified $P(1)$ (as well as $P(2)$, $P(3)$, and $P(4)$ for good measure). We now assume the inductive hypothesis $P(k)$, which is the equation displayed above with k substituted for n, and must derive $P(k+1)$, which is the equation

$$\frac{1}{2}+\frac{1}{4}+\frac{1}{8}+\cdots+\frac{1}{2^k}+\frac{1}{2^{k+1}} = \frac{2^{k+1}-1}{2^{k+1}}.$$

The "obvious" thing to try is to add $1/2^{k+1}$ to both sides of the inductive hypothesis and see whether the algebra works out as we hope it will. We obtain

$$\left(\frac{1}{2}+\frac{1}{4}+\frac{1}{8}+\cdots+\frac{1}{2^k}\right)+\frac{1}{2^{k+1}} = \frac{2^k-1}{2^k}+\frac{1}{2^{k+1}} = \frac{2\cdot 2^k - 2\cdot 1+1}{2^{k+1}} = \frac{2^{k+1}-1}{2^{k+1}},$$

as desired.

13. The base case of the statement $P(n): 1^2-2^2+3^2-\cdots+(-1)^{n-1}n^2 = (-1)^{n-1}n(n+1)/2$, when $n=1$, is $1^2 = (-1)^0\cdot 1\cdot 2/2$, which is certainly true. Assume the inductive hypothesis $P(k)$, and try to derive $P(k+1)$:

$$1^2-2^2+3^2-\cdots+(-1)^{k-1}k^2+(-1)^k(k+1)^2 = (-1)^k\frac{(k+1)(k+2)}{2}.$$

Starting with the left-hand side of $P(k+1)$, we have

$$\begin{aligned}&(1^2-2^2+3^2-\cdots+(-1)^{k-1}k^2)+(-1)^k(k+1)^2\\ &\qquad= (-1)^{k-1}\frac{k(k+1)}{2}+(-1)^k(k+1)^2 \quad \text{(by the inductive hypothesis)}\\ &\qquad= (-1)^k(k+1)((-k/2)+k+1)\\ &\qquad= (-1)^k(k+1)\left(\frac{k}{2}+1\right) = (-1)^k\frac{(k+1)(k+2)}{2},\end{aligned}$$

the right-hand side of $P(k+1)$.

15. The base case of the statement $P(n) : 1 \cdot 2 + 2 \cdot 3 + \cdots + n(n+1) = n(n+1)(n+2)/3$, when $n = 1$, is $1 \cdot 2 = 1 \cdot 2 \cdot 3/3$, which is certainly true. We assume the inductive hypothesis $P(k)$, and try to derive $P(k+1)$:

$$1 \cdot 2 + 2 \cdot 3 + \cdots + k(k+1) + (k+1)(k+2) = \frac{(k+1)(k+2)(k+3)}{3}$$

Starting with the left-hand side of $P(k+1)$, we have

$$\begin{aligned}
&\big(1 \cdot 2 + 2 \cdot 3 + \cdots + k(k+1)\big) + (k+1)(k+2) \\
&\qquad = \frac{k(k+1)(k+2)}{3} + (k+1)(k+2) \quad \text{(by the inductive hypothesis)} \\
&\qquad = (k+1)(k+2)\left(\frac{k}{3} + 1\right) = \frac{(k+1)(k+2)(k+3)}{3},
\end{aligned}$$

the right-hand side of $P(k+1)$.

17. This proof follows the basic pattern of the solution to Exercise 3, but the algebra gets more complex. The statement $P(n)$ that we wish to prove is

$$1^4 + 2^4 + 3^4 + \cdots + n^4 = \frac{n(n+1)(2n+1)(3n^2+3n-1)}{30},$$

where n is a positive integer. The basis step, $n = 1$, is true, since $1 \cdot 2 \cdot 3 \cdot 5/30 = 1$. Assume the displayed statement as the inductive hypothesis, and proceed as follows to prove $P(n+1)$:

$$\begin{aligned}
(1^4 + 2^4 + \cdots + n^4) + (n+1)^4 &= \frac{n(n+1)(2n+1)(3n^2+3n-1)}{30} + (n+1)^4 \\
&= \frac{n+1}{30}\big(n(2n+1)(3n^2+3n-1) + 30(n+1)^3\big) \\
&= \frac{n+1}{30}(6n^4 + 39n^3 + 91n^2 + 89n + 30) \\
&= \frac{n+1}{30}(n+2)(2n+3)(3(n+1)^2 + 3(n+1) - 1)
\end{aligned}$$

The last equality is straightforward to check; it was obtained not by attempting to factor the next to last expression from scratch but rather by knowing exactly what we expected the simplified expression to be.

19. **a)** $P(2)$ is the statement that $1 + \frac{1}{4} < 2 - \frac{1}{2}$.

b) This is true because $5/4$ is less than $6/4$.

c) The inductive hypothesis is the statement that

$$1 + \frac{1}{4} + \cdots + \frac{1}{k^2} < 2 - \frac{1}{k}.$$

d) For the inductive step, we want to show for each $k \geq 2$ that $P(k)$ implies $P(k+1)$. In other words, we want to show that assuming the inductive hypothesis (see part **(c)**) we can show

$$1 + \frac{1}{4} + \cdots + \frac{1}{k^2} + \frac{1}{(k+1)^2} < 2 - \frac{1}{k+1}.$$

e) Assume the inductive hypothesis. Then we have

$$\begin{aligned}
1 + \frac{1}{4} + \cdots + \frac{1}{k^2} + \frac{1}{(k+1)^2} &< 2 - \frac{1}{k} + \frac{1}{(k+1)^2} \\
&= 2 - \left(\frac{1}{k} - \frac{1}{(k+1)^2}\right) \\
&= 2 - \left(\frac{k^2 + 2k + 1 - k}{k(k+1)^2}\right) \\
&= 2 - \frac{k^2 + k}{k(k+1)^2} - \frac{1}{k(k+1)^2} \\
&= 2 - \frac{1}{k+1} - \frac{1}{k(k+1)^2} < 2 - \frac{1}{k+1}.
\end{aligned}$$

f) We have completed both the basis step and the inductive step, so by the principle of mathematical induction, the statement is true for every positive integer n greater than 1.

21. Let $P(n)$ be the proposition $2^n > n^2$. We want to show that $P(n)$ is true for all $n > 4$. The base case is therefore $n = 5$, and we check that $2^5 = 32 > 25 = 5^2$. Now we assume the inductive hypothesis that $2^k > k^2$ and want to derive the statement that $2^{k+1} > (k+1)^2$. Working from the right-hand side, we have $(k+1)^2 = k^2 + 2k + 1 < k^2 + 2k + k = k^2 + 3k < k^2 + k^2$ (since $k > 3$). Thus we have $(k+1)^2 < 2k^2 < 2 \cdot 2^k$ (by the inductive hypothesis), which in turn equals 2^{k+1}, as desired.

23. We compute the values of $2n+3$ and 2^n for the first few values of n and come to the immediate conjecture that $2n + 3 \le 2^n$ for $n \ge 4$ but for no other nonnegative integer values of n. The negative part of this statement is just the fact that $3 > 1$, $5 > 2$, $7 > 4$, and $9 > 8$. We must prove by mathematical induction that $2n + 3 \le 2^n$ for all $n \ge 4$. The base case is $n = 4$, in which we see that, indeed, $11 \le 16$. Next assume the inductive hypothesis that $2n + 3 \le 2^n$, and consider $2(n+1) + 3$. This equals $2n + 3 + 2$, which by the inductive hypothesis is less than or equal to $2^n + 2$. But since $n \ge 1$, this in turn is at most $2^n + 2^n = 2^{n+1}$, precisely the statement we wished to prove.

25. We can assume that $h > -1$ is fixed, and prove the proposition by induction on n. Let $P(n)$ be the proposition $1 + nh \le (1+h)^n$. The base case is $n = 0$, in which case $P(0)$ is simply $1 \le 1$, certainly true. Now we assume the inductive hypothesis, that $1 + kh \le (1+h)^k$; we want to show that $1 + (k+1)h \le (1+h)^{k+1}$. Since $h > -1$, it follows that $1 + h > 0$, so we can multiply both sides of the inductive hypothesis by $1+h$ to obtain $(1+h)(1+kh) \le (1+h)^{k+1}$. Thus to complete the proof it is enough to show that $1+(k+1)h \le (1+h)(1+kh)$. But the right-hand side of this inequality is the same as $1 + h + kh + kh^2 = 1 + (k+1)h + kh^2$, which is greater than or equal to $1 + (k+1)h$ because $kh^2 \ge 0$.

27. This exercise involves some messy algebra, but the logic is the usual logic for proofs using the principle of mathematical induction. The basis step ($n = 1$) is true, since 1 is greater than $2(\sqrt{2} - 1) \approx 0.83$. We assume that

$$1 + \frac{1}{\sqrt{2}} + \cdots + \frac{1}{\sqrt{n}} > 2(\sqrt{n+1} - 1)$$

and try to derive the corresponding statement for $n+1$:

$$1 + \frac{1}{\sqrt{2}} + \cdots + \frac{1}{\sqrt{n}} + \frac{1}{\sqrt{n+1}} > 2(\sqrt{n+2} - 1)$$

Since by the inductive hypothesis we know that

$$1 + \frac{1}{\sqrt{2}} + \cdots + \frac{1}{\sqrt{n}} + \frac{1}{\sqrt{n+1}} > 2(\sqrt{n+1} - 1) + \frac{1}{\sqrt{n+1}},$$

we will be finished if we can show that the inequality

$$2(\sqrt{n+1} - 1) + \frac{1}{\sqrt{n+1}} > 2(\sqrt{n+2} - 1)$$

holds. By canceling the -2 from both sides and rearranging, we obtain the equivalent inequality

$$2(\sqrt{n+2} - \sqrt{n+1}) < \frac{1}{\sqrt{n+1}},$$

which in turn is equivalent to

$$2(\sqrt{n+2} - \sqrt{n+1})(\sqrt{n+2} + \sqrt{n+1}) < \frac{\sqrt{n+1}}{\sqrt{n+1}} + \frac{\sqrt{n+2}}{\sqrt{n+1}}.$$

This last inequality simplifies to

$$2 < 1 + \frac{\sqrt{n+2}}{\sqrt{n+1}},$$

which is clearly true. Therefore the original inequality is true, and our proof is complete.

29. Recall that $H_k = 1/1 + 1/2 + \cdots + 1/k$. We want to prove that $H_{2^n} \le 1 + n$ for all natural numbers n. We proceed by mathematical induction, noting that the basis step $n = 0$ is the trivial statement $H_1 = 1 \le 1 + 0$. Therefore we assume that $H_{2^n} \le 1 + n$; we want to show that $H_{2^{n+1}} \le 1 + (n+1)$. We have

$$\begin{aligned}
H_{2^{n+1}} &= H_{2^n} + \frac{1}{2^n+1} + \frac{1}{2^n+2} + \cdots + \frac{1}{2^{n+1}} \quad \text{(by definition; there are } 2^n \text{ fractions here)} \\
&\le (1+n) + \frac{1}{2^n+1} + \frac{1}{2^n+2} + \cdots + \frac{1}{2^{n+1}} \quad \text{(by the inductive hypothesis)} \\
&\le (1+n) + \frac{1}{2^n+1} + \frac{1}{2^n+1} + \cdots + \frac{1}{2^n+1} \quad \text{(we made the denominators smaller)} \\
&= 1 + n + \frac{2^n}{2^n+1} < 1 + n + 1 = 1 + (n+1).
\end{aligned}$$

31. This is easy to prove without mathematical induction, because we can observe that $n^2 + n = n(n+1)$, and either n or $n+1$ is even. If we want to use the principle of mathematical induction, we can proceed as follows. The basis step is the observation that $1^2 + 1 = 2$ is divisible by 2. Assume the inductive hypothesis, that $k^2 + k$ is divisible by 2; we must show that $(k+1)^2 + (k+1)$ is divisible by 2. But $(k+1)^2 + (k+1) = k^2 + 2k + 1 + k + 1 = (k^2 + k) + 2(k+1)$. But now $k^2 + k$ is divisible by 2 by the inductive hypothesis, and $2(k+1)$ is divisible by 2 by definition, so this sum of two multiples of 2 must be divisible by 2.

33. To prove that $P(n) : 5 \,|\, (n^5 - n)$ holds for all nonnegative integers n, we first check that $P(0)$ is true; indeed $5 \,|\, 0$. Next assume that $5 \,|\, (n^5 - n)$, so that we can write $n^5 - n = 5t$ for some integer t. Then we want to prove $P(n+1)$, namely that $5 \,|\, \big((n+1)^5 - (n+1)\big)$. We expand and then factor the right-hand side to obtain

$$\begin{aligned}
(n+1)^5 - (n+1) &= n^5 + 5n^4 + 10n^3 + 10n^2 + 5n + 1 - n - 1 \\
&= (n^5 - n) + 5(n^4 + 2n^3 + 2n^2 + n) \\
&= 5t + 5(n^4 + 2n^3 + 2n^2 + n) \quad \text{(by the inductive hypothesis)} \\
&= 5(t + n^4 + 2n^3 + 2n^2 + n).
\end{aligned}$$

Thus we have shown that $(n+1)^5 - (n+1)$ is also a multiple of 5, and our proof by induction is complete. (Note that here we have used n as the dummy variable in the inductive step, rather than k. It really makes no difference.)

We should point out that using mathematical induction is not the only way to prove this proposition; it can also be proved by considering the five cases determined by the value of n **mod** 5. The reader is encouraged to write down such a proof.

35. First let us rewrite this proposition so that it is a statement about all nonnegative integers, rather than just the odd positive integers. An odd positive integer can be written as $2n - 1$, so let us prove the proposition $P(n)$ that $(2n-1)^2 - 1$ is divisible by 8 for all positive integers n. We first check that $P(1)$ is true; indeed $8 \,|\, 0$. Next assume that $8 \,|\, ((2n-1)^2 - 1)$. Then we want to prove $P(n+1)$, namely that $8 \,|\, ((2n+1)^2 - 1)$. Let us look at the difference of these two expressions: $(2n+1)^2 - 1 - ((2n-1)^2 - 1)$. A little algebra reduces this to $8n$, which is certainly a multiple of 8. But if this difference is a multiple of 8, and if, by the inductive hypothesis, $(2n-1)^2 - 1$ is a multiple of 8, then $(2n+1)^2 - 1$ must be a multiple of 8, and our proof by induction is complete.

37. It is not easy to stumble upon the trick needed in the inductive step in this exercise, so do not feel bad if you did not find it. The form is straightforward. For the basis step ($n = 1$), we simply observe that $11^{1+1} + 12^{2\cdot 1 - 1} = 121 + 12 = 133$, which is divisible by 133. Then we assume the inductive hypothesis, that

$11^{n+1} + 12^{2n-1}$ is divisible by 133, and let us look at the expression when $n+1$ is plugged in for n. We want somehow to manipulate it so that the expression for n appears. We have

$$\begin{aligned} 11^{(n+1)+1} + 12^{2(n+1)-1} &= 11 \cdot 11^{n+1} + 144 \cdot 12^{2n-1} \\ &= 11 \cdot 11^{n+1} + (11 + 133) \cdot 12^{2n-1} \\ &= 11(11^{n+1} + 12^{2n-1}) + 133 \cdot 12^{2n-1} . \end{aligned}$$

Looking at the last line, we see that the expression in parentheses is divisible by 133 by the inductive hypothesis, and obviously the second term is divisible by 133, so the entire quantity is divisible by 133, as desired.

39. The basis step is trivial, as usual: $A_1 \subseteq B_1$ implies that $\bigcap_{j=1}^{1} A_j \subseteq \bigcap_{j=1}^{1} B_j$ because the intersection of one set is itself. Assume the inductive hypothesis that if $A_j \subseteq B_j$ for $j = 1, 2, \ldots, k$, then $\bigcap_{j=1}^{k} A_j \subseteq \bigcap_{j=1}^{k} B_j$. We want to show that if $A_j \subseteq B_j$ for $j = 1, 2, \ldots, k+1$, then $\bigcap_{j=1}^{k+1} A_j \subseteq \bigcap_{j=1}^{k+1} B_j$. To show that one set is a subset of another we show that an arbitrary element of the first set must be an element of the second set. So let $x \in \bigcap_{j=1}^{k+1} A_j = \left(\bigcap_{j=1}^{k} A_j\right) \cap A_{k+1}$. Because $x \in \bigcap_{j=1}^{k} A_j$, we know by the inductive hypothesis that $x \in \bigcap_{j=1}^{k} B_j$; because $x \in A_{k+1}$, we know from the given fact that $A_{k+1} \subseteq B_{k+1}$ that $x \in B_{k+1}$. Therefore $x \in \left(\bigcap_{j=1}^{k} B_j\right) \cap B_{k+1} = \bigcap_{j=1}^{k+1} B_j$.

This is really easier to do directly than by using the principle of mathematical induction. For a noninductive proof, suppose that $x \in \bigcap_{j=1}^{n} A_j$. Then $x \in A_j$ for each j from 1 to n. Since $A_j \subseteq B_j$, we know that $x \in B_j$. Therefore by definition, $x \in \bigcap_{j=1}^{n} B_j$.

41. In order to prove this statement, we need to use one of the distributive laws from set theory: $(X \cup Y) \cap Z = (X \cap Z) \cup (Y \cap Z)$ (see Section 2.2). Indeed, the proposition at hand is the generalization of this distributive law, from two sets in the union to n sets in the union. We will also be using implicitly the associative law for set union.

The basis step, $n = 1$, is the statement $A_1 \cap B = A_1 \cap B$, which is obviously true. Therefore we assume the inductive hypothesis that

$$(A_1 \cup A_2 \cup \cdots \cup A_n) \cap B = (A_1 \cap B) \cup (A_2 \cap B) \cup \cdots \cup (A_n \cap B) .$$

We wish to prove the similar statement for $n+1$, namely

$$(A_1 \cup A_2 \cup \cdots \cup A_n \cup A_{n+1}) \cap B = (A_1 \cap B) \cup (A_2 \cap B) \cup \cdots \cup (A_n \cap B) \cup (A_{n+1} \cap B) .$$

Starting with the left-hand side, we apply the distributive law for two sets:

$$\begin{aligned} ((A_1 \cup A_2 \cup \cdots \cup A_n) \cup A_{n+1}) \cap B &= ((A_1 \cup A_2 \cup \cdots \cup A_n) \cap B) \cup (A_{n+1} \cap B) \\ &= ((A_1 \cap B) \cup (A_2 \cap B) \cup \cdots \cup (A_n \cap B)) \cup (A_{n+1} \cap B) \\ &\qquad \text{(by the inductive hypothesis)} \\ &= (A_1 \cap B) \cup (A_2 \cap B) \cup \cdots \cup (A_n \cap B) \cup (A_{n+1} \cap B) \end{aligned}$$

43. In order to prove this statement, we need to use one of De Morgan's laws from set theory: $\overline{(A \cup B)} = \overline{A} \cap \overline{B}$ (see Section 2.2). Indeed, the proposition at hand is the generalization of this law, from two sets in the union to n sets in the union. We will also be using implicitly the associative laws for set union and intersection.

The basis step, $n = 1$, is the statement $\overline{A_1} = \overline{A_1}$ (since the union or intersection of just one set is the set itself), and this proposition is obviously true. Therefore we assume the inductive hypothesis that

$$\overline{\bigcup_{k=1}^{n} A_k} = \bigcap_{k=1}^{n} \overline{A_k} .$$

We wish to prove the similar statement for $n+1$, namely

$$\overline{\bigcup_{k=1}^{n+1} A_k} = \bigcap_{k=1}^{n+1} \overline{A_k}\,.$$

Starting with the left-hand side, we group, apply De Morgan's law for two sets, and then the inductive hypothesis:

$$\begin{aligned}
\overline{\bigcup_{k=1}^{n+1} A_k} &= \overline{(\bigcup_{k=1}^{n} A_k) \cup A_{n+1}} \\
&= \overline{\bigcup_{k=1}^{n} A_k} \cap \overline{A_{n+1}} \quad \text{(by DeMorgan's law)} \\
&= (\bigcap_{k=1}^{n} \overline{A_k}) \cap \overline{A_{n+1}} \quad \text{(by the inductive hypothesis)} \\
&= \bigcap_{k=1}^{n+1} \overline{A_k}
\end{aligned}$$

45. This proof will be similar to the proof in Example 9. The basis step is clear, since for $n = 2$, the set has exactly one subset containing exactly two elements, and $2(2-1)/2 = 1$. Assume the inductive hypothesis, that a set with n elements has $n(n-1)/2$ subsets with exactly two elements; we want to prove that a set S with $n+1$ elements has $(n+1)n/2$ subsets with exactly two elements. Fix an element a in S, and let T be the set of elements of S other than a. There are two varieties of subsets of S containing exactly two elements. First there are those that do not contain a. These are precisely the two-element subsets of T, and by the inductive hypothesis, there are $n(n-1)/2$ of them. Second, there are those that contain a together with one element of T. Since T has n elements, there are exactly n subsets of this type. Therefore the total number of subsets of S containing exactly two elements is $(n(n-1)/2) + n$, which simplifies algebraically to $(n+1)n/2$, as desired.

47. The one and only flaw in this proof is in this statement, which is part of the inductive step: "the set of the first n horses and the set of the last n horses [in the collection of $n+1$ horses being considered] overlap." The only assumption made about the number n in this argument is that n is a positive integer. When $n = 1$, so that $n+1 = 2$, the statement quoted is obviously nonsense: the set of the first one horse and the set of the last one horse, in this set of two horses, are disjoint.

49. The mistake is in applying the inductive hypothesis to look at $\max(x-1, y-1)$, because even though x and y are positive integers, $x-1$ and $y-1$ need not be (one or both could be 0). In fact, that is what happens if we let $x = 1$ and $y = 2$ when $k = 1$.

51. We use the notation (i,j) to mean the square in row i and column j, where we number from the left and from the bottom, starting at $(0,0)$ in the lower left-hand corner. We use induction on $i+j$ to show that every square can be reached by the knight. There are six base cases, for the cases when $i+j \le 2$. The knight is already at $(0,0)$ to start, so the empty sequence of moves reaches that square. To reach $(1,0)$, the knight moves successively from $(0,0)$ to $(2,1)$ to $(0,2)$ to $(1,0)$. Similarly, to reach $(0,1)$, the knight moves successively from $(0,0)$ to $(1,2)$ to $(2,0)$ to $(0,1)$. Note that the knight has reached $(2,0)$ and $(0,2)$ in the process. For the last basis step, note this path to $(1,1)$: $(0,0)$ to $(1,2)$ to $(2,0)$ to $(0,1)$ to $(2,2)$ to $(0,3)$ to $(1,1)$. We now assume the inductive hypothesis, that the knight can reach any square (i,j) for which $i+j = k$, where k is an integer greater than 1, and we want to show how the knight can reach each square (i,j) when $i+j = k+1$. Since $k+1 \ge 3$, at least one of i and j is at least 2. If $i \ge 2$, then by the inductive hypothesis, there is a sequence of moves ending at $(i-2, j+1)$, since $i-2+j+1=i+j-1=k$; from there it is just one step to (i,j). Similarly, if $j \geq 2$, then by the inductive hypothesis, there is a sequence of moves ending at $(i+1, j-2)$, since $i+1+j-2=i+j-1=k$; from there it is again just one step to (i,j).

53. The base cases are $n=0$ and $n=1$, and it is a simple matter to evaluate, directly from the "limit of difference quotient" definition, the derivatives of $x^0=1$ and $x^1=x$:

$$\frac{d}{dx}x^0 = \lim_{h\to 0}\frac{(x+h)^0-x^0}{h} = \lim_{h\to 0}\frac{1-1}{h} = 0 = 0\cdot x^{-1}$$

$$\frac{d}{dx}x^1 = \lim_{h\to 0}\frac{(x+h)^1-x^1}{h} = \lim_{h\to 0}\frac{h}{h} = 1 = 1\cdot x^0$$

We are told to assume that the product rule holds:

$$\frac{d}{dx}(f(x)\cdot g(x)) = f(x)\cdot g'(x) + g(x)\cdot f'(x)$$

So we work as follows, invoking the inductive hypothesis and the base cases:

$$\begin{aligned}\frac{d}{dx}x^{n+1} &= \frac{d}{dx}(x\cdot x^n) = x\cdot\frac{d}{dx}x^n + x^n\cdot\frac{d}{dx}x\\ &= x\cdot nx^{n-1} + x^n\cdot 1 = nx^n + x^n = (n+1)x^n\end{aligned}$$

55. We prove this by induction on k. The basis step $k=0$ is the trivial statement that $1\equiv 1 \pmod{m}$. Suppose that the statement is true for k. We must show it for $k+1$. So let $a\equiv b \pmod{m}$. By the inductive hypothesis we know that $a^k\equiv b^k \pmod{m}$. Then we apply Theorem 5 from Section 3.4 to conclude that $a\cdot a^k\equiv b\cdot b^k \pmod{m}$, which by definition says that $a^{k+1}\equiv b^{k+1} \pmod{m}$, as desired.

57. Let $P(n)$ be the proposition

$$[(p_1\to p_2)\wedge(p_2\to p_3)\wedge\cdots\wedge(p_{n-1}\to p_n)]\to[(p_1\wedge p_2\wedge\cdots\wedge p_{n-1})\to p_n].$$

We want to prove this proposition for all $n\geq 2$. The basis step, $(p_1\to p_2)\to(p_1\to p_2)$, is clearly true (a tautology), since every proposition implies itself. Now we assume $P(n)$ and want to show $P(n+1)$, namely

$$\begin{aligned}&[(p_1\to p_2)\wedge(p_2\to p_3)\wedge\cdots\wedge(p_{n-1}\to p_n)\wedge(p_n\to p_{n+1})]\to\\ &\qquad[(p_1\wedge p_2\wedge\cdots\wedge p_{n-1}\wedge p_n)\to p_{n+1}].\end{aligned}$$

To show this, we will assume that the hypothesis (everything in the first square brackets) is true and show that the conclusion (the conditional statement in the second square brackets) is also true.

So we assume $(p_1\to p_2)\wedge(p_2\to p_3)\wedge\cdots\wedge(p_{n-1}\to p_n)\wedge(p_n\to p_{n+1})$. By the associativity of $\wedge$, we can group this as $((p_1\to p_2)\wedge(p_2\to p_3)\wedge\cdots\wedge(p_{n-1}\to p_n))\wedge(p_n\to p_{n+1})$. By the simplification rule, we can conclude that the first group, $(p_1\to p_2)\wedge(p_2\to p_3)\wedge\cdots\wedge(p_{n-1}\to p_n)$, must be true. Now the inductive hypothesis allows us to conclude that $(p_1\wedge p_2\wedge\cdots\wedge p_{n-1})\to p_n$. This together with the rest of the assumption, namely $p_n\to p_{n+1}$, yields, by the hypothetical syllogism rule, $(p_1\wedge p_2\wedge\cdots\wedge p_{n-1})\to p_{n+1}$.

That is almost what we wanted to prove, but not quite. We wanted to prove that $(p_1\wedge p_2\wedge\cdots\wedge p_{n-1}\wedge p_n)\to p_{n+1}$. In order to prove this, let us assume its hypothesis, $p_1\wedge p_2\wedge\cdots\wedge p_{n-1}\wedge p_n$. Again using the simplification rule we obtain $p_1\wedge p_2\wedge\cdots\wedge p_{n-1}$. Now by modus ponens with the proposition $(p_1\wedge p_2\wedge\cdots\wedge p_{n-1})\to p_{n+1}$, which we proved above, we obtain p_{n+1}. Thus we have proved $(p_1\wedge p_2\wedge\cdots\wedge p_{n-1}\wedge p_n)\to p_{n+1}$, as desired.

59. This exercise, as the double star indicates, is quite hard. The trick is to induct not on n itself, but rather on $\log_2 n$. In other words, we write $n=2^k$ and prove the statement by induction on k. This will prove the statement for every n that is a power of 2; a separate argument is needed to extend to the general case.

We take the basis step to be $k=1$ (the case $k=0$ is trivially true, as well), so that $n=2^1=2$. In this case the trick is to start with the true inequality $(\sqrt{a_1}-\sqrt{a_2})^2 \geq 0$. Expanding, we have $a_1 - 2\sqrt{a_1a_2}+a_2 \geq 0$, whence $(a_1+a_2)/2 \geq (a_1a_2)^{1/2}$, as desired. For the inductive step, we assume that the inequality holds for $n=2^k$ and prove that it also holds for $2n=2^{k+1}$. What we need to show, then, is that

$$\frac{a_1+a_2+\cdots+a_{2n}}{2n} \geq (a_1a_2\cdots a_{2n})^{1/(2n)}.$$

First we observe that

$$\frac{a_1+a_2+\cdots+a_{2n}}{2n} = \left(\frac{a_1+a_2+\cdots+a_n}{n} + \frac{a_{n+1}+a_{n+2}+\cdots+a_{2n}}{n}\right)\Big/2$$

and

$$(a_1a_2\cdots a_{2n})^{1/(2n)} = \left((a_1a_2\cdots a_n)^{1/n}(a_{n+1}a_{n+2}\cdots a_{2n})^{1/n}\right)^{1/2}.$$

Now to simplify notation, let $A(x,y,\ldots)$ denote the arithmetic mean and $G(x,y,\ldots)$ denote the geometric mean of the numbers $x,y,\ldots$. It is clear that if $x\leq x'$, $y \leq y'$, and so on, then $A(x,y,\ldots)\leq A(x',y',\ldots)$ and $G(x,y,\ldots)\leq G(x',y',\ldots)$. Now we have

$$\begin{aligned}
A(a_1,\ldots,a_{2n}) &= A\big(A(a_1,\ldots,a_n),A(a_{n+1},\ldots,a_{2n})\big) \quad \text{(by the first observation above)}\\
&\geq A\big(G(a_1,\ldots,a_n),G(a_{n+1},\ldots,a_{2n})\big) \quad \text{(by the inductive hypothesis)}\\
&\geq G\big(G(a_1,\ldots,a_n),G(a_{n+1},\ldots,a_{2n})\big) \quad \text{(this was the case } n=2)\\
&= G(a_1,\ldots,a_{2n}) \quad \text{(by the second observation above)}.
\end{aligned}$$

Having proved the inequality in the case in which n is a power of 2, we now turn to the case of n that is not a power of 2. Let m be the smallest power of 2 bigger than m. (For instance, if $n=25$, then $m=32$.) Denote the arithmetic mean $A(a_1,\ldots,a_n)$ by a, and set $a_{n+1}=a_{n+2}=\cdots=a_m$ all equal to a. One effect of this is that then $A(a_1,\ldots,a_m)=a$. Now we have

$$\left(\Big(\prod_{i=1}^n a_i\Big)a^{m-n}\right)^{1/m} \leq A(a_1,\ldots,a_m)$$

by the case we have already proved, since m is a power of 2. Using algebra on the left-hand side and the observation that $A(a_1,\ldots,a_m)=a$ on the right, we obtain

$$\Big(\prod_{i=1}^n a_i\Big)^{1/m} a^{1-n/m} \leq a$$

or

$$\Big(\prod_{i=1}^n a_i\Big)^{1/m} \leq a^{n/m}.$$

Finally we raise both sides to the power m/n to give

$$\Big(\prod_{i=1}^n a_i\Big)^{1/n} \leq a,$$

as desired.

61. Let us check the cases $n=1$ and $n=2$, both to establish the basis and to try to see what is going on. For $n=1$ there is only one nonempty subset of $\{1\}$, so the left-hand side is just $\frac{1}{1}$, and that equals 1. For $n=2$ there are three nonempty subsets: $\{1\}$, $\{2\}$, and $\{1,2\}$, so the left-hand side is $\frac{1}{1}+\frac{1}{2}+\frac{1}{1\cdot 2}=2$. To prove the inductive step, assume that the statement is true for n, and consider it for $n+1$. Now the set of the first $n+1$ positive integers has many nonempty subsets, but they fall into three categories: a nonempty subset of the first n positive integers, a nonempty subset of the first n positive integers together with $n+1$, or just $\{n+1\}$.

So we need to sum over these three categories. By the inductive hypothesis, the sum over the first category is n. For the second category, we can factor out $\frac{1}{n+1}$ from each term of the sum, note that the remaining factor again gives n by the inductive hypothesis, and so conclude that this part of the sum is $\frac{n}{n+1}$. Finally, the third category simply yields the value $\frac{1}{n+1}$. Therefore the entire summation is $n + \frac{n}{n+1} + \frac{1}{n+1} = n+1$, as desired.

63. The basis step ($n = 2$) is clear, because if $A_1 \subseteq A_2$, then A_1 satisfies the condition of being a subset of each set in the collection, and otherwise A_2 does, because in that case, A_2 must be a subset of A_1 (by the stated assumptions). For the inductive step, assume the inductive hypothesis, that the conditional statement is true for k sets, and suppose we are given $k+1$ sets that satisfy the given conditions. By the inductive hypothesis, there must be a set A_i for some $i \le k$ such that $A_i \subseteq A_j$ for $1 \le j \le k$. If $A_i \subseteq A_{k+1}$, then we are done. Otherwise, we know that $A_{k+1} \subseteq A_i$, and this tells us that A_{k+1} satisfies the condition of being a subset of A_j for $1 \le j \le k+1$.

65. Number the people 1, 2, 3, and 4, and let s_i be the scandal originally known only to person i. It is clear that $G(1) = 0$ and $G(2) = 1$. For three people, without loss of generality assume that 1 calls 2 first and 1 calls 3 next. At this point 1 and 3 know all three scandals, but it takes one more call to let 2 know s_3. Thus $G(3) = 3$. For four people, without loss of generality assume that 1 calls 2 first. If now 3 calls 4, then after two calls 1 and 2 both know s_1 and s_2, while 3 and 4 both know s_3 and s_4. It is clear that two more calls (between 1 and 3, and between 2 and 4, say) are necessary and sufficient to complete the exchange. This makes a total of four calls. The only other case to consider (to see whether $G(4)$ might be less than 4) is when the second call, without loss of generality, occurs between 1 and 3. At this point, both 2 and 4 still need to learn s_3, and talking to each other won't give them that information, so at least two more calls would be required. Thus $G(4) = 4$.

67. We need to show that $2n - 4$ calls are both necessary and sufficient to exchange all the gossip. Sufficiency is easier. Select four of the people, say 1, 2, 3, and 4, to be the central committee. Every person outside the central committee calls one person on the central committee. This takes $n - 4$ calls, and at this point the central committee members *as a group* know all the scandals. They then exchange information among themselves by making the calls 12, 34, 13, and 24 in that order (of course the first two can be done in parallel and the last two can be done in parallel). At this point, *every* central committee member knows all the scandals, and we have used $n - 4 + 4 = n$ calls. Finally, again every person outside the central committee calls one person on the central committee, at which point everyone knows all the scandals. This takes $n - 4$ more calls, for a total of $2n - 4$ calls.

That this cannot be done with fewer than $2n - 4$ calls is much harder to prove, and the proof will not be presented here. See the following website for details:

`www.cs.cornell.edu/vogels/Epidemics/gossips-telephones.pdf`

69. We prove this by mathematical induction. The basis step ($n = 2$) is true tautologically (if $I_1 \cap I_2 \ne \emptyset$ then $I_1 \cap I_2 \ne \emptyset$). The heart of the argument occurs with three sets, so we will give the proof for $n = 3$ explicitly. Recall the notation $(u, v) = \{x \mid u < x < v\}$. Suppose that the intervals are (a, b), (c, d), and (e, f), where without loss of generality we can assume that $a \le c \le e$. Because $(a, b) \cap (e, f) \ne \emptyset$, we must have $e < b$; for a similar reason, $e < d$. It follows that the number halfway between e and the smaller of b and d is common to all three intervals. Now for the inductive step, assume that whenever we have k intervals that have pairwise nonempty intersections then there is a point common to all the intervals, and suppose that we are given intervals $I_1, I_2, \ldots, I_{k+1}$ that have pairwise nonempty intersections. For each i from 1 to k, let $J_i = I_i \cap I_{k+1}$. We claim that the collection $J_1, J_2, \ldots, J_k$ satisfies the inductive hypothesis, that is, that $J_{i_1} \cap J_{i_2} \ne \emptyset$ for each choice of subscripts i_1 and i_2. This follows from the $n = 3$ case proved above, using the sets I_{i_1}, I_{i_2}, and I_{k+1}. We can now invoke the inductive hypothesis to conclude that there is a number common to all of the sets J_i for $i = 1, 2, \ldots, k$, which perforce is in the intersection of all the sets I_i for $i = 1, 2, \ldots, k+1$.

71. Pair up the people. Have the people stand at mutually distinct small distances from their partners but far away from everyone else. Then each person throws a pie at his or her partner, so everyone gets hit.

73. The proof in Example 13 guides us to one solution (it is certainly not unique). We begin by placing a right triomino in the center, with its gap in the same quadrant as the missing square in the upper left corner of the board (this piece is distinctively shaded in our solution below). This reduces the problem to four problems on 4×4 boards. Then we place triominoes in the centers of these four quadrants, using the same principle (shaded somewhat differently below). Finally, we place pieces in the remaining squares to fill up each quadrant.

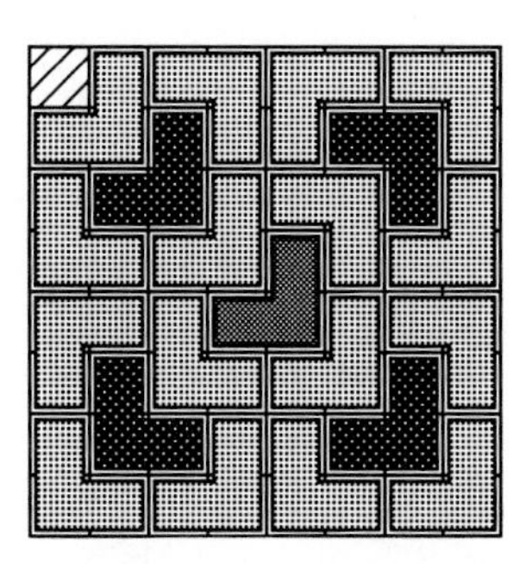

75. This problem is very similar to Example 13; the only difficulty is in visualizing what's happening in three dimensions. The basis step ($n = 1$) is trivial, since one tile coincides with the solid to be tiled. To make this read a little easier, let us call a $1 \times 1 \times 1$ cube a "cubie"; and let us call the object we are tiling with, namely the $2 \times 2 \times 2$ cube with one cubie removed, a tile. For the inductive step, assume the inductive hypothesis, that the $2^n \times 2^n \times 2^n$ cube with one cubie removed can be covered with tiles, and suppose that a $2^{n+1} \times 2^{n+1} \times 2^{n+1}$ cube with one cubie removed is given. We must show how to cover it with tiles. Think of this large object as split into eight octants through its center, by planes parallel to the faces. The missing cubie occurs in one of these octants. Now position one tile with its center at the center of the large object, so that the missing cubie in the tile lies in the octant in which the large object is missing its cubie. This creates eight $2^n \times 2^n \times 2^n$ cubes, each missing exactly one cubie—one in each octant. By the inductive hypothesis, we can fill each of these smaller objects with tiles. Putting these tilings together gives us the desired tiling of the $2^{n+1} \times 2^{n+1} \times 2^{n+1}$ cube with one cubie removed, as desired.

77. For this specific tiling, the most straightforward proof consists of producing the desired picture. It can be discovered by playing around with the tiles (either make a set, or use paper and pencil).

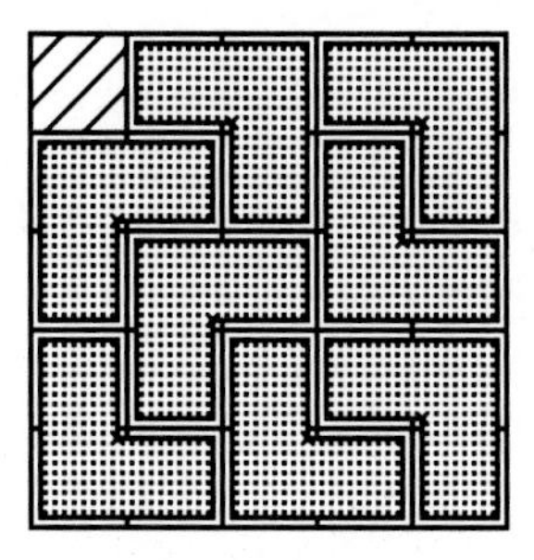

79. Let $Q(n)$ be $P(n+b-1)$. Thus $Q(1)$ is $P(b)$, $Q(2)$ is $P(b+1)$, and so on. Therefore the statement that $P(n)$ is true for $n = b, b+1, b+2, \ldots$ is the same as the statement that $Q(m)$ is true for all positive integers m.

We are given that $P(b)$ is true (i.e., that $Q(1)$ is true), and that $P(k) \to P(k+1)$ for all $k \geq b$ (i.e., that $Q(m) \to Q(m+1)$ for all positive integers m). Therefore by the principle of mathematical induction, $Q(m)$ is true for all positive integers m, as desired.

SECTION 4.2 Strong Induction and Well-Ordering

In this section we extend the technique of proof by mathematical induction by using a stronger inductive hypothesis. The inductive step is now to prove the proposition for $k+1$, assuming the inductive hypothesis that the proposition is true for all values less than or equal to k. Aside from that, the two methods are the same.

1. Let $P(n)$ be the statement that you can run n miles. We want to prove that $P(n)$ is true for all positive integers n. For the basis step we note that the given conditions tell us that $P(1)$ and $P(2)$ are true. For the inductive step, fix $k \geq 2$ and assume that $P(j)$ is true for all $j \leq k$. We want to show that $P(k+1)$ is true. Since $k \geq 2$, $k-1$ is a positive integer less than or equal to k, so by the inductive hypothesis, we know that $P(k-1)$ is true. That is, we know that you can run $k-1$ miles. We were told that "you can always run two more miles once you have run a specified number of miles," so we know that you can run $(k-1)+2 = k+1$ miles. This is $P(k+1)$.

Note that we didn't use strong induction exactly as stated on page 284. Instead, we considered both $n=1$ and $n=2$ as part of the basis step. We could have more formally included $n=2$ in the inductive step as a special case. Writing our proof this way, the basis step is just to note that we are told we can run one mile, so $P(1)$ is true. For the inductive step, if $k=1$ then we are already told that we can run two miles. If $k>1$, then the inductive hypothesis tells us that we can run $k-1$ miles, so we can run $(k-1)+2=k+1$ miles.

3. a) $P(8)$ is true, because we can form 8 cents of postage with one 3-cent stamp and one 5-cent stamp. $P(9)$ is true, because we can form 9 cents of postage with three 3-cent stamps. $P(10)$ is true, because we can form 10 cents of postage with two 5-cent stamps.

b) The inductive hypothesis is the statement that using just 3-cent and 5-cent stamps we can form j cents postage for all j with $8 \leq j \leq k$, where we assume that $k \geq 10$.

c) In the inductive step we must show, assuming the inductive hypothesis, that we can form $k+1$ cents postage using just 3-cent and 5-cent stamps.

d) We want to form $k+1$ cents of postage. Since $k \geq 10$, we know that $P(k-2)$ is true, that is, that we can form $k-2$ cents of postage. Put one more 3-cent stamp on the envelope, and we have formed $k+1$ cents of postage, as desired.

e) We have completed both the basis step and the inductive step, so by the principle of strong induction, the statement is true for every integer n greater than or equal to 8.

5. a) We can form the following amounts of postage as indicated: $4=4$, $8=4+4$, $11=11$, $12=4+4+4$, $15=11+4$, $16=4+4+4+4$, $19=11+4+4$, $20=4+4+4+4+4$, $22=11+11$, $23=11+4+4+4$, $24=4+4+4+4+4+4$, $26=11+11+4$, $27=11+4+4+4+4$, $28=4+4+4+4+4+4+4$, $30=11+11+4+4$, $31=11+4+4+4+4+4$, $32=4+4+4+4+4+4+4+4$, $33=11+11+11$. By having considered all the combinations, we know that the gaps in this list cannot be filled. We claim that we can form all amounts of postage greater than or equal to 30 cents using just 4-cent and 11-cent stamps.

b) Let $P(n)$ be the statement that we can form n cents of postage using just 4-cent and 11-cent stamps. We want to prove that $P(n)$ is true for all $n \geq 30$. The basis step, $n=30$, is handled above. Assume that we can form k cents of postage (the inductive hypothesis); we will show how to form $k+1$ cents of postage. If the k cents included an 11-cent stamp, then replace it by three 4-cent stamps $(3 \cdot 4 = 11+1)$. Otherwise,

k cents was formed from just 4-cent stamps. Because $k \geq 30$, there must be at least eight 4-cent stamps involved. Replace eight 4-cent stamps by three 11-cent stamps, and we have formed $k+1$ cents in postage $(3 \cdot 11 = 8 \cdot 4 + 1)$.

c) $P(n)$ is the same as in part **(b)**. To prove that $P(n)$ is true for all $n \geq 30$, we note for the basis step that from part **(a)**, $P(n)$ is true for $n = 30, 31, 32, 33$. Assume the inductive hypothesis, that $P(j)$ is true for all j with $30 \leq j \leq k$, where k is a fixed integer greater than or equal to 33. We want to show that $P(k+1)$ is true. Because $k - 3 \geq 30$, we know that $P(k-3)$ is true, that is, that we can form $k-3$ cents of postage. Put one more 4-cent stamp on the envelope, and we have formed $k+1$ cents of postage, as desired. In this proof our inductive hypothesis included all values between 30 and k inclusive, and that enabled us to jump back four steps to a value for which we knew how to form the desired postage.

7. We can form the following amounts of money as indicated: $2 = 2$, $4 = 2+2$, $5 = 5$, $6 = 2+2+2$. By having considered all the combinations, we know that the gaps in this list ($1 and $3) cannot be filled. We claim that we can form all amounts of money greater than or equal to 5 dollars. Let $P(n)$ be the statement that we can form n dollars using just 2-dollar and 5-dollar bills. We want to prove that $P(n)$ is true for all $n \geq 5$. We already observed that the basis step is true for $n = 5$ and 6. Assume the inductive hypothesis, that $P(j)$ is true for all j with $5 \leq j \leq k$, where k is a fixed integer greater than or equal to 6. We want to show that $P(k+1)$ is true. Because $k - 1 \geq 5$, we know that $P(k-1)$ is true, that is, that we can form $k-1$ dollars. Add another 2-dollar bill, and we have formed $k+1$ dollars, as desired.

9. Following the hint, we let $P(n)$ be the statement that there is no positive integer b such that $\sqrt{2} = n/b$. For the basis step, $P(1)$ is true because $\sqrt{2} > 1 \geq 1/b$ for all positive integers b. For the inductive step, assume that $P(j)$ is true for all $j \leq k$, where k is an arbitrary positive integer; we must prove that $P(k+1)$ is true. So assume the contrary, that $\sqrt{2} = (k+1)/b$ for some positive integer b. Squaring both sides and clearing fractions, we have $2b^2 = (k+1)^2$. This tells us that $(k+1)^2$ is even, and so $k+1$ is even as well (the square of an odd number is odd, by Example 1 in Section 1.6). Therefore we can write $k+1 = 2t$ for some positive integer t. Substituting, we have $2b^2 = 4t^2$, so $b^2 = 2t^2$. By the same reasoning as before, b is even, so $b = 2s$ for some positive integer s. Then we have $\sqrt{2} = (k+1)/b = (2t)/(2s) = t/s$. But $t \leq k$, so this contradicts the inductive hypothesis, and our proof of the inductive step is complete.

11. There are four base cases. If $n = 1 = 4 \cdot 0 + 1$, then clearly the first player is doomed, so the second player wins. If there are two, three, or four matches ($n = 4 \cdot 0 + 2$, $n = 4 \cdot 0 + 3$, or $n = 4 \cdot 1$), then the first player can win by removing all but one match. Now assume the strong inductive hypothesis, that in games with k or fewer matches, the first player can win if $k \equiv 0$, 2 or, 3 (mod 4) and the second player can win if $k \equiv 1 \pmod 4$. Suppose we have a game with $k+1$ matches, with $k \geq 4$. If $k+1 \equiv 0 \pmod 4$, then the first player can remove three matches, leaving $k-2$ matches for the other player. Since $k - 2 \equiv 1 \pmod 4$, by the inductive hypothesis, this is a game that the second player at that point (who is the first player in our game) can win. Similarly, if $k+1 \equiv 2 \pmod 4$, then the first player can remove one match, leaving k matches for the other player. Since $k \equiv 1 \pmod 4$, by the inductive hypothesis, this is a game that the second player at that point (who is the first player in our game) can win. And if $k+1 \equiv 3 \pmod 4$, then the first player can remove two matches, leaving $k-1$ matches for the other player. Since $k - 1 \equiv 1 \pmod 4$, by the inductive hypothesis, this is again a game that the second player at that point (who is the first player in our game) can win. Finally, if $k+1 \equiv 1 \pmod 4$, then the first player must leave k, $k-1$, or $k-2$ matches for the other player. Since $k \equiv 0 \pmod 4$, $k - 1 \equiv 3 \pmod 4$, and $k - 2 \equiv 2 \pmod 4$, by the inductive hypothesis, this is a game that the first player at that point (who is the second player in our game) can win. Thus the first player in our game is doomed, and the proof is complete.

13. Let $P(n)$ be the statement that exactly $n-1$ moves are required to assemble a puzzle with n pieces. Now $P(1)$ is trivially true. Assume that $P(j)$ is true for all $j < n$, and consider a puzzle with n pieces. The final move must be the joining of two blocks, of size k and $n-k$ for some integer k, $1 \leq k \leq n-1$. By the inductive hypothesis, it required $k-1$ moves to construct the one block, and $n-k-1$ moves to construct the other. Therefore $1+(k-1)+(n-k-1)=n-1$ moves are required in all, so $P(n)$ is true. Notice that for variety here we proved $P(n)$ under the assumption that $P(j)$ was true for $j<n$; so n played the role that $k+1$ plays in the statement of strong induction given in the text. It is worthwhile to understand how all of these forms are saying the same thing and to be comfortable moving between them.

15. Let the Chomp board have n rows and n columns. We claim that the first player can win the game by making the first move to leave just the top row and left-most column. (He does this by selecting the cookie in the second column of the second row.) Let $P(n)$ be the statement that if a player has presented his opponent with a Chomp configuration consisting of just n cookies in the top row and n cookies in the left-most column (both of these including the poisoned cookie in the upper left corner), then he can win the game. We will prove $\forall n P(n)$ by strong induction. We know that $P(1)$ is true, because the opponent is forced to take the poisoned cookie at his first turn. Fix $k \geq 1$ and assume that $P(j)$ is true for all $j \leq k$. We claim that $P(k+1)$ is true. It is the opponent's turn to move. If she picks the poisoned cookie, then the game is over and she loses. Otherwise, assume that she picks the cookie in the top row in column j, or the cookie in the left column in row j, for some j with $2 \leq j \leq k+1$. The first player now picks the cookie in the left column in row j, or the cookie in the top row in column j, respectively. This leaves the position covered by $P(j-1)$ for his opponent, so by the inductive hypothesis, he can win.

17. Let $P(n)$ be the statement that if a simple polygon with n sides is triangulated, then at least two of the triangles in the triangulation have two sides that border the exterior of the polygon. We will prove $\forall n \geq 4\, P(n)$. The statement is clearly true for $n=4$, because there is only one diagonal, leaving two triangles with the desired property. Fix $k \geq 4$ and assume that $P(j)$ is true for all j with $4 \leq j \leq k$. Consider a polygon with $k+1$ sides, and some triangulation of it. Pick one of the diagonals in this triangulation. First suppose that this diagonal divides the polygon into one triangle and one polygon with k sides. Then the triangle has two sides that border the exterior. Furthermore, the k-gon has, by the inductive hypothesis, two triangles that have two sides that border the exterior of that k-gon, and only one of these triangles can fail to be a triangle that has two sides that border the exterior of the original polygon. The only other case is that this diagonal divides the polygon into two polygons with j sides and $k+3-j$ sides for some j with $4 \leq j \leq k-1$. By the inductive hypothesis, each of these two polygons has two triangles that have two sides that border their exterior, and in each case only one of these triangles can fail to be a triangle that has two sides that border the exterior of the original polygon.

19. Let $P(n)$ be the statement that the area of a simple polygon with n sides and vertices all at lattice points is given by $I+B/2-1$, where I and B are as defined in the exercise. We will prove $\forall n \geq 3\, P(n)$. We begin by proving an additivity lemma. If P is a simple polygon with all vertices at the lattice points, divided into polygons P_1 and P_2 by a diagonal, then $I(P)+B(P)/2-1=(I(P_1)+B(P_1)/2-1)+(I(P_2)+B(P_2)/2-1)$. To see this, suppose there are k lattice points on the diagonal, not counting its endpoints. Then $I(P)=I(P_1)+I(P_2)+k$ and $B(P)=B(P_1)+B(P_2)-2k-2$; and the result follows by simple algebra. What this says in particular is that if Pick's formula gives the correct area for P_1 and P_2, then it must give the correct formula for P, whose area is the sum of the areas for P_1 and P_2; and similarly if Pick's formula gives the correct area for P and one of the P_i's, then it must give the correct formula for the other P_i.

Next we prove the theorem for rectangles whose sides are parallel to the coordinate axes. Such a rectangle necessarily has vertices at (a,b), (a,c), (d,b), and (d,c), where a, b, c, and d are integers with $b<c$ and $a<d$. Its area is clearly $(c-b)(d-a)$. By looking at the perimeter, we see that it has $B=2(c-b+d-a)$, and we see also that it has $I=(c-b-1)(d-a-1)=(c-b)(d-a)-(c-b)-(d-a)+1$. Therefore

$I + B/2 - 1 = (c-b)(d-a) - (c-b) - (d-a) + 1 + (c-b+d-a) - 1 = (c-b)(d-a)$, which is the desired area.

Next consider a right triangle whose legs are parallel to the coordinate axes. This triangle is half a rectangle of the type just considered, for which Pick's formula holds, so by the additivity lemma, it holds for the triangle as well. (The values of B and I are the same for each of the two triangles, so if Pick's formula gave an answer that was either too small or too large, then it would give a correspondingly wrong answer for the rectangle.)

For the next step, consider an arbitrary triangle with vertices at the lattice points that is not of the type already considered. Embed it in as small a rectangle as possible. There are several possible ways this can happen, but in any case (and adding one more edge in one case), the rectangle will have been partitioned into the given triangle and two or three right triangles with sides parallel to the coordinate axes. See the figure for a typical case. Again by the additivity lemma, we are guaranteed that Pick's formula gives the correct area for the central triangle.

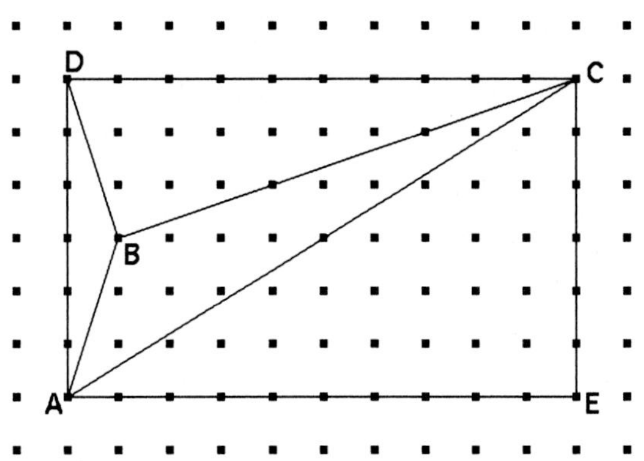

Note that we have now proved $P(3)$, the basis step in our strong induction proof. For the inductive step, given an arbitrary polygon, use Lemma 1 in the text to split it into two polygons. Then by the additivity lemma above and the inductive hypothesis, we know that Pick's formula gives the correct area for this polygon.

Here are some good websites for more details:

`http://planetmath.org/encyclopedia/ProofOfPicksTheorem.html`
`http://www.cut-the-knot.org/ctk/Pick_proof.shtml`
`http://mathforum.org/library/drmath/view/65670.html`

21. **a)** Use the left figure. Angle abp is smallest for p, but the segment bp is not an interior diagonal.

b) Use the right figure. The vertex other than b with smallest x-coordinate is d, but the segment bd is not an interior diagonal.

c) Use the right figure. The vertex closest to b is d, but the segment bd is not an interior diagonal.

23. **a)** When we try to prove the inductive step and find a triangle in each subpolygon with at least two sides bordering the exterior, it may happen in each case that the triangle we are guaranteed in fact borders the diagonal (which is part of the boundary of that polygon). This leaves us with no triangles guaranteed to touch the boundary of the *original* polygon.

b) We proved $\forall n \geq 4\, T(n)$ in Exercise 17. Since we can always find two triangles that satisfy the property, perforce, at least one triangle does. Thus we have proved $\forall n \geq 4\, E(n)$.

25. **a)** The inductive step here allows us to conclude that $P(3)$, $P(5)$, ... are all true, but we can conclude nothing about $P(2)$, $P(4)$, ...

b) We can conclude that $P(n)$ is true for all positive integers n, using strong induction.

c) The inductive step here allows us to conclude that $P(2)$, $P(4)$, $P(8)$, $P(16)$, ... are all true, but we can conclude nothing about $P(n)$ when n is not a power of 2.

d) This is mathematical induction; we can conclude that $P(n)$ is true for all positive integers n.

27. Suppose, for a proof by contradiction, that there is some positive integer n such that $P(n)$ is not true. Let m be the smallest positive integer greater than n for which $P(m)$ *is* true; we know that such an m exists because $P(m)$ is true for infinitely many values of m, and therefore true for more than just $1, 2, \ldots, n-1$. But we are given that $P(m) \to P(m-1)$, so $P(m-1)$ is true. Thus $m-1$ cannot be greater than n, so $m-1=n$ and $P(n)$ is in fact true. This contradiction shows that $P(n)$ is true for all n.

29. The error is in going from the basis step $n=0$ to the next value, $n=1$. We cannot write 1 as the sum of two smaller natural numbers, so we cannot invoke the inductive hypothesis. In the notation of the "proof," when $k=0$, we cannot write $0+1=i+j$ where $0 \le i \le 0$ and $0 \le j \le 0$.

31. To show that strong induction is valid, let us suppose that we have a proposition $\forall n P(n)$ which has been proved using it. We must show that in fact $\forall n P(n)$ is true (to say that a principle of proof is valid means that it proves only true propositions). Let S be the set of counterexamples, i.e., $S = \{\, n \mid \neg P(n) \,\}$. We want to show that $S = \emptyset$. We argue by contradiction. Assume that $S \ne \emptyset$. Then by the well-ordering property, S has a smallest element. Since part of the method of strong induction is to show that $P(1)$ is true, this smallest counterexample must be greater than 1. Let us call it $k+1$. Since $k+1$ is the smallest element of S, it must be the case that $P(1) \wedge P(2) \wedge \cdots \wedge P(k)$ is true. But the rest of the proof using strong induction involved showing that $P(1) \wedge P(2) \wedge \cdots \wedge P(k)$ implied $P(k+1)$; therefore since the hypothesis is true, the conclusion must be true as well, i.e., $P(k+1)$ is true. This contradicts our assumption that $k+1 \in S$. Therefore we conclude that $S = \emptyset$, so $\forall n P(n)$ is true.

33. In each case we will argue on the basis of a "smallest counterexample."

a) Suppose that there is a counterexample, that is, that there are values of n and k such that $P(n,k)$ is not true. Choose a counterexample with $n+k$ as small as possible. We cannot have $n=1$ and $k=1$, because we are given that $P(1,1)$ is true. Therefore either $n>1$ or $k>1$. In the former case, by our choice of counterexample, we know that $P(n-1,k)$ is true. But the inductive step then forces $P(n,k)$ to be true, a contradiction. The latter case is similar. So our supposition that there is a counterexample must be wrong, and $P(n,k)$ is true in all cases.

b) Suppose that there is a counterexample, that is, that there are values of n and k such that $P(n,k)$ is not true. Choose a counterexample with n as small as possible. We cannot have $n=1$, because we are given that $P(1,k)$ is true for all k. Therefore $n>1$. By our choice of counterexample, we know that $P(n-1,k)$ is true. But the inductive step then forces $P(n,k)$ to be true, a contradiction. So our supposition that there is a counterexample must be wrong, and $P(n,k)$ is true in all cases.

c) Suppose that there is a counterexample, that is, that there are values of n and k such that $P(n,k)$ is not true. Choose a counterexample with k as small as possible. We cannot have $k=1$, because we are given that $P(n,1)$ is true for all n. Therefore $k>1$. By our choice of counterexample, we know that $P(n,k-1)$ is true. But the inductive step then forces $P(n,k)$ to be true, a contradiction. So our supposition that there is a counterexample must be wrong, and $P(n,k)$ is true in all cases.

35. We want to calculate the product $a_1 a_2 \cdots a_n$ by inserting parentheses to express the calculation as a sequence of multiplications of two quantities. For example, we can insert parentheses into $a_1a_2a_3a_4a_5$ to render it $(a_1(a_2a_3))(a_4a_5)$, and then the four multiplications are $a_2 \cdot a_3$, $a_4 \cdot a_5$, $a_1 \cdot (a_2a_3)$, and finally the product of these last two quantities. We must show that no matter how the parentheses are inserted, $n-1$ multiplications will be required. If $n=1$, then clearly 0 multiplications are required, so the basis step is trivial. Now assume the strong inductive hypothesis, that for all $k < n$, no matter how parentheses are inserted into the product of k numbers, $k-1$ multiplications are required to compute the answer. Consider a parenthesized product of a_1 through a_n, and look at the last multiplication. Thus we have $(a_1a_2\cdots a_r)\cdot(a_{r+1}\cdots a_n)$, where we do not care how the parentheses are distributed within the pairs shown here. By the inductive hypothesis, it requires $r-1$ multiplications to obtain the first product in parentheses and $n-r-1$ to obtain the second (that second product has $n-r$ factors). Furthermore, 1 additional multiplication is needed to multiply these two answers together. This gives a total of $(r-1)+(n-r-1)+1 = n-1$ multiplications for the given problem, exactly what we needed to show.

37. Suppose that we have two such pairs, say (q,r) and (q',r'), so that $a = dq+r = dq'+r'$, with $0 \le r,r' < d$. We will show that the pairs are really the same, that is, that $q = q'$ and $r = r'$. From $dq+r = dq'+r'$ we obtain $d(q-q') = r'-r$. Therefore $d\,|(r'-r)$. But $|r'-r| < d$ (since both r' and r are nonnegative integers less than d). The only multiple of d in that range is 0, so we are forced to conclude that $r' = r$. Then it easily follows that $q = q'$ as well, since $q = (a-r)/d = (a-r')/d = q'$.

39. This problem deals with a paradox caused by self-reference. First of all, the answer to the question is clearly "no," because there are a finite number of English words, and so only a finite number of strings of fifteen words or fewer; therefore only a finite number of positive integers can be so described, not all of them. On the other hand, we might offer the following "proof" that every positive integer *can* be so expressed. Clearly 1 can be so expressed (e.g., "one" or "the cardinality of the power set of the empty set"). By the well-ordering principle, if there is a positive integer that cannot be expressed in fifteen words or fewer, then there is a smallest such, say s. Then the phrase "the smallest positive integer that cannot be described using no more than fifteen English words" is a description of s using no more than fifteen English words, a contradiction. Therefore no such s exists, and we seem to have proved that every positive integer can be so expressed, in obvious violation to common sense (and the argument presented above). Paradoxes like this are likely to arise whenever we try to use language to talk about itself; the use of language in this way, while seeming to be meaningful, is in fact nonsense.

41. We will prove this by contradiction. Suppose that the well-ordering property were false. Let S be a counterexample: a nonempty set of nonnegative integers that contains no smallest element. Let $P(n)$ be the statement "$i \notin S$ for all $i \le n$." We will show that $P(n)$ is true for all n (which will contradict the assertion that S is nonempty). Now $P(0)$ must be true, because if $0 \in S$ then clearly S would have a smallest element, namely 0. Suppose now that $P(n)$ is true, so that $i \notin S$ for $i = 0,1,\ldots,n$. We must show that $P(n+1)$ is true, which amounts to showing that $n+1 \notin S$. If $n+1 \in S$, then $n+1$ would be the smallest element of S, and this would contradict our assumption. Therefore $n+1 \notin S$. Thus we have shown by the principle of mathematical induction that $P(n)$ is true for all n, which means that there can be no elements of S. This contradicts our assumption that $S \ne \emptyset$, and our proof by contradiction is complete.

43. Strong induction clearly implies the principle of mathematical induction, for if one has shown that $P(k) \to P(k+1)$ is true, then one has also shown that $[P(1)\wedge\cdots\wedge P(k)] \to P(k+1)$ is true. By Exercise 41, the principle of mathematical induction implies the well-ordering property. Therefore by assuming either the principle of mathematical induction or strong induction as an axiom, we can prove the well-ordering property.

SECTION 4.3 Recursive Definitions and Structural Induction

The best way to approach a recursive definition is first to compute several instances. For example, if you are given a recursive definition of a function f, then compute $f(0)$ through $f(8)$ to get a feeling for what is happening. Most of the time it is necessary to prove statements about recursively defined objects using structural induction (or mathematical induction or strong induction), and the induction practically takes care of itself, mimicking the recursive definition.

1. In each case, we compute $f(1)$ by using the recursive part of the definition with $n = 0$, together with the given fact that $f(0) = 1$. Then we compute $f(2)$ by using the recursive part of the definition with $n = 1$, together with the given value of $f(1)$. We continue in this way to obtain $f(3)$ and $f(4)$.

a) $f(1) = f(0)+2 = 1+2 = 3$; $f(2) = f(1)+2 = 3+2 = 5$; $f(3) = f(2)+2 = 5+2 = 7$; $f(4) = f(3)+2 = 7+2 = 9$

b) $f(1) = 3f(0) = 3\cdot 1 = 3$; $f(2) = 3f(1) = 3\cdot 3 = 9$; $f(3) = 3f(2) = 3\cdot 9 = 27$; $f(4) = 3f(3) = 3\cdot 27 = 81$

c) $f(1) = 2^{f(0)} = 2^1 = 2$; $f(2) = 2^{f(1)} = 2^2 = 4$; $f(3) = 2^{f(2)} = 2^4 = 16$; $f(4) = 2^{f(3)} = 2^{16} = 65{,}536$

d) $f(1) = f(0)^2+f(0)+1 = 1^2+1+1 = 3$; $f(2) = f(1)^2+f(1)+1 = 3^2+3+1 = 13$; $f(3) = f(2)^2+f(2)+1 = 13^2+13+1 = 183$; $f(4) = f(3)^2+f(3)+1 = 183^2+183+1 = 33{,}673$

3. In each case we compute the subsequent terms by plugging into the recursive formula, using the previously given or computed values.

a) $f(2) = f(1)+3f(0) = 2+3(-1) = -1$; $f(3) = f(2)+3f(1) = -1+3\cdot 2 = 5$; $f(4) = f(3)+3f(2) = 5+3(-1) = 2$; $f(5) = f(4)+3f(3) = 2+3\cdot 5 = 17$

b) $f(2) = f(1)^2 f(0) = 2^2\cdot(-1) = -4$; $f(3) = f(2)^2 f(1) = (-4)^2\cdot 2 = 32$; $f(4) = f(3)^2 f(2) = 32^2\cdot(-4) = -4096$; $f(5) = f(4)^2 f(3) = (-4096)^2\cdot 32 = 536{,}870{,}912$

c) $f(2) = 3f(1)^2 - 4f(0)^2 = 3\cdot 2^2 - 4\cdot(-1)^2 = 8$; $f(3) = 3f(2)^2 - 4f(1)^2 = 3\cdot 8^2 - 4\cdot 2^2 = 176$; $f(4) = 3f(3)^2 - 4f(2)^2 = 3\cdot 176^2 - 4\cdot 8^2 = 92{,}672$; $f(5) = 3f(4)^2 - 4f(3)^2 = 3\cdot 92672^2 - 4\cdot 176^2 = 25{,}764{,}174{,}848$

d) $f(2) = f(0)/f(1) = (-1)/2 = -1/2$; $f(3) = f(1)/f(2) = 2/(-\frac{1}{2}) = -4$; $f(4) = f(2)/f(3) = (-\frac{1}{2})/(-4) = 1/8$; $f(5) = f(3)/f(4) = (-4)/\frac{1}{8} = -32$

5. a) This is not valid, since letting $n = 1$ we would have $f(1) = 2f(-1)$, but $f(-1)$ is not defined.

b) This is valid. The basis step tells us what $f(0)$ is, and the recursive step tells us how each subsequent value is determined from the one before. It is not hard to look at the pattern and conjecture that $f(n) = 1-n$. We prove this by induction. The basis step is $f(0) = 1 = 1-0$; and if $f(k) = 1-k$, then $f(k+1) = f(k)-1 = 1-k-1 = 1-(k+1)$.

c) The basis conditions specify $f(0)$ and $f(1)$, and the recursive step gives $f(n)$ in terms of $f(n-1)$ for $n \ge 2$, so this is a valid definition. If we compute the first several values, we conjecture that $f(n) = 4-n$ if $n > 0$, but $f(0) = 2$. That is our "formula." To prove it correct by induction we need two basis steps: $f(0) = 2$, and $f(1) = 3 = 4-1$. For the inductive step (with $k \ge 1$), $f(k+1) = f(k)-1 = (4-k)-1 = 4-(k+1)$.

d) The basis conditions specify $f(0)$ and $f(1)$, and the recursive step gives $f(n)$ in terms of $f(n-2)$ for $n \ge 2$, so this is a valid definition. The sequence of function values is 1, 2, 2, 4, 4, 8, 8, $\ldots$, and we can fit a formula to this if we use the floor function: $f(n) = 2^{\lfloor (n+1)/2 \rfloor}$. For a proof, we check the base cases: $f(0) = 1 = 2^{\lfloor (0+1)/2 \rfloor}$ and $f(1) = 2 = 2^{\lfloor (1+1)/2 \rfloor}$. For the inductive step: $f(k+1) = 2f(k-1) = 2\cdot 2^{\lfloor k/2 \rfloor} = 2^{\lfloor k/2 \rfloor + 1} = 2^{\lfloor ((k+1)+1)/2 \rfloor}$.

e) The definition tells us explicitly what $f(0)$ is. The recursive step specifies $f(1)$, $f(3)$, $\ldots$ in terms of $f(0)$, $f(2)$, $\ldots$; and it also gives $f(2)$, $f(4)$, $\ldots$ in terms of $f(0)$, $f(2)$, $\ldots$. So the definition is valid. We compute that $f(1) = 3$, $f(2) = 9$, $f(3) = 27$, and so conjecture that $f(n) = 3^n$. The basis step of the inductive proof is clear. For odd n greater than 0 we have $f(n) = 3f(n-1) = 3\cdot 3^{n-1} = 3^n$, and for even n greater than 1 we have $f(n) = 9f(n-2) = 9\cdot 3^{n-2} = 3^n$. Note that we used a slightly different notation here, letting n be the new value, rather than $k+1$, but the logic is the same.

7. There are many correct answers for these sequences. We will give what we consider to be the simplest ones.

a) Clearly each term in this sequence is 6 greater than the preceding term. Thus we can define the sequence by setting $a_1 = 6$ and declaring that $a_{n+1} = a_n + 6$ for all $n \geq 1$.

b) This is just like part **(a)**, in that each term is 2 more than its predecessor. Thus we have $a_1 = 3$ and $a_{n+1} = a_n + 2$ for all $n \geq 1$.

c) Each term is 10 times its predecessor. Thus we have $a_1 = 10$ and $a_{n+1} = 10a_n$ for all $n \geq 1$.

d) Just set $a_1 = 5$ and declare that $a_{n+1} = a_n$ for all $n \geq 1$.

9. We need to write $F(n+1)$ in terms of $F(n)$. Since $F(n)$ is the sum of the first n positive integers (namely 1 through n), and $F(n+1)$ is the sum of the first $n+1$ positive integers (namely 1 through $n+1$), we can obtain $F(n+1)$ from $F(n)$ by adding $n+1$. Therefore the recursive part of the definition is $F(n+1) = F(n)+n+1$. The initial condition is a specification of the value of $F(0)$; the sum of no positive integers is clearly 0, so we set $F(0) = 0$. (Alternately, if we assume that the argument for F is intended to be strictly positive, then we set $F(1) = 1$, since the sum of the first one positive integer is 1.)

11. We need to see how $P_m(n+1)$ relates to $P_m(n)$. Now $P_m(n+1) = m(n+1) = mn + m = P_m(n) + m$. Thus the recursive part of our definition is just $P_m(n+1) = P_m(n) + m$. The basis step is $P_m(0) = 0$, since $m \cdot 0 = 0$, no matter what value m has.

13. We prove this using the principle of mathematical induction. The base case is $n = 1$, and in that case the statement to be proved is just $f_1 = f_2$; this is true since both values are 1. Next we assume the inductive hypothesis, that

$$f_1 + f_3 + \cdots + f_{2n-1} = f_{2n} \,,$$

and try to prove the corresponding statement for $n+1$, namely,

$$f_1 + f_3 + \cdots + f_{2n-1} + f_{2n+1} = f_{2n+2} \,.$$

We have

$$\begin{aligned} f_1 + f_3 + \cdots + f_{2n-1} + f_{2n+1} &= f_{2n} + f_{2n+1} \quad \text{(by the inductive hypothesis)} \\ &= f_{2n+2} \quad \text{(by the definition of the Fibonacci numbers).} \end{aligned}$$

15. We prove this using the principle of mathematical induction. The basis step is for $n = 1$, and in that case the statement to be proved is just $f_0 f_1 + f_1 f_2 = f_2^2$; this is true since $0 \cdot 1 + 1 \cdot 1 = 1^2$. Next we assume the inductive hypothesis, that

$$f_0 f_1 + f_1 f_2 + \cdots + f_{2n-1} f_{2n} = f_{2n}^2 \,,$$

and try to prove the corresponding statement for $n+1$, namely,

$$f_0 f_1 + f_1 f_2 + \cdots + f_{2n-1} f_{2n} + f_{2n} f_{2n+1} + f_{2n+1} f_{2n+2} = f_{2n+2}^2 \,.$$

Note that *two* extra terms were added, since the final subscript has to be even. We have

$$\begin{aligned} f_0 f_1 + f_1 f_2 + \cdots + f_{2n-1} f_{2n} + f_{2n} f_{2n+1} + f_{2n+1} f_{2n+2} &= f_{2n}^2 + f_{2n} f_{2n+1} + f_{2n+1} f_{2n+2} \\ &\qquad \text{(by the inductive hypothesis)} \\ &= f_{2n}(f_{2n} + f_{2n+1}) + f_{2n+1} f_{2n+2} \\ &\qquad \text{(by factoring)} \\ &= f_{2n} f_{2n+2} + f_{2n+1} f_{2n+2} \\ &\qquad \text{(by the definition of the Fibonacci numbers)} \\ &= (f_{2n} + f_{2n+1}) f_{2n+2} \\ &= f_{2n+2} f_{2n+2} = f_{2n+2}^2. \end{aligned}$$

17. Let d_n be the number of divisions used by Algorithm 6 in Section 3.6 (the Euclidean algorithm) to find $\gcd(f_{n+1}, f_n)$. We write the calculation in this order, since $f_{n+1} \geq f_n$. We begin by finding the values of d_n for the first few values of n, in order to find a pattern and make a conjecture as to what the answer is. For $n = 0$ we are computing $\gcd(f_1, f_0) = \gcd(1, 0)$. Without performing any divisions, we know immediately that the answer is 1, so $d_0 = 0$. For $n = 1$ we are computing $\gcd(f_2, f_1) = \gcd(1, 1)$. One division is used to show that $\gcd(1, 1) = \gcd(1, 0)$, so $d_1 = 1$. For $n = 2$ we are computing $\gcd(f_3, f_2) = \gcd(2, 1)$. One division is used to show that $\gcd(2, 1) = \gcd(1, 0)$, so $d_2 = 1$. For $n = 3$, the computation gives successively $\gcd(f_4, f_3) = \gcd(3, 2) = \gcd(2, 1) = \gcd(1, 0)$, for a total of 2 divisions; thus $d_3 = 2$. For $n = 4$, we have $\gcd(f_5, f_4) = \gcd(5, 3) = \gcd(3, 2) = \gcd(2, 1) = \gcd(1, 0)$, for a total of 3 divisions; thus $d_4 = 3$. At this point we see that each increase of 1 in n seems to add one more division, in order to reduce $\gcd(f_{n+1}, f_n)$ to $\gcd(f_n, f_{n-1})$. Perhaps, then, for $n \geq 2$, we have $d_n = n - 1$. Let us make that conjecture. We have already verified the basis step when we computed that $d_2 = 1$. Now assume the inductive hypothesis, that $d_n = n - 1$. We must show that $d_{n+1} = n$. Now d_{n+1} is the number of divisions used in computing $\gcd(f_{n+2}, f_{n+1})$. The first step in the algorithm is to divide f_{n+1} into f_{n+2}. Since $f_{n+2} = f_{n+1} + f_n$ (this is the key point) and $f_n < f_{n+1}$, we get a quotient of 1 and a remainder of f_n. Thus we have, after one division, $\gcd(f_{n+2}, f_{n+1}) = \gcd(f_{n+1}, f_n)$. Now by the inductive hypothesis we need exactly $d_n = n - 1$ more divisions, since the algorithm proceeds from this point exactly as it proceeded given the inputs for the case of n. Therefore $1 + (n - 1) = n$ divisions are used in all, and our proof is complete. The answer, then, is that $d_0 = 0$, $d_1 = 1$, and $d_n = n - 1$ for $n \geq 2$. (If we interpreted the problem as insisting that we compute $\gcd(f_n, f_{n+1})$, with that order of the arguments, then the analysis and the answer are slightly different: $d_0 = 1$, and $d_n = n$ for $n \geq 1$.)

19. The determinant of the matrix $\mathbf{A} = \begin{bmatrix} a & b \\ c & d \end{bmatrix}$, written $|\mathbf{A}|$, is by definition $ad - bc$; and the determinant has the multiplicative property that $|\mathbf{AB}| = |\mathbf{A}||\mathbf{B}|$. Therefore the determinant of the matrix $\mathbf{A} = \begin{bmatrix} 1 & 1 \\ 1 & 0 \end{bmatrix}$ in Exercise 16 is $1 \cdot 0 - 1 \cdot 1 = -1$, and $|\mathbf{A}^n| = |\mathbf{A}|^n = (-1)^n$. On the other hand, the determinant of the matrix $\begin{bmatrix} f_{n+1} & f_n \\ f_n & f_{n-1} \end{bmatrix}$ is by definition $f_{n+1}f_{n-1} - f_n^2$. In Exercise 18 we showed that $\mathbf{A}^n$ is this latter matrix. The identity in Exercise 14 follows.

21. Assume that the definitions given in Exercise 20 were as follows: the max or min of one number is itself; $\max(a_1, a_2) = a_1$ if $a_1 \geq a_2$ and a_2 if $a_1 < a_2$, whereas $\min(a_1, a_2) = a_2$ if $a_1 \geq a_2$ and a_1 if $a_1 < a_2$; and for $n \geq 2$,

$$\max(a_1, a_2, \ldots, a_{n+1}) = \max(\max(a_1, a_2, \ldots, a_n), a_{n+1})$$

and

$$\min(a_1, a_2, \ldots, a_{n+1}) = \min(\min(a_1, a_2, \ldots, a_n), a_{n+1}).$$

We can then prove the three statements here by induction on n.

a) For $n = 1$, both sides of the equation equal $-a_1$. For $n = 2$, we must show that $\max(-a_1, -a_2) = -\min(a_1, a_2)$. There are two cases, depending on the relationship between a_1 and a_2. If $a_1 \leq a_2$, then $-a_1 \geq -a_2$, so by our definition, $\max(-a_1, -a_2) = -a_1$. On the other hand our definition implies that $\min(a_1, a_2) = a_1$ in this case. Therefore $\max(-a_1, -a_2) = -a_1 = -\min(a_1, a_2)$. The other case, $a_1 > a_2$, is similar: $\max(-a_1, -a_2) = -a_2 = -\min(a_1, a_2)$. Now we are ready for the inductive step. Assume the inductive hypothesis, that

$$\max(-a_1, -a_2, \ldots, -a_n) = -\min(a_1, a_2, \ldots, a_n).$$

We need to show the corresponding equality for $n+1$. We have

$$\begin{aligned}
&\max(-a_1, -a_2, \ldots, -a_n, -a_{n+1}) \\
&\quad = \max(\max(-a_1, -a_2, \ldots, -a_n), -a_{n+1}) \quad \text{(by definition)} \\
&\quad = \max(-\min(a_1, a_2, \ldots, a_n), -a_{n+1}) \quad \text{(by the inductive hypothesis)} \\
&\quad = -\min(\min(a_1, a_2, \ldots, a_n), a_{n+1}) \quad \text{(by the already proved case } n=2\text{)} \\
&\quad = -\min(a_1, a_2, \ldots, a_n, a_{n+1}) \quad \text{(by definition)}.
\end{aligned}$$

b) For $n = 1$, the equation is simply the identity $a_1 + b_1 = a_1 + b_1$. For $n = 2$, the situation is a little messy. Let us consider first the case that $a_1 + b_1 \geq a_2 + b_2$. Then $\max(a_1 + b_1, a_2 + b_2) = a_1 + b_1$. Also note that $a_1 \leq \max(a_1, b_1)$, and $b_1 \leq \max(b_1, b_2)$, so that $a_1 + b_1 \leq \max(a_1, a_2) + \max(b_1, b_2)$. Therefore we have $\max(a_1 + b_1, a_2 + b_2) = a_1 + b_1 \leq \max(a_1, a_2) + \max(b_1, b_2)$. The other case, in which $a_1 + b_1 < a_2 + b_2$, is similar. Now for the inductive step, we first need a lemma: if $u \leq v$, then $\max(u, w) \leq \max(v, w)$; this is easy to prove by looking at the three cases determined by the size of w relative to the sizes of u and v. Now assuming the inductive hypothesis, we have

$$\begin{aligned}
&\max(a_1 + b_1, a_2 + b_2, \ldots, a_n + b_n, a_{n+1} + b_{n+1}) \\
&\quad = \max(\max(a_1 + b_1, a_2 + b_2, \ldots, a_n + b_n), a_{n+1} + b_{n+1}) \quad \text{(by definition)} \\
&\quad \leq \max(\max(a_1, a_2, \ldots, a_n) + \max(b_1, b_2, \ldots, b_n), a_{n+1} + b_{n+1}) \\
&\qquad \text{(by the inductive hypothesis and the lemma)} \\
&\quad \leq \max(\max(a_1, a_2, \ldots, a_n), a_{n+1}) + \max(\max(b_1, b_2, \ldots, b_n), b_{n+1}) \\
&\qquad \text{(by the already proved case } n=2\text{)} \\
&\quad = \max(a_1, a_2, \ldots, a_n, a_{n+1}) + \max(b_1, b_2, \ldots, b_n, b_{n+1}) \quad \text{(by definition)}.
\end{aligned}$$

c) The proof here is exactly dual to the proof in part **(b)**. We replace every occurrence of "max" by "min," and invert each inequality. The proof then reads as follows. For $n = 1$, the equation is simply the identity $a_1 + b_1 = a_1 + b_1$. For $n = 2$, the situation is a little messy. Let us consider first the case that $a_1 + b_1 \leq a_2 + b_2$. Then $\min(a_1 + b_1, a_2 + b_2) = a_1 + b_1$. Also note that $a_1 \geq \min(a_1, a_2)$, and $b_1 \geq \min(b_1, b_2)$, so that $a_1 + b_1 \geq \min(a_1, a_2) + \min(b_1, b_2)$. Therefore we have $\min(a_1 + b_1, a_2 + b_2) = a_1 + b_1 \geq \min(a_1, a_2) + \min(b_1, b_2)$. The other case, in which $a_1 + b_1 > a_2 + b_2$, is similar. Now for the inductive step, we first need a lemma: if $u \geq v$, then $\min(u, w) \geq \min(v, w)$; this is easy to prove by looking at the three cases determined by the size of w relative to the sizes of u and v. Now assuming the inductive hypothesis, we have

$$\begin{aligned}
&\min(a_1 + b_1, a_2 + b_2, \ldots, a_n + b_n, a_{n+1} + b_{n+1}) \\
&\quad = \min(\min(a_1 + b_1, a_2 + b_2, \ldots, a_n + b_n), a_{n+1} + b_{n+1}) \quad \text{(by definition)} \\
&\quad \geq \min(\min(a_1, a_2, \ldots, a_n) + \min(b_1, b_2, \ldots, b_n), a_{n+1} + b_{n+1}) \\
&\qquad \text{(by the inductive hypothesis and the lemma)} \\
&\quad \geq \min(\min(a_1, a_2, \ldots, a_n), a_{n+1}) + \min(\min(b_1, b_2, \ldots, b_n), b_{n+1}) \\
&\qquad \text{(by the already proved case } n=2\text{)} \\
&\quad = \min(a_1, a_2, \ldots, a_n, a_{n+1}) + \min(b_1, b_2, \ldots, b_n, b_{n+1}) \quad \text{(by definition)}.
\end{aligned}$$

23. We can define the set $S = \{ x \mid x \text{ is a positive integer and } x \text{ is a multiple of } 5 \}$ by the basis step requirement that $5 \in S$ and the recursive requirement that if $n \in S$, then $n+5 \in S$. Alternately we can mimic Example 7, making the recursive part of the definition that $x + y \in S$ whenever x and y are in S.

25. a) Since we can generate all the even integers by starting with 0 and repeatedly adding or subtracting 2, a simple recursive way to define this set is as follows: $0 \in S$; and if $x \in S$ then $x + 2 \in S$ and $x - 2 \in S$.

b) The smallest positive integer congruent to 2 modulo 3 is 2, so we declare $2 \in S$. All the others can be obtained by adding multiples of 3, so our inductive step is that if $x \in S$, then $x+3 \in S$.

c) The positive integers not divisible by 5 are the ones congruent to 1, 2, 3, or 4 modulo 5. Therefore we can proceed just as in part **(b)**, setting $1 \in S$, $2 \in S$, $3 \in S$, and $4 \in S$ as the base cases, and then declaring that if $x \in S$, then $x+5 \in S$.

27. **a)** If we apply each of the recursive step rules to the only element given in the basis step, we see that $(0,1)$, $(1,1)$, and $(2,1)$ are all in S. If we apply the recursive step to these we add $(0,2)$, $(1,2)$, $(2,2)$, $(3,2)$, and $(4,2)$. The next round gives us $(0,3)$, $(1,3)$, $(2,3)$, $(3,3)$, $(4,3)$, $(5,3)$, and $(6,3)$. And a fourth set of applications adds $(0,4)$, $(1,4)$, $(2,4)$, $(3,4)$, $(4,4)$, $(5,4)$, $(6,4)$, $(7,4)$, and $(8,4)$.

b) Let $P(n)$ be the statement that $a \le 2b$ whenever $(a,b) \in S$ is obtained by n applications of the recursive step. For the basis step, $P(0)$ is true, since the only element of S obtained with no applications of the recursive step is $(0,0)$, and indeed $0 \le 2 \cdot 0$. Assume the strong inductive hypothesis that $a \le 2b$ whenever $(a,b) \in S$ is obtained by k or fewer applications of the recursive step, and consider an element obtained with $k+1$ applications of the recursive step. Since the final application of the recursive step to an element (a,b) must be applied to an element obtained with fewer applications of the recursive step, we know that $a \le 2b$. So we just need to check that this inequality implies $a \le 2(b+1)$, $a+1 \le 2(b+1)$, and $a+2 \le 2(b+1)$. But this is clear, since we just add $0 \le 2$, $1 \le 2$, and $2 \le 2$, respectively, to $a \le 2b$ to obtain these inequalities.

c) This holds for the basis step, since $0 \le 0$. If this holds for (a,b), then it also holds for the elements obtained from (a,b) in the recursive step, since adding $0 \le 2$, $1 \le 2$, and $2 \le 2$, respectively, to $a \le 2b$ yields $a \le 2(b+1)$, $a+1 \le 2(b+1)$, and $a+2 \le 2(b+1)$.

29. **a)** Since we are working with positive integers, the smallest pair in which the sum of the coordinates is even is $(1,1)$. So our basis step is $(1,1) \in S$. If we start with a point for which the sum of the coordinates is even and want to maintain this parity, then we can add 2 to the first coordinate, or add 2 to the second coordinate, or add 1 to each coordinate. Thus our recursive step is that if $(a,b) \in S$, then $(a+2,b) \in S$, $(a,b+2) \in S$, and $(a+1,b+1) \in S$. To prove that our definition works, we note first that $(1,1)$ has an even sum of coordinates, and if (a,b) has an even sum of coordinates, then so do $(a+2,b)$, $(a,b+2)$, and $(a+1,b+1)$, since we added 2 to the sum of the coordinates in each case. Conversely, we must show that if $a+b$ is even, then $(a,b) \in S$ by our definition. We do this by induction on the sum of the coordinates. If the sum is 2, then $(a,b)=(1,1)$, and the basis step put (a,b) into S. Otherwise the sum is at least 4, and at least one of $(a-2,b)$, $(a,b-2)$, and $(a-1,b-1)$ must have positive integer coordinates whose sum is an even number smaller than $a+b$, and therefore must be in S by our definition. Then one application of the recursive step shows that $(a,b) \in S$ by our definition.

b) Since we are working with positive integers, the smallest pairs in which there is an odd coordinate are $(1,1)$, $(1,2)$, and $(2,1)$. So our basis step is that these three points are in S. If we start with a point for which a coordinate is odd and want to maintain this parity, then we can add 2 to that coordinate. Thus our recursive step is that if $(a,b) \in S$, then $(a+2,b) \in S$ and $(a,b+2) \in S$. To prove that our definition works, we note first that $(1,1)$, $(1,2)$, and $(2,1)$ all have an odd coordinate, and if (a,b) has an odd coordinate, then so do $(a+2,b)$ and $(a,b+2)$, since adding 2 does not change the parity. Conversely (and this is the harder part), we must show that if (a,b) has at least one odd coordinate, then $(a,b) \in S$ by our definition. We do this by induction on the sum of the coordinates. If $(a,b)=(1,1)$ or $(a,b)=(1,2)$ or $(a,b)=(2,1)$, then the basis step put (a,b) into S. Otherwise either a or b is at least 3, so at least one of $(a-2,b)$ and $(a,b-2)$ must have positive integer coordinates whose sum is smaller than $a+b$, and therefore must be in S by our definition, since we haven't changed the parities. Then one application of the recursive step shows that $(a,b) \in S$ by our definition.

c) We use two basis steps here, $(1,6) \in S$ and $(2,3) \in S$. If we want to maintain the parity of $a+b$ and the fact that b is a multiple of 3, then we can add 2 to a (leaving b alone), or we can add 6 to b (leaving a

alone). So our recursive step is that if $(a,b) \in S$, then $(a+2,b) \in S$ and $(a,b+6) \in S$. To prove that our definition works, we note first that $(1,6)$ and $(2,3)$ satisfy the condition, and if (a,b) satisfies the condition, then so do $(a+2,b)$ and $(a,b+6)$, since adding 2 or 6 does not change the parity of the sum, and adding 6 maintains divisibility by 3. Conversely (and this is the harder part), we must show that if (a,b) satisfies the condition, then $(a,b) \in S$ by our definition. We do this by induction on the sum of the coordinates. The smallest sums of coordinates satisfying the condition are 5 and 7, and the only points are $(1,6)$, which the basis step put into S, $(2,3)$, which the basis step put into S, and $(4,3) = (2+2,3)$, which is in S by one application of our recursive definition. For a sum greater than 7, either $a \geq 3$, or $a \leq 2$ and $b \geq 9$ (since $2+6$ is not odd). This implies that either $(a-2,b)$ or $(a,b-6)$ must have positive integer coordinates whose sum is smaller than $a+b$ and satisfy the condition for being in S, and hence are in S by our definition. Then one application of the recursive step shows that $(a,b) \in S$ by our definition.

31. The answer depends on whether we require fully parenthesized expressions. Assuming that we do not, then the following definition is the most straightforward. Let F be the required collection of formulae. The basis step is that all specific sets and all variables representing sets are to be in F. The recursive part of the definition is that if α and β are in F, then so are $\overline{\alpha}$, (α), $\alpha \cup \beta$, $\alpha \cap \beta$, and $\alpha - \beta$. If we insist on parentheses, then the recursive part of the definition is that if α and β are in F, then so are $\overline{\alpha}$, $(\alpha \cup \beta)$, $(\alpha \cap \beta)$, and $(\alpha - \beta)$.

33. Let $D = \{0,1,2,3,4,5,6,7,8,9\}$ be the set of decimal digits. We think of a string as either being an element of D or else coming from a shorter string by appending an element of D, as in Definition 2. This problem is like Example 9.

a) The basis step is for a string of length 1, i.e., an element of D. If $x \in D$, then $m(x) = x$. For the recursive step, if the string $s = tx$, where $t \in D^*$ and $x \in D$, then $m(s) = \min(m(s), x)$. In other words, if the last digit in the string is smaller than the minimum digit in the rest of the string, then the last digit is the smallest digit in the string; otherwise the smallest digit in the rest of the string is the smallest digit in the string.

b) Recall the definition of concatenation (Definition 3). The basis step does not apply, since s and t here must be nonempty. Let $t = wx$, where $w \in D^*$ and $x \in D$. If $w = \lambda$, then $m(st) = m(sx) = \min(m(s), x) = \min(m(s), m(x))$ by the recursive step and the basis step of the definition of m in part **(a)**. Otherwise, $m(st) = m((sw)x) = \min(m(sw), x)$ by the definition of m in part **(a)**. But $m(sw) = \min(m(s), m(w))$ by the inductive hypothesis of our structural induction, so $m(st) = \min(\min(m(s), m(w)), x) = \min(m(s), \min(m(w), x))$ by the meaning of min. But $\min(m(w), x) = m(wx) = m(t)$ by the recursive step of the definition of m in part **(a)**. Thus $m(st) = \min(m(s), m(t))$.

35. The string of length 0, namely the empty string, is its own reversal, so we define $\lambda^R = \lambda$. A string w of length $n+1$ can always be written as vy, where v is a string of length n (the first n symbols of w), and y is a symbol (the last symbol of w). To reverse w, we need to start with y, and then follow it by the first part of w (namely v), reversed. Thus we define $w^R = y(v^R)$. (Note that the parentheses are for our benefit—they are not part of the string.)

37. We set $w^0 = \lambda$ (the concatenation of no copies of w should be defined to be the empty string). For $i \geq 0$, we define $w^{i+1} = ww^i$, where this notation means that we first write down w and then follow it with w^i.

39. The recursive part of this definition tells us that the only way to modify a string in A to obtain another string in A is to tack a 0 onto the front and a 1 onto the end. Starting with the empty string, then, the only strings we get are λ, 01, 0011, 000111, ... In other words, $A = \{0^n1^n \mid n \geq 0\}$.

41. The basis step is $i = 0$, where we need to show that the length of w^0 is 0 times the length of w. This is true, no matter what w is, since $l(w^0) = l(\lambda) = 0$. Assume the inductive hypothesis that $l(w^i) = i \cdot l(w)$. Then

$l(w^{i+1}) = l(ww^i) = l(w) + l(w^i)$, this latter equality having been shown in Example 14. Now by the inductive hypothesis we have $l(w) + l(w^i) = l(w) + i \cdot l(w) = (i+1) \cdot l(w)$, as desired.

43. This is similar to Theorem 2. For the full binary tree consisting of just the root r the result is true since $n(T) = 1$ and $h(T) = 0$, and $1 \geq 2 \cdot 0 + 1$. For the inductive hypothesis we assume that $n(T_1) \geq 2h(T_1) + 1$ and $n(T_2) \geq 2h(T_2) + 1$ where T_1 and T_2 are full binary trees. By the recursive definitions of $n(T)$ and $h(T)$, we have $n(T) = 1 + n(T_1) + n(T_2)$ and $h(T) = 1 + \max(h(T_1), h(T_2))$. Therefore $n(T) = 1 + n(T_1) + n(T_2) \geq 1 + 2h(T_1) + 1 + 2h(T_2) + 1 \geq 1 + 2 \cdot \max(h(T_1), h(T_2)) + 2$ since the sum of two nonnegative numbers is at least as large as the larger of the two. But this equals $1 + 2(\max(h(T_1), h(T_2)) + 1) = 1 + 2h(T)$, and our proof is complete.

45. The basis step requires that we show that this formula holds when $(m, n) = (0, 0)$. The inductive step requires that we show that if the formula holds for all pairs smaller than (m, n) in the lexicographic ordering of $\mathbf{N} \times \mathbf{N}$, then it also holds for (m, n). For the basis step we have $a_{0,0} = 0 = 0 + 0$. For the inductive step, assume that $a_{m',n'} = m' + n'$ whenever (m', n') is less than (m, n) in the lexicographic ordering of $\mathbf{N} \times \mathbf{N}$. By the recursive definition, if $n = 0$ then $a_{m,n} = a_{m-1,n} + 1$; since $(m-1, n)$ is smaller than (m, n), the inductive hypothesis tells us that $a_{m-1,n} = m - 1 + n$, so $a_{m,n} = m - 1 + n + 1 = m + n$, as desired. Now suppose that $n > 0$, so that $a_{m,n} = a_{m,n-1} + 1$. Again we have $a_{m,n-1} = m + n - 1$, so $a_{m,n} = m + n - 1 + 1 = m + n$, and the proof is complete.

47. a) It is clear that $P_{mm} = P_m$, since a number exceeding m can never be used in a partition of m.

b) We need to verify all five lines of this definition, show that the recursive references are to a smaller value of m or n, and check that they take care of all the cases and are mutually compatible. Let us do the last of these first. The first two lines take care of the case in which either m or n is equal to 1. They are consistent with each other in case $m = n = 1$. The last three lines are mutually exclusive and take care of all the possibilities for m and n if neither is equal to 1, since, given any two numbers, either they are equal or one is greater than the other. Note finally that the third line allows $m = 1$; in that case the value is defined to be P_{11}, which is consistent with line one, since $P_{1n} = 1$.

Next let us make sure that the logic of the definition is sound, specifically that P_{mn} is being defined in terms of P_{ij} for $i \leq m$ and $j \leq n$, with at least one of the inequalities strict. There is no problem with the first two lines, since these are not recursive. The third line is okay, since $m < n$, and P_{mn} is being defined in terms of P_{mm}. The fourth line is also okay, since here P_{mm} is being defined in terms of $P_{m,m-1}$. Finally, the last line is okay, since the subscripts satisfy the desired inequalities.

Finally, we need to check the content of each line. (Note that so far we have hardly even discussed what P_{mn} means!) The first line says that there is only one way to write the number 1 as the sum of positive integers, none of which exceeds n, and that is patently true, namely as $1 = 1$. The second line says that there is only one way to write the number m as the sum of positive integers, none of which exceeds 1, and that, too, is obvious, namely $m = 1 + 1 + \cdots + 1$. The third line says that the number of ways to write m as the sum of integers not exceeding n is the same as the number of ways to write m as the sum of integers not exceeding m as long as $m < n$. This again is true, since we could never use a number from $\{m+1, m+2, \ldots, n\}$ in such a sum anyway. Now we begin to get to the meat. The fourth line says that the number of ways to write m as the sum of positive integers not exceeding m is 1 plus the number of ways to write m as the sum of positive integers not exceeding $m - 1$. Indeed, there is exactly one way to write m as the sum of positive integers not exceeding m that actually uses m, namely $m = m$; all the rest use only numbers less than or equal to $m - 1$. This verifies line four. The real heart of the matter is line five. How can we write m as the sum of positive integers not exceeding n? We may use an n, or we may not. There are exactly $P_{m,n-1}$ ways to form the sum without using n, since in that case each summand is less than or equal to $n - 1$. If we do use at least one n, then we have $m = n + (m - n)$. The number of ways this can be done, then, is the same as the number of ways to complete the partition by writing $(m - n)$ as the sum of positive integers not exceeding n. Thus there are $P_{m-n,n}$ ways to write m as the sum of numbers not exceeding n, at least one of which equals n. By the sum rule (see Chapter 5), we have $P_{mn} = P_{m,n-1} + P_{m-n,n}$, as desired.

c) We expand each P_{mn} according to the definition. For the first problem we have $P_5 = P_{55} = 1 + P_{54} = 1+P_{53}+P_{14} = 1+P_{52}+P_{23}+1 = 1+P_{51}+P_{32}+P_{22}+1 = 1+1+P_{31}+P_{12}+1+P_{21}+1 = 1+1+1+1+1+1+1 = 7$. For the second problem we have $P_6 = P_{66} = 1+P_{65} = 1+P_{64}+P_{15} = 1+P_{63}+P_{24}+1 = 1+P_{62}+P_{33}+P_{22}+1 = 1+P_{61}+P_{42}+1+P_{32}+1+P_{21}+1 = 1+1+P_{41}+P_{22}+1+P_{31}+P_{12}+1+1+1 = 1+1+1+1+P_{21}+1+1+1+1+1+1 = 1+1+1+1+1+1+1+1+1+1+1 = 11$.

49. We prove this by induction on m. The basis step is $m = 1$, so we need to compute $A(1,2)$. Line four of the definition tells us that $A(1,2) = A(0, A(1,1))$. Since $A(1,1) = 2$, by line three, we see that $A(1,2) = A(0,2)$. Now line one of the definition applies, and we see that $A(1,2) = A(0,2) = 2 \cdot 2 = 4$, as desired. For the inductive step, assume that $A(m-1,2) = 4$, and consider $A(m,2)$. Applying first line four of the definition, then line three, and then the inductive hypothesis, we have $A(m,2) = A(m-1, A(m,1)) = A(m-1,2) = 4$.

51. a) We use the results of Exercises 49 and 50: $A(2,3) = A(1, A(2,2)) = A(1,4) = 2^4 = 16$.

b) We have $A(3,3) = A(2, A(3,2)) = A(2,4)$ by Exercise 49. Now one can show by induction (using the result of Exercise 50) that $A(2,n)$ is equal to $2^{2^{\cdot^{\cdot^{2}}}}$, with n 2's in the tower. Therefore the answer is $2^{2^{2^{2}}} = 2^{16} = 65{,}536$.

53. It is often the case in proofs by induction that you need to prove something stronger than the given proposition, in order to have a stronger inductive hypothesis to work with. This is called **inductive loading** (see the preamble to Exercise 70 in Section 4.1). That is the case with our proof here. We will prove the statement "$A(m,k) > A(m,l)$ if $k > l$ for all m, k, and l," and we will use **double induction**, inducting first on m, and then within the inductive step for that induction, inducting on k (using strong induction). Note that this stronger statement implies the statement we are trying to prove—just take $k = l + 1$.

The basis step is $m = 0$, in which the statement at hand reduces (by line one of the definition) to the true conditional statement that if $k > l$, then $2k > 2l$. Next we assume the inductive hypothesis on m, namely that $A(m,x) > A(m,y)$ for *all* values of x and y with $x > y$. We want now to show that if $k > l$, then $A(m+1,k) > A(m+1,l)$. This we will do by induction on k. For the basis step, $k = 0$, there is nothing to prove, since the condition $k > l$ is vacuous. Similarly, if $k = 1$, then $A(m+1,k) = 2$ and $A(m+1,l) = 0$ (since necessarily $l = 0$), so the desired inequality holds. So assume the inductive hypothesis (using strong induction), that $A(m+1,r) > A(m+1,s)$ whenever $k > r > s$, where $k \geq 2$. We need to show that $A(m+1,k) > A(m+1,l)$ if $k > l$. Now $A(m+1,k) = A(m, A(m+1,k-1))$ by line four of the definition. Since $k - 1 \geq l$, we apply the inductive hypothesis on k to yield $A(m+1,k-1) > A(m+1,l-1)$, and therefore by the inductive hypothesis on m, we have $A(m, A(m+1,k-1)) > A(m, A(m+1,l-1))$. But this latter value equals $A(m+1,l)$, as long as $l \geq 2$. Thus we have shown that $A(m+1,k) > A(m+1,l)$ as long as $l \geq 2$. On the other hand, if $l = 0$ or 1, then $A(m+1,l) \leq 2$ (by lines two and three of the definition), whereas $A(m+1,2) = 4$ by Exercise 49. Therefore $A(m+1,k) \geq A(m+1,2) > A(m+1,l)$. This completes the proof.

55. We repeatedly invoke the result of Exercise 54, which says that $A(m+1,j) \geq A(m,j)$. Indeed, we have $A(i,j) \geq A(i-1,j) \geq \cdots \geq A(0,j) = 2j \geq j$.

57. Let $P(n)$ be the statement "F is well-defined at n; i.e., $F(n)$ is a well-defined number." We need to show that $P(n)$ is true for all n. We do this by strong induction. First $P(0)$ is true, since $F(0)$ is well-defined by the specification of $F(0)$. Next assume that $P(k)$ is true for all $k < n$. We want to show that $P(n)$ is also true, in other words that $F(n)$ is well-defined. Since the definition gave $F(n)$ in terms of $F(0)$ through $F(n-1)$, and since we are assuming that these are all well-defined (our inductive hypothesis), we conclude that $F(n)$ is well-defined, as desired.

59. **a)** This would be a proper definition if the recursive part were stated to hold for $n \geq 2$. As it stands, however, $F(1)$ is ambiguous.

b) This definition makes no sense as it stands; $F(2)$ is not defined, since $F(0)$ isn't.

c) For $n = 4$, the recursive part makes no sense, since we would have to know $F(4/3)$. Also, $F(3)$ is ambiguous.

d) The definition is ambiguous about $n = 1$, since both the second clause and the third clause seem to apply. If the second clause is restricted to odd $n \geq 3$, then the sequence is well-defined and begins 1, 2, 2, 3, 3, 3, 4, 4, 5, 4.

e) We note that $F(1)$ is defined explicitly, but we run into problems trying to compute $F(2)$:

$$F(2) = 1 + F(F(1)) = 1 + F(2).$$

This not only leaves us begging the question as to what $F(2)$ is, but is a contradiction, since $0 \neq 1$.

61. In each case we will apply the definition to compute $\log^{(0)}$, then $\log^{(1)}$, then $\log^{(2)}$, then $\log^{(3)}$ and so on. As soon as we get an answer no larger than 1 we stop; the last "exponent" is the answer. In other words $\log^* n$ is the number of times we need to apply the log function until we get a value less than or equal to 1. Note that $\log^{(1)} n = \log n$ for $n > 0$. Similarly, $\log^{(2)} n = \log(\log n)$ as long as it is defined ($n > 1$), $\log^{(3)} n = \log(\log(\log n))$ as long as it is defined ($n > 2$), and so on. Normally the parentheses are understood and omitted.

a) $\log^{(0)} 2 = 2$, $\log^{(1)} 2 = \log 2 = 1$; therefore $\log^* 2 = 1$, the last "exponent".

b) $\log^{(0)} 4 = 4$, $\log^{(1)} 4 = \log 4 = 2$, $\log^{(2)} 4 = \log 2 = 1$; therefore $\log^* 4 = 2$, the last "exponent". We had to take the log twice to get from 4 down to 1.

c) $\log^{(0)} 8 = 8$, $\log^{(1)} 8 = \log 8 = 3$, $\log^{(2)} 8 = \log 3 \approx 1.585$, $\log^{(3)} 8 \approx \log 1.585 \approx 0.664$; therefore $\log^* 8 = 3$, the last "exponent". We had to take the log three times to get from 8 down to something no bigger than 1.

d) $\log^{(0)} 16 = 16$, $\log^{(1)} 16 = \log 16 = 4$, $\log^{(2)} 16 = \log 4 = 2$, $\log^{(3)} 16 = \log 2 = 1$; therefore $\log^* 16 = 3$, the last "exponent". We had to take the log three times to get from 16 down to 1.

e) $\log^{(0)} 256 = 256$, $\log^{(1)} 256 = \log 256 = 8$; by part **(c)**, we need to take the log three more times in order to get from 8 down to something no bigger than 1, so we have to take the log four times in all to get from 256 down to something no bigger than 1. Thus $\log^* 256 = 4$.

f) $\log 65536 = 16$; by part **(d)**, we need to take the log three more times in order to get from 16 down to 1, so we have to take the log four times in all to get from 65536 down to 1. Thus $\log^* 65536 = 4$.

g) $\log 2^{2048} = 2048$; taking log four more times gives us, successively, 11, approximately 3.46, approximately 1.79, approximately 0.84. So $\log^* 2^{2048} = 5$.

63. Each application of the function f subtracts another a from the argument. Therefore iterating this function k times (which is what $f^{(k)}$ does) has the effect of subtracting ka. Therefore $f^{(k)}(n) = n - ka$. Now $f_0^*(n)$ is the smallest k such that $f^{(k)}(n) \leq 0$, i.e., $n - ka \leq 0$. Solving this for k easily yields $k \geq n/a$. Thus $f_0^*(n) = \lceil n/a \rceil$ (we need to take the ceiling function because k must be an integer).

65. Each application of the function f takes the square root of its argument. Therefore iterating this function k times (which is what $f^{(k)}$ does) has the effect of taking the $(2^k)^{\text{th}}$ root. Therefore $f^{(k)}(n) = n^{1/2^k}$. Now $f_2^*(n)$ is the smallest k such that $f^{(k)}(n) \leq 2$, that is, $n^{1/2^k} \leq 2$. Solving this for n easily yields $n \leq 2^{2^k}$, so

$k \ge \log\log n$, where logarithm is taken to the base 2. Thus $f_2^*(n) = \lceil \log\log n \rceil$ for $n \ge 2$ (we need to take the ceiling function because k must be an integer) and $f_2^*(1) = 0$.

SECTION 4.4 Recursive Algorithms

Recursive algorithms are important theoretical entities, but they often cause a lot of grief on first encounter. Sometimes it is helpful to "play computer" very carefully to see how a recursive algorithm works. Ironically, however, it is good to avoid doing that after you get the idea. Instead, convince yourself that if the recursive algorithm handles the base case correctly, and handles the recursive step correctly (gives the correct answer assuming that the correct answer was obtained on the recursive call), then the algorithm works. Let the computer worry about actually "recursing all the way down to the base case"!

1. First, we use the recursive step to write $5! = 5 \cdot 4!$. We then use the recursive step repeatedly to write $4! = 4 \cdot 3!$, $3! = 3 \cdot 2!$, $2! = 2 \cdot 1!$, and $1! = 1 \cdot 0!$. Inserting the value of $0! = 1$, and working back through the steps, we see that $1! = 1 \cdot 1 = 1$, $2! = 2 \cdot 1! = 2 \cdot 1 = 2$, $3! = 3 \cdot 2! = 3 \cdot 2 = 6$, $4! = 4 \cdot 3! = 4 \cdot 6 = 24$, and $5! = 5 \cdot 4! = 5 \cdot 24 = 120$.

3. First, because $n = 11$ is odd, we use the **else** clause to see that

$$mpower(3, 11, 5) = (mpower(3, 5, 5)^2 \ \mathbf{mod}\ 5 \cdot 3 \ \mathbf{mod}\ 5) \ \mathbf{mod}\ 5\,.$$

We next use the **else** clause again to see that

$$mpower(3, 5, 5) = (mpower(3, 2, 5)^2 \ \mathbf{mod}\ 5 \cdot 3 \ \mathbf{mod}\ 5) \ \mathbf{mod}\ 5\,.$$

Then we use the **else if** clause to see that

$$mpower(3, 2, 5) = mpower(3, 1, 5)^2 \ \mathbf{mod}\ 5\,.$$

Using the **else** clause again, we have

$$mpower(3, 1, 5) = (mpower(3, 0, 5)^2 \ \mathbf{mod}\ 5 \cdot 3 \ \mathbf{mod}\ 5) \ \mathbf{mod}\ 5\,.$$

Finally, using the **if** clause, we see that $mpower(3, 0, 5) = 1$. Now we work backward: $mpower(3, 1, 5) = (1^2 \ \mathbf{mod}\ 5 \cdot 3 \ \mathbf{mod}\ 5) \ \mathbf{mod}\ 5 = 3$, $mpower(3, 2, 5) = 3^2 \ \mathbf{mod}\ 5 = 4$, $mpower(3, 5, 5) = (4^2 \ \mathbf{mod}\ 5 \cdot 3 \ \mathbf{mod}\ 5) \ \mathbf{mod}\ 5 = 3$, and finally that $mpower(3, 11, 5) = (3^2 \ \mathbf{mod}\ 5 \cdot 3 \ \mathbf{mod}\ 5) \ \mathbf{mod}\ 5 = 2$. We conclude that $3^{11} \ \mathbf{mod}\ 5 = 2$.

5. With this input, the algorithm uses the **else** clause to find that $\gcd(8, 13) = \gcd(13 \ \mathbf{mod}\ 8, 8) = \gcd(5, 8)$. It uses this clause again to find that $\gcd(5, 8) = \gcd(8 \ \mathbf{mod}\ 5, 5) = \gcd(3, 5)$, then to get $\gcd(3, 5) = \gcd(5 \ \mathbf{mod}\ 3, 3) = \gcd(2, 3)$, then $\gcd(2, 3) = \gcd(3 \ \mathbf{mod}\ 2, 2) = \gcd(1, 2)$, and once more to get $\gcd(1, 2) = \gcd(2 \ \mathbf{mod}\ 1, 1) = \gcd(0, 1)$. Finally, to find $\gcd(0, 1)$ it uses the first step with $a = 0$ to find that $\gcd(0, 1) = 1$. Consequently, the algorithm finds that $\gcd(8, 13) = 1$.

7. The key idea for this recursive procedure is that $nx = (n-1)x + x$. Thus we compute nx by calling the procedure recursively, with n replaced by $n - 1$, and adding x. The base case is $1 \cdot x = x$.

```
procedure product(n : positive integer, x : integer)
if n = 1 then product(n, x) := x
else product(n, x) := product(n − 1, x) + x
```

9. If we have already found the sum of the first $n - 1$ odd positive integers, then we can find the sum of the first n positive integers simply by adding on the value of the nth odd positive integer. We need to realize that the nth odd positive integer is $2n - 1$, and we need to note that the base case (in which $n = 1$) gives us a sum of 1. The algorithm is then straightforward.

procedure *sum of odds*(n : positive integer)
if $n = 1$ **then** *sum of odds*(n) := 1
else *sum of odds*(n) := *sum of odds*($n - 1$) + $2n - 1$

11. We recurse on the size of the list. If there is only one element, then it is the smallest. Otherwise, we find the smallest element in the list consisting of all but the last element of our original list, and compare it with the last element of the original list. Whichever is smaller is the answer. (We assume that there is already a function min, defined for two arguments, which returns the smaller.)

procedure *smallest*($a_1, a_2, \ldots, a_n$: integers)
if $n = 1$ **then** *smallest*($a_1, a_2, \ldots, a_n$) := a_1
else *smallest*($a_1, a_2, \ldots, a_n$) := min(*smallest*($a_1, a_2, \ldots, a_{n-1}$), a_n)

13. We basically just take the recursive algorithm for finding $n!$ and apply the **mod** operation at each step. Note that this enables us to calculate $n!$ **mod** m without using excessively large numbers, even if $n!$ is very large.

procedure *modfactorial*(n, m : positive integers)
if $n = 1$ **then** *modfactorial*(n, m) := 1
else *modfactorial*(n, m) := ($n \cdot$ (*modfactorial*($n - 1, m$)) **mod** m

15. We need to worry about which of our arguments is the larger. Since we are given $a < b$ as input, we need to make sure always to call the algorithm with the first argument less than the second. There need to be two stopping conditions: when $a = 0$ (in which case the answer is b), and when the two arguments have become equal (in which case the answer is their common value). Otherwise we use the recursive condition that $\gcd(a, b) = \gcd(a, b - a)$, taking care to reverse the arguments if necessary.

procedure *gcd*(a, b : nonnegative integers with $a < b$)
if $a = 0$ **then** *gcd*(a, b) := b
else if $a = b - a$ **then** *gcd*(a, b) := a
else if $a < b - a$ **then** *gcd*(a, b) := *gcd*($a, b - a$)
else *gcd*(a, b) := *gcd*($b - a, a$)

17. We build the recursive steps into the algorithm.

procedure *multiply*(x, y : nonnegative integers)
if $y = 0$ **then** *multiply*(x, y) := 0
else if y is even **then** *multiply*(x, y) := $2 \cdot$ *multiply*($x, y/2$)
else *multiply*(x, y) := $2 \cdot$ *multiply*($x, (y - 1)/2$) + x

19. We use strong induction on a, starting at $a = 0$. If $a = 0$, we know that $\gcd(0, b) = b$ for all $b > 0$, so the **if** clause handles this basis case correctly. Now fix $k > 0$ and assume the inductive hypothesis—that the algorithm works correctly for all values of its first argument less than k. Consider what happens with input (k, b), where $k < b$. Since $k > 0$, the **else** clause is executed, and the answer is whatever the algorithm gives as output for inputs (b **mod** k, k). Note that b **mod** $k < k$, so the input pair is valid. By our inductive hypothesis, this output is in fact gcd(b **mod** k, k). By Lemma 1 in Section 3.6 (when the Euclidean algorithm was introduced), we know that $\gcd(k, b) =$ gcd(b **mod** k, k), and our proof is complete.

21. For the basis step, if $n = 1$, then $nx = x$, and the algorithm correctly returns x. For the inductive step, assume that the algorithm correctly computes kx, and consider what it does to compute $(k + 1)x$. The recursive clause applies, and it recursively computes the product of $k + 1 - 1 = k$ and x, and then adds x. By the inductive hypothesis, it computes that product correctly, so the answer returned is $kx + x = (k + 1)x$, which is the correct answer.

23. As usual with recursive algorithms, the algorithm practically writes itself.

```
procedure square(n : nonnegative integer)
if n = 0 then square(n) := 0
else square(n) := square(n − 1) + 2(n − 1) + 1
```

The proof of correctness, by mathematical induction, practically writes itself as well. Let $P(n)$ be the statement that this algorithm correctly computes n^2. Since $0^2 = 0$, the algorithm works correctly (using the **if** clause) if the input is 0. Assume that the algorithm works correctly for input k. Then for input $k+1$ it gives as output (because of the **else** clause) its output when the input is k, plus $2(k+1-1)+1$. By the inductive hypothesis, its output at k is k^2, so its output at $k+1$ is $k^2+2(k+1-1)+1 = k^2+2k+1 = (k+1)^2$, exactly what it should be.

25. Algorithm 2 uses 2^n multiplications by a, one for each factor of a in the product a^{2^n}. The algorithm in Exercise 24, based on squaring, uses only n multiplications (each of which is a multiplication of a number by itself). For instance, to compute $a^{2^4} = a^{16}$, this algorithm will compute $a \cdot a = a^2$ (one multiplication), then $a^2 \cdot a^2 = a^4$ (a second multiplication), then $a^4 \cdot a^4 = a^8$ (a third), and finally $a^8 \cdot a^8 = a^{16}$ (a fourth multiplication).

27. Algorithm 2 uses n multiplications by a, one for each factor of a in the product a^n. The algorithm in Exercise 26 will use $O(\log n)$ multiplications as it computes squares. Furthermore, in addition to squaring, sometimes a multiplication by a is needed; this will add at most another $O(\log n)$ multiplications. Thus a total of $O(\log n)$ multiplications are used altogether.

29. This is very similar to the recursive procedure for computing the Fibonacci numbers. Note that we can combine the two base cases (stopping rules) into one.

```
procedure sequence(n : nonnegative integer)
if n < 2 then sequence(n) := n + 1
else sequence(n) := sequence(n − 1) · sequence(n − 2)
```

31. The iterative version is much more efficient. The analysis is exactly the same as that for the Fibonacci sequence given in this section. Indeed, the nth term in this sequence is actually just 2^{f_n}, as is easily shown by induction.

33. This is essentially just Algorithm 8, with a different operation and with different (and more) initial conditions.

```
procedure iterative(n : nonnegative integer)
if n = 0 then z := 1
else if n = 1 then z := 2
else
begin
        x := 1
        y := 2
        z := 3
        for i := 1 to n − 2
        begin
                w := x + y + z
                x := y
                y := z
                z := w
        end
end
{z is the nth term of the sequence}
```

35. These algorithms are very similar to the procedures for computing the Fibonacci numbers. Note that for the recursive version, we can combine the three base cases (stopping rules) into one.

procedure *recursive*(n : nonnegative integer)
if $n < 3$ **then** $recursive(n) := 2n + 1$
else $recursive(n) := recursive(n-1) \cdot (recursive(n-2))^2 \cdot (recursive(n-3))^3$

procedure *iterative*(n : nonnegative integer)
if $n = 0$ **then** $z := 1$
else if $n = 1$ **then** $z := 3$
else
begin
  $x := 1$
  $y := 3$
  $z := 5$
  for $i := 1$ **to** $n - 2$
  begin
    $w := z \cdot y^2 \cdot x^3$
    $x := y$
    $y := z$
    $z := w$
  end
end
{z is the nth term of the sequence}

The recursive version is much easier to write, but the iterative version is much more efficient. In doing the computation for the iterative version, we just need to go through the loop $n-2$ times in order to compute a_n, so it requires $O(n)$ steps. In doing the computation for the recursive version, we are constantly recalculating previous values that we've already calculated, just as was the case with the recursive version of the algorithm to calculate the Fibonacci numbers.

37. We use the recursive definition of the reversal of a string given in Exercise 35 of Section 4.3, namely that $(vy)^R = y(v^R)$, where y is the last symbol in the string and v is the substring consisting of all but the last symbol. The right-hand side of the last statement in this procedure means that we concatenate b_n with the output of the recursive call.

procedure *reverse*($b_1 b_2 \ldots b_n$: bit string)
if $n = 0$ **then** $reverse(b_1 b_2 \ldots b_n) := \lambda$
else $reverse(b_1 b_2 \ldots b_n) := b_n reverse(b_1 b_2 \ldots b_{n-1})$

39. The procedure correctly gives the reversal of λ as λ (the basis step of our proof by mathematical induction on n), and because the reversal of a string consists of its last character followed by the reversal of its first $n-1$ characters (see Exercise 35 in Section 4.3), the algorithm behaves correctly when $n > 0$ by the inductive hypothesis.

41. The algorithm merely implements the idea of Example 13 in Section 4.1. If $n = 1$ (the basis step here), we simply place the one right triomino so that its armpit corresponds to the hole in the 2×2 board. If $n > 1$, then we divide the board into four boards, each of size $2^{n-1} \times 2^{n-1}$, notice which quarter the hole occurs in, position one right triomino at the center of the board with its armpit in the quarter where the missing square is (see Figure 8 in Section 4.1), and invoke the algorithm recursively four times—once on each of the $2^{n-1} \times 2^{n-1}$ boards, each of which has one square missing (either because it was missing to begin with, or because it is covered by the central triomino).

43. Essentially all we do is write down the definition as a procedure.

procedure *ackermann*(m, n : nonnegative integers)
if $m = 0$ **then** $ackermann(m, n) := 2 \cdot n$
else if $n = 0$ **then** $ackermann(m, n) := 0$
else if $n = 1$ **then** $ackermann(m, n) := 2$
else $ackermann(m, n) := ackermann(m - 1, ackermann(m, n - 1))$

45. We assume that sorting is to be done into alphabetical order. First the list is split into the two lists b, d, a, f, g and h, z, p, o, k, and each of these is sorted by merge sort. Let us assume for a moment that this has been done, so the two lists are a, b, d, f, g and h, k, o, p, z. Then these two lists are merged into one sorted list, as follows. We compare a with h and find that a is smaller; thus a comes first in the merged list, and we pass on to b. Comparing b with h, we find that b is smaller, so b comes next in the merged list, and we pass on to d. We repeat this process (using Algorithm 10) until the lists are merged into one sorted list, $a, b, d, f, g, h, k, o, p, z$. (It was just a coincidence that every element in the first of these two lists came before every element in the second.)

Let us return to the question of how each of the 5-element lists was sorted. For the list b, d, a, f, g, we divide it into the sublists b, d, a and f, g. Again we sort each piece by the same algorithm, producing a, b, d and f, g, and we merge them into the sorted list a, b, d, f, g. Going one level deeper into the recursion, we see that sorting b, d, a was accomplished by splitting it into b, d and a, and sorting each piece by the same algorithm. The first of these required further splitting into b and d. One element lists are already sorted, of course. Similarly, the other 5-element list was sorted by a similar recursive process. A tree diagram for this problem is displayed below. The top half of the picture is a tree showing the splitting part of the algorithm. The bottom half shows the merging part as an upside-down tree.

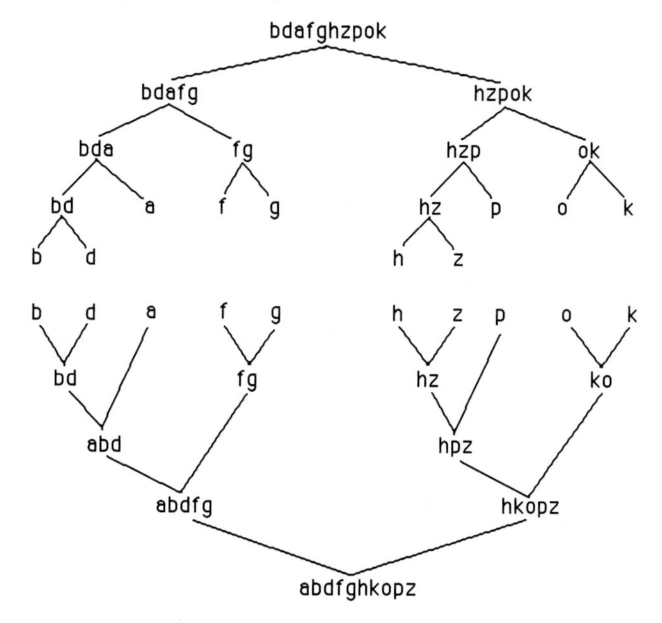

47. All we have to do is to make sure that one of the lists is exhausted only when the other list has only one element left in it. In this case, a comparison is needed to place every element of the merged list into place, except for the last element. Clearly this condition is met if and only if the largest element in the combined list is in one of the initial lists and the second largest element is in the other. One such pair of lists is $\{1, 2, \ldots, m - 1, m + n\}$ and $\{m, m + 1, \ldots, m + n - 1\}$.

49. We use strong induction on n, showing that the algorithm works correctly if $n = 1$, and that if it works correctly for $n = 1$ through $n = k$, then it also works correctly for $n = k + 1$. If $n = 1$, then the algorithm does nothing, which is correct, since a list with one element is already sorted. If $n = k + 1$, then the list is split into two lists, L_1 and L_2. By the inductive hypothesis, *mergesort* correctly sorts each of these sublists, and so it remains only to show that *merge* correctly merges two sorted lists into one. This is clear, from the code given in Algorithm 10, since with each comparison, the smallest element in $L_1 \cup L_2$ not yet put into L is put there.

51. We need to compare every other element with a_1. Thus at least $n-1$ comparisons are needed (we will assume for Exercises 53 and 55 that the answer is exactly $n-1$). The actual number of comparisons depends on the actual encoding of the algorithm. With any reasonable encoding, it should be $O(n)$.

53. In our analysis we assume that a_1 is considered to be put between the two sublists, not as the last element of the first sublist (which would require an extra pass in some cases). In the worst case, the original list splits into lists of length 3 and 0 (with a_1 between them); by Exercise 51, this requires $4-1=3$ comparisons. No comparisons are needed to sort the second of these lists (since it is empty). To sort the first, we argue in the same way: the worst case is for a splitting into lists of length 2 and 0, requiring $3-1=2$ comparisons. Similarly, $2-1=1$ comparison is needed to split the list of length 2 into lists of length 1 and 0. In all, then, $3+2+1=6$ comparisons are needed in this worst case. (One can prove that this discussion really does deal with the worst case by looking at what happens in the various other cases.)

55. In our analysis we assume that a_1 is considered to be put between the two sublists, not as the last element of the first sublist (which would require an extra pass in some cases). We claim that the worst case complexity is $n(n-1)/2$ comparisons, and we prove this by induction on n. This is certainly true for $n=1$, since no comparisons are needed. Otherwise, suppose that the initial split is into lists of size k and $n-k-1$. By the inductive hypothesis, it will require $(k(k-1)/2)+((n-k-1)(n-k-2)/2)$ comparisons in the worst case to finish the algorithm. This quadratic function of k attains its maximum value if $k=0$ (or $k=n-1$), namely the value $(n-1)(n-2)/2$. Also, it took $n-1$ comparisons to perform the first splitting. If we add these two quantities ($(n-1)(n-2)/2$ and $n-1$) and do the algebra, then we obtain $n(n-1)/2$, as desired. Thus in the worst case the complexity is $O(n^2)$.

SECTION 4.5 Program Correctness

Entire books have been written on program verification; obviously here we barely scratch the surface. In some sense program verification is just a very careful stepping through a program to prove that it works correctly. There should be no problem verifying anything except loops; indeed it may seem that there is nothing to prove. Loops are harder to deal with. The trick is to find the right invariant. Once you have the invariant, it again is really a matter of stepping through one pass of the loop to make sure that the invariant is satisfied at the end of that pass, given that it was satisfied at the beginning. Analogous to our remark in the comments for Section 1.6, there is a deep theorem of logic (or theoretical computer science) that says essentially that there is no algorithm for proving correct programs correct, so the task is much more an art than a science.

Your proofs must be valid proofs, of course. You may use the rules of inference discussed in Section 1.5. What is special about proofs of program correctness is the addition of some special rules for this setting. These exercises (and the examples in the text) may seem overly simple, but unfortunately it is extremely hard to prove all but the simplest programs correct.

1. We suppose that initially $x = 0$. The segment causes two things to happen. First y is assigned the value of 1. Next the value of $x+y$ is computed to be $0+1=1$, and so z is assigned the value 1. Therefore at the end z has the value 1, so the final assertion is satisfied.

3. We suppose that initially $y = 3$. The effect of the first two statements is to assign x the value 2 and z the value $2+3=5$. Next, when the **if**...**then** statement is encountered, since the value of y is 3, and $3 > 0$ is true, the statement $z := z+1$ assigns the value $5+1=6$ to z (and the **else** clause is not executed). Therefore at the end, z has the value 6, so the final assertion $z = 6$ is true.

5. We generalize the rule of inference for the **if**...**then**...**else** statement, given before Example 3. Let p be the initial assertion, and let q be the final assertion. If *condition* 1 is true, then S_1 will get executed, so we need $(p \wedge \textit{condition } 1)\{S_1\}q$. Similarly, if *condition* 2 is true, but *condition* 1 is false, then S_2 will get executed, so we need $(p \wedge \neg(\textit{condition } 1) \wedge \textit{condition } 2)\{S_2\}q$. This pattern continues, until the last statement we need is $(p \wedge \neg(\textit{condition } 1) \wedge \neg(\textit{condition } 2) \wedge \cdots \wedge \neg(\textit{condition } n-1))\{S_n\}q$. Given all of these, we can conclude that pTq, where T is the entire statement. In symbols, we have

$$\begin{array}{l}
(p \wedge \textit{condition } 1)\{S_1\}q \\
(p \wedge \neg(\textit{condition } 1) \wedge \textit{condition } 2)\{S_2\}q \\
\quad\vdots \\
(p \wedge \neg(\textit{condition } 1) \wedge \neg(\textit{condition } 2) \wedge \cdots \wedge \neg(\textit{condition } n-1))\{S_n\}q \\
\hline
\therefore\ p\{\textbf{if } \textit{condition } 1 \textbf{ then } S_1 \textbf{ else if } \textit{condition } 2 \textbf{ then } S_2 \ \ldots \textbf{else } S_n\}q\,.
\end{array}$$

7. The problem is similar to Example 4. We will use the loop invariant p: "$\textit{power} = x^{i-1}$ and $i \le n+1$." Now p is true initially, since before the loop starts, $i = 1$ and $\textit{power} = 1 = x^0 = x^{1-1}$. (There is a technicality here: we define 0^0 to equal 1 in order for this to be correct if $x = 0$. There is no harm in this, since $n > 0$, so if $x = 0$, then the program certainly computes the correct answer $x^n = 0$.) We must now show that if p is true and $i \le n$ before some pass through the loop, then p remains true after that pass. The loop increments i by one. Hence since $i \le n$ before this pass through the loop, $i \le n+1$ after this pass. Also the loop assigns $\textit{power} \cdot x$ to $\textit{power}$. By the inductive hypothesis, $\textit{power}$ started with the value x^{i-1} (the old value of i). Therefore its new value is $x^{i-1} \cdot x = x^i = x^{(i+1)-1}$. But since $i+1$ is the new value of i, the statement $\textit{power} = x^{i-1}$ is true at the completion of this pass through the loop. Hence p remains true, so p is a loop invariant. Furthermore, the loop terminates after n traversals, with $i = n+1$, since i is assigned the value 1 prior to entering the loop, i is incremented by 1 on each pass, and the loop terminates when $i > n$. At termination we have $(i \le n+1) \wedge \neg(i \le n)$, so $i = n+1$. Hence $\textit{power} = x^{(n+1)-1} = x^n$, as desired.

9. We will break the problem up into the various statements that are left unproved in Example 5.

We must show that $p\{S_1\}q$, where p is the assertion that m and n are integers, and q is the proposition $p \wedge (a = |n|)$. This follows from Example 3 and the fact that the values of m and n have not been tampered with.

We must show that $q\{S_2\}r$, where r is the proposition $q \wedge (k = 0) \wedge (x = 0)$. This is clear, since in S_2, none of m, n, or a have been altered, but k and x have been assigned the value 0.

We must show that $(x = mk) \wedge (k \le a)$ is an invariant for the loop in S_3. Assume that this statement is true before a pass through the loop. In the loop, k is incremented by 1. If $k < a$ (the condition of the loop), then at the end of the pass, $k < a+1$, so $k \le a$. Furthermore, since $x = mk$ at the beginning of the pass, and x is incremented by m inside the loop, we have $x = mk + m = m(k+1)$ at the end of the pass, which is precisely the statement that $x = mk$ for the updated value of k. When the loop terminates (which it clearly does after a iterations), clearly $k = a$ (since $(k \le a) \wedge \neg(k < a)$), and so $x = ma$ at this point.

Finally we must show that $s\{S_4\}t$, where s is the proposition $(x = ma) \wedge (a = |n|)$, and t is the proposition $product = mn$. The program segment S_4 assigns the value x or $-x$ to *product*, depending on the sign of n. We need to consider the two cases. If $n < 0$, then since $a = |n|$ we know that $a = -n$. Therefore $product = -x = -(ma) = -m(-n) = mn$. If $n \not< 0$, then since $a = |n|$ we know that $a = n$. Therefore $product = x = ma = mn$.

11. To say that the program assertion $p\{S\}q_1$ is true is to say that if p is true before S is executed, then q_1 is true after S is executed. To prove this, let us assume that p is true and then S is executed. Since $p\{S\}q_0$, we know that after S is executed q_0 is true. By modus ponens, since q_0 is true and $q_0 \to q_1$ is true, we know that q_1 is true, as desired.

13. Our loop invariant p is the proposition "$\gcd(a,b) = \gcd(x,y)$ and $y \geq 0$." First note that p is true before the loop is entered, since at that point $x = a$, $y = b$, and y is a positive integer (we use the initial assertion here). Now assume that p is true and $y > 0$; then the loop will be executed again. Within the loop x and y are replaced by y and x **mod** y, respectively. According to Lemma 1 in Section 3.6, $\gcd(x,y) = \gcd(y, x \textbf{ mod } y)$. Therefore after the execution of the loop, the value of $\gcd(x,y)$ remains what it was before. Furthermore, since y is a remainder, it is still greater than or equal to 0. Hence p remains true—it is a loop invariant. Furthermore, if the loop terminates, then it must be the case that $y = 0$. In this case we know that $\gcd(x,y) = x$, the desired final assertion. Therefore the program, which gives x as its output, has correctly computed $\gcd(a,b)$. Finally (although this is not necessary to establish *partial* correctness), we can prove that the loop must terminate, since each iteration causes the value of y to decrease by at least 1 (by the definition of **mod**). Thus the loop can be iterated at most b times.

GUIDE TO REVIEW QUESTIONS FOR CHAPTER 4

1. a) no

b) Sometimes yes. If the given formula is correct, then it is often possible to prove it using the principle of mathematical induction (although it would be wishful thinking to believe that *every* such true formula could be so proved). If the formula is incorrect, then induction would not work, of course; thus an incorrect formula could not be shown to be incorrect using the principle.

c) See Exercise 9 in Section 4.1.

2. a) $n \geq 7$

b) For the basis step we just check that $11 \cdot 7 + 17 \leq 2^7$. Fix $n \geq 7$, and assume the inductive hypothesis, that $11n + 17 \leq 2^n$. Then $11(n+1) + 17 = (11n + 17) + 11 \leq 2^n + 11 < 2^n + 2^n = 2^{n+1}$. The strict inequality here follows from the fact that $n \geq 4$.

3. a) Carefully considering all the possibilities shows that the amounts of postage less than 32 cents that can be achieved are 0, 5, 9, 10, 14, 15, 18, 19, 20, 23, 24, 25, 27, 28, 29, and 30. All amounts greater than or equal to 32 cents can be achieved.

b) To prove this latter statement, we check the basis step by noting that $32 = 9 + 9 + 9 + 5$. Assume that we can achieve n cents, and consider $n + 1$ cents, where $n \geq 32$. If the stamps used for n cents included a 9-cent stamp, then replacing it by two 5-cent stamps gives us $n + 1$ cents, as desired. Otherwise only 5-cent stamps were used to achieve n cents, and since $n > 30$, there must be at least seven such stamps. Replace seven of the 5-cent stamps by four 9-cent stamps; this increases the amount of postage by $4 \cdot 9 - 7 \cdot 5 = 1$ cent, again as desired.

c) We check the base cases $32 = 3 \cdot 9 + 5$, $33 = 2 \cdot 9 + 3 \cdot 5$, $34 = 9 + 5 \cdot 5$, $35 = 7 \cdot 5$, and $36 = 4 \cdot 9$. Fix $n \geq 37$ and assume that all amounts from 32 to $n - 1$ can be achieved. To achieve n cents postage, take the stamps used for $n-5$ cents (since $n \geq 37$, $n-5 \geq 32$, so the inductive hypothesis applies) and adjoin a 5-cent stamp.

d) Let n be an integer greater than or equal to 32. We want to express n as a sum of a nonnegative multiple of 5 and a nonnegative multiple of 9. Divide n by 5 to obtain a quotient q and remainder r such that $n = 5q + r$ and $0 \leq r \leq 4$. Note that since $n \geq 32$, $q \geq 6$. If $r = 0$, then we already have n expressed in the desired form. If $r = 1$, then $n \geq 36$, so $q \geq 7$; thus we can write $n = 5q + 1 = 5(q-7) + 4 \cdot 9$ to get the desired decomposition. If $r = 2$, then we rewrite $n = 5q + 2 = 5(q-5) + 3 \cdot 9$. If $r = 3$, then we rewrite $n = 5q + 3 = 5(q-3) + 2 \cdot 9$. And if $r = 4$, then we rewrite $n = 5q + 4 = 5(q-1) + 9$. In each case we have the desired sum.

4. See Examples 2 and 3 in Section 4.2.

5. **a)** See p. 278 and Appendix 1 (Axiom 4 for the positive integers).

b) Let S be the set of positive integers that cannot be written as the product of primes. If $S \neq \emptyset$, then S has a least element, c. Clearly $c \neq 1$, since 1 is the product of no primes. Thus c is greater than 1. Now c cannot be prime, since as such it would already be written as the product of primes (namely itself). Therefore c is a composite number, say $c = ab$, where a and b are both positive integers less than c. Since c is the smallest element of S, neither a nor b is in S. Therefore both a and b *can* be written as the product of primes. But multiplying these products together patently shows that c is the product of primes. This is a contradiction to the choice of c. Therefore our assumption that $S \neq \emptyset$ was wrong, and the theorem is proved.

6. **a)** See Exercise 56 in Section 4.3. **b)** $f(1) = 2$, and $f(n) = (n+1)f(n-1)$ for all $n \geq 2$

7. **a)** See p. 297. **b)** See Example 6 in Section 4.3.

8. **a)** See Exercise 57 in Section 4.3. **b)** $a_n = 3 \cdot 2^{n-3}$

9. See Examples 10 and 11 in Section 4.3.

10. **a)** See Example 9 in Section 4.3. **b)** See Example 14 in Section 4.3.

11. **a)** See p. 311.

b) Call the sequence a_1, a_2, $\ldots$, a_n. If $n = 1$, then the $sum(a_1) = a_1$. Otherwise $sum(a_1, a_2, \ldots, a_n) = a_n + sum(a_1, a_2, \ldots, a_{n-1})$.

12. See Example 4 in Section 4.4.

13. **a)** See p. 317.

b) We split the list into the two halves: $4, 10, 1, 5, 3$ and $8, 7, 2, 6, 9$. We then merge sort each half by applying this algorithm recursively and merging the results. For the first half, for example, this means splitting $4, 10, 1, 5, 3$ into the two halves $4, 10, 1$ and $5, 3$, recursively sorting each half, and merging. For the second half of this, for example, it means splitting into 5 and 3, recursively sorting each half, and merging. Since these two halves are already sorted, we just merge, into the sorted list $3, 5$. Similarly, we will get $1, 4, 10$ for the result of merge sort applied to $4, 10, 1$. When we merge $1, 4, 10$ and $3, 5$, we get $1, 3, 4, 5, 10$. Finally, we merge this with the sorted second half, $2, 6, 7, 8, 9$, to obtain the completely sorted list $1, 2, 3, 4, 5, 6, 7, 8, 9, 10$.

c) $O(n \log n)$; see pp. 319–321

14. **a)** no

b) No—you also need to show that it halts for all inputs, and the initial and final assertions for which you provide a proof of partial correctness need to be appropriate ones (i.e., relevant to the question of whether the program produces the correct output).

15. See the rules displayed in Section 4.5.

16. See p. 326.

SUPPLEMENTARY EXERCISES FOR CHAPTER 4

1. Let $P(n)$ be the statement that this equation holds. The basis step consists of verifying that $P(1)$ is true, which is trivial because $2/3 = 1 - (1/3^1)$. For the inductive step we assume that $P(k)$ is true and try to prove $P(k+1)$. We have

$$\begin{aligned}\frac{2}{3}+\frac{2}{9}+\frac{2}{27}+\cdots+\frac{2}{3^n}+\frac{2}{3^{n+1}} &= 1-\frac{1}{3^n}+\frac{2}{3^{n+1}} \quad \text{(by the inductive hypothesis)}\\ &= 1-\frac{3}{3^{n+1}}+\frac{2}{3^{n+1}}\\ &= 1-\frac{1}{3^{n+1}},\end{aligned}$$

as desired.

3. We prove this by induction on n. If $n = 1$ (basis step), then the equation reads $1 \cdot 2^0 = (1-1) \cdot 2^1 + 1$, which is the true statement $1 = 1$. Assume that the statement is true for n:

$$1 \cdot 2^0 + 2 \cdot 2^1 + 3 \cdot 2^2 + \cdots + n \cdot 2^{n-1} = (n-1) \cdot 2^n + 1$$

We must show that it is true for $n+1$. Thus we have

$$\begin{aligned}1 \cdot 2^0 + 2 \cdot 2^1 &+ 3 \cdot 2^2 + \cdots + n \cdot 2^{n-1} + (n+1) \cdot 2^n\\ &= (n-1) \cdot 2^n + 1 + (n+1) \cdot 2^n \quad \text{(by the inductive hypothesis)}\\ &= (2n) \cdot 2^n + 1\\ &= n \cdot 2^{n+1} + 1\\ &= \big((n+1)-1\big) \cdot 2^{n+1} + 1,\end{aligned}$$

exactly as desired.

5. We prove this by induction on n. If $n = 1$ (basis step), then the equation reads $1/(1 \cdot 4) = 1/4$, which is true. Assume that the statement is true for n:

$$\frac{1}{1 \cdot 4}+\frac{1}{4 \cdot 7}+\cdots+\frac{1}{(3n-2)(3n+1)} = \frac{n}{3n+1}$$

We must show that it is true for $n+1$. Thus we have

$$\begin{aligned}\frac{1}{1 \cdot 4}+\frac{1}{4 \cdot 7}+\cdots+\frac{1}{(3n-2)(3n+1)}&+\frac{1}{\big(3(n+1)-2\big)\big(3(n+1)+1\big)}\\ &= \frac{n}{3n+1}+\frac{1}{\big(3(n+1)-2\big)\big(3(n+1)+1\big)} \quad \text{(by the inductive hypothesis)}\\ &= \frac{n}{3n+1}+\frac{1}{(3n+1)(3n+4)}\\ &= \frac{1}{3n+1}\left(n+\frac{1}{3n+4}\right)\\ &= \frac{1}{3n+1}\left(\frac{3n^2+4n+1}{3n+4}\right)\\ &= \frac{1}{3n+1}\left(\frac{(3n+1)(n+1)}{3n+4}\right)\\ &= \frac{n+1}{3n+4}\\ &= \frac{n+1}{3(n+1)+1},\end{aligned}$$

exactly as desired.

7. Let $P(n)$ be the statement $2^n > n^3$. We want to prove that $P(n)$ is true for all $n > 9$. The basis step is $n = 10$, in which we have $2^{10} = 1024 > 1000 = 10^3$. Assume $P(n)$; we want to show $P(n+1)$. Then we have

$$\begin{aligned}(n+1)^3 &= n^3 + 3n^2 + 3n + 1 \\ &\le n^3 + 3n^2 + 3n^2 + 3n^2 \quad (\text{since } n \ge 1) \\ &= n^3 + 9n^2 \\ &< n^3 + n^3 \quad (\text{since } n > 9) \\ &= 2n^3 < 2 \cdot 2^n \quad (\text{by the inductive hypothesis}) \\ &= 2^{n+1},\end{aligned}$$

as desired.

9. This problem deals with factors in algebra. We have to be just a little clever. Let $P(n)$ be the statement that $a - b$ is a factor of $a^n - b^n$. We want to show that $P(n)$ is true for all positive integers n, and of course we will do so by induction. If $n = 1$, then we have the trivial statement that $a - b$ is a factor of $a - b$. Next assume the inductive hypothesis, that $P(n)$ is true. We want to show $P(n+1)$, that $a - b$ is a factor of $a^{n+1} - b^{n+1}$. The trick is to rewrite $a^{n+1} - b^{n+1}$ by subtracting and adding ab^n (and hence not changing its value). We obtain $a^{n+1} - b^{n+1} = a^{n+1} - ab^n + ab^n - b^{n+1} = a(a^n - b^n) + b^n(a - b)$. Now this expression contains two terms. By the inductive hypothesis, $a - b$ is a factor of the first term. Obviously $a - b$ is a factor of the second. Therefore $a - b$ is a factor of the entire expression, and we are done.

11. Let $P(n)$ be the given equation. It is certainly true for $n = 0$, since it reads $a = a$ in that case. Assume that $P(n)$ is true:

$$a + (a + d) + \cdots + (a + nd) = \frac{(n+1)(2a + nd)}{2}$$

Then

$$\begin{aligned}&a + (a + d) + \cdots + (a + nd) + \big(a + (n+1)d\big) \\ &\qquad = \frac{(n+1)(2a+nd)}{2} + \big(a + (n+1)d\big) \quad (\text{by the inductive hypothesis}) \\ &\qquad = \frac{(n+1)(2a+nd) + 2\big(a + (n+1)d\big)}{2} \\ &\qquad = \frac{(n+1)(2a+nd) + 2a + nd + nd + 2d}{2} \\ &\qquad = \frac{(n+2)(2a+nd) + (n+2)d}{2} \\ &\qquad = \frac{(n+2)\big(2a + (n+1)d\big)}{2},\end{aligned}$$

which is exactly $P(n+1)$.

13. We use induction. If $n = 1$, then the left-hand side has just one term, namely $5/6$, and the right-hand side is $10/12$, which is the same number. Assume that the equation holds for $n = k$, and consider $n = k + 1$. Then we have

$$\begin{aligned}\sum_{i=1}^{k+1} \frac{i+4}{i(i+1)(i+2)} &= \sum_{i=1}^{k} \frac{i+4}{i(i+1)(i+2)} + \frac{k+5}{(k+1)(k+2)(k+3)} \\ &= \frac{k(3k+7)}{2(k+1)(k+2)} + \frac{k+5}{(k+1)(k+2)(k+3)} \quad \text{(by the inductive hypothesis)} \\ &= \frac{1}{(k+1)(k+2)} \cdot \left(\frac{k(3k+7)}{2} + \frac{k+5}{k+3}\right) \\ &= \frac{1}{2(k+1)(k+2)(k+3)} \cdot (k(3k+7)(k+3) + 2(k+5)) \\ &= \frac{1}{2(k+1)(k+2)(k+3)} \cdot (3k^3 + 16k^2 + 23k + 10) \\ &= \frac{1}{2(k+1)(k+2)(k+3)} \cdot (3k+10)(k+1)^2 \\ &= \frac{1}{2(k+2)(k+3)} \cdot (3k+10)(k+1) \\ &= \frac{(k+1)(3(k+1)+7)}{2((k+1)+1)((k+1)+2)},\end{aligned}$$

as desired.

15. When $n = 1$, we are looking for the derivative of $g(x) = xe^x$, which, by the product rule, is $x \cdot e^x + e^x = (x+1)e^x$, so the statement is true for $n = 1$. Assume that the statement is true for $n = k$, that is, the kth derivative is given by $g^{(k)} = (x+k)e^x$. Differentiating by the product rule gives us the $(k+1)^{\text{st}}$ derivative: $g^{(k+1)} = (x+k)e^x + e^x = (x+(k+1))e^x$, as desired.

17. We look at the first few Fibonacci numbers to see if there is a pattern: $f_0 = 0$ (even), $f_1 = 1$ (odd), $f_2 = 1$ (odd), $f_3 = 2$ (even), $f_4 = 3$ (odd), $f_5 = 5$ (odd), ... The pattern seems to be even-odd-odd, repeated forever. Since the pattern has period 3, we can formulate our conjecture as follows: f_n is even if $n \equiv 0 \pmod 3$, and is odd in the other two cases. Let us prove this by mathematical induction. There are two base cases, $n = 0$ and $n = 1$. The conjecture is certainly true in each of them, since $0 \equiv 0 \pmod 3$ and f_0 is even, and $1 \not\equiv 0 \pmod 3$ and f_0 is odd. So we assume the inductive hypothesis and consider a given $n+1$. There are three cases to consider, depending on the value of $(n+1)$ **mod** 3. If $n+1 \equiv 0 \pmod 3$, then $n-1$ and n are congruent to 1 and 2 modulo 3, respectively. By the inductive hypothesis, both f_{n-1} and f_n are odd. Therefore f_{n+1}, which is the sum of these two numbers, is even, as desired. The other two cases are similar. If $n+1 \equiv 1 \pmod 3$, then $n-1$ and n are congruent to 2 and 0 modulo 3, respectively. By the inductive hypothesis, f_{n-1} is odd and f_n is even. Therefore f_{n+1}, which is the sum of these two numbers, is odd, as desired. On the other hand, if $n+1 \equiv 2 \pmod 3$, then $n-1$ and n are congruent to 0 and 1 modulo 3, respectively. By the inductive hypothesis, f_{n-1} is even and f_n is odd. Therefore f_{n+1}, which is the sum of these two numbers, is odd, as desired.

19. The important point to note here is that k can be thought of as a universally quantified variable for each n. Thus the statement we wish to prove is $P(n)$: for every k, $f_k f_n + f_{k+1} f_{n+1} = f_{n+k+1}$. We use mathematical induction. If $n = 0$ (the first base case), then we want to prove $P(0)$: for every k, $f_k f_0 + f_{k+1} f_1 = f_{0+k+1}$, which reduces to the identity $f_{k+1} = f_{k+1}$, since $f_0 = 0$ and $f_1 = 1$. If $n = 1$ (the second base case), then we want to prove $P(1)$: for every k, $f_k f_1 + f_{k+1} f_2 = f_{1+k+1}$, which reduces to the defining recurrence $f_k + f_{k+1} = f_{k+2}$, since $f_1 = 1$ and $f_2 = 1$. Now we assume the inductive hypothesis $P(n)$ and try to prove $P(n+1)$. It is a straightforward calculation, using the inductive hypothesis and the recursive definition of the

Fibonacci numbers:

$$\begin{aligned} f_k f_{n+1} + f_{k+1} f_{n+2} &= f_k(f_{n-1} + f_n) + f_{k+1}(f_n + f_{n+1}) \\ &= f_k f_{n-1} + f_k f_n + f_{k+1} f_n + f_{k+1} f_{n+1} \\ &= (f_k f_{n-1} + f_{k+1} f_n) + (f_k f_n + f_{k+1} f_{n+1}) \\ &= f_{n-1+k+1} + f_{n+k+1} = f_{n+k+2}\,, \end{aligned}$$

as desired.

21. Let $P(n)$ be the statement $l_0^2 + l_1^2 + \cdots + l_n^2 = l_n l_{n+1} + 2$. We easily verify the two base cases, $P(0)$ and $P(1)$, since $2^2 = 2 \cdot 1 + 2$ and $2^2 + 1^2 = 1 \cdot 3 + 2$. Next assume the inductive hypothesis and consider $P(n+1)$. We have

$$\begin{aligned} l_0^2 + l_1^2 + \cdots + l_n^2 + l_{n+1}^2 &= l_n l_{n+1} + 2 + l_{n+1}^2 \\ &= l_{n+1}(l_n + l_{n+1}) + 2 \\ &= l_{n+1} l_{n+2} + 2\,, \end{aligned}$$

which is exactly what we wanted.

23. The identity is clearly true for $n = 1$. Let us expand the right-hand side for $n+1$, invoking the inductive hypothesis at the appropriate point (and using the suggested trigonometric identities as well as the fact that $i^2 = -1$):

$$\begin{aligned} \cos(n+1)x + i\sin(n+1)x &= \cos(nx + x) + i\sin(nx + x) \\ &= \cos nx \cos x - \sin nx \sin x + i(\sin nx \cos x + \cos nx \sin x) \\ &= \cos x(\cos nx + i\sin nx) + \sin x(-\sin nx + i\cos nx) \\ &= \cos x(\cos nx + i\sin nx) + i\sin x(i\sin nx + \cos nx) \\ &= (\cos nx + i\sin nx)(\cos x + i\sin x) \\ &= (\cos x + i\sin x)^n(\cos x + i\sin x) \\ &= (\cos x + i\sin x)^{n+1} \end{aligned}$$

25. First let's rewrite the right-hand side to make it simpler to work with, namely as $2^{n+1}(n^2 - 2n + 3) - 6$. We use induction. If $n = 1$, then the left-hand side has just one term, namely 2, and the right-hand side is $4 \cdot 2 - 6 = 2$ as well. Assume that the equation holds for $n = k$, and consider $n = k+1$. Then we have

$$\begin{aligned} \sum_{j=1}^{k+1} j^2 2^j &= \sum_{j=1}^{k} j^2 2^j + (k+1)^2 2^{k+1} \\ &= 2^{k+1}(k^2 - 2k + 3) - 6 + (k^2 + 2k + 1)2^{k+1} \quad \text{(by the inductive hypothesis)} \\ &= 2^{k+1}(2k^2 + 4) - 6 \\ &= 2^{k+2}(k^2 + 2) - 6 \\ &= 2^{k+2}((k+1)^2 - 2(k+1) + 3) - 6, \end{aligned}$$

as desired.

27. One solution here is to use partial fractions and telescoping. First note that

$$\frac{1}{j^2 - 1} = \frac{1}{2}\left(\frac{1}{j-1} - \frac{1}{j+1}\right).$$

Therefore when summing from 1 to n, the terms being added and the terms being subtracted all cancel out except for $1/(j-1)$ when $j = 2$ and 3, and $1/(j+1)$ when $j = n-1$ and n. Thus the sum is just

$$\frac{1}{2}\left(\frac{1}{1} + \frac{1}{2} - \frac{1}{n} - \frac{1}{n+1}\right).$$

which simplifies, with a little algebra, to the expression given on the right-hand side of the formula in the exercise.

In the spirit of this chapter, however, we also give a proof by mathematical induction. Let $P(n)$ be the formula in the exercise. The basis step is for $n=2$, in which case both sides reduce to $1/3$. For the inductive step assume that the equation holds for $n=k$, and consider $n=k+1$. Then we have

$$\begin{aligned}
\sum_{j=1}^{k+1}\frac{1}{j^2-1} &= \sum_{j=1}^{k}\frac{1}{j^2-1}+\frac{1}{(k+1)^2-1}\\
&= \frac{(k-1)(3k+2)}{4k(k+1)}+\frac{1}{(k+1)^2-1} \qquad \text{(by the inductive hypothesis)}\\
&= \frac{(k-1)(3k+2)}{4k(k+1)}+\frac{1}{k^2+2k} = \frac{(k-1)(3k+2)}{4k(k+1)}+\frac{1}{k(k+2)}\\
&= \frac{(k-1)(3k+2)(k+2)+4(k+1)}{4k(k+1)(k+2)}\\
&= \frac{3k^3+5k^2}{4k(k+1)(k+2)} = \frac{3k^2+5k}{4(k+1)(k+2)}\\
&= \frac{k(3k+5)}{4(k+1)(k+2)} = \frac{((k+1)-1)(3(k+1)+2)}{4(k+1)(k+2)},
\end{aligned}$$

which is exactly what $P(k+1)$ asserts.

29. Let $P(n)$ be the assertion that at least $n+1$ lines are needed to cover the lattice points in the given triangular region. Clearly $P(0)$ is true, because we need at least one line to cover the one point at $(0,0)$. Assume the inductive hypothesis, that at least $k+1$ lines are needed to cover the lattice points with $x\geq 0$, $y\geq 0$, and $x+y\leq k$. Consider the triangle of lattice points defined by $x\geq 0$, $y\geq 0$, and $x+y\leq k+1$. Because this set includes the previous set, at least $k+1$ lines are required just to cover the smaller set (by the inductive hypothesis). By way of contradiction, assume that $k+1$ lines could cover this larger set as well. Then these lines must also cover the $k+2$ points on the line $x+y=k+1$, namely $(0,k+1)$, $(1,k)$, $(2,k-1)$, ..., $(k,1)$, $(k+1,0)$. But only the line $x+y=k+1$ itself can cover more than one of these points, because two distinct lines intersect in at most one point, and this line does nothing toward covering the lattice points in the smaller triangle. Therefore none of the $k+1$ lines that are needed to cover the lattice points in the smaller triangle can cover more than one of the points on the line $x+y=k+1$, and this leaves at least one point uncovered. Therefore our assumption that $k+1$ line could cover the larger set is wrong, and our proof is complete.

31. The basis step is the given statement defining $\mathbf{B}$. Assume the inductive hypothesis, that $\mathbf{B}^k=\mathbf{MA}^k\mathbf{M}^{-1}$. We want to prove that $\mathbf{B}^{k+1}=\mathbf{MA}^{k+1}\mathbf{M}^{-1}$. By definition $\mathbf{B}^{k+1}=\mathbf{BB}^k=\mathbf{MAM}^{-1}\mathbf{B}^k=\mathbf{MAM}^{-1}\mathbf{MA}^k\mathbf{M}^{-1}$ by the inductive hypothesis. But this simplifies, using rules for matrices, to $\mathbf{MAIA}^k\mathbf{M}^{-1}=\mathbf{MAA}^k\mathbf{M}^{-1}=\mathbf{MA}^{k+1}\mathbf{M}^{-1}$, as desired.

33. It takes some luck to be led to the solution here. We see that we can write $3!=3+2+1$. We also have a recursive definition of factorial, that $(n+1)!=(n+1)n!$, so we might hope to multiply each of the divisors we got at the previous stage by $n+1$ to get divisors at this stage. Thus we would have $4!=4\cdot 3!=4(3+2+1)=12+8+4$, but that gives us only three divisors in the sum, and we want four. That last divisor, which is $n+1$ can, however, be rewritten as the sum of n and 1, so our sum for $4!$ is $12+8+3+1$. Let's see if we can continue this. We have $5!=5\cdot 4!=5(12+8+3+1)=50+40+15+5=50+40+15+4+1$. It seems to be working. The basis step $n=3$ is already done, so let's see if we can prove the inductive step. Assume that we can write $k!$ as a sum of the desired form, say $k!=a_1+a_2+\cdots+a_k$, where each a_i is a divisor of $k!$ and the divisors are listed in strictly decreasing order, and consider $(k+1)!$. Then we have $(k+1)!=(k+1)k!=$

$(k+1)(a_1+a_2+\cdots+a_k) = (k+1)a_1+(k+1)a_2+\cdots+(k+1)a_k = (k+1)a_1+(k+1)a_2+\cdots+k\cdot a_k+a_k$. Because each a_i was a divisor of $k!$, each $(k+1)a_i$ is a divisor of $(k+1)!$, but what about those last two terms? We don't seem to have any way to know that $k\cdot a_k$ is a factor of $(k+1)!$. Hold on, in our exploration we always had the last divisor in our sum being 1. If so, then $k\cdot a_k = k$, which is a divisor of $(k+1)!$, and $a_k = 1$, so the new last summand is again 1. (Notice also that our list of summands is still in strictly decreasing order.) So our proof by mathematical induction needs to be of the following stronger result: For every $n \geq 3$, we can write $n!$ as the sum of n of its distinct positive divisors, one of which is 1. The argument we have just given proves this by mathematical induction.

35. When $n=1$ the statement is vacuously true. If $n=2$ there must be a woman first and a man second, so the statement is true. Assume that the statement is true for $n=k$, where $k\geq 2$, and consider $k+1$ people standing in a line, with a woman first and a man last. If the kth person is a woman, then we have that woman standing in front of the man at the end, and we are done. If the kth person is a man, then the first k people in line satisfy the conditions of the inductive hypothesis for the first k people in line, so again we can conclude that there is a woman directly in front of a man somewhere in the line.

37. (It will be helpful for the reader to draw a diagram to help in following this proof.) When $n=1$ there is one circle, and we can color the inside blue and the outside red to satisfy the conditions. Assume the inductive hypothesis that if there are k circles, then the regions can be 2-colored such that no regions with a common boundary have the same color, and consider a situation with $k+1$ circles. Remove one of the circles, producing a picture with k circles, and invoke the inductive hypothesis to color it in the prescribed manner. Then replace the removed circle and change the color of every region inside this circle (from red to blue, and from blue to red). It is clear that the resulting figure satisfies the condition, since if two regions have a common boundary, then either that boundary was an arc of the new circle, in which case the regions on either side used to be the same region and now the inside portion is colored differently from the outside, or else the boundary did not involve the new circle, in which case the regions are colored differently because they were colored differently before the new circle was restored.

39. We use induction. If $n=1$ then the equation reads $1\cdot 1 = 1\cdot 2/2$, which is true. Assume that the equation is true for n and consider it for $n+1$. (We use the letter n rather than k, because k is used for something else here.) Then we have, with some messy algebra,

$$\begin{aligned}
\sum_{j=1}^{n+1}(2j-1)\left(\sum_{k=j}^{n+1}\frac{1}{k}\right) &= \sum_{j=1}^{n}(2j-1)\left(\sum_{k=j}^{n+1}\frac{1}{k}\right) + (2(n+1)-1)\cdot\frac{1}{n+1} \\
&= \sum_{j=1}^{n}(2j-1)\left(\frac{1}{n+1}+\sum_{k=j}^{n}\frac{1}{k}\right) + \frac{2n+1}{n+1} \\
&= \left(\frac{1}{n+1}\sum_{j=1}^{n}(2j-1)\right) + \left(\sum_{j=1}^{n}(2j-1)\sum_{k=j}^{n}\frac{1}{k}\right) + \frac{2n+1}{n+1} \\
&= \left(\frac{1}{n+1}\cdot n^2\right) + \frac{n(n+1)}{2} + \frac{2n+1}{n+1} \quad \text{(by the inductive hypothesis)} \\
&= \frac{2n^2+n(n+1)^2+(4n+2)}{2(n+1)} \\
&= \frac{2(n+1)^2+n(n+1)^2}{2(n+1)} \\
&= \frac{(n+1)(n+2)}{2},
\end{aligned}$$

as desired.

41. **a)** $M(102) = 102 - 10 = 92$ **b)** $M(101) = 101 - 10 = 91$

c) $M(99) = M(M(99+11)) = M(M(110)) = M(100) = M(M(111)) = M(101) = 91$

d) $M(97) = M(M(108)) = M(98) = M(M(109)) = M(99) = 91$ (using part **(c)**)

e) This one is too long to show in its entirety here, but here is what is involved. First, $M(87) = M(M(98)) = M(91)$, using part **(d)**. Then $M(91) = M(M(102)) = M(92)$ from part **(a)**. In a similar way, we find that $M(92) = M(93)$, and so on, until it equals $M(97)$, which we found in part **(d)** to be 91. Hence the answer is 91.

f) Using what we learned from part **(e)**, we have $M(76) = M(M(87)) = M(91) = 91$.

43. The basis step is wrong. The statement makes no sense for $n = 1$, since the last term on the left-hand side would then be $1/(0 \cdot 1)$, which is undefined. The first n for which it makes sense is $n = 2$, when it reads

$$\frac{1}{1 \cdot 2} = \frac{3}{2} - \frac{1}{2}.$$

Of course this statement is false, since $\frac{1}{2} \neq 1$. Therefore the basis step fails, and so the "theorem" is not true.

45. We will prove by induction that n circles divide the plane into $n^2 - n + 2$ regions. One circle certainly divides the plane into two regions (the inside and the outside), and $1^2 - 1 + 2 = 2$. Thus the statement is correct for $n = 1$. We assume that the statement is true for n circles, and consider it for $n + 1$ circles. Let us imagine an arrangement of $n + 1$ circles in the plane, each pair intersecting in exactly two points, no point common to three circles. If we remove one circle, then we are left with n circles, and by the inductive hypothesis they divide the plane into $n^2 - n + 2$ regions. Now let us draw the circle that we removed, starting at a point at which it intersects another circle. As we proceed around the circle, every time we encounter a point on one of the circles that was already there, we cut off a new region (in other words, we divide one old region into two). Therefore the number of regions that are added on account of this circle is equal to the number of points of intersection of this circle with the other n circles. We are told that each other circle intersects this one in exactly two points. Therefore there are a total of $2n$ points of intersection, and hence $2n$ new regions. Therefore the number of regions determined by $n+1$ circles is $n^2 - n + 2 + 2n = n^2 + n + 2 = (n+1)^2 - (n+1) + 2$ (the last equality is just algebra). Thus we have derived that the statement is also true for $n + 1$, and our proof is complete.

47. We will give a proof by contradiction. Let us consider the set $B = \{\, b\sqrt{2} \mid b \text{ and } b\sqrt{2} \text{ are positive integers} \,\}$. Clearly B is a subset of the set of positive integers. Now if $\sqrt{2}$ is rational, say $\sqrt{2} = p/q$, then $B \neq \emptyset$, since $q\sqrt{2} = p \in B$. Therefore by the well-ordering property, B contains a smallest element, say $a = b\sqrt{2}$. Then $a\sqrt{2} - a = a\sqrt{2} - b\sqrt{2} = (a - b)\sqrt{2}$. Since $a\sqrt{2} = 2b$ and a are both integers, so is this quantity. Furthermore, it is a positive integer, since it equals $a(\sqrt{2} - 1)$ and $\sqrt{2} - 1 > 0$. Therefore $a\sqrt{2} - a \in B$. But clearly $a\sqrt{2} - a < a$, since $\sqrt{2} < 2$. This contradicts our choice of a to be the smallest element of B. Therefore our original assumption that $\sqrt{2}$ is rational is false.

49. **a)** We use the following lemma: A positive integer d is a common divisor of $a_1, a_2, \ldots, a_n$ if and only if d is a divisor of $\gcd(a_1, a_2, \ldots, a_n)$. [Proof: The prime factorization of $\gcd(a_1, a_2, \ldots, a_n)$ is $\prod p_i^{e_i}$, where e_i is the minimum exponent of p_i among $a_1, a_2, \ldots, a_n$. Clearly d divides every a_j if and only if the exponent of p_i in the prime factorization of d is less than or equal to e_i for every i, which happens if and only if $d \mid \gcd(a_1, a_2, \ldots, a_n)$.] Now let $d = \gcd(a_1, a_2, \ldots, a_n)$. Then d must be a divisor of each a_i, and hence must be a divisor of $\gcd(a_{n-1}, a_n)$ as well. Therefore d is a common divisor of $a_1, a_2, \ldots, a_{n-2}, \gcd(a_{n-1}, a_n)$. To show that it is the greatest common divisor of these numbers, suppose that e is any common divisor of these numbers. Then e is a divisor of each a_i for $1 \le i \le n-2$, and, being a divisor of $\gcd(a_{n-1}, a_n)$, it is also a divisor of a_{n-1} and a_n. Therefore e is a common divisor of *all* the a_i and hence a divisor of their common divisor, d. This shows that d is the greatest common divisor of $a_1, a_2, \ldots, a_{n-2}, \gcd(a_{n-1}, a_n)$.

b) If $n = 2$, then we just apply the Euclidean algorithm to a_1 and a_2. Otherwise, we apply the Euclidean algorithm to a_{n-1} and a_n, obtaining an answer d, and then apply this algorithm recursively to $a_1, a_2, \ldots, a_{n-2}, d$. Note that this last sequence has only $n-1$ numbers in it.

51. We begin by computing $f(n)$ for the first few values of n, using the recursive definition. Thus we have $f(1) = 1$, $f(2) = f(1)+4-1 = 1+4-1 = 4$, $f(3) = f(2)+6-1 = 4+6-1 = 9$, $f(4) = f(3)+8-1 = 9+8-1 = 16$. The pattern seems clear, so we conjecture that $f(n) = n^2$. Now we prove this by induction. The base case we have already verified. So assume that $f(n) = n^2$. All we have to do is show that $f(n+1) = (n+1)^2$. By the recursive definition we have $f(n+1) = f(n) + 2(n+1) - 1$. This equals $n^2 + 2(n+1) - 1$ by the inductive hypothesis, and by algebra we have $n^2 + 2(n+1) - 1 = (n+1)^2$, as desired.

53. The recursive definition says that we can "grow" strings in S by appending 0's on the left and 1's on the right, as many as we wish.

a) The only string of length 0 in S is λ. There are two strings of length 1 in S, obtained either by appending a 0 to the front of λ or a 1 to the end of λ, namely the strings 0 and 1. The strings of length 2 in S come from the strings of length 1 by appending either a 0 to the front or a 1 to the end; they are 00, 01, and 11. Similarly, we can append a 0 to the front or a 1 to the end of any of these strings to get the strings of length 3 in S, namely 000, 001, 011, and 111. Continuing in this manner, we see that the other strings in S of length less than or equal to 5 are 0000, 0001, 0011, 0111, 1111, 00000, 00001, 00011, 00111, 01111, and 11111.

b) The simplest way to describe these strings is $\{\, 0^m 1^n \mid m \text{ and } n \text{ are nonnegative integers} \,\}$.

55. Applying the first recursive step once to λ tells us that $() \in B$. Then applying the second recursive step to this string tells us that $()() \in B$. Finally, we apply the first recursive step once more to get $(()()) \in B$. To see that $(()))$ is not in B, we invoke Exercise 58. Since the number of left parentheses does not equal the number of right parentheses, this string is not balanced.

57. There is of course the empty string, with 0 symbols. By the first recursive rule, we get the string $()$. If we apply the first recursive rule to this string, then we get $(())$, and if we apply the second recursive rule, then we get $()()$. These are the only strings in B with four or fewer symbols.

59. The definition simply says that N of a string is a count of the parentheses, with each left parenthesis counting $+1$ and each right parenthesis counting -1.

a) There is one left parenthesis and one right parenthesis, so $N\big(()\big) = 1 - 1 = 0$.

b) There are 3 left parentheses and 5 right parentheses, so $N()))())(() = 3 - 5 = -2$.

c) There are 4 left parentheses and 2 right parentheses, so $N\big(((()(()\big) = 4 - 2 = 2$.

d) There are 6 left parentheses and 6 right parentheses, so $N\big(()((()))(())\big) = 6 - 6 = 0$.

61. The basic idea, of course, is to turn the definition into a procedure. The recursive part of the definition tells us how to find elements of B from shorter elements of B. The naive approach, however, is not very good, because we end up adding to B strings that already are there. For example, the string $()()()$ occurs in two different ways from the rule "$xy \in B$ if $x, y \in B$": by letting $x = ()()$ and $y = ()$, and by letting $x = ()$ and $y = ()()$.

To avoid this problem, we will keep two lists of strings, whose union is the set $B(n)$ of balanced strings of parentheses of length not exceeding n. The set $S(n)$ will be those balanced strings w of length at most n such that $w = uv$, where $u, v \neq \lambda$ and u and v are balanced. The set $T(n)$ will be all other balanced strings of length at most n. Note that, for example, $\lambda \in T$, $() \in T$, $(()) \in T$, but $()() \in S$. Since all the strings in B are of even length, we really only need to work with even values of n, dragging the odd values along for the ride.

procedure *generate*(n : nonnegative integer)
if n is odd **then**
begin
  $S := S(n-1)$ {the S constructed by *generate*$(n-1)$}
  $T := T(n-1)$ {the T constructed by *generate*$(n-1)$}
end
else if $n = 0$ **then**
begin
  $S := \emptyset$
  $T := \{\lambda\}$
end
else
begin
  $S' := S(n-2)$ {the S constructed by *generate*$(n-2)$}
  $T' := T(n-2)$ {the T constructed by *generate*$(n-2)$}
  $T := T' \cup \{\, (x) \mid x \in T' \cup S' \wedge \text{length}(x) = n-2 \,\}$
  $S := S' \cup \{\, xy \mid x \in T' \wedge y \in T' \cup S' \wedge \text{length}(xy) = n \,\}$
end
{$T \cup S$ is the set of balanced strings of length at most n}

63. There are two cases. If $x \le y$ initially, then the statement $x := y$ is not executed, so x and y remain unchanged and $x \le y$ is a true final assertion. If $x > y$ initially, then the statement $x := y$ is executed, so $x = y$ at the end, and thus $x \le y$ is again a true final assertion. These are the only two possibilities associated with the initial condition **T** (true), so our proof is complete.

65. If the list has just one element in it, then the number of 0's is 1 if the element is 0 and is 0 otherwise. That forms the basis step of our algorithm. For the recursive step, the number of occurrences of 0 in a list L is the same as the number of occurrences in the list L without its last term if that last term is not a 0, and is one more than this value if it is. We can write this in pseudocode as follows.

procedure *zerocount*($a_1, a_2, \ldots, a_n$: list of integers)
if $n = 1$ **then**
  if $a_1 = 0$ **then** *zerocount*$(a_1, a_2, \ldots, a_n) := 1$
  else *zerocount*$(a_1, a_2, \ldots, a_n) := 0$
else
  if $a_n = 0$ **then** *zerocount*$(a_1, a_2, \ldots, a_n) :=$ *zerocount*$(a_1, a_2, \ldots, a_{n-1}) + 1$
  else *zerocount*$(a_1, a_2, \ldots, a_n) :=$ *zerocount*$(a_1, a_2, \ldots, a_{n-1})$

67. From the numerical evidence in Exercise 66, it appears that $a(n)$ is a natural number and $a(n) \le n$ for all n. We prove that a is well-defined by showing that this observation is in fact true. Obviously the proof is by mathematical induction. The basis step is $n = 0$, for which the statement is obviously true, since $a(0) = 0$. Now assume that $a(n-1)$ is a natural number and $a(n-1) \le n-1$. Then $a(a(n-1))$ is a applied to some natural number less than or equal to $n-1$; by the inductive hypothesis this value is some natural number less than or equal to $n-1$. Therefore $a(a(n-1))$ is also some natural number less than or equal to $n-1$ (again by the inductive hypothesis). Therefore $n - a(a(n-1))$ is n minus some natural number less than or equal to $n-1$, which is some natural number less than or equal to n, and we are done.

69. From Exercise 68 we know that $a(n) = \lfloor (n+1)\mu \rfloor$ and that $a(n-1) = \lfloor n\mu \rfloor$. Since μ is less than 1, these two values are either equal or they differ by 1. First suppose that $\mu n - \lfloor \mu n \rfloor < 1 - \mu$. This is equivalent to $\mu(n+1) < 1 + \lfloor \mu n \rfloor$. If this is true, then clearly $\lfloor \mu(n+1) \rfloor = \lfloor \mu n \rfloor$. On the other hand, if $\mu n - \lfloor \mu n \rfloor \ge 1 - \mu$, then $\mu(n+1) \ge 1 + \lfloor \mu n \rfloor$, so $\lfloor \mu(n+1) \rfloor = \lfloor \mu n \rfloor + 1$, as desired.

71. We apply the definition:

$$\begin{aligned}
m(0) &= 0 & f(0) &= 1\\
m(1) &= 1 - f(m(0)) = 1 - f(0) = 1 - 1 = 0 & f(1) &= 1 - m(f(0)) = 1 - m(1) = 1 - 0 = 1\\
m(2) &= 2 - f(m(1)) = 2 - f(0) = 2 - 1 = 1 & f(2) &= 2 - m(f(1)) = 2 - m(1) = 2 - 0 = 2\\
m(3) &= 3 - f(m(2)) = 3 - f(1) = 3 - 1 = 2 & f(3) &= 3 - m(f(2)) = 3 - m(2) = 3 - 1 = 2\\
m(4) &= 4 - f(m(3)) = 4 - f(2) = 4 - 2 = 2 & f(4) &= 4 - m(f(3)) = 4 - m(2) = 4 - 1 = 3\\
m(5) &= 5 - f(m(4)) = 5 - f(2) = 5 - 2 = 3 & f(5) &= 5 - m(f(4)) = 5 - m(3) = 5 - 2 = 3\\
m(6) &= 6 - f(m(5)) = 6 - f(3) = 6 - 2 = 4 & f(6) &= 6 - m(f(5)) = 6 - m(3) = 6 - 2 = 4\\
m(7) &= 7 - f(m(6)) = 7 - f(4) = 7 - 3 = 4 & f(7) &= 7 - m(f(6)) = 7 - m(4) = 7 - 2 = 5\\
m(8) &= 8 - f(m(7)) = 8 - f(4) = 8 - 3 = 5 & f(8) &= 8 - m(f(7)) = 8 - m(5) = 8 - 3 = 5\\
m(9) &= 9 - f(m(8)) = 9 - f(5) = 9 - 3 = 6 & f(9) &= 9 - m(f(8)) = 9 - m(5) = 9 - 3 = 6
\end{aligned}$$

73. By Exercise 72 the sequence starts out 1, 2, 2, 3, 3, 4, 4, 4, 5, 5, 5, 6, 6, 6, 6, 7, 7, 7, 7, 8, ..., and we see that $f(1) = 1$ (since the last occurrence of 1 is in position 1), $f(2) = 3$ (since the last occurrence of 2 is in position 3), $f(3) = 5$ (since the last occurrence of 3 is in position 5), $f(4) = 8$ (since the last occurrence of 4 is in position 8), and so on. Since the sequence is nondecreasing, the last occurrence of n must be in the position for which the total number of 1s, 2s, 3s, ..., n's all together is that position number. But since a_k gives the number of occurrences of k, this is just $\sum_{k=1}^{n} a_k$, as desired. For example,

$$\sum_{k=1}^{6} a_k = 1 + 2 + 2 + 3 + 3 + 4 = 15 = f(6),$$

the position where the last 6 occurs.

Since we just saw that $f(n)$ is the sum of the first n terms of the sequence, $f(f(n))$ must be the sum of the first $f(n)$ terms of the sequence. But since $f(n)$ is the last term whose value is n, this means the sum of all the terms of the sequence whose value is at most n. Since there are a_k terms of the sequence whose value is k, this sum must be $\sum_{k=1}^{n} k \cdot a_k$, as desired. For example,

$$\sum_{k=1}^{3} k \cdot a_k = 1 \cdot 1 + 2 \cdot 2 + 2 \cdot 3 = 11 = f(f(3)) = f(5),$$

the position where the last 5 occurs.

WRITING PROJECTS FOR CHAPTER 4

Books and articles indicated by bracketed symbols below are listed near the end of this manual. You should also read the general comments and advice you will find there about researching and writing these essays.

1. Start with the historical footnote in the text. The standard history of mathematics references, such as [Bo4] and [Ev3], might have something or might provide a hint of where to look next.

2. There is a nice chapter on this in [He]. A Web search should also turn up useful pages, such as this one: `http://www-cgrl.cs.mcgill.ca/~godfried/teaching/cg-projects/97/Octavian/compgeom.html`

3. There are several textbooks on computational geometry, such as [O'R]. A comprehensive website on the subject can be found here: `http://compgeom.cs.uiuc.edu/~jeffe/compgeom/`

4. The ratio of successive Fibonacci numbers approaches a value known as the "golden ratio," so it would be useful to search for this topic as well. A recent and somewhat controversial book on the subject [Li] debunks some of the more outrageous claimed applications. This website seems to have a wealth of information on applications of Fibonacci numbers: `http://www.cs.rit.edu/~pga/Fibo/fact_sheet.html`

5. You can find some references, as well as an historical discussion of the Ackermann function and an iterative algorithm for computing it, in [GrZe].

6. Try searching your library's on-line catalog or the Web under keywords like *program correctness* or *verification.* Or look at [Ba1], [Di], or [Ho1].

7. As in Writing Project 6, a key-word search might turn up something. One book to look at is [De1].

CHAPTER 5
Counting

SECTION 5.1 The Basics of Counting

The secret to solving counting problems is to look at the problem the right way and apply the correct rules (usually the product rule or the sum rule), often with some common sense and cleverness thrown in. This is usually easier said than done, but it gets easier the more problems you do. Sometimes you need to count more than you want and then subtract the overcount. (This notion is made more precise in Section 7.5, where the inclusion–exclusion principle is discussed explicitly.) At other times you compensate by dividing; see Exercise 61, for example. Counting problems are sometimes ambiguous, so it is possible that your answer, although different from the answer we obtain, is the correct answer to a different interpretation of the problem; try to figure out whether that is the case.

If you have trouble with a problem, simplify the parameters to make them more manageable (if necessary) and try to list the set in question explicitly. This will often give you an idea of what is going on and suggest a general method of attack that will solve the problem as given. For example, in Exercise 11 you are asked about bit strings of length 10. If you are having difficulty, investigate the analogous question about bit strings of length 2 or 3, where you can write down the entire set, and see if a pattern develops. (Some people define mathematics as the study of patterns.) Sometimes tree diagrams make the analysis in these small cases easier to keep track of.

See the solution to Exercise 41 for a discussion of the powerful tool of **symmetry**, *which you will often find helpful. Another useful trick is* **gluing**; *see the solution to Exercise 53.*

Finally, do not be misled, if you find these exercises easy, into thinking that combinatorial problems are a piece of cake. It is very easy to ask combinatorial questions that look just like the ones asked here but in fact are extremely difficult, if not impossible. For example, try your hand at the following problem: how many strings are there, using 10 A's, 12 B's, 11 C's, and 15 D's, such that no A is followed by a B, and no C is followed by a D?

1. This problem illustrates the difference between the product rule and the sum rule. If we must make one choice *and* then another choice, the product rule applies, as in part **(a)**. If we must make one choice *or* another choice, the sum rule applies, as in part **(b)**. We assume in this problem that there are no double majors.

a) The product rule applies here, since we want to do two things, one after the other. First, since there are 18 mathematics majors, and we are to choose one of them, there are 18 ways to choose the mathematics major. Then we must choose the computer science major from among the 325 computer science majors, and that can clearly be done in 325 ways. Therefore there are, by the product rule, $18 \cdot 325 = 5850$ ways to pick the two representatives.

b) The sum rule applies here, since we want to do one of two mutually exclusive things. Either we can choose a mathematics major to be the representative, or we can choose a computer science major. There are 18 ways to choose a mathematics major, and there are 325 ways to choose a computer science major. Since these two actions are mutually exclusive (no one is both a mathematics major and a computer science major), and since we want to do one of them or the other, there are $18 + 325 = 343$ ways to pick the representative.

3. **a)** The product rule applies, since the student will perform each of 10 tasks, one after the other. There are 4 ways to do each task. Therefore there are $4 \cdot 4 \cdots 4 = 4^{10} = 1{,}048{,}576$ ways to answer the questions on the test.

b) This is identical to part **(a)**, except that now there are 5 ways to answer each question—give any of the 4 answers or give no answer at all. Therefore there are $5^{10} = 9{,}765{,}625$ ways to answer the questions on the test.

5. The product rule applies here, since a flight is determined by choosing an airline for the flight from New York to Denver (which can be done in 6 ways) and then choosing an airline for the flight from Denver to San Francisco (which can be done in 7 ways). Therefore there are $6 \cdot 7 = 42$ different possibilities for the entire flight.

7. Three-letter initials are determined by specifying the first initial (26 ways), then the second initial (26 ways), and then the third initial (26 ways). Therefore by the product rule there are $26 \cdot 26 \cdot 26 = 26^3 = 17{,}576$ possible three-letter initials.

9. There is only one way to specify the first initial, but as in Exercise 7, there are 26 ways to specify each of the other initials. Therefore there are, by the product rule, $1 \cdot 26 \cdot 26 = 26^2 = 676$ possible three-letter initials beginning with A.

11. A bit string is determined by choosing the bits in the string, one after another, so the product rule applies. We want to count the number of bit strings of length 10, except that we are not free to choose either the first bit or the last bit (they are mandated to be 1's). Therefore there are 8 choices to make, and each choice can be made in 2 ways (the bit can be either a 1 or a 0). Thus the product rule tells us that there are $2^8 = 256$ such strings.

13. This is a trick question, since it is easier than one might expect. Since the string is given to consist entirely of 1's, there is nothing to choose except the length. Since there are $n+1$ possible lengths not exceeding n (if we include the empty string, of length 0), the answer is simply $n+1$. Note that the empty string consists—vacuously—entirely of 1's.

15. By the sum rule we can count the number of strings of length 4 or less by counting the number of strings of length i, for $0 \le i \le 4$, and then adding the results. Now there are 26 letters to choose from, and a string of length i is specified by choosing its characters, one after another. Therefore, by the product rule there are 26^i strings of length i. The answer to the question is thus $\sum_{i=0}^{4} 26^i = 1 + 26 + 676 + 17576 + 456976 = 475{,}255$.

17. The easiest way to count this is to find the total number of ASCII strings of length five and then subtract off the number of such strings that do not contain the @ character. Since there are 128 characters to choose from in each location in the string, the answer is $128^5 - 127^5 = 34{,}359{,}738{,}368 - 33{,}038{,}369{,}407 = 1{,}321{,}368{,}961$.

19. Because neither 100 nor 50 is divisible by either 7 or 11, whether the ranges are meant to be inclusive or exclusive of their endpoints is moot.

a) There are $\lfloor 100/7 \rfloor = 14$ integers less than 100 that are divisible by 7, and $\lfloor 50/7 \rfloor = 7$ of them are less than 50 as well. This leaves $14 - 7 = 7$ numbers between 50 and 100 that are divisible by 7. They are 56, 63, 70, 77, 84, 91, and 98.

b) There are $\lfloor 100/11 \rfloor = 9$ integers less than 100 that are divisible by 11, and $\lfloor 50/11 \rfloor = 4$ of them are less than 50 as well. This leaves $9 - 4 = 5$ numbers between 50 and 100 that are divisible by 11. They are 55, 66, 77, 88, and 99.

c) A number is divisible by both 7 and 11 if and only if it is divisible by their least common multiple, which is 77. Obviously there is only one such number between 50 and 100, namely 77. We could also work this out as we did in the previous parts: $\lfloor 100/77 \rfloor - \lfloor 50/77 \rfloor = 1 - 0 = 1$. Note also that the intersection of the sets we found in the previous two parts is precisely what we are looking for here.

21. This problem deals with the set of positive integers between 100 and 999, inclusive. Note that there are exactly $999 - 100 + 1 = 900$ such numbers. A second way to see this is to note that to specify a three-digit number, we need to choose the first digit to be nonzero (which can be done in 9 ways) and then the second and third digits (which can each be done in 10 ways), for a total of $9 \cdot 10 \cdot 10 = 900$ ways, by the product rule. A third way to see this (perhaps most relevant for this problem) is to note that a number of the desired form is a number less than or equal to 999 (and there are 999 such numbers) but not less than or equal to 99 (and there are 99 such numbers); therefore there are $999 - 99 = 900$ numbers in the desired range.

a) Every seventh number—7, 14, and so on—is divisible by 7. Therefore the number of positive integers less than or equal to n and divisible by 7 is $\lfloor n/7 \rfloor$ (the floor function is used—we have to round down—because the first six positive integers are not multiples of 7; for example there are only $\lfloor 20/7 \rfloor = 2$ multiples of 7 less than or equal to 20). So we find that there are $\lfloor 999/7 \rfloor = 142$ multiples of 7 not exceeding 999, of which $\lfloor 99/7 \rfloor = 14$ do not exceed 99. Therefore there are exactly $142 - 14 = 128$ numbers in the desired range divisible by 7.

b) This is similar to part **(a)**, with 7 replaced by 2, but with the added twist that we want to count the numbers *not* divisible by 2. Mimicking the analysis in part **(a)**, we see that there are $\lfloor 999/2 \rfloor = 499$ even numbers not exceeding 999, and therefore $999 - 499 = 500$ odd ones; there are similarly $99 - \lfloor 99/2 \rfloor = 50$ odd numbers less than or equal to 99. Therefore there are $500 - 50 = 450$ odd numbers between 100 and 999 inclusive.

c) There are just 9 possible digits that a three-digit number can start with. If all of its digits are to be the same, then there is no choice after the leading digit has been specified. Therefore there are 9 such numbers.

d) This is similar to part **(b)**, except that 2 is replaced by 4. Following the analysis there, we find that there are $999 - \lfloor 999/4 \rfloor = 750$ positive integers less than or equal to 999 not divisible by 4, and $99 - \lfloor 99/4 \rfloor = 75$ such positive integers less than or equal to 99. Therefore there are $750 - 75 = 675$ three-digit integers not divisible by 4.

e) The method is similar to that used in the earlier parts. There are $\lfloor 999/3 \rfloor - \lfloor 99/3 \rfloor = 300$ three-digit numbers divisible by 3, and $\lfloor 999/4 \rfloor - \lfloor 99/4 \rfloor = 225$ three-digit numbers divisible by 4. Moreover there are $\lfloor 999/12 \rfloor - \lfloor 99/12 \rfloor = 75$ numbers divisible by both 3 and 4, i.e., divisible by 12. In order to count each number divisible by 3 or 4 once and only once, we need to add the number of numbers divisible by 3 to the number of numbers divisible by 4, and then subtract the number of numbers divisible by both 3 and 4 so as not to count them twice. Therefore the answer is $300 + 225 - 75 = 450$.

f) In part **(e)** we found that there were 450 three-digit integers that *are* divisible by either 3 or 4. The others, of course, are not. Therefore there are $900 - 450 = 450$ three-digit integers that are not divisible by either 3 or 4.

g) We saw in part **(e)** that there are 300 three-digit numbers divisible by 3 and that 75 of them are also divisible by 4. The remainder of those 300 numbers, therefore, are not divisible by 4. Thus the answer is $300 - 75 = 225$.

h) We saw in part **(e)** that there are 75 three-digit numbers divisible by both 3 and 4.

23. This problem involves 1000 possible strings, since there is a choice of 10 digits for each of the three positions in the string.

a) This is most easily done by subtracting from the total number of strings the number of strings that violate the condition. Clearly there are 10 strings that consist of the same digit three times (000, 111, ..., 999). Therefore there are $1000 - 10 = 990$ strings that do not.

b) If we must begin our string with an odd digit, then we have only 5 choices for this digit. We still have 10 choices for each of the remaining digits. Therefore there are $5 \cdot 10 \cdot 10 = 500$ such strings. Alternatively, we note that by symmetry exactly half the strings begin with an odd digit (there being the same number of odd digits as even ones). Therefore half of the 1000 strings, or 500 of them, begin with an odd digit.

c) Here we need to choose the position of the digit that is not a 4 (3 ways) and choose that digit (9 ways). Therefore there are $3 \cdot 9 = 27$ such strings.

25. There are 50 choices to make, each of which can be done in 3 ways, namely by choosing the governor, choosing the senior senator, or choosing the junior senator. By the product rule the answer is therefore $3^{50} \approx 7.2 \times 10^{23}$.

27. By the sum rule we need to add the number of license plates of the first type and the number of license plates of the second type. By the product rule there are $26 \cdot 26 \cdot 10 \cdot 10 \cdot 10 \cdot 10 = 6{,}760{,}000$ license plates consisting of 2 letters followed by 4 digits; and there are $10 \cdot 10 \cdot 26 \cdot 26 \cdot 26 \cdot 26 = 45{,}697{,}600$ license plates consisting of 2 digits followed by 4 letters. Therefore the answer is $6{,}760{,}000 + 45{,}697{,}600 = 52{,}457{,}600$.

29. First let us compute the number of ways to choose the letters. By the sum rule this is the sum of the number of ways to use two letters and the number of ways to use three letters. By the product rule there are 26^2 ways to choose two letters and 26^3 ways to choose three letters. Therefore there are $26^2 + 26^3$ ways to choose the letters. By similar reasoning there are $10^2 + 10^3$ ways to choose the digits. Thus the answer to the question is $(26^2 + 26^3)(10^2 + 10^3) = 18252 \cdot 1100 = 20{,}077{,}200$.

31. We take as known that there are 26 letters including 5 vowels in the English alphabet.

a) There are 8 slots, each of which can be filled with one of the $26 - 5 = 21$ nonvowels (consonants), so by the product rule the answer is $21^8 = 37{,}822{,}859{,}361$.

b) There are 21 choices for the first slot in our string, but only 20 choices for the second slot, 19 for the third, and so on. So the answer is $21 \cdot 20 \cdot 19 \cdot 18 \cdot 17 \cdot 16 \cdot 15 \cdot 14 = 8{,}204{,}716{,}800$.

c) There are 26 choices for each slot except the first, for which there are 5 choices, so the answer is $5 \cdot 26^7 = 40{,}159{,}050{,}880$.

d) This is similar to **(b)**, except that there are only five choice in the first slot, and we are free to choose from all the letters not used so far, rather than just the consonants. Thus the answer is $5 \cdot 25 \cdot 24 \cdot 23 \cdot 22 \cdot 21 \cdot 20 \cdot 19 = 12{,}113{,}640{,}000$.

e) We subtract from the total number of strings (26^8) the number that do not contain at least one vowel (21^8, the answer to **(a)**), obtaining the answer $26^8 - 21^8 = 208{,}827{,}064{,}576 - 37{,}822{,}859{,}361 = 171{,}004{,}205{,}215$.

f) The best way to do this is first to decide where the vowel goes (8 choices), then to decide what the vowel is to be (A, E, I, O, or U—5 choices), and then to fill the remaining slots with any consonants (21^7 choices, since one slot has already been filled). Therefore the answer is $8 \cdot 5 \cdot 21^7 = 72{,}043{,}541{,}640$.

g) We can ignore the first slot, since there is no choice. Now the problem is almost identical to **(e)**, except that there are only 7 slots to fill. So the answer is $26^7 - 21^7 = 8{,}031{,}810{,}176 - 1{,}801{,}088{,}541 = 6{,}230{,}721{,}635$.

h) The problem is almost identical to **(g)**, except that there are only 6 slots to fill. So the answer is $26^6 - 21^6 = 308{,}915{,}776 - 85{,}766{,}121 = 223{,}149{,}655$.

33. For each part of this problem, we need to find the number of one-to-one functions from a set with 5 elements to a set with k elements. To specify such a function, we need to make 5 choices, in succession, namely the values of the function at each of the 5 elements in its domain. Therefore the product rule applies. The first choice can be made in k ways, since any element of the codomain can be the image of the first element of the domain. After that choice has been made, there are only $k - 1$ elements of the codomain available to be the image of the second element of the domain, since images must be distinct for the function to be one-to-one. Similarly, for the third element of the domain, there are $k - 2$ possible choices for a function value. Continuing in this way, and applying the product rule, we see that there are $k(k-1)(k-2)(k-3)(k-4)$ one-to-one functions from a set with 5 elements to a set with k elements.

a) By the analysis above the answer is $4 \cdot 3 \cdot 2 \cdot 1 \cdot 0 = 0$, what we would expect since there are no one-to-one functions from a set to a strictly smaller set.

b) By the analysis above the answer is $5 \cdot 4 \cdot 3 \cdot 2 \cdot 1 = 120$.

c) By the analysis above the answer is $6 \cdot 5 \cdot 4 \cdot 3 \cdot 2 = 720$.

d) By the analysis above the answer is $7 \cdot 6 \cdot 5 \cdot 4 \cdot 3 = 2520$.

35. **a)** There can clearly be no one-to-one function from $\{1, 2, \ldots, n\}$ to $\{0, 1\}$ if $n > 2$. If $n = 1$, then there are 2 such functions, the one that sends 1 to 0, and the one that sends 1 to 1. If $n = 2$, then there are again 2 such functions, since once it is determined where 1 is sent, the function must send 2 to the other value in the codomain.

b) If the function assigns 0 to both 1 and n, then there are $n-2$ function values free to be chosen. Each can be chosen in 2 ways. Therefore, by the product rule (since we have to choose values for all the elements of the domain) there are 2^{n-2} such functions, as long as $n > 1$. If $n = 1$, then clearly there is just one such function.

c) If $n = 1$, then there are no such functions, since there are no positive integers less than n. So assume $n > 1$. In order to specify such a function, we have to decide which of the numbers from 1 to $n-1$, inclusive, will get sent to 1. There are $n-1$ ways to make this choice. There is no choice for the remaining numbers from 1 to $n-1$, inclusive, since they all must get sent to 0. Finally, we are free to specify the value of the function at n, and this may be done in 2 ways. Hence, by the product rule the final answer is $2(n-1)$.

37. The easiest way to view a partial function in terms of counting is to add an additional element to the codomain of the function—let's call it u for "undefined"— and then imagine that the function assigns a value to *all* elements of the domain. If the original function f had previously been undefined at x, we now say that $f(x) = u$. Thus all we have done is to increase the size of the codomain from n elements to $n+1$ elements. By Example 6 we conclude that there are $(n+1)^m$ such partial functions.

39. The trick here is to realize that a palindrome of length n is completely determined by its first $\lceil n/2 \rceil$ bits. This is true because once these bits are specified, the remaining bits, read from right to left, must be identical to the first $\lfloor n/2 \rfloor$ bits, read from left to right. Furthermore, these first $\lceil n/2 \rceil$ bits can be specified at will, and by the product rule there are $2^{\lceil n/2 \rceil}$ ways to do so.

41. **a)** Here is a good way (but certainly not the only way) to approach this problem. Since the bride and groom must stand next to each other, let us treat them as one unit. Then the question asks for the number of ways to arrange five units in a row (the bride-and-groom unit and the four other people). We can think of filling five positions one at a time, so by the product rule there are $5 \cdot 4 \cdot 3 \cdot 2 \cdot 1 = 120$ ways to make these choices. This is not quite the answer, however, since there are also two ways to decide on which side of the groom the bride will stand. Therefore the final answer is $120 \cdot 2 = 240$.

b) There are clearly $6 \cdot 5 \cdot 4 \cdot 3 \cdot 2 \cdot 1 = 720$ arrangements in all. We just determined in part **(a)** that 240 of them involve the bride standing next to the groom. Therefore there are $720 - 240 = 480$ ways to arrange the people with the bride not standing next to the groom.

c) Of the 720 arrangements of these people (see part **(b)**), exactly half must have the bride somewhere to the left of the groom. (We are invoking **symmetry** here—a useful tool for solving some combinatorial problems.) Therefore the answer is $720/2 = 360$.

43. There are 2^7 bit strings of length 10 that begin with three 0's, since each of the remaining seven bits has two possible values. Similarly, there are 2^8 bit strings of length 10 that end with two 0's. Furthermore, there are 2^5 bit strings of length 10 that both begin with three 0's and end with two 0's, since each of the five "middle" bits can be specified in two ways. Using the principle of inclusion–exclusion, we conclude that there are $2^7 + 2^8 - 2^5 = 352$ such strings. The idea behind this principle here is that the strings that both begin with three 0's and end with two 0's were counted twice when we added 2^7 and 2^8, so we need to subtract for the overcounting. It is definitely *not* the case that we are subtracting because we do not want to count such strings at all.

45. First let us count the number of 8-bit strings that contain three consecutive 0's. We will organize our count by looking at the leftmost bit that contains a 0 followed by at least two more 0's. If this is the first bit, then the second and third bits are determined (namely, they are both 0), but bits 4 through 8 are free to be specified, so there are $2^5 = 32$ such strings. If it is the second, third, or fourth bits, then the bit preceding it must be a 1 and the two bits after it must be 0's, but the other four bits are free. Therefore there are $2^4 = 16$ such strings in each of these three cases, or 48 in all. Next let us suppose that the substring of three or more 0's starts at bit 5. Then bit 4 must be a 1. Bits 1 through 3 can be anything other than three 0's (if it were three 0's, then we already counted this string); thus there are $2^3 - 1 = 7$ ways to specify them. Bit 8 is free. Therefore there are $7 \cdot 2 = 14$ such strings. Finally, suppose that the substring 000 starts in bit 6, so that bit 5 is a 1. There are $2^4 = 16$ possibilities for the first four bits, but three of them contain three consecutive 0's (0000, 0001, and 1000); therefore there are only 13 such strings. Adding up all the cases we have discussed, we obtain the final answer of $32 + 48 + 14 + 13 = 107$ for the number of 8-bit strings that contain three consecutive 0's.

Next we need to compute the number of 8-bit strings that contain four consecutive 1's. The analysis is very similar to what we have just done. There are $2^4 = 16$ such strings starting with 1111; there are $2^3 = 8$ starting 01111; there are $2^3 = 8$ starting $x01111$ (where x is either 0 or 1); and there are similarly 8 starting each of $xy01111$ and $xyz01111$. This gives us a total of $16 + 4 \cdot 8 = 48$ strings containing four consecutive 1's.

Finally, we need to count the number of strings that contain both three consecutive 0's and four consecutive 1's. It is easy enough to just list them all: 00001111, 00011111, 00011110, 10001111, 11110000, 11111000, 01111000, and 11110001, eight in all. Now applying the principle of inclusion–exclusion to what we have calculated above, we obtain the answer to the entire problem: $107 + 48 - 8 = 147$. There are only 256 bit strings of length eight altogether, so this answer is somewhat surprising, in that more than half of them satisfy the stated condition.

47. We can solve this problem by computing the number of positive integers not exceeding 100 that are divisible by 4, the number of them divisible by 6, and the number divisible by both 4 and 6; and then applying the principle of inclusion–exclusion. There are clearly $100/4 = 25$ multiples of 4 in this range, since every fourth number is divisible by 4. The number of integers in this range divisible by 6 is $\lfloor 100/6 \rfloor = 16$; we needed to round down because the multiples of 6 occur at the end of each consecutive block of 6 integers (6, 12, 18, etc.). Furthermore a number is divisible by both 4 and 6 if and only if it is divisible by their least common multiple, namely 12. Therefore there are $\lfloor 100/12 \rfloor = 8$ numbers divisible by both 4 and 6 in this range. Finally, applying inclusion–exclusion, we see that the number of positive integers not exceeding 100 that are divisible either by 4 or by 6 is $25 + 16 - 8 = 33$.

49. **a)** We are told that there are $26 + 26 + 10 + 6 = 68$ available characters. A password of length k using these characters can be formed in 68^k ways. Therefore the number of passwords with the specified length restriction is $68^8 + 68^9 + 68^{10} + 68^{11} + 68^{12} = 9{,}920{,}671{,}339{,}261{,}325{,}541{,}376$, which is about 9.9×10^{21} or about ten sextillion.

b) For a password *not* to contain one of the special characters, it must be constructed from the other 62 characters. There are $62^8 + 62^9 + 62^{10} + 62^{11} + 62^{12} = 3{,}279{,}156{,}377{,}874{,}257{,}103{,}616$ of these. Thus there are 6,641,514,961,387,068,437,760 $\approx 6.6 \times 10^{21}$ (about seven sextillion) passwords that do contain at least one occurrence of one of the special symbols.

c) Assuming no restrictions, it will take one nanosecond (one billionth of a second, or 10^{-9} sec) for each password. We just multiply this by our answer in part **(a)** to find the number of seconds the hacker will require. We can convert this to years by dividing by $60 \cdot 60 \cdot 24 \cdot 365.2425$ (the average number of seconds in a year). It will take about 314,374 years.

51. By the result of Example 8, there are $C =$ 6,400,000,000 possible numbers of the form *NXX-NXX-XXXX*. To determine the number of different telephone numbers worldwide, then, we need to determine how many country codes there are and multiply by C. There are clearly 10 country codes of length 1, 100 country codes of length 2, and 1000 country codes of length 3. Thus there are $10 + 100 + 1000 = 1110$ country codes in all, by the sum rule. Our final answer is $1110 \cdot C =$ 7,104,000,000,000.

53. We assume that what is intended is that each of the 4 letters is to be used exactly once. There are at least two ways to do this problem. First let us break it into two cases, depending on whether the a comes at the end of the string or not. If a is not at the end, then there are 3 places to put it. After we have placed the a, there are only 2 places to put the b, since it cannot go into the position occupied by the a and it cannot go into the position following the a. Then there are 2 positions in which the c can go, and only 1 position for the d. Therefore, there are by the product rule $3 \cdot 2 \cdot 2 \cdot 1 = 12$ allowable strings in which the a does not come last. Second, if the a comes last, then there are $3 \cdot 2 \cdot 1 = 6$ ways to arrange the letters b, c, and d in the first three positions. The answer, by the sum rule, is therefore $12 + 6 = 18$.

Here is another approach. Ignore for a moment the restriction that a b cannot follow an a. Then we need to choose the letter that comes first (which can be done in 4 ways), then the letter that comes second (which can be done in 3 ways, since one letter has already been used), then the letter that comes third (which can be done in 2 ways, since two of the letters have already been used), and finally the letter that comes last (which can only be done in 1 way, since there is only one unused letter at that point). Therefore there are, by the product rule, $4 \cdot 3 \cdot 2 \cdot 1 = 24$ such strings. Now we need to subtract from this total the number of strings in which the a is immediately followed by the b. To count these, let us imagine the a and b glued together into one superletter, ab. (This **gluing** technique often comes in handy.) Now there are 3 things to arrange. We can choose any of them (the letters c or d or the superletter ab) to come first, and that can be done in 3 ways. We can choose either of the other two to come second (which can be done in 2 ways), and we are forced to choose the remaining one to come third. By the product rule there are $3 \cdot 2 \cdot 1 = 6$ ways to make these choices. Therefore our final answer is $24 - 6 = 18$.

55. There are at least two approaches that are effective here. In our first tree, we let each branching point represent a decision as to whether to include the next element in the set (starting with the largest element). At the top of the tree, for example, we can either choose to include 24 or to exclude it (denoted $\neg 24$). We branch one way for each possibility. In the first figure below, the entire subtree to the right represents those sets that do not include 24, and the subtree to the left represents those that do. At the point below and to the left of the 24, we have only one branch, $\neg 11$, since after we have included 24 in our set, we cannot include 11, because the sum would not be less than 28 if we did. At the point below and to the right of the $\neg 24$, however, we again branch twice, since we can choose either to include 11 or to exclude it. To answer the question, we look at the points in the last row of the tree. Each represents a set whose sum is less than 28. For example, the sixth point from the right represents the set $\{11, 3\}$. Since there are 17 such points, the answer to the problem is 17.

Our other solution is more compact. In the tree below we show branches from a point only for the inclusion of new numbers in the set. The set formed by including no more numbers is represented by the point itself. This time *every* point represents a set. For example, the point at the top represents the empty set, the point below and to the right of the number 11 represents the set $\{11\}$, and the left-most point on the bottom row represents the set $\{3,7,9\}$. In general the set that a point represents is the set of numbers found on the path up the tree from that point to the root of the tree (the point at the top). We only included branches when the sum would be less than 28. Since there are 17 points altogether in this figure, the answer to the problem is 17.

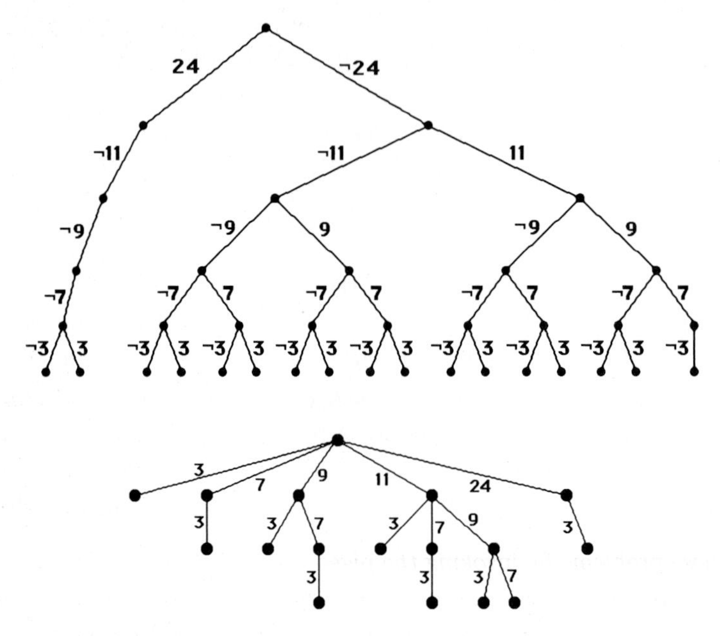

57. **a)** The tree shown here enumerates the possible outcomes. First we branch on gender, then on size, and finally on color. There are 22 ends, so the answer to the question is 22.

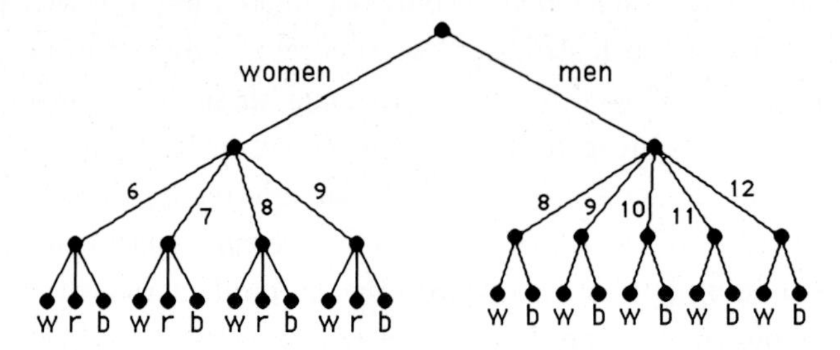

b) First we apply the sum rule: the number of shoes is the sum of the number of men's shoes and the number of women's shoes. Next we apply the product rule. For a woman's shoe we need to specify size (4 choices) and then for each choice of size, we need to specify color (3 choices). Therefore there are $4 \cdot 3 = 12$ possible women's models. Similarly, there are $5 \cdot 2 = 10$ men's models. Therefore the answer is $12 + 10 = 22$.

59. We want to prove $P(m)$, the sum rule for m tasks, which says that if tasks T_1, T_2, $\ldots$, T_m can be done in n_1, n_2, $\ldots$, n_m ways, respectively, and no two of them can be done at the same time, then there are $n_1 + n_2 + \cdots + n_m$ ways to do one of the tasks. The basis step of our proof by mathematical induction is $m = 2$, and that has already been given. Now assume that $P(m)$ is true, and we want to prove $P(m+1)$. There are $m+1$ tasks, no two of which can be done at the same time; we want to do one of them. Either we choose one from among the first m, or we choose the task T_{m+1}. By the sum rule for two tasks, the number of ways we can do this is $n + n_{m+1}$, where n is the number of ways we can do one of the tasks among the first m. But by the inductive hypothesis $n = n_1 + n_2 + \cdots + n_m$. Therefore the number of ways we can do one of the $m+1$ tasks is $n_1 + n_2 + \cdots + n_m + n_{m+1}$, as desired.

61. A diagonal joins two vertices of the polygon, but they must be vertices that are not already joined by a side of the polygon. Thus there are $n-3$ diagonals emanating from each vertex of the polygon (we've excluded two of the $n-1$ other vertices as possible targets for diagonals). If we multiply $n-3$ by n, the number of vertices, we will have counted each diagonal exactly twice—once for each endpoint. We compensate for this overcounting by dividing by 2. Therefore there are $n(n-3)/2$ diagonals. (Note that the convexity of the polygon had nothing to do with the problem—we were counting the diagonals, whether or not we could be sure that they all lay inside the polygon.)

SECTION 5.2 The Pigeonhole Principle

The pigeonhole principle seems so trivial that it is difficult to realize how powerful it is in solving some mathematical problems. As usual with combinatorial problems, the trick is to look at things the right way, which usually means coming up with the clever insight, perhaps after hours of agonizing and frustrating exploration with a problem.

Try to solve these problems by invoking the pigeonhole principle explicitly, even if you can see other ways of doing them; you will gain some insights by formulating the problem and your solution in this way. The trick, of course, is to figure out what should be the pigeons and what should be the pigeonholes. Many of the the hints of Section 5.1 apply here, as well as general problem-solving techniques, especially the willingness to play with a problem for a long time before giving up.

Many of the elegant applications are quite subtle and difficult, and there are even more subtle and difficult applications not touched on here. Not every problem, of course, fits into the model of one of the examples in the text. In particular, Exercise 38 looks deceptively like a problem amenable to the technique discussed in Example 10. Keep in mind that the process of grappling with problems such as these is worthwhile and educational in itself, even if you never find the solution.

1. There are six classes: these are the pigeons. There are five days on which classes may meet (Monday through Friday): these are the pigeonholes. Each class must meet on a day (each pigeon must occupy a pigeonhole). By the pigeonhole principle at least one day must contain at least two classes.

3. **a)** There are two colors: these are the pigeonholes. We want to know the least number of pigeons needed to insure that at least one of the pigeonholes contains two pigeons. By the pigeonhole principle the answer is 3. If three socks are taken from the drawer, at least two must have the same color. On the other hand two socks are not enough, because one might be brown and the other black. Note that the number of socks was irrelevant (assuming that it was at least 3).

b) He needs to take out 14 socks in order to insure at least two black socks. If he does so, then at most 12 of them are brown, so at least two are black. On the other hand, if he removes 13 or fewer socks, then 12 of them could be brown, and he might not get his pair of black socks. This time the number of socks did matter.

5. There are four possible remainders when an integer is divided by 4 (these are the pigeonholes here): 0, 1, 2, or 3. Therefore, by the pigeonhole principle at least two of the five given remainders (these are the pigeons) must the same.

7. Let the n consecutive integers be denoted $x+1$, $x+2$, ..., $x+n$, where x is some integer. We want to show that exactly one of these is divisible by n. There are n possible remainders when an integer is divided by n, namely 0, 1, 2, ..., $n-1$. There are two possibilities for the remainders of our collection of n numbers: either they cover all the possible remainders (in which case exactly one of our numbers has a remainder of 0 and is therefore divisible by n), or they do not. If they do not, then by the pigeonhole principle, since there are then fewer than n pigeonholes (remainders) for n pigeons (the numbers in our collection), at least one remainder must occur twice. In other words, it must be the case that $x+i$ and $x+j$ have the same remainder when divided by n for some pair of numbers i and j with $0 < i < j \le n$. Since $x+i$ and $x+j$ have the same remainder when divided by n, if we subtract $x+i$ from $x+j$, then we will get a number divisible by n. This means that $j-i$ is divisible by n. But this is impossible, since $j-i$ is a positive integer strictly less than n. Therefore the first possibility must hold, that exactly one of the numbers in our collection is divisible by n.

9. The generalized pigeonhole principle applies here. The pigeons are the students (no slur intended), and the pigeonholes are the states, 50 in number. By the generalized pigeonhole principle if we want there to be at least 100 pigeons in at least one of the pigeonholes, then we need to have a total of N pigeons so that $\lceil N/50 \rceil \ge 100$. This will be the case as long as $N \ge 99 \cdot 50 + 1 = 4951$. Therefore we need at least 4951 students to guarantee that at least 100 come from a single state.

11. We must recall from analytic geometry that the midpoint of the segment whose endpoints are (a,b,c) and (d,e,f) is $((a+d)/2, (b+e)/2, (c+f)/2)$. We are concerned only with integer values of the original coordinates. Clearly the coordinates of these fractions will be integers as well if and only if a and d have the same parity (both odd or both even), b and e have the same parity, and c and f have the same parity. Thus what matters in this problem is the parities of the coordinates. There are eight possible triples of parities: $(\text{odd}, \text{odd}, \text{odd})$, $(\text{odd}, \text{odd}, \text{even})$, $(\text{odd}, \text{even}, \text{odd})$, ..., $(\text{even}, \text{even}, \text{even})$. Since we are given nine points, the pigeonhole principle guarantees that at least two of them will have the same triple of parities. The midpoint of the segment joining these two points will therefore have integer coordinates.

13. **a)** We can group the first eight positive integers into four subsets of two integers each, each subset adding up to 9: $\{1,8\}$, $\{2,7\}$, $\{3,6\}$, and $\{4,5\}$. If we select five integers from this set, then by the pigeonhole principle (at least) two of them must come from the same subset. These two integers have a sum of 9, as desired.

b) No. If we select one element from each of the subsets specified in part **(a)**, then no sum will be 9. For example, we can select 1, 2, 3, and 4.

15. We can apply the pigeonhole principle by grouping the numbers cleverly into pairs (subsets) that add up to 7, namely $\{1,6\}$, $\{2,5\}$, and $\{3,4\}$. If we select four numbers from the set $\{1,2,3,4,5,6\}$, then at least two of them must fall within the same subset, since there are only three subsets. Two numbers in the same subset are the desired pair that add up to 7. We also need to point out that choosing three numbers is not enough, since we could choose $\{1,2,3\}$, and no pair of them add up to more than 5.

17. The given information tells us that there are $50 \cdot 85 \cdot 5 = 21{,}250$ bins. If we had this many products, then each could be stored in a separate bin. By the pigeonhole principle, however, if there are at least 21,251 products, then at least two of them must be stored in the same bin. This is the number the problem is asking us for.

19. **a)** If this statement were not true, then there would be at most 8 from each class standing, for a total of at most 24 students. This contradicts the fact that there are 25 students in the class.

b) If this statement were not true, then there would be at most 2 freshmen, at most 18 sophomores, and at most 4 juniors, for a total of at most 24 students. This contradicts the fact that there are 25 students in the class.

21. One way to do this is to have the sequence contain four groups of four numbers each, so that the numbers within each group are decreasing, and so that the numbers between groups are increasing. For example, we could take the sequence to be 4, 3, 2, 1; 8, 7, 6, 5; 12, 11, 10, 9; 16, 15, 14, 13. There can be no increasing subsequence of five terms, because any increasing subsequence can have only one element from each of the four groups. There can be no decreasing subsequence of five terms, because any decreasing subsequence cannot have elements from more than one group.

23. This is actually a fairly hard problem, in terms of what we need to keep track of. Call the given sequence $a_1, a_2, \ldots, a_n$. We will keep track of the lengths of long increasing or decreasing subsequences by assigning values i_k and d_k for each k from 1 to n, indicating the length of the longest increasing subsequence ending with a_k and the length of the longest decreasing subsequence ending with a_k, respectively. Thus $i_1 = d_1 = 1$, since a_1 is both an increasing and a decreasing subsequence of length 1. If $a_2 < a_1$, then $i_2 = 1$ and $d_2 = 2$, since the longest increasing subsequence ending at a_2 is just a_2, but the longest decreasing subsequence ending at a_2 is a_1, a_2. If $a_2 > a_1$, then it is the other way around: $i_2 = 2$ and $d_2 = 1$. In general, we can determine i_k in the following manner (the determination of d_k is similar, with the roles of i and d, and the roles of $<$ and $>$, reversed). We look at the numbers $a_1, a_2, \ldots, a_{k-1}$. For each a_j that is less than a_k, we know that the value of d_k is at least $i_j + 1$, since the increasing subsequence of length i_j ending at a_j can be extended by following it by a_k, resulting in an increasing subsequence of length $i_j + 1$, ending at a_k. Furthermore, there is no other way of producing an increasing subsequence ending at a_k, other than the subsequence of length 1. Thus we set i_k equal to either 1 or the maximum of the numbers $i_j + 1$ for those values of $j < k$ for which $a_j < a_k$. Finally, once we have determined all the values i_k and d_k, we choose the largest of these $2n$ numbers as our answer.

The procedure just described, however, does not keep track of what the longest subsequence is, so we need to use two more sets of variables, *iprev*$_k$ and *dprev*$_k$. These will point back to the terms in the sequence that caused the values of i_k and d_k to be what they are. To retrieve the longest increasing or decreasing subsequence, once we know which type it is and where it ends, we follow these pointers, thereby exhibiting the subsequence backwards. We will not write the pseudocode for this final phase of the algorithm.

```
procedure long subsequence(a_1, a_2, ..., a_n : distinct integers)
for k := 1 to n
begin
    i_k := 1; d_k := 1
    iprev_k := k; dprev_k := k
    for j := 1 to k - 1
    begin
        if a_j < a_k and i_j + 1 > i_k then
        begin
            i_k := i_j + 1
            iprev_k := j
        end
        if a_j > a_k and d_j + 1 > d_k then
        begin
            d_k := d_j + 1
            dprev_k := j
        end
    end
end {at this point correct values of i_k and d_k have all been assigned}
longest := 1
for k := 2 to n
begin
    if i_k > longest then longest := i_k
    if d_k > longest then longest := d_k
```

end
{ *longest* is the length of the longest increasing or decreasing subsequence}

25. We can prove these statements using both the result and the method of Example 13. First note that the role of "mutual friend" and "mutual enemy" is symmetric, so it is really enough to prove one of these statements; the other will follow by interchanging the roles. So let us prove that in every group of 10 people, either there are 3 mutual friends or 4 mutual enemies. Consider one person; call this person A. Of the 9 other people, either there must be 6 enemies of A, or there must be 4 friends of A (if there were 5 or fewer enemies and 3 or fewer friends, that would only account for 8 people). We need to consider the two cases separately. First suppose that A has 6 enemies. Apply the result of Example 13 to these 6 people: among them either there are 3 mutual friends or there are 3 mutual enemies. If there are 3 mutual friends, then we are done. If there are 3 mutual enemies, then these 3 people, together with A, form a group of 4 mutual enemies, and again we are done. That finishes the first case. The second case was that A had 4 friends. If some pair of these people are friends, then they, together with A, form the desired group of 3 mutual friends. Otherwise, these 4 people are the desired group of 4 mutual enemies. Thus in either case we have found either 3 mutual friends or 4 mutual enemies.

27. We need to show two things: that if we have a group of n people, then among them we must find either a pair of friends or a subset of n of them all of whom are mutual enemies; and that there exists a group of $n-1$ people for which this is not possible. For the first statement, if there is any pair of friends, then the condition is satisfied, and if not, then every pair of people are enemies, so the second condition is satisfied. For the second statement, if we have a group of $n-1$ people all of whom are enemies of each other, then there is neither a pair of friends nor a subset of n of them all of whom are mutual enemies.

29. First we need to figure out how many distinct combinations of initials and birthdays there are. The product rule tells us that since there are 26 ways to choose each of the 3 initials and 366 ways to choose the birthday, there are $26 \cdot 26 \cdot 26 \cdot 366 = 6{,}432{,}816$ such combinations. By the generalized pigeonhole principle, with these combinations as the pigeonholes and the 36 million people as the pigeons, there must be at least $\lceil 36{,}000{,}000/6{,}432{,}816 \rceil = 6$ people with the same combination.

31. The 38 time periods are the pigeonholes, and the 677 classes are the pigeons. By the generalized pigeonhole principle there is some time period in which at least $\lceil 677/38 \rceil = 18$ classes are meeting. Since each class must meet in a different room, we need 18 rooms.

33. Let c_i be the number of computers that the ith computer is connected to. Each of these integers is between 0 and 5, inclusive. The pigeonhole principle does not allow us to conclude immediately that two of these numbers must be the same, since there are six numbers (pigeons) and six possible values (pigeonholes). However, if not all of the values are used, then the pigeonhole principle would allow us to draw the desired conclusion. Let us therefore show that not all of the numbers can be used. The only way that the value 5 can appear as one of the c_i's is if one computer is connected to each of the others. In that case, the number 0 cannot appear, since no computer could be connected to none of the others. Thus not both 5 and 0 can appear in our list, and the above argument is valid.

35. This is similar to Example 9. Label the computers C_1 through C_{100}, and label the printers P_1 through P_{20}. If we connect C_k to P_k for $k = 1, 2, \ldots, 20$ and connect each of the computers C_{21} though C_{100} to *all* the printers, then we have used a total of $20 + 80 \cdot 20 = 1620$ cables. Clearly this is sufficient, because if computers C_1 through C_{20} need printers, then they can use the printers with the same subscripts, and if any computers with higher subscripts need a printer instead of one or more of these, then they can use the printers that are not being used, since they are connected to all the printers. Now we must show that 1619 cables are not enough. Since there are 1619 cables and 20 printers, the average number of computers per printer is $1619/20$, which is less than 81. Therefore some printer must be connected to fewer than 81 computers (the average of a set of numbers cannot be bigger than each of the numbers in the set). That means it is connected to 80 or fewer computers, so there are at least 20 computers that are not connected to it. If those 20 computers all needed a printer simultaneously, then they would be out of luck, since they are connected to at most the 19 other printers.

37. This problem is similar to Example 10, so we follow the method of solution suggested there. Let a_j be the number of matches held during or before the jth hour. Then $a_1, a_2, \ldots, a_{75}$ is an increasing sequence of distinct positive integers, since there was at least one match held every hour. Furthermore $1 \le a_j \le 125$, since there were only 125 matches altogether. Moreover, $a_1 + 24, a_2 + 24, \ldots, a_{75} + 24$ is also an increasing sequence of distinct positive integers, with $25 \le a_i + 24 \le 149$.

Now the 150 positive integers $a_1, a_2, \ldots, a_{75}, a_1 + 24, a_2 + 24, \ldots, a_{75} + 24$ are all less than or equal to 149. Hence by the pigeonhole principle two of these integers are equal. Since the integers $a_1, a_2, \ldots, a_{75}$ are all distinct, and the integers $a_1 + 24, a_2 + 24, \ldots, a_{75} + 24$ are all distinct, there must be distinct indices i and j such that $a_j = a_i + 24$. This means that exactly 24 matches were held from the beginning of hour $i + 1$ to the end of hour j, precisely the occurrence we wanted to find.

39. This is exactly a restatement of the generalized pigeonhole principle. The pigeonholes are the elements in the codomain (the elements of the set T), and the pigeons are the elements of the domain (the elements of the set S). To say that a pigeon s is in pigeonhole t is just to say that $f(s) = t$. The elements $s_1, s_2, \ldots, s_m$ are just the m pigeons guaranteed by the generalized pigeonhole principle to occupy a common pigeonhole.

41. Let d_j be $jx - N(jx)$, where $N(jx)$ is the integer closest to jx, for $1 \le j \le n$. We want to show that $|d_j| < 1/n$ for some j. Note that each d_j is an irrational number strictly between $-1/2$ and $1/2$ (since jx is irrational and every irrational number is closer than $1/2$ to the nearest integer). The proof is slightly messier if n is odd, so let us assume that n is even. Consider the n intervals

$$\left(0, \frac{1}{n}\right), \left(\frac{1}{n}, \frac{2}{n}\right), \ldots, \left(\frac{(n/2)-1}{n}, \frac{1}{2}\right), \left(-\frac{1}{n}, 0\right), \left(-\frac{2}{n}, -\frac{1}{n}\right), \ldots, \left(-\frac{1}{2}, -\frac{(n/2)-1}{n}\right).$$

The intervals are the pigeonholes and the d_j's are the pigeons. If the interval $(0, 1/n)$ or $(-1/n, 0)$ is occupied, then we are done, since the d_j in that interval tells us which j makes $|d_j| < 1/n$. If not, then there are n pigeons for at most $n-2$ pigeonholes, so by the pigeonhole principle there is some interval, say $((k-1)/n, k/n)$, with two pigeons in it, say d_r and d_s, with $r < s$. Now we will consider $sx - rx$ and show that it is within $1/n$ of its nearest integer; that will complete our proof, since $sx - rx = (s-r)x$, and $s - r$ is a positive integer less than n.

We can write $rx = N(rx) + d_r$ and $sx = N(sx) + d_s$, where $(k-1)/n \le d_r < k/n$ and $(k-1)/n \le d_s < k/n$. Subtracting, we have that $sx - rx = [N(sx) - N(rx)] + [d_s - d_r]$. Now the quantity in the first pair of square brackets is an integer. Furthermore the quantity in the second pair of square brackets is the difference of two numbers in the interval $((k-1)/n, k/n)$ and hence has absolute value less than $1/n$ (the extreme case would be when one of them is very close to $(k-1)/n$ and the other is very close to k/n). Therefore, by definition of "closest integer" $sx - rx$ is at most a distance $1/n$ from its closest integer, i.e., $|(sx - rx) - N(sx - rx)| < 1/n$, as desired. (The case in which n is odd is similar, but we need to extend our intervals slightly past $\pm 1/2$, using $n + 1$ intervals rather than n. This is okay, since when we subtract 2 from $n + 1$ we still have more pigeons than pigeonholes.)

43. a) Assuming that each $i_k \le n$, there are only n pigeonholes (namely $1, 2, \ldots, n$) for the $n^2 + 1$ numbers

$i_1, i_2, \ldots, i_{n^2+1}$. Hence, by the generalized pigeonhole principle at least $\lceil (n^2+1)/n \rceil = n+1$ of the numbers are in the same pigeonhole, i.e., equal.

b) If $a_{k_j} < a_{k_{j+1}}$, then the subsequence consisting of a_{k_j} followed by a maximal increasing subsequence of length $i_{k_{j+1}}$ starting at $a_{k_{j+1}}$ contradicts the fact that $i_{k_j} = i_{k_{j+1}}$. Hence $a_{k_j} > a_{k_{j+1}}$.

c) If there is no increasing subsequence of length greater than n, then parts **(a)** and **(b)** apply. Therefore we have $a_{k_{n+1}} > a_{k_n} > \cdots > a_{k_2} > a_{k_1}$, a decreasing subsequence of length $n+1$.

SECTION 5.3 Permutations and Combinations

In this section we look at counting problems more systematically than in Section 5.1. We have some formulae that apply in many instances, and the trick is to recognize the instances. If an ordered arrangement without repetitions is asked for, then the formula for permutations usually applies; if an unordered selection without repetition is asked for, then the formula for combinations usually applies. Of course the product rule and the sum rule (and common sense and cleverness) are still needed to solve some of these problems—having formulae for permutations and combinations definitely does not reduce the solving of counting problems to a mechanical algorithm.

Again the general comments of Section 5.1 apply. Try to solve problems more than one way and come up with the same answer—you will learn from the process of looking at the same problem from two or more different angles, and you will be (almost) sure that your answer is correct.

1. Permutations are ordered arrangements. Thus we need to list all the ordered arrangements of all 3 of these letters. There are 6 such: a, b, c; a, c, b; b, a, c; b, c, a; c, a, b; and c, b, a. Note that we have listed them in alphabetical order. Algorithms for generating permutations and combinations are discussed in Section 5.6.

3. If we want the permutation to end with a, then we may as well forget about the a, and just count the number of permutations of $\{b, c, d, e, f, g\}$. Each permutation of these 6 letters, followed by a, will be a permutation of the desired type, and conversely. Therefore the answer is $P(6,6) = 6! = 720$.

5. We simply plug into the formula $P(n,r) = n(n-1)(n-2)\cdots(n-r+1)$, given in Theorem 1. Note that there are r terms in this product, starting with n. This is the same as $P(n,r) = n!/(n-r)!$, but the latter formula is not as nice for computation, since it ignores the fact that each of the factors in the denominator cancels one factor in the numerator. Thus to compute $n!$ and $(n-r)!$ and then to divide is to do a lot of extra arithmetic. Of course if the denominator is 1, then there is no extra work, so we note that $P(n,n) = P(n,n-1) = n!$.

a) $P(6,3) = 6 \cdot 5 \cdot 4 = 120$ **b)** $P(6,5) = 6! = 720$ **c)** $P(8,1) = 8$
d) $P(8,5) = 8 \cdot 7 \cdot 6 \cdot 5 \cdot 4 = 6720$ **e)** $P(8,8) = 8! = 40{,}320$ **f)** $P(10,9) = 10! = 3{,}628{,}800$

7. This is $P(9,5) = 9 \cdot 8 \cdot 7 \cdot 6 \cdot 5 = 15{,}120$ by Theorem 1.

9. We need to pick 3 horses from the 12 horses in the race, and we need to arrange them in order (first, second, and third), in order to specify the win, place, and show. Thus there are $P(12,3) = 12 \cdot 11 \cdot 10 = 1320$ possibilities.

11. **a)** To specify a bit string of length 10 that contains exactly four 1's, we simply need to choose the four positions that contain the 1's. There are $C(10,4) = 210$ ways to do that.

b) To contain at most four 1's means to contain four 1's, three 1's, two 1's, one 1, or no 1's. Reasoning as in part **(a)**, we see that there are $C(10,4)+C(10,3)+C(10,2)+C(10,1)+C(10,0) = 210+120+45+10+1 = 386$ such strings.

c) To contain at least four 1's means to contain four 1's, five 1's, six 1's, seven 1's, eight 1's, nine 1's, or ten 1's. Reasoning as in part **(b)**, we see that there are $C(10,4)+C(10,5)+C(10,6)+C(10,7)+C(10,8)+C(10,9)+C(10,10) = 210+252+210+120+45+10+1 = 848$ such strings. A simpler approach would be to figure out the number of ways not to have at least four 1's (i.e., to have three 1's, two 1's, one 1, or no 1's) and then subtract that from 2^{10}, the total number of bit strings of length 10. This way we get $1024 - (120+45+10+1) = 848$, fortunately the same answer as before. Solving a combinatorial problem in more than one way is a useful check on the correctness of the answer.

d) To have an equal number of 0's and 1's in this case means to have five 1's. Therefore the answer is $C(10,5) = 252$. Incidentally, this gives us another way to do part **(b)**. If we don't have an equal number of 0's and 1's, then we have either at most four 1's or at least six 1's. By symmetry, having at most four 1's occurs in half of these cases. Therefore the answer to part **(b)** is $(2^{10} - C(10,5))/2 = 386$, as above.

13. We assume that the row has a distinguished head. Consider the order in which the men appear relative to each other. There are n men, and all of the $P(n,n) = n!$ arrangements is allowed. Similarly, there are $n!$ arrangements in which the women can appear. Now the men and women must alternate, and there are the same number of men and women; therefore there are exactly two possibilities: either the row starts with a man and ends with a woman ($MWMW\ldots MW$) or else it starts with a woman and ends with a man ($WMWM\ldots WM$). We have three tasks to perform, then: arrange the men among themselves, arrange the women among themselves, and decide which sex starts the row. By the product rule there are $n!\cdot n!\cdot 2 = 2(n!)^2$ ways in which this can be done.

15. We assume that a combination is called for, not a permutation, since we are told to *select a set*, not *form an arrangement*. We need to choose 5 things from 26, so there are $C(26,5) = 26\cdot 25\cdot 24\cdot 23\cdot 22/5! = 65{,}780$ ways to do so.

17. We know that there are 2^{100} subsets of a set with 100 elements. All of them have more than two elements except the empty set, the 100 subsets consisting of one element each, and the $C(100,2) = 4950$ subsets with two elements. Therefore the answer is $2^{100} - 5051 \approx 1.3\times 10^{30}$.

19. **a)** Each flip can be either heads or tails, so there are $2^{10} = 1024$ possible outcomes.

b) To specify an outcome that has exactly two heads, we simply need to choose the two flips that came up heads. There are $C(10,2) = 45$ such outcomes.

c) To contain at most three tails means to contain three tails, two tails, one tail, or no tails. Reasoning as in part **(b)**, we see that there are $C(10,3)+C(10,2)+C(10,1)+C(10,0) = 120+45+10+1 = 176$ such outcomes.

d) To have an equal number of heads and tails in this case means to have five heads. Therefore the answer is $C(10,5) = 252$.

21. **a)** If BCD is to be a substring, then we can think of that block of letters as one superletter, and the problem is to count permutations of five items—the letters A, E, F, and G, and the superletter BCD. Therefore the answer is $P(5,5) = 5! = 120$.

b) Reasoning as in part **(a)**, we see that the answer is $P(4,4) = 4! = 24$.

c) As in part **(a)**, we glue BA into one item and glue GF into one item. Therefore we need to permute five items, and there are $P(5,5) = 5! = 120$ ways to do it.

d) This is similar to part **(c)**. Glue ABC into one item and glue DE into one item, producing four items, so the answer is $P(4,4) = 4! = 24$.

e) If both ABC and CDE are substrings, then $ABCDE$ has to be a substring. So we are really just permuting three items: $ABCDE$, F, and G. Therefore the answer is $P(3,3) = 3! = 6$.

f) There are no permutations with both of these substrings, since B cannot be followed by both A and E at the same time.

23. First position the men relative to each other. Since there are eight men, there are $P(8,8)$ ways to do this. This creates nine slots where a woman (but not more than one woman) may stand: in front of the first man, between the first and second men, $\ldots$, between the seventh and eight men, and behind the eighth man. We need to choose five of these positions, in order, for the first through fifth woman to occupy (order matters, because the women are distinct people). This can be done is $P(9,5)$ ways. Therefore the answer is $P(8,8) \cdot P(9,5) = 8! \cdot 9!/4! = 609{,}638{,}400$.

25. a) Since the prizes are different, we want an ordered arrangement of four numbers from the set of the first 100 positive integers. Thus there are $P(100,4) = 94{,}109{,}400$ ways to award the prizes.

b) If the grand prize winner is specified, then we need to choose an ordered set of three tickets to win the other three prizes. This can be done is $P(99,3) = 941{,}094$ ways.

c) We can first determine which prize the person holding ticket 47 will win (this can be done in 4 ways), and then we can determine the winners of the other three prizes, exactly as in part **(b)**. Therefore the answer is $4P(99,3) = 3{,}764{,}376$.

d) This is the same calculation as in part **(a)**, except that there are only 99 viable tickets. Therefore the answer is $P(99,4) = 90{,}345{,}024$. Note that this answer plus the answer to part **(c)** equals the answer to part **(a)**, since the person holding ticket 47 either wins the grand prize or does not win the grand prize.

e) This is similar to part **(c)**. There are $4 \cdot 3 = 12$ ways to determine which prizes these two lucky people will win, after which there are $P(98,2) = 9506$ ways to award the other two prizes. Therefore the answer is $12 \cdot 9506 = 114{,}072$.

f) This is like part **(e)**. There are $P(4,3) = 24$ ways to choose the prizes for the three people mentioned, and then 97 ways to choose the other winner. This gives $24 \cdot 97 = 2328$ ways in all.

g) Here it is just a matter of ordering the prizes for these four people, so the answer is $P(4,4) = 24$.

h) This is similar to part **(d)**, except that this time the pool of viable numbers has only 96 numbers in it. Therefore the answer is $P(96,4) = 79{,}727{,}040$.

i) There are four ways to determine the grand prize winner under these conditions. Then there are $P(99,3)$ ways to award the remaining prizes. This gives an answer of $4P(99,3) = 3{,}764{,}376$.

j) First we need to choose the prizes for the holder of 19 and 47. Since there are four prizes, there are $P(4,2) = 12$ ways to do this. Then there are 96 people who might win the remaining prizes, and there are $P(96,2) = 9120$ ways to award these prizes. Therefore the answer is $12 \cdot 9120 = 109{,}440$.

27. a) Since the order of choosing the members is not relevant (the offices are not differentiated), we need to use a combination. The answer is clearly $C(25,4) = 12{,}650$.

b) In contrast, here we need a permutation, since the order matters (we choose first a president, then a vice president, then a secretary, then a treasurer). The answer is clearly $P(25,4) = 303{,}600$.

29. a) In this part the permutation $5, 6, 32, 7$, for example, is to be counted, since it contains the consecutive numbers 5, 6, and 7 in their correct order (even though separated by the 32). In order to specify such a

4-permutation, we first need to choose the 3 consecutive integers. They can be anything from $\{1, 2, 3\}$ to $\{98, 99, 100\}$; thus there are 98 possibilities. Next we need to decide which slot is to contain a number not in this set; there are 4 possibilities. Finally, we need to decide which of the 97 other positive integers not exceeding 100 is to fill this slot, and there are of course 97 choices. Thus our first attempt at an answer gives us, by the product rule, $98 \cdot 4 \cdot 97$.

Unfortunately, this answer is not correct, because we have counted some 4-permutations more than once. Consider the 4-permutation $4, 5, 6, 7$, for example. We cannot tell whether it arose from choosing 4, 5, and 6 as the consecutive numbers, or from choosing 5, 6, and 7. (These are the only two ways it could have arisen.) In fact, every 4-permutation consisting of 4 consecutive numbers, in order, has been double counted. Therefore to correct our count, we need to subtract the number of such 4-permutations. Clearly there are 97 of them (they can begin with any number from 1 to 97). Further thought shows that every other 4-permutation in our collection arises in a unique way (in other words, there is a unique subsequence of three consecutive integers). Thus our final answer is $98 \cdot 4 \cdot 97 - 97 = 37{,}927$.

b) In this part we are insisting that the consecutive numbers be consecutive in the 4-permutation as well. The analysis in part **(a)** works here, except that there are only 2 places to put the fourth number—in slot 1 or in slot 4. Therefore the answer is $98 \cdot 2 \cdot 97 - 97 = 18{,}915$.

31. We need to be careful here, because strings can have repeated letters.

a) We need to choose the position for the vowel, and this can be done in 6 ways. Next we need to choose the vowel to use, and this can be done in 5 ways. Each of the other five positions in the string can contain any of the 21 consonants, so there are 21^5 ways to fill the rest of the string. Therefore the answer is $6 \cdot 5 \cdot 21^5 = 122{,}523{,}030$.

b) We need to choose the position for the vowels, and this can be done in $C(6, 2) = 15$ ways (we need to choose two positions out of six). We need to choose the two vowels (5^2 ways). Each of the other four positions in the string can contain any of the 21 consonants, so there are 21^4 ways to fill the rest of the string. Therefore the answer is $15 \cdot 5^2 \cdot 21^4 = 72{,}930{,}375$.

c) The best way to do this is to count the number of strings with no vowels and subtract this from the total number of strings. We obtain $26^6 - 21^6 = 223{,}149{,}655$.

d) As in part **(c)**, we will do this by subtracting from the total number of strings, the number of strings with no vowels and the number of strings with one vowel (this latter quantity having been computed in part **(a)**). We obtain $26^6 - 21^6 - 6 \cdot 5 \cdot 21^5 = 223149655 - 122523030 = 100{,}626{,}625$.

33. We are told that we must select three of the 10 men and three of the 15 women. This can be done is $C(10, 3)C(15, 3) = 54{,}600$ ways.

35. To implement the condition that every 0 be immediately followed by a 1, let us think of "gluing" a 1 to the right of each 0. Then the objects we have to work with are eight blocks consisting of the string 01 and two 1's. The question is, then, how many strings are there consisting of these ten objects? This is easy to calculate, for we simply have to choose two of the "positions" in the string to contain the 1's and fill the remaining "positions" with the 01 blocks. Therefore the answer is $C(10, 2) = 45$.

37. Perhaps the most straightforward way to do this is to look at the several cases. The string might contain three 1's and seven 0's, four 1's and six 0's, five of each, six 1's and four 0's, or seven 1's and three 0's. In each case we can determine the number of strings by calculating a binomial coefficient, since we simply need to choose the positions for the 1's. Therefore the answer is $C(10, 3) + C(10, 4) + C(10, 5) + C(10, 6) + C(10, 7) = 120 + 210 + 252 + 210 + 120 = 912$.

39. To specify such a license plate we need to write down a 3-permutation of the set of 26 letters and follow it by a 3-permutation of the set of 10 digits. By the product rule the answer is therefore $P(26,3) \cdot P(10,3) = 26 \cdot 25 \cdot 24 \cdot 10 \cdot 9 \cdot 8 = 11{,}232{,}000$.

41. If there are no ties, then there are $3! = 6$ possible finishes. If two of the horses tie and the third has a different time, then there are 3 ways to decide which horse is not tied and then 2 ways to decide whether that horse finishes first or last. That gives $3 \cdot 2 = 6$ possibilities. Finally, all three horses can tie. So the answer is $6 + 6 + 1 = 13$.

43. We can solve this problem by breaking it down into cases depending on the ties. There are four basic cases. (1) If there are unique gold and silver winners, then we can choose these winners in $6 \cdot 5 = 30$ ways. Any nonempty subset of the remaining four runners can win the bronze medal. There are $2^4 - 1 = 15$ ways to choose these people, giving us $30 \cdot 15 = 450$ ways in all for this case. (2) If there is a 2-way tie for first place, then there are $C(6,2) = 15$ ways to choose the gold medalists. Any nonempty subset of the remaining four runners can win the bronze medal, so there are $2^4 - 1 = 15$ ways to choose these people, giving us $15 \cdot 15 = 225$ ways in all for this case. (3) If there is a k-way tie for first with $k \geq 3$, then there are $C(6,k)$ ways to choose the gold medalists (there are no other medals in this case). This gives us $C(6,3) + C(6,4) + C(6,5) + C(6,6) = 20 + 15 + 6 + 1 = 42$ more possibilities. (4) The only other case is that there is a single gold medal winner and a k-way tie for second with $k \geq 2$. We can choose the winner in 6 ways and the silver medalists in $2^5 - C(5,1) - C(5,0) = 32 - 5 - 1 = 26$ ways. This gives us $6 \cdot 26 = 156$ possibilities. Putting this all together, the answer is $450 + 225 + 42 + 156 = 873$.

SECTION 5.4 Binomial Coefficients

In this section we usually write the binomial coefficients using the $\binom{n}{r}$ notation, rather than the $C(n,r)$ notation. These numbers tend to come up in many parts of discrete mathematics.

1. a) When $(x+y)^4 = (x+y)(x+y)(x+y)(x+y)$ is expanded, all products of a term in the first sum, a term in the second sum, a term in the third sum, and a term in the fourth sum are added. Terms of the form x^4, x^3y, x^2y^2, xy^3 and y^4 arise. To obtain a term of the form x^4, an x must be chosen in each of the sums, and this can be done in only one way. Thus, the x^4 term in the product has a coefficient of 1. (We can think of this coefficient as $\binom{4}{4}$.) To obtain a term of the form x^3y, an x must be chosen in three of the four sums (and consequently a y in the other sum). Hence, the number of such terms is the number of 3-combinations of four objects, namely $\binom{4}{3} = 4$. Similarly, the number of terms of the form x^2y^2 is the number of ways to pick two of the four sums to obtain x's (and consequently take a y from each of the other two factors). This can be done in $\binom{4}{2} = 6$ ways. By the same reasoning there are $\binom{4}{1} = 4$ ways to obtain the xy^3 terms, and only one way (which we can think of as $\binom{4}{0}$) to obtain a y^4 term. Consequently, the product is $x^4 + 4x^3y + 6x^2y^2 + 4xy^3 + y^4$.

b) This is explained in Example 2. The expansion is $\binom{4}{0}x^4 + \binom{4}{1}x^3y + \binom{4}{2}x^2y^2 + \binom{4}{3}xy^3 + \binom{4}{4}y^4 = x^4 + 4x^3y + 6x^2y^2 + 4xy^3 + y^4$. Note that it does not matter whether we think of the bottom of the binomial coefficient expression as corresponding to the exponent on x, as we did in part **(a)**, or the exponent on y, as we do here.

3. The coefficients are the binomial coefficients $\binom{6}{i}$, as i runs from 0 to 6, namely 1, 6, 15, 20, 15, 6, 1. Therefore $(x+y)^6 = \sum_{i=0}^{6} \binom{6}{i} x^{6-i}y^i = x^6 + 6x^5y + 15x^4y^2 + 20x^3y^3 + 15x^2y^4 + 6xy^5 + y^6$.

5. There is one term for each i from 0 to 100, so there are 101 terms.

7. By the Binomial Theorem the term involving x^9 in the expansion of $(2+(-x))^{19}$ is $\binom{19}{9}2^{10}(-x)^9$. Therefore the coefficient is $\binom{19}{9}2^{10}(-1)^9 = -2^{10}\binom{19}{9} = -94{,}595{,}072$.

9. Using the Binomial Theorem, we see that the term involving x^{101} in the expansion of $((2x)+(-3y))^{200}$ is $\binom{200}{99}(2x)^{101}(-3y)^{99}$. Therefore the coefficient is $\binom{200}{99}2^{101}(-3)^{99} = -2^{101}3^{99}C(200,99)$.

11. Let us apply the Binomial Theorem to the given binomial:

$$(x^2 - x^{-1})^{100} = \sum_{j=0}^{100}\binom{100}{j}(x^2)^{100-j}(-x^{-1})^j$$
$$= \sum_{j=0}^{100}\binom{100}{j}(-1)^j x^{200-2j-j} = \sum_{j=0}^{100}\binom{100}{j}(-1)^j x^{200-3j}$$

Thus the only nonzero coefficients are those of the form $200-3j$ where j is an integer between 0 and 100, inclusive, namely 200, 197, 194, ..., 2, -1, -4, ..., -100. If we denote $200-3j$ by k, then we have $j = (200-k)/3$. This gives us our answer. The coefficient of x^k is zero for k not in the list just given (namely those vales of k between -100 and 200, inclusive, that are congruent to 2 modulo 3), and for those values of k in the list, the coefficient is $(-1)^{(200-k)/3}\binom{100}{(200-k)/3}$.

13. We are asked simply to display these binomial coefficients. Each can be computed from the formula in Theorem 2 in Section 5.3. Alternatively, we can apply Pascal's Identity to the last row of Figure 1(b), adding successive numbers in that row to produce the desired row. We thus obtain

$$1 \quad 9 \quad 36 \quad 84 \quad 126 \quad 126 \quad 84 \quad 36 \quad 9 \quad 1\,.$$

15. There are many ways to see why this is true. By Corollary 1 the sum of *all* the positive numbers $\binom{n}{k}$, as k runs from 0 to n, is 2^n, so certainly each one of them is no bigger than this sum. Another way to see this is to note that $\binom{n}{k}$ counts the number of subsets of an n-set having k elements, and 2^n counts even more—the number of subsets of an n-set with no restriction as to size; so certainly the former is smaller than the latter.

17. We know that

$$\binom{n}{k} = \frac{n(n-1)(n-2)\cdots(n-k+1)}{k(k-1)(k-2)\cdots 2}.$$

Now if we make the numerator of the right-hand side larger by raising each factor up to n, and make the denominator smaller by lowering each factor to 2, then we have certainly not decreased the value, so the left-hand side is less than or equal to this altered expression. But the result is precisely $n^k/2^{k-1}$, as desired.

19. Using the formula (Theorem 2 in Section 5.3) we have

$$\binom{n}{k-1} + \binom{n}{k} = \frac{n!}{(k-1)!(n-(k-1))!} + \frac{n!}{k!(n-k)!}$$
$$= \frac{n!k + n!(n-k+1)}{k!(n-k+1)!} \quad \text{(having found a common denominator)}$$
$$= \frac{(n+1)n!}{k!((n+1)-k)!} = \frac{(n+1)!}{k!((n+1)-k)!} = \binom{n+1}{k}.$$

21. **a)** We show that each side counts the number of ways to choose from a set with n elements a subset with k elements and a distinguished element of that set. For the left-hand side, first choose the k-set (this can be done in $\binom{n}{k}$ ways) and then choose one of the k elements in this subset to be the distinguished element (this can be done in k ways). For the right-hand side, first choose the distinguished element out of the entire n-set (this can be done in n ways), and then choose the remaining $k-1$ elements of the subset from the remaining $n-1$ elements of the set (this can be done in $\binom{n-1}{k-1}$ ways).

b) This is straightforward algebra:

$$k\binom{n}{k} = k \cdot \frac{n!}{k!(n-k)!} = \frac{n\cdot(n-1)!}{(k-1)!(n-k)!} = n\binom{n-1}{k-1}.$$

23. This identity can be proved algebraically or combinatorially. Algebraically, we compute as follows, starting with the right-hand side (we use twice the fact that $(x+1)x! = (x+1)!$):

$$\begin{aligned}\frac{(n+1)\binom{n}{k-1}}{k} &= \frac{(n+1)n!}{(k-1)!(n-(k-1))!k}\\ &= \frac{(n+1)!}{k!(n-(k-1))!}\\ &= \frac{(n+1)!}{k!((n+1)-k)!}\\ &= \binom{n+1}{k}\end{aligned}$$

For a combinatorial argument, we need to construct a situation in which both sides count the same thing. Suppose that we have a set of $n+1$ people, and we wish to choose k of them. Clearly there are $\binom{n+1}{k}$ ways to do this. On the other hand, we can choose our set of k people by first choosing one person to be in the set (there are $n+1$ choices), and then choosing $k-1$ additional people to be in the set, from the n people remaining. This can be done in $\binom{n}{k-1}$ ways. Therefore apparently there are $(n+1)\binom{n}{k-1}$ ways to choose the set of k people. However, we have overcounted: there are k ways that every such set can be chosen, since once we have the set, we realize that any of the k people could have been chosen first. Thus we have overcounted by a factor of k, and the real answer is $(n+1)\binom{n}{k-1}/k$ (we correct for the overcounting by dividing by k). Comparing our two approaches, one yielding the answer $\binom{n+1}{k}$, and the other yielding the answer $(n+1)\binom{n}{k-1}/k$, we conclude that $\binom{n+1}{k} = (n+1)\binom{n}{k-1}/k$.

Finally, we are asked to use this identity to give a recursive definition of the $\binom{n}{k}$'s. Note that this identity expresses $\binom{n}{k}$ in terms of $\binom{i}{j}$ for values of i and j less than n and k, respectively (namely $i = n-1$ and $j = k-1$). Thus the identity will be the recursive part of the definition. We need the base cases to handle $n = 0$ or $k = 0$. Our full definition becomes

$$\binom{n}{k} = \begin{cases} 1 & \text{if } k = 0\\ 0 & \text{if } k > 0 \text{ and } n = 0\\ n\binom{n-1}{k-1}/k & \text{if } n > 0 \text{ and } k > 0.\end{cases}$$

Actually, if we assume (as we usually do) that $k \le n$, then we do not need the second line of the definition. Note that $\binom{n}{k} = 0$ for $n < k$ under the definition given here, which is consistent with the combinatorial definition, since there are no ways to choose k different elements from a set with fewer than k elements.

25. We use Pascal's Identity twice (Theorem 2 of this section) and Corollary 1 of the previous section:

$$\begin{aligned}\binom{2n}{n+1} + \binom{2n}{n} = \binom{2n+1}{n+1} &= \frac{1}{2}\left(\binom{2n+1}{n+1} + \binom{2n+1}{n+1}\right)\\ &= \frac{1}{2}\left(\binom{2n+1}{n+1} + \binom{2n+1}{n}\right) = \frac{1}{2}\left(\binom{2n+2}{n+1}\right)\end{aligned}$$

27. **a)** We need to find something to count so that the left-hand side of the equation counts it in one way and the right-hand side counts it in a different way. After much thought, we might try the following. We will count the number of bit strings of length $n+r+1$ containing exactly r 0's and $n+1$ 1's. There are $\binom{n+r+1}{r}$ such strings, since a string is completely specified by deciding which r of the bits are to be the 0's. To see that the left-hand side of the identity counts the same thing, let $l+1$ be the position of the last 1 in the string. Since there are $n+1$ 1's, we know that l cannot be less than n. Thus there are disjoint cases for each l from n to $n+r$. For each such l, we completely determine the string by deciding which of the l positions in the string before the last 1 are to be 0's. Since there are n 1's in this range, there are $l-n$ 0's. Thus there are $\binom{l}{l-n}$ ways to choose the positions of the 0's. Now by the sum rule the total number of bit strings will be $\sum_{l=n}^{n+r} \binom{l}{l-n}$. By making the change of variable $k=l-n$, this transforms into the left-hand side, and we are finished.

b) We need to prove this by induction on r; Pascal's Identity will enter at the crucial step. We let $P(r)$ be the statement to be proved. The basis step is clear, since the equation reduces to $\binom{n}{0} = \binom{n+1}{0}$, which is the true proposition $1=1$. Assuming the inductive hypothesis, we derive $P(r+1)$ in the usual way:

$$\begin{aligned}\sum_{k=0}^{r+1} \binom{n+k}{k} &= \left(\sum_{k=0}^{r} \binom{n+k}{k}\right) + \binom{n+r+1}{r+1} \\ &= \binom{n+r+1}{r} + \binom{n+r+1}{r+1} \quad \text{(by the inductive hypothesis)} \\ &= \binom{n+(r+1)+1}{r+1} \quad \text{(by Pascal's Identity)}\end{aligned}$$

29. We will follow the hint and count the number of ways to choose a committee with leader from a set of n people. Note that the size of the committee is not specified, although it clearly needs to have at least one person (its leader). On the one hand, we can choose the leader first, in any of n ways. We can then choose the rest of the committee, which can be any subset of the remaining $n-1$ people; this can be done in 2^{n-1} ways since there are this many subsets. Therefore the right-hand side of the proposed identity counts this. On the other hand, we can organize our count by the size of the committee. Let k be the number of people who will serve on the committee. The number of ways to select a committee with k people is clearly $\binom{n}{k}$, and once we have chosen the committee, there are clearly k ways in which to choose its leader. By the sum rule the left-hand side of the proposed identity therefore also counts the number of such committees. Since the two sides count the same quantity, they must be equal.

31. Corollary 2 says that $\binom{n}{0} - \binom{n}{1} + \binom{n}{2} - \cdots \pm \binom{n}{n} = 0$. If we put all the negative terms on the other side, we obtain $\binom{n}{0} + \binom{n}{2} + \binom{n}{4} + \cdots = \binom{n}{1} + \binom{n}{3} + \cdots$ (one side ends at $\binom{n}{n}$ and the other side ends at $\binom{n}{n-1}$—which is which depends on whether n is even or odd). Now the left-hand side counts the number of subsets with even cardinality of a set with n elements, and the right-hand side counts the number of subsets with odd cardinality of the same set. That these two quantities are equal is precisely what we wanted to prove.

33. **a)** Clearly a path of the desired type must consist of m moves to the right and n moves up. Therefore each such path can be represented by a bit string consisting of m 0's and n 1's, with the 0's representing moves to the right and the 1's representing moves up. Note that the total length of this bit string is $m+n$.

b) We know from this section that the number of bit strings of length $m+n$ containing exactly n 1's is $\binom{m+n}{n}$, since one need only specify the positions of the 1's. Note that this is the same as $\binom{m+n}{m}$.

35. We saw in Exercise 33 that the number of paths of length n was the same as the number of bit strings of length n, which we know to be 2^n, the right-hand side of the identity. On the other hand, a path of length n must end up at some point the sum of whose coordinates is n, say at $(n-k,k)$ for some k from 0 to n. We saw in Exercise 33 that the number of paths ending up at $(n-k,k)$ was equal to $\binom{n-k+k}{k} = \binom{n}{k}$. Therefore the left-hand side of the identity counts the number of such paths, too.

37. The right-hand side of the identity we are asked to prove counts, by Exercise 33, the number of paths from $(0,0)$ to $(n+1,r)$. Now let us count these paths by breaking them down into $r+1$ cases, depending on how many steps upward they begin with. Let k be the number of steps upward they begin with before taking a step to the right. Then k can take any value from 0 to r. The number of paths from $(0,0)$ to $(n+1,r)$ that begin with exactly k steps upward before turning to the right is clearly the same as the number of paths from $(1,k)$ to $(n+1,r)$, since after these k upward steps and the move to the right we have reached $(1,k)$. This latter quantity is the same as the number of paths from $(0,0)$ to $(n+1-1, r-k) = (n, r-k)$, since we can relabel our diagram to make $(1,k)$ the origin. From Exercise 33, this latter quantity is $\binom{n+r-k}{r-k}$. Therefore the total number of paths is the desired sum

$$\sum_{k=0}^{r} \binom{n+r-k}{r-k} = \sum_{k=0}^{r} \binom{n+k}{k},$$

where the equality comes from changing the dummy variable from k to $r-k$. Since both sides count the same thing, they are equal.

39. **a)** This looks like the third negatively sloping diagonal of Pascal's triangle, starting with the leftmost entry in the second row and reading down and to the right. In other words, the nth term of this sequence is $\binom{n+1}{n-1} = \binom{n+1}{2}$.

b) This looks like the fourth negatively sloping diagonal of Pascal's triangle, starting with the leftmost entry in the third row and reading down and to the right. In other words, the nth term of this sequence is $\binom{n+2}{n-1} = \binom{n+2}{3}$.

c) These seem to be the entries reading straight down the middle of the Pascal's triangle. Only every other row has a middle element. The first entry in the sequence is $\binom{0}{0}$, the second is $\binom{2}{1}$, the third is $\binom{4}{2}$, the fourth is $\binom{6}{3}$, and so on. In general, then, the nth term is $\binom{2n-2}{n-1}$.

d) These seem to be the entries reading down the middle of the Pascal's triangle. Only every other row has an exact middle element, but in the other rows, there are two elements sharing the middle. The first entry in the sequence is $\binom{0}{0}$, the second is $\binom{1}{0}$, the third is $\binom{2}{1}$, the fourth is $\binom{3}{1}$, the fifth is $\binom{4}{2}$, the fourth is $\binom{5}{2}$, and so on. In general, then, the nth term is $\binom{n-1}{\lfloor (n-1)/2 \rfloor}$.

e) One pattern here is a wandering through Pascal's triangle according to the following rule. The nth term is in the nth row of the triangle. We increase the position in that row as we go down, but as soon as we are about to reach the middle of the row, we jump back to the start of the next row. For example, the fifth term is the first entry in row 5; the sixth term is the second entry in row 6; the seventh term is the third entry in row 7; the eighth term is the fourth entry in row 8. The next entry following that pattern would take us to the middle of the ninth row, so instead we jump back to the beginning, and the ninth term is the first entry of row 9. To come up with a formula here, we see that the nth entry is $\binom{n-1}{k-1}$ for a particular k; let us determine k as a function of n. A little playing around with the pattern reveals that k is n minus the largest power of 2 less than n (where for this purpose we consider 0 to be the largest power of 2 less than 1). For example, the 14th term has $k = 14 - 8 = 6$, so it is $\binom{13}{5} = 1287$.

f) The terms seem to come from every third row, so the nth term is $\binom{3n-3}{k}$ for some k. A little observation indicates that in fact these terms are $\binom{0}{0}$, $\binom{3}{1}$, $\binom{6}{2}$, $\binom{9}{3}$, $\binom{12}{4}$, and so on. Thus the nth term is $\binom{3n-3}{n-1}$.

SECTION 5.5 Generalized Permutations and Combinations

As in Section 5.3, we have formulae that give us the answers to some combinatorial problems, if we can figure out which formula applies to which problem, and in what way it applies. Here, even more than in previous sections, the ability to see a problem from the right perspective is the key to solving it. Expect to spend several minutes staring at each problem before any insight comes. Reread the examples in the section and try to imagine yourself going through the thought processes explained there. Gradually your mind will begin to think in the same terms. In particular, ask yourself what is being selected from what, whether ordered or unordered selections are to be made, and whether repetition is allowed. In most cases, after you have answered these questions, you can find the appropriate formula from Table 1.

1. Since order is important here, and since repetition is allowed, this is a simple application of the product rule. There are 3 ways in which the first element can be selected, 3 ways in which the second element can be selected, and so on, with finally 3 ways in which the fifth element can be selected, so there are $3^5 = 243$ ways in which the 5 elements can be selected. The general formula is that there are n^k ways to select k elements from a set of n elements, in order, with unlimited repetition allowed.

3. Since we are considering strings, clearly order matters. The choice for each position in the string is from the set of 26 letters. Therefore, using the same reasoning as in Exercise 1, we see that there are $26^6 = 308{,}915{,}776$ strings.

5. We assume that the jobs and the employees are distinguishable. For each job, we have to decide which employee gets that job. Thus there are 5 ways in which the first job can be assigned, 5 ways in which the second job can be assigned, and 5 ways in which the third job can be assigned. Therefore, by the multiplication principle (just as in Exercise 1) there are $5^3 = 125$ ways in which the assignments can be made. (Note that we do not require that every employee get at least one job.)

7. Since the selection is to be an unordered one, Theorem 2 applies. We want to choose $r = 3$ items from a set with $n = 5$ elements. Theorem 2 tells us that there are $C(5+3-1,3) = C(7,3) = 7\cdot 6\cdot 5/(3\cdot 2) = 35$ ways to do so. (Equivalently, this problem is asking us to count the number of nonnegative integer solutions to $x_1+x_2+x_3+x_4+x_5 = 3$, where x_i represents the number of times that the ith element of the 5-element set gets selected.)

9. Let b_1, b_2, ..., b_8 be the number of bagels of the 8 types listed (in the order listed) that are selected. Order does not matter: we are presumably putting the bagels into a bag to take home, and the order in which we put them there is irrelevant.

a) If we want to choose 6 bagels, then we are asking for the number of nonnegative solutions to the equation $b_1 + b_2 + \cdots + b_8 = 6$. Theorem 2 applies, with $n = 8$ and $r = 6$, giving us the answer $C(8+6-1,6) = C(13,6) = 1716$.

b) This is the same as part **(a)**, except that $r = 12$ rather than 6. Thus there are $C(8+12-1,12) = C(19,12) = C(19,7) = 50{,}388$ ways to make the selection. (Note that $C(19,7)$ was easier to compute than $C(19,12)$, and since they are equal, we chose the latter form.)

c) This is the same as part **(a)**, except that $r = 24$ rather than 6. Thus there are $C(8+24-1,24) = C(31,24) = C(31,7) = 2{,}629{,}575$ ways to make the selection.

d) This one is more complicated. Here we want to solve the equation $b_1 + b_2 + \cdots + b_8 = 12$, subject to the constraint that each $b_i \geq 1$. We reduce this problem to the form in which Theorem 2 is applicable with the following trick. Let $b_i' = b_i - 1$; then b_i' represents the number of bagels of type i, in excess of the required 1, that are selected. If we substitute $b_i = b_i' + 1$ into the original equation, we obtain $(b_1'+1)+(b_2'+1)+\cdots+(b_8'+1) = 12$, which reduces to $b_1'+b_2'+\cdots+b_8' = 4$. In other words, we are asking

how many ways are there to choose the 4 extra bagels (in excess of the required 1 of each type) from among the 8 types, repetitions allowed. By Theorem 2 the number of solutions is $C(8+4-1,4) = C(11,4) = 330$.
e) This final part is even trickier. First let us ignore the restriction that there can be no more than 2 salty bagels (i.e., that $b_4 \le 2$). We will take into account, however, the restriction that there must be at least 3 egg bagels (i.e., that $b_3 \ge 3$). Thus we want to count the number of solutions to the equation $b_1+b_2+\cdots+b_8 = 12$, subject to the condition that $b_i \ge 0$ for all i and $b_3 \ge 3$. As in part **(d)**, we use the trick of choosing the 3 egg bagels at the outset, leaving only 9 bagels free to be chosen; equivalently, we set $b_3' = b_3 - 3$, to represent the extra egg bagels, above the required 3, that are chosen. Now Theorem 2 applies to the number of solutions of $b_1 + b_2 + b_3' + b_4 + \cdots + b_8 = 9$, so there are $C(8+9-1,9) = C(16,9) = C(16,7) = 11{,}440$ ways to make this selection.

Next we need to worry about the restriction that $b_4 \le 2$. We will impose this restriction by subtracting from our answer so far the number of ways to violate this restriction (while still obeying the restriction that $b_3 \ge 3$). The difference will be the desired answer. To violate the restriction means to have $b_4 \ge 3$. Thus we want to count the number of solutions to $b_1 + b_2 + \cdots + b_8 = 12$, with $b_3 \ge 3$ and $b_4 \ge 3$. Using the same technique as we have just used, this is equal to the number of nonnegative solutions to the equation $b_1 + b_2 + b_3' + b_4' + b_5 + \cdots + b_8 = 6$ (the 6 on the right being $12 - 3 - 3$). By Theorem 2 there are $C(8+6-1,6) = C(13,6) = 1716$ ways to make this selection. Therefore our final answer is $11440 - 1716 = 9724$.

11. This can be solved by common sense. Since the pennies are all identical and the nickels are all identical, all that matters is the number of each type of coin selected. We can select anywhere from 0 to 8 pennies (and the rest nickels); since there are nine numbers in this range, the answer is 9. (The number of pennies and nickels is irrelevant, as long as each is at least eight.) If we wanted to use a high-powered theorem for this problem, we could observe that Theorem 2 applies, with $n = 2$ (there are two types of coins) and $r = 8$. The formula gives $C(2+8-1,8) = C(9,8) = 9$.

13. Assuming that the warehouses are distinguishable, let w_i be the number of books stored in warehouse i. Then we are asked for the number of solutions to the equation $w_1 + w_2 + w_3 = 3000$. By Theorem 2 there are $C(3+3000-1,3000) = C(3002,3000) = C(3002,2) = 4{,}504{,}501$ of them.

15. a) Let $x_1 = x_1' + 1$; thus x_1' is the value that x_1 has in excess of its required 1. Then the problem asks for the number of nonnegative solutions to $x_1' + x_2 + x_3 + x_4 + x_5 = 20$. By Theorem 2 there are $C(5+20-1,20) = C(24,20) = C(24,4) = 10{,}626$ of them.
b) Substitute $x_i = x_i' + 2$ into the equation for each i; thus x_i' is the value that x_i has in excess of its required 2. Then the problem asks for the number of nonnegative solutions to $x_1' + x_2' + x_3' + x_4' + x_5' = 11$. By Theorem 2 there are $C(5+11-1,11) = C(15,11) = C(15,4) = 1365$ of them.
c) There are $C(5+21-1,21) = C(25,21) = C(25,4) = 12650$ solutions with no restriction on x_1. The restriction on x_1 will be violated if $x_1 \ge 11$. Following the procedure in part **(a)**, we find that there are $C(5+10-1,10) = C(14,10) = C(14,4) = 1001$ solutions in which the restriction is violated. Therefore there are $12650 - 1001 = 11{,}649$ solutions of the equation with its restriction.
d) First let us impose the restrictions that $x_3 \ge 15$ and $x_2 \ge 1$. Then the problem is equivalent to counting the number of solutions to $x_1 + x_2' + x_3' + x_4 + x_5 = 5$, subject to the constraints that $x_1 \le 3$ and $x_2' \le 2$ (the latter coming from the original restriction that $x_2 < 4$). Note that these two restrictions cannot be violated simultaneously. Thus if we count the number of solutions to $x_1 + x_2' + x_3' + x_4 + x_5 = 5$, subtract the number of its solutions in which $x_1 \ge 4$, and subtract the numbers of its solutions in which $x_2' \ge 3$, then we will have the answer. By Theorem 2 there are $C(5+5-1,5) = C(9,5) = 126$ solutions of the unrestricted equation. Applying the first restriction reduces the equation to $x_1' + x_2' + x_3' + x_4 + x_5 = 1$,

which has $C(5+1-1,1) = C(5,1) = 5$ solutions. Applying the second restriction reduces the equation to $x_1 + x_2'' + x_3' + x_4 + x_5 = 2$, which has $C(5+2-1,2) = C(6,2) = 15$ solutions. Therefore the answer is $126 - 5 - 15 = 106$.

17. Theorem 3 applies here, with $n = 10$ and $k = 3$. The answer is therefore

$$\frac{10!}{2!3!5!} = 2520\,.$$

19. Theorem 3 applies here, with $n = 14$, $n_1 = n_2 = 3$ (the triplets), $n_3 = n_4 = n_5 = 2$ (the twins), and $n_6 = n_7 = 1$. The answer is therefore

$$\frac{14!}{3!3!2!2!2!1!1!} = 302{,}702{,}400\,.$$

21. If we think of the balls as doing the choosing, then this is asking for the number of ways to choose six bins from the nine given bins, with repetition allowed. (The number of times each bin is chosen is the number of balls in that bin.) By Theorem 2 with $n = 9$ and $r = 6$, this choice can be made in $C(9+6-1,6) = C(14,6) = 3003$ ways.

23. There are several ways to count this. We can first choose the two objects to go into box #1 ($C(12,2)$ ways), then choose the two objects to go into box #2 ($C(10,2)$ ways, since only 10 objects remain), then choose the two objects to go into box #3 ($C(8,2)$ ways), and so on. So the answer is $C(12,2) \cdot C(10,2) \cdot C(8,2) \cdot C(6,2) \cdot C(4,2) \cdot C(2,2) = (12 \cdot 11/2)(10 \cdot 9/2)(8 \cdot 7/2)(6 \cdot 5/2)(4 \cdot 3/2)(2 \cdot 1/2) = 12!/2^6 = 7{,}484{,}400$. Alternatively, just line up the 12 objects in a row (12! ways to do that), and put the first two into box #1, the next two into box #2, and so on. This overcounts by a factor of 2^6, since there are that many ways to swap objects in the permutation without affecting the result (swap the first and second objects or not, and swap the third and fourth objects or not, and so on). So this results in the same answer. Here is a third way to get this answer. First think of pairing the objects. Think of the objects as ordered (a first, a second, and so on). There are 11 ways to choose a mate for the first object, then 9 ways to choose a mate for the first unused object, then 7 ways to choose a mate for the first still unused object, and so on. This gives $11 \cdot 9 \cdot 7 \cdot 5 \cdot 3$ ways to do the pairing. Then there are 6! ways to choose the boxes for the pairs. So the answer is the product of these two quantities, which is again 7,484,400.

25. Let d_1, d_2, ..., d_6 be the digits of a natural number less than 1,000,000; they can each be anything from 0 to 9 (in particular, we may as well assume that there are leading 0's if necessary to make the number exactly 6 digits long). If we want the sum of the digits to equal 19, then we are asking for the number of solutions to the equation $d_1 + d_2 + \cdots + d_6 = 19$ with $0 \le d_i \le 9$ for each i. Ignoring the upper bound restriction, there are, by Theorem 2, $C(6+19-1,19) = C(24,19) = C(24,5) = 42504$ of them. We must subtract the number of solutions in which the restriction is violated. If the digits are to add up to 19 and one or more of them is to exceed 9, then exactly one of them will have to exceed 9, since $10 + 10 > 19$. There are 6 ways to choose the digit that will exceed 9. Once we have made that choice (without loss of generality assume it is d_1 that is to be made greater than or equal to 10), then we count the number of solutions to the equation by counting the number of solutions to $d_1' + d_2 + \cdots + d_6 = 19 - 10 = 9$; by Theorem 2 there are $C(6+9-1,9) = C(14,9) = C(14,5) = 2002$ of them. Thus there are $6 \cdot 2002 = 12012$ solutions that violate the restriction. Subtracting this from the 42504 solutions altogether, we find that $42504 - 12012 = 30{,}492$ is the answer to the problem.

27. We assume that each problem is worth a whole number of points. Then we want to find the number of integer solutions to $x_1 + x_2 + \cdots + x_{10} = 100$, subject to the constraint that each $x_i \geq 5$. Letting x_i' be the number of points assigned to problem i in excess of its required 5, and substituting $x_i = x_i' + 5$ into the equation, we obtain the equivalent equation $x_1' + x_2' + \cdots + x_{10}' = 50$. By Theorem 2 the number of solutions is given by $C(10+50-1, 50) = C(59, 50) = C(59, 9) = 12{,}565{,}671{,}261$.

29. There are at least two good ways to do this problem. First we present a solution in the spirit of this section. Let us place the 1's and some gaps in a row. A 1 will come first, followed by a gap, followed by another 1, another gap, a third 1, a third gap, a fourth 1, and a fourth gap. Into the gaps we must place the 12 0's that are in this string. Let g_1, g_2, g_3, and g_4 be the numbers of 0's placed in gaps 1 through 4, respectively. The only restriction is that each $g_i \geq 2$. Thus we want to count the number of solutions to the equation $g_1+g_2+g_3+g_4 = 12$, with $g_i \geq 2$ for each i. Letting $g_i = g_i'+2$, we want to count, equivalently, the number of nonnegative solutions to $g_1'+g_2'+g_3'+g_4' = 4$. By Theorem 2 there are $C(4+4-1, 4) = C(7,4) = C(7,3) = 35$ solutions. Thus our answer is 35.

Here is another way to solve the problem. Since each 1 must be followed by two 0's, suppose we glue 00 to the right end of each 1. This uses up 8 of the 0's, leaving 4 unused 0's. Now we have 8 objects, namely 4 0's and 4 100's. We want to find the number of strings we can form with these 8 objects, starting with a 100. After placing the 100 first, there are 7 places left for objects, 3 of which have to be 100's. Clearly there are $C(7,3) = 35$ ways to choose the positions for the 100's, so our answer is 35.

31. This is a direct application of Theorem 3, with $n = 11$, $n_1 = 5$, $n_2 = 2$, $n_3 = n_4 = 1$, and $n_5 = 2$ (where n_1 represents the number of A's, etc.). Thus the answer is $11!/(5!2!1!1!2!) = 83{,}160$.

33. We need to use the sum rule at the outermost level here, adding the number of strings using each subset of letters. There are quite a few cases. First, there are 3 strings of length 1, namely O, R, and N. There are several strings of length 2. If the string uses no O's, then there are 2; if it uses 1 O, then there are 2 ways to choose the other letter, and 2 ways to permute the letters in the string, so there are 4; and of course there is just 1 string of length 2 using 2 O's. Strings of length 3 can use 1, 2, or 3 O's. A little thought shows that the number of such strings is $3! = 6$, $2 \cdot 3 = 6$, and 1, respectively. There are 3 possibilities of the choice of letters for strings of length 4. If we omit an O, then there are $4!/2! = 12$ strings; if we omit either of the other letters (2 ways to choose the letter), then there are 4 strings. Finally, there are $5!/3! = 20$ strings of length 5. This gives a total of $3+2+4+1+6+6+1+12+2\cdot 4+20 = 63$ strings using some or all of the letters.

35. We need to consider the three cases determined by the number of characters used in the string: 7, 8, or 9. If all nine letters are to be used, then Theorem 3 applies and we get

$$\frac{9!}{4!2!1!1!1!} = 7560$$

strings. If only eight letters are used, then we need to consider which letter is left out. In each of the cases in which the V, G, or N is omitted, Theorem 3 tells us that there are

$$\frac{8!}{4!2!1!1!} = 840$$

strings, for a total of 2520 for these cases. If an R is left out, then Theorem 3 tells us that there are

$$\frac{8!}{4!1!1!1!} = 1680$$

strings, and if an E is left out, then Theorem 3 tells us that there are

$$\frac{8!}{3!2!1!1!} = 3360$$

strings. This gives a total of $2520+1680+3360=7560$ strings of length 8. (It was not an accident that there are as many strings of length 8 as there are of length 9, since there is a one-to-one correspondence between these two sets, given by associating with a string of length 9 its first 8 characters.) For strings of length 7 there are even more cases. We tabulate them here:

omitting VG	$7!/(4!2!1!) = 105$ strings
omitting VN	$7!/(4!2!1!) = 105$ strings
omitting GN	$7!/(4!2!1!) = 105$ strings
omitting VR	$7!/(4!1!1!1!) = 210$ strings
omitting GR	$7!/(4!1!1!1!) = 210$ strings
omitting NR	$7!/(4!1!1!1!) = 210$ strings
omitting RR	$7!/(4!1!1!1!) = 210$ strings
omitting EV	$7!/(3!2!1!1!) = 420$ strings
omitting EG	$7!/(3!2!1!1!) = 420$ strings
omitting EN	$7!/(3!2!1!1!) = 420$ strings
omitting ER	$7!/(3!1!1!1!1!) = 840$ strings
omitting EE	$7!/(2!2!1!1!1!) = 1260$ strings

Adding up these numbers we see that there are 4515 strings of length 7. Thus the answer is $7560+7560+4515=19{,}635$.

37. We assume that all the fruit is to be eaten; in other words, this process ends after 7 days. This is a permutation problem since the order in which the fruit is consumed matters (indeed, there is nothing else that matters here). Theorem 3 applies, with $n=7$, $n_1=3$, $n_2=2$, and $n_3=2$. The answer is therefore $7!/(3!2!2!)=210$.

39. We can describe any such travel in a unique way by a sequence of 4 x's, 3 y's, and 5 z's. By Theorem 3 there are

$$\frac{12!}{4!3!5!}=27720$$

such sequences.

41. This is like Example 9. If we approach it as is done there, we see that the answer is

$$C(52,7)C(45,7)C(38,7)C(31,7)C(24,7)=\frac{52!}{7!45!}\cdot\frac{45!}{7!38!}\cdot\frac{38!}{7!31!}\cdot\frac{31!}{7!24!}\cdot\frac{24!}{7!17!}=\frac{52!}{7!7!7!7!7!17!}\approx 7.0\times 10^{34}.$$

Applying Theorem 4 will yield the same answer; in this approach we think of the five players and the undealt cards as the six distinguishable boxes.

43. We assume that we are to care about which player gets which cards. For example, a deal in which Laurel gets a royal flush in spades and Blaine gets a royal flush in hearts will be counted as different from a deal in which Laurel gets a royal flush in hearts and Blaine gets a royal flush in spades (and the other four players get the same cards each time). The order in which a player receives his or her cards is not relevant, however, so we are dealing with combinations. We can look at one player at a time. There are $C(48,5)$ ways to choose the cards for the first player, then $C(43,5)$ ways to choose the cards for the second player (because five of the cards are gone), and so on. So the answer, by the multiplication principle, is $C(48,5)\cdot C(43,5)\cdot C(38,5)\cdot C(33,5)\cdot C(28,5)\cdot C(23,5)=649{,}352{,}163{,}073{,}816{,}339{,}512{,}038{,}979{,}194{,}880\approx 6.5\times 10^{32}$.

45. **a)** All that matters is how many copies of the book get placed on each shelf. Letting x_i be the number of copies of the book placed on shelf i, we are asking for the number of solutions to the equation $x_1+x_2+\cdots+x_k=n$, with each x_i a nonnegative integer. By Theorem 2 this is $C(k+n-1,n)$.

b) No generality is lost if we number the books b_1, b_2, ..., b_n and think of placing book b_1, then placing b_2, and so on. There are clearly k ways to place b_1, since we can put it as the first book (for now) on any of the shelves. After b_1 is placed, there are $k+1$ ways to place b_2, since it can go to the right of b_1 or it can be the first book on any of the shelves. We continue in this way: there are $k+2$ ways to place b_3 (to the right of b_1, to the right of b_2, or as the first book on some shelf), $k+3$ ways to place b_4, ..., $k+n-1$ ways to place b_n. Therefore the answer is the product of these numbers, which can more easily be expressed as $(k+n-1)!/(k-1)!$.

Another, perhaps easier, way to obtain this answer is to think of first choosing the locations for the books, which is what we counted in part **(a)**, and then choose a permutation of the n books to put into those locations (shelf by shelf, from the top down, and from left to right on each shelf). Thus the answer is $C(k+n-1,n)\cdot n!$, which evaluates to the same thing we obtained with our other analysis.

47. The first box holds n_1 objects, and there are $C(n,n_1)$ ways to choose those objects from among the n objects in the collection. Once these objects are chosen, we can choose the objects to be placed in the second box in $C(n-n_1,n_2)$ ways, since there are $n-n_1$ objects not yet placed, and we need to put n_2 of them into the second box. Similarly, there are then $C(n-n_1-n_2,n_3)$ ways to choose objects for the third box. We continue in this way, until finally there are $C(n-n_1-n_2-\cdots-n_{k-1},n_k)$ ways to choose the objects to put in the last (k^{th}) box. Note that this last expression equals $C(n_k,n_k)=1$, since $n_1+n_2+\cdots+n_k=n$. Now by the product rule the number of ways to make the entire assignment is

$$C(n,n_1)\cdot C(n-n_1,n_2)\cdot C(n-n_1-n_2,n_3)\cdots C(n-n_1-n_2-\cdots-n_{k-1},n_k).$$

We use the formula for combinations to write this as

$$\frac{n!}{n_1!(n-n_1)!}\cdot\frac{(n-n_1)!}{n_2!(n-n_1-n_2)!}\cdot\frac{(n-n_1-n_2)!}{n_3!(n-n_1-n_2-n_3)!}\cdots\frac{(n-n_1-n_2-\cdots-n_{k-1})!}{n_k!(n-n_1-n_2-\cdots-n_{k-1}-n_k)!},$$

which simplifies after the telescoping cancellation to

$$\frac{n!}{n_1!n_2!\cdots n_k!}$$

(we use the fact that $n-n_1-n_2-\cdots-n_{k-1}=n_k$, since $n_1+n_2+\cdots+n_k=n$), as desired.

49. a) The sequence was nondecreasing to begin with. By adding $k-1$ to the kth term, we are adding more to each term than was added to the previous term. Hence even if two successive terms in the sequence were originally equal, the second term must strictly exceed the first after this addition is completed. Therefore the sequence is made up of distinct numbers. The smallest can be no smaller than $1+(1-1)=1$, and the largest can be no larger than $n+(r-1)=n+r-1$; therefore the terms all come from T.
b) If we are given an increasing sequence of r terms from T, then by subtracting $k-1$ from the kth term we have a nondecreasing sequence of r terms from S, repetitions allowed. (The kth term in the original sequence must be between k and $n+r-1-(r-k)=n+(k-1)$, so subtracting $k-1$ leaves a number between 1 and n, inclusive. Furthermore, only 1 more is subtracted from a term than is subtracted from the previous term; thus no term can become strictly smaller than its predecessor, since it exceeded it by at least 1 to start with.) This operation exactly inverts the operation described in part **(a)**, so the correspondence is one-to-one.
c) The first two parts show that there are exactly as many r-combinations with repetition allowed from S as there are r-combinations (without repetition) from T. Since T has $n+r-1$ elements, this latter quantity is clearly $C(n+r-1,r)$.

51. We use the formula given on page 378 for the number of ways to distribute n distinguishable objects into j indistinguishable boxes with no box empty:

$$S(n,j)=\frac{1}{j!}\sum_{i=0}^{j-1}(-1)^i\binom{j}{i}(j-i)^n$$

In this case, $n = 6$ and $j = 4$, so we have

$$S(6,4) = \frac{1}{4!}\left(\binom{4}{0}4^6 - \binom{4}{1}3^6 + \binom{4}{2}2^6 - \binom{4}{3}1^6\right) = \frac{1}{4!}(4096 - 2916 + 384 - 4) = 65\,.$$

If we want to work this out from scratch, we can argue as follows. There are two patterns possible. We can put three of the objects into one box and each of the remaining objects into a separate box; there are $C(6,3) = 20$ ways to choose the objects for the crowded box. Alternatively, we can choose a pair of objects for one box ($C(6,2) = 15$ ways) and a pair of remaining objects for the second box ($C(4,2) = 6$ ways) and put the other two objects into separate boxes, but divide by 2 because of the overcounting caused by the indistinguishability of the first two boxes, for a total of 45 ways. Therefore the answer is $20 + 45 = 65$.

53. We assume that people are distinguishable, so this problem is identical to Exercise 51. There are 65 ways to place the employees.

55. Since each box has to contain at least one object, we might as well put one object into each box to begin with. This leaves us with just two more objects, and there are only two choices: we can put them both into the same box (so that the partition we end up with is $6 = 3 + 1 + 1 + 1$), or we can put them into different boxes (so that the partition we end up with is $6 = 2 + 2 + 1 + 1$). So the answer is 2.

57. Since each box has to contain at least two DVDs, we might as well put two DVDs into each box to begin with. This leaves us with just three more DVDs, and there are only three choices: we can put them all into the same box (so that the partition we end up with is $9 = 5 + 2 + 2$), or we can put two into one box and one into another (so that the partition we end up with is $9 = 4 + 3 + 2$), or we can put them all into different boxes (so that the partition we end up with is $9 = 3 + 3 + 3$). So the answer is 3.

59. To begin, notice that because each box must have at least one ball, there are only two basic arrangements: to put three balls into one box and one ball into each of the other two boxes (denoted 3-1-1), or to put one ball into one box and two balls into each of the other two boxes (denoted 1-2-2).

a) For the 3-1-1 arrangement, there are 3 ways to choose the crowded box, $C(5,3) = 10$ ways to choose the balls to be put there, and 2 ways to decide where the other balls go, for a total of $3 \cdot 10 \cdot 2 = 60$ possibilities. For the 1-2-2 arrangement, there are 3 ways to choose the box that will have just one ball, 5 ways to choose which ball goes there, and $C(4,2) = 6$ ways to decide which two balls go into the lower-numbered remaining box, for a total of $3 \cdot 5 \cdot 6 = 90$ possibilities. Thus the answer is $60 + 90 = 150$.

b) There are $C(5,3) = 10$ ways to choose the balls for the crowded box in the 3-1-1 arrangement. For the 1-2-2 arrangement there are 5 ways to choose the lonely ball and 3 ways to choose the partner of the lowest-numbered remaining ball. Therefore the answer is $10 + 5 \cdot 3 = 25$.

c) There are 3 ways to choose the crowded box for the 3-1-1 arrangement, and there are 3 ways to choose the solo box for the 1-2-2 arrangement. Therefore the answer is $3 + 3 = 6$.

d) There are just the 2 possibilities we have been discussing: 3-1-1 and 1-2-2.

61. Without the restriction on site X, we are simply asking for the number of ways to order the ten symbols V, V, W, W, X, X, Y, Y, Z, Z (the ordering will give us the visiting schedule). By Theorem 3 this can be done in $10!/(2!)^5 = 113{,}400$ ways. If the inspector visits site X on consecutive days, then in effect we are ordering nine symbols (including only one X), where now the X means to visit site X twice in a row. There are $9!/(2!)^4 = 22{,}680$ ways to do this. Therefore the answer is $113{,}400 - 22{,}680 = 90{,}720$.

63. When $(x_1 + x_2 + \cdots + x_m)^n$ is expanded, each term will clearly be of the form $Cx_1^{n_1}x_2^{n_2}\cdots x_m^{n_m}$, for some constants C that depend on the exponents, where the exponents sum to n. Thus the form of the given formula is correct, and the only question is whether the constants are correct. We need to count the number of ways in which a product of one term from each of the n factors can be $x_1^{n_1}x_2^{n_2}\cdots x_m^{n_m}$. In order for this to happen, we must choose n_1 x_1's, n_2 x_2's, ..., n_m x_m's. By Theorem 3 this can be done in

$$C(n; n_1, n_2, \ldots, n_m) = \frac{n!}{n_1!n_2!\cdots n_m!}$$

ways.

65. By the Multinomial Theorem, given in Exercise 63, the coefficient is

$$C(10; 3, 2, 5) = \frac{10!}{3!2!5!} = \frac{10\cdot 9\cdot 8\cdot 7\cdot 6}{12} = 2520\,.$$

SECTION 5.6 Generating Permutations and Combinations

This section is quite different from the rest of this chapter. It is really about algorithms and programming. These algorithms are not easy, and it would be worthwhile to "play computer" with them to get a feeling for how they work. In constructing such algorithms yourself, try assuming that you will list the permutations or combinations in a nice order (such as lexicographic order); then figure out how to find the "next" one in this order.

1. Lexicographic order is the same as numerical order in this case, so the ordering from smallest to largest is 14532, 15432, 21345, 23451, 23514, 31452, 31542, 43521, 45213, 45321.

3. We use Algorithm 1 to find the next permutation. Our notation follows that algorithm, with j being the largest subscript such that $a_j < a_{j+1}$ and k being the subscript of the smallest number to the right of a_j that is larger than a_j.

a) Since $4 > 3 > 2$, we know that the 1 is our a_j. The smallest integer to the right of 1 and greater than 1 is 2, so $k = 4$. We interchange a_j and a_k, giving the permutation 2431, and then we reverse the entire substring to the right of the position now occupied by the 2, giving the answer 2134.

b) The first integer from the right that is less than its right neighbor is the 2 in position 4. Therefore $j = 4$ here, and of course k has to be 5. The next permutation is the one that we get by interchanging the fourth and fifth numbers, 54132. (Note that the last phase of the algorithm, reversing the end of the string, was vacuous this time—there was only one element to the right of position 4, so no reversing was necessary.)

c) Since $5 > 3$, we know that the 4 is our a_j. The smallest integer to the right of 4 and greater than 4 is $a_k = 5$. We interchange a_j and a_k, giving the permutation 12543, and then we reverse the entire substring to the right of the position now occupied by the 5, giving the answer 12534.

d) Since $3 > 1$, we know that the 2 is our a_j. The smallest integer to the right of 2 and greater than 2 is $a_k = 3$. We interchange a_j and a_k, giving the permutation 45321, and then we reverse the entire substring to the right of the position now occupied by the 3, giving the answer 45312.

e) The first integer from the right that is less than its right neighbor is the 3 in position 6. Therefore $j = 6$ here, and of course k has to be 7. The next permutation is the one that we get by interchanging the sixth and seventh numbers, 6714253. As in part **(b)**, no reversing was necessary.

f) Since $8 > 7 > 6 > 4$, we know that the 2 is our a_j, so $j = 4$. The smallest integer to the right of 2 and greater than 2 is $a_8 = 4$. We interchange a_4 and a_8, giving the permutation 31548762, and then we reverse the entire substring to the right of the position now occupied by the 4, giving the answer 31542678.

5. We begin with the permutation 1234. Then we apply Algorithm 1 23 times in succession, giving us the other 23 permutations in lexicographic order: 1243, 1324, 1342, 1423, 1432, 2134, 2143, 2314, 2341, 2413, 2431, 3124, 3142, 3214, 3241, 3412, 3421, 4123, 4132, 4213, 4231, 4312, and 4321. The last permutation is the one entirely in decreasing order. Each application of Algorithm 1 follows the pattern in Exercise 3.

7. We begin with the first 3-combination, namely $\{1,2,3\}$. Let us trace through Algorithm 3 to find the next. Note that $n=5$ and $r=3$; also $a_1=1$, $a_2=2$, and $a_3=3$. We set i equal to 3 and then decrease i until $a_i \neq 5-3+i$. This inequality is already satisfied for $i=3$, since $a_3 \neq 5$. At this point we increment a_i by 1 (so that now $a_3=4$), and fill the remaining spaces with consecutive integers following a_i (in this case there are no more remaining spaces). Thus our second 3-combination is $\{1,2,4\}$. The next call to Algorithm 3 works the same way, producing the third 3-combination, namely $\{1,2,5\}$. To get the fourth 3-combination, we again call Algorithm 3. This time the i that we end up with is 2, since $5=a_3=5-3+3$. Therefore the second element in the list is incremented, namely goes from a 2 to a 3, and the third element is the next larger element after 3, namely 4. Thus this 3-combination is $\{1,3,4\}$. Another call to the algorithm gives us $\{1,3,5\}$, and another call gives us $\{1,4,5\}$. Now when we call the algorithm, we find $i=1$ at the end of the **while** loop, since in this case the last two elements are the two largest elements in the set. Thus a_1 is increased to 2, and the remainder of the list is filled with the next two consecutive integers, giving us $\{2,3,4\}$. Continuing in this manner, we get the rest of the 3-combinations: $\{2,3,5\}$, $\{2,4,5\}$, $\{3,4,5\}$.

9. Clearly the next larger r-combination must differ from the old one in position i, since positions $i+1$, $i+2$, $\ldots$, r are occupied by the largest possible numbers (namely $i+n-r+1$ to n). Also a_i+1 is the smallest possible number that can be put in position i if we want an r-combination greater than the given one, and then similarly a_i+2, a_i+3, $\ldots$, $a_i+r-i+1$ are the smallest allowable numbers for positions $i+1$ to r. Therefore there is no r-combination between the given one and the one that Algorithm 3 produces, which is exactly what we had to prove.

11. One way to do this problem (and to have done Exercise 10) is to generate the r-combinations using Algorithm 3, and then to find all the permutations of each, using Algorithm 1 (except that now the elements to be permuted are not the integers from 1 to r, but are instead the r elements of the r-combination currently being used). Thus we start with the first 3-combination, $\{1,2,3\}$, and we list all 6 of its permutations: 123, 132, 213, 231, 312, 321. Next we find the next 3-combination, namely $\{1,2,4\}$, and list all of its permutations: 124, 142, 214, 241, 412, 421. We continue in this manner to generate the remaining 48 3-permutations of $\{1,2,3,4,5\}$: 125, 152, 215, 251, 512, 521; 134, 143, 314, 341, 413, 431; 135, 153, 315, 351, 513, 531; 145, 154, 415, 451, 514, 541; 234, 243, 324, 342, 423, 432; 235, 253, 325, 352, 523, 532; 245, 254, 425, 452, 524, 542; 345, 354, 435, 453, 534, 543. There are of course $P(5,3)=5\cdot 4\cdot 3=60$ items in our list.

13. One way to show that a function is a bijection is to find its inverse, since only bijections can have inverses. Note that the sizes of the two sets in question are the same, since there are $n!$ nonnegative integers less than $n!$, and there are $n!$ permutations of $\{1,2,\ldots,n\}$. In this case, since Cantor expansions are unique, we need to take the digits a_1, a_2, $\ldots$, a_{n-1} of the Cantor expansion of a nonnegative integer m less than $n!$ (so that $m=a_1 1!+a_2 2!+\cdots+a_{n-1}(n-1)!$), and produce a permutation with these a_k's satisfying the definition given before Exercise 12—indeed the only such permutation.

We will fill the positions in the permutation one at a time. First we put n into position $n-a_{n-1}$; clearly a_{n-1} will be the number of integers less than n that follow n in the permutation, since exactly a_{n-1} positions remain empty to the right of where we put the n. Next we renumber the free positions (the ones other than the one into which we put n), from left to right, as $1,2,\ldots,n-1$. Under this numbering, we put $n-1$ into position $(n-1)-a_{n-2}$. Again it is clear that a_{n-2} will be the number of integers less than $n-1$ that follow $n-1$ in the permutation. We continue in this manner, renumbering the free positions, from left to right, as $1,2,\ldots,n-k+1$, and then placing $n-k+1$ in position $(n-k+1)-a_{n-k}$, for $k=1,2,\ldots,n-1$. Finally we place 1 in the only remaining position.

15. The algorithm is really given in our solution to Exercise 13. To produce all the permutations, we find the permutation corresponding to i, where $0\leq i<n!$, under the correspondence given in Exercise 13. To do

this, we need to find the digits in the Cantor expansion of i, starting with the digit a_{n-1}. In what follows, that digit will be called c. We use k to keep track of which digit we are working on; as k goes from 1 to $n-1$, we will be computing the digit a_{n-k} in the Cantor expansion and inserting $n-k+1$ into the proper location in the permutation. (At the end, we need to insert 1 into the only remaining position.) We will call the positions in the permutation $p_1, p_2, \ldots, p_n$. We write only the procedure that computes the permutation corresponding to i; obviously to get all the permutations we simply call this procedure for each i from 0 to $n!-1$.

```
procedure Cantor permutation(n, i : integers with n ≥ 1 and 0 ≤ i < n!)
x := n {to help in computing Cantor digits}
for j := 1 to n {initialize permutation to be all 0's}
        p_j := 0
for k := 1 to n − 1 {figure out where to place n − k + 1}
begin
        c := ⌊x/(n − k)!⌋ {the Cantor digit}
        x := x − c(n − k)! {what's left of x}
        h := n {now find the (c + 1)^th free position from the right}
        while p_h ≠ 0
                h := h − 1
        for j := 1 to c
        begin
                h := h − 1
                while p_h ≠ 0
                        h := h − 1
        end
        p_h := n − k + 1 {here is the key step}
end
h := 1 {now find the last free position}
while p_h ≠ 0
        h := h + 1
p_h := 1
{p_1, p_2, ..., p_n is the permutation corresponding to i}
```

GUIDE TO REVIEW QUESTIONS FOR CHAPTER 5

1. $1+2+2\cdot2+2\cdot2\cdot2+\cdots+2\cdot2\cdots2=2047$

2. Subtract 11 from the answer to the previous review question, since λ, 1, 11, ..., 11...1 are the bit strings that do not have at least one 0 bit.

3. **a)** See Example 6 in Section 5.1. **b)** 10^5 **c)** See Example 7 in Section 5.1. **d)** $10\cdot9\cdot8\cdot7\cdot6$ **e)** 0

4. with a tree diagram; see Example 20 in Section 5.1 (extended to larger tree)

5. Using the inclusion–exclusion principle, we get $2^7+2^7-2^4$; see Example 17 in Section 5.1.

6. **a)** See p. 347. **b)** 11 pigeons, 10 holes (digits)

7. **a)** See p. 349. **b)** $N=91$, $k=10$

8. **a)** Permutations are ordered arrangements; combinations are unordered (or just arbitrarily ordered for convenience) selections.
b) $P(n,r)=C(n,r)\cdot r!$ (see the proof of Theorem 2 in Section 5.3) **c)** $C(25,6)$ **d)** $P(25,6)$

9. **a)** See pp. 366–367. **b)** by adding the two numbers above each number in the new row

10. A combinatorial proof is a proof of an algebraic identity that shows that both sides count the same thing (in some application). An algebraic proof is totally different—it shows that the two sides are equal by doing formal manipulations with the unknowns, with no reference to what the expressions might mean in an application.

11. See p. 366.

12. **a)** See p. 363. **b)** See p. 363. **c)** $2^{100}5^{101}C(201,101)$

13. **a)** See Theorem 2 in Section 5.5. **b)** $C(5+12-1,12)$ **c)** $C(5+9-1,9)$
d) $C(5+12-1,12)-C(5+7-1,7)$ **e)** $C(5+10-1,10)-C(5+6-1,6)$

14. **a)** See Example 5 in Section 5.5. **b)** $C(4+17-1,17)$
c) $C(4+13-1,13)$ (see Exercise 15a in Section 5.5)

15. **a)** See Theorem 3 in Section 5.5. **b)** $14!/(2!2!1!3!1!1!3!1!)$

16. See pp. 382–384.

17. **a)** $C(52,5)\cdot C(47,5)\cdot C(42,5)\cdot C(37,5)\cdot C(32,5)\cdot C(27,5)$ **b)** See Theorem 4 in Section 5.5.

18. See p. 385.

SUPPLEMENTARY EXERCISES FOR CHAPTER 5

1. In each part of this problem we have $n=10$ and $r=6$.

a) If the items are to be chosen in order with no repetition allowed, then we need a simple permutation. Therefore the answer is $P(10,6)=10\cdot 9\cdot 8\cdot 7\cdot 6\cdot 5=151{,}200$.

b) If repetition is allowed, then this is just a simple application of the product rule, with 6 tasks, each of which can be done in 10 ways. Therefore the answer is $10^6=1{,}000{,}000$.

c) If the items are to be chosen without regard to order but with no repetition allowed, then we need a simple combination. Therefore the answer is $C(10,6)=C(10,4)=10\cdot 9\cdot 8\cdot 7\cdot/(4\cdot 3\cdot 2)=210$.

d) Unordered choices with repetition allowed are counted by $C(n+r-1,r)$, which in this case is $C(15,6)=5005$.

3. The student has 3 choices for each question: true, false, and no answer. There are 100 questions, so by the product rule there are $3^{100}\approx 5.2\times 10^{47}$ ways to answer the test.

5. We will apply the inclusion–exclusion principle from Section 5.1. First let us calculate the number of these strings with exactly three a's. To specify such a string we need to choose the positions for the a's, which can be done in $C(10,3)$ ways. Then we need to choose either a b or a c to fill each of the other 7 positions in the string, which can be done in 2^7 ways. Therefore there are $C(10,3)\cdot 2^7=15360$ such strings. Similarly, there are $C(10,4)\cdot 2^6=13440$ strings with exactly four b's. Next we need to compute the number of strings satisfying both of these conditions. To specify a string with exactly three a's and exactly four b's, we need to choose the positions for the a's, which can be done in $C(10,3)$ ways, and then choose the positions for the b's, which can be done in $C(7,4)$ ways (only seven slots remain after the a's are placed). Therefore there are $C(10,3)\cdot C(7,4)=4200$ such strings. Finally, by the inclusion–exclusion principle the number of strings having either exactly three a's or exactly four b's is $15360+13440-4200=24{,}600$.

7. a) We want a combination with repetition allowed, with $n = 28$ and $r = 3$. By Theorem 2 of Section 5.5, there are $C(28+3-1,3) = C(30,3) = 4060$ possibilities.

b) This is just a simple application of the product rule. There are 28 ways to choose the ice cream, 8 ways to choose the sauce, and 12 ways to choose the topping, so there are $28 \cdot 8 \cdot 12 = 2688$ possible small sundaes.

c) By the product rule we have to multiply together the number of ways to choose the ice cream, the number of ways to choose the sauce, and the number of ways to choose the topping. There are $C(28+3-1,3)$ ways to choose the ice cream, just as in part **(a)**. There are $C(8,2)$ ways to choose the sauce, since repetition is not allowed. There are similarly $C(12,3)$ ways to choose the toppings. Multiplying these numbers together, we find that the answer is $4060 \cdot 28 \cdot 220 = 25{,}009{,}600$ different large sundaes.

9. We can solve this problem by counting the number of numbers that have the given digit in 1, 2, or 3 places.

a) The digit 0 appears in 1 place in some two-digit numbers and in some three-digit numbers. There are clearly 9 two-digit numbers in which 0 appears, namely 10, 20, ..., 90. We can count the number of three-digit numbers in which 0 appears exactly once as follows: first choose the place in which it is to appear (2 ways, since it cannot be the leading digit), then choose the left-most of the remaining digits (9 ways, since it cannot be a 0), then choose the final digit (also 9 ways). Therefore there are $9 + 2 \cdot 9 \cdot 9 = 171$ numbers in which the 0 appears exactly once, accounting for 171 appearances of the digit 0. Finally there are another 9 numbers in which the digit 0 appears twice, namely 100, 200, ..., 900. This accounts for 18 more 0's. And of course the number 1000 contributes 3 0's. Therefore our final answer is $171 + 18 + 3 = 192$.

b) The analysis for the digit 1 is not the same as for the digit 0, since we can have leading 1's but not leading 0's. One 1 appears in the one-digit numbers. Two-digit numbers can have a 1 in the ones place (and there are 9 of these, namely 11, 21, ..., 91), or in the tens place (and there are 10 of these, namely 10 through 19). Of course the number 11 is counted in both places, but that is proper, since we want to count each appearance of a 1. Therefore there are $10 + 9 = 19$ 1's appearing in two-digit numbers. Similarly, the three-digit numbers have 90 1's appearing in the ones place (every tenth number, and there are 900 numbers), 90 1's in the tens place (10 per decade, and there are 9 decades), and 100 1's in the hundreds place (100 through 199); therefore there are 280 ones appearing in three-digit numbers. Finally there is a 1 in 1000, so the final answer is $1 + 19 + 280 + 1 = 301$.

c) The analysis for the digit 2 is exactly the same as for the digit 1, with the exception that we do not get any 2's in 1000. Therefore the answer is $301 - 1 = 300$.

d) The analysis for the digit 9 is exactly the same as for the digit 2, so the answer is again 300.

Let us check all of the answers to this problem simultaneously. There are 300 each of the digits 2 through 9, for a total of 2400 digits. There are 192 0's and 301 1's. Therefore $2400+192+301 = 2893$ digits are used altogether. Let us count this another way. There are 9 one-digit numbers, 90 two-digit numbers, 900 three-digit numbers, and 1 four-digit number, so the total number of digits is $9\cdot1+90\cdot2+900\cdot3+1\cdot4 = 2893$. This agreement tends to confirm our analysis.

11. This is a negative instance of the generalized pigeonhole principle. The worst case would be if the student gets each fortune 3 times, for a total of $3 \times 213 = 639$ meals. If the student ate 640 or more meals, then the student will get the same fortune at least $\lceil 640/213 \rceil = 4$ times.

13. We have no guarantee ahead of time that this will work, but we will try applying the pigeonhole principle. Let us count the number of different possible sums. If the numbers in the set do not exceed 50, then the largest possible sum of a 5-element subset will be $50 + 49 + 48 + 47 + 46 = 240$. The smallest possible sum will be $1 + 2 + 3 + 4 + 5 = 15$. Therefore the sum has to be a number between 15 and 240, inclusive, and there are $240 - 15 + 1 = 226$ such numbers. Now let us count the number of different subsets. That is of course $C(10,5) = 252$. Since there are more subsets (pigeons) than sums (pigeonholes), we know that there must be two subsets with the same sum.

15. We assume that the drawings of the cards is done without replacement (i.e., no repetition allowed).

a) The worst case would be that we drew 1 ace and the 48 cards that are not aces, a total of 49 cards. Therefore we need to draw 50 cards to guarantee at least 2 aces (and it is clear that 50 is sufficient, since at worst 2 of the 4 aces would then be left in the deck).

b) The same analysis as in part **(a)** applies, so again the answer is 50.

c) In this problem we can use the pigeonhole principle. If we drew 13 cards, then they might all be of different kinds (ranks). If we drew 14 cards, however, then since there are only 13 kinds we would be assured of having at least two of the same kind. (The drawn cards are the pigeons and the kinds are the pigeonholes.)

d) If we drew 4 cards, they might all be of the same kind. However, if we draw 5 cards, then since there are only 4 of one kind, we are assured of seeing at least two different kinds.

17. This problem can be solved using the pigeonhole principle if we look at it correctly. Let s_i be the sum of the first i of these numbers, where $1 \le i \le m$. Now if $s_i \equiv 0 \pmod{m}$ for some i, then we have our desired consecutive terms whose sum is divisible by m. Otherwise the numbers $s_1 \mathbf{~mod~} m$, $s_2 \mathbf{~mod~} m$, ..., $s_m \mathbf{~mod~} m$ are all integers in the set $\{1, 2, \ldots, m-1\}$. By the pigeonhole principle we know that two of them are the same, say $s_i \mathbf{~mod~} m = s_j \mathbf{~mod~} m$ with $i < j$. Then $s_j - s_i$ is divisible by m. But $s_j - s_i$ is just the sum of the $(i+1)^{\text{th}}$ through j^{th} terms in the sequence, and we are done.

19. The decimal expansion of a rational number a/b (we can assume that a and b are positive integers) can be obtained by long division of the number b into the number a, where a is written with a decimal point and an arbitrarily long string of 0's following it. The basic step in long division is finding the next digit of the quotient, which is just $\lfloor r/b \rfloor$, where r is the current remainder with the next digit of the dividend brought down. Now in our case, eventually the dividend has nothing but 0's to bring down. Furthermore there are only b possible remainders, namely the numbers from 0 to $b-1$. Thus at some point in our calculation after we have passed the decimal point, we will, by the pigeonhole principle, be looking at exactly the same situation as we had previously. From that point onward, then, the calculation must follow the same pattern as it did previously. In particular, the digits of the quotient will repeat.

For example, to compute the decimal expansion of the rational number $349/11$, we divide 11 into $349.00\ldots$. The first digit of the quotient is 3, and the remainder is 1. The next digit of the quotient is 1 and the remainder is 8. At this point there are only 0's left to bring down. The next digit of the quotient is a 7 with a remainder of 3, and then a quotient digit of 2 with a remainder of 8. We are now in exactly the same situation as at the previous appearance of a remainder of 8, so the quotient digits 72 repeat forever. Thus $349/11 = 31.\overline{72}$.

21. a) This is a simple combination, so the answer is $C(20, 12) = 125{,}970$.

b) The only choice is the choice of a variety, so the answer is 20.

c) We assume that order does not matter (all the donuts will go into a bag). Therefore, since repetitions are allowed, Theorem 2 of Section 5.5 applies, and the answer is $C(20+12-1, 12) = C(31, 12) = 141{,}120{,}525$.

d) We can simply subtract from our answer to part **(c)** our answer to part **(b)**, which asks for the number of ways this restriction can be violated. Therefore the answer is 141,120,505.

e) We put the 6 blueberry filled donuts into our bag, and the problem becomes one of choosing 6 donuts with no restrictions. In analogy with part **(c)**, we obtain the answer $C(20+6-1, 6) = C(25, 6) = 177{,}100$.

f) There are $C(20+5-1, 5) = C(24, 5) = 42504$ ways to choose at least 7 blueberry donuts among our dozen (the calculation is essentially the same as that in part **(e)**). Our answer is therefore 42504 less than our unrestricted answer to part **(c)**: $141120525 - 42504 = 141{,}078{,}021$.

23. a) The given equation is equivalent to $n(n-1)/2 = 45$, which reduces to $n^2 - n - 90 = 0$. The quadratic formula (or factoring) tells us that the roots are $n = 10$ and $n = -9$. Since n is assumed to be nonnegative, the only relevant solution is $n = 10$.

b) The given equation is equivalent to $n(n-1)(n-2)/6 = n(n-1)$. Since $P(n,2)$ is not defined for $n < 2$, we know that neither n nor $n-1$ is 0, so we can divide both sides by these factors, obtaining $n - 2 = 6$, whence $n = 8$.

c) Recall the identity $C(n,k) = C(n, n-k)$. The given equation fits that model if $n = 7$ and $k = 5$. Hence $n = 7$ is a solution. That there are no more solutions follows from the fact that $C(n,k)$ is an increasing function in k for $0 \le k \le n/2$, and hence there are no other numbers i and j for which $C(n,i) = C(n,j)$.

25. Following the hint, we see that each element of S falls into exactly one of three categories: either it is an element of A, or else it is not an element of A but is an element of B (in other words, is an element of $B - A$), or else it is not an element of B either (in other words, is an element of $S - B$). So the number of ways to choose sets A and B to satisfy these conditions is the same as the number of ways to place each element of S into one of these three categories. Therefore the answer is 3^n. For example, if $n = 2$ and $S = \{x, y\}$, then there are 9 pairs: $(\emptyset, \emptyset)$, $(\emptyset, \{x\})$, $(\emptyset, \{y\})$, $(\emptyset, \{x,y\})$, $(\{x\}, \{x\})$, $(\{x\}, \{x,y\})$, $(\{y\}, \{y\})$, $(\{y\}, \{x,y\})$, $(\{x,y\}, \{x,y\})$.

27. We start with the right-hand side and use Pascal's Identity three times to obtain the left-hand side:

$$\begin{aligned}
&C(n+2, r+1) - 2C(n+1, r+1) + C(n, r+1) \\
&\qquad = C(n+1, r+1) + C(n+1, r) - 2C(n+1, r+1) + C(n, r+1) \\
&\qquad = C(n+1, r) - C(n+1, r+1) + C(n, r+1) \\
&\qquad = [C(n,r) + C(n, r-1)] - [C(n, r+1) + C(n,r)] + C(n, r+1) \\
&\qquad = C(n, r-1)
\end{aligned}$$

29. Substitute $x = 1$ and $y = 3$ into the Binomial Theorem (Theorem 1 in Section 5.4) and we obtain exactly this identity.

31. The trick to the analysis here is to imagine what such a string has to look like. Every string of 0's and 1's can be thought of as consisting of alternating blocks—a block of 1's (possibly empty) followed by a block of 0's followed by a block of 1's followed by a block of 0's, and so on, ending with a block of 0's (again, possibly empty). If we want there to be exactly two occurrences of 01, then in fact there must be exactly six such blocks, the middle four all being nonempty (the transitions from 0's to 1's create the 01's) and the outer two possibly being empty. In other words, the string must look like this:

$$x_1 \text{ 1's} — x_2 \text{ 0's} — x_3 \text{ 1's} — x_4 \text{ 0's} — x_5 \text{ 1's} — x_6 \text{ 0's},$$

where $x_1 + x_2 + \cdots + x_6 = n$ and $x_1 \ge 0$, $x_6 \ge 0$, and $x_i \ge 1$ for $i = 2, 3, 4, 5$. Clearly such a string is totally specified by the values of the x_i's. Therefore we are simply asking for the number of solutions to the equation $x_1 + x_2 + \cdots + x_6 = n$ subject to the stated constraints. This kind of problem is solved in Section 5.5 (Example 5 and several exercises). The stated problem is equivalent to finding the number of solutions to $x_1 + x_2' + x_3' + x_4' + x_5' + x_6 = n - 4$ where each variable here is nonnegative (we let $x_i = x_i' + 1$ for $i = 2, 3, 4, 5$ in order to insure that these x_i's are strictly positive). The number of such solutions is, by the results just cited, $C(6 + n - 4 - 1, n - 4)$, which simplifies to $C(n+1, n-4)$ or $C(n+1, 5)$.

33. An answer key is just a permutation of 8 a's, 3 b's, 4 c's, and 5 d's. We know from Theorem 3 in Section 5.5 that there are

$$\frac{20!}{8!3!4!5!} = 3{,}491{,}888{,}400$$

such permutations.

35. We assume that each student is to get one advisor, that there are no other restrictions, and that the students and advisors are to be considered distinct. Then there are 5 ways to assign each student, so by the product rule there are $5^{24} \approx 6.0 \times 10^{16}$ ways to assign all of them.

37. For all parts of this problem, Theorem 2 in Section 5.5 is used.

a) We let $x_1 = x_1' + 2$, $x_2 = x_2' + 3$, and $x_3 = x_3' + 4$. Then the restrictions are equivalent to requiring that each of the x_i''s be nonnegative. Therefore we want the number of nonnegative integer solutions to the equation $x_1' + x_2' + x_3' = 8$. There are $C(3+8-1,8) = C(10,8) = C(10,2) = 45$ of them.

b) The number of solutions with $x_3 > 5$ is the same as the number of solutions to $x_1 + x_2 + x_3' = 11$, where $x_3 = x_3' + 6$. There are $C(3+11-1,11) = C(13,11) = C(13,2) = 78$ of these. Now we want to subtract the number of solutions for which also $x_1 \geq 6$. This is equivalent to the number of solutions to $x_1' + x_2 + x_3' = 5$, where $x_1 = x_1' + 6$. There are $C(3+5-1,5) = C(7,5) = C(7,2) = 21$ of these. Therefore the answer to the problem is $78 - 21 = 57$.

c) Arguing as in part **(b)**, we know that there are 78 solutions to the equation $x_1 + x_2 + x_3' = 11$, which is equivalent to the number of solutions to $x_1 + x_2 + x_3 = 17$ with $x_3 > 5$. We now need to subtract the number of these solutions that violate one or both of the restrictions $x_1 < 4$ and $x_2 < 3$. The number of solutions with $x_1 \geq 4$ is the number of solutions to $x_1' + x_2 + x_3' = 7$, namely $C(3+7-1,7) = C(9,7) = C(9,2) = 36$. The number of solutions with $x_2 \geq 3$ is the number of solutions to $x_1 + x_2' + x_3' = 8$, namely $C(3+8-1,7) = C(10,8) = C(10,2) = 45$. However, there are also solutions in which both restrictions are violated, namely the solutions to $x_1' + x_2' + x_3' = 4$. There are $C(3+4-1,4) = C(6,4) = C(6,2) = 15$ of these. Therefore the number of solutions in which one or both conditions are violated is $36 + 45 - 15 = 66$; we needed to subtract the 15 so as not to count these solutions twice. Putting this all together, we see that there are $78 - 66 = 12$ solutions of the given problem.

39. a) We want to find the number of r-element subsets for $r = 0,1,2,3,4$ and add. Therefore the answer is $C(10,0) + C(10,1) + C(10,2) + C(10,3) + C(10,4) = 1 + 10 + 45 + 120 + 210 = 386$.

b) This time we want $C(10,8) + C(10,9) + C(10,10) = C(10,2) + C(10,1) + C(10,0) = 45 + 10 + 1 = 56$.

c) This time we want $C(10,1) + C(10,3) + C(10,5) + C(10,7) + C(10,9) = C(10,1) + C(10,3) + C(10,5) + C(10,3) + C(10,1) = 10 + 120 + 252 + 120 + 10 = 512$. We can also solve this problem by using the fact from Exercise 31 in Section 5.4 that a set has the same number of subsets with an even number of elements as it has subsets with an odd number of elements. Since the set has $2^{10} = 1024$ subsets altogether, half of these—512 of them—must have an odd number of elements.

41. Since the objects are identical, all that matters is the number of objects put into each container. If we let x_i be the number of objects put into the ith container, then we are asking for the number of solutions to the equation $x_1 + x_2 + \cdots + x_m = n$ with the restriction that each $x_i \geq 1$. By the usual trick this is equivalent to asking for the number of nonnegative integer solutions to $x_1' + x_2' + \cdots + x_m' = n - m$, where we have set $x_i = x_i' + 1$ to insure that each container gets at least one object. By Theorem 2 in Section 5.5, there are $C(m + (n-m) - 1, n-m) = C(n-1, n-m)$ solutions. This can also be written as $C(n-1, m-1)$, since $(n-1) - (n-m) = m-1$. (Of course if $n < m$, then there are no solutions, since it would be impossible to put at least one object in each container. Our answer is consistent with this observation if we think of $C(x,y)$ as being 0 if $y > x$.)

43. a) This can be done with the multiplication principle. There are five choices for each ball, so the answer is $5^6 = 15{,}625$.

b) This is like Example 10 in Section 5.5, and we can use the formula on page 378:

$$\sum_{j=1}^{k} \frac{1}{j!} \sum_{i=0}^{j-1} (-1)^i \binom{j}{i} (j-i)^n$$

with $n = 6$ and $k = 5$. We get

$$\frac{1}{1!}1^6+\frac{1}{2!}(1\cdot 2^6-2\cdot 1^6)+\frac{1}{3!}(1\cdot 3^6-3\cdot 2^6+3\cdot 1^6)+\frac{1}{4!}(1\cdot 4^6-4\cdot 3^6+6\cdot 2^6-4\cdot 1^6)+\frac{1}{5!}(1\cdot 5^6-5\cdot 4^6+10\cdot 3^6-10\cdot 2^6+5\cdot 1^6),$$

which is $1 + 31 + 90 + 65 + 15 = 202$. The command in *Maple* for this, using the "combinat" package, is `sum(stirling2(6,j),j=1..5);`.

c) We saw in the discussion surrounding Example 9 in Section 5.5 that the number of ways to distribute n unlabeled objects into k labeled boxes is $C(n+k-1,k-1)$, because this is really the same as the problem of choosing an n-combination from the set of k boxes, with repetitions allowed. In this case we have $n = 6$ and $k = 5$, so the answer is $C(10,4) = 210$.

d) Since both the boxes and the objects are indistinguishable, what is really being asked is how many different ways there are to write 6 as the sum of five nonnegative integers, with order ignored. We will just enumerate the possibilities and count them. We have $6 = 6+0+0+0+0$; $6 = 5+1+0+0+0$; $6 = 4+2+0+0+0$; $6 = 4+1+1+0+0$; $6 = 3+3+0+0+0$; $6 = 3+2+1+0+0$; $6 = 3+1+1+1+0$; $6 = 2+2+2+0+0$; $6 = 2+2+1+1+0$; and $6 = 2+1+1+1+1$. There are ten ways in all. Notice that we are allowing some of the boxes to be empty.

45. For convenience let us assume that the finite set is $\{1,2,\ldots,n\}$. If we call a permutation $a_1a_2\ldots a_r$, then we simply need to allow each of the variables a_i to take on all n of the values from 1 to n. This is essentially just counting in base n, so our algorithm will be similar to Algorithm 2 in Section 5.6. The procedure shown here generates the next permutation. To get all the permutations, we just start with $11\ldots 1$ and call this procedure $r^n - 1$ times.

procedure *next permutation*(n : positive integer, $a_1, a_2, \ldots, a_r$: positive integers $\le n$)
{this procedure replaces the input with the next permutation, repetitions allowed,
  in lexicographic order; assume that there is a next permutation, i.e., $a_1a_2\ldots a_r \ne nn\ldots n$}
$i := r$
while $a_i = n$
begin
  $a_i := 1$
  $i := i - 1$
end
$a_i := a_i + 1$
{$a_1a_2\ldots a_r$ is the next permutation in lexicographic order}

47. We must show that if there are $R(m,n-1)+R(m-1,n)$ people at a party, then there must be at least m mutual friends or n mutual enemies. Consider one person; let's call him Jerry. Then there are $R(m-1,n)+R(m,n-1)-1$ other people at the party, and by the pigeonhole principle there must be at least $R(m-1,n)$ friends of Jerry or $R(m,n-1)$ enemies of Jerry among these people. First let's suppose there are $R(m-1,n)$ friends of Jerry. By the definition of R, among these people we are guaranteed to find either $m-1$ mutual friends or n mutual enemies. In the former case these $m-1$ mutual friends together with Jerry are a set of m mutual friends; and in the latter case we have the desired set of n mutual enemies. The other situation is similar: Suppose there are $R(m,n-1)$ enemies of Jerry; we are guaranteed to find among them either m mutual friends or $n-1$ mutual enemies. In the former case we have the desired set of m mutual friends, and in the latter case these $n-1$ mutual enemies together with Jerry are a set of n mutual enemies.

WRITING PROJECTS FOR CHAPTER 5

Books and articles indicated by bracketed symbols below are listed near the end of this manual. You should also read the general comments and advice you will find there about researching and writing these essays.

1. You might start with the standard history of mathematics books, such as [Bo4] or [Ev3].

2. To learn about telephone numbers in North America, refer to books on telecommunications, such [Fr]. The term to look for in an index is the North American Numbering Plan.

3. A lot of progress has been made recently by research mathematicians such as Herbert Wilf in finding general methods of proving essentially *all* true combinatorial identities, more or less mechanically. See whether you can find some of this work by looking in *Mathematical Reviews* (MathSciNet on the Web) or the book [PeWi]. There is also some discussion of this in [Wi2], a book on generating functions. Also, a classical book on combinatorial identities is [Ri2].

4. Students who have had an advanced physics course will be at an advantage here. Maybe you have a friend who is a physics major! In any case, it should not be hard to find a fairly elementary textbook on this subject.

5. More advanced combinatorics textbooks usually deal with Stirling numbers, at least in the exercises. See [Ro1], for instance. Other sources here are a chapter in [MiRo] and the amazing [GrKn].

6. See the comments for Writing Project 5.

7. There are entire books devoted to Ramsey theory, dealing not only with the classical Ramsey numbers, but also with applications to number theory, graph theory, geometry, linear algebra, etc. For a fairly advanced such book, see [GrRo]; for a gentler introduction, see the relevant sections of [Ro1] or the chapter in [MiRo]. Up-to-the-minute results can be found with a Web search.

8. Try books with titles such as "combinatorial algorithms"—that's what methods of generating permutations are, after all. See [Ev1] or [ReNi], for example. Another fascinating source (which deals with combinatorial algorithms as well as many other topics relevant to this text) is [GrKn]. Volume 2 of Knuth's classic [Kn] should have some relevant material. There is also an older article you might want to check out, [Le1]. An interesting related problem is to generate a *random* permutation; this is needed, for example, when using a computer to simulate the shuffling of a deck of cards for playing card games.

9. See the comments for Writing Project 8.

CHAPTER 6
Discrete Probability

SECTION 6.1 An Introduction to Discrete Probability

Calculating probabilities is one of the most immediate applications of combinatorics. Many people play games in which discrete probability plays a role, such as card games like poker or bridge, board games, casino games, and state lotteries. Probability is also important in making decisions in such areas as business and medicine—for example, in deciding how high a deductible to have on your automobile insurance. This section only scratches the surface, of course, but it is surprising how many useful calculations can be made using just the techniques discussed in this textbook.

The process is basically the same in each problem. First count the number of possible, equally likely, outcomes; this is the denominator of the probability you are seeking. Then count the number of ways that the event you are looking for can happen; this is the numerator. We have given approximate decimal (or percentage) answers to many of the problems, since the human mind can comprehend the magnitude of a number much better this way than by looking at a fraction with large numerator and denominator.

1. There are 52 equally likely cards to be selected, and 4 of them are aces. Therefore the probability is $4/52 = 1/13 \approx 7.7\%$.

3. Among the first 100 positive integers there are exactly 50 odd ones. Therefore the probability is $50/100 = 1/2$.

5. One way to do this is to look at the 36 equally likely outcomes of the roll of two dice, which we can represent by the set of ordered pairs (i,j) with $1 \leq i,j \leq 6$. A better way is to argue as follows. Whatever the number of spots showing on the first die, the sum will be even if and only if the number of spots showing on the second die has the same parity (even or odd) as the first. Since there are 3 even faces (2, 4, and 6) and 3 odd faces (1, 3, and 5), the probability is $3/6 = 1/2$.

7. There are $2^6 = 64$ possible outcomes, represented by all the sequences of length 6 of H's and T's. Only one of those sequences, $HHHHHH$, represents the event under consideration, so the probability is $1/64 \approx 0.016$.

9. We saw in Example 11 of Section 5.3 that there are $C(52,5)$ possible poker hands, and we assume by symmetry that they are all equally likely. In order to solve this problem, we need to compute the number of poker hands that do not contain the queen of hearts. Such a hand is simply an unordered selection from a deck with 51 cards in it (all cards except the queen of hearts), so there are $C(51,5)$ such hands. Therefore the answer to the question is the ratio

$$\frac{C(51,5)}{C(52,5)} = \frac{47}{52} \approx 90.4\% .$$

11. This question completely specifies the poker hand, so there is only one hand satisfying the conditions. Since there are $C(52,5)$ equally likely poker hands (see Example 11 of Section 5.3), the probability of drawing this one is $1/C(52,5)$, which is about 1 out of 2.5 million.

13. Let us compute the probability that the hand contains no aces and then subtract from 1 (invoking Theorem 1). A hand with no aces must be drawn from the 48 nonace cards, so there are $C(48,5)$ such hands. Therefore the probability of drawing such a hand is $C(48,5)/C(52,5)$, which works out to about 66%. Thus the probability of holding a hand with at least one ace is $1-(C(48,5)/C(52,5))$, or about 34%.

15. We need to compute the number of ways to hold two pairs. To specify the hand we first choose the kinds (ranks) the pairs will be (such as kings and fives); there are $C(13,2)=78$ ways to do this, since we need to choose 2 kinds from the 13 possible kinds. Then we need to decide which 2 cards of each of the kinds of the pairs we want to include. There are 4 cards of each kind (4 suits), so there are $C(4,2)=6$ ways to make each of these two choices. Finally, we need to decide which card to choose for the fifth card in the hand. We cannot choose any card in either of the 2 kinds that are already represented (we do not want to construct a full house by accident), so there are $52-8=44$ cards to choose from and hence $C(44,1)=44$ ways to make the choice. Putting this all together by the product rule, there are $78\cdot 6\cdot 6\cdot 44 = 123{,}552$ different hands classified as "two pairs."

Since each hand is equally likely, and since there are $C(52,5)=2{,}598{,}960$ different hands (see Example 11 in Section 5.3), the probability of holding two pairs is $123552/2598960 = 198/4165 \approx 0.0475$.

17. First we need to compute the number of ways to hold a straight. We can specify the hand by first choosing the starting (lowest) kind for the straight. Since the straight can start with any card from the set $\{A,2,3,4,5,6,7,8,9,10\}$, there are $C(10,1)=10$ ways to do this. Then we need to decide which card of each of the kinds in the straight we want to include. There are 4 cards of each kind (4 suits), so there are $C(4,1)=4$ ways to make each of these 5 choices. Putting this all together by the product rule, there are $10\cdot 4^5 = 10{,}240$ different hands containing a straight. (For poker buffs, it should be pointed out that a hand is classified as a "straight" in poker if it contains a straight but does not contain a straight flush, which is a straight in which all of the cards are in the same suit. Since there are $10\cdot 4 = 40$ straight flushes, we would need to subtract 40 from our answer above in order to find the number of hands classified as a "straight." Also, some poker books do not count $A,2,3,4,5$ as a straight.)

Since each hand is equally likely, and since there are $C(52,5)=2{,}598{,}960$ different hands (see Example 11 in Section 5.3), the probability of holding a hand containing a straight is $10240/2598960 = 128/32487 \approx 0.00394$.

19. First we can calculate the number of hands that contain cards of five different kinds (this is Exercise 14). All that is required to specify such a hand is to choose the five kinds (which can be done in $C(13,5)$ ways, since there are 13 kinds in all), and then for each of those cards to specify a suit (which can be done in 4^5 ways, since there are four possible suits for each card). Thus there are $C(13,5)\cdot 4^5 = 1317888$ hands of this type. Now we need to figure out how many of them violate the conditions—in other words, how many of them contain a flush or a straight. To obtain a flush (this is Exercise 16) we need to choose a suit and then choose 5 cards from the 13 in this suit, so there are $4\cdot C(13,5) = 5148$ different flushes. We solved the problem of straights in Exercise 17; there are 10240 straights. Furthermore, there are 40 hands that are both straights and flushes (such a hand can have its lowest card be any of the ten kinds A, 2, ..., 10, and be in any of the four suits). Now we are ready to put this all together. The number of hands that are flushes or straights is, by the principle of inclusion–exclusion, $5148+10240-40=15348$. Subtracting this from the number of hands containing five different kinds, we see that there are $1317888-15348=1302540$ hands of the desired type. Therefore the probability of drawing such a hand is $1302540/C(52,5) = 1302540/2598960 = 1277/2548$, which works out to just a hair over 50%.

21. Looked at properly, this is the same as Exercise 7. There are 2 equally likely outcomes for the parity on the roll of a die—even and odd. Of the $2^6=64$ parity outcomes in the roll of a die 6 times, only one consists of

6 odd numbers. Therefore the probability is $1/64$.

23. We need to count the number of positive integers not exceeding 100 that are divisible by 5 or 7. Using an analysis similar to Exercise 21e in Section 5.1, we see that there are $\lfloor 100/5 \rfloor = 20$ numbers in that range divisible by 5 and $\lfloor 100/7 \rfloor = 14$ divisible by 7. However, we have counted the numbers 35 and 70 twice, since they are divisible by both 5 and 7 (i.e., divisible by 35). Therefore there are $20 + 14 - 2 = 32$ such numbers. (We needed to subtract 2 to compensate for the double counting.) Now since there are 100 equally likely numbers in the set, the probability of choosing one of these 32 numbers is $32/100 = 8/25 = 0.32$.

25. In each case, if the numbers are chosen from the integers from 1 to n, then there are $C(n, 6)$ possible entries, only one of which is the winning one, so the answer is $1/C(n, 6)$.
a) $1/C(50, 6) = 1/15890700 \approx 6.3 \times 10^{-8}$ **b)** $1/C(52, 6) = 1/20358520 \approx 4.9 \times 10^{-8}$
c) $1/C(56, 6) = 1/32468436 \approx 3.1 \times 10^{-8}$ **d)** $1/C(60, 6) = 1/50063860 \approx 2.0 \times 10^{-8}$

27. In each case, there are $C(n, 6)$ possible choices of winning numbers. If we want to choose exactly one of them correctly, then we have 6 ways to specify which number it is to be, and then $C(n - 6, 5)$ ways to pick five losing numbers from the $n - 6$ losing numbers. Thus the probability is $6C(n - 6, 5)/C(n, 6)$ in each case. We will calculate these numbers for the various values of n. Note that the probability decreases as n increases (it gets harder to choose one of the winning numbers as the pool of numbers grows).
a) $6C(40 - 6, 5)/C(40, 6) = 6 \cdot C(34, 5)/C(40, 6) = 6 \cdot 278256/3838380 = 139128/319865 \approx 0.435$
b) $6C(48 - 6, 5)/C(48, 6) = 6 \cdot C(42, 5)/C(48, 6) = 6 \cdot 850668/12271512 = 212667/511313 \approx 0.416$
c) $6C(56 - 6, 5)/C(56, 6) = 6 \cdot C(50, 5)/C(56, 6) = 6 \cdot 2118760/32468436 = 151340/386529 \approx 0.392$
d) $6C(64 - 6, 5)/C(64, 6) = 6 \cdot C(58, 5)/C(64, 6) = 6 \cdot 4582116/74974368 = 163647/446276 \approx 0.367$

29. There is only one winning choice of numbers, namely the same 8 numbers the computer chooses. Therefore the probability of winning is $1/C(100, 8) \approx 1/(1.86 \times 10^{11})$.

31. Since the drawing is done at random, and there are three winners and 97 losers, Michelle's chance of winning is $3/100$. To see this more formally, we reason as follows, thinking of the winners as drawn in order (first, second, third). There are $100 \cdot 99 \cdot 98$ equally likely outcomes of the drawing. The number of possible outcomes in which Michelle wins first prize is $1 \cdot 99 \cdot 98$. The number of possible outcomes in which Michelle wins second prize is $99 \cdot 1 \cdot 98$ (99 people other than Michelle could have won first prize). The number of possible outcomes in which Michelle wins third prize is $99 \cdot 98 \cdot 1$. Adding, we see that there are $3 \cdot 99 \cdot 98$ ways for Michelle to win a prize. Therefore the probability we seek is $(3 \cdot 99 \cdot 98)/(100 \cdot 99 \cdot 98) = 3/100$.

33. a) There are $200 \cdot 199 \cdot 198$ equally likely outcomes of the drawings. In only one of these do Abby, Barry, and Sylvia win the first, second, and third prizes, respectively. Therefore the probability is $1/(200 \cdot 199 \cdot 198) = 1/7880400$.
b) There are $200 \cdot 200 \cdot 200$ equally likely outcomes of the drawings. In only one of these do Abby, Barry, and Sylvia win the first, second, and third prizes, respectively. Therefore the probability is $1/(200 \cdot 200 \cdot 200) = 1/8000000$.

35. a) There are 18 red numbers and 38 numbers in all, so the probability is $18/38 = 9/19 \approx 0.474$.
b) There are 38^2 equally likely outcomes for two spins, since each spin can result in 38 different outcomes. Of these, 18^2 are a pair of black numbers. Therefore the probability is $18^2/38^2 = 81/361 \approx 0.224$.
c) There are 2 outcomes being considered here, so the probability is $2/38 = 1/19$.
d) There are 38^5 equally likely outcomes in five spins of the wheel. Since 36 outcomes on each spin are not 0 or 00, there are 36^5 outcomes being considered. Therefore the probability is $36^5/38^5 = 1889568/2476099 \approx 0.763$.

e) There are 38^2 equally likely outcomes for two spins. The number of outcomes that meet the conditions specified here is $6 \cdot (38 - 6) = 192$ (by the product rule). Therefore the probability is $192/38^2 = 48/361 \approx 0.133$.

37. Reasoning as in Example 2, we see that there are 4 ways to get a total of 9 when two dice are rolled: $(6,3)$, $(5,4)$, $(4,5)$, and $(3,6)$. There are $6^2 = 36$ equally likely possible outcomes of the roll of two dice, so the probability of getting a total of 9 when two dice are rolled is $4/36 \approx 0.111$. For three dice, there are $6^3 = 216$ equally likely possible outcomes, which we can represent as ordered triples (a,b,c). We need to enumerate the possibilities that give a total of 9. This is done in a more systematic way in Section 5.5, but we will do it here by brute force. The first die could turn out to be a 6, giving rise to the 2 triples $(6,2,1)$ and $(6,1,2)$. The first die could be a 5, giving rise to the 3 triples $(5,3,1)$, $(5,2,2)$, and $(5,1,3)$. Continuing in this way, we see that there are 4 triples giving a total of 9 when the first die shows a 4, 5 triples when it shows a 3, 6 triples when it shows a 2, and 5 triples when it shows a 1 (namely $(1,6,2)$, $(1,5,3)$, $(1,4,4)$, $(1,3,5)$, and $(1,2,6)$). Therefore there are $2+3+4+5+6+5 = 25$ possible outcomes giving a total of 9. This tells us that the probability of rolling a 9 when three dice are thrown is $25/216 \approx 0.116$, slightly larger than the corresponding value for two dice. Thus rolling a total of 9 is more likely when using three dice than when using only two.

39. It's hard to know how to respond to this argument, other than to say that the claim—that the probabilities that the prize is behind each of the doors are equal—is nonsense. For example, if one flips a thumbtack, then it would be silly to say that the probability that it lands with the point up must equal the probability that it lands with the point down—there is no symmetry in the situation to justify such a conclusion. (It is as absurd as claiming that every time you enter a contest you have a 50% probability of winning, since there are only two outcomes—either you win or you lose.) Here, too, there is a lack of symmetry, since the door you chose was chosen by you at random, without knowing where the prize was, and the door chosen by the host was not chosen at random—he carefully avoided opening the door with the prize. (In fact, if the host's algorithm were to choose one of the other two doors at random, then whenever he opens a nonprize door, the probability that the prize is behind your door *does* become $\frac{1}{2}$. Of course in this case, sometimes the game won't advance that far, since the door he chooses might contain the prize.)

41. a) There are 6^4 possible outcomes when a die is rolled four times. There are 5^4 outcomes in which a 6 does not appear, so the probability of not rolling a 6 is $5^4/6^4$. Therefore the probability that at least one 6 does appear is $1 - 5^4/6^4 = 671/1296$, which is about 0.518.

b) There are 36^{24} possible outcomes when a pair of dice is rolled 24 times. There are 35^{24} outcomes in which a double 6 does not appear, so the probability of not rolling a double 6 is $35^{24}/36^{24}$. Therefore the probability that at least one double 6 does appear is $1 - 35^{24}/36^{24}$, which is about 0.491. No, the probability is not greater than $1/2$.

c) From our answers above we see that the answer is yes, since $0.518 > 0.491$.

SECTION 6.2 Probability Theory

This section introduced several basic concepts from probability. You should be able to apply the definitions to do the kinds of computations shown in the examples. Students interested in further work in probability, especially as it applies to statistics, should consult an elementary probability and statistics textbook. It is advisable (almost imperative) for all mathematics, engineering, and computer science majors to take a good statistics course.

1. We are told that $p(H) = 3p(T)$. We also know that $p(H) + p(T) = 1$, since heads and tails are the only two outcomes. Solving these simultaneous equations, we find that $p(T) = 1/4$ and $p(H) = 3/4$. An interesting example of an experiment in which intuition would tell you that outcomes should be equally likely but in fact they are not is to spin a penny on its edge on a smooth table and let it fall. Repeat this experiment 50 times or so, and you will be amazed at the outcomes. Make sure to count only those trials in which the coin spins freely for a second or more, not bumping into any objects or falling off the table.

3. Let us denote by t the probability that a 2 or 4 appears. Then the given information tells us that $t = 3(1-t)$, since $1-t$ is the probability that some other number appear. Solving this equation gives $t = 3/4$. We assume from the statement of the problem that 2 and 4 are equally likely. Since together they have probability $3/4$, each of them must have probability $3/8$. Similarly, each of the other numbers (1, 3, 5, or 6) must have probability $(1-t)/4$, which works out to $1/16$.

5. There are six ways to roll a sum of 7. We can denote them as $(1,6)$, $(2,5)$, $(3,4)$, $(4,3)$, $(5,2)$, and $(6,1)$, where (i,j) means rolling i on the first die and j on the second. We need to compute the probability of each of these outcomes and then add them to find the probability of rolling a 7. The two dice are independent, so we can argue as follows, using the given information about the probability of each outcome on each die: $p((1,6)) = \frac{1}{7}\cdot\frac{1}{7} = \frac{1}{49}$; $p((2,5)) = \frac{1}{7}\cdot\frac{1}{7} = \frac{1}{49}$; $p((3,4)) = \frac{1}{7}\cdot\frac{1}{7} = \frac{1}{49}$; $p((4,3)) = \frac{2}{7}\cdot\frac{2}{7} = \frac{4}{49}$; $p((5,2)) = \frac{1}{7}\cdot\frac{1}{7} = \frac{1}{49}$; $p((6,1)) = \frac{1}{7}\cdot\frac{1}{7} = \frac{1}{49}$. Adding, we find that the probability of rolling a 7 as the sum is $9/49$.

7. We exploit symmetry in answering many of these.

a) Since 1 has either to precede 4 or to follow it, and there is no reason that one of these should be any more likely than the other, we immediately see that the answer is $1/2$. We could also use brute force here, list all 24 permutations, and count that 12 of them have 1 preceding 4.

b) By the same reasoning as in part **(a)**, the answer is again $1/2$.

c) We could list all 24 permutations, and count that 8 of them have 4 preceding both 1 and 2. But here is a better argument. Among the numbers 1, 2, and 4, each is just as likely as the others to occur first. Thus by symmetry the answer is $1/3$.

d) We could list all 24 permutations, and count that 6 of them have 4 preceding 1, 2, and 3 (i.e., 4 occurring first); or we could argue that there are $3! = 6$ ways to write down the rest of a permutation beginning with 4. But here is a better argument. Each of the four numbers is just as likely as the others to occur first. Thus by symmetry the answer is $1/4$.

e) We could list all 24 permutations, and count that 6 of them have 4 preceding 3, and 2 preceding 1. But here is a better argument. Between 4 and 3, each is just as likely to precede the other, so the probability that 4 precedes 3 is $1/2$. Similarly, the probability that 2 precedes 1 is $1/2$. The relative position of 4 and 3 is independent of the relative position of 2 and 1, so the probability that both happen is the product $(1/2)(1/2) = 1/4$.

9. Note that there are 26! permutations of the letters, so the denominator in all of our answers is 26!. To find the numerator, we have to count the number of ways that the given event can happen. Alternatively, in some cases we may be able to exploit symmetry.

a) There is only one way for this to happen, so the answer is $1/26!$.

b) There are 25! ways to choose the rest of the permutation after the first letter has been specified to be z. Therefore the answer is $25!/26! = 1/26$. Alternatively, each of the 26 letters is equally likely to be first, so the probability that z is first is $1/26$.

c) Since z has either to precede a or to follow it, and there is no reason that one of these should be any more likely than the other, we immediately see that the answer is $1/2$.

d) In effect we are forming a permutation of 25 items—the letters b through y and the double letter combination az. There are 25! ways to do this, so the answer is $25!/26! = 1/26$. Here is another way to reason. For a to immediately precede z in the permutation, we must first make sure that z does not occur in the first spot (since nothing precedes it), and the probability of that is clearly $25/26$. Then the probability that a is the letter immediately preceding z given that z is not first is $1/25$, since each of the 25 other letters is equally likely to be in the position in front of z. Therefore the desired probability is $(25/26)(1/25) = 1/26$. Note that this "product rule" is essentially just the definition of conditional probability.

e) We solve this by the first technique used in part **(d)**. In effect we are forming a permutation of 24 items, one of which is the triple letter combination amz. There are 24! ways to do this, so the answer is $24!/26! = 1/650$.

f) If m, n, and o are specified to be in their original positions, then there are only 23 letters to permute, and there are 23! ways to do this. Therefore the probability is $23!/26! = 1/15600$.

11. Clearly $p(E \cup F) \geq p(E) = 0.7$. Also, $p(E \cup F) \leq 1$. If we apply Theorem 2 from Section 6.1, we can rewrite this as $p(E) + p(F) - p(E \cap F) \leq 1$, or $0.7 + 0.5 - p(E \cap F) \leq 1$. Solving for $p(E \cap F)$ gives $p(E \cap F) \geq 0.2$.

13. The items in this inequality suggest that it may have something to do with the formula for the probability of the union of two events given in this section:

$$p(E \cup F) = p(E) + p(F) - p(E \cap F)$$

We know that $p(E \cup F) \leq 1$, since no event can have probability exceeding 1. Thus we have

$$1 \geq p(E) + p(F) - p(E \cap F).$$

A little algebraic manipulation easily transforms this to the desired inequality.

15. Let us start with the simplest nontrivial case, namely $n = 2$. We want to show that

$$p(E_1 \cup E_2) \leq p(E_1) + p(E_2).$$

We know from the formula for the probability of the union of two events given in this section that

$$p(E_1 \cup E_2) = p(E_1) + p(E_2) - p(E_1 \cap E_2).$$

Since $p(E_1 \cap E_2) \geq 0$, the desired inequality follows immediately. We can use this as the basis step of a proof by mathematical induction. (Technically, we should point out that for $n = 1$ there is nothing to prove, since $p(E_1) \leq p(E_1)$.) For the inductive step, assume the stated inequality for n. Then

$$\begin{aligned} p(E_1 \cup E_2 \cup \cdots \cup E_n \cup E_{n+1}) &\leq p(E_1 \cup E_2 \cup \cdots \cup E_n) + p(E_{n+1}) \\ &\leq p(E_1) + p(E_2) + \cdots + p(E_n) + p(E_{n+1}), \end{aligned}$$

as desired. The first inequality follows from the case $n = 2$, and the second follows from the inductive hypothesis.

17. There are various ways to prove this algebraically. Here is one. Since $E \cup \overline{E}$ is the entire sample space S, we can break the event F up into two disjoint events, $F = S \cap F = (E \cup \overline{E}) \cap F = (E \cap F) \cup (\overline{E} \cap F)$, using the distributive law. Therefore $p(F) = p((E \cap F) \cup (\overline{E} \cap F)) = p(E \cap F) + p(\overline{E} \cap F)$, since these two events are disjoint. Subtracting $p(E \cap F)$ from both sides, using the fact that $p(E \cap F) = p(E) \cdot p(F)$ (our hypothesis that E and F are independent), and factoring, we have $p(F)(1 - p(E)) = p(\overline{E} \cap F)$. Since $1 - p(E) = p(\overline{E})$, this says that $p(\overline{E} \cap F) = p(\overline{E}) \cdot p(F)$, as desired.

19. As instructed, we are assuming here that births are independent and the probability of a birth in each month is $1/12$. Although this is clearly not exactly true (for example, the months do not all have the same lengths), it is probably close enough for our answers to be approximately accurate.

a) The probability that the second person has the same birth month as the first person (whatever that was) is $1/12$.

b) We proceed as in Example 13. The probability that all the birth months are different is

$$p_n = \frac{11}{12} \cdot \frac{10}{12} \cdots \frac{13-n}{12}$$

since each person after the first must have a different birth month from all the previous people in the group. Note that if $n \geq 13$, then $p_n = 0$ since the 12th fraction is 0 (this also follows from the pigeonhole principle). The probability that at least two are born in the same month is therefore $1 - p_n$.

c) We compute $1 - p_n$ for $n = 2, 3, \ldots$ and find that the first time this exceeds $1/2$ is when $n = 5$, so that is our answer. With five people, the probability that at least two will share a birth month is about 62%.

21. If n people are chosen at random, then the probability that all of them were born on a day other than April 1 is $(365/366)^n$. To compute the probability that exactly one of them is born on April 1, we note that this can happen in n different ways (it can be any of the n people), and the probability that it happens for each particular person is $(1/366)(365/366)^{n-1}$, since the other $n-1$ people must be born on some other day. Putting this all together, the probability that two of them were born on April 1 is $1 - (365/366)^n - n(1/366)(365/366)^{n-1}$. Using a calculator or computer algebra system, we find that this first exceeds $1/2$ when $n = 614$. Interestingly, if the problem asked about *exactly* two April 1 birthdays, then the probability is $C(n,2)(1/366)^2(365/366)^{n-2}$, which never exceeds $1/2$.

23. There are 16 equally likely outcomes of flipping a fair coin five times in which the first flip comes up heads (each of the other flips can be either heads or tails). Of these only four will result in four heads appearing, namely $HHHHT$, $HHHTH$, $HHTHH$, and $HTHHH$. Therefore by the definition of conditional probability the answer is $4/16$, or $1/4$.

25. There are 16 equally likely bit strings of length 4, but only 8 of them start with a 1. Three of these contain at least two consecutive 0's, namely 1000, 1001, and 1100. Therefore by the definition of conditional probability the answer is $3/8$.

27. In each case we need to compute $p(E)$, $p(F)$, and $p(E \cap F)$. Then we need to compare $p(E) \cdot p(F)$ to $p(E \cap F)$; if they are equal, then by definition the events are independent, and otherwise they are not. We assume that boys and girls are equally likely, and that successive births are independent. (Medical science suggests that neither of these assumptions is exactly correct, although both are reasonably good approximations.)

a) If the family has only two children, then there are four equally likely outcomes: BB, BG, GB, and GG. There are two ways to have children of both sexes, so $p(E) = 2/4$. There are three ways to have at most one boy, so $p(F) = 3/4$. There are two ways to have children of both sexes and at most one boy, so $p(E \cap F) = 2/4$. Since $p(E) \cdot p(F) = 3/8 \neq 2/4$, the events are not independent.

b) If the family has four children, then there are 16 equally likely outcomes, since there are 16 strings of length 4 consisting of B's and G's. All but two of these outcomes give children of both sexes, so $p(E) = 14/16$. Only five of them result in at most one boy, so $p(F) = 5/16$. There are four ways to have children of both sexes and at most one boy, so $p(E \cap F) = 4/16$. Since $p(E) \cdot p(F) = 35/128 \neq 4/16$, the events are not independent.

c) If the family has five children, then there are 32 equally likely outcomes, since there are 32 strings of length 5 consisting of B's and G's. All but two of these outcomes give children of both sexes, so $p(E) = 30/32$. Only six of them result in at most one boy, so $p(F) = 6/32$. There are five ways to have children of both sexes and at most one boy, so $p(E \cap F) = 5/32$. Since $p(E) \cdot p(F) = 45/256 \neq 5/32$, the events are not independent.

29. We can model this problem using the binomial distribution. We have here $n = 6$ Bernoulli trials (the six coins being flipped), with $p = 1/2$ (the probability of heads, which we will arbitrarily call success). There are two ways in which there could be an odd person out. Either there could be five heads and one tail, or there could be one head and five tails. Thus we want to know the probability that the number of successes is either $k = 5$ or $k = 1$. According to the formula developed in this section,

$$b(5; 6, \frac{1}{2}) = C(6,5)\left(\frac{1}{2}\right)^5\left(1-\frac{1}{2}\right)^1 = \frac{3}{32},$$

and by a similar calculation, $b(1; 6, 1/2) = 3/32$. Therefore the probability that there is an odd man out is $2 \cdot (3/32) = 3/16$, or about one in five.

31. In each case we need to calculate the probability of having five girls. By the independence assumption, this is just the product of the probabilities of having a girl on each birth.

a) Since the probability of a girl is $1/2$, the answer is $(1/2)^5 = 1/32 \approx 0.031$.

b) The is the same as part **(a)**, except that the probability of a girl is 0.49. Therefore the answer is $0.49^5 \approx 0.028$.

c) Plugging in $i = 1, 2, 3, 4, 5$, we see that the probability of having boys on the successive births are 0.50, 0.49, 0.48, 0.47, and 0.46. Therefore the probability of having girls on the successive births are 0.50, 0.51, 0.52, 0.53, and 0.54. The answer is thus $0.50 \cdot 0.51 \cdot 0.52 \cdot 0.53 \cdot 0.54 \approx 0.038$.

33. In each case we need to calculate the probability that the first child is a boy (call this $p(E)$) and the probability that the last two children are girls (call this $p(F)$). Then the desired answer is $p(E \cup F)$, which equals $p(E) + p(F) - p(E) \cdot p(F)$. This last product comes from the fact that events E and F are independent.

a) Clearly $p(E) = 1/2$ and $p(F) = (1/2) \cdot (1/2) = 1/4$. Therefore the answer is

$$\frac{1}{2} + \frac{1}{4} - \frac{1}{2} \cdot \frac{1}{4} = \frac{5}{8}.$$

b) Clearly $p(E) = 0.51$ and $p(F) = 0.49 \cdot 0.49 = 0.2401$. Therefore the answer is $0.51 + 0.2401 - 0.51 \cdot 0.2401 = 0.627649$.

c) Plugging in $i = 1, 2, 3, 4, 5$, we see that the probability of having boys on the successive births are 0.50, 0.49, 0.48, 0.47, and 0.46. Thus $p(E) = 0.50$ and $p(F) = 0.53 \cdot 0.54 = 0.2862$. Therefore the answer is $0.50 + 0.2862 - 0.50 \cdot 0.2862 = 0.6431$.

35. We need to use the binomial distribution, which tells us that the probability of k successes is

$$b(k; n, p) = C(n,k)p^k(1-p)^{n-k}.$$

a) Here $k = n$, since we want all the trials to result in success. Plugging in and computing, we have $b(n; n, p) = 1 \cdot p^n \cdot (1-p)^0 = p^n$.

b) There is at least one failure if and only if it is not the case that there are no failures. Thus we obtain the answer by subtracting the probability in part **(a)** from 1, namely $1 - p^n$.

c) There are two ways in which there can be at most one failure: no failures or one failure. We already computed that the probability of no failures is p^n. Plugging in $k = n - 1$ (one failure means $n - 1$ successes), we compute that the probability of exactly one failure is $b(n-1; n, p) = n \cdot p^{n-1} \cdot (1-p)$. Therefore the answer is $p^n + np^{n-1}(1-p)$. This formula only makes sense if $n > 0$, of course; if $n = 0$, then the answer is clearly 1.

d) Since this event is just that the event in part **(c)** does not happen, the answer is $1 - [p^n + np^{n-1}(1-p)]$. Again, this is for $n > 0$; the probability is clearly 0 if $n = 0$.

37. By Definition 2, the probability of an event is the sum of the probabilities of the outcomes in that event. Thus $p(\bigcup_{i=1}^{\infty} E_i)$ is the sum of $p(s)$ for each outcome s in $\bigcup_{i=1}^{\infty} E_i$. Since the E_i's are pairwise disjoint, this is the sum of the probabilities of all the outcomes in any of the E_i's, which is what $\sum_{i=1}^{\infty} p(E_i)$ is. Thus the issue is really whether one can rearrange the summands in an infinite sum of positive numbers and still get the same answer. Note that the series converges absolutely, because all terms are positive and all partial sums are at most 1. From calculus, we know that rearranging terms is legitimate in this case. An alternative proof could be based on the hint, since $\lim_{i\to\infty} p(E_i)$ is necessarily 0.

39. **a)** Since E is the event that for every set S with k players there is a player who has beaten all k of them, $\overline{E}$ is the event that for some set S with k players there is no player who has beaten all k of them. Thus for $\overline{E}$ to happen, some F_j must happen, so $\overline{E} = \bigcup_{j=1}^{C(m,k)} F_j$. The given inequality now follows from Boole's inequality (Exercise 15).

b) The probability that a particular player not in the jth set beats all k of the players in the jth set is $(1/2)^k = 2^{-k}$. Therefore the probability that this player does not have such a perfect record is $1 - 2^{-k}$, so the probability that all $m - k$ of the players not in the jth set are unable to boast of a perfect record is $(1 - 2^{-k})^{m-k}$. That is precisely $p(F_j)$.

c) The first inequality follows immediately, since all the summands are the same and there are $C(m,k) = \binom{m}{k}$ of them. If this probability is less than 1, then it must be possible that $\overline{E}$ fails, i.e., that E happens. So there is a tournament that meets the conditions of the problem as long as the second inequality holds.

d) We ask a computer algebra system to compute $C(m,2)(1-2^{-2})^{m-2}$ and $C(m,3)(1-2^{-3})^{m-3}$ for successive values of m to determine which values of m make the expression less than 1. We conclude that such a tournament exists for $k = 2$ when $m \geq 21$, and for $k = 3$ when $m \geq 91$. In fact, however, these are not the smallest values of m for which such a tournament exists. Indeed, for $k = 2$, we can take the tournament in which Barb beats Laurel, Laurel beats David, and David beats Barb ($m = 3$), and according to *The Probabilistic Method*, second edition, by Noga Alon and Joel Spencer (Wiley, 2000), there is a tournament with seven players meeting the condition when $k = 3$.

41. The input to this algorithm is the integer $n > 1$ to be tested for primality and the number k of iterations desired. If the output is "composite" then we know for sure that n is composite. If the output is "probably prime" then we do not know whether or not n is prime, and of course it makes no sense to talk about the probability that n is prime (either it is or it isn't!—there is no chance involved). What we do know is that the chance that a composite number would produce the output "probably prime" is at most $1/4^k$.

```
procedure probabilistic prime(n, k)
composite := false
i := 0
while composite = false and i < k
begin
        i := i + 1
        choose b uniformly at random with 1 < b < n
        apply Miller's test to base b
        if n fails the test then composite = true
end
if composite = true then print ("composite")
else print ("probably prime")
```

SECTION 6.3 Bayes' Theorem

Bayes' Theorem is extremely useful, as you have seen in applications featured in this section. Most of the exercises are similar to examples in the textbook. Make sure to write everything down precisely and follow the formulae exactly—this is one area in which the notation can be your salvation! We have rounded all numerical results to three decimal places.

1. We know that $p(F \mid E) = p(F \cap E)/p(E)$, so we need to find those two quantities. We are given $p(E) = 1/3$. To compute $p(F \cap E)$, we can use the fact that $p(F \cap E) = p(F)p(E \mid F)$. We are given that $p(F) = 1/2$ and that $p(E \mid F) = 2/5$; therefore $p(F \cap E) = (1/2)(2/5) = 1/5$. Putting this together, we have $p(F \mid E) = (1/5)/(1/3) = 3/5$.

3. Let F be the event that Frida picks the first box. Thus we know that $p(F) = p(\overline{F}) = 1/2$. Let B be the event that Frida picks a blue ball. Because of the contents of the boxes, we know that $p(B \mid F) = 3/5$ (three of the five balls in the first box are blue) and $p(B \mid \overline{F}) = 1/5$. We are asked for $p(F \mid B)$. We use Bayes' Theorem:

$$p(F \mid B) = \frac{p(B \mid F)p(F)}{p(B \mid F)p(F) + p(B \mid \overline{F})p(\overline{F})} = \frac{(3/5)(1/2)}{(3/5)(1/2) + (1/5)(1/2)} = \frac{3}{4}$$

5. Let S be the event that a randomly chosen racer uses steroids. We know that $p(S) = 0.08$ and therefore $p(\overline{S}) = 0.92$. Let P be the event that a randomly chosen person tests positive for steroid use. We are told that $p(P \mid S) = 0.96$ and $p(P \mid \overline{S}) = 0.09$ (this is a "false positive" test result). We are asked for $p(S \mid P)$. We use Bayes' Theorem:

$$p(S \mid P) = \frac{p(P \mid S)p(S)}{p(P \mid S)p(S) + p(P \mid \overline{S})p(\overline{S})} = \frac{(0.96)(0.08)}{(0.96)(0.08) + (0.09)(0.92)} \approx 0.481$$

7. Let O be the event that a randomly chosen person uses opium. We are told that $p(O) = 0.01$ and therefore $p(\overline{O}) = 0.99$. Let P be the event that a randomly chosen person tests positive for opium use. We are told that $p(P \mid \overline{O}) = 0.02$ ("false positive") and $p(\overline{P} \mid O) = 0.05$ ("false negative"). From these we can conclude that $p(\overline{P} \mid \overline{O}) = 0.98$ ("true negative") and $p(P \mid O) = 0.95$ ("true positive").

a) We are asked for $p(\overline{O} \mid \overline{P})$. We use Bayes' Theorem:

$$p(\overline{O} \mid \overline{P}) = \frac{p(\overline{P} \mid \overline{O})p(\overline{O})}{p(\overline{P} \mid \overline{O})p(\overline{O}) + p(\overline{P} \mid O)p(O)} = \frac{(0.98)(0.99)}{(0.98)(0.99) + (0.05)(0.01)} \approx 0.999$$

b) We are asked for $p(O \mid P)$. We use Bayes' Theorem:

$$p(O \mid P) = \frac{p(P \mid O)p(O)}{p(P \mid O)p(O) + p(P \mid \overline{O})p(\overline{O})} = \frac{(0.95)(0.01)}{(0.95)(0.01) + (0.02)(0.99)} \approx 0.324$$

9. Let H be the event that a randomly chosen person in the clinic is infected with HIV. We are told that $p(H) = 0.08$ and therefore $p(\overline{H}) = 0.92$. Let P be the event that a randomly chosen person tests positive for HIV on the blood test. We are told that $p(P \mid H) = 0.98$ and $p(P \mid \overline{H}) = 0.03$ ("false positive"). From these we can conclude that $p(\overline{P} \mid H) = 0.02$ ("false negative") and $p(\overline{P} \mid \overline{H}) = 0.97$.

a) We are asked for $p(H \mid P)$. We use Bayes' Theorem:

$$p(H \mid P) = \frac{p(P \mid H)p(H)}{p(P \mid H)p(H) + p(P \mid \overline{H})p(\overline{H})} = \frac{(0.98)(0.08)}{(0.98)(0.08) + (0.03)(0.92)} \approx 0.740$$

b) In part **(a)** we found $p(H \mid P)$. Here we are asked for the probability of the complementary event (given a positive test result). Therefore we have simply $p(\overline{H} \mid P) = 1 - p(H \mid P) \approx 1 - 0.740 = 0.260$.

c) We are asked for $p(H \mid \overline{P})$. We use Bayes' Theorem:

$$p(H \mid \overline{P}) = \frac{p(\overline{P} \mid H)p(H)}{p(\overline{P} \mid H)p(H) + p(\overline{P} \mid \overline{H})p(\overline{H})} = \frac{(0.02)(0.08)}{(0.02)(0.08) + (0.97)(0.92)} \approx 0.002$$

d) In part **(c)** we found $p(H \mid \overline{P})$. Here we are asked for the probability of the complementary event (given a negative test result). Therefore we have simply $p(\overline{H} \mid \overline{P}) = 1 - p(H \mid \overline{P}) \approx 1 - 0.002 = 0.998$.

11. Let S be the event that a randomly chosen product actually is a success. We are told that $p(S) = 0.6$ and therefore $p(\overline{S}) = 0.4$. Let P be the event that a randomly chosen product is predicted to be successful. We are told that $p(P \mid S) = 0.7$ and $p(P \mid \overline{S}) = 0.4$. We are asked for $p(S \mid P)$. We use Bayes' Theorem:

$$p(S \mid P) = \frac{p(P \mid S)p(S)}{p(P \mid S)p(S) + p(P \mid \overline{S})p(\overline{S})} = \frac{(0.7)(0.6)}{(0.7)(0.6) + (0.4)(0.4)} \approx 0.724$$

13. By Generalized Bayes' Theorem,

$$\begin{aligned} p(F_1 \mid E) &= \frac{p(E \mid F_1)p(F_1)}{p(E \mid F_1)p(F_1) + p(E \mid F_2)p(F_2) + p(E \mid F_3)p(F_3)} \\ &= \frac{(1/8)(1/4)}{(1/8)(1/4) + (1/4)(1/4) + (1/6)(1/2)} = \frac{3}{17}. \end{aligned}$$

15. a) Because the winning door was chosen uniformly at random, your chance of winning, $p(W = i)$, is $1/3$, no matter which door you chose.

b) If you have chosen a door other than the winning door (i.e., $i \neq k$), then Monty opens the other non-winning door. If you have chosen the winning door (i.e., $i = k$), then Monty opens the other two doors with equal likelihood (we are told that he selects "at random"). Thus $p(M = j \mid W = k) = 1$ if i, j, and k are distinct; $p(M = j \mid W = k) = 0$ if $j = k$ or $j = i$; and $p(M = j \mid W = k) = 1/2$ if $i = k$ and $j \neq i$.

c) Without loss of generality, assume that $i = 1$, $j = 2$, and $k = 3$ (we would get the same answer for any combination of distinct values). By Bayes' Theorem,

$$\begin{aligned} &p(W = 3 \mid M = 2) \\ &= \frac{p(M = 2 \mid W = 3)p(W = 3)}{p(M = 2 \mid W = 1)p(W = 1) + p(M = 2 \mid W = 2)p(W = 2) + p(M = 2 \mid W = 3)p(W = 3)} \\ &= \frac{1 \cdot (1/3)}{(1/2) \cdot (1/3) + 0 \cdot (1/3) + 1 \cdot (1/3)} = \frac{1}{3/2} = \frac{2}{3}. \end{aligned}$$

d) You should change doors, because you now have a $2/3$ chance to win by switching. (A similar calculation to that in part **(c)** shows that your chance of winning if you do not switch doors is $1/3$.)

17. The proof is basically the same as the proof of Theorem 1. The definition of conditional probability tells us that $p(F_j \mid E) = p(E \cap F_j)/p(E)$. For the numerator, again using the definition of conditional probability, we have $p(E \cap F_j) = p(E \mid F_j)p(F_j)$, as desired. For the denominator we show that $p(E) = \sum_{i=1}^{n} p(E \mid F_i)p(F_i)$. Just as in the proof of Theorem 1, the events $E \cap F_i$ partition the event E; that is $(E \cap F_{i_1}) \cap (E \cap F_{i_2}) = \emptyset$ when $i_i \neq i_2$ (because the F_i's are mutually exclusive), and $\bigcup_{i=1}^{n} (E \cap F_{i_1}) = E$ (because the $\bigcup_{i=1}^{n} F_i = S$). Therefore $p(E) = \sum_{i=1}^{n} p(E \cap F_i) = \sum_{i=1}^{n} p(E \mid F_i)p(F_i)$, and our proof is complete.

19. We follow the procedure in Example 3. We first compute that $p(\text{opportunity}) = 175/1000 = 0.175$ and $q(\text{opportunity}) = 20/400 = 0.05$. Then we compute that

$$r(\text{opportunity}) = \frac{p(\text{opportunity})}{p(\text{opportunity}) + q(\text{opportunity})} = \frac{0.175}{0.175 + 0.05} \approx 0.778.$$

Because $r(\text{opportunity})$ is less than the threshold 0.9, an incoming message containing "opportunity" would not be rejected.

21. We follow the procedure in Example 4. We first compute that $p(\text{enhancement}) = 1500/10000 = 0.15$, $q(\text{enhancement}) = 20/5000 = 0.004$, $p(\text{herbal}) = 800/10000 = 0.08$, and $q(\text{herbal}) = 200/5000 = 0.04$. Then, assuming the necessary independence, we compute

$$\begin{aligned} r(\text{enhancement}, \text{herbal}) &= \frac{p(\text{enhancement})p(\text{herbal})}{p(\text{enhancement})p(\text{herbal}) + q(\text{enhancement})q(\text{herbal})} \\ &= \frac{(0.15)(0.08)}{(0.15)(0.08) + (0.004)(0.04)} \approx 0.987. \end{aligned}$$

Because $r(\text{enhancement}, \text{herbal})$ is greater than the threshold 0.9, an incoming message containing "enhancement" and "herbal" will be rejected.

23. First, by Bayes' Theorem, we have

$$p(S \mid E_1 \cap E_2) = \frac{p(E_1 \cap E_2 \mid S)p(S)}{p(E_1 \cap E_2 \mid S)p(S) + p(E_1 \cap E_2 \mid \overline{S})p(\overline{S})}.$$

Next, because we are assuming no prior knowledge about whether a message is or is not spam, we set $p(S) = p(\overline{S}) = 0.5$, and so the equation above simplifies to

$$p(S \mid E_1 \cap E_2) = \frac{p(E_1 \cap E_2 \mid S)}{p(E_1 \cap E_2 \mid S) + p(E_1 \cap E_2 \mid \overline{S})}.$$

Finally, because of the assumed independence of E_1, E_2, and S, we have $p(E_1 \cap E_2 \mid S) = p(E_1 \mid S) \cdot p(E_2 \mid S)$, and similarly for the $\overline{S}$. Thus the equation is equivalent to what we were asked to show.

SECTION 6.4 Expected Value and Variance

This section concludes the discussion of probability theory. The linearity of expectation is a surprisingly powerful tool, as Exercises 21 and 37, for example, illustrate.

1. By Theorem 2 with $p = 1/2$ and $n = 5$, we see that the expected number of heads is 2.5.

3. By Theorem 2 the expected number of successes for n Bernoulli trials is np. In the present problem we have $n = 10$ and $p = 1/6$. Therefore the expected number of successes (i.e., appearances of a 6) is $10 \cdot (1/6) = 1\frac{2}{3}$.

5. This problem involves a lot of computation. It is similar to Example 3, which relied only the results of Example 12 in Section 6.2. We need to compute the probability of each outcome in order to be able to apply Definition 1 (expected value). It is easy to see that the given information implies that for one roll of such a die $p(3) = 2/7$ and $p(1) = p(2) = p(4) = p(5) = p(6) = 1/7$ (this was Exercise 2 in Section 6.2). Next we need to do several computations similar to those required in Exercise 5 in Section 6.2, in order to compute $p(X = k)$ for each k from 2 to 12, where the random variable X represents the sum (for example, $X(3, 5) = 8$). The probability of a sum of 2, $p(X = 2)$, is $\frac{1}{7} \cdot \frac{1}{7} = \frac{1}{49}$, since the only way to achieve a sum of 2 is to roll a 1 on each die, and the two dice are independent. Similarly, $p(X = 3) = \frac{1}{7} \cdot \frac{1}{7} + \frac{1}{7} \cdot \frac{1}{7} = \frac{2}{49}$, since both of the outcomes $(1, 2)$ and $(2, 1)$ give a sum of 3. We perform similar calculations for the other outcomes of the sum. Here is the entire set of values:

$$p(X=2)=\frac{1}{7}\cdot\frac{1}{7}=\frac{1}{49}$$

$$p(X=3)=\frac{1}{7}\cdot\frac{1}{7}+\frac{1}{7}\cdot\frac{1}{7}=\frac{2}{49}$$

$$p(X=4)=\frac{1}{7}\cdot\frac{2}{7}+\frac{1}{7}\cdot\frac{1}{7}+\frac{2}{7}\cdot\frac{1}{7}=\frac{5}{49}$$

$$p(X=5)=\frac{1}{7}\cdot\frac{1}{7}+\frac{1}{7}\cdot\frac{2}{7}+\frac{2}{7}\cdot\frac{1}{7}+\frac{1}{7}\cdot\frac{1}{7}=\frac{6}{49}$$

$$p(X=6)=\frac{1}{7}\cdot\frac{1}{7}+\frac{1}{7}\cdot\frac{1}{7}+\frac{2}{7}\cdot\frac{2}{7}+\frac{1}{7}\cdot\frac{1}{7}+\frac{1}{7}\cdot\frac{1}{7}=\frac{8}{49}$$

$$p(X=7)=\frac{1}{7}\cdot\frac{1}{7}+\frac{1}{7}\cdot\frac{1}{7}+\frac{2}{7}\cdot\frac{1}{7}+\frac{2}{7}\cdot\frac{1}{7}+\frac{1}{7}\cdot\frac{1}{7}+\frac{1}{7}\cdot\frac{1}{7}=\frac{8}{49}$$

$$p(X=8)=\frac{1}{7}\cdot\frac{1}{7}+\frac{2}{7}\cdot\frac{1}{7}+\frac{1}{7}\cdot\frac{1}{7}+\frac{1}{7}\cdot\frac{2}{7}+\frac{1}{7}\cdot\frac{1}{7}=\frac{7}{49}$$

$$p(X=9)=\frac{2}{7}\cdot\frac{1}{7}+\frac{1}{7}\cdot\frac{1}{7}+\frac{1}{7}\cdot\frac{1}{7}+\frac{1}{7}\cdot\frac{2}{7}=\frac{6}{49}$$

$$p(X=10)=\frac{1}{7}\cdot\frac{1}{7}+\frac{1}{7}\cdot\frac{1}{7}+\frac{1}{7}\cdot\frac{1}{7}=\frac{3}{49}$$

$$p(X=11)=\frac{1}{7}\cdot\frac{1}{7}+\frac{1}{7}\cdot\frac{1}{7}=\frac{2}{49}$$

$$p(X=12)=\frac{1}{7}\cdot\frac{1}{7}=\frac{1}{49}$$

A check on our calculation is that the sum of the probabilities is 1. Finally, we need to add the values of X times the corresponding probabilities:

$$E(X)=2\cdot\frac{1}{49}+3\cdot\frac{2}{49}+4\cdot\frac{5}{49}+5\cdot\frac{6}{49}+6\cdot\frac{8}{49}+7\cdot\frac{8}{49}+8\cdot\frac{7}{49}+9\cdot\frac{6}{49}+10\cdot\frac{3}{49}+11\cdot\frac{2}{49}+12\cdot\frac{1}{49}=\frac{336}{49}\approx 6.86$$

This is a reasonable answer (you should always ask yourself if the answer is reasonable!), since the dice are not very different from ordinary dice, and with ordinary dice the expectation is 7.

7. By Theorem 3 we know that the expectation of a sum is the sum of the expectations. In the current exercise we can let X be the random variable giving the score on the true-false questions and let Y be the random variable giving the score on the multiple choice questions. In order to compute the expectation of X and of Y, let us for a moment ignore the point values, and instead just look at the number of true-false or multiple choice questions that Linda gets right. The expected number of true-false questions she gets right is the expectation of the number of successes when 50 Bernoulli trials are performed with $p=0.9$. By Theorem 2 the expectation for the number of successes is $np=50\cdot 0.9=45$. Since each problem counts 2 points, the expectation of X is $45\cdot 2=90$. Similarly, the expected number of multiple choice questions she gets right is the expectation of the number of successes when 25 Bernoulli trials are performed with $p=0.8$, namely $25\cdot 0.8=20$. Since each problem counts 4 points, the expectation of Y is $20\cdot 4=80$. Therefore her expected score on the exam is $E(X+Y)=E(X)+E(Y)=90+80=170$.

9. In Example 8 we found that the answer to this question when the probability that x is in the list is p is $p(n+2)+(2n+2)(1-p)$. Plugging in $p=2/3$ we have

$$\frac{2}{3}\cdot(n+2)+(2n+2)\cdot\frac{1}{3}=\frac{4n+6}{3}.$$

11. There are 10 different outcomes of our experiment (it really doesn't matter whether we get a 6 on the last roll or not). Let the random variable X be the number of times we roll the die. For $i = 1, 2, \ldots, 9$, the probability that $X = i$ is $(5/6)^{i-1}(1/6)$, since to roll the die exactly i times requires that we obtain something other than a 6 exactly $i-1$ times followed by a 6 on the ith roll. Furthermore $p(X = 10) = (5/6)^9$. We need to compute $\sum_{i=1}^{10} i \cdot p(X = i)$. A computer algebra system gives the answer as $50700551/10077696 \approx 5.03$. Note that this is reasonable, since if there were no cut-off then the expected number of rolls would be 6.

13. The random variable that counts the number of rolls has a geometric distribution with $p = 1/6$, since the probability of getting a sum of 7 when a pair of dice is rolled is $1/6$. According to Theorem 4 the expected value is $1/(1/6) = 6$.

15. For a geometric distribution $p(X = k) = (1-p)^{k-1}p$ for $k = 1, 2, 3, \ldots$. Therefore

$$p(X \geq j) = \sum_{k=j}^{\infty} p(X = k) = \sum_{k=j}^{\infty} (1-p)^{k-1}p = p(1-p)^{j-1}\sum_{k=0}^{\infty}(1-p)^k = p(1-p)^{j-1}\frac{1}{1-(1-p)} = (1-p)^{j-1},$$

where we have used the formula for the sum of a geometric series at the end.

17. The random variable that counts the number of integers we need to select has a geometric distribution with $p = 1/2302$. According to Theorem 4 the expected value is $1/(1/2302) = 2302$.

19. We know from Examples 1 and 4 that $E(X) = 7/2$ and $E(Y) = 7$. To compute $E(XY)$ we need to find the value of XY averaged over the 36 equally likely possible outcomes. The following table shows the value of XY. For example, when the outcome is $(3, 4)$, then $X = 3$ and $Y = 7$, so $XY = 21$.

	1	2	3	4	5	6
1	2	3	4	5	6	7
2	6	8	10	12	14	16
3	12	15	18	21	24	27
4	20	24	28	32	36	40
5	30	35	40	45	50	55
6	42	48	54	60	66	72

We compute the average to be $E(XY) = 329/12$; and $(7/2) \cdot 7 \neq 329/12$.

21. We follow the hint. Let $I_j = 1$ if a run begins at the jth Bernoulli trial and $I_j = 0$ otherwise. Note that for I_j to equal 1, we must have the jth trial result in S and either the $(j-1)^{\text{st}}$ trial result in F or $j = 1$. Clearly the number of runs is the sum $R = \sum_{j=1}^{n} I_j$, since exactly one I_j is 1 for each run. Now $E(I_1) = p$, the probability of S on the first trial, and $E(I_j) = p(1-p)$ for $1 < j \leq n$ since we need success on the jth trial and failure on the $(j-1)^{\text{st}}$ trial. By linearity (Theorem 3, which applies even when these random variables are not independent, which they certainly are not here) we have $E(R) = p + \sum_{j=2}^{n} p(1-p) = p + (n-1)p(1-p)$.

23. In Example 18 we saw that the variance of the number of successes in n Bernoulli trials is npq. Here $n = 10$ and $p = q = 1/2$. Therefore the variance is $5/2$. Note that the unit for stating variance is not flips here, but $(\text{flips})^2$. To restate this in terms of flips, we must take the square root to compute the standard deviation. The standard deviation is $\sqrt{2.5} \approx 1.6$ flips.

25. The question is asking about the signed difference. For example, if $n = 6$ and we get five tails and one head, then $X_6 = 4$, whereas if we get five heads and one tail, then $X_6 = -4$. The key here is to notice that X_n is just n minus twice the number of heads.

a) The expected number of heads is $n/2$. Therefore the expected value of twice the number of heads is twice this, or n, and the expected value of n minus this is $n - n = 0$. This is not surprising; if it were not zero, then there would be a bias favoring heads or favoring tails.

b) Since the expected value is 0, the variance is the expected value of the square of X_n, which is the same as the square of twice the number of heads. This is clearly four times the square of the number of heads, so its expected value is $4 \cdot n/4 = n$, from Example 18, since $p = q = 1/2$. Therefore the answer is n. We are implicitly using here the result of Exercise 25 in the supplementary exercises for this chapter.

27. **a)** The probabilities that $(X_1, X_2, X_3) = (x_1, x_2, x_3)$ are as follows, because with probability 1 we have that $X_3 = (X_1 + X_2) \textbf{ mod } 2$:

$$p(0,0,0) = \frac{1}{2} \cdot \frac{1}{2} \cdot 1 = \frac{1}{4}$$
$$p(0,0,1) = \frac{1}{2} \cdot \frac{1}{2} \cdot 0 = 0$$
$$p(0,1,0) = \frac{1}{2} \cdot \frac{1}{2} \cdot 0 = 0$$
$$p(0,1,1) = \frac{1}{2} \cdot \frac{1}{2} \cdot 1 = \frac{1}{4}$$
$$p(1,0,0) = \frac{1}{2} \cdot \frac{1}{2} \cdot 0 = 0$$
$$p(1,0,1) = \frac{1}{2} \cdot \frac{1}{2} \cdot 1 = \frac{1}{4}$$
$$p(1,1,0) = \frac{1}{2} \cdot \frac{1}{2} \cdot 1 = \frac{1}{4}$$
$$p(1,1,1) = \frac{1}{2} \cdot \frac{1}{2} \cdot 0 = 0$$

We must show that X_1 and X_2 are independent, that X_1 and X_3 are independent, and that X_2 and X_3 are independent. We are told that X_1 and X_2 are independent. To see that X_1 and X_3 are independent, we note from the list above that $p(X_1 = 0 \wedge X_3 = 0) = 1/4 + 0 = 1/4$, that $p(X_1 = 0) = 1/2$, and that $p(X_3 = 0) = 1/2$, so it is true that $p(X_1 = 0 \wedge X_3 = 0) = p(X_1 = 0)p(X_3 = 0)$. Essentially the same calculation shows that $p(X_1 = 0 \wedge X_3 = 1) = p(X_1 = 0)p(X_3 = 1)$, $p(X_1 = 1 \wedge X_3 = 0) = p(X_1 = 1)p(X_3 = 0)$, and $p(X_1 = 1 \wedge X_3 = 1) = p(X_1 = 1)p(X_3 = 1)$. Therefore by definition, X_1 and X_3 are independent. The same reasoning shows that X_2 and X_3 are independent. To see that X_3 and $X_1 + X_2$ are not independent, we observe that $p(X_3 = 1 \wedge X_1 + X_2 = 2) = 0$, because $2 \textbf{ mod } 2 \neq 1$. But $p(X_3 = 1)p(X_1 + X_2 = 2) = (1/2)(1/4) = 1/8$.

b) We see from the table in part **(a)** that X_1, X_2, and X_3 are all Bernoulli random variables, so the variance of each is $(1/2)(1/2) = 1/4$ by Example 14. Therefore $V(X_1) + V(X_2) + V(X_3) = 3/4$. We see from the table above that $p(X_1 + X_2 + X_3 = 0) = 1/4$, $p(X_1 + X_2 + X_3 = 1) = 0$, $p(X_1 + X_2 + X_3 = 2) = 3/4$, and $p(X_1 + X_2 + X_3 = 3) = 0$. Therefore the expected value of $X_1 + X_2 + X_3$ is $(1/4)(0) + (3/4)(2) = 3/2$, and the variance of $X_1 + X_2 + X_3$ is $(1/4)(0 - 3/2)^2 + (3/4)(2 - 3/2)^2 = 3/4$.

c) If we attempt to prove this by mathematical induction, then presumably we would like the inductive step to be $V((X_1 + X_2 + \cdots + X_k) + X_{k+1}) = V(X_1 + X_2 + \cdots + X_k) + V(X_{k+1})$ (by the $n = 2$ case of Theorem 7, which was proved in the text), which then equals $(V(X_1) + V(X_2) + \cdots + V(X_k)) + V(X_{k+1})$ by the inductive hypothesis. However, in order to invoke Theorem 7, we must have that $X_1 + X_2 + \cdots + X_k$ and X_{k+1} are independent, and we see from part **(a)** that this is not a valid conclusion if all we know is the *pairwise* independence of the variables. Notice that the conclusion (that the variance of the sum is the sum of variances) *is* true assuming only pairwise independence; it's just that we cannot prove it in this manner. See Exercise 28.

29. We proceed as in Example 19, applying Chebyshev's inequality with $V(X) = n/4$ by Example 18 and $r = 5\sqrt{n}$. We have $p(|X(s) - E(X)| \geq 5\sqrt{n}) \leq V(X)/r^2 = (n/4)/(5\sqrt{n})^2 = 1/100$.

31. For simplicity we suppress the argument and write simply X for $X(s)$. As in the proof of Theorem 5, $E(X) = \sum_r r \cdot p(X = r)$. Dividing both sides by a we obtain $E(X)/a = \sum_r (r/a) \cdot p(X = r)$. Now this sum is at least as great as the subsum restricted to those values of $r \geq a$, and for those values, $r/a \geq 1$. Thus we have $E(X)/a \geq \sum_{r \geq a} 1 \cdot p(X = r)$. But this latter expression is just $p(X \geq a)$, as desired.

33. It is interesting to note that Markov was Chebyshev's student in Russia. One note of caution—the variance is not 2,500 cans; it is 2,500 square cans (the units for the variance of X are the square of the units for X). So a measure of how much the number of cans recycled per day varies is about the square root of this, or about 50 cans.

a) We have $E(X) = 50{,}000$ and we take $a = 55{,}000$. Then $p(X \geq 55{,}000) \leq 50{,}000/55{,}000 = 10/11$. This is not a terribly good estimate.

b) We apply Theorem 8, with $r = 10{,}000$. The probability that the number of cans recycled will differ from the expectation of 50,000 by at least 10,000 is at most $10{,}000/2{,}500^2 = 0.0016$. Therefore the probability is at least 0.9984 that the center will recycle between 40,000 and 60,000 cans. This is also not a very good estimate, since if the number of cans recycled per day usually differs by only about 50 from the mean of 50,000, it is virtually impossible that the difference would ever be over 200 times this amount—the probability is much, much less than 1.6 in 1000.

35. a) Each of the $n!$ permutations occurs with probability $1/n!$, so clearly $E(X)$ is the average number of comparisons, averaged over all these permutations.

b) In Example 5 of Section 3.3, we noted that the version of bubble sort that continues $n-1$ rounds regardless of whether new changes were made uses $n(n-1)/2$ comparisons, so X in this problem is always at most $n(n-1)/2$. It follows from the formula for expectation that $E(X) \leq n(n-1)/2$.

c) An inversion in a permutation is a pair of integers a_j and a_k with $j < k$ (so that $a_j < a_k$) such that a_k precedes a_j in the permutation (the elements are out of order). Because the bubble sort works by comparing adjacent elements and then swapping them if they are out of order, the only way that these elements can end up in their correct positions is if they are swapped, and therefore they must be compared.

d) For each permutation P, we know from part **(c)** that $X(P) \geq I(P)$. It follows from the definition of expectation that $E(X) \geq E(I)$.

e) This summation just counts 1 for every instance of an inversion.

f) This follows from the linearity of expectation (Theorem 3).

g) By Theorem 2 with $n = 1$, the expectation of $I_{j,k}$ is the probability that a_k precedes a_j in the permutation. But by symmetry, since the permutation is randomly chosen, this is clearly $1/2$. (There are $n!/2$ permutations in which a_k precedes a_j and $n!/2$ permutations in which a_j precedes a_k out of the $n!$ permutations in all.)

h) The summation in part **(f)** consists of $C(n,2) = n(n-1)/2$ terms, each equal to $1/2$, so the sum is $(n(n-1)/2)(1/2) = n(n-1)/4$.

i) From part **(b)** we know that $E(X)$, the object of interest, is at most $n(n-1)/2$, and from part **(d)** and part **(h)** we know that $E(X)$ is at least $n(n-1)/4$. Since both of these are $\Theta(n^2)$, the result follows.

37. Following the hint, we let $X = X_1 + X_2 + \cdots + X_n$, where $X_i = 1$ if the permutation fixes the ith element and $X_i = 0$ otherwise. Then X is the number of fixed elements, so we are being asked to compute $V(X)$. By Theorem 6, it suffices to compute $E(X^2) - E(X)^2$. Now $E(X) = E(X_1) + E(X_2) + \cdots + E(X_n) = n \cdot (1/n) = 1$, since the probability that the ith element stays in its original position is clearly $1/n$ by symmetry. To compute $E(X^2)$ we first multiply out:

$$X^2 = (X_1 + X_2 + \cdots + X_n)^2 = \sum_{i=1}^{n} X_i^2 + \sum_{i \neq j} X_i X_j$$

Now since $X_i^2 = X_i$, $E(X_i^2) = 1/n$. Next note that $X_i X_j = 1$ if both the ith and the jth elements are left fixed, and since there are $(n-2)!$ permutations that do this, the probability that $X_i X_j = 1$ is

$(n-2)!/n! = 1/(n(n-1))$. Of course $X_iX_j = 0$ otherwise. Therefore $E(X_iX_j) = 1/(n(n-1))$. Note also that there are n summands in the first sum of the displayed equation and $n(n-1)$ in the second. So $E(X^2) = n(1/n) + n(n-1) \cdot 1/(n(n-1)) = 1 + 1 = 2$. Therefore $V(X) = E(X^2) - E(X)^2 = 2 - 1^2 = 1$.

39. We can prove this by doing some algebra, using Theorems 3 and 6 and Exercise 38:

$$\begin{aligned} V(X+Y) &= E((X+Y)^2) - E(X+Y)^2 \\ &= E(X^2 + 2X \cdot Y + Y^2) - (E(X) + E(Y))^2 \\ &= E(X^2) + 2E(X \cdot Y) + E(Y^2) - E(X)^2 - 2E(X)E(Y) - E(Y)^2 \\ &= (E(X^2) - E(X)^2) + 2(E(X \cdot Y) - E(X)E(Y)) + (E(Y^2) - E(Y)^2) \\ &= V(X) + 2\,\mathrm{Cov}(X,Y) + V(Y) \end{aligned}$$

41. The probability that a particular ball fails to go into the first bin is $(n-1)/n$. Since these choices are assumed to be made independently, the probability that the first bin remains empty is then $((n-1)/n)^m$.

43. Let $X = X_1 + X_2 + \cdots + X_n$, where $X_i = 1$ if the ith bin remains empty and $X_i = 0$ otherwise. Then X is the number of bins that remain empty, so we are being asked to compute $E(X)$. From Exercise 41 we know that $p(X_i = 1) = ((n-1)/n)^m$ if m balls are distributed, so $E(X_i) = ((n-1)/n)^m$. By linearity of expectation (Theorem 3), the expected number of bins that remain empty is therefore $n((n-1)/n)^m = (n-1)^m/n^{m-1}$.

GUIDE TO REVIEW QUESTIONS FOR CHAPTER 6

1. a) See p. 394. **b)** $1/C(50,6)$

2. a) $\forall i(0 \le p(x_i) \le 1)$ and $\sum_{i=1}^{n} p(x_i) = 1$ **b)** $p(H) = 3/4$, $p(T) = 1/4$

3. a) See p. 404. **b)** $1/3$

4. a) See p. 405. **b)** yes

5. a) See p. 408. **b)** 1, 2, 3, 4, 5, 6

6. a) See p. 426. **b)** $1 \cdot \frac{1}{36} + 2 \cdot \frac{3}{36} + 3 \cdot \frac{5}{36} + 4 \cdot \frac{7}{36} + 5 \cdot \frac{9}{36} + 6 \cdot \frac{11}{36} = \frac{161}{36} \approx 4.47$

7. a) See p. 431. **b)** $(5n+6)/3$ (see Example 8 in Section 6.4)

8. a) See p. 406. **b)** See Theorem 2 in Section 6.2. **c)** See Theorem 2 in Section 6.4.

9. a) See p. 429. **b)** See Example 6 in Section 6.4.

10. a) See the discussion of Monte Carlo algorithms on p. 411. **b)** See Example 16 in Section 6.2.

11. See p. 418; $p(F \mid E) = \dfrac{p(E \mid F)p(F)}{p(E \mid F)p(F) + p(E \mid \overline{F})p(\overline{F})} = \dfrac{(1/3)(2/3)}{(1/3)(2/3) + (1/4)(1/3)} = \dfrac{8}{11}$

12. a) See pp. 433–434. **b)** See Theorem 4 in Section 6.4.

13. a) See p. 436. **b)** See Example 14 in Section 6.4.

14. a) See Theorem 7 in Section 6.4. **b)** See Example 18 in Section 6.4.

15. See pp. 438–439.

SUPPLEMENTARY EXERCISES FOR CHAPTER 6

1. There are 35 outcomes in which the numbers chosen are consecutive, since the first of these numbers can be anything from 1 to 35. There are $C(40,6) = 3{,}838{,}380$ possible choices in all. Therefore the answer is $35/3838380 = 1/109668$.

3. Each probability is of the form s/t where s is the number of hands of the described type and t is the total number of hands, which is clearly $C(52,13) = 635{,}013{,}559{,}600$. Hence in each case we will count the number of hands and divide by this value.

a) There is only one hand with all 13 hearts, so the probability is $1/t$, which is about 1.6×10^{-12}.

b) There are four such hands, since there are four ways to choose the suit, so the answer is $4/t$, which is about 6.3×10^{-12}.

c) To specify such a hand we need to choose 7 spades from the 13 spades available and then choose 6 clubs from the 13 clubs available. Thus there are $C(13,7)C(13,6) = 2944656$ such hands. The probability is therefore $2944656/t \approx 4.6 \times 10^{-6}$.

d) This event is 12 times more likely than the event in part **(c)**, since there are $P(4,2) = 12$ ways to choose the two suits. Thus the answer is $35335872/t \approx 5.6 \times 10^{-5}$.

e) This is similar to part **(c)**, but with four choices to make. The answer is $C(13,4)C(13,6)C(13,2)$ $C(13,1)/t = 1244117160/635013559600 \approx 2.0 \times 10^{-3}$.

f) There are $P(4,4) = 24$ ways to specify the suits, and then there are $C(13,4)C(13,6)C(13,2)C(13,1)$ ways to choose the cards from these suits to construct the desired hand. Therefore the answer is 24 times as big as the answer to part **(e)**, namely $29858811840/635013559600 \approx 0.047$.

5. a) Each of the outcomes 1 through 8 occurs with probability $1/8$, so the expectation is $(1/8)(1+2+3+\cdots+8) = 9/2$.

b) We compute $V(X) = E(X^2) - E(X)^2 = (1/8)(1^2+2^2+3^2+\cdots+8^2) - (9/2)^2 = (51/2) - (81/4) = 21/4$.

7. a) Since expected value is linear, the expected value of the sum is the sum of the expected values, each of which is $9/2$ by Exercise 5a. Therefore the answer is 9.

b) Since variance is linear for independent random variables, and clearly these variables are independent, the variance of the sum is the sum of the variances, each of which is $21/4$ by Exercise 5b. Therefore the answer is $21/2$.

9. a) Since expected value is linear, the expected value of the sum is the sum of the expected values, which are $7/2$ by Example 1 in Section 6.4 and $9/2$ by Exercise 5a. Therefore the answer is $(7/2) + (9/2) = 8$.

b) Since variance is linear for independent random variables, and clearly these variables are independent, the variance of the sum is the sum of the variances, which are $35/12$ by Example 15 in Section 6.4 and $21/4$ by Exercise 5b. Therefore the answer is $(35/12) + (21/4) = 49/6$.

11. a) There are 2^n possible outcomes of the flips. In order for the odd person out to be decided, we must have one head and $n-1$ tails, or one tail and $n-1$ heads. The number of ways for this to happen is clearly $2n$ (choose the odd person and choose whether it is heads or tails). Therefore the probability that there is an odd person out is $2n/2^n = n/2^{n-1}$. Call this value p.

b) Clearly the number of flips has a geometric distribution with parameter $p = n/2^{n-1}$, from part **(a)**. Therefore the probability that the odd person out is decided with the kth flip is $p(1-p)^{k-1}$.

c) By Theorem 4 in Section 6.4, the expectation is $1/p = 2^{n-1}/n$.

13. We start by counting the number of positive integers less than mn that *are* divisible by either m or n. Certainly all the integers m, $2m$, $3m$, ..., nm are divisible by m. There are n numbers in this list. All but one of them are less than mn. Therefore $n-1$ positive integers less than mn are divisible by m. Similarly, $m-1$ positive integers less than mn are divisible by n. Next we need to see how many numbers are divisible by both m and n. A number is divisible by both m and n if and only if it is divisible by the least common multiple of m and n. Let $L = \text{lcm}(m,n)$. Thus the numbers divisible by both m and n are L, $2L$, ..., mn. This list has $\gcd(m,n)$ numbers in it, since we know that $\text{lcm}(m,n)\cdot\gcd(m,n) = mn$. Therefore $\gcd(m,n)-1$ positive integers strictly less than mn are divisible by both m and n. Using the inclusion–exclusion principle, we deduce that $(n-1)+(m-1)-(\gcd(m,n)-1) = n+m-\gcd(m,n)-1$ positive integers less than mn are divisible by either m or n. Therefore $(mn-1)-(n+m-\gcd(m,n)-1) = mn-n-m+\gcd(m,n) = (m-1)(n-1)+\gcd(m,n)-1$ numbers in this range are not divisible by either m or n. This gives us our answer:

$$\frac{(m-1)(n-1)+\gcd(m,n)-1}{mn-1}$$

15. **a)** Label the faces of the cards $F1$, $B1$, $F2$, $B2$, $F3$, and $B3$ (here F stands for front, B stands for back, and the numeral stands for the card number). Without loss of generality, assume that $F1$, $B1$, and $F2$ are the black faces. There are six equally likely outcomes of this experiment, namely that we are looking at each of these faces. Then the event that we are looking at a black face is the event $E_1 = \{F1, B1, F2\}$. The event that the other side is also black is the event $E_2 = \{F1, B1\}$. We are asked for $p(E_2 \mid E_1)$, which is by definition $p(E_2 \cap E_1)/p(E_1) = p(\{F1, B1\})/p(\{F1, B1, F2\}) = (2/6)/(3/6) = 2/3$.
b) The argument in part **(a)** works for red as well, so the answer is again $2/3$. This seeming paradox comes up in other contexts, such as the Law of Restricted Choice in the game of bridge.

17. There are 2^{10} bit strings. There are 2^5 palindromic bit strings, since once the first five bits are specified arbitrarily, the remaining five bits are forced. If a bit string is picked at random, then, the probability that it is a palindrome is $2^5/2^{10} = 1/32$.

19. **a)** We assume that the coin is fair, so the probability of a head is $1/2$ on each flip, and the flips are independent. The probability that one wins 2^n dollars (i.e., $p(X = 2^n)$) is $1/2^n$, since that happens precisely when the player gets $n-1$ tails followed by a head. The expected value of the winnings is therefore the sum of 2^n times $1/2^n$ as n goes from 1 to infinity. Since each of these terms is 1, the sum is infinite. In other words, one should be willing to wager any amount of money and expect to come out ahead in the long run. The catch, of course, and that is partly why it is a paradox, is that the long run is too long, and the bank could not actually pay 2^n dollars for large n anyway (it would exceed the world's money supply). It would not make sense for someone to pay a million dollars to play this game just once.
b) Now the expectation is $(1/2)(2^1)+(1/2^2)(2^2)+(1/2^3)(2^3)+(1/2^4)(2^4)+(1/2^5)(2^5)+(1/2^6)(2^6)+(1/2^7)(2^7)+(1/2^7)(2^8) = 9$. Therefore a fair wager would be \$9.

21. **a)** The intersection of two sets is a subset of each of them, so the largest $p(A \cap B)$ could be would occur when the smaller is a subset of the larger. In this case, that would mean that we want $B \subseteq A$, in which case $A \cap B = B$, so $p(A \cap B) = p(B) = 1/3$. To construct an example, we find a common denominator of the fractions involved, namely 12, and let the sample space consist of 12 equally likely outcomes, say numbered 1 through 12. We let $B = \{1,2,3,4\}$ and $A = \{1,2,3,4,5,6,7,8,9\}$. The smallest intersection would occur when $A \cup B$ is as large as possible, since $p(A \cup B) = p(A) + p(B) - p(A \cap B)$. The largest $A \cup B$ could ever be is the entire sample space, whose probability is 1, and that certainly can occur here. So we have $1 = (3/4)+(1/3)-p(A \cap B)$, which gives $p(A \cap B) = 1/12$. To construct an example, again we find a common denominator of these fractions, namely 12, and let the sample space consist of 12 equally likely outcomes, say numbered 1 through 12. We let $B = \{1,2,3,4\}$ and $A = \{4,5,6,7,8,9,10,11,12\}$. Then $A \cap B = \{4\}$, and $p(A \cap B) = 1/12$.

b) The largest $p(A \cup B)$ could ever be is 1, which occurs when $A \cup B$ is the entire sample space. As we saw in part **(a)**, that is possible here, using the second example above. The union of two sets is a superset of each of them, so the smallest $p(A \cup B)$ could be would occur when the smaller is a subset of the larger. In this case, that would mean that we want $B \subseteq A$, in which case $A \cup B = A$, so $p(A \cup B) = p(A) = 3/4$. This occurs in the first example given above.

23. a) We need three conditions for two of the events at once and one condition for all three:

$$\begin{aligned} p(E_1 \cap E_2) &= p(E_1)p(E_2) \\ p(E_1 \cap E_3) &= p(E_1)p(E_3) \\ p(E_2 \cap E_3) &= p(E_2)p(E_3) \\ p(E_1 \cap E_2 \cap E_3) &= p(E_1)p(E_2)p(E_3) \end{aligned}$$

b) Intuitively, it is clear that these three events are independent, since successive flips do not depend on the results of previous flips. Mathematically, we need to look at the various events. There are 8 possible outcomes of this experiment. In four of them the first flip comes up heads, so $p(E_1) = 4/8 = 1/2$. Similarly, $p(E_2) = 1/2$ and $p(E_3) = 1/2$. In two of these outcomes the first flip is a head and the second flip is a tail, so $p(E_1 \cap E_2) = 2/8 = 1/4$. Similarly, $p(E_1 \cap E_3) = 1/4$ and $p(E_2 \cap E_3) = 1/4$. Only one outcome has all three events happening, so $p(E_1 \cap E_2 \cap E_3) = 1/8$. We now need to plug these numbers into the four equations displayed in part **(a)** and check the they are satisfied:

$$\begin{aligned} p(E_1 \cap E_2) &= \frac{1}{4} = \frac{1}{2} \cdot \frac{1}{2} = p(E_1)p(E_2) \\ p(E_1 \cap E_3) &= \frac{1}{4} = \frac{1}{2} \cdot \frac{1}{2} = p(E_1)p(E_3) \\ p(E_2 \cap E_3) &= \frac{1}{4} = \frac{1}{2} \cdot \frac{1}{2} = p(E_2)p(E_3) \\ p(E_1 \cap E_2 \cap E_3) &= \frac{1}{8} = \frac{1}{2} \cdot \frac{1}{2} \cdot \frac{1}{2} = p(E_1)p(E_2)p(E_3) \end{aligned}$$

The first three lines show that E_1, E_2, and E_3 are pairwise independent, and these together with the last line show that they are mutually independent.

c) We need to compute the following quantities, which we do by counting outcomes. $p(E_1) = 4/8 = 1/2$, $p(E_2) = 4/8 = 1/2$, $p(E_3) = 4/8 = 1/2$, $p(E_1 \cap E_2) = 2/8 = 1/4$, $p(E_1 \cap E_3) = 2/8 = 1/4$, $p(E_2 \cap E_3) = 2/8 = 1/4$, and $p(E_1 \cap E_2 \cap E_3) = 1/8$. Note that these are the same values obtained in part **(b)**. Therefore when we plug them into the defining equations for independence, we must again get true statements, so these events are both pairwise and mutually independent.

d) We need to compute the following quantities, which we do by counting outcomes. $p(E_1) = 4/8 = 1/2$, $p(E_2) = 4/8 = 1/2$, $p(E_3) = 4/8 = 1/2$, $p(E_1 \cap E_2) = 2/8 = 1/4$, $p(E_1 \cap E_3) = 2/8 = 1/4$, $p(E_2 \cap E_3) = 2/8 = 1/4$, and $p(E_1 \cap E_2 \cap E_3) = 0/8$. Note that these are the same values obtained in part **(b)**, except that now $p(E_1 \cap E_2 \cap E_3) = 0/8$. Therefore when we plug them into the defining equations for independence, we again get true statements for the first three, but not for the last one. Therefore, these events are pairwise independent, but they are not mutually independent.

e) There will be one condition for each subset of the set of events, other than subsets consisting of no events or just one event. There are 2^n subsets of a set with n elements, and $n+1$ of them have fewer than two elements. Therefore there are $2^n - n - 1$ subsets of interest and that many conditions to check.

25. Using Theorems 3 and 6 of Section 6.4 and the fact that the expectation of a constant is itself (this is easy to prove from the definition), we have

$$\begin{aligned} V(aX+b) &= E((aX+b)^2) - E(aX+b)^2 \\ &= E(a^2X^2 + 2abX + b^2) - (aE(X)+b)^2 \\ &= E(a^2X^2) + E(2abX) + E(b^2) - (a^2E(X)^2 + 2abE(X) + b^2) \\ &= a^2E(X^2) + 2abE(X) + b^2 - a^2E(X)^2 - 2abE(X) - b^2) \\ &= a^2(E(X^2) - E(X)^2) = a^2V(X). \end{aligned}$$

27. This is essentially an application of inclusion–exclusion (Section 7.5). To count every element in the sample space exactly once, we want to include every element in each of the sets and then take away the double counting of the elements in the intersections. Thus $p(E_1 \cup E_2 \cup \cdots \cup E_m) = p(E_1) + p(E_2) + \cdots + p(E_m) - p(E_1 \cap E_2) - p(E_1 \cap E_3) - \cdots - p(E_1 \cap E_m) - p(E_2 \cap E_3) - p(E_2 \cap E_4) - \cdots - p(E_2 \cap E_m) - \cdots - p(E_{m-1} \cap E_m) = qm - (m(m-1)/2)r$, since $C(m,2)$ terms are being subtracted. But $p(E_1 \cup E_2 \cup \cdots \cup E_m) = 1$, so we have $qm - (m(m-1)/2)r = 1$. Since $r \ge 0$, this equation tells us that $qm \ge 1$, so $q \ge 1/m$. Since $q \le 1$, this equation also implies that $(m(m-1)/2)r = qm - 1 \le m - 1$, from which it follows that $r \le 2/m$.

29. **a)** We purchase the cards until we have gotten one of each type. That means we have purchased X cards in all. On the other hand, that also means that we purchased X_0 cards until we got the first type we got (of course $X_0 = 1$ in all cases), and then purchased X_1 more cards until we got the second type we got, and so on. Thus X is the sum of the X_j's.
b) Once j distinct types have been obtained, there are $n - j$ new types available out of a total of n types available. Since it is equally likely that we get each type, the probability of success on the next purchase (getting a new type) is $(n-j)/n$.
c) This follows immediately from the definition of geometric distribution, the definition of X_j, and part **(b)**.
d) From part **(c)** it follows that $E(X_j) = n/(n-j)$. Thus by linearity of expectation from part **(a)** we have

$$E(X) = E(X_0) + E(X_1) + \cdots + E(X_{n-1}) = \frac{n}{n} + \frac{n}{n-1} + \cdots + \frac{n}{1} = n\left(\frac{1}{n} + \frac{1}{n-1} + \cdots + \frac{1}{1}\right).$$

e) If $n = 50$, then

$$E(X) = n\sum_{j=1}^{n} \frac{1}{j} \approx n(\ln n + \gamma) \approx 50(\ln 50 + 0.57721) \approx 224.46.$$

We can compute the exact answer using a computer algebra system:

$$\frac{13943237577224054960759}{61980890084919934128} \approx 224.96$$

31. We see from Exercise 42 in Section 5.5 (applying the idea in Example 8 of that section) that there are $52!/13!^4$ possible ways to deal the cards. In order to answer this question, we need to find the number of ways to deal them so that each player gets an ace. There are $4! = 24$ ways to distribute the aces so that each player receives one. Once this is done, there are 48 cards left, 12 to be dealt to each player, so using the idea in Example 8 in Section 5.5 again, there are $48!/12!^4$ possible ways to deal these cards. Taking the quotient of these two quantities will give us the desired probability:

$$\frac{24 \cdot (48!/12!^4)}{52!/13!^4} = \frac{24 \cdot 13^4}{52 \cdot 51 \cdot 50 \cdot 49} = \frac{2197}{20825} \approx \frac{1}{10}$$

WRITING PROJECTS FOR CHAPTER 6

Books and articles indicated by bracketed symbols below are listed near the end of this manual. You should also read the general comments and advice you will find there about researching and writing these essays.

1. A general mathematics history text should cover this topic well. Introductory probability books might also have a few words on the subject.

2. It will be instructive to see whether the advice given in popular gambling books (which is where to go for this project) is correct! Your university library might not be a good place to look for this project; try your local bookstore instead.

3. As in the previous project, you should consult popular books on this subject. James Thorpe was one of the first persons to realize that the player can win against the house in blackjack by using the right strategy (which involves keeping track of the cards that have already been used, as well as doing the right thing in terms of drawing additional cards on each hand).

4. See the comments for Writing Project 2.

5. Google gives about five million hits on the phrase "spam filter" (in quotation marks). You might also want to include the word "successful" to narrow the search.

6. There is an article on this in an old issue of *The American Statistician* [Sc].

7. See the classical book on the probabilistic method, in which Erdős has written an appendix, [AlSp].

8. Modern books on algorithms, such as [CoLe], are good sources here.

CHAPTER 7
Advanced Counting Techniques

SECTION 7.1 Recurrence Relations

This section is related to Section 4.3, in that recurrence relations are in some sense really recursive or inductive definitions. Many of the exercises in this set provide practice in setting up such relations from a given applied situation. In each problem of this type, ask yourself how the nth term of the sequence can be related to one or more previous terms; the answer is the desired recurrence relation.

Some of these exercises deal with solving recurrence relations by the iterative approach. The trick here is to be precise and patient. First write down a_n in terms of a_{n-1}. Then use the recurrence relation with $n-1$ plugged in for n to rewrite what you have in terms of a_{n-2}; simplify algebraically. Continue in this manner until a pattern emerges. Then write down what the expression is in terms of a_0 (or a_1, depending on the initial condition), following the pattern that developed in the first few terms. Usually at this point either the answer is what you have just written down, or else the answer can be obtained from what you have by summing a series. The iterative approach is not usually effective for recurrence relations of degree greater than 1 (i.e., those in which a_n depends on previous terms other than just a_{n-1}).

Exercise 37 is interesting and challenging, and shows that the inductive step may be quite nontrivial. Exercise 45 deals with onto functions; another—totally different—approach to counting onto functions is given in Section 7.6. Exercises 46–61 deal with additional interesting applications.

1. We need to compute the terms of the sequence one at a time, since each term is dependent upon one or more of the previous terms.

a) We are given $a_0 = 2$. Then by the recurrence relation $a_n = 6a_{n-1}$ we see (by letting $n = 1$) that $a_1 = 6a_0 = 6\cdot 2 = 12$. Similarly $a_2 = 6a_1 = 6\cdot 12 = 72$, then $a_3 = 6a_2 = 6\cdot 72 = 432$, and $a_4 = 6a_3 = 6\cdot 432 = 2592$.
b) $a_1 = 2$ (given), $a_2 = a_1^2 = 2^2 = 4$, $a_3 = a_2^2 = 4^2 = 16$, $a_4 = a_3^2 = 16^2 = 256$, $a_5 = a_4^2 = 256^2 = 65536$
c) This time each term depends on the two previous terms. We are given $a_0 = 1$ and $a_1 = 2$. To compute a_2 we let $n = 2$ in the recurrence relation, obtaining $a_2 = a_1 + 3a_0 = 2 + 3\cdot 1 = 5$. Then we have $a_3 = a_2 + 3a_1 = 5 + 3\cdot 2 = 11$ and $a_4 = a_3 + 3a_2 = 11 + 3\cdot 5 = 26$.
d) $a_0 = 1$ (given), $a_1 = 1$ (given), $a_2 = 2a_1 + 2^2a_0 = 2\cdot 1 + 4\cdot 1 = 6$, $a_3 = 3a_2 + 3^2a_1 = 3\cdot 6 + 9\cdot 1 = 27$, $a_4 = 4a_3 + 4^2a_2 = 4\cdot 27 + 16\cdot 6 = 204$
e) We are given $a_0 = 1$, $a_1 = 2$, and $a_2 = 0$. Then $a_3 = a_2 + a_0 = 0 + 1 = 1$ and $a_4 = a_3 + a_1 = 1 + 2 = 3$.

3. a) We simply plug in $n = 0$, $n = 1$, $n = 2$, $n = 3$, and $n = 4$. Thus we have $a_0 = 2^0 + 5\cdot 3^0 = 1 + 5\cdot 1 = 6$, $a_1 = 2^1 + 5\cdot 3^1 = 2 + 5\cdot 3 = 17$, $a_2 = 2^2 + 5\cdot 3^2 = 4 + 5\cdot 9 = 49$, $a_3 = 2^3 + 5\cdot 3^3 = 8 + 5\cdot 27 = 143$, and $a_4 = 2^4 + 5\cdot 3^4 = 16 + 5\cdot 81 = 421$.
b) Using our data from part **(a)**, we see that $49 = 5\cdot 17 - 6\cdot 6$, $143 = 5\cdot 49 - 6\cdot 17$, and $421 = 5\cdot 143 - 6\cdot 49$.
c) This is algebra. The messiest part is factoring out a large power of 2 and a large power of 3. If we substitute $n-1$ for n in the definition we have $a_{n-1} = 2^{n-1} + 5\cdot 3^{n-1}$; similarly $a_{n-2} = 2^{n-2} + 5\cdot 3^{n-2}$. We start with the right-hand side of our desired identity:

$$\begin{aligned}5a_{n-1} - 6a_{n-2} &= 5(2^{n-1} + 5 \cdot 3^{n-1}) - 6(2^{n-2} + 5 \cdot 3^{n-2}) \\ &= 2^{n-2}(10 - 6) + 3^{n-2}(75 - 30) \\ &= 2^{n-2} \cdot 4 + 3^{n-2} \cdot 9 \cdot 5 \\ &= 2^n + 3^n \cdot 5 = a_n\end{aligned}$$

5. In each case we have to substitute the given equation for a_n into the recurrence relation $a_n = 8a_{n-1} - 16a_{n-2}$ and see if we get a true statement. Remember to make the appropriate substitutions for n (either $n-1$ or $n-2$) on the right-hand side. What we are really doing here is performing the inductive step in a proof by mathematical induction: if the formula is correct for a_{n-1} (and also for a_{n-2}, etc., in some cases), then the formula is also correct for a_n.

a) Plugging $a_n = 0$ into the equation $a_n = 8a_{n-1} - 16a_{n-2}$, we obtain the true statement that $0 = 0$. Therefore $a_n = 0$ is a solution of the recurrence relation.

b) Plugging $a_n = 1$ into the equation $a_n = 8a_{n-1} - 16a_{n-2}$, we obtain the false statement $1 = 8 \cdot 1 - 16 \cdot 1 = -8$. Therefore $a_n = 1$ is not a solution.

c) Plugging $a_n = 2^n$ into the equation $a_n = 8a_{n-1} - 16a_{n-2}$, we obtain the statement $2^n = 8 \cdot 2^{n-1} - 16 \cdot 2^{n-2}$. By algebra, the right-hand side equals $2^{n-2}(8 \cdot 2 - 16) = 0$. Since this is not equal to the left-hand side, we conclude that $a_n = 2^n$ is not a solution.

d) Plugging $a_n = 4^n$ into the equation $a_n = 8a_{n-1} - 16a_{n-2}$, we obtain the statement $4^n = 8 \cdot 4^{n-1} - 16 \cdot 4^{n-2}$. By algebra, the right-hand side equals $4^{n-2}(8 \cdot 4 - 16) = 4^{n-2} \cdot 16 = 4^{n-2} \cdot 4^2 = 4^n$. Since this is the left-hand side, we conclude that $a_n = 4^n$ is a solution.

e) Plugging $a_n = n4^n$ into the equation $a_n = 8a_{n-1} - 16a_{n-2}$, we obtain the statement $n4^n = 8(n-1)4^{n-1} - 16(n-2)4^{n-2}$. By algebra, the right-hand side equals $4^{n-2}\big(8(n-1) \cdot 4 - 16(n-2)\big) = 4^{n-2}(32n - 32 - 16n + 32) = 4^{n-2}(16n) = 4^{n-2} \cdot 4^2 n = n4^n$. Since this is the left-hand side, we conclude that $a_n = n4^n$ is a solution.

f) Plugging $a_n = 2 \cdot 4^n + 3n4^n$ into the equation $a_n = 8a_{n-1} - 16a_{n-2}$, we obtain the statement $2 \cdot 4^n + 3n4^n = 8\big(2 \cdot 4^{n-1} + 3(n-1)4^{n-1}\big) - 16\big(2 \cdot 4^{n-2} + 3(n-2)4^{n-2}\big)$. By algebra, the right-hand side equals $4^{n-2}\big(8 \cdot 2 \cdot 4 + 8 \cdot 3(n-1) \cdot 4 - 16 \cdot 2 - 16 \cdot 3(n-2)\big) = 4^{n-2}(64 + 96n - 96 - 32 - 48n + 96) = 4^{n-2}(48n + 32) = 4^{n-2} \cdot 4^2(3n + 2) = (2 + 3n)4^n$. Since this is the same as the left-hand side, we conclude that $a_n = 2 \cdot 4^n + 3n4^n$ is a solution.

g) Plugging $a_n = (-4)^n$ into the equation $a_n = 8a_{n-1} - 16a_{n-2}$, we obtain the statement $(-4)^n = 8 \cdot (-4)^{n-1} - 16 \cdot (-4)^{n-2}$. By algebra the right-hand side equals $(-4)^{n-2}(8 \cdot (-4) - 16) = (-4)^{n-2}(-48) = -3(-4)^n$. Since this is not equal to the left-hand side, we conclude that $a_n = (-4)^n$ is not a solution.

h) Plugging $a_n = n^2 4^n$ into the equation $a_n = 8a_{n-1} - 16a_{n-2}$, we obtain the statement $n^2 4^n = 8(n-1)^2 4^{n-1} - 16(n-2)^2 4^{n-2}$. By algebra, the right-hand side equals $4^{n-2}\big(8(n-1)^2 \cdot 4 - 16(n-2)^2\big) = 4^{n-2}\big(32(n^2 - 2n + 1) - 16(n^2 - 4n + 4)\big) = 4^{n-2}(32n^2 - 64n + 32 - 16n^2 + 64n - 64) = 4^{n-2}(16n^2 - 32) = 4^{n-2} \cdot 4^2(n^2 - 2) = 4^n(n^2 - 2)$. Since this is not equal to the left-hand side, we conclude that $a_n = n^2 4^n$ is not a solution.

7. In each case we have to plug the purported solution into the right-hand side of the recurrence relation and see if it simplifies to the left-hand side. The algebra can get tedious, and it is easy to make a mistake.

a) We have

$$\begin{aligned}a_{n-1} + 2a_{n-2} + 2n - 9 &= -(n-1) + 2 + 2(-(n-2) + 2) + 2n - 9 \\ &= -n + 2 = a_n.\end{aligned}$$

b) We have

$$\begin{aligned}a_{n-1} + 2a_{n-2} + 2n - 9 &= 5(-1)^{n-1} - (n-1) + 2 + 2(5(-1)^{n-2} - (n-2) + 2) + 2n - 9 \\ &= 5(-1)^{n-2}(-1 + 2) - n + 2 = a_n.\end{aligned}$$

Note that we had to factor out $(-1)^{n-2}$ and that this is the same as $(-1)^n$ since $(-1)^2 = 1$.

c) We have

$$\begin{aligned} a_{n-1} + 2a_{n-2} + 2n - 9 &= 3(-1)^{n-1} + 2^{n-1} - (n-1) + 2 + 2(3(-1)^{n-2} + 2^{n-2} - (n-2) + 2) + 2n - 9 \\ &= 3(-1)^{n-2}(-1+2) + 2^{n-2}(2+2) - n + 2 = a_n. \end{aligned}$$

Note that we had to factor out 2^{n-2} and that $2^{n-2} \cdot 4 = 2^n$.

d) We have

$$\begin{aligned} a_{n-1} + 2a_{n-2} + 2n - 9 &= 7 \cdot 2^{n-1} - (n-1) + 2 + 2(7 \cdot 2^{n-2} - (n-2) + 2) + 2n - 9 \\ &= 2^{n-2}(7 \cdot 2 + 2 \cdot 7) - n + 2 = a_n. \end{aligned}$$

9. In the iterative approach, we write a_n in terms of a_{n-1}, then write a_{n-1} in terms of a_{n-2} (using the recurrence relation with $n-1$ plugged in for n), and so on. When we reach the end of this procedure, we use the given initial value of a_0. This will give us an explicit formula for the answer or it will give us a finite series, which we then sum to obtain an explicit formula for the answer.

a) We write

$$\begin{aligned} a_n &= 3a_{n-1} \\ &= 3(3a_{n-2}) = 3^2 a_{n-2} \\ &= 3^2(3a_{n-3}) = 3^3 a_{n-3} \\ &\ \ \vdots \\ &= 3^n a_{n-n} = 3^n a_0 = 3^n \cdot 2. \end{aligned}$$

Note that we figured out the last line by following the pattern that had developed in the first few lines. Therefore the answer is $a_n = 2 \cdot 3^n$.

b) We write

$$\begin{aligned} a_n &= 2 + a_{n-1} \\ &= 2 + (2 + a_{n-2}) = (2+2) + a_{n-2} = (2 \cdot 2) + a_{n-2} \\ &= (2 \cdot 2) + (2 + a_{n-3}) = (3 \cdot 2) + a_{n-3} \\ &\ \ \vdots \\ &= (n \cdot 2) + a_{n-n} = (n \cdot 2) + a_0 = (n \cdot 2) + 3 = 2n + 3. \end{aligned}$$

Again we figured out the last line by following the pattern that had developed in the first few lines. Therefore the answer is $a_n = 2n + 3$.

c) We write (note that it is more convenient to put the a_{n-1} at the end)

$$\begin{aligned} a_n &= n + a_{n-1} \\ &= n + \big((n-1) + a_{n-2}\big) = \big(n + (n-1)\big) + a_{n-2} \\ &= \big(n + (n-1)\big) + \big((n-2) + a_{n-3}\big) = \big(n + (n-1) + (n-2)\big) + a_{n-3} \\ &\ \ \vdots \\ &= \big(n + (n-1) + (n-2) + \cdots + (n - (n-1))\big) + a_{n-n} \\ &= \big(n + (n-1) + (n-2) + \cdots + 1\big) + a_0 \\ &= \frac{n(n+1)}{2} + 1 = \frac{n^2 + n + 2}{2}. \end{aligned}$$

Therefore the answer is $a_n = (n^2 + n + 2)/2$. The formula used to obtain the last line—for the sum of the first n positive integers—was developed in Example 1 of Section 4.1.

d) We write

$$\begin{aligned}
a_n &= 3 + 2n + a_{n-1} \\
&= 3 + 2n + \big(3 + 2(n-1) + a_{n-2}\big) = \big(2\cdot 3 + 2n + 2(n-1)\big) + a_{n-2} \\
&= \big(2\cdot 3 + 2n + 2(n-1)\big) + \big(3 + 2(n-2) + a_{n-3}\big) \\
&= \big(3\cdot 3 + 2n + 2(n-1) + 2(n-2)\big) + a_{n-3} \\
&\ \vdots \\
&= \big(n\cdot 3 + 2n + 2(n-1) + 2(n-2) + \cdots + 2(n-(n-1))\big) + a_{n-n} \\
&= \big(n\cdot 3 + 2n + 2(n-1) + 2(n-2) + \cdots + 2\cdot 1\big) + a_0 \\
&= 3n + 2\cdot \frac{n(n+1)}{2} + 4 = n^2 + 4n + 4\,.
\end{aligned}$$

Therefore the answer is $a_n = n^2 + 4n + 4$. Again we used the formula for the sum of the first n positive integers developed in Example 1 of Section 4.1.

e) We write

$$\begin{aligned}
a_n &= -1 + 2a_{n-1} \\
&= -1 + 2(-1 + 2a_{n-2}) = -3 + 4a_{n-2} \\
&= -3 + 4(-1 + 2a_{n-3}) = -7 + 8a_{n-3} \\
&= -7 + 8(-1 + 2a_{n-4}) = -15 + 16a_{n-4} \\
&= -15 + 16(-1 + 2a_{n-5}) = -31 + 32a_{n-5} \\
&\ \vdots \\
&= -(2^n - 1) + 2^n a_{n-n} = -2^n + 1 + 2^n\cdot 1 = 1\,.
\end{aligned}$$

This time it was somewhat harder to figure out the pattern developing in the coefficients, but it became clear after we carried out the computation far enough. The answer, namely that $a_n = 1$ for all n, it is clear in retrospect, after we found it, since $2\cdot 1 - 1 = 1$.

f) We write

$$\begin{aligned}
a_n &= 1 + 3a_{n-1} \\
&= 1 + 3(1 + 3a_{n-2}) = (1+3) + 3^2 a_{n-2} \\
&= (1+3) + 3^2(1 + 3a_{n-3}) = (1 + 3 + 3^2) + 3^3 a_{n-3} \\
&\ \vdots \\
&= (1 + 3 + 3^2 + \cdots + 3^{n-1}) + 3^n a_{n-n} \\
&= 1 + 3 + 3^2 + \cdots + 3^{n-1} + 3^n \\
&= \frac{3^{n+1} - 1}{3 - 1} \qquad \text{(a geometric series)} \\
&= \frac{3^{n+1} - 1}{2}\,.
\end{aligned}$$

Thus the answer is $a_n = (3^{n+1} - 1)/2$.

g) We write

$$\begin{aligned}
a_n &= na_{n-1} = n(n-1)a_{n-2} \\
&= n(n-1)(n-2)a_{n-3} = n(n-1)(n-2)(n-3)a_{n-4} \\
&\ \vdots \\
&= n(n-1)(n-2)(n-3)\cdots(n-(n-1))\,a_{n-n} \\
&= n(n-1)(n-2)(n-3)\cdots 1\cdot a_0 \\
&= n!\cdot 5 = 5n!\,.
\end{aligned}$$

h) We write

$$\begin{aligned}
a_n &= 2na_{n-1} \\
&= 2n\big(2(n-1)a_{n-2}\big) = 2^2\big(n(n-1)\big)a_{n-2} \\
&= 2^2\big(n(n-1)\big)\big(2(n-2)a_{n-3}\big) = 2^3\big(n(n-1)(n-2)\big)a_{n-3} \\
&\ \vdots \\
&= 2^n n(n-1)(n-2)(n-3)\cdots\big(n-(n-1)\big)a_{n-n} \\
&= 2^n n(n-1)(n-2)(n-3)\cdots 1\cdot a_0 \\
&= 2^n n!\,.
\end{aligned}$$

11. **a)** Since the number of bacteria triples every hour, the recurrence relation should say that the number of bacteria after n hours is 3 times the number of bacteria after $n-1$ hours. Letting b_n denote the number of bacteria after n hours, this statement translates into the recurrence relation $b_n = 3b_{n-1}$.
b) The given statement is the initial condition $b_0 = 100$ (the number of bacteria at the beginning is the number of bacteria after no hours have elapsed). We solve the recurrence relation by iteration: $b_n = 3b_{n-1} = 3^2b_{n-2} = \cdots = 3^nb_{n-n} = 3^nb_0$. Letting $n = 10$ and knowing that $b_0 = 100$, we see that $b_{10} = 3^{10}\cdot 100 = 5{,}904{,}900$.

13. **a)** Let c_n be the number of cars produced in the first n months. The initial condition could be taken to be $c_0 = 0$ (no cars are made in the first 0 months). Since n cars are made in the nth month, and since c_{n-1} cars are made in the first $n-1$ months, we see that $c_n = c_{n-1} + n$.
b) The number of cars produced in the first year is c_{12}. To compute this we will solve the recurrence relation and initial condition, then plug in $n = 12$ (alternately, we could just compute the terms c_1, c_2, $\ldots$, c_{12} directly from the definition). We proceed by iteration exactly as we did in Exercise 9c:

$$\begin{aligned}
c_n &= n + c_{n-1} \\
&= n + \big((n-1) + c_{n-2}\big) = \big(n + (n-1)\big) + c_{n-2} \\
&= \big(n + (n-1)\big) + \big((n-2) + c_{n-3}\big) = \big(n + (n-1) + (n-2)\big) + c_{n-3} \\
&\ \vdots \\
&= \big(n + (n-1) + (n-2) + \cdots + (n-(n-1))\big) + c_{n-n} \\
&= \big(n + (n-1) + (n-2) + \cdots + 1\big) + c_0 \\
&= \frac{n(n+1)}{2} + 0 = \frac{n^2+n}{2}
\end{aligned}$$

Therefore the number of cars produced in the first month is $(12^2 + 12)/2 = 78$.
c) We found the formula in our solution to part **(b)**.

15. Each month our account accrues some interest that must be paid. Since the balance the previous month is $B(k-1)$, the amount of interest we owe is $(0.07/12)B(k-1)$. After paying this interest, the rest of the \$100 payment we make each month goes toward reducing the principle. Therefore we have $B(k) = B(k-1) - (100 - (0.07/12)B(k-1))$. This can be simplified to $B(k) = (1 + (0.07/12))B(k-1) - 100$. The initial condition is $B(0) = 5000$. If one calculates this as k goes from 0 to 60, we see the balance gradually decrease and finally become negative when $k = 60$ (i.e., after five years).

17. We want to show that $H_n = 2^n - 1$ is a solution to the recurrence relation $H_n = 2H_{n-1} + 1$ with initial condition $H_1 = 1$. For $n = 1$ (the base case), this is simply the calculation that $2^1 - 1 = 1$. Assume that $H_n = 2^n - 1$. Then by the recurrence relation we have $H_{n+1} = 2H_n + 1$, whereupon if we substitute on the basis of the inductive hypothesis we obtain $2(2^n - 1) + 1 = 2^{n+1} - 2 + 1 = 2^{n+1} - 1$, exactly the formula for the case of $n+1$. Thus we have shown that if the formula is correct for n, then it is also correct for $n+1$, and our proof by induction is complete.

19. **a)** Let a_n be the number of ways to deposit n dollars in the vending machine. We must express a_n in terms of earlier terms in the sequence. If we want to deposit n dollars, we may start with a dollar coin and then deposit $n-1$ dollars. This gives us a_{n-1} ways to deposit n dollars. We can also start with a dollar bill and then deposit $n-1$ dollars. This gives us a_{n-1} more ways to deposit n dollars. Finally, we can deposit a five-dollar bill and follow that with $n-5$ dollars; there are a_{n-5} ways to do this. Therefore the recurrence relation is $a_n = 2a_{n-1} + a_{n-5}$. Note that this is valid for $n \geq 5$, since otherwise a_{n-5} makes no sense.

b) We need initial conditions for all subscripts from 0 to 4. It is clear that $a_0 = 1$ (deposit nothing) and $a_1 = 2$ (deposit either the dollar coin or the dollar bill). It is also not hard to see that $a_2 = 2^2 = 4$, $a_3 = 2^3 = 8$, and $a_4 = 2^4 = 16$, since each sequence of n C's and B's corresponds to a way to deposit n dollars—a C meaning to deposit a coin and a B meaning to deposit a bill.

c) We will compute a_5 through a_{10} using the recurrence relation:

$$\begin{aligned} a_5 &= 2a_4 + a_0 = 2\cdot 16 + 1 = 33 \\ a_6 &= 2a_5 + a_1 = 2\cdot 33 + 2 = 68 \\ a_7 &= 2a_6 + a_2 = 2\cdot 68 + 4 = 140 \\ a_8 &= 2a_7 + a_3 = 2\cdot 140 + 8 = 288 \\ a_9 &= 2a_8 + a_4 = 2\cdot 288 + 16 = 592 \\ a_{10} &= 2a_9 + a_5 = 2\cdot 592 + 33 = 1217 \end{aligned}$$

Thus there are 1217 ways to deposit $10.

21. Since this problem concerns a bill of 17 pesos, we can ignore all denominations greater than 17. Therefore we assume that we have coins for 1, 2, 5, and 10 pesos, and bills for 5 and 10 pesos. Then we proceed as in Exercise 19 to write down a recurrence relation and initial conditions for a_n, the number of ways to pay a bill of n pesos (order mattering). If we want to achieve a total of n pesos, we can start with a 1-peso coin and then pay out $n-1$ pesos. This gives us a_{n-1} ways to pay n pesos. Similarly, there are a_{n-2} ways to pay starting with a 2-peso coin, a_{n-5} ways to pay starting with a 5-peso coin, a_{n-10} ways to pay starting with a 10-peso coin, a_{n-5} ways to pay starting with a 5-peso bill, and a_{n-10} ways to pay starting with a 10-peso bill. This gives the recurrence relation $a_n = a_{n-1} + a_{n-2} + 2a_{n-5} + 2a_{n-10}$, valid for all $n \geq 10$. As for initial conditions, we see immediately that $a_0 = 1$ (there is one way to pay nothing, namely by using no coins or bills), $a_1 = 1$ (use a 1-peso coin), $a_2 = 2$ (use a 2-peso coin or two 1-peso coins), $a_3 = 3$ (use only 1-peso coins, or use a 2-peso coin either first or second), and $a_4 = 5$ (the bill can be paid using the schemes 1111, 112, 121, 211, or 22, with the obvious notation). For $n = 5$ through $n = 9$, we can iterate the recurrence relation $a_n = a_{n-1} + a_{n-2} + 2a_{n-5}$, since no 10-peso bills are involved. This yields:

$$\begin{aligned} a_5 &= a_4 + a_3 + 2a_0 = 5 + 3 + 2\cdot 1 = 10 \\ a_6 &= a_5 + a_4 + 2a_1 = 10 + 5 + 2\cdot 1 = 17 \\ a_7 &= a_6 + a_5 + 2a_2 = 17 + 10 + 2\cdot 2 = 31 \\ a_8 &= a_7 + a_6 + 2a_3 = 31 + 17 + 2\cdot 3 = 54 \\ a_9 &= a_8 + a_7 + 2a_4 = 54 + 31 + 2\cdot 5 = 95 \end{aligned}$$

Next we iterate the full recurrence relation to get up to $n = 17$:

$$a_{10} = a_9 + a_8 + 2a_5 + 2a_0 = 95 + 54 + 2 \cdot 10 + 2 \cdot 1 = 171$$
$$a_{11} = a_{10} + a_9 + 2a_6 + 2a_1 = 171 + 95 + 2 \cdot 17 + 2 \cdot 1 = 302$$
$$a_{12} = a_{11} + a_{10} + 2a_7 + 2a_2 = 302 + 171 + 2 \cdot 31 + 2 \cdot 2 = 539$$
$$a_{13} = a_{12} + a_{11} + 2a_8 + 2a_3 = 539 + 302 + 2 \cdot 54 + 2 \cdot 3 = 955$$
$$a_{14} = a_{13} + a_{12} + 2a_9 + 2a_4 = 955 + 539 + 2 \cdot 95 + 2 \cdot 5 = 1694$$
$$a_{15} = a_{14} + a_{13} + 2a_{10} + 2a_5 = 1694 + 955 + 2 \cdot 171 + 2 \cdot 10 = 3011$$
$$a_{16} = a_{15} + a_{14} + 2a_{11} + 2a_6 = 3011 + 1694 + 2 \cdot 302 + 2 \cdot 17 = 5343$$
$$a_{17} = a_{16} + a_{15} + 2a_{12} + 2a_7 = 5343 + 3011 + 2 \cdot 539 + 2 \cdot 31 = 9494$$

Thus the final answer is that there are 9494 ways to pay a 17-peso debt using the coins and bills described here, assuming that order matters.

23. a) Let a_n be the number of bit strings of length n containing a pair of consecutive 0's. In order to construct a bit string of length n containing a pair of consecutive 0's we could start with 1 and follow with a string of length $n-1$ containing a pair of consecutive 0's, or we could start with a 01 and follow with a string of length $n-2$ containing a pair of consecutive 0's, or we could start with a 00 and follow with any string of length $n-2$. These three cases are mutually exclusive and exhaust the possibilities for how the string might start. From this analysis we can immediately write down the recurrence relation, valid for all $n \geq 2$: $a_n = a_{n-1} + a_{n-2} + 2^{n-2}$. (Recall that there are 2^k bit strings of length k.)
b) There are no bit strings of length 0 or 1 containing a pair of consecutive 0's, so the initial conditions are $a_0 = a_1 = 0$.
c) We will compute a_2 through a_7 using the recurrence relation:

$$a_2 = a_1 + a_0 + 2^0 = 0 + 0 + 1 = 1$$
$$a_3 = a_2 + a_1 + 2^1 = 1 + 0 + 2 = 3$$
$$a_4 = a_3 + a_2 + 2^2 = 3 + 1 + 4 = 8$$
$$a_5 = a_4 + a_3 + 2^3 = 8 + 3 + 8 = 19$$
$$a_6 = a_5 + a_4 + 2^4 = 19 + 8 + 16 = 43$$
$$a_7 = a_6 + a_5 + 2^5 = 43 + 19 + 32 = 94$$

Thus there are 94 bit strings of length 7 containing two consecutive 0's.

25. a) This problem is very similar to Example 6, with the recurrence required to go one level deeper. Let a_n be the number of bit strings of length n that do not contain three consecutive 0's. In order to construct a bit string of length n of this type we could start with 1 and follow with a string of length $n-1$ not containing three consecutive 0's, or we could start with a 01 and follow with a string of length $n-2$ not containing three consecutive 0's, or we could start with a 001 and follow with a string of length $n-2$ not containing three consecutive 0's. These three cases are mutually exclusive and exhaust the possibilities for how the string might start, since it cannot start 000. From this analysis we can immediately write down the recurrence relation, valid for all $n \geq 3$: $a_n = a_{n-1} + a_{n-2} + a_{n-3}$.
b) The initial conditions are $a_0 = 1$, $a_1 = 2$, and $a_2 = 4$, since all strings of length less than 3 satisfy the conditions (recall that the empty string has length 0).

c) We will compute a_3 through a_7 using the recurrence relation:

$$a_3 = a_2 + a_1 + a_0 = 4 + 2 + 1 = 7$$
$$a_4 = a_3 + a_2 + a_1 = 7 + 4 + 2 = 13$$
$$a_5 = a_4 + a_3 + a_2 = 13 + 7 + 4 = 24$$
$$a_6 = a_5 + a_4 + a_3 = 24 + 13 + 7 = 44$$
$$a_7 = a_6 + a_5 + a_4 = 44 + 24 + 13 = 81$$

Thus there are 81 bit strings of length 7 that do not contain three consecutive 0's.

27. a) Let a_n be the number of ways to climb n stairs. In order to climb n stairs, a person must either start with a step of one stair and then climb $n-1$ stairs (and this can be done in a_{n-1} ways) or else start with a step of two stairs and then climb $n-2$ stairs (and this can be done in a_{n-2} ways). From this analysis we can immediately write down the recurrence relation, valid for all $n \geq 2$: $a_n = a_{n-1} + a_{n-2}$.

b) The initial conditions are $a_0 = 1$ and $a_1 = 1$, since there is one way to climb no stairs (do nothing) and clearly only one way to climb one stair. Note that the recurrence relation is the same as that for the Fibonacci sequence, and the initial conditions are that $a_0 = f_1$ and $a_1 = f_2$, so it must be that $a_n = f_{n+1}$ for all n.

c) Each term in our sequence $\{a_n\}$ is the sum of the previous two terms, so the sequence begins $a_0 = 1$, $a_1 = 1$, $a_2 = 2$, $a_3 = 3$, $a_4 = 5$, $a_5 = 8$, $a_6 = 13$, $a_7 = 21$, $a_8 = 34$. Thus a person can climb a flight of 8 stairs in 34 ways under the restrictions in this problem.

29. a) Let a_n be the number of ternary strings of length n that do not contain two consecutive 0's. In order to construct a bit string of length n of this type we could start with a 1 or a 2 and follow with a string of length $n-1$ not containing two consecutive 0's, or we could start with 01 or 02 and follow with a string of length $n-2$ not containing two consecutive 0's. There are clearly $2a_{n-1}$ possibilities in the first case and $2a_{n-2}$ possibilities in the second. These two cases are mutually exclusive and exhaust the possibilities for how the string might start, since it cannot start 00. From this analysis we can immediately write down the recurrence relation, valid for all $n \geq 2$: $a_n = 2a_{n-1} + 2a_{n-2}$.

b) The initial conditions are $a_0 = 1$ (for the empty string) and $a_1 = 3$ (all three strings of length 1 fail to contain two consecutive 0's).

c) We will compute a_2 through a_6 using the recurrence relation:

$$a_2 = 2a_1 + 2a_0 = 2 \cdot 3 + 2 \cdot 1 = 8$$
$$a_3 = 2a_2 + 2a_1 = 2 \cdot 8 + 2 \cdot 3 = 22$$
$$a_4 = 2a_3 + 2a_2 = 2 \cdot 22 + 2 \cdot 8 = 60$$
$$a_5 = 2a_4 + 2a_3 = 2 \cdot 60 + 2 \cdot 22 = 164$$
$$a_6 = 2a_5 + 2a_4 = 2 \cdot 164 + 2 \cdot 60 = 448$$

Thus there are 448 ternary strings of length 6 that do not contain two consecutive 0's.

31. a) Let a_n be the number of ternary strings of length n that do not contain two consecutive 0's or two consecutive 1's. In order to construct a bit string of length n of this type we could start with a 2 and follow with a string of length $n-1$ not containing two consecutive 0's or two consecutive 1's, or we could start with 02 or 12 and follow with a string of length $n-2$ not containing two consecutive 0's or two consecutive 1's, or we could start with 012 or 102 and follow with a string of length $n-3$ not containing two consecutive 0's or two consecutive 1's, or we could start with 0102 or 1012 and follow with a string of length $n-4$ not containing two consecutive 0's or two consecutive 1's, and so on. In other words, once we encounter a 2, we can, in effect, start fresh, but the first 2 may not appear for a long time. Before the first 2 there are always two possibilities—the sequence must alternate between 0's and 1's, starting with either a 0 or a 1.

Furthermore, there is one more possibility—that the sequence contains no 2's at all, and there are two cases in which this can happen: 0101... and 1010.... Putting this all together we can write down the recurrence relation, valid for all $n \geq 2$:

$$a_n = a_{n-1} + 2a_{n-2} + 2a_{n-3} + 2a_{n-4} + \cdots + 2a_0 + 2$$

(It turns out that the sequence also satisfies the recurrence relation $a_n = 2a_{n-1} + a_{n-2}$, which can be derived algebraically from the recurrence relation we just gave by subtracting the recurrence for a_{n-1} from the recurrence for a_n. Can you find a direct argument for it?)

b) The initial conditions are that $a_0 = 1$ (the empty string satisfies the conditions) and $a_1 = 3$ (the condition cannot be violated in so short a string).

c) We will compute a_2 through a_6 using the recurrence relation:

$$a_2 = a_1 + 2a_0 + 2 = 3 + 2 \cdot 1 + 2 = 7$$

$$a_3 = a_2 + 2a_1 + 2a_0 + 2 = 7 + 2 \cdot 3 + 2 \cdot 1 + 2 = 17$$

$$a_4 = a_3 + 2a_2 + 2a_1 + 2a_0 + 2 = 17 + 2 \cdot 7 + 2 \cdot 3 + 2 \cdot 1 + 2 = 41$$

$$a_5 = a_4 + 2a_3 + 2a_2 + 2a_1 + 2a_0 + 2 = 41 + 2 \cdot 17 + 2 \cdot 7 + 2 \cdot 3 + 2 \cdot 1 + 2 = 99$$

$$a_6 = a_5 + 2a_4 + 2a_3 + 2a_2 + 2a_1 + 2a_0 + 2 = 99 + 2 \cdot 41 + 2 \cdot 17 + 2 \cdot 7 + 2 \cdot 3 + 2 \cdot 1 + 2 = 239$$

Thus there are 239 ternary strings of length 6 that do not contain two consecutive 0's or two consecutive 1's.

33. **a)** Let a_n be the number of ternary strings that do not contain consecutive symbols that are the same. By symmetry we know that $a_n/3$ of these must start with each of the symbols 0, 1, and 2. Now to construct such a string, we can begin with any symbol (3 choices), but we must follow it with a string of length $n-1$ not containing two consecutive symbols that are the same and not beginning with the symbol with which we began ($\frac{2}{3}a_{n-1}$ choices). This tells us that $a_n = 3 \cdot \frac{2}{3}a_{n-1}$, or more simply $a_n = 2a_{n-1}$, valid for every $n \geq 2$.

b) The initial condition is clearly that $a_1 = 3$. (We could also mention that $a_0 = 1$, but the recurrence only goes one level deep.)

c) Here it is easy to compute the terms in the sequence, since each is just 2 times the previous one. Thus $a_6 = 2a_5 = 2^2a_4 = 2^3a_3 = 2^4a_2 = 2^5a_1 = 2^5 \cdot 3 = 96$.

35. **a)** This problem is really the same as ("isomorphic to," as a mathematician would say) Exercise 27, since a sequence of signals exactly corresponds to a sequence of steps in that exercise. Therefore the recurrence relation is $a_n = a_{n-1} + a_{n-2}$ for all $n \geq 2$.

b) The initial conditions are again the same as in Exercise 27, namely $a_0 = 1$ (the empty message) and $a_1 = 1$.

c) Continuing where we left off our calculation in Exercise 27, we find that $a_9 = a_8 + a_7 = 34 + 21 = 55$ and then $a_{10} = a_9 + a_8 = 55 + 34 = 89$. (If we allow only part of the time period to be used, and if we rule out the empty message, then the answer will be $1 + 2 + 3 + 5 + 8 + 13 + 21 + 34 + 55 + 89 = 231$.)

37. **a)** This problem is related to Exercise 58 in Section 4.1. Consider the plane already divided by $n-1$ lines into R_{n-1} regions. The nth line is now added, intersecting each of the other $n-1$ lines in exactly one point, $n-1$ intersections in all. Think of drawing that line, beginning at one of its ends (out at "infinity"). (You should be drawing a picture as you read these words!) As we move toward the first point of intersection, we are dividing the unbounded region of the plane through which it is passing into two regions; the division is complete when we reach the first point of intersection. Then as we draw from the first point of intersection to the second, we cut off another region (in other words we divide another of the regions that were already there into two regions). This process continues as we encounter each point of intersection. By the time we have reached the last point of intersection, the number of regions has increased by $n-1$ (one for each point of intersection). Finally, as we move off to infinity, we divide the unbounded region through which we pass into two regions, increasing the count by yet 1 more. Thus there are exactly n more regions than there were before the nth line was added. The analysis we have just completed shows that the recurrence relation we seek is $R_n = R_{n-1} + n$. The initial condition is $R_0 = 1$ (since there is just one region—the whole plane—when there are no lines). Alternately, we could specify $R_1 = 2$ as the initial condition.

b) The recurrence relation and initial condition we have are precisely those in Exercise 9c, so the solution is $R_n = (n^2 + n + 2)/2$.

39. This problem is intimately related to Exercise 46 in the supplementary set of exercises in Chapter 4. It also will use the result of Exercise 37 of the present section.

a) Imagine $n-1$ planes meeting the stated conditions, dividing space into S_{n-1} solid regions. (This may be hard to visualize once n gets to be more than 2 or 3, but you should try to see it in your mind, even if the picture is blurred.) Now a new plane is drawn, intersecting each of the previous $n-1$ planes in a line. Look at the pattern these lines form on the new plane. There are $n-1$ lines, each two of which intersect and no three of which pass through the same point (because of the requirement on the "general position" of the planes). According to the result of Exercise 37b, they form $((n-1)^2 + (n-1) + 2)/2 = (n^2 - n + 2)/2$ regions in the new plane. Now each of these planar regions is actually splitting a former solid region into two. Thus the number of new solid regions this new plane creates is $(n^2 - 2n + 2)/2$. In other words, we have our recurrence relation: $S_n = S_{n-1} + (n^2 - n + 2)/2$. The initial condition is $S_0 = 1$ (if there are no planes, we get one region). Let us verify this for some small values of n. If $n = 1$, then the recurrence relation gives $S_1 = S_0 + (1^2 - 1 + 2)/2 = 1 + 1 = 2$, which is correct (one plane divides space into two half-spaces). Next $S_2 = S_1 + (2^2 - 2 + 2)/2 = 2 + 2 = 4$, and again it is easy to see that this is correct. Similarly, $S_3 = S_2 + (3^2 - 3 + 2)/2 = 4 + 4 = 8$, and we know that this is right from our familiarity with 3-dimensional graphing (space has eight octants). The first surprising case is $n = 4$, when we have $S_4 = S_3 + (4^2 - 4 + 2)/2 = 8 + 7 = 15$. This takes some concentration to see (consider the plane $x + y + z = 1$ passing through space. It splits each octant into two parts except for the octant in which all coordinates are negative, because it does not pass through that octant. Thus 7 regions become 14, and the additional region makes a total of 15.

b) The iteration here gets a little messy. We need to invoke two summation formulae from Table 2 in Section 2.4: $1 + 2 + 3 + \cdots + n = n(n+1)/2$ and $1^2 + 2^2 + 3^2 + \cdots + n^2 = n(n+1)(2n+1)/6$. We proceed as follows:

$$\begin{aligned}
S_n &= \frac{n^2}{2} - \frac{n}{2} + 1 + S_{n-1} \\
&= \frac{n^2}{2} - \frac{n}{2} + 1 + \left(\frac{(n-1)^2}{2} - \frac{(n-1)}{2} + 1\right) + S_{n-2} \\
&= \frac{n^2}{2} - \frac{n}{2} + 1 + \left(\frac{(n-1)^2}{2} - \frac{(n-1)}{2} + 1\right) + \left(\frac{(n-2)^2}{2} - \frac{(n-2)}{2} + 1\right) + S_{n-3} \\
&\ \ \vdots \\
&= \frac{n^2}{2} - \frac{n}{2} + 1 + \left(\frac{(n-1)^2}{2} - \frac{(n-1)}{2} + 1\right) + \left(\frac{(n-2)^2}{2} - \frac{(n-2)}{2} + 1\right) + \cdots \\
&\qquad + \left(\frac{1^2}{2} - \frac{1}{2} + 1\right) + S_0 \\
&= \frac{1}{2}\left((n^2 + (n-1)^2 + \cdots + 1^2) - (n + (n-1) + \cdots + 1)\right) + (1 + 1 + \cdots + 1) + 1 \\
&= \frac{1}{2}\left(\frac{n(n+1)(2n+1)}{6} - \frac{n(n+1)}{2}\right) + n + 1 \\
&= \frac{n^3 + 5n + 6}{6} \qquad \text{(after a little algebra).}
\end{aligned}$$

Note that this answer agrees with that given in supplementary Exercise 46 of Chapter 4.

41. The easy way to do this problem is to invoke symmetry. A bit string of length 7 has an even number of 0's if and only if it has an odd number of 1's, since the sum of the number of 0's and the number of 1's, namely 7, is odd. Because of the symmetric role of 0 and 1, there must be just as many 7-bit strings with an even number of 0's as there are with an odd number of 0's, each therefore being $2^7/2$ (since there are 2^7 bit strings altogether). Thus the answer is $2^{7-1} = 64$.

The solution can also be found using recurrence relations. Let e_n be the number of bit strings of length n with an even number of 0's. A bit string of length n with an even number of 0's is either a bit string that starts with a 1 and is then followed by a bit string of length $n-1$ with an even number of 0's (of which there are e_{n-1}), or else it starts with a 0 and is then followed by a bit string of length $n-1$ with an odd number of 0's (of which there are $2^{n-1}-e_{n-1}$). Therefore we have the recurrence relation $e_n = e_{n-1}+(2^{n-1}-e_{n-1}) = 2^{n-1}$. In other words, it is a recurrence relation that is its own solution. In our case, $n = 7$, so there are $2^{7-1} = 64$ such strings. (See also Exercise 31 in Section 5.4.)

43. We assume that the walkway is one tile in width and n tiles long, from start to finish. Thus we are talking about ternary sequences of length n that do not contain two consecutive 0's, say. This was studied in Exercise 29, so the answers obtained there apply. We let a_n represent the desired quantity.

a) As in Exercise 29, we find the recurrence relation to be $a_n = 2a_{n-1} + 2a_{n-2}$.

b) As in Exercise 29, the initial conditions are $a_0 = 1$ and $a_1 = 3$.

c) Continuing the computation started in the solution to Exercise 29, we find

$$a_7 = 2a_6 + 2a_5 = 2 \cdot 448 + 2 \cdot 164 = 1224\,.$$

Thus there are 1224 such colored paths.

45. If the codomain has only one element, then there is only one function (namely the function that takes each element of the domain to the unique element of the codomain). Therefore when $n = 1$ we have $S(m,n) = S(m,1) = 1$, the initial condition we are asked to verify. Now assume that $m \geq n > 1$, and we want to count $S(m,n)$, the number of functions from a domain with m elements *onto* a codomain with n elements. The form of the recurrence relation we are supposed to verify suggests that what we want to do is to look at the non-onto functions. There are n^m functions from the m-set to the n-set altogether (by the product rule, since we need to choose an element from the n-set, which can be done in n ways, a total of m times). Therefore we must show that there are $\sum_{k=1}^{n-1} C(n,k)S(m,k)$ functions from the domain to the codomain that are *not* onto. First we use the sum rule and break this count down into the disjoint cases determined by the number of elements—let us call it k—in the range of the function. Since we want the function not to be onto, k can have any value from 1 to $n-1$, but k cannot equal n. Once we have specified k, in order to specify a function we need to first specify the actual range, and this can be done in $C(n,k)$ ways, namely choosing the subset of k elements from the codomain that are to constitute the range; and second choose an *onto* function from the domain to this set of k elements. This latter task can be done in $S(m,k)$ ways, since (and here is the key recursive point) we are defining $S(m,k)$ to be precisely this number. Therefore by the product rule there are $C(n,k)S(m,k)$ different functions with our original domain having a range of k elements, and so by the sum rule there are $\sum_{k=1}^{n-1} C(n,k)S(m,k)$ non-onto functions from our original domain to our original codomain. Note that this two-dimensional recurrence relation can be used to compute $S(m,n)$ for any desired positive integers m and n. Using it is much easier than trying to list all onto functions.

47. We will see that the answer is too large for us to list all the possibilities by hand with a reasonable amount of effort.

a) We know from Example 8 that $C_0 = 1$, $C_1 = 1$, and $C_3 = 5$. It is also easy to see that $C_2 = 2$, since there are only two ways to parenthesize the product of three numbers. We know from Exercise 46 that $C_4 = 14$. Therefore the recurrence relation tells us that $C_5 = C_0C_4 + C_1C_3 + C_2C_2 + C_3C_1 + C_4C_0 = 1 \cdot 14 + 1 \cdot 5 + 2 \cdot 2 + 5 \cdot 1 + 14 \cdot 1 = 42$.

b) Here $n = 5$, so the formula gives $\frac{1}{6}C(10,5) = \frac{1}{6} \cdot 10 \cdot 9 \cdot 8 \cdot 7 \cdot 6/5! = 42$.

49. Obviously $J(1) = 1$. When $n = 2$, the second person is killed, so $J(2) = 1$. When $n = 3$, person 2 is killed off, then person 3 is skipped, so person 1 is killed, making $J(3) = 3$. When $n = 4$, the order of death is 2, 4, 3; so $J(4) = 1$. For $n = 5$, the order of death is 2, 4, 1, 5; so $J(5) = 3$. With pencil and paper (or a computer, if we feel like writing a little program), we find the remaining values:

n	$J(n)$	n	$J(n)$
1	1	9	3
2	1	10	5
3	3	11	7
4	1	12	9
5	3	13	11
6	5	14	13
7	7	15	15
8	1	16	1

51. If the number of players is even (call it $2n$), then after we have gone around the circle once we are back at the beginning, with two changes. First, the number assigned to every player has been changed, since all the even numbers are now missing. The first remaining player is 1, the second remaining player is 3, the third remaining player is 5, and so on. In general, the player in location i at this point is the player whose original number was $2i - 1$. Second, the number of players is half of what it used to be; it's now n. Therefore we know that the survivor will be the player currently occupying spot $J(n)$, namely $2J(n) - 1$. Thus we have shown that $J(2n) = 2J(n) - 1$. The argument when there are an odd number of players is similar. If there are $2n+1$ players, then after we have gone around the circle once and then killed off player 1, we will have n players left. The first remaining spot is occupied by player 3, the second remaining player is 5, and so on—the ith remaining player is $2i + 1$. Therefore we know that the survivor will be the player currently occupying spot $J(n)$, namely $2J(n) + 1$. Thus we have shown that $J(2n + 1) = 2J(n) + 1$. The base case is clearly $J(1) = 1$.

53. We use the conjecture from Exercise 50: If $n = 2^m + k$, where $k < 2^m$, then $J(n) = 2k + 1$. Thus $J(100) = J(2^6 + 36) = 2 \cdot 36 + 1 = 73$; $J(1000) = J(2^9 + 488) = 2 \cdot 488 + 1 = 977$; and $J(10000) = J(2^{13} + 1808) = 2 \cdot 1808 + 1 = 3617$.

55. It is not too hard to find the winning moves (here $a \xrightarrow{b} c$ means to move disk b from peg a to peg c, where we label the smallest disk 1 and the largest disk 4): $1 \xrightarrow{1} 2$, $1 \xrightarrow{2} 3$, $2 \xrightarrow{1} 3$, $1 \xrightarrow{3} 2$, $1 \xrightarrow{4} 4$, $2 \xrightarrow{3} 4$, $3 \xrightarrow{1} 2$, $3 \xrightarrow{2} 4$, $2 \xrightarrow{1} 4$. We can argue that at least seven moves are required no matter how many pegs we have: three to unstack the disks, one to move disk 4, and three more to restack them. We need to show that at least two additional moves are required because of the congestion caused by there being only four pegs. Note that in order to move disk 4 from peg 1 to peg 4, the other three disks must reside on pegs 2 and 3. That requires at least one move to restack them and one move to unstack them. This completes the argument.

57. It is helpful to do Exercise 56 first to get a feeling for what is going on. The base cases are obvious. If $n > 1$, then the algorithm consists of three stages. In the first stage, by the inductive hypothesis, it takes $R(n - k)$ moves to transfer the smallest $n - k$ disks to peg 2. Then by the usual Tower of Hanoi algorithm, it takes

$2^k - 1$ moves to transfer the largest k disks (i.e., the rest of them) to peg 4, avoiding peg 2. Then again by induction, it takes $R(n-k)$ moves to transfer the smallest $n-k$ disks to peg 4; all the pegs are available for this work, since the largest disks, now residing on peg 4, do not interfere. The recurrence relation is therefore established.

59. First write $R(n) = \sum_{j=1}^{n}(R(j) - R(j-1))$, which is immediate from the telescoping nature of the sum (and the fact that $R(0) = 0$. By the result from Exercise 58, this is the sum of $2^{k'-1}$ for this range of values of j (here j is playing the role that n played in Exercise 58, and k' is the value selected by the algorithm for j). But k' is constant (call this constant i) for i successive values of j. Therefore this sum is $\sum_{i=1}^{k} i2^{i-1}$, except that if n is not a triangular number, then the last few values when $i = k$ are missing, and that is what the final term in the given expression accounts for.

61. By Exercise 59, $R(n)$ is no greater than $\sum_{i=1}^{k} i2^{i-1}$. By using algebra and calculus, we can show that this equals $(k+1)2^k - 2^{k+1} + 1$, so it is no greater than $(k+1)2^k$. (The proof is to write the formula for a geometric series $\sum_{i=0}^{k} x^i = (1-x^{k+1})/(1-x)$, differentiate both sides, and simplify.) Since $n > k(k-1)/2$, we see from the quadratic formula that $k < \frac{1}{2} + \sqrt{2n + \frac{1}{4}} < 1 + \sqrt{2n}$ for all $n > 1$. Therefore $R(n)$ is bounded above by $(1 + \sqrt{2n} + 1)2^{1+\sqrt{2n}} \le 8 \cdot \sqrt{n}\, 2^{\sqrt{2n}}$ for all $n > 2$. This shows that $R(n)$ is $O(\sqrt{n}\, 2^{\sqrt{2n}})$, as desired.

63. We have to do Exercise 62 before we can do this exercise.

a) We found that the first differences were $\nabla a_n = 0$. Therefore the second differences are given by $\nabla^2 a_n = 0 - 0 = 0$.

b) We found that the first differences were $\nabla a_n = 2n - 2(n-1) = 2$. Therefore the second differences are given by $\nabla^2 a_n = 2 - 2 = 0$.

c) We found that the first differences were $\nabla a_n = n^2 - (n-1)^2 = 2n - 1$. Therefore the second differences are given by $\nabla^2 a_n = (2n-1) - (2(n-1) - 1) = 2$.

d) We found that the first differences were $\nabla a_n = 2^n - 2^{n-1} = 2^{n-1}$. Therefore the second differences are given by $\nabla^2 a_n = 2^{n-1} - 2^{n-2} = 2^{n-2}$.

65. This is just an exercise in algebra. The right-hand side of the given expression is by definition $a_n - 2\nabla a_n + \nabla a_n - \nabla a_{n-1} = a_n - \nabla a_n - \nabla a_{n-1} = a_n - (a_n - a_{n-1}) - (a_{n-1} - a_{n-2})$. Everything in this expression cancels except the last term, yielding a_{n-2}, as desired.

67. In order to express the recurrence relation $a_n = a_{n-1} + a_{n-2}$ in terms of a_n, ∇a_n, and $\nabla^2 a_n$, we use the results of Exercise 64 (that $a_{n-1} = a_n - \nabla a_n$) and Exercise 65 (that $a_{n-2} = a_n - 2\nabla a_n + \nabla^2 a_n$). Thus the given recurrence relation is equivalent to $a_n = (a_n - \nabla a_n) + (a_n - 2\nabla a_n + \nabla^2 a_n)$, which simplifies algebraically to $a_n = 3\nabla a_n - \nabla^2 a_n$.

SECTION 7.2 Solving Linear Recurrence Relations

In many ways this section is extremely straightforward. Theorems 1–6 give an algorithm for solving linear homogeneous recurrence relations with constant coefficients. The only difficulty that sometimes occurs is that the algebra involved becomes messy or impossible. (Although the Fundamental Theorem of Algebra says that every n^{th} degree polynomial equation has exactly n roots (counting multiplicities), there is in general no way to find their exact values. For example, there is nothing analogous to the quadratic formula for equations of degree 5. Also, the roots may be irrational, as we saw in Example 4, or complex, as is discussed in Exercises 38 and 39. Patience is required with the algebra in such cases.) Many other techniques are available in other special cases, in analogy to the situation with differential equations; see Exercises 48–50, for example. If you have access to a computer algebra package, you should investigate how good it is at solving recurrences. See the solution to Exercise 49 for the kind of command to use in Maple.

1. a) This is linear (the terms a_i all appear to the first power), has constant coefficients (3, 4, and 5), and is homogeneous (no terms are functions of just n). It has degree 3, since a_n is expressed in terms of a_{n-1}, a_{n-2}, and a_{n-3}.

b) This does not have constant coefficients, since the coefficient of a_{n-1} is the nonconstant $2n$.

c) This is linear, homogeneous, with constant coefficients. It has degree 4, since a_n is expressed in terms of a_{n-1}, a_{n-2}, a_{n-3} and a_{n-4} (the fact that the coefficient of a_{n-2}, for example, is 0 is irrelevant—the degree is the largest k such that a_{n-k} is present).

d) This is not homogeneous because of the 2.

e) This is not linear, since the term a_{n-1}^2 appears.

f) This is linear, homogeneous, with constant coefficients. It has degree 2.

g) This is linear but not homogeneous because of the n.

3. a) We can solve this problem by iteration (or even by inspection), but let us use the techniques in this section instead. The characteristic equation is $r-2=0$, so the only root is $r=2$. Therefore the general solution to the recurrence relation, by Theorem 3 (with $k=1$), is $a_n=\alpha 2^n$ for some constant α. We plug in the initial condition to solve for α. Since $a_0=3$ we have $3=\alpha 2^0$, whence $\alpha=3$. Therefore the solution is $a_n=3\cdot 2^n$.

b) Again this is trivial to solve by inspection, but let us use the algorithm. The characteristic equation is $r-1=0$, so the only root is $r=1$. Therefore the general solution to the recurrence relation, by Theorem 3 (with $k=1$), is $a_n=\alpha 1^n=\alpha$ for some constant α. In other words, the sequence is constant. We plug in the initial condition to solve for α. Since $a_0=2$ we have $\alpha=2$. Therefore the solution is $a_n=2$ for all n.

c) The characteristic equation is $r^2-5r+6=0$, which factors as $(r-2)(r-3)=0$, so the roots are $r=2$ and $r=3$. Therefore by Theorem 1 the general solution to the recurrence relation is $a_n=\alpha_1 2^n+\alpha_2 3^n$ for some constants α_1 and α_2. We plug in the initial condition to solve for the α's. Since $a_0=1$ we have $1=\alpha_1+\alpha_2$, and since $a_1=0$ we have $0=2\alpha_1+3\alpha_2$. These linear equations are easily solved to yield $\alpha_1=3$ and $\alpha_2=-2$. Therefore the solution is $a_n=3\cdot 2^n-2\cdot 3^n$.

d) The characteristic equation is $r^2-4r+4=0$, which factors as $(r-2)^2=0$, so there is only one root, $r=2$, which occurs with multiplicity 2. Therefore by Theorem 2 the general solution to the recurrence relation is $a_n=\alpha_1 2^n+\alpha_2 n2^n$ for some constants α_1 and α_2. We plug in the initial conditions to solve for the α's. Since $a_0=6$ we have $6=\alpha_1$, and since $a_1=8$ we have $8=2\alpha_1+2\alpha_2$. These linear equations are easily solved to yield $\alpha_1=6$ and $\alpha_2=-2$. Therefore the solution is $a_n=6\cdot 2^n-2\cdot n2^n=(6-2n)2^n$. Incidentally, there is a good way to check a solution to a recurrence relation problem, namely by calculating the next term in two ways. In this exercise, the recurrence relation tells us that $a_2=4a_1-4a_0=4\cdot 8-4\cdot 6=8$, whereas the solution tells us that $a_2=(6-2\cdot 2)2^2=8$. Since these answers agree, we are somewhat confident that our solution is correct. We could calculate a_3 in two ways for another confirmation.

e) This time the characteristic equation is $r^2+4r+4=0$, which factors as $(r+2)^2=0$, so again there is only one root, $r=-2$, which occurs with multiplicity 2. Therefore by Theorem 2 the general solution to the recurrence relation is $a_n = \alpha_1(-2)^n + \alpha_2 n(-2)^n$ for some constants α_1 and α_2. We plug in the initial conditions to solve for the α's. Since $a_0 = 0$ we have $0 = \alpha_1$, and since $a_1 = 1$ we have $1 = -2\alpha_1 - 2\alpha_2$. These linear equations are easily solved to yield $\alpha_1 = 0$ and $\alpha_2 = -1/2$. Therefore the solution is $a_n = (-1/2)n(-2)^n = n(-2)^{n-1}$.

f) The characteristic equation is $r^2 - 4 = 0$, so the roots are $r = 2$ and $r = -2$. Therefore the solution is $a_n = \alpha_1 2^n + \alpha_2(-2)^n$ for some constants α_1 and α_2. We plug in the initial conditions to solve for the α's. We have $0 = \alpha_1 + \alpha_2$, and $4 = 2\alpha_1 - 2\alpha_2$. These linear equations are easily solved to yield $\alpha_1 = 1$ and $\alpha_2 = -1$. Therefore the solution is $a_n = 2^n - (-2)^n$.

g) The characteristic equation is $r^2 - 1/4 = 0$, so the roots are $r = 1/2$ and $r = -1/2$. Therefore the solution is $a_n = \alpha_1(1/2)^n + \alpha_2(-1/2)^n$ for some constants α_1 and α_2. We plug in the initial conditions to solve for the α's. We have $1 = \alpha_1 + \alpha_2$, and $0 = \alpha_1/2 - \alpha_2/2$. These linear equations are easily solved to yield $\alpha_1 = \alpha_2 = 1/2$. Therefore the solution is $a_n = (1/2)(1/2)^n + (1/2)(-1/2)^n = (1/2)^{n+1} - (-1/2)^{n+1}$.

5. The recurrence relation found in Exercise 35 of Section 7.1 was $a_n = a_{n-1} + a_{n-2}$, with initial conditions $a_0 = a_1 = 1$. To solve this, we look at the characteristic equation $r^2 - r - 1 = 0$ (exactly as in Example 4) and obtain, by the quadratic formula, the roots $r_1 = (1+\sqrt{5})/2$ and $r_2 = (1-\sqrt{5})/2$. Therefore from Theorem 1 we know that the solution is given by

$$a_n = \alpha_1\left(\frac{1+\sqrt{5}}{2}\right)^n + \alpha_2\left(\frac{1-\sqrt{5}}{2}\right)^n,$$

for some constants α_1 and α_2. The initial conditions $a_0 = 1$ and $a_1 = 1$ allow us to determine these constants. We plug them into the equation displayed above and obtain

$$1 = a_0 = \alpha_1 + \alpha_2$$
$$1 = a_1 = \alpha_1\left(\frac{1+\sqrt{5}}{2}\right) + \alpha_2\left(\frac{1-\sqrt{5}}{2}\right).$$

By algebra we solve these equations (one way is to solve the first for α_2 in terms of α_1, and plug that into the second equation to get one equation in α_1, which can then be solved—the fact that these coefficients are messy irrational numbers involving $\sqrt{5}$ does not change the rules of algebra, of course). The solutions are $\alpha_1 = (5+\sqrt{5})/10$ and $\alpha_2 = (5-\sqrt{5})/10$. Therefore the specific solution is given by

$$a_n = \frac{5+\sqrt{5}}{10}\left(\frac{1+\sqrt{5}}{2}\right)^n + \frac{5-\sqrt{5}}{10}\left(\frac{1-\sqrt{5}}{2}\right)^n.$$

Alternatively, by not rationalizing the denominators when we solve for α_1 and α_2, we get $\alpha_1 = (1+\sqrt{5})/(2\sqrt{5})$ and $\alpha_2 = -(1-\sqrt{5})/(2\sqrt{5})$. With these expressions, we can write our solution as

$$a_n = \frac{1}{\sqrt{5}}\left(\frac{1+\sqrt{5}}{2}\right)^{n+1} - \frac{1}{\sqrt{5}}\left(\frac{1-\sqrt{5}}{2}\right)^{n+1}.$$

7. First we need to find a recurrence relation and initial conditions for the problem. Let t_n be the number of ways to tile a $2 \times n$ board with 1×2 and 2×2 pieces. To obtain the recurrence relation, imagine what tiles are placed at the left-hand end of the board. We can place a 2×2 tile there, leaving a $2 \times (n-2)$ board to be tiled, which of course can be done in t_{n-2} ways. We can place a 1×2 tile at the edge, oriented vertically, leaving a $2 \times (n-1)$ board to be tiled, which of course can be done in t_{n-1} ways. Finally, we can place two 1×2 tiles horizontally, one above the other, leaving a $2 \times (n-2)$ board to be tiled, which of course can be done in t_{n-2} ways. These three possibilities are disjoint. Therefore our recurrence relation is $t_n = t_{n-1} + 2t_{n-2}$. The initial conditions are $t_0 = t_1 = 1$, since there is only one way to tile a 2×0 board (the way that uses no tiles) and only one way to tile a 2×1 board. This recurrence relation is the same one that appeared in Example 3; it has characteristic roots 2 and -1, so the general solution is

$$t_n = \alpha_1 2^n + \alpha_2(-1)^n .$$

To determine the coefficients we plug in the initial conditions, giving us the equations

$$1 = t_0 = \alpha_1 + \alpha_2$$

$$1 = t_1 = 2\alpha_1 - \alpha_2 .$$

Solving these yields $\alpha_1 = 2/3$ and $\alpha_2 = 1/3$, so our final solution is $t_n = 2^{n+1}/3 + (-1)^n/3$.

9. a) The amount P_n in the account at the end of the nth year is equal to the amount at the end of the previous year (P_{n-1}), plus the 20% dividend on that amount ($0.2P_{n-1}$) plus the 45% dividend on the amount at the end of the year before that ($0.45P_{n-2}$). Thus we have $P_n = 1.2P_{n-1} + 0.45P_{n-2}$. We need two initial conditions, since the equation has degree 2. Clearly $P_0 = 100000$. The other initial condition is that $P_1 = 120000$, since there is only one dividend at the end of the first year.

b) Solving this recurrence relation requires looking at the characteristic equation $r^2 - 1.2r - 0.45 = 0$. By the quadratic formula, the roots are $r_1 = 1.5$ and $r_2 = -0.3$. Therefore the general solution of the recurrence relation is $P_n = \alpha_1(1.5)^n + \alpha_2(-0.3)^n$. Plugging in the initial conditions gives us the equations $100000 = \alpha_1 + \alpha_2$ and $120000 = 1.5\alpha_1 - 0.3\alpha_2$. These are easily solved to give $\alpha_1 = 250000/3$ and $\alpha_2 = 50000/3$. Therefore the solution of our problem is

$$P_n = \frac{250000}{3}(1.5)^n + \frac{50000}{3}(-0.3)^n .$$

11. a) We prove this by induction on n. We need to verify two base cases. For $n = 1$ we have $L_1 = 1 = 0 + 1 = f_0 + f_2$; and for $n = 2$ we have $L_2 = 3 = 1 + 2 = f_1 + f_3$. Assume the inductive hypothesis that $L_k = f_{k-1} + f_{k+1}$ for $k < n$. We must show that $L_n = f_{n-1} + f_{n+1}$. To do this, we let $k = n - 1$ and $k = n - 2$:

$$L_{n-1} = f_{n-2} + f_n$$

$$L_{n-2} = f_{n-3} + f_{n-1} .$$

If we add these two equations, we obtain

$$L_{n-1} + L_{n-2} = (f_{n-2} + f_{n-3}) + (f_n + f_{n-1}) ,$$

which is the same as

$$L_n = f_{n-1} + f_{n+1}$$

as desired, using the recurrence relations for the Lucas and Fibonacci numbers.

b) To find an explicit formula for the Lucas numbers, we need to solve the recurrence relation and initial conditions. Since the recurrence relation is the same as that of the Fibonacci numbers, we get the same general solution as in Example 4, namely

$$L_n = \alpha_1 \left(\frac{1+\sqrt{5}}{2}\right)^n + \alpha_2 \left(\frac{1-\sqrt{5}}{2}\right)^n ,$$

for some constants α_1 and α_2. The initial conditions are different, though. When we plug them in we get the system

$$2 = L_0 = \alpha_1 + \alpha_2$$

$$1 = L_1 = \alpha_1 \left(\frac{1+\sqrt{5}}{2}\right) + \alpha_2 \left(\frac{1-\sqrt{5}}{2}\right) .$$

By algebra we solve these equations, yielding $\alpha_1 = \alpha_1 = 1$. Therefore the specific solution is given by

$$L_n = \left(\frac{1+\sqrt{5}}{2}\right)^n + \left(\frac{1-\sqrt{5}}{2}\right)^n .$$

13. This is a third degree equation. The characteristic equation is $r^3 - 7r - 6 = 0$. Assuming the composer of the problem has arranged that the roots are nice numbers, we use the rational root test, which says that rational roots must be of the form $\pm p/q$, where p is a factor of the constant term (6 in this case) and q is a factor of the coefficient of the leading term (the coefficient of r^3 is 1 in this case). Hence the possible rational roots are $\pm 1, \pm 2, \pm 3, \pm 6$. We find that $r = -1$ is a root, so one factor of $r^3 - 7r + 6$ is $r + 1$. Dividing $r + 1$ into $r^3 - 7r - 6$ by long (or synthetic) division, we find that $r^3 - 7r - 6 = (r + 1)(r^2 - r - 6)$. By inspection we factor the rest, obtaining $r^3 - 7r - 6 = (r+1)(r-3)(r+2)$. Hence the roots are -1, 3, and -2, so the general solution is $a_n = \alpha_1(-1)^n + \alpha_2 3^n + \alpha_3(-2)^n$. To find these coefficients, we plug in the initial conditions:

$$9 = a_0 = \alpha_1 + \alpha_2 + \alpha_3$$
$$10 = a_1 = -\alpha_1 + 3\alpha_2 - 2\alpha_3$$
$$32 = a_2 = \alpha_1 + 9\alpha_2 + 4\alpha_3 .$$

Solving this system of equations (by elimination, for instance), we get $\alpha_1 = 8$, $\alpha_2 = 4$, and $\alpha_3 = -3$. Therefore the specific solution is $a_n = 8(-1)^n + 4 \cdot 3^n - 3(-2)^n$.

15. This is a third degree recurrence relation. The characteristic equation is $r^3 - 2r^2 - 5r + 6 = 0$. By the rational root test, the possible rational roots are $\pm 1, \pm 2, \pm 3, \pm 6$. We find that $r = 1$ is a root. Dividing $r - 1$ into $r^3 - 2r^2 - 5r + 6$, we find that $r^3 - 2r^2 - 5r + 6 = (r - 1)(r^2 - r - 6)$. By inspection we factor the rest, obtaining $r^3 - 2r^2 - 5r + 6 = (r - 1)(r - 3)(r + 2)$. Hence the roots are 1, 3, and -2, so the general solution is $a_n = \alpha_1 1^n + \alpha_2 3^n + \alpha_3(-2)^n$, or more simply $a_n = \alpha_1 + \alpha_2 3^n + \alpha_3(-2)^n$. To find these coefficients, we plug in the initial conditions:

$$7 = a_0 = \alpha_1 + \alpha_2 + \alpha_3$$
$$-4 = a_1 = \alpha_1 + 3\alpha_2 - 2\alpha_3$$
$$8 = a_2 = \alpha_1 + 9\alpha_2 + 4\alpha_3 .$$

Solving this system of equations, we get $\alpha_1 = 5$, $\alpha_2 = -1$, and $\alpha_3 = 3$. Therefore the specific solution is $a_n = 5 - 3^n + 3(-2)^n$.

17. We almost follow the hint and let a_{n+1} be the right-hand side of the stated identity. Clearly $a_1 = C(0,0) = 1$ and $a_2 = C(1,0) = 1$. Thus $a_1 = f_1$ and $a_2 = f_2$. Now if we can show that the sequence $\{a_n\}$ satisfies the same recurrence relation that the Fibonacci numbers do, namely $a_{n+1} = a_n + a_{n-1}$, then we will know that $a_n = f_n$ for all $n \geq 1$ (precisely what we want to show), since the solution of a second degree recurrence relation with two initial conditions is unique.

To show that $a_{n+1} = a_n + a_{n-1}$, we start with the right-hand side, which is, by our definition, $C(n-1,0) + C(n-2,1) + \cdots + C(n-1-k,k) + C(n-2,0) + C(n-3,1) + \cdots + C(n-2-l,l)$, where $k = \lfloor (n-1)/2 \rfloor$ and $l = \lfloor (n-2)/2 \rfloor$. Note that $k = l$ if n is even, and $k = l + 1$ if n is odd. Let us first take the case in which $k = l = (n-2)/2$. By Pascal's Identity, we regroup the sum above and rewrite it as

$$\begin{aligned}
C(n-1,0) &+ [C(n-2,0) + C(n-2,1)] + [C(n-3,1) + C(n-3,2)] + \cdots \\
&\qquad + [C(n-2-((n-2)/2-1), (n-2)/2-1) + C(n-1-(n-2)/2, (n-2)/2)] \\
&\qquad + C(n-2-(n-2)/2, (n-2)/2) \\
&= C(n-1,0) + C(n-1,1) + C(n-2,2) + \cdots \\
&\qquad + C(n-(n-2)/2, (n-2)/2) + C(n-2-(n-2)/2, (n-2)/2) \\
&= 1 + C(n-1,1) + C(n-2,2) + \cdots + C(n-(n-2)/2, (n-2)/2) + 1 \\
&= C(n,0) + C(n-1,1) + C(n-2,2) + \cdots + C(n-(n-2)/2, (n-2)/2) + C(n-n/2, n/2) \\
&= C(n,0) + C(n-1,1) + C(n-2,2) + \cdots + C(n-j,j) ,
\end{aligned}$$

where $j = n/2 = \lfloor n/2 \rfloor$. This is precisely a_{n+1}, as desired. In case n is odd, so that $k = (n-1)/2$ and $l = (n-3)/2$, we have a similar calculation (in this case the sum involving k has one more term than the sum involving l):

$$\begin{aligned}
C(n-1,0) &+ [C(n-2,0)+C(n-2,1)] + [C(n-3,1)+C(n-3,2)] + \cdots \\
&\quad + [C(n-2-(n-3)/2,(n-3)/2) + C(n-1-(n-1)/2,(n-1)/2)] \\
&= C(n-1,0)+C(n-1,1)+C(n-2,2)+\cdots+C(n-(n-1)/2,(n-1)/2) \\
&= 1+C(n-1,1)+C(n-2,2)+\cdots+C(n-(n-1)/2,(n-1)/2) \\
&= C(n,0)+C(n-1,1)+C(n-2,2)+\cdots+C(n-j,j)\,,
\end{aligned}$$

where $j=(n-1)/2=\lfloor n/2\rfloor$. Again, this is precisely a_{n+1}, as desired.

19. This is a third degree recurrence relation. The characteristic equation is $r^3+3r^2+3r+1=0$. We easily recognize this polynomial as $(r+1)^3$. Hence the only root is -1, with multiplicity 3, so the general solution is (by Theorem 4) $a_n=\alpha_1(-1)^n+\alpha_2 n(-1)^n+\alpha_3 n^2(-1)^n$. To find these coefficients, we plug in the initial conditions:

$$\begin{aligned}
5 &= a_0 = \alpha_1 \\
-9 &= a_1 = -\alpha_1-\alpha_2-\alpha_3 \\
15 &= a_2 = \alpha_1+2\alpha_2+4\alpha_3
\end{aligned}$$

Solving this system of equations, we get $\alpha_1=5$, $\alpha_2=3$, and $\alpha_3=1$. Therefore the answer is $a_n=5(-1)^n+3n(-1)^n+n^2(-1)^n$. We could also write this in factored form, of course, as $a_n=(n^2+3n+5)(-1)^n$. As a check of our answer, we can calculate a_3 both from the recurrence and from our formula, and we find that it comes out to be -23 in both cases.

21. This is similar to Example 6. We can immediately write down the general solution using Theorem 4. In this case there are four distinct roots, so $t=4$. The multiplicities are 4, 3, 2, and 1. So the general solution is $a_n=(\alpha_{1,0}+\alpha_{1,1}n+\alpha_{1,2}n^2+\alpha_{1,3}n^3)+(\alpha_{2,0}+\alpha_{2,1}n+\alpha_{2,2}n^2)(-2)^n+(\alpha_{3,0}+\alpha_{3,1}n)3^n+\alpha_{4,0}(-4)^n$.

23. Theorem 5 tells us that the general solution to the inhomogeneous linear recurrence relation

$$a_n=c_1a_{n-1}+c_2a_{n-2}+\cdots+c_ka_{n-k}+F(n)$$

can be found by finding one particular solution of this recurrence relation and adding it to the general solution of the corresponding homogeneous recurrence relation

$$a_n=c_1a_{n-1}+c_2a_{n-2}+\cdots+c_ka_{n-k}\,.$$

If we let f_n be the particular solution to the inhomogeneous recurrence relation and g_n be the general solution to the homogeneous recurrence relation (which will have some unspecified parameters α_1, α_2, ..., α_k), then the general solution to the inhomogeneous recurrence relation is f_n+g_n (so it, too, will have some unspecified parameters α_1, α_2, ..., α_k).

a) To show that $a_n=-2^{n+1}$ is a solution to $a_n=3a_{n-1}+2^n$, we simply substitute it in and see if we get a true statement. Upon substituting into the right-hand side we get $3a_{n-1}+2^n=3(-2^n)+2^n=2^n(-3+1)=-2^{n+1}$, which is precisely the left-hand side.

b) By Theorem 5 and the comments above, we need to find the general solution to the corresponding homogeneous recurrence relation $a_n=3a_{n-1}$. This is easily seen to be $a_n=\alpha 3^n$ (either by the iterative method or by the method of this section with a linear characteristic equation). Putting these together as discussed above, we find the general solution to the given recurrence relation: $a_n=\alpha 3^n-2^{n+1}$.

c) To find the solution with $a_0=1$, we need to plug this initial condition (where $n=0$) into our answer to part **(b)**. Doing so gives the equation $1=\alpha-2$, whence $\alpha=3$. Therefore the solution to the given recurrence relation and initial condition is $a_n=3\cdot 3^n-2^{n+1}=3^{n+1}-2^{n+1}$.

25. See the introductory remarks to Exercise 23, which apply here as well.

a) We solve this problem by wishful thinking. Suppose that $a_n = An + B$, and substitute into the given recurrence relation. This gives us $An+B = 2\big(A(n-1)+B\big)+n+5$, which simplifies to $(A+1)n+(-2A+B+5) = 0$. Now if this is going to be true for all n, then both of the quantities in parentheses will have to be 0. In other words, we need to solve the simultaneous equations $A+1=0$ and $-2A+B+5=0$. The solution is $A=-1$ and $B=-7$. Therefore a solution to the recurrence relation is $a_n = -n-7$.

b) By Theorem 5 and the comments at the beginning of Exercise 23, we need to find the general solution to the corresponding homogeneous recurrence relation $a_n = 2a_{n-1}$. This is easily seen to be $a_n = \alpha 2^n$ (either by the iterative method or by the method of this section with a linear characteristic equation). Putting these together as discussed above, we find the general solution to the given recurrence relation: $a_n = \alpha 2^n - n - 7$.

c) To find the solution with $a_0 = 4$, we need to plug this initial condition (where $n=0$) into our answer to part **(b)**. Doing so gives the equation $4 = \alpha - 7$, whence $\alpha = 11$. Therefore the solution to the given recurrence relation and initial condition is $a_n = 11\cdot 2^n - n - 7$.

27. We need to use Theorem 6, and so we need to find the roots of the characteristic polynomial of the associated homogeneous recurrence relation. The characteristic equation is $r^4-8r^2+16=0$, and as we saw in Exercise 20, $r=\pm 2$ are the only roots, each with multiplicity 2.

a) Since 1 is not a root of the characteristic polynomial of the associated homogeneous recurrence relation, Theorem 6 tells us that the particular solution will be of the form $p_3n^3+p_2n^2+p_1n+p_0$. Note that $s=1$ here, in the notation of Theorem 6.

b) Since -2 is a root with multiplicity 2 of the characteristic polynomial of the associated homogeneous recurrence relation, Theorem 6 tells us that the particular solution will be of the form $n^2p_0(-2)^n$.

c) Since 2 is a root with multiplicity 2 of the characteristic polynomial of the associated homogeneous recurrence relation, Theorem 6 tells us that the particular solution will be of the form $n^2(p_1n+p_0)2^n$.

d) Since 4 is not a root of the characteristic polynomial of the associated homogeneous recurrence relation, Theorem 6 tells us that the particular solution will be of the form $(p_2n^2+p_1n+p_0)4^n$.

e) Since -2 is a root with multiplicity 2 of the characteristic polynomial of the associated homogeneous recurrence relation, Theorem 6 tells us that the particular solution will be of the form $n^2(p_2n^2+p_1n+p_0)(-2)^n$. Note that we needed a second degree polynomial inside the parenthetical expression because the polynomial in $F(n)$ was second degree.

f) Since 2 is a root with multiplicity 2 of the characteristic polynomial of the associated homogeneous recurrence relation, Theorem 6 tells us that the particular solution will be of the form $n^2(p_4n^4+p_3n^3+p_2n^2+p_1n+p_0)2^n$.

g) Since 1 is not a root of the characteristic polynomial of the associated homogeneous recurrence relation, Theorem 6 tells us that the particular solution will be of the form p_0. Note that $s=1$ here, in the notation of Theorem 6.

29. a) The associated homogeneous recurrence relation is $a_n = 2a_{n-1}$. We easily solve it to obtain $a_n^{(h)} = \alpha 2^n$. Next we need a particular solution to the given recurrence relation. By Theorem 6 we want to look for a function of the form $a_n = c\cdot 3^n$. We plug this into our recurrence relation and obtain $c\cdot 3^n = 2c\cdot 3^{n-1}+3^n$. We divide through by 3^{n-1} and simplify, to find easily that $c=3$. Therefore the particular solution we seek is $a_n^{(p)} = 3\cdot 3^n = 3^{n+1}$. So the general solution is the sum of the homogeneous solution and this particular solution, namely $a_n = \alpha 2^n + 3^{n+1}$.

b) We plug the initial condition into our solution from part **(a)** to obtain $5 = a_1 = 2\alpha + 9$. This tells us that $\alpha = -2$. So the solution is $a_n = -2\cdot 2^n + 3^{n+1} = -2^{n+1}+3^{n+1}$. At this point it would be very useful to check our answer. One method is to let a computer do the work; a computer algebra package such as *Maple* will solve equations of this type (see Exercise 49 for the syntax of the command). Alternatively, we can compute the next term of the sequence in two ways and verify that we obtain the same answer in each case. From the recurrence relation, we expect that $a_2 = 2 \cdot a_1 + 3^2 = 2 \cdot 5 + 9 = 19$. On the other hand, our solution tells us that $a_2 = -2^{2+1} + 3^{2+1} = -8 + 27 = 19$. Since the values agree, we can be fairly confident that our solution is correct.

31. The associated homogeneous recurrence relation is $a_n = 5a_{n-1} - 6a_{n-2}$. To solve it we find the characteristic equation $r^2 - 5r + 6 = 0$, find that $r = 2$ and $r = 3$ are its solutions, and therefore obtain the homogeneous solution $a_n^{(h)} = \alpha 2^n + \beta 3^n$. Next we need a particular solution to the given recurrence relation. By using the idea in Theorem 6 twice (or following the hint), we want to look for a function of the form $a_n = cn \cdot 2^n + dn + e$. (The reason for the factor n in front of 2^n is that 2^n was already a solution of the homogeneous equation. The reason for the term $dn + e$ is the first degree polynomial $3n$.) We plug this into our recurrence relation and obtain $cn \cdot 2^n + dn + e = 5c(n-1) \cdot 2^{n-1} + 5d(n-1) + 5e - 6c(n-2) \cdot 2^{n-2} - 6d(n-2) - 6e + 2^n + 3n$. In order for this equation to be true, the exponential parts must be equal, and the polynomial parts must be equal. Therefore we have $c \cdot 2^n = 5c(n-1) \cdot 2^{n-1} - 6c(n-2) \cdot 2^{n-2} + 2^n$ and $dn + e = 5d(n-1) + 5e - 6d(n-2) - 6e + 3n$. To solve the first of these equations, we divide through by 2^{n-1}, obtaining $2c = 5c(n-1) - 3c(n-2) + 2$, whence a little algebra yields $c = -2$. To solve the second equation, we note that the coefficients of n as well as the constant terms must be equal on each side, so we know that $d = 5d - 6d + 3$ and $e = -5d + 5e + 12d - 6e$. This tells us that $d = 3/2$ and $e = 21/4$. Therefore the particular solution we seek is $a_n^{(p)} = -2n \cdot 2^n + 3n/2 + 21/4$. So the general solution is the sum of the homogeneous solution and this particular solution, namely $a_n = \alpha 2^n + \beta 3^n - 2n \cdot 2^n + 3n/2 + 21/4 = \alpha 2^n + \beta 3^n - n \cdot 2^{n+1} + 3n/2 + 21/4$.

33. The associated homogeneous recurrence relation is $a_n = 4a_{n-1} - 4a_{n-2}$. To solve it we find the characteristic equation $r^2 - 4r + 4 = 0$, find that $r = 2$ is a repeated root, and therefore obtain the homogeneous solution $a_n^{(h)} = \alpha 2^n + \beta n \cdot 2^n$. Next we need a particular solution to the given recurrence relation. By Theorem 6 we want to look for a function of the form $a_n = n^2(cn + d)2^n$. (The reason for the factor $cn + d$ is that there is a linear polynomial factor in front of 2^n in the nonhomogeneous term; the reason for the factor n^2 is that the root $r = 2$ already appears twice in the associated homogeneous relation.) We plug this into our recurrence relation and obtain $n^2(cn+d)2^n = 4(n-1)^2(cn-c+d)2^{n-1} - 4(n-2)^2(cn-2c+d)2^{n-2} + (n+1)2^n$. We divide through by 2^n, obtaining $n^2(cn + d) = 2(n-1)^2(cn - c + d) - (n-2)^2(cn - 2c + d) + (n + 1)$. Some algebra transforms this into $cn^3 + dn^2 = cn^3 + dn^2 + (-6c+1)n + (6c - 2d + 1)$. Equating like powers of n tells us that $c = 1/6$ and $d = 1$. Therefore the particular solution we seek is $a_n^{(p)} = n^2(n/6 + 1)2^n$. So the general solution is the sum of the homogeneous solution and this particular solution, namely $a_n = (\alpha + \beta n + n^2 + n^3/6)2^n$.

35. The associated homogeneous recurrence relation is $a_n = 4a_{n-1} - 3a_{n-2}$. To solve it we find the characteristic equation $r^2 - 4r + 3 = 0$, find that $r = 1$ and $r = 3$ are its solutions, and therefore obtain the homogeneous solution $a_n^{(h)} = \alpha + \beta 3^n$. Next we need a particular solution to the given recurrence relation. By using the idea in Theorem 6 twice, we want to look for a function of the form $a_n = c \cdot 2^n + n(dn + e) = c \cdot 2^n + dn^2 + en$. (The factor n in front of $(dn + e)$ is needed since 1 is already a root of the characteristic polynomial.) We plug this into our recurrence relation and obtain $c \cdot 2^n + dn^2 + en = 4c \cdot 2^{n-1} + 4d(n-1)^2 + 4e(n-1) - 3c \cdot 2^{n-2} - 3d(n-2)^2 - 3e(n-2) + 2^n + n + 3$. A lot of messy algebra transforms this into the following equation, where we group by function of n: $2^{n-2}(-c - 4) + n^2 \cdot 0 + n(-4d - 1) + (8d - 2e - 3) = 0$. The coefficients must therefore all be 0, whence $c = -4$, $d = -1/4$, and $e = -5/2$. Therefore the particular solution we seek is $a_n^{(p)} = -4 \cdot 2^n - n^2/4 - 5n/2$. So the general solution is the sum of the homogeneous solution and this particular solution, namely $a_n = -4 \cdot 2^n - n^2/4 - 5n/2 + \alpha + \beta 3^n$. Next we plug in the initial conditions to obtain $1 = a_0 = -4 + \alpha + \beta$ and $4 = a_1 = -8 - 11/4 + \alpha + 3\beta$. We solve this system of equations to obtain $\alpha = 1/8$ and $\beta = 39/8$. So the final solution is $a_n = -4 \cdot 2^n - n^2/4 - 5n/2 + 1/8 + (39/8)3^n$. As a check of our work (it would be too much to hope that we could always get this far without making an algebraic error), we can compute a_2 both from the recurrence and from the solution, and we find that $a_2 = 22$ both ways.

37. Obviously the nth term of the sequence comes from the $(n-1)^{\text{st}}$ term by adding the nth triangular number; in symbols, $a_{n-1} + n(n+1)/2 = \left(\sum_{k=1}^{n-1} k(k+1)/2\right) + n(n+1)/2 = \sum_{k=1}^{n} k(k+1)/2 = a_n$. Also, the sum of the first triangular number is clearly 1. To solve this recurrence relation, we easily see that the homogeneous solution is $a_n^{(h)} = \alpha$, so since the nonhomogeneous term is a second degree polynomial, we need a particular solution of the form $a_n = cn^3 + dn^2 + en$. Plugging this into the recurrence relation gives $cn^3 + dn^2 + en = c(n-1)^3 + d(n-1)^2 + e(n-1) + n(n+1)/2$. Expanding and collecting terms, we have $(3c - \frac{1}{2})n^2 + (-3c + 2d - \frac{1}{2})n + (c - d + e) = 0$, whence $c = \frac{1}{6}$, $d = \frac{1}{2}$, and $e = \frac{1}{3}$. Thus $a_n^{(p)} = \frac{1}{6}n^3 + \frac{1}{2}n^2 + \frac{1}{3}n$. So the general solution is $a_n = \alpha + \frac{1}{6}n^3 + \frac{1}{2}n^2 + \frac{1}{3}n$. It is now a simple matter to plug in the initial condition $a_1 = 1$ to see that $\alpha = 0$. Note that we can find a common denominator and write our solution in the nice form $a_n = n(n+1)(n+2)/6$, which is the binomial coefficient $C(n+2, 3)$.

39. Nothing in the discussion of solving recurrence relations by the methods of this section relies on the roots of the characteristic equation being real numbers. Sometimes the roots are complex numbers (involving $i = \sqrt{-1}$). The situation is analogous to the fact that we sometimes get irrational numbers when solving the characteristic equation (for example, for the Fibonacci numbers), even though the coefficients are all integers and the terms in the sequence are all integers. It is just that we need irrational numbers in order to write down an algebraic solution. Here we need complex numbers in order to write down an algebraic solution, even though all the terms in the sequence are real.

a) The characteristic equation is $r^4 - 1 = 0$. This factors as $(r-1)(r+1)(r^2+1) = 0$, so the roots are $r = 1$ and $r = -1$ (from the first two factors) and $r = i$ and $r = -i$ (from the third factor).

b) By our work in part **(a)**, the general solution to the recurrence relation is $a_n = \alpha_1 + \alpha_2(-1)^n + \alpha_3 i^n + \alpha_4(-i)^n$. In order to figure out the α's we plug in the initial conditions, yielding the following system of linear equations:

$$1 = a_0 = \alpha_1 + \alpha_2 + \alpha_3 + \alpha_4$$

$$0 = a_1 = \alpha_1 - \alpha_2 + i\alpha_3 - i\alpha_4$$

$$-1 = a_2 = \alpha_1 + \alpha_2 - \alpha_3 - \alpha_4$$

$$1 = a_3 = \alpha_1 - \alpha_2 - i\alpha_3 + i\alpha_4$$

Remembering that i is just a constant, we solve this system by elimination or other means. For instance, we could begin by subtracting the third equation from the first, to give $2 = 2\alpha_3 + 2\alpha_4$ and subtracting the fourth from the second to give $-1 = 2i\alpha_3 - 2i\alpha_4$. This gives us two equation in two unknowns. Solving them yields $\alpha_3 = (2i-1)/(4i)$ which can be put into nicer form by multiplying by i/i, so $\alpha_3 = (2+i)/4$; and then $\alpha_4 = 1 - \alpha_3 = (2-i)/4$. We plug these values back into the first and fourth equations, obtaining $\alpha_1 + \alpha_2 = 0$ and $\alpha_1 - \alpha_2 = 1/2$. These tell us that $\alpha_1 = 1/4$ and $\alpha_2 = -1/4$. Therefore the answer to the problem is

$$a_n = \frac{1}{4} - \frac{1}{4}(-1)^n + \frac{2+i}{4}i^n + \frac{2-i}{4}(-i)^n.$$

41. a) To say that f_n is the integer closest to $\frac{1}{\sqrt{5}}\left(\frac{1+\sqrt{5}}{2}\right)^n$ is to say that the absolute difference between these two numbers is less than $\frac{1}{2}$. But the difference is just $\left|\frac{1}{\sqrt{5}}\left(\frac{1-\sqrt{5}}{2}\right)^n\right|$. Thus we are asked to show that this latter number is less than $\frac{1}{2}$. The value within the parentheses is about -0.62. When raised to the nth power, for $n \geq 0$, we get a number of absolute value less than or equal to 1. When we then divide by $\sqrt{5}$ (which is greater than 2), we get a number less than $\frac{1}{2}$, as desired.

b) Clearly the second term in the formula for f_n alternates sign as n increases: a positive number is being subtracted for n even, and a negative number is being subtracted for n odd. Therefore f_n is less than $\frac{1}{\sqrt{5}}\left(\frac{1+\sqrt{5}}{2}\right)^n$ for even n and greater than $\frac{1}{\sqrt{5}}\left(\frac{1+\sqrt{5}}{2}\right)^n$ for odd n.

43. We follow the hint and let $b_n = a_n + 1$, or, equivalently, $a_n = b_n - 1$. Then the recurrence relation becomes $b_n - 1 = b_{n-1} - 1 + b_{n-2} - 1 + 1$, or $b_n = b_{n-1} + b_{n-2}$; and the initial conditions become $b_0 = a_0 + 1 = 0 + 1 = 1$ and $b_1 = a_1 + 1 = 1 + 1 = 2$. We now apply the result of Exercise 42, with b playing the role of a, and $s = 1$ and $t = 2$, to get $b_n = f_{n-1} + 2f_n$. Therefore $a_n = f_{n-1} + 2f_n - 1$. We can check this with a few small values of n: for $n = 2$, our solution predicts that $a_2 = f_1 + 2f_2 - 1 = 1 + 2 \cdot 1 - 1 = 2$; similarly, $a_3 = f_2 + 2f_3 - 1 = 1 + 2 \cdot 2 - 1 = 4$ and $a_4 = 7$. These are precisely the values we would get by applying directly the recurrence relation defining a_n in this problem. A reality check like this is a good way to increase the chances that we haven't made a mistake.

An alternative answer is $a_n = f_{n+2} - 1$. We can prove this as follows:

$$f_{n-1} + 2f_n - 1 = f_{n-1} + f_n + f_n - 1 = f_{n+1} + f_n - 1 = f_{n+2} - 1$$

45. Let a_n be the desired quantity, the number of pairs of rabbits on the island after n months. So $a_0 = 1$, since one pair is there initially. We need to read the problem carefully and decide how we will interpret what it says. Since a pair produces two new pairs "at the age of one month" and six new pairs "at the age of two months" and every month thereafter, the original pair has already produced two new pairs at the end of one month, so $a_1 = 3$ (the original pair plus two new pairs), and $a_2 = 3 + 6 + 4 = 13$ (the three pairs that were already there, six new pairs produced by the original inhabitants, and two new pairs produced by each of the two pairs born at the end of the first month). If you interpret the wording to imply that births do not occur until after the month has finished, then naturally you will get different answers from those we are about to find.

a) We already have stated the initial conditions $a_0 = 1$ and $a_1 = 3$. To obtain a recurrence relation for a_n, the number of pairs of rabbits present at the end of the nth month, we observe (as was the case in analyzing Fibonacci's example) that all the rabbit pairs who were present at the end of the $(n-2)^{\text{nd}}$ month will give rise to six new ones, giving us $6a_{n-2}$ new pairs; and all the rabbit pairs who were present at the end of the $(n-1)^{\text{st}}$ month but not at the end of the $(n-2)^{\text{nd}}$ month will give rise to two new ones, namely $2(a_{n-1} - a_{n-2})$ new pairs. Of course, the a_{n-1} pairs who were there stay around as well. Thus our recurrence relation is $a_n = 6a_{n-2} + 2(a_{n-1} - a_{n-2}) + a_{n-1}$, or, more simply, $a_n = 3a_{n-1} + 4a_{n-2}$. As a check, we compute that $a_2 = 3 \cdot 3 + 4 \cdot 1 = 13$, which is the number we got above.

b) We proceed by the method of this section, as we did in, say, Exercise 3. The characteristic equation is $r^2 - 3r - 4 = 0$, which factors as $(r-4)(r+1) = 0$, so we get roots 4 and -1. Thus the general solution is $a_n = \alpha_1 4^n + \alpha_2(-1)^n$. Plugging in the initial conditions $a_0 = 1$ and $a_1 = 3$, we find $1 = \alpha_1 + \alpha_2$ and $3 = 4\alpha_1 - \alpha_2$, which are easily solved to yield $\alpha_1 = 4/5$ and $\alpha_2 = 1/5$. Therefore the number of pairs of rabbits on the island after n months is $a_n = 4 \cdot 4^n/5 + (-1)^n/5 = \big(4^{n+1} + (-1)^n\big)/5$. As a check, we see that $a_2 = (4^3 + 1)/5 = 65/5 = 13$, the number we found above.

47. Let a_n be the employee's salary for the nth year of employment, in tens of thousands of dollars (this makes the numbers easier to work with). Thus we are told that $a_1 = 5$, and applying the given rule for raises, we have $a_2 = 2 \cdot 5 + 1 = 11$, $a_3 = 2 \cdot 11 + 2 = 24$, and so on.

a) For her n^{th} year of employment, she has $n-1$ years of experience, so the raise rule says that $a_n = 2a_{n-1} + (n-1)$. (Remember that we are using $10,000 as the unit of pay here.)

b) The associated homogeneous recurrence relation is $a_n = 2a_{n-1}$, which clearly has the solution $a_n^{(h)} = \alpha 2^n$. For the particular solution of the given relation, we note that the nonhomogeneous term is a linear function of n and try $a_n = cn + d$. Plugging into the relation yields $cn + d = 2c(n-1) + 2d + n - 1$, which, upon grouping like terms, becomes $(-c-1)n + (2c - d + 1) = 0$. Therefore $c = -1$ and $d = -1$, so $a_n^{(p)} = -n - 1$. Therefore the general solution is $a_n = \alpha 2^n - n - 1$. Plugging in the initial condition gives $5 = a_1 = 2\alpha - 2$, whence $\alpha = 7/2$. Our solution is therefore $a_n = 7 \cdot 2^{n-1} - n - 1$. We can check that this gives the correct salary for the first few years, as computed above.

49. Using the notation of Exercise 48 we have $f(n) = n+1$, $g(n) = n+3$, $h(n) = n$, and $C = 1$. Therefore

$$Q(n) \cdot n = \frac{(2 \cdot 3 \cdot 4 \cdots n) \cdot n}{4 \cdot 5 \cdot 6 \cdots (n+3)} = \frac{6n}{(n+1)(n+2)(n+3)} = \frac{-3}{n+1} + \frac{12}{n+2} + \frac{-9}{n+3}.$$

The last decomposition was by standard partial fractions techniques from calculus (write the fraction as $A/(n+1) + B/(n+2) + C/(n+3)$ and solve for A, B, and C by multiplying it out and equating like powers of n with the original fraction). Now we can give a closed form for $\sum_{i=1}^{n} Q(i)i$, since almost all the terms cancel out in a telescoping manner:

$$\begin{aligned}\sum_{i=1}^{n} Q(i)i &= \sum_{i=1}^{n} \frac{-3}{i+1} + \frac{12}{i+2} + \frac{-9}{i+3} \\ &= -\frac{3}{2} + \frac{12}{3} - \frac{9}{4} - \frac{3}{3} + \frac{12}{4} - \frac{9}{5} - \frac{3}{4} + \frac{12}{5} - \frac{9}{6} - \frac{3}{5} + \frac{12}{6} - \frac{9}{7} + \cdots \\ &\quad - \frac{3}{n-1} + \frac{12}{n} - \frac{9}{n+1} - \frac{3}{n} + \frac{12}{n+1} - \frac{9}{n+2} - \frac{3}{n+1} + \frac{12}{n+2} - \frac{9}{n+3} \\ &= -\frac{3}{2} + \frac{12}{3} - \frac{3}{3} - \frac{9}{n+2} + \frac{12}{n+2} - \frac{9}{n+3} = \frac{3}{2} - \frac{6n+9}{(n+2)(n+3)}\end{aligned}$$

This plus 1 gives us the numerator for a_n, according to the formula given in Exercise 48. For the denominator, we need

$$g(n+1)Q(n+1) = \frac{(n+4) \cdot 2 \cdot 3 \cdots (n+1)}{4 \cdot 5 \cdots (n+4)} = \frac{6}{(n+2)(n+3)}.$$

Putting this all together algebraically, we obtain $a_n = (5n^2 + 13n + 12)/12$. We can (and should!) check that this conforms to the recurrence when we calculate a_1, a_2, and so on. Indeed, we get $a_1 = 5/2$ and $a_2 = 29/6$ both ways. It is interesting to note that asking *Maple* to do this with the command

```
rsolve({a(n) = ((n + 3) * a(n - 1) + n)/(n + 1), a(0) = 1}, a(n));
```

produces the correct answer.

51. A proof of this theorem can be found in textbooks such as *Discrete Mathematics with Applications* by H. E. Mattson, Jr. (Wiley, 1993), Chapter 11.

53. We follow the hint, letting $n = 2^k$ and $a_k = \log T(n) = \log T(2^k)$. We take the log (base 2, of course) of both sides of the given recurrence relation and use the properties of logarithms to obtain

$$\log T(n) = \log n + 2 \log T(n/2),$$

so we have

$$\log T(2^k) = k + 2 \log T(2^{k-1})$$

or

$$a_k = k + 2a_{k-1}.$$

The initial condition becomes $a_0 = \log 6$. Using the techniques in this section, we find that the general solution of the recurrence relation is $a_k = c \cdot 2^k - k - 2$. Plugging in the initial condition leads to $c = 2 + \log 6$. Now we have to translate this back into terms involving T. Since $T(n) = 2^{a_k}$ and $n = 2^k$ we have

$$T(n) = 2^{(2+\log 6) \cdot 2^k - k - 2} = (2^{\log 6})^{2^k} (2^{2 \cdot 2^k - 2})(2^{-k}) = \frac{6^n \cdot 4^{n-1}}{n}.$$

SECTION 7.3 Divide-and-Conquer Algorithms and Recurrence Relations

Many of these exercises are fairly straightforward applications of Theorem 2 (or its special case, Theorem 1). The messiness of the algebra and analysis in this section is indicative of what often happens when trying to get reasonably precise estimates for the efficiency of complicated or clever algorithms.

1. Let $f(n)$ be the number of comparisons needed in a binary search of a list of n elements. From Example 1 we know that f satisfies the divide-and-conquer recurrence relation $f(n) = f(n/2) + 2$. Also, 2 comparisons are needed for a list with one element, i.e., $f(1) = 2$ (see Example 3 in Section 3.3 for further discussion). Thus $f(64) = f(32) + 2 = f(16) + 4 = f(8) + 6 = f(4) + 8 = f(2) + 10 = f(1) + 12 = 2 + 12 = 14$.

3. In the notation of Example 4 (all numerals in base 2), we want to multiply $a = 1110$ by $b = 1010$. Note that $n = 2$. Therefore $A_1 = 11$, $A_0 = 10$, $B_1 = 10$ and $B_0 = 10$. We need to form $A_1 - A_0 = 11 - 10 = 01$ and $B_0 - B_1 = 00$. Then we need the following three products: $A_1B_1 = (11)(10)$, $(A_1 - A_0)(B_0 - B_1) = (01)(00)$, and $A_0B_0 = (10)(10)$. In order to from these products, the algorithm would in fact recurse, but let us not worry about that, assuming instead that we have these answers, namely $A_1B_1 = 0110$, $(A_1 - A_0)(B_0 - B_1) = 0000$, and $A_0B_0 = 0100$. Now we need to shift these products various numbers of places to the left. We shift A_1B_1 $2n = 4$ places and also $n = 2$ places, obtaining 01100000 and 011000; we shift $(A_1 - A_0)(B_0 - B_1)$ $n = 2$ places, obtaining 000000, and we shift A_0B_0 $n = 2$ places and also no places, obtaining 010000 and 0100. Finally we add all five of these binary numbers, obtaining 10001100.

5. This problem is asking us to estimate the number of bit operations needed to do the shifts, additions, and subtractions in multiplying two $2n$-bit integers by the algorithm in Example 4. First recall from Example 8 in Section 3.6 that the number of bit operations needed for an addition of two k-bit numbers is at most $3k$; the same bound holds for subtraction. Let us assume that to shift a number k bits also requires k bit operations. Thus we need to count the number of additions and shifts of various sizes that occur in the fast multiplication algorithm. First, we need to perform two subtractions of n-bit numbers to get $A_1 - A_0$ and $B_0 - B_1$; these will take up to $6n$ bit operations altogether. We need to shift A_1B_1 $2n$ places (requiring $2n$ bit operations), and also n places (requiring n bit operations); we need to shift $(A_1 - A_0)(B_0 - B_1)$ n places (requiring n bit operations); and we need to shift A_0B_0 n places, also requiring n bit operations. This makes a total of $5n$ bit operations for the shifting. Finally we need to worry about the additions (which actually might include a subtraction if the middle term is negative). If we are clever, we can add the four terms that involve at most $3n$ bits first (that is, everything except the $2^{2n}A_1B_1$). Three additions are required, each taking $9n$ bit operations, for a total of $27n$ bit operations. Finally we need to perform one addition involving a $4n$-bit number, taking $12n$ operations. This makes a total of $39n$ bit operations for the additions.

Putting all these operations together, we need perhaps a total of $6n + 5n + 39n = 50n$ bit operations to perform all the additions, subtractions, and shifts. Obviously this bound is not exact; it depends on the actual implementation of these binary operations.

Using $C = 50$ as estimated above, the recurrence relation for fast multiplication is $f(2n) = 3f(n) + 50n$, with $f(1) = 1$ (one multiplication of bits is all that is needed if we have 1-bit numbers). Thus we can compute $f(64)$ as follows: $f(2) = 3 \cdot 1 + 50 = 53$; $f(4) = 3 \cdot 53 + 100 = 259$; $f(8) = 3 \cdot 259 + 200 = 977$; $f(16) = 3 \cdot 977 + 400 = 3331$; $f(32) = 3 \cdot 3331 + 800 = 10793$; and finally $f(64) = 3 \cdot 10793 + 1600 = 33979$. Thus about 34,000 bit operations are needed.

7. We compute these from the bottom up. (In fact, it is easy to see by induction that $f(3^k) = k + 1$, so no computation is really needed at all.)

a) $f(3) = f(1) + 1 = 1 + 1 = 2$ **b)** $f(9) = f(3) + 1 = 2 + 1 = 3$; $f(27) = f(9) + 1 = 3 + 1 = 4$

c) $f(81) = f(27) + 1 = 4 + 1 = 5$; $f(243) = f(81) + 1 = 5 + 1 = 6$; $f(729) = f(243) + 1 = 6 + 1 = 7$

9. We compute these from the bottom up.

a) $f(5) = f(1) + 3 \cdot 5^2 = 4 + 75 = 79$

b) $f(25) = f(5) + 3 \cdot 25^2 = 79 + 1875 = 1954$; $f(125) = f(25) + 3 \cdot 125^2 = 1954 + 46875 = 48{,}829$

c) $f(625) = f(125) + 3 \cdot 625^2 = 48829 + 1171875 = 1220704$; $f(3125) = f(625) + 3 \cdot 3125^2 = 1220704 + 29296875 = 30{,}517{,}579$

11. We apply Theorem 2, with $a = 1$, $b = 2$, $c = 1$, and $d = 0$. Since $a = b^d$, we have that $f(n)$ is $O(n^d \log n) = O(\log n)$.

13. We apply Theorem 2, with $a = 2$, $b = 3$, $c = 4$, and $d = 0$. Since $a > b^d$, we have that $f(n)$ is $O(n^{\log_b a}) = O(n^{\log_3 2}) \approx O(n^{0.63})$.

15. After 1 round, there are 16 teams left; after 2 rounds, 8 teams; after 3 rounds, 4 teams; after 4 rounds, 2 teams; and after 5 rounds, only 1 team remains, so the tournament is over. In general, k rounds are needed if there are 2^k teams (easily proved by induction).

17. a) Our recursive algorithm will take a sequence of names and determine whether one name occurs as more than half of the elements of the sequence, and if so, which name that is. If the sequence has just one element, then the one person on the list is the winner. For the recursive step, divide the list into two parts—the first half and the second half—as equally as possible. As is pointed out in the hint, no one could have gotten a majority of the votes on this list without having a majority in one half or the other, since if a candidate got less than or equal to half the votes in each half, then he got less than or equal to half the votes in all (this is essentially just the distributive law). Apply the algorithm recursively to each half to come up with at most two names. Then run through the entire list to count the number of occurrences of each of those names to decide which, if either, is the winner. This requires at most $2n$ additional comparisons for a list of length n.
b) We apply the Master Theorem with $a = 2$, $b = 2$, $c = 2$, and $d = 1$. Since $a = b^d$, we know that the number of comparisons is $O(n^d \log n) = O(n \log n)$.

19. a) We compute x^n when n is even by first computing $y = x^{n/2}$ recursively and then doing one multiplication, namely $y \cdot y$. When n is odd, we first compute $y = x^{(n-1)/2}$ recursively and then do two multiplications, namely $y \cdot y \cdot x$. So if $f(n)$ is the number of multiplications required, assuming the worst, then we have essentially $f(n) = f(n/2) + 2$.
b) By the Master Theorem, with $a = 1$, $b = 2$, $c = 2$, and $d = 0$, we see that $f(n)$ is $O(n^0 \log n) = O(\log n)$.

21. a) $f(16) = 2f(4) + 1 = 2(2f(2) + 1) + 1 = 2(2 \cdot 1 + 1) + 1 = 7$
b) Let $m = \log n$, so that $n = 2^m$. Also, let $g(m) = f(2^m)$. Then our recurrence becomes $f(2^m) = 2f(2^{m/2}) + 1$, since $\sqrt{2^m} = (2^m)^{1/2} = 2^{m/2}$. Rewriting this in terms of g we have $g(m) = 2g(m/2) + 1$. Theorem 1 now tells us that $g(m)$ is $O(m^{\log_2 2}) = O(m)$. Since $m = \log n$, this says that our function is $O(\log n)$.

23. a) The messiest part of this is just the bookkeeping. Note that we start with *best* set to 0, since the empty subsequence has a sum of 0, and this is the best we can do if all the terms are negative. Also note that it would be an easy improvement to keep track of where the subsequence is located, as well as what its sum is.

```
procedure largest sum(a_1, ..., a_n : real numbers)
best := 0 {empty subsequence has sum 0}
for i := 1 to n
begin
        sum := 0
        for j := i to n
        begin
                sum := sum + a_j
                if sum > best then best := sum
        end
end
{ best is the maximum possible sum of numbers in the list}
```

b) One sum and one comparison are made inside the inner loop. This loop is executed $C(n+1,2)$ times—once for each choice of a pair (i,j) of endpoints of the sequence of consecutive terms being examined (this is a combination with repetition allowed, since $i=j$ when we are examining one term by itself). Since $C(n+1,2)=n(n+1)/2$ the computational complexity is $O(n^2)$.

c) We divide the list into a first half and a second half and apply the algorithm recursively to find the largest sum of consecutive terms for each half. The largest sum of consecutive terms in the entire sequence is either one of these two numbers or the sum of a sequence of consecutive terms that crosses the middle of the list. To find the largest possible sum of a sequence of consecutive terms that crosses the middle of the list, we start at the middle and move forward to find the largest possible sum in the second half of the list, and move backward to find the largest possible sum in the first half of the list; the desired sum is the sum of these two quantities. The final answer is then the largest of this sum and the two answers obtained recursively. The base case is that the largest sum of a sequence of one term is the larger of that number and 0.

d) (i) Split the list into the first half, $-2,4,-1,3$, and the second half, $5,-6,1,2$. Apply the algorithm recursively to each half (we omit the details of this step) to find that the largest sum in the first half is 6 and the largest sum in the second half is 5. Now find the largest sum of a sequence of consecutive terms that crosses the middle of the list. Moving forward, the best we can do is 5; moving backward, the best we can do is 6. Therefore we can get a sum of 11 by adding the second through fifth terms. This is better than either recursive answer, so the desired answer is 11. (ii) Split the list into the first half, $4,1,-3,7$, and the second half, $-1,-5,3,-2$. Apply the algorithm recursively to each half (we omit the details of this step) to find that the largest sum in the first half is 9 and the largest sum in the second half is 3. Now find the largest sum of a sequence of consecutive terms that crosses the middle of the list. Moving forward, the best we can do is -1; moving backward, the best we can do is 9. Therefore we can get a sum of 8 by crossing the middle. The best of these three possibilities is 9, which we get from the first through fourth terms. (iii) Split the list into the first half, $-1,6,3,-4$, and the second half, $-5,8,-1,7$. Apply the algorithm recursively to each half (we omit the details of this step) to find that the largest sum in the first half is 9 and the largest sum in the second half is 14. Now find the largest sum of a sequence of consecutive terms that crosses the middle of the list. Moving forward, the best we can do is 9; moving backward, the best we can do is 5. Therefore we can get a sum of 14 by adding the second through eighth terms. The best of these three is actually a tie, between the second through eighth terms and the sixth through eighth terms, with a sum of 14 in each case.

e) Let $S(n)$ be the number of sums and $C(n)$ the number of comparisons used. Since the "conquer" step requires n sums and $n+2$ comparisons (two extra comparisons to determine the winner among the three possible largest sums), we have $S(n)=2S(n/2)+n$ and $C(n)=2C(n/2)+n+2$. The basis step is $C(1)=1$ and $S(1)=0$.

f) By the Master Theorem with $a=b=2$ and $d=1$, we see that we need only $O(n\log n)$ of each type of operation. This is a significant improvement over the $O(n^2)$ complexity we found in part **(b)** for the algorithm in part **(a)**.

25. To carry this down to its base level would require applying the algorithm seven times, so to keep things within reason, we will show only the outermost step. The points are already sorted for us, and so we divide them into two groups, using x coordinate. The left side will have the first eight points listed in it (they all have x coordinates less than 4.5), and the right side will have the rest, all of which have x coordinates greater than 4.5. Thus our vertical line will be taken to be $x = 4.5$. Now assume that we have already applied the algorithm recursively to find the minimum distance between two points on the left, and the minimum distance on the right. It turns out that $d = d_L = d_R = 2$. This is achieved, for example, by the points $(1,6)$ and $(3,6)$. Thus we want to concentrate on the strip from $x = 2.5$ to $x = 6.5$ of width $2d$. The only points in this strip are $(3,1)$, $(3,6)$, $(3,10)$, $(4,3)$, $(5,1)$, $(5,5)$, $(5,9)$, and $(6,7)$, Working from the bottom up, we compute distances from these points to points as much as $d = 2$ vertical units above them. According to the discussion in the text, there can never be more than seven such computations for each point in the strip. The distances we need, then, are $\overline{(3,1)(5,1)}$, $\overline{(3,1)(4,3)}$, $\overline{(5,1)(4,3)}$, $\overline{(4,3)(5,5)}$, $\overline{(5,5)(3,6)}$, $\overline{(5,5)(6,7)}$, $\overline{(3,6)(6,7)}$, $\overline{(6,7)(5,9)}$, and $\overline{(5,9)(3,10)}$. The smallest of these turns out to be 2, so the minimum distance $d = 2$ in fact is the smallest distance among all the points. (Actually we did not need to compute the distances between points that were already on the same side of the dividing line, since their distance had already been computed in the recursive step, but checking whether they are on opposite sides of the vertical line would entail additional computation anyway.)

27. The algorithm is essentially the same as the algorithm given in Example 12. The only difference is in constructing the boxes centered on the vertical line that divides the two halves of the set of points. In this variation, our strip still has width $2d$ (i.e., d units to the left and d units to the right of the vertical line), because it would be possible for two points within this box, one on each side of the line, to lie at a distance less than d from each other, but no point outside this strip has a chance to contribute to a small "across the line" distance. In this variation, however, we do not need to construct eight boxes of size $(d/2) \times (d/2)$, but rather just two boxes of size $d \times d$. The reason for this is that there can be at most one point in each of those boxes using the distance formula given in this exercise—two points within such a box (which is on the same side of the dividing line) can be at most d units apart and so would already have been considered in the recursive step. Thus the recurrence relation is the same as the recurrence relation in Example 12, except that the coefficient 7 is replaced by 1. The analysis via the Master Theorem remains the same, and again we get a $O(n \log n)$ algorithm.

29. Suppose $n = b^k$, so that $k = \log_b n$. We will prove by induction on k that $f(b^k) = f(1)(b^k)^d + c(b^k)^d k$, which is what we are asked to prove, translated into this notation. If $k = 0$, then the equation reduces to $f(1) = f(1)$, which is certainly true. We assume the inductive hypothesis, that $f(b^k) = f(1)(b^k)^d + c(b^k)^d k$, and we try to prove that $f(b^{k+1}) = f(1)(b^{k+1})^d + c(b^{k+1})^d(k+1)$. By the recurrence relation for $f(n)$ in terms of $f(n/b)$, we have $f(b^{k+1}) = b^d f(b^k) + c(b^{k+1})^d$. Then we invoke the inductive hypothesis and work through the algebra:

$$\begin{aligned} b^d f(b^k) + c(b^{k+1})^d &= b^d\big(f(1)(b^k)^d + c(b^k)^d k\big) + c(b^{k+1})^d \\ &= f(1)b^{kd+d} + cb^{kd+d}k + c(b^{k+1})^d \\ &= f(1)(b^{k+1})^d + c(b^{k+1})^d k + c(b^{k+1})^d \\ &= f(1)(b^{k+1})^d + c(b^{k+1})^d(k+1) \end{aligned}$$

31. The algebra is quite messy, but this is a straightforward proof by induction on $k = \log_b n$. If $k = 0$, so that $n = 1$, then we have the true statement

$$f(1) = C_1 + C_2 = \frac{b^d c}{b^d - a} + f(1) + \frac{b^d c}{a - b^d}$$

(since the fractions cancel each other out). Assume the inductive hypothesis, that for $n = b^k$ we have

$$f(n) = \frac{b^d c}{b^d - a} n^d + \left(f(1) + \frac{b^d c}{a - b^d} \right) n^{\log_b a} .$$

Then for $n = b^{k+1}$ we apply first the recurrence relation, then the inductive hypothesis, and finally some algebra:

$$\begin{aligned} f(n) &= af\left(\frac{n}{b}\right) + cn^d \\ &= a\left(\frac{b^d c}{b^d - a}\left(\frac{n}{b}\right)^d + \left(f(1) + \frac{b^d c}{a - b^d}\right)\left(\frac{n}{b}\right)^{\log_b a}\right) + cn^d \\ &= \frac{b^d c}{b^d - a} \cdot n^d \cdot \frac{a}{b^d} + \left(f(1) + \frac{b^d c}{a - b^d}\right) n^{\log_b a} + cn^d \\ &= n^d \left(\frac{ac}{b^d - a} + \frac{c(b^d - a)}{b^d - a}\right) + \left(f(1) + \frac{b^d c}{a - b^d}\right) n^{\log_b a} \\ &= \frac{b^d c}{b^d - a} \cdot n^d + \left(f(1) + \frac{b^d c}{a - b^d}\right) n^{\log_b a} \end{aligned}$$

Thus we have verified that the equation holds for $k + 1$, and our induction proof is complete.

33. The equation given in Exercise 31 says that $f(n)$ is the sum of a constant times n^d and a constant times $n^{\log_b a}$. Therefore we need to determine which term dominates, i.e., whether d or $\log_b a$ is larger. But we are given $a > b^d$; hence $\log_b a > \log_b b^d = d$. It therefore follows (we are also using the fact that f is increasing) that $f(n)$ is $O(n^{\log_b a})$.

35. We use the result of Exercise 33, since $a = 5 > 4^1 = b^d$. Therefore $f(n)$ is $O(n^{\log_b a}) = O(n^{\log_4 5}) \approx O(n^{1.16})$.

37. We use the result of Exercise 33, since $a = 8 > 2^2 = b^d$. Therefore $f(n)$ is $O(n^{\log_b a}) = O(n^{\log_2 8}) = O(n^3)$.

SECTION 7.4 Generating Functions

Generating functions are an extremely powerful tool in mathematics (not just in discrete mathematics). This section, as well as some material introduced in these exercises, gives you an introduction to them. The algebra in many of these exercises gets very messy, and you probably want to check your answers, either by computing values when solving recurrence relation problems, or by using a computer algebra package. See the solution to Exercise 11, for example, to learn how to get Maple *to produce the sequence for a given generating function. For more information on generating functions, consult reference [Wi2] given at the end of this* Guide *(in the List of References for the Writing Projects).*

1. By definition we want the function $f(x) = 2 + 2x + 2x^2 + 2x^3 + 2x^4 + 2x^5 = 2(1 + x + x^2 + x^3 + x^4 + x^5)$. From Example 2, we see that the expression in parentheses can also be written as $(x^6 - 1)/(x - 1)$. Thus we can write the answer as $f(x) = 2(x^6 - 1)/(x - 1)$.

3. We will use Table 1 in much of this solution.

a) Apparently all the terms are 0 except for the six 2's shown. Thus $f(x) = 2x + 2x^2 + 2x^3 + 2x^4 + 2x^5 + 2x^6$. This is already in closed form, but we can also write it more compactly as $f(x) = 2x(1 - x^6)/(1 - x)$ by factoring out $2x$, or as $f(x) = 2(1 - x^7)/(1 - x) - 2$ by subtracting away the missing term. In each case we use the identity from Example 2.

b) Apparently all the terms beyond the first three are 1's. Since $1/(1 - x) = 1 + x + x^2 + x^3 + \cdots$, we can write this generating function as $1/(1 - x) - 1 - x - x^2$, or we can write it as $x^3/(1 - x)$, depending on whether we want to simplify by adding back the missing terms or by factoring out x^3.

c) This generating function is the sequence $x + x^4 + x^7 + x^{10} + \cdots$. If we factor out an x, we have $x(1 + (x^3) + (x^3)^2 + (x^3)^3 + \cdots) = x/(1 - x^3)$, from Table 1.

d) We factor out a 2 and then include the remaining factors of 2 along with the x terms. Thus our generating function is $2(1 + (2x) + (2x)^2 + (2x)^3 + \cdots) = 2/(1 - 2x)$, again using Table 1.

e) By the Binomial Theorem (see also Table 1), the generating function is $(1 + x)^7$.

f) From Table 1 we know that $1/(1 - ax) = 1 + ax + a^2x^2 + a^3x^3 + \cdots$. That is what we have here, with $a = -1$ (and a factor of 2 in front of it all). Therefore the generating function is $2/(1 + x)$.

g) This sequence is all 1's except for a 0 where the x^2 coefficient should be. Therefore the generating function is $(1/(1 - x)) - x^2$.

h) If we factor out x^3, then we can use a formula from Table 1: $x^3 + 2x^4 + 3x^5 + \cdots = x^3(1 + 2x + 3x^2 + \cdots) = x^3/(1 - x)^2$.

5. As in Exercise 3, we make extensive use of Table 1.

a) Since the sequence with $a_n = 1$ for all n has generating function $1/(1 - x)$, this sequence has generating function $5/(1 - x)$.

b) By Table 1 the answer is $1/(1 - 3x)$.

c) We can either subtract the missing terms and write this generating function as $(2/(1 - x)) - 2 - 2x - 2x^2$, or we can factor out x^3 and write it as $2x^3/(1 - x)$. Note that these two algebraic expressions are equivalent.

d) We need to split this into two parts. Since we know that the generating function for the sequence $\{n + 1\}$ is $1/(1 - x)^2$, we write $2n + 3 = 2(n + 1) + 1$. Therefore the generating function is $2/(1 - x)^2 + 1/(1 - x)$. We can combine terms and write this function as $(3 - x)/(1 - x)^2$, but there is no particular reason to prefer that form in general.

e) By Table 1 the answer is $(1 + x)^8$. Note that $C(8, n) = 0$ by definition for all $n > 8$.

f) By Table 1 the generating function is $1/(1 - x)^5$.

7. a) We can rewrite this as $(-4(1 - \frac{3}{4}x))^3 = -64(1 - \frac{3}{4}x)^3$ and then apply the Binomial Theorem (the second line of Table 1) to get $a_n = -64C(3, n)(-\frac{3}{4})^n$. Explicitly, this says that $a_0 = -64$, $a_1 = 144$, $a_2 = -108$, $a_3 = 27$, and $a_n = 0$ for all $n \geq 4$. Alternatively, we could (by hand or with *Maple*) just multiply out this finite polynomial and note the coefficients.

b) This is like part **(a)**. By the Binomial Theorem (the third line of Table 1) we get $a_{3n} = C(3, n)$, and the other coefficients are all 0. Alternatively, we could just multiply out this finite polynomial and note the nonzero coefficients: $a_0 = 1$, $a_3 = 3$, $a_6 = 3$, $a_9 = 1$.

c) By Table 1, $a_n = 5^n$.

d) Note that $x^3/(1 + 3x) = x^3 \sum_{n=0}^{\infty}(-3)^n x^n = \sum_{n=0}^{\infty}(-3)^n x^{n+3} = \sum_{n=3}^{\infty}(-3)^{n-3}x^n$. So $a_n = (-3)^{n-3}$ for $n \geq 3$, and $a_0 = a_1 = a_2 = 0$.

e) We know what the coefficients are for the power series of $1/(1 - x^2)$, namely 0 for the odd ones and 1 for the even ones. The first three terms of this function force us to adjust the values of a_0, a_1 and a_2. So we have $a_0 = 7 + 1 = 8$, $a_1 = 3 + 0 = 3$, $a_2 = 1 + 1 = 2$, $a_n = 0$ for odd n greater than 2, and $a_n = 1$ for even n greater than 2.

f) Perhaps this is easiest to see if we write it out: $x^4(1 + x^4 + x^8 + x^{12} + \cdots) - x^3 - x^2 - x - 1 = x^4 + x^8 + x^{12} + \cdots - x^3 - x^2 - x - 1$. Therefore we have $a_n = 1$ if n is a positive multiple of 4; $a_n = -1$ if $n < 4$, and $a_n = 0$ otherwise.

g) We know that $x^2/(1-x)^2 = x^2 \sum_{n=0}^{\infty}(n+1)x^n = \sum_{n=0}^{\infty}(n+1)x^{n+2} = \sum_{n=2}^{\infty}(n-1)x^n$. Therefore $a_n = n-1$ for $n \geq 2$ and $a_0 = a_1 = 0$.

h) We know that $2e^{2x} = 2\sum_{n=0}^{\infty}(2x)^n/n! = \sum_{n=0}^{\infty}(2^{n+1}/n!)x^n$. Therefore $a_n = 2^{n+1}/n!$.

9. Different approaches are possible for obtaining these answers. One can use brute force algebra and just multiply everything out, either by hand or with computer algebra software such as *Maple*. One can view the problem as asking for the solution to a particular combinatorial problem and solve the problem by other means (e.g., listing all the possibilities). Or one can get a closed form expression for the coefficients, using the generating function theory developed in this section.

a) First we view this combinatorially. To obtain a term x^{10} when multiplying out these three factors, we could either take two x^5's and one x^0, or we could take two x^0's and one x^{10}. In each case there are $C(3,1)=3$ choices for the factor from which to pick the single value. Therefore the answer is $3+3=6$. Second, it is clear that we can view this problem as asking for the coefficient of x^2 in $(1+x+x^2+x^3+\cdots)^3$, since each x^5 in the original is playing the role of x here. Since $(1+x+x^2+x^3+\cdots)^3 = 1/(1-x)^3 = \sum_{n=0}^{\infty} C(n+2,2)x^n$, the answer is clearly $C(2+2,2)=C(4,2)=6$. A third way to get the answer is to ask *Maple* to compute $(1+x^5+x^{10})^3$ and look at the coefficient of x^{10}, which will turn out to be 6. Note that we don't have to go beyond x^{10} in each factor, because the higher terms can't contribute to an x^{10} term in the answer.

b) If we factor out x^3 from each factor, we can write this as $x^9(1+x+x^2+\cdots)^3$. Thus we are seeking the coefficient of x in $(1+x+x^2+\cdots)^3 = \sum_{n=0}^{\infty} C(n+2,2)x^n$, so the answer is $C(1+2,2)=3$. The other two methods explained in part **(a)** work here as well.

c) If we factor out as high a power of x from each factor as we can, then we can write this as

$$x^7(1+x+x^2)(1+x+x^2+x^3+x^4)(1+x+x^2+x^3+\cdots),$$

and so we seek the coefficient of x^3 in $(1+x+x^2)(1+x+x^2+x^3+x^4)(1+x+x^2+x^3+\cdots)$. By brute force we can list the nine ways to obtain x^3 in this product (where "ijk" means choose an x^i term from the first factor, an x^j term from the second factor, and an x^k term from the third factor): 003, 012, 021, 030, 102, 111, 120, 201, 210. If we want to do this more analytically, let us write our expression in closed form as

$$\frac{1-x^3}{1-x}\cdot\frac{1-x^5}{1-x}\cdot\frac{1}{1-x} = \frac{1-x^3-x^5+x^8}{(1-x)^3} = \frac{1}{(1-x)^3} - x^3\cdot\frac{1}{(1-x)^3} + \text{irrelevant terms}.$$

Now the coefficient of x^n in $1/(1-x)^3$ is $C(n+2,2)$. Furthermore, the coefficient of x^3 in this power series comes either from the coefficient of x^3 in the first term in the final expression displayed above, or from the coefficient of x^0 in the second factor of the second term of that expression. Therefore our answer is $C(3+2,2)-C(0+2,2)=10-1=9$.

d) Note that only even powers appear in the first and third factor, so to get x^{10} when we multiply this out, we can only choose the x^6 term in the second factor. But this would require terms from the first and third factors with a total exponent of 4, and clearly that is not possible. Therefore the desired coefficient is 0.

e) The easiest approach here might be brute force. Using the same notation as explained in part **(c)** above, the ways to get x^{10} are 046, 280, 406, 640, and (10)00. Therefore the answer is 5. We can check this with *Maple*. An analytic approach would be rather messy for this problem.

11. a) By Table 1 the coefficient of x^n in this power series is 2^n. Therefore the answer is $2^{10}=1024$.

b) By Table 1 the coefficient of x^n in this power series is $(-1)^n C(n+1,1)$. Therefore the answer is $(-1)^{10}C(10+1,1)=11$.

c) By Table 1 the coefficient of x^n in this power series is $C(n+2,2)$. Therefore the answer is $C(10+2,2)=66$.

d) By Table 1 the coefficient of x^n in this power series is $(-2)^n C(n+3,3)$. Therefore the answer is $(-2)^{10}C(10+3,3)=292{,}864$. Incidentally, *Maple* can do this kind of problem as well. Typing

```
series(1/(1 + 2 * x)^4, x = 0, 11);
```

will cause *Maple* to give the terms of the power series for this function, including all terms less than x^{11}. The output looks like

$$1-8x+40x^2-160x^3+560x^4-1792x^5+5376x^6-15360x^7+42240x^8-112640x^9+292864x^{10}+O(x^{11}).$$

(You might wonder why *Maple* says that the terms involving x^{11}, x^{12}, and so on are big-O of x^{11}. That seems backward! The reason is that one thinks of x as approaching 0 here, rather than infinity. Then, indeed, each term with a higher power of x (greater than 11) is smaller than x^{11}, up to a constant multiple.)

e) This is really asking for the coefficient of x^6 in $1/(1-3x)^3$. Following the same idea as in part **(d)**, we see that the answer is $3^6C(6+2,2) = 20{,}412$.

13. Each child will correspond to a factor in our generating function. We can give any number of balloons to the child, as long as it is at least 2; therefore the generating function for each child is $x^2+x^3+x^4+\cdots$. We want to find the coefficient of x^{10} in the expansion of $(x^2+x^3+x^4+\cdots)^4$. This function is the same as $x^8(1+x+x^2+x^3+\cdots)^4 = x^8/(1-x)^4$. Therefore we want the coefficient of x^2 in the generating function for $1/(1-x)^4$, which we know from Table 1 is $C(2+3,3) = 10$. Alternatively, to find the coefficient of x^2 in $(1+x+x^2+x^3+\cdots)^4$, we can multiply out $(1+x+x^2)^4$ (perhaps with a computer algebra package such as *Maple*), and the coefficient of x^2 turns out to be 10. Note that we truncated the series to be multiplied out, since terms higher than x^2 can't contribute to the x^2 term.

15. Each child will correspond to a factor in our generating function. We can give 1, 2, or 3 animals to the child; therefore the generating function for each child is $x+x^2+x^3$. We want to find the coefficient of x^{15} in the expansion of $(x+x^2+x^3)^6$. Factoring out an x from each term, we see that this is the same as the coefficient of x^9 in $(1+x+x^2)^6$. We can multiply this out (preferably with a computer algebra package such as *Maple*), and the coefficient of x^9 turns out to be 50. To solve it analytically, we write our generating function $(1+x+x^2)^6$ as

$$\left(\frac{1-x^3}{1-x}\right)^6 = \frac{1-6x^3+15x^6-20x^9+\text{higher order terms}}{(1-x)^6}.$$

There are four contributions to the coefficient of x^9, one for each listed term in the numerator, from the power series for $1/(1-x)^6$. Since the coefficient of x^n in $1/(1-x)^6$ is $C(n+5,5)$, our answer is $C(9+5,5) - 6C(6+5,5) + 15C(3+5,5) - 20C(0+5,5) = 2002 - 2772 + 840 - 20 = 50$.

17. The factor in the generating function for choosing the donuts for each policeman is $x^3+x^4+x^5+x^6+x^7$. Therefore the generating function for this problem is $(x^3+x^4+x^5+x^6+x^7)^4$. We want to find the coefficient of x^{25}, since we want 25 donuts in all. This is equivalent to finding the coefficient of x^{13} in $(1+x+x^2+x^3+x^4)^4$, since we can factor out $(x^3)^4 = x^{12}$. At this point, we could multiply it out (perhaps with *Maple*), and see that the desired coefficient is 20. Alternatively, we can write our generating function as

$$\left(\frac{1-x^5}{1-x}\right)^4 = \frac{1-4x^5+6x^{10}+\text{higher order terms}}{(1-x)^4}.$$

There are three contributions to the coefficient of x^{13}, one for each term in the numerator, from the power series for $1/(1-x)^4$. Since the coefficient of x^n in $1/(1-x)^4$ is $C(n+3,3)$, our answer is $C(13+3,3) - 4C(8+3,3) + 6C(3+3,3) = 560 - 660 + 120 = 20$.

19. We want the coefficient of x^k to be the number of ways to make change for k dollars. One-dollar bills contribute 1 each to the exponent of x. Thus we can model the choice of the number of one-dollar bills by the choice of a term from $1+x+x^2+x^3+\cdots$. Two-dollar bills contribute 2 each to the exponent of x. Thus we can model the choice of the number of two-dollar bills by the choice of a term from $1+x^2+x^4+x^6+\cdots$. Similarly, five-dollar bills contribute 5 each to the exponent of x, so we can model the choice of the number of five-dollar bills by the choice of a term from $1+x^5+x^{10}+x^{15}+\cdots$. Similar reasoning applies to ten-dollar bills. Thus the generating function is $f(x) = (1+x+x^2+x^3+\cdots)(1+x^2+x^4+x^6+\cdots)(1+x^5+x^{10}+x^{15}+\cdots)(1+x^{10}+x^{20}+x^{30}+\cdots)$, which can also be written (see Table 1) as

$$f(x) = \frac{1}{(1-x)(1-x^2)(1-x^5)(1-x^{10})}.$$

21. Let e_i, for $i = 1, 2, 3$, be the exponent of x taken from the ith factor in forming a term x^4 in the expansion. Thus $e_1 + e_2 + e_3 = 4$. The coefficient of x^4 is therefore the number of ways to solve this equation with nonnegative integers, which, from Section 5.5, is $C(3+4-1, 4) = C(6,4) = 15$.

23. a) The restriction on x_1 gives us the factor $x^2 + x^3 + x^4 + \cdots$. The restriction on x_2 gives us the factor $1 + x + x^2 + x^3$. The restriction on x_3 gives us the factor $x^2 + x^3 + x^4 + x^5$. Thus the answer is the product of these: $(x^2 + x^3 + x^4 + \cdots)(1 + x + x^2 + x^3)(x^2 + x^3 + x^4 + x^5)$. We can use algebra and Table 1 to rewrite this in closed form as $x^4(1 + x + x^2 + x^3)^2/(1 - x)$.

b) We want the coefficient of x^6 in this series, which is the same as the coefficient of x^2 in the series for

$$\frac{(1 + x + x^2 + x^3)^2}{1 - x} = \frac{1 + 2x + 3x^2 + \text{higher order terms}}{1 - x}.$$

Since the coefficient of x^n in $1/(1 - x)$ is 1, our answer is $1 + 2 + 3 = 6$.

25. This problem reinforces the point that "and" corresponds to multiplication and "or" corresponds to addition.

a) The only issue is how many stamps of each denomination we choose. The exponent on x will be the number of cents. So the generating function for choosing 3-cent stamps is $1 + x^3 + x^6 + x^9 + \cdots$, the generating function for 4-cent stamps is $1 + x^4 + x^8 + x^{12} + \cdots$, and the generating function for 20-cent stamps is $1 + x^{20} + x^{40} + x^{60} + \cdots$. In closed form this is $1/((1 - x^3)(1 - x^4)(1 - x^{20}))$. The coefficient of x^r gives the answer—the number of ways to choose stamps totaling r cents of postage.

b) Again the exponent on x will be the number of cents, but this time we paste one stamp at a time. For the first pasting, we can choose a 3-cent stamp, a 4-cent stamp, or a 20-cent stamp. Hence the generating function for the number of ways to paste one stamp is $x^3 + x^4 + x^{20}$. For the second pasting, we can make these same choices, so the generating function for the number of ways to paste two stamps is $(x^3 + x^4 + x^{20})^2$. In general, if we use n stamps, the generating function is $(x^3 + x^4 + x^{20})^n$. Since a pasting consists of a pasting of zero or more stamps, the entire generating function will be

$$\sum_{n=0}^{\infty} (x^3 + x^4 + x^{20})^n = \frac{1}{1 - x^3 - x^4 - x^{20}}.$$

c) We seek the coefficient of x^{46} in the power series for our answer to part **(a)**, $1/((1 - x^3)(1 - x^4)(1 - x^{20}))$. Other than working this out by brute force (enumerating the combinations), the best way to get the answer is probably asking *Maple* or another computer algebra package to multiply out these series. If we do so, the answer turns out to be 7. (The choices are $2 \cdot 20 + 2 \cdot 3$, $20 + 5 \cdot 4 + 2 \cdot 3$, $20 + 2 \cdot 4 + 6 \cdot 3$, $10 \cdot 4 + 2 \cdot 3$, $7 \cdot 4 + 6 \cdot 3$, $4 \cdot 4 + 10 \cdot 3$, and $1 \cdot 4 + 14 \cdot 3$.)

d) We seek the coefficient of x^{46} in the power series for our answer to part **(b)**, $1/(1 - x^3 - x^4 - x^{20})$. The best way to get the answer is probably asking *Maple* or another computer algebra package to find this power series using calculus. If we do so, the answer turns out to be 3224. Alternatively, for each of the seven combinations in our answer to part **(c)**, we can find the number of ordered arrangements, as in Section 5.5. Thus the answer is

$$\frac{4!}{2!2!} + \frac{8!}{1!5!2!} + \frac{9!}{1!2!6!} + \frac{12!}{10!2!} + \frac{13!}{7!6!} + \frac{14!}{4!10!} + \frac{15!}{1!14!} = 6 + 168 + 252 + 66 + 1716 + 1001 + 15 = 3224.$$

27. We will write down the generating function in each case and then use a computer algebra package to find the desired coefficients. As a check, one could carefully enumerate these by hand. In making change, one usually considers order irrelevant.

a) The generating function for the dimes is $1 + x^{10} + x^{20} + x^{30} + \cdots = 1/(1 - x^{10})$, and the generating function for the quarters is $1 + x^{25} + x^{50} + x^{75} + \cdots = 1/(1 - x^{25})$, so the generating function for the whole problem is $1/((1 - x^{10})(1 - x^{25}))$. The coefficient of x^k gives the number of ways to make change for k cents, so we seek the coefficient of x^{100}. If we ask a computer algebra system to find this coefficient (it uses calculus to get the power series), we find that the answer is 3. In fact, this is correct, since we can use four quarters, two quarters, or no quarters (and the number of dimes is uniquely determined by this choice).

b) This is identical to part **(a)** except for a factor for the nickels. Thus we seek the coefficient of x^{100} in $1/((1-x^5)(1-x^{10})(1-x^{25}))$, which turns out to be 29. (If we wanted to list these systematically, we could organize our work by the number of quarters, and within that by the number of dimes.)

c) This is identical to part **(a)** except for a factor for the pennies. Thus we seek the coefficient of x^{100} in $1/((1-x)(1-x^{10})(1-x^{25}))$, which turns out to be 29 again. (In retrospect, this is obvious. The only difference between parts **(b)** and **(c)** is that five pennies are substituted for each nickel.)

d) This is identical to part **(a)** except for factors for the pennies and nickels. Thus we seek the coefficient of x^{100} in $1/((1-x)(1-x^5)(1-x^{10})(1-x^{25}))$, which turns out to be 242.

29. We will write down the generating function in each case and then use a computer algebra package to find the desired coefficients. In making change, one usually considers order irrelevant.

a) The generating function for the \$10 bills is $1+x^{10}+x^{20}+x^{30}+\cdots = 1/(1-x^{10})$, the generating function for the \$20 bills is $1+x^{20}+x^{40}+x^{60}+\cdots = 1/(1-x^{20})$, and the generating function for the \$50 bills is $1+x^{50}+x^{100}+x^{150}+\cdots = 1/(1-x^{50})$, so the generating function for the whole problem is $1/((1-x^{10})(1-x^{20})(1-x^{50}))$. The coefficient of x^k gives the number of ways to make change for k dollars, so we seek the coefficient of x^{100}. If we ask a computer algebra system to find this coefficient (it uses calculus to get the power series), we find that the answer is 10. In fact, this is correct, since there is one way in which we can use two \$50 bills, three ways in which we use one \$50 bill (using either two, one, or no \$20 bills), and six ways to use no \$50 bills (using zero through five \$20's).

b) This is identical to part **(a)** except for a factor for the \$5 bills. Thus we seek the coefficient of x^{100} in $1/((1-x^5)(1-x^{10})(1-x^{20})(1-x^{50}))$, which turns out to be 49.

c) In part **(b)** we saw that the generating function for this problem is $1/((1-x^5)(1-x^{10})(1-x^{20})(1-x^{50}))$. If at least one of each bill must be used, let us assume that this $\$50+\$20+\$10+\$5=\$85$ has already been dispersed. Then we seek the coefficient of x^{15}. The computer algebra package tells us that the answer is 2, but it is trivial to see that there are only two ways to make \$15 with these bills.

d) This time the generating function is $(x^5+x^{10}+x^{15}+x^{20})(x^{10}+x^{20}+x^{30}+x^{40})(x^{20}+x^{40}+x^{60}+x^{80})$. When the computer multiplies this out, it tells us that the coefficient of x^{100} is 4, so that is the answer. (In retrospect, we see that the only solutions are $4\cdot\$20+1\cdot\$10+2\cdot\$5$, $3\cdot\$20+3\cdot\$10+2\cdot\$5$, $3\cdot\$20+2\cdot\$10+4\cdot\$5$, and $2\cdot\$20+4\cdot\$10+4\cdot\$5$.)

31. a) The terms involving a_0, a_1, and a_2 are missing; $G(x)-a_0-a_1x-a_2x^2 = a_3x^3+a_4x^4+\cdots$. That is the generating function for precisely the sequence we are given. Thus the answer is $G(x)-a_0-a_1x-a_2x^2$.

b) Every other term is missing, and the old coefficient of x^n is now the coefficient of x^{2n}. This suggests that maybe x^2 should be used in place of x. Indeed, this works; the answer is $G(x^2)=a_0+a_1x^2+a_2x^4+\cdots$.

c) If we want a_0 to be the coefficient of x^4 (and similarly for the other powers), we must throw in an extra factor. Thus the answer is $x^4G(x)$. Note that $x^4(a_0+a_1x+a_2x^2+\cdots)=a_0x^4+a_1x^5+a_2x^6+\cdots$.

d) Extra factors of 2 are applied to each term, with the power of 2 matching the subscript (which, of course, gives us the power of x). Thus the answer must be $G(2x)=a_0+a_1(2x)+a_2(2x)^2+a_3(2x)^3\cdots = a_0+2a_1x+4a_2x^2+8a_3x^3\cdots$.

e) Following the hint, we integrate $G(t)=\sum_{n=0}^{\infty}a_nt^n$ from 0 to x, to obtain $\int_0^x G(t)\,dt=\sum_{n=0}^{\infty}a_n\int_0^x t^n\,dt = \sum_{n=0}^{\infty}a_nx^{n+1}/(n+1)$. (If we had tried differentiating first, we'd see that that didn't work. It is a theorem of advanced calculus that it is legal to integrate inside the summation within the open interval of convergence.) This is the series

$$a_0x+\frac{a_1}{2}x^2+\frac{a_2}{3}x^3+\cdots,$$

precisely the sequence we are given (note that the constant term is 0). Thus $\int_0^x G(t)\,dt$ is the generating function for this sequence.

f) If we look at Theorem 1, it is not hard to see that the sequence shown here is precisely the coefficients of $G(x)\cdot(1+x+x^2+\cdots) = G(x)/(1-x)$.

33. This problem is like Example 16. First let $G(x)=\sum_{k=0}^{\infty} a_k x^k$. Then $xG(x)=\sum_{k=0}^{\infty} a_k x^{k+1} = \sum_{k=1}^{\infty} a_{k-1}x^k$ (by changing the name of the variable from k to $k+1$). Thus

$$G(x)-3xG(x)=\sum_{k=0}^{\infty} a_k x^k - \sum_{k=1}^{\infty} 3a_{k-1}x^k = a_0 + \sum_{k=1}^{\infty}(a_k - 3a_{k-1})x^k = a_0 + \sum_{k=1}^{\infty} 2x^k$$
$$= 1 + \frac{2}{1-x} - 2 = \frac{1+x}{1-x},$$

because of the given recurrence relation, the initial condition, and the fact from Table 1 that $\sum_{k=0}^{\infty} 2x^k = 2/(1-x)$. Thus $G(x)(1-3x) = (1+x)/(1-x)$, so $G(x) = (1+x)/((1-3x)(1-x))$. At this point we must use partial fractions to break up the denominator. Setting

$$\frac{1+x}{(1-3x)(1-x)} = \frac{A}{1-3x} + \frac{B}{1-x},$$

multiplying through by the common denominator, and equating coefficients, we find that $A=2$ and $B=-1$. Thus

$$G(x) = \frac{2}{1-3x} + \frac{-1}{1-x} = \sum_{k=0}^{\infty}(2\cdot 3^k - 1)x^k$$

(the last equality came from using Table 1). Therefore $a_k = 2\cdot 3^k - 1$.

35. Let $G(x)=\sum_{k=0}^{\infty} a_k x^k$. Then $xG(x)=\sum_{k=0}^{\infty} a_k x^{k+1} = \sum_{k=1}^{\infty} a_{k-1}x^k$ (by changing the name of the variable from k to $k+1$), and similarly $x^2G(x) = \sum_{k=0}^{\infty} a_k x^{k+2} = \sum_{k=2}^{\infty} a_{k-2}x^k$. Thus

$$G(x)-5xG(x)+6x^2G(x)=\sum_{k=0}^{\infty} a_k x^k - \sum_{k=1}^{\infty} 5a_{k-1}x^k + \sum_{k=2}^{\infty} 6a_{k-2}x^k = a_0 + a_1x - 5a_0x + \sum_{k=2}^{\infty} 0\cdot x^k = 6,$$

because of the given recurrence relation and the initial conditions. Thus $G(x)(1-5x+6x^2)=6$, so $G(x) = 6/((1-3x)(1-2x))$. At this point we must use partial fractions to break up the denominator. Setting

$$\frac{6}{(1-3x)(1-2x)} = \frac{A}{1-3x} + \frac{B}{1-2x},$$

multiplying through by the common denominator, and equating coefficients, we find that $A=18$ and $B=-12$. Thus

$$G(x) = \frac{18}{1-3x} + \frac{-12}{1-2x} = \sum_{k=0}^{\infty}(18\cdot 3^k - 12\cdot 2^k)x^k$$

(the last equality came from using Table 1). Therefore $a_k = 18\cdot 3^k - 12\cdot 2^k$. Incidentally, it would be wise to check our answers, either with a computer algebra package (see the solution to Exercise 37 for the syntax in *Maple*) or by computing the next term of the sequence from both the recurrence and the formula (here $a_2 = 114$ both ways).

37. Let $G(x)=\sum_{k=0}^{\infty} a_k x^k$. Then $xG(x)=\sum_{k=0}^{\infty} a_k x^{k+1} = \sum_{k=1}^{\infty} a_{k-1}x^k$ (by changing the name of the variable from k to $k+1$), and similarly $x^2G(x) = \sum_{k=0}^{\infty} a_k x^{k+2} = \sum_{k=2}^{\infty} a_{k-2}x^k$. Thus

$$G(x)-4xG(x)+4x^2G(x)=\sum_{k=0}^{\infty} a_k x^k - \sum_{k=1}^{\infty} 4a_{k-1}x^k + \sum_{k=2}^{\infty} 4a_{k-2}x^k = a_0 + a_1x - 4a_0x + \sum_{k=2}^{\infty} k^2\cdot x^k$$
$$= 2 - 3x + \frac{2}{(1-x)^3} - \frac{3}{(1-x)^2} + \frac{1}{1-x} - x$$
$$= 2 - 4x + \frac{2}{(1-x)^3} - \frac{3}{(1-x)^2} + \frac{1}{1-x},$$

because of the given recurrence relation, the initial conditions, Table 1, and a calculation of the generating function for $\{k^2\}$ (the last "$-x$" comes from the fact that the k^2 sum starts at 2). (To find the generating function for $\{k^2\}$, start with the fact that $1/(1-x)^3$ is the generating function for $\{C(k,2)=(k+2)(k+1)/2\}$, that $1/(1-x)^2$ is the generating function for $\{k+1\}$, and that $1/(1-x)$ is the generating function for $\{1\}$, and take an appropriate linear combination of these to get the generating function for $\{k^2\}$.) Thus

$$G(x)(1-4x+4x^2) = 2-4x+\frac{2}{(1-x)^3}-\frac{3}{(1-x)^2}+\frac{1}{1-x},$$

so

$$G(x) = \frac{2-4x}{(1-2x)^2}+\frac{2}{(1-2x)^2(1-x)^3}-\frac{3}{(1-2x)^2(1-x)^2}+\frac{1}{(1-2x)^2(1-x)}.$$

At this point we must use partial fractions to break up the denominators. Setting the previous expression equal to

$$\frac{A}{1-x}+\frac{B}{(1-x)^2}+\frac{C}{(1-x)^3}+\frac{D}{1-2x}+\frac{E}{(1-2x)^2},$$

multiplying through by the common denominator, and equating coefficients, we find (after a lot of algebra) that $A=13$, $B=5$, $C=2$, $D=-24$, and $E=6$. (Alternatively, one can ask *Maple* to produce the partial fraction decomposition, with the command

$$\texttt{convert}(\textit{expression}, \texttt{parfrac}, \texttt{x});$$

where the expression is $G(x)$.) Thus

$$\begin{aligned} G(x) &= \frac{13}{1-x}+\frac{5}{(1-x)^2}+\frac{2}{(1-x)^3}+\frac{-24}{1-2x}+\frac{6}{(1-2x)^2} \\ &= \sum_{k=0}^{\infty}\left(13+5(k+1)+2(k+2)(k+1)/2-24\cdot 2^k+6(k+1)2^k\right)x^k \end{aligned}$$

(from Table 1). Therefore $a_k = k^2+8k+20+(6k-18)2^k$. Incidentally, it would be most wise to check our answers, either with a computer algebra package, or by computing the next term of the sequence from both the recurrence and the formula (here $a_2 = 16$ both ways). The command in *Maple* for solving this recurrence is this:

$$\texttt{rsolve(\{a(k) = 4 * a(k-1) - 4 * a(k-2) + k\^{}2, a(0) = 2, a(1) = 5\}, a(k));}$$

39. In principle this exercise is similar to the examples and previous exercises. In fact, the algebra is quite a bit messier. We want to solve the recurrence relation $f_k = f_{k-1}+f_{k-2}$, with initial conditions $f_0 = 0$ and $f_1 = 1$. Let G be the generating function for f_k, so that $G(x) = \sum_{k=0}^{\infty} f_k x^k$. We look at $G(x)-xG(x)-x^2G(x)$ in order to take advantage of the recurrence relation:

$$\begin{aligned} G(x)-xG(x)-x^2G(x) &= \sum_{k=0}^{\infty} f_k x^k - \sum_{k=1}^{\infty} f_{k-1}x^k - \sum_{k=2}^{\infty} f_{k-2}x^k \\ &= f_0 + f_1x - f_0x + \sum_{k=2}^{\infty}(f_k - f_{k-1} - f_{k-2})x^k \\ &= 0 + x - 0 + 0 = x \end{aligned}$$

Thus G satisfies the equation

$$G(x) = \frac{x}{1-x-x^2}.$$

To write this more usefully, we need to use partial fractions. The roots of the denominator are $r_1 = (-1+\sqrt{5})/2$ and $r_2 = (-1-\sqrt{5})/2$. We want to find constants A and B such that

$$\frac{x}{1-x-x^2} = \frac{-x}{x^2+x-1} = \frac{A}{x-r_1}+\frac{B}{x-r_2}.$$

This means that A and B have to satisfy the simultaneous equations $A+B=-1$ and $r_2A+r_1B=0$ (multiply the last displayed equation through by the denominator and equate like powers of x). Solving, we obtain $A=(1-\sqrt{5})/(2\sqrt{5})$ and $B=(-1-\sqrt{5})/(2\sqrt{5})$. Now we have

$$\begin{aligned} G(x) &= \frac{A}{x-r_1}+\frac{B}{x-r_2} \\ &= \frac{-A}{r_1}\frac{1}{1-(x/r_1)}+\frac{-B}{r_2}\frac{1}{1-(x/r_2)} \\ &= \frac{-A}{r_1}\sum_{k=0}^{\infty}\left(\frac{1}{r_1}\right)^k x^k+\frac{-B}{r_2}\sum_{k=0}^{\infty}\left(\frac{1}{r_2}\right)^k x^k . \end{aligned}$$

Therefore

$$\begin{aligned} f_k &= -A\left(\frac{1}{r_1}\right)^{k+1}-B\left(\frac{1}{r_2}\right)^{k+1} \\ &= \frac{1}{\sqrt{5}}\left(\frac{2}{-1+\sqrt{5}}\right)^k-\frac{1}{\sqrt{5}}\left(\frac{2}{-1-\sqrt{5}}\right)^k . \end{aligned}$$

We can check our answer by computing the first few terms with a calculator, and indeed we find that $f_2=1$, $f_3=2$, $f_4=3$, $f_5=5$, and so on.

41. a) Let $G(x)=\sum_{n=0}^{\infty}C_nx^n$ be the generating function for the sequence of Catalan numbers. Then by Theorem 1 a change of variable in the middle, and the recurrence relation for the Catalan numbers,

$$G(x)^2=\sum_{n=0}^{\infty}\left(\sum_{k=0}^{n}C_kC_{n-k}\right)x^n=\sum_{n=1}^{\infty}\left(\sum_{k=0}^{n-1}C_kC_{n-1-k}\right)x^{n-1}=\sum_{n=1}^{\infty}C_nx^{n-1} .$$

So $xG(x)^2=\sum_{n=1}^{\infty}C_nx^n$. Therefore,

$$xG(x)^2-G(x)+1=\left(\sum_{n=1}^{\infty}C_nx^n\right)-\left(\sum_{n=0}^{\infty}C_nx^n\right)+1=-C_0+1=0 .$$

We now apply the quadratic formula to solve for $G(x)$:

$$G(x)=\frac{1\pm\sqrt{1-4x}}{2x}$$

We must decide whether to use the plus sign or the minus sign. If we use the plus sign, then trying to calculate $G(0)$, which, after all, is supposed to be C_0, gives us the undefined value $2/0$. Therefore we must use the minus sign, and indeed one can find using calculus that the indeterminate form $0/0$ equals 1 here, since $\lim_{x\to 0}G(x)=1$.

b) By Exercise 40 we know that

$$(1-4x)^{-1/2}=\sum_{n=0}^{\infty}\binom{2n}{n}x^n ,$$

so by integrating term by term (which is valid) we have

$$\int_0^x(1-4t)^{-1/2}\,dt=\frac{1-\sqrt{1-4x}}{2}=x\cdot\frac{1-\sqrt{1-4x}}{2x}=\sum_{n=0}^{\infty}\frac{1}{n+1}\binom{2n}{n}x^{n+1}=x\sum_{n=0}^{\infty}\frac{1}{n+1}\binom{2n}{n}x^n .$$

Since $G(x)=(1-\sqrt{1-4x})/(2x)$, equating coefficients of the power series tells us that $C_n=\frac{1}{n+1}\binom{2n}{n}$.

43. Following the hint, we note that $(1+x)^{m+n}=(1+x)^m(1+x)^n$. Then applying the Binomial Theorem, we have

$$\sum_{k=0}^{m+n}C(m+n,r)x^r=\sum_{r=0}^{m}C(m,r)x^r\cdot\sum_{r=0}^{m}C(n,r)x^r=\sum_{r=0}^{m+n}\left(\sum_{k=0}^{r}C(m,r-k)C(n,k)\right)x^r$$

by Theorem 1. Comparing coefficients gives us the desired identity.

45. We will make heavy use of the identity $e^x = \sum_{n=0}^{\infty} \frac{1}{n!}x^n$.

a) $\sum_{n=0}^{\infty} \frac{2}{n!}x^n = 2\sum_{n=0}^{\infty} \frac{1}{n!}x^n = 2e^x$

b) $\sum_{n=0}^{\infty} \frac{(-1)^n}{n!}x^n = \sum_{n=0}^{\infty} \frac{1}{n!}(-x)^n = e^{-x}$

c) $\sum_{n=0}^{\infty} \frac{3^n}{n!}x^n = \sum_{n=0}^{\infty} \frac{1}{n!}(3x)^n = e^{3x}$

d) This generating function can be obtained either with calculus or without. To do it without calculus, write

$$\begin{aligned}\sum_{n=0}^{\infty} \frac{n+1}{n!}x^n = \sum_{n=0}^{\infty} \frac{n}{n!}x^n + \sum_{n=0}^{\infty} \frac{1}{n!}x^n = \sum_{n=1}^{\infty} \frac{1}{(n-1)!}x^n + e^x &= x\sum_{n=1}^{\infty} \frac{1}{(n-1)!}x^{n-1} + e^x \\ &= x\sum_{n=0}^{\infty} \frac{1}{n!}x^n + e^x = xe^x + e^x \,.\end{aligned}$$

To do it with calculus, differentiate both sides of $xe^x = \sum_{n=0}^{\infty} \frac{x^{n+1}}{n!}$ to obtain $xe^x + e^x = \sum_{n=0}^{\infty} (n+1)\frac{x^n}{n!}$.

e) This generating function can be obtained either with calculus or without. To do it without calculus, write

$$\sum_{n=0}^{\infty} \frac{1}{n+1}\cdot\frac{x^n}{n!} = \sum_{n=0}^{\infty} \frac{x^n}{(n+1)!} = \frac{1}{x}\sum_{n=0}^{\infty} \frac{x^{n+1}}{(n+1)!} = \frac{1}{x}\sum_{n=1}^{\infty} \frac{x^n}{n!} = \frac{1}{x}(e^x - 1)\,.$$

To do it with calculus, integrate $e^t = \sum_{n=0}^{\infty} \frac{t^n}{n!}$ from 0 to x to obtain

$$e^x - 1 = \sum_{n=0}^{\infty} \frac{x^{n+1}}{n+1}\cdot\frac{1}{n!} = x\sum_{n=0}^{\infty} \frac{1}{(n+1)}\frac{x^n}{n!}\,.$$

Therefore $\sum_{n=0}^{\infty} \frac{1}{(n+1)}\frac{x^n}{n!} = (e^x - 1)/x$.

47. In many of these cases, it's a matter of plugging the exponent of e into the generating function for e^x. We let a_n denote the nth term of the sequence whose generating function is given.

a) The generating function is $e^{-x} = \sum_{n=0}^{\infty} \frac{(-x)^n}{n!} = \sum_{n=0}^{\infty} (-1)^n\frac{x^n}{n!}$, so the sequence is $a_n = (-1)^n$.

b) The generating function is $3e^{2x} = 3\sum_{n=0}^{\infty} \frac{(2x)^n}{n!} = \sum_{n=0}^{\infty} (3\cdot 2^n)\frac{x^n}{n!}$, so the sequence is $a_n = 3\cdot 2^n$.

c) The generating function is $e^{3x} - 3e^{2x} = \sum_{n=0}^{\infty} \frac{(3x)^n}{n!} - 3\sum_{n=0}^{\infty} \frac{(2x)^n}{n!} = \sum_{n=0}^{\infty} (3^n - 3\cdot 2^n)\frac{x^n}{n!}$, so the sequence is $a_n = 3^n - 3\cdot 2^n$.

d) The sequence whose exponential generating function is e^{-2x} is clearly $\{(-2)^n\}$, as in the previous parts of this exercise. Since

$$1 - x = \frac{1}{0!}x^0 + \frac{-1}{1!}x^1 + \sum_{n=2}^{\infty} \frac{0}{n!}x^n\,,$$

we know that $a_n = (-2)^n$ for $n \geq 2$, with $a_1 = (-2)^1 - 1 = -3$ and $a_0 = (-2)^0 + 1 = 2$.

e) We know that

$$\frac{1}{1-x} = \sum_{n=0}^{\infty} x^n = \sum_{n=0}^{\infty} \frac{n!}{n!}x^n\,,$$

so the sequence for which $1/(1-x)$ is the exponential generating function is $\{n!\}$. Combining this with the rest of the function (similar to previous parts of this exercise), we have $a_n = (-2)^n + n!$.

f) This is similar to part **(e)**. The three functions being added here are the exponential generating functions for $\{(-3)^n\}$, $(-1,-1,0,0,0,\ldots)$, and $\{n!\cdot 2^n\}$. Therefore $a_n = (-3)^n + n!\cdot 2^n$ for $n \geq 2$, with $a_0 = (-3)^0 - 1 + 0!\cdot 2^0 = 1$ and $a_1 = (-3)^1 - 1 + 1!\cdot 2^1 = -2$.

g) First we note that

$$e^{x^2} = \sum_{n=0}^{\infty} \frac{(x^2)^n}{n!} = 1 + \frac{x^2}{1!} + \frac{x^4}{2!} + \frac{x^6}{3!} + \cdots$$

$$= \frac{x^0}{0!}\cdot\frac{0!}{0!} + \frac{x^2}{2!}\cdot\frac{2!}{1!} + \frac{x^4}{4!}\cdot\frac{4!}{2!} + \frac{x^6}{6!}\cdot\frac{6!}{3!} + \cdots.$$

Therefore we see that $a_n = 0$ if n is odd, and $a_n = n!/(n/2)!$ if n is even.

49. a) Let a_n be the number of codewords of length n. There are 8^n strings of length n in all, and only those that contain an even number of 7's are code words. The initial condition is clearly $a_0 = 1$ (the empty string has an even number of 7's); if that seems too obscure, one can take $a_1 = 7$, since one of the eight strings of length 1 (namely the string 7) is disallowed. To write down a recurrence, we observe that a valid string of length n consists either of a valid string of length $n-1$ followed by a digit other than 7 (so that there will still be an even number of 7's), and there are $7a_{n-1}$ of these; or of an invalid string of length $n-1$ followed by a 7 (so that there will still be an even number of 7's), and there are $8^{n-1} - a_{n-1}$ of these. Putting these together, we have the recurrence relation $a_n = 7a_{n-1} + 8^{n-1} - a_{n-1} = 6a_{n-1} + 8^{n-1}$. For example, $a_2 = 6\cdot 7 + 8 = 50$.

b) Using the techniques of Section 7.2, we note that the general solution to the associated homogeneous recurrence relation is $a_n^{(h)} = \alpha 6^n$, and then we seek a particular solution of the form $a_n = c\cdot 8^n$. Plugging this into the recurrence relation, we have $c\cdot 8^n = 6c\cdot 8^{n-1} + 8^{n-1}$, which is easily solved to yield $c = \frac{1}{2}$ (first divide through by 8^{n-1}). This gives $a_n^{(p)} = \frac{1}{2}\cdot 8^n$, so the general solution is $a_n = \alpha 6^n + \frac{1}{2}\cdot 8^n$. Next we plug in the initial condition and easily find that $\alpha = \frac{1}{2}$. Therefore the solution is $a_n = (6^n + 8^n)/2$. We can check that this gives the correct answer when $n = 2$.

c) We proceed as in Example 17. Let $G(x) = \sum_{k=0}^{\infty} a_k x^k$. Then $xG(x) = \sum_{k=0}^{\infty} a_k x^{k+1} = \sum_{k=1}^{\infty} a_{k-1} x^k$ (by a change of variable). Thus

$$G(x) - 6xG(x) = \sum_{k=0}^{\infty} a_k x^k - \sum_{k=1}^{\infty} 6a_{k-1}x^k = a_0 + \sum_{k=1}^{\infty} (a_k - 6a_{k-1})x^k = 1 + \sum_{k=1}^{\infty} 8^{k-1}x^k$$

$$= 1 + x\sum_{k=1}^{\infty} 8^{k-1}x^{k-1} = 1 + x\sum_{k=0}^{\infty} 8^k x^k = 1 + x\cdot\frac{1}{1-8x} = \frac{1-7x}{1-8x}.$$

Thus $G(x)(1-6x) = (1-7x)/(1-8x)$, so $G(x) = (1-7x)/((1-6x)(1-8x))$. At this point we need to use partial fractions to break this up (see, for example, Exercise 35):

$$G(x) = \frac{1-7x}{(1-6x)(1-8x)} = \frac{1/2}{(1-6x)} + \frac{1/2}{(1-8x)}$$

Therefore, with the help of Table 1, $a_n = (6^n + 8^n)/2$, as we found in part **(b)**.

51. To form a partition of n, we must choose some 1's, some 2's, some 3's, and so on. The generating function for choosing 1's is

$$1 + x + x^2 + x^3 + \cdots = \frac{1}{1-x}$$

(the exponent gives the number so obtained). Similarly, the generating function for choosing 2's is

$$1 + x^2 + x^4 + x^6 + \cdots = \frac{1}{1-x^2}$$

(again the exponent gives the number so obtained). The other choices have analogous generating functions. Therefore the generating function for the entire problem, so that the coefficient of x^n will give $p(n)$, the number of partitions of n, is the infinite product

$$\frac{1}{1-x}\cdot\frac{1}{1-x^2}\cdot\frac{1}{1-x^3}\cdots.$$

53. This is similar to Exercise 51. Since all the parts have to be of different sizes, we can choose only no 1's or one 1; thus the generating function for choosing 1's is $1+x$ (the exponent gives the number so obtained). Similarly the generating function for choosing 2's is $1+x^2$, and analogously for higher choices. Therefore the generating function for the entire problem, so that the coefficient of x^n will give $p_d(n)$, the number of partitions of n into distinct-sized parts, is the infinite product

$$(1+x)(1+x^2)(1+x^3)\cdots.$$

55. It suffices to show that the generating functions obtained in Exercises 53 and 52 are equal, that is, that

$$(1+x)(1+x^2)(1+x^3)\cdots=\frac{1}{1-x}\cdot\frac{1}{1-x^3}\cdot\frac{1}{1-x^5}\cdots.$$

Assuming that the symbol-pushing we are about to do with infinite products is valid, we simply rewrite the left-hand side using the trivial algebraic identity $(1-x^{2r})/(1-x^r)=1+x^r$ and cancel common factors:

$$\begin{aligned}(1+x)(1+x^2)(1+x^3)\cdots&=\frac{1-x^2}{1-x}\cdot\frac{1-x^4}{1-x^2}\cdot\frac{1-x^6}{1-x^3}\cdot\frac{1-x^8}{1-x^4}\cdots\\&=\frac{1}{1-x}\cdot\frac{1}{1-x^3}\cdot\frac{1}{1-x^5}\cdots\end{aligned}$$

57. These follow fairly easily from the definitions.

a) $G_X(1)=\sum_{k=0}^{\infty}p(X=k)\cdot 1^k=\sum_{k=0}^{\infty}p(X=k)=1$, since X has to take on some nonnegative integer value. (That the sum of the probabilities is 1 is one of the axioms of a sample space; see Section 6.2.)

b) $G_X'(1)=\frac{d}{dx}\sum_{k=0}^{\infty}p(X=k)\cdot x^k\big|_{x=1}=\sum_{k=0}^{\infty}p(X=k)\cdot k\cdot x^{k-1}\big|_{x=1}=\sum_{k=0}^{\infty}p(X=k)\cdot k=E(X)$, by the definition of expected value from Section 6.4.

c) $G_X''(1)=\frac{d^2}{dx^2}\sum_{k=0}^{\infty}p(X=k)\cdot x^k\big|_{x=1}=\sum_{k=0}^{\infty}p(X=k)\cdot k(k-1)\cdot x^{k-2}\big|_{x=1}=\sum_{k=0}^{\infty}p(X=k)\cdot(k^2-k)=V(X)+E(X)^2-E(X)$, since, by Theorem 6 in Section 6.4, $V(X)=E(X^2)-E(X)^2$. Combining this with the result of part **(b)** gives the desired equality.

59. a) In order to have the mth success on the $(m+n)^{\text{th}}$ trial, where $n\geq 0$, we must have $m-1$ successes and n failures in any order among the first $m+n-1$ trials, followed by a success. The probability of each such ordered arrangement is clearly q^np^m, where p is the probability of success and $q=1-p$ is the probability of failure; and there are $C(n+m-1,n)$ such orders. Therefore $p(X=n)=C(n+m-1,n)q^np^m$. (This was Exercise 28 in the Supplementary Exercises for Chapter 6.) Therefore the probability generating function is

$$G(x)=\sum_{n=0}^{\infty}C(n+m-1,n)q^np^mx^n=p^m\sum_{n=0}^{\infty}C(n+m-1,n)(qx)^n=p^m\frac{1}{(1-qx)^m}$$

by Table 1.

b) By Exercise 57, $E(X)$ is the derivative of $G(x)$ at $x=1$. Here we have

$$G'(x)=\frac{p^mmq}{(1-qx)^{m+1}},\quad\text{so}\quad G'(1)=\frac{p^mmq}{(1-q)^{m+1}}=\frac{p^mmq}{p^{m+1}}=\frac{mq}{p}.$$

From the same exercise, we know that the variance is $G''(1) + G'(1) - G'(1)^2$; so we compute:

$$G''(x) = \frac{p^m m(m+1)q^2}{(1-qx)^{m+2}}, \quad \text{so} \quad G''(1) = \frac{p^m m(m+1)q^2}{(1-q)^{m+2}} = \frac{m(m+1)q^2}{p^2},$$

and therefore

$$V(X) = G''(1) + G'(1) - G'(1)^2 = \frac{m(m+1)q^2}{p^2} + \frac{mq}{p} - \left(\frac{mq}{p}\right)^2 = \frac{mq}{p^2}.$$

SECTION 7.5 Inclusion–Exclusion

Inclusion–exclusion is not a nice compact formula in practice, but it is often the best that is available. In Exercise 19, for example, the answer contains over 30 terms. The applications in this section are somewhat contrived, but much more interesting applications are presented in Section 7.6. The inclusion–exclusion principle in some sense gives a methodical way to apply common sense. Presumably anyone could solve a problem such as Exercise 9 by trial and error or other ad hoc techniques, given enough time; the inclusion–exclusion principle makes the solution straightforward. Be careful when using the inclusion–exclusion principle to get the signs right—some terms need to be subtracted and others need to be added. In general the sign changes when the size of the expression changes.

1. In all cases we use the fact that $|A_1 \cup A_2| = |A_1| + |A_2| - |A_1 \cap A_2| = 12 + 18 - |A_1 \cap A_2| = 30 - |A_1 \cap A_2|$.
a) Here $|A_1 \cap A_2| = 0$, so the answer is $30 - 0 = 30$.
b) This time we are told that $|A_1 \cap A_2| = 1$, so the answer is $30 - 1 = 29$.
c) This time we are told that $|A_1 \cap A_2| = 6$, so the answer is $30 - 6 = 24$.
d) If $A_1 \subseteq A_2$, then $A_1 \cap A_2 = A_1$, so $|A_1 \cap A_2| = |A_1| = 12$. Therefore the answer is $30 - 12 = 18$.

3. We may as well treat percentages as if they were cardinalities—as if the population were exactly 100. Let V be the set of households with television sets, and let P be the set of households with phones. Then we are given $|V| = 96$, $|P| = 98$, and $|V \cap P| = 95$. Therefore $|V \cup P| = 96 + 98 - 95 = 99$, so only 1% of the households have neither telephones nor televisions.

5. For all parts we need to use the formula $|A_1 \cup A_2 \cup A_3| = |A_1| + |A_2| + |A_3| - |A_1 \cap A_2| - |A_1 \cap A_3| - |A_2 \cap A_3| + |A_1 \cap A_2 \cap A_3|$.
a) If the sets are pairwise disjoint, then the cardinality of the union is the sum of the cardinalities, namely 300, since all but the first three terms on the right-hand side of the formula are equal to 0.
b) Using the formula, we have $100 + 100 + 100 - 50 - 50 - 50 + 0 = 150$.
c) Using the formula, we have $100 + 100 + 100 - 50 - 50 - 50 + 25 = 175$.
d) In this case the answer is obviously 100. By the formula, the cardinality of each set on the right-hand side is 100, so we can arrive at this answer through the computation $100+100+100-100-100-100+100 = 100$.

7. We need to use the formula $|P \cup F \cup C| = |P| + |F| + |C| - |P \cap F| - |P \cap C| - |F \cap C| + |P \cap F \cap C|$, where, for example, P is the set of students who have taken a course in Pascal. Thus we have $|P \cup F \cup C| = 1876 + 999 + 345 - 876 - 290 - 231 + 189 = 2012$. Therefore, since there are 2504 students altogether, we know that $2504 - 2012 = 492$ have taken none of these courses.

9. We need to use the inclusion–exclusion formula for four sets, C (the students taking calculus), D (the students taking discrete mathematics), S (those taking data structures), and L (those taking programming languages). The formula says $|C \cup D \cup S \cup L| = |C| + |D| + |S| + |L| - |C \cap D| - |C \cap S| - |C \cap L| - |D \cap S| - |D \cap L| - |S \cap L| + |C \cap D \cap S| + |C \cap D \cap L| + |C \cap S \cap L| + |D \cap S \cap L| - |C \cap D \cap S \cap L|$. Plugging the given information into this formula gives us a total of $507 + 292 + 312 + 344 - 0 - 14 - 213 - 211 - 43 - 0 + 0 + 0 + 0 + 0 - 0 = 974$.

11. There are clearly 50 odd positive integers not exceeding 100 (half of these 100 numbers are odd), and there are 10 squares (from 1^2 to 10^2). Furthermore, half of these squares are odd. Thus we compute the cardinality of the set in question to be $50 + 10 - 5 = 55$.

13. Let us count the strings that have 6 or more consecutive 0's. There are 4 strings that have 0's in the first six places, since there are $2 \cdot 2 = 4$ ways to specify the last two bits. Similarly, there are 4 strings that have 0's in bits 2 through 7, and there are 4 strings that have 0's in bits 3 through 8. We have overcounted, though. There are 2 strings that have 0's in bits 1 through 7 (the intersection of the first two sets mentioned above); 2 strings that have 0's in bits 2 through 8 (the intersection of the last two sets mentioned above); and 1 string that has 0's in all bits (the intersection of the first and last sets mentioned above). Moreover, there is 1 string with 0's in bits 1 through 8, the intersection of all three sets mentioned above. Putting this all together, we know that the number of strings with 6 consecutive 0's is $4+4+4-2-2-1+1 = 8$. Since there are $2^8 = 256$ strings in all, there must be $256 - 8 = 248$ that do not contain 6 consecutive 0's.

15. We need to use inclusion–exclusion with three sets. There are 7! permutations that begin 987, since there are 7 digits free to be permuted among the last 7 spaces (we are assuming that it is meant that the permutations are to start with 987 *in that order*, not with 897, for instance). Similarly, there are 8! permutations that have 45 in the fifth and sixth positions, and there are 7! that end with 123. (We assume that the intent is that these digits are to appear in the order given.) There are 5! permutations that begin with 987 and have 45 in the fifth and sixth positions; 4! that begin with 987 and end with 123; and 5! that have 45 in the fifth and sixth positions and end with 123. Finally, there are 2! permutations that begin with 987, have 45 in the fifth and sixth positions, and end with 123 (since only the 0 and the 6 are left to place). Therefore the total number of permutations meeting any of these conditions is $7! + 8! + 7! - 5! - 4! - 5! + 2! = 50{,}138$.

17. By inclusion–exclusion, the answer is $50+60+70+80-6\cdot5+4\cdot1-0 = 234$. Note that there were $C(4,2) = 6$ pairs to worry about (each with 5 elements in common) and $C(4,1) = 4$ triples to worry about (each with 1 element in common).

19. $|A_1 \cup A_2 \cup A_3 \cup A_4 \cup A_5| = |A_1|+|A_2|+|A_3|+|A_4|+|A_5|-|A_1\cap A_2|-|A_1\cap A_3|-|A_1\cap A_4|-|A_1\cap A_5|-|A_2\cap A_3|-|A_2\cap A_4|-|A_2\cap A_5|-|A_3\cap A_4|-|A_3\cap A_5|-|A_4\cap A_5|+|A_1\cap A_2\cap A_3|+|A_1\cap A_2\cap A_4|+|A_1\cap A_2\cap A_5|+|A_1\cap A_3\cap A_4|+|A_1\cap A_3\cap A_5|+|A_1\cap A_4\cap A_5|+|A_2\cap A_3\cap A_4|+|A_2\cap A_3\cap A_5|+|A_2\cap A_4\cap A_5|+|A_3\cap A_4\cap A_5|-|A_1\cap A_2\cap A_3\cap A_4|-|A_1\cap A_2\cap A_3\cap A_5|-|A_1\cap A_2\cap A_4\cap A_5|-|A_1\cap A_3\cap A_4\cap A_5|-|A_2\cap A_3\cap A_4\cap A_5|+|A_1\cap A_2\cap A_3\cap A_4\cap A_5|$

21. Since no three of the sets have a common intersection, we need only carry our expression out as far as pairs. Thus we have $|A_1 \cup A_2 \cup A_3 \cup A_4 \cup A_5 \cup A_6| = |A_1| + |A_2| + |A_3| + |A_4| + |A_5| + |A_6| - |A_1 \cap A_2| - |A_1 \cap A_3| - |A_1 \cap A_4| - |A_1 \cap A_5| - |A_1 \cap A_6| - |A_2 \cap A_3| - |A_2 \cap A_4| - |A_2 \cap A_5| - |A_2 \cap A_6| - |A_3 \cap A_4| - |A_3 \cap A_5| - |A_3 \cap A_6| - |A_4 \cap A_5| - |A_4 \cap A_6| - |A_5 \cap A_6|$.

23. Since the probability of an event (i.e., a set) E is proportional to the number of elements in the set E, this problem is just asking about cardinalities, and so inclusion–exclusion gives us the answer. Thus $p(E_1 \cup E_2 \cup E_3) = p(E_1) + p(E_2) + p(E_3) - p(E_1 \cap E_2) - p(E_1 \cap E_3) - p(E_2 \cap E_3) + p(E_1 \cap E_2 \cap E_3)$.

25. We can do this problem either by working directly with probabilities or by counting ways to satisfy the condition. We choose to do the former. First we need to determine the probability that all the numbers are odd. There are $C(100,4)$ ways to choose the numbers, and there are $C(50,4)$ ways to choose them all to be odd (since there are 50 odd numbers in the given interval). Therefore the probability that they are all odd is $C(50,4)/C(100,4)$. Similarly, since there are 33 multiples of 3 in the given interval, the probability of having all four numbers divisible by 3 is $C(33,4)/C(100,4)$. Finally, the probability that all four are divisible by 5 is $C(20,4)/C(100,4)$.

Next we need to know the probabilities that two of these events occur simultaneously. A number is both odd and divisible by 3 if and only if it is divisible by 3 but not by 6; therefore, since there are $\lfloor 100/6 \rfloor = 16$ multiples of 6 in the given interval, there are $33 - 16 = 17$ numbers that are both odd and divisible by 3. Thus the probability is $C(17,4)/C(100,4)$. Similarly there are 10 odd numbers divisible by 5, so the probability that all four numbers meet those conditions is $C(10,4)/C(100,4)$. Finally, the probability that all four numbers are divisible by both 3 and 5 is $C(6,4)/C(100,4)$, since there are only $\lfloor 100/15 \rfloor = 6$ such numbers.

Finally, the only numbers satisfying all three conditions are the odd multiplies of 15, namely 15, 45, and 75. Since there are only 3 such numbers, it is impossible that all chosen four numbers are divisible by 2, 3, and 5; in other words, the probability of that event is 0. We are now ready to apply the result of Exercise 23 (i.e., inclusion–exclusion viewed in terms of probabilities). We get

$$\frac{C(50,4)}{C(100,4)} + \frac{C(33,4)}{C(100,4)} + \frac{C(20,4)}{C(100,4)} - \frac{C(17,4)}{C(100,4)} - \frac{C(10,4)}{C(100,4)} - \frac{C(6,4)}{C(100,4)} + 0$$
$$= \frac{230300 + 40920 + 4845 - 2380 - 210 - 15}{3921225}$$
$$= \frac{273460}{3921225} = \frac{4972}{71295} \approx 0.0697 .$$

27. We are asked to write down inclusion–exclusion for five sets, just as in Exercise 19, except that intersections of more than three sets can be omitted. Furthermore, we are to use event notation, rather than set notation. Thus we have $p(E_1 \cup E_2 \cup E_3 \cup E_4 \cup E_5) = p(E_1) + p(E_2) + p(E_3) + p(E_4) + p(E_5) - p(E_1 \cap E_2) - p(E_1 \cap E_3) - p(E_1 \cap E_4) - p(E_1 \cap E_5) - p(E_2 \cap E_3) - p(E_2 \cap E_4) - p(E_2 \cap E_5) - p(E_3 \cap E_4) - p(E_3 \cap E_5) - p(E_4 \cap E_5) + p(E_1 \cap E_2 \cap E_3) + p(E_1 \cap E_2 \cap E_4) + p(E_1 \cap E_2 \cap E_5) + p(E_1 \cap E_3 \cap E_4) + p(E_1 \cap E_3 \cap E_5) + p(E_1 \cap E_4 \cap E_5) + p(E_2 \cap E_3 \cap E_4) + p(E_2 \cap E_3 \cap E_5) + p(E_2 \cap E_4 \cap E_5) + p(E_3 \cap E_4 \cap E_5)$.

29. We are simply asked to rephrase Theorem 1 in terms of probabilities of events. Thus we have

$$p(E_1 \cup E_2 \cup \cdots \cup E_n) = \sum_{1 \le i \le n} p(E_i) - \sum_{1 \le i < j \le n} p(E_i \cap E_j) + \sum_{1 \le i < j < k \le n} p(E_i \cap E_j \cap E_k)$$
$$\cdots + (-1)^{n+1} p(E_1 \cap E_2 \cap \cdots \cap E_n) .$$

SECTION 7.6 Applications of Inclusion–Exclusion

Some of these applications are quite subtle and not easy to understand on first encounter. They do point out the power of the inclusion–exclusion principle. Many of the exercises are closely tied to the examples, so additional study of the examples should be helpful in doing the exercises. It is often helpful, in organizing your work, to write down (in complete English sentences) exactly what the properties of interest are, calling them the P_i's. To find the number of elements lacking all the properties (as you need to do in Exercise 2, for example), use the formula above Example 1.

1. We want to find the number of apples that have neither of the properties of having worms or of having bruises. By inclusion–exclusion, we know that this is equal to the number of apples, minus the numbers with each of the properties, plus the number with both properties. In this case, this is $100 - 20 - 15 + 10 = 75$.

3. We need first to find the number of solutions with no restrictions. By the results of Section 5.5, there are $C(3+13-1, 13) = C(15,13) = C(15,2) = 105$. Next we need to find the number of solutions in which each restriction is violated. There are three variables that can fail to be less than 6, and the situation is symmetric, so the total number of solutions in which each restriction is violated is 3 times the number of solutions in which $x_1 \geq 6$. By the trick we used in Section 5.5, this is the same as the number of nonnegative integer solutions to $x_1' + x_2 + x_3 = 7$, where $x_1 = x_1' + 6$. This of course is $C(3+7-1, 7) = C(9,7) = C(9,2) = 36$. Therefore there are $3 \cdot 36 = 108$ solutions in which at least one of the restrictions is violated (with some of these counted more than once).

Next we need to find the number of solutions with at least two of the restrictions violated. There are $C(3,2) = 3$ ways to choose the pair to be violated, so the number we are seeking is 3 times the number of solutions in which $x_1 \geq 6$ and $x_2 \geq 6$. Again by the trick we used in Section 5.5, this is the same as the number of nonnegative integer solutions to $x_1' + x_2' + x_3 = 1$, where $x_1 = x_1' + 6$ and $x_2 = x_2' + 6$. This of course is $C(3+1-1, 1) = C(3,1) = 3$. Therefore there are $3 \cdot 3 = 9$ solutions in which two of the restrictions are violated. Finally, we note that there are no solutions in which all three of the solutions are violated, since if each of the variables is at least 6, then their sum is at least 18, and hence cannot equal 13.

Thus by inclusion–exclusion, we see that there are $105 - 108 + 9 = 6$ solutions to the original problem. (We can check this on an ad hoc basis. The only way the sum of three numbers, not as big as 6, can be 13, is to have either two 5's and one 3, or else one 5 and two 4's. There are three variables that can be the "odd man out" in each case, for a total of 6 solutions.)

5. We follow the procedure described in the text. There are 198 positive integers less than 200 and greater than 1. The ones that are not prime are divisible by at least one of the primes in the set $\{2, 3, 5, 7, 11, 13\}$. The number of integers in the given range divisible by the prime p is given by $\lfloor 199/p \rfloor$. Therefore we apply inclusion–exclusion and obtain the following number of integers from 2 to 199 that are not divisible by at least one of the primes in our set. (We have only listed those terms that contribute to the result, deleting all those that equal 0.)

$$\begin{aligned}
198 &- \left\lfloor \frac{199}{2} \right\rfloor - \left\lfloor \frac{199}{3} \right\rfloor - \left\lfloor \frac{199}{5} \right\rfloor - \left\lfloor \frac{199}{7} \right\rfloor - \left\lfloor \frac{199}{11} \right\rfloor - \left\lfloor \frac{199}{13} \right\rfloor + \left\lfloor \frac{199}{2 \cdot 3} \right\rfloor \\
&+ \left\lfloor \frac{199}{2 \cdot 5} \right\rfloor + \left\lfloor \frac{199}{2 \cdot 7} \right\rfloor + \left\lfloor \frac{199}{2 \cdot 11} \right\rfloor + \left\lfloor \frac{199}{2 \cdot 13} \right\rfloor + \left\lfloor \frac{199}{3 \cdot 5} \right\rfloor + \left\lfloor \frac{199}{3 \cdot 7} \right\rfloor + \left\lfloor \frac{199}{3 \cdot 11} \right\rfloor \\
&+ \left\lfloor \frac{199}{3 \cdot 13} \right\rfloor + \left\lfloor \frac{199}{5 \cdot 7} \right\rfloor + \left\lfloor \frac{199}{5 \cdot 11} \right\rfloor + \left\lfloor \frac{199}{5 \cdot 13} \right\rfloor + \left\lfloor \frac{199}{7 \cdot 11} \right\rfloor + \left\lfloor \frac{199}{7 \cdot 13} \right\rfloor + \left\lfloor \frac{199}{11 \cdot 13} \right\rfloor \\
&- \left\lfloor \frac{199}{2 \cdot 3 \cdot 5} \right\rfloor - \left\lfloor \frac{199}{2 \cdot 3 \cdot 7} \right\rfloor - \left\lfloor \frac{199}{2 \cdot 3 \cdot 11} \right\rfloor - \left\lfloor \frac{199}{2 \cdot 3 \cdot 13} \right\rfloor - \left\lfloor \frac{199}{2 \cdot 5 \cdot 7} \right\rfloor - \left\lfloor \frac{199}{2 \cdot 5 \cdot 11} \right\rfloor \\
&- \left\lfloor \frac{199}{2 \cdot 5 \cdot 13} \right\rfloor - \left\lfloor \frac{199}{2 \cdot 7 \cdot 11} \right\rfloor - \left\lfloor \frac{199}{2 \cdot 7 \cdot 13} \right\rfloor - \left\lfloor \frac{199}{3 \cdot 5 \cdot 7} \right\rfloor - \left\lfloor \frac{199}{3 \cdot 5 \cdot 11} \right\rfloor - \left\lfloor \frac{199}{3 \cdot 5 \cdot 13} \right\rfloor \\
&= 198 - 99 - 66 - 39 - 28 - 18 - 15 + 33 + 19 + 14 + 9 + 7 + 13 + 9 + 6 + 5 \\
&\quad + 5 + 3 + 3 + 2 + 2 + 1 - 6 - 4 - 3 - 2 - 2 - 1 - 1 - 1 - 1 - 1 - 1 - 1 = 40
\end{aligned}$$

These 40 numbers are therefore all prime, as are the 6 numbers in our set. Therefore there are exactly 46 prime numbers less than 200.

7. We can apply inclusion–exclusion if we reason as follows. First, we restrict ourselves to numbers greater than 1. If the number N is the power of an integer, then it is certainly the prime power of an integer, since if

$N = x^k$, where $k = mp$, with p prime, then $N = (x^m)^p$. Thus we need to count the number of perfect second powers, the number of perfect third powers, the number of perfect fifth powers, etc., less than 10,000. Let us first determine how many positive integers greater than 1 and less than 10,000 are the square of an integer. Since $\lfloor\sqrt{9999}\rfloor = 99$, there must be $99 - 1 = 98$ such numbers (namely 2^2 through 99^2). Similarly, since $\lfloor\sqrt[3]{9999}\rfloor - 1 = 20$, there are 20 cubes of integers less than 10,000. Similarly, there are $\lfloor\sqrt[5]{9999}\rfloor - 1 = 5$ fifth powers, $\lfloor\sqrt[7]{9999}\rfloor - 1 = 2$ seventh powers, $\lfloor\sqrt[11]{9999}\rfloor - 1 = 1$ eleventh power, and $\lfloor\sqrt[13]{9999}\rfloor - 1 = 1$ thirteenth power. There are no higher prime powers, since $\lfloor\sqrt[17]{9999}\rfloor - 1 = 0$ (and indeed, $2^{17} = 131072 > 9999$).

Now we need to account for the double counting. There are $\lfloor\sqrt[6]{9999}\rfloor - 1 = 3$ sixth powers, and these were counted as both second powers and third powers. Similarly, there is $\lfloor\sqrt[10]{9999}\rfloor - 1 = 1$ tenth power ($10 = 2\cdot 5$). These are the only two cases of double counting, since all other combinations give a count of 0. Therefore among the 9998 numbers from 2 to 9999, inclusive, we found that there were $98+20+5+2+1+1-3-1 = 123$ powers. Therefore there are $9998 - 123 = 9875$ numbers that are not powers.

9. This exercise is just asking for the number of onto functions from a set with 6 elements (the toys) to a set with 3 elements (the children), since each toy is assigned a unique child. By Theorem 1 there are $3^6 - C(3,1)2^6 + C(3,2)1^6 = 540$ such functions.

11. Here is one approach. Let us ignore temporarily the stipulation about the most difficult job being assigned to the best employee (we assume that this language uniquely specifies a job and an employee). Then we are looking for the number of onto functions from the set of 7 jobs to the set of 4 employees. By Theorem 1 there are $4^7 - C(4,1)3^7 + C(4,2)2^7 - C(4,1)1^7 = 8400$ such functions. Now by symmetry, in exactly one fourth of those assignments should the most difficult job be given to the best employee, as opposed to one of the other three employees. Therefore the answer is $8400/4 = 2100$.

13. We simply apply Theorem 2:

$$\begin{aligned} D_7 &= 7!\left(1 - \frac{1}{1!} + \frac{1}{2!} - \frac{1}{3!} + \frac{1}{4!} - \frac{1}{5!} + \frac{1}{6!} - \frac{1}{7!}\right) \\ &= 5040 - 5040 + 2520 - 840 + 210 - 42 + 7 - 1 = 1854 \end{aligned}$$

15. **a)** An arrangement in which no letter is put into the correct envelope is a derangement. There are by definition D_{100} derangements. Since there are $P(100,100) = 100!$ equally likely permutations altogether, the probability of a derangement is $D_{100}/100!$. Numerically, this is almost exactly equal to $1/e$, which is about 0.368.

b) We need to count the number of ways to put exactly one letter into the correct envelope. First, there are $C(100,1) = 100$ ways to choose the letter that is to be correctly stuffed. Then there are D_{99} ways to insert the remaining 99 letters so that none of them go into their correct envelopes. By the product rule, there are $100D_{99}$ such arrangements. As in part **(a)** the denominator is $P(100,100) = 100!$. Therefore the answer is $100D_{99}/100! = D_{99}/99!$. Again this is almost exactly $1/e \approx 0.368$.

c) This time, to count the number of ways that exactly 98 letters can be put into their correct envelopes, we need simply to choose the two letters that are to be misplaced, since there is only one way to misplace them. There are of course $C(100,2) = 4950$ ways to do this. As in part **(a)** the denominator is $P(100,100) = 100!$. Therefore the answer is $4950/100!$. This is substantially less than 10^{-100}, so for all practical purposes, the answer is 0.

d) There is no way that exactly 99 letters can be inserted into their correct envelopes, since as soon as 99 letters have been correctly inserted, there is only one envelope left for the remaining letter, and it is the correct one. Therefore the answer is exactly 0. (The probability of an event that cannot happen is 0.)

e) Only one of the 100! permutations is the correct stuffing, so the answer is $1/100!$. As in part **(c)** this is 0 for all practical purposes.

17. We can derive this answer by mimicking the derivation of the formula for the number of derangements, but worrying only about the even digits. There are 10! permutations altogether. Let e be one of the 5 even digits. The number of permutations in which e is in its original position is 9! (the other 9 digits need to be permuted). Therefore we need to subtract from 10! the $5 \cdot 9!$ ways in which the even digits can end up in their original positions. However, we have overcounted, since there are $C(5,2)8!$ ways in which 2 of the even digits can end up in their original positions, $C(5,3)7!$ ways in which 3 of them can, $C(5,4)6!$ ways in which 4 of them can, and $C(5,5)5!$ ways in which they can all retain their original positions. Applying inclusion–exclusion, we therefore have the answer

$$10! - 5 \cdot 9! + 10 \cdot 8! - 10 \cdot 7! + 5 \cdot 6! - 5! = 2{,}170{,}680\,.$$

19. We want to show that $D_n - nD_{n-1} = (-1)^n$. We will use an iterative approach, taking advantage of the result of Exercise 18, which can be rewritten algebraically as $D_k - kD_{k-1} = -\big(D_{k-1} - (k-1)D_{k-2}\big)$ for all $k \geq 2$. We have

$$\begin{aligned} D_n - nD_{n-1} &= -\big(D_{n-1} - (n-1)D_{n-2}\big) \\ &= -\big(-(D_{n-2} - (n-2)D_{n-3})\big) \\ &= (-1)^2\big(D_{n-2} - (n-2)D_{n-3}\big) \\ &\;\;\vdots \\ &= (-1)^{n-2}(D_2 - 2D_1) \\ &= (-1)^n \end{aligned}$$

since $D_2 = 1$ and $D_1 = 0$, and since $(-1)^{n-2} = (-1)^n$.

21. We can solve this problem by looking at the explicit formula we have for D_n from Theorem 2 (multiplying through by $n!$):

$$D_n = n! - n! + \frac{n!}{2} - \frac{n!}{3!} + \cdots + (-1)^{n-1}\frac{n!}{(n-1)!} + (-1)^n\frac{n!}{n!}$$

Now all of these terms are even except possibly for the last two, since (after being reduced to natural numbers) they all contain the factors n and $n-1$, at least one of which must be even. Therefore to determine whether D_n is even or odd, we need only look at these last two terms, which are $\pm n \mp 1$. If n is even, then this difference is odd; but if n is odd, then this difference is even. Therefore D_n is even precisely when n is odd.

23. Recall that $\phi(n)$, for a positive integer $n > 1$, denotes the number of positive integers less than (or, vacuously, equal to) n and relatively prime to n (in other words, that have no common prime factors with n). We will derive a formula for $\phi(n)$ using inclusion–exclusion. We are given that the prime factorization of n is $n = p_1^{a_1}p_2^{a_2}\cdots p_m^{a_m}$. Let P_i be the property that a positive integer less than or equal to n has p_i as a factor. Then $\phi(n)$ is precisely the number of positive integers less than or equal to n that have none of the properties P_i. By the alternative form of the principle of inclusion–exclusion, we have the following formula for this quantity:

$$\begin{aligned} N(P_1'P_2'\cdots P_m') = n - \sum_{1\leq i\leq m} N(P_i) + \sum_{1\leq i<j\leq m} N(P_iP_j) - \sum_{1\leq i<j<k\leq m} N(P_iP_jP_k) \\ + \cdots + (-1)^m N(P_1P_2\cdots P_m) \end{aligned}$$

Our only remaining task is to find a formula for each of these sums. This is not hard. First $N(P_i)$, the number of positive integers less than or equal to n divisible by p_i, is equal to n/p_i, just as in the discussion of the sieve of Eratosthenes (we need no floor function symbols since n/p_i is necessarily an integer). Similarly, $N(P_iP_j)$, the number of positive integers less than or equal to n divisible by both p_i and p_j, i.e., by the product p_ip_j, is equal to $n/(p_ip_j)$, and so on. Making these substitutions, we can rewrite the formula displayed above as

$$N(P_1'P_2'\cdots P_m') = n - \sum_{1\le i\le m} \frac{n}{p_i} + \sum_{1\le i<j\le m} \frac{n}{p_ip_j} - \sum_{1\le i<j<k\le m} \frac{n}{p_ip_jp_k} + \cdots + (-1)^m \frac{n}{p_1p_2\cdots p_m}.$$

This formula can be written in a more useful form. If we factor out the n from every term, then it is not hard to see that what remains is the product $(1-1/p_1)(1-1/p_2)\cdots(1-1/p_m)$. Therefore our answer is

$$n\prod_{i=1}^{m}\left(1-\frac{1}{p_i}\right).$$

25. A permutation meeting these conditions must be a derangement of 123 followed by a derangement of 456 in positions 4, 5, and 6. Since there are $D_3 = 2$ derangements of the first 3 elements to choose from for the first half of our permutation and $D_3 = 2$ derangements of the last 3 elements to choose from for the second half, there are, by the product rule, $2\cdot 2 = 4$ derangements satisfying the given conditions. Indeed, these 4 derangements are 231564, 231645, 312564, and 312645.

27. Let P_i be the property that a function from a set with m elements to a set with n elements does not have the ith element of the codomain included in its range. We want to compute $N(P_1'P_2'\cdots P_n')$. In order to use the principle of inclusion–exclusion we need to determine $\sum N(P_i)$, $\sum N(P_iP_j)$, etc. By the product rule, there are n^m functions from the set with m elements to the set with n elements. If we want the function not to have the ith element of the codomain in its range, then there are only $n-1$ choices at each stage, rather than n, to assign to each element of the domain; therefore $N(P_i) = (n-1)^m$, for each i. Furthermore, there are $C(n,1)$ different i's. Therefore $\sum N(P_i) = C(n,1)(n-1)^m$. Similarly, to compute $\sum N(P_iP_j)$, we note that there are $C(n,2)$ ways to specify i and j, and that once we have determined which 2 elements are to be omitted from the codomain, there are $(n-2)^m$ different functions with this smaller codomain. Therefore $\sum N(P_iP_j) = C(n,2)(n-2)^m$. We continue in this way, until finally we need to find $N(P_1P_2\cdots P_n)$, which is clearly equal to 0, since the function must have at least one element in its range. The formula given in the statement of Theorem 1 therefore follows from the inclusion–exclusion principle.

GUIDE TO REVIEW QUESTIONS FOR CHAPTER 7

1. **a)** See pp. 449–450. **b)** $\$1{,}000{,}000\cdot 1.09^n$

2. See Example 4 in Section 7.1.

3. See Example 5 in Section 7.1.

4. **a)** See Example 6 in Section 7.1 (interchange the roles of 0 and 1). **b)** See Exercise 27 in Section 7.1.

5. an equation of the form $a_n = c_1a_{n-1} + c_2a_{n-2} + \cdots + c_ka_{n-k}$

6. **a)** See Theorem 1 and Example 3 in Section 7.2 if the roots of the characteristic equation are distinct; otherwise see Theorem 2 and Example 5.

b) The characteristic equation is $r^2 - 13r + 22 = 0$, leading to roots 2 and 11. This gives the general solution $a_n = \alpha_1 2^n + \alpha_2 11^n$. Substituting in the initial conditions gives $\alpha_1 = 2$ and $\alpha_2 = 1$. Therefore the solution is $a_n = 2^{n+1} + 11^n$.

c) The characteristic equation is $r^2 - 14r + 49 = 0$, leading to the repeated root 7. This gives the general solution $a_n = \alpha_1 7^n + \alpha_2 n7^n$. Substituting in the initial conditions gives $\alpha_1 = 3$ and $\alpha_2 = 2$. Therefore the solution is $a_n = (3+2n)7^n$.

7. a) See p. 476. The exact solution is $f(b^k) = a^k f(1) + \sum_{j=0}^{k-1} a^j g(b^{k-j})$. **b)** 1442

8. a) See Example 1 in Section 7.3. **b)** $O(\log n)$

9. a) See p. 502. **b)** See pp. 501–502.

c) Note that the number of integers not exceeding 1000 that are divisible by a and b is $\lfloor 1000/\text{lcm}(a,b) \rfloor$. Thus the answer is

$$\left\lfloor \frac{1000}{6} \right\rfloor + \left\lfloor \frac{1000}{10} \right\rfloor + \left\lfloor \frac{1000}{15} \right\rfloor - \left\lfloor \frac{1000}{\text{lcm}(6,10)} \right\rfloor - \left\lfloor \frac{1000}{\text{lcm}(6,15)} \right\rfloor - \left\lfloor \frac{1000}{\text{lcm}(10,15)} \right\rfloor + \left\lfloor \frac{1000}{\text{lcm}(6,10,15)} \right\rfloor$$
$$= 166 + 100 + 66 - 33 - 33 - 33 + 33 = 266\,.$$

This is similar to (but slightly harder than) the discussion on p. 507.

d) For a similar problem, see Example 1 in Section 7.6. The solution is $C(4+22-1,22) - C(4+14-1,14) - C(4+16-1,16) - C(4+17-1,17) + C(4+8-1,8) + C(4+9-1,9) + C(4+11-1,11) - C(4+3-1,3)$.

10. a) See Example 5 in Section 7.5. **b)** $4 \cdot 25 - 6 \cdot 5 + 4 \cdot 2 - 1 = 77$

11. See Theorem 1 in Section 7.5.

12. See p. 509.

13. a) Count the number of onto functions from an m-set to an n-set, using Theorem 1 in Section 7.6; see Example 3 in Section 7.6.

b) $3^7 - 3 \cdot 2^7 + 3 \cdot 1^7 = 1806$

14. See the discussion of the Sieve of Eratosthenes on p. 507.

15. a) See p. 510.

b) Think of the hats permuted among the heads (which are the positions for the objects being permuted).

c) See Theorem 2 in Section 7.6.

SUPPLEMENTARY EXERCISES FOR CHAPTER 7

1. Let L_n be the number of chain letters sent at the nth stage.

a) Since each person receiving a letter sends it to 4 new people, there will be 4 times as many letters sent at the nth stage as were sent at the $(n-1)^{\text{st}}$ stage. Therefore the recurrence relation is $L_n = 4L_{n-1}$.

b) The initial condition is that at the first stage 40 letters are sent (each of the original 10 people sent it to 4 others), i.e., $L_1 = 40$.

c) We need to solve this recurrence relation. We do so easily by iteration, since $L_n = 4L_{n-1} = 4^2 L_{n-2} = \cdots = 4^{n-1} L_1 = 4^{n-1} \cdot 40$, or more simply $L_n = 10 \cdot 4^n$.

3. Let M_n be the amount of money (in dollars) that the government prints in the nth hour.

a) According to the given information, the amount of money printed in the nth hour is \$10,000 in \$1 bills, \$20,000 in \$5 bills, \$30,000 in \$10 bills, \$50,000 in \$20 bills, and \$50,000 in \$50 bills, for a total of \$160,000. Therefore our recurrence relation is $M_n = M_{n-1} + 160000$.

b) Since 1000 of each bill was produced in the first hour, we know that $M_1 = 1000(1+5+10+20+50+100) = 186000$.

c) We solve the recurrence relation by iteration:

$$\begin{aligned} M_n &= 160000 + M_{n-1} \\ &= 160000 + 160000 + M_{n-2} = 2 \cdot 160000 + M_{n-2} \\ &\vdots \\ &= (n-1) \cdot 160000 + M_1 \\ &= 160000(n-1) + 186000 = 160000n + 26000 \end{aligned}$$

d) Let T_n be the total amount of money produced in the first n hours. Then $T_n = T_{n-1} + M_n$, since the total amount of money produced in the first n hours is the same as the total amount of money produced in the first $n-1$ hours, plus the amount of money produced in the nth hour. Thus, from our result in part **(c)**, the recurrence relation is $T_n = T_{n-1} + 160000n + 26000$, with initial condition $T_0 = 0$ (no money is produced in 0 hours).

e) We solve the recurrence relation from part **(d)** by iteration:

$$\begin{aligned} T_n &= 26000 + 160000n + T_{n-1} \\ &= 26000 + 160000n + 26000 + 160000(n-1) + T_{n-2} \\ &= 2 \cdot 26000 + \big(n + (n-1)\big) \cdot 160000 + T_{n-2} \\ &\vdots \\ &= n \cdot 26000 + 160000 \cdot \big(n + (n-1) + \cdots + 1\big) + T_0 \\ &= 26000n + 160000 \cdot \frac{n(n+1)}{2} = 80000n^2 + 106000n \end{aligned}$$

5. This problem is similar to Exercise 35 in Section 7.1. Let m_n be the number of messages that can be sent in n microseconds.

a) A message must begin with either the two-microsecond signal or the three-microsecond signal. If it begins with the two-microsecond signal, then the rest of the message is of length $n-2$; if it begins with the three-microsecond signal, then the message continues as a message of length $n-3$. Therefore the recurrence relation is $m_n = m_{n-2} + m_{n-3}$.

b) We need initial conditions for $n = 0$, 1, and 2, since the recurrence relation has degree 3. Clearly $m_0 = 1$, since the empty message is the one and only message of length 0. Also $m_1 = 0$, since every nonempty message contains at least one signal, and the shortest signal has length 2. Finally $m_2 = 1$, since there is only one message of length 2, namely the one that uses one of the shorter signals and none of the longer signals.

c) There are two approaches here. One is to solve the recurrence relation, using the methods of Section 7.2. Unfortunately, the characteristic equation is $r^3 - r - 1 = 0$, and it has no rational roots. It is possible to find real roots, but the formula for solving third degree equations is messy, and the algebra in completing the solution this way would not be pleasant. (Alternatively, one could get approximations to the roots, then get approximations to the coefficients in the solution, plug in $n = 12$, and round to the nearest integer; again the calculation involved would be unpleasant.)

The other approach is simply to use the recurrence relation to compute m_3, m_4, ..., m_{12}. First $m_3 = m_1 + m_0 = 0 + 1 = 1$; then $m_4 = m_2 + m_1 = 1 + 0 = 1$, then $m_5 = m_3 + m_2 = 1 + 1 = 2$, and so on. Starting with m_6, the sequence continues 2, 3, 4, 5, 7, 9, 12. Therefore there are 12 different messages that can be sent in exactly 12 microseconds. (If we wanted to find the number of nonempty messages that could be sent in at most 12 microseconds—which is certainly one interpretation of the question—then we would add m_1 through m_{12}, obtaining 47 as our answer.)

7. The recurrence relation found in Exercise 6 is of degree 10, namely $a_n = a_{n-4} + a_{n-6} + a_{n-10}$. It needs 10 initial conditions, namely $a_0 = 1$, $a_1 = a_2 = a_3 = a_5 = a_7 = a_9 = 0$, and $a_4 = a_6 = a_8 = 1$.

a) $a_{12} = a_8 + a_6 + a_2 = 1 + 1 + 0 = 2$ (Indeed, the 2 ways to affix 12 cents postage is either to use 3 4-cent stamps or to use 2 6-cent stamps.)

b) First we need to compute $a_{10} = a_6 + a_4 + a_0 = 1 + 1 + 1 = 3$. Then $a_{14} = a_{10} + a_8 + a_4 = 3 + 1 + 1 = 5$.

c) We use the results of previous parts here: $a_{18} = a_{14} + a_{12} + a_8 = 5 + 2 + 1 = 8$.

d) First we need to compute $a_{16} = a_{12} + a_{10} + a_6 = 2 + 3 + 1 = 6$. Using this (and previous parts), we have $a_{22} = a_{18} + a_{16} + a_{12} = 8 + 6 + 2 = 16$.

9. Following the hint, let $b_n = \log a_n$ (remember that we mean log base 2). Then using the property that the log of a quotient is the difference of the logs and the log of a power is the multiple of the log, we take the logarithm of both sides of the recurrence relation for a_n to obtain $b_n = 2b_{n-1} - b_{n-2}$. The initial conditions translate into $b_0 = \log a_0 = \log 1 = 0$ and $b_1 = \log a_1 = \log 2 = 1$. Thus we have transformed our problem into a linear, homogeneous, second degree recurrence relation with constant coefficients.

To solve $b_n = 2b_{n-1} - b_{n-2}$, we form the characteristic equation $r^2 - 2r + 1 = 0$, which has the repeated root $r = 1$. By Theorem 2 in Section 7.2, the general solution is $b_n = \alpha_1 1^n + \alpha_2 n 1^n = \alpha_1 + \alpha_2 n$. Plugging in the initial conditions gives the equations $\alpha_1 = 0$ and $\alpha_1 + \alpha_2 = 1$, whence $\alpha_2 = 1$. Therefore the solution is $b_n = n$. Finally, $b_n = \log a_n$ implies that $a_n = 2^{b_n}$. Therefore our solution to the original problem is $a_n = 2^n$.

11. The characteristic equation of the associated homogeneous equation is $r^3 - 3r^2 + 3r - 1 = 0$. This factors as $(r-1)^3 = 0$, so there is only one root, 1, and its multiplicity is 3. Therefore the general solution is $a_n^{(h)} = \alpha + \beta n + \gamma n^2$. Since the nonhomogeneous term is 1, Theorem 6 in Section 7.2 tells us to look for a particular solution of the form $a_n = c \cdot n^3$. Plugging this into the recurrence gives $c \cdot n^3 = 3c(n-1)^3 - 3c(n-2)^3 + (n-3)^3 + 1$. Simplifying this by multiplying it out and collecting like powers of n gives us $6c = 1$ (all the other terms cancel out), so $c = 1/6$. Thus $a_n^{(p)} = n^3/6$. Plugging in the initial conditions to the general solution $a_n = \alpha + \beta n + \gamma n^2 + n^3/6$ gives us $2 = \alpha$, $4 = \alpha + \beta + \gamma + 1/6$, and $8 = \alpha + 2\beta + 4\gamma + 4/3$. Solving yields $\alpha = 2$, $\beta = 4/3$, and $\gamma_3 = 1/2$. Therefore the solution is $a_n = 2 + 4n/3 + n^2/2 + n^3/6$. As a check we can compute a_3 both from the recurrence and from the formula, and we get 15 in both cases.

13. One way to approach this problem is by temporarily using three variables. We assume that rabbits are born at the beginning of the month. Let a_n be the number of pairs of $\frac{1}{2}$-month-old rabbits present in the middle of the nth month, let b_n be the number of pairs of $1\frac{1}{2}$-month-old rabbits present in the middle of the nth month, and let c_n be the number of pairs of $2\frac{1}{2}$-month-old rabbits present in the middle of the nth month. All the older rabbits have left the island, by the conditions of the exercise. Let us see how each of these depends on previous values. First note that $b_n = a_{n-1}$, since these rabbits are one month older. Similarly $c_n = b_{n-1}$. Combining these two equations gives $c_n = a_{n-2}$. Finally, $a_n = b_{n-1} + c_{n-1}$, since newborns come from these two groups of rabbits. Writing this last equation totally in terms of a_n (using the previous equations) gives $a_n = a_{n-2} + a_{n-3}$.

Now we are interested in $T_n = a_n + b_n + c_n$, the total number of pairs of rabbits in the middle of the nth month. Since we have seen that the sequences $\{b_n\}$ and $\{c_n\}$ are the same as the sequence $\{a_n\}$, just shifted by one or two months, they must satisfy the same recurrence relation, so we have $b_n = b_{n-2} + b_{n-3}$ and $c_n = c_{n-2} + c_{n-3}$. If we add these three recurrence relations, we obtain $T_n = T_{n-2} + T_{n-3}$. We can take as the initial conditions $T_1 = T_2 = 1$ and $T_3 = 2$.

(We are not asked to solve this recurrence relation, and fortunately so. The characteristic equation, $r^3 - r - 1 = 0$ has no nice roots—one is irrational and two are complex. The roots are distinct, however, so let us call them r_1, r_2, and r_3. Then the general solution to the recurrence relation is $T_n = \alpha_1 r_1^n + \alpha_2 r_2^n + \alpha_3 r_3^n$.

We could in principle determine the values of the α's by plugging in the initial conditions, thereby obtaining an explicit solution. We will not do this.)

15. We use Theorem 2 in Section 7.3, with $a = 3$, $b = 5$, $c = 2$ and $d = 4$. Since $a < b^d$, we have that $f(n)$ is $O(n^d) = O(n^4)$.

17. In the algorithm in Exercise 16, we need 2 comparisons to determine the largest and second largest elements of the sequence, knowing the largest and second largest elements of the first half and the second half. Thus letting $f(n)$ be the number of comparisons needed for a list with n elements, and assuming that n is even, we have $f(n) = 2f(n/2) + 2$. Now by Theorem 2 in Section 7.3, with $a = 2$, $b = 2$, $c = 2$ and $d = 0$, we know that $f(n)$ is $O(n^{\log_b a}) = O(n^1) = O(n)$.

19. First we have to find Δa_n. By definition we have $\Delta a_n = a_{n+1} - a_n = 3(n+1)^3 + (n+1) + 2 - (3n^3 + n + 2) = 9n^2 + 9n + 4$.

a) By definition $\Delta^2 a_n = \Delta a_{n+1} - \Delta a_n = 9(n+1)^2 + 9(n+1) + 4 - (9n^2 + 9n + 4) = 18n + 18$.

b) By definition $\Delta^3 a_n = \Delta^2 a_{n+1} - \Delta^2 a_n = 18(n+1) + 18 - (18n + 18) = 18$.

c) By definition $\Delta^4 a_n = \Delta^3 a_{n+1} - \Delta^3 a_n = 18 - 18 = 0$.

21. We apply the definition, starting with the right-hand side:

$$\begin{aligned} a_{n+1}(\Delta b_n) + b_n(\Delta a_n) &= a_{n+1}(b_{n+1} - b_n) + b_n(a_{n+1} - a_n) \\ &= a_{n+1}b_{n+1} - a_n b_n \quad \text{(by algebra)} \\ &= \Delta(a_n b_n) \quad \text{(by definition)} \end{aligned}$$

23. a) Let $G(x) = \sum_{n=0}^{\infty} a_n x^n$. Then $G'(x) = \sum_{n=1}^{\infty} n a_n x^{n-1} = \sum_{n=0}^{\infty} (n+1)a_{n+1}x^n$. Therefore

$$G'(x) - G(x) = \sum_{n=0}^{\infty} ((n+1)a_{n+1} - a_n)x^n = \sum_{n=0}^{\infty} \frac{x^n}{n!} = e^x ,$$

as desired. That $G(0) = a_0 = 1$ is given.

b) We compute the indicated derivative:

$$(e^{-x}G(x))' = e^{-x}G'(x) - e^{-x}G(x) = e^{-x}(G'(x) - G(x)) = e^{-x} \cdot e^x = 1$$

This means that $e^{-x}G(x)$ is x plus a constant, say $x + c$. So $G(x) = xe^x + ce^x$. Plugging in the initial condition shows that $c = 1$, and we are done.

c) We work with the generating function for the exponential function:

$$G(x) = \sum_{n=0}^{\infty} \frac{x^{n+1}}{n!} + \sum_{n=0}^{\infty} \frac{x^n}{n!} = \sum_{n=1}^{\infty} \frac{x^n}{(n-1)!} + \sum_{n=0}^{\infty} \frac{x^n}{n!}$$

Therefore $a_n = 1/(n-1)! + 1/n!$ for all $n \geq 1$ (and $a_0 = 1$). As a check we can compute the first few terms of the sequence both from this solution and from the recurrence, and in each case we find the sequence $a_0 = 1$, $a_1 = 2$, $a_2 = 3/2$, $a_3 = 2/3$, $a_4 = 5/24$, ...

25. Let H, C, and S stand for the sets of farms that have horses, cows, and sheep, respectively. We are told that $|H \cup C \cup S| = 323$, $|H| = 224$, $|C| = 85$, $|S| = 57$, and $|H \cap C \cap S| = 18$. We are asked to find $|H \cap C| + |H \cap S| + |C \cap S| - 3|H \cap C \cap S|$ (the reason for the subtraction is that the indicated sum counts the farms with all three animals 3 times, and we wish to count it no times). By the principle of inclusion–exclusion we know that $|H \cup C \cup S| = |H| + |C| + |S| - |H \cap C| - |H \cap S| - |C \cap S| + |H \cap C \cap S|$. Solving for the expression we are interested in, we get $|H \cap C| + |H \cap S| + |C \cap S| - 3|H \cap C \cap S| = |H| + |C| + |S| - |H \cup C \cup S| - 2|H \cap C \cap S| = 224 + 85 + 57 - 323 - 2 \cdot 18 = 7$. Thus 7 farms have exactly two of the three types of animals.

27. We apply the principle of inclusion–exclusion: $|AM \cup PM \cup OR \cup CS| = 23 + 17 + 44 + 63 - 5 - 8 - 4 - 6 - 5 - 14 + 2 + 2 + 1 + 1 - 1 = 110$.

29. Since the largest possible value for $x_1 + x_2 + x_3$ under these constraints is $5+9+4 = 18$, there are no solutions to the given equation.

31. **a)** We solve this problem in the same manner as we solved Exercise 7 in Section 7.6. As explained in our solution there, we need only look at prime powers. Let us restrict ourselves to integers greater than 1, and add 1 at the end. There are $\lfloor\sqrt{199}\rfloor - 1 = 13$ perfect second powers in the given range, namely 2^2 through 14^2. There are $\lfloor\sqrt[3]{199}\rfloor - 1 = 4$ perfect third powers, $\lfloor\sqrt[5]{199}\rfloor - 1 = 1$ perfect fifth power, and $\lfloor\sqrt[7]{199}\rfloor - 1 = 1$ perfect seventh power. Furthermore, there is $\lfloor\sqrt[6]{199}\rfloor - 1 = 1$ perfect sixth power, which is both a perfect square and a perfect cube. Therefore by inclusion–exclusion, the number of numbers between 2 and 199 inclusive that are powers greater than the first power of an integer is $13 + 4 + 1 + 1 - 1 = 18$; adding on the number 1 itself (since $1 = 1^2$), we get the answer 19.

b) We saw in Exercise 5 in Section 7.6 that there are 46 primes less than 200, and we just saw above that there are 19 powers. Since these two sets are disjoint, we just add the cardinalities, obtaining $19 + 46 = 65$.

c) Solving this problem is like counting prime numbers, except that the squares of primes play the role of the primes themselves. The squares of primes relevant to the problem are 4, 9, 25, 49, 121, and 169. The number of positive integers less than 200 divisible by p^2 is $\lfloor 199/p^2 \rfloor$. There is overcounting, however, since a number divisible by a number like $36 = 2^2 \cdot 3^2$ is counted in both $\lfloor 199/2^2 \rfloor$ and $\lfloor 199/3^2 \rfloor$; hence we need to subtract $\lfloor 199/6^2 \rfloor$. The number of numbers divisible by squares of primes is therefore

$$\left\lfloor\frac{199}{2^2}\right\rfloor + \left\lfloor\frac{199}{3^2}\right\rfloor + \left\lfloor\frac{199}{5^2}\right\rfloor + \left\lfloor\frac{199}{7^2}\right\rfloor + \left\lfloor\frac{199}{11^2}\right\rfloor + \left\lfloor\frac{199}{13^2}\right\rfloor - \left\lfloor\frac{199}{6^2}\right\rfloor - \left\lfloor\frac{199}{10^2}\right\rfloor - \left\lfloor\frac{199}{14^2}\right\rfloor,$$

which is just $49 + 22 + 7 + 4 + 1 + 1 - 5 - 1 - 1 = 77$. Therefore there are $199 - 77 = 122$ positive integers less than 200 that are not divisible by the square of an integer greater than 1.

d) This is similar to part **(c)**, with cubes in place of squares. Reasoning the same way, we get

$$199 - \left(\left\lfloor\frac{199}{2^3}\right\rfloor + \left\lfloor\frac{199}{3^3}\right\rfloor + \left\lfloor\frac{199}{5^3}\right\rfloor\right) = 199 - (24 + 7 + 1) = 167.$$

e) For each set of three prime numbers $\{p, q, r\}$, the number of positive integers less than 200 divisible by p, q, and r is given by $\lfloor 200/(pqr) \rfloor$. There is no overcounting to worry about in this problem, since no number less than 200 is divisible by four primes (the smallest such number is $2 \cdot 3 \cdot 5 \cdot 7 = 210$). Therefore the number of positive integers less than 200 divisible by three primes is the sum of $\lfloor 200/(pqr) \rfloor$ over all triples of distinct primes whose product is at most 200. A tedious listing shows that there are 19 such triples, and when we form the sum we get 31. Therefore there are $199 - 31 = 168$ positive integers less than 200 that are not divisible by three or more primes.

33. There are n ways to choose which person is to receive the correct hat, and there are D_{n-1} ways to have the remaining hats returned totally incorrectly (where D_{n-1} is the number of derangements of $n-1$ objects). On the other hand there are $n!$ possible ways to return the hats. Therefore the probability is $nD_{n-1}/n! = D_{n-1}/(n-1)!$. Note that this happens to be the same as the probability that none of $n-1$ people is given the correct hat; therefore it is approximately $1/e \approx 0.368$ for large n.

35. There are $2^6 = 64$ bit strings of length 6. We need to find the number that contain at least 4 1's. The number that contain exactly i 1's is $C(6, i)$, since such a string is determined by choosing i of the 6 positions to contain the 1's. Therefore there are $C(6,4)+C(6,5)+C(6,6) = C(6,2)+C(6,1)+C(6,0) = 15+6+1 = 22$ strings with at least 4 1's. Hence the probability in question is $22/64 = 11/32$.

WRITING PROJECTS FOR CHAPTER 7

Books and articles indicated by bracketed symbols below are listed near the end of this manual. You should also read the general comments and advice you will find there about researching and writing these essays.

1. Obviously you will need to find a translated version if you want to read what Fibonacci actually said. The search technique of gradually working your way backwards usually works: If you can't find what you want in the place you start (here, for example, maybe with a standard mathematics history textbook), then search the references provided by that work, then check the references in the references, and so on backwards.

2. Articles and books at all levels have dealt with this subject. You might find something in, say, *Scientific American* (which is indexed in hard-copy and electronic versions of *Readers' Guide*); you might find some articles in materials for high-school students (see, for example, *Mathematics Teacher*, a magazine for high school teachers); and just browsing through the mathematics section of a public library or popular bookstore might yield something on this topic. Talk to someone who teaches a "math for poets" course at your school (i.e., a course with almost no mathematical prerequisite that deals with appreciating the beauty or applications of mathematics); some of the textbooks for that kind of course have material on this topic, as well as references to other sources of information.

3. Paul Stockmeyer, a professor at the College of William and Mary, describes the Tower of Hanoi problem and its variations as his "main professional hobby." Consult his website (`http://www.cs.wm.edu/~pkstoc`) for some of his papers on the subject, as well as some great links.

4. There are many articles about the Catalan numbers, as well as treatments in textbooks. One article to start with might be [HiPe2].

5. Obviously, consult the reference mentioned in that exercise.

6. Andrzej Pelc has written several papers on this topic; search for his Web page.

7. Numerous websites discuss this, and several of them have working demos, as well. Search for the key words.

8. See [Ro3] for a brief introduction. An extensive discussion, as well as a long list of references, can be found in [Gu].

9. One book on sieve methods is [HaRi].

10. See the article [Da2]. There is also relevant material in the chapter on arrangements with forbidden positions in [MiRo].

11. A wonderful book on generating functions is [Wi2], and [GrKn] also has a lot of relevant material. There are sections on generating functions in [Gr2] as well as any of the advanced combinatorics books mentioned in the general suggestions at the back of this *Guide*.

12. See advanced combinatorics texts, such as [Ro1] or [Tu1]. For another writing project, find out about George Polyá, a fascinating figure in 20th century mathematics and mathematics education.

13. See [BoDo], which should also have pointers to historical sources.

14. Advanced combinatorics texts, such as [Br2] or [Ro1], discuss this topic.

CHAPTER 8
Relations

SECTION 8.1 Relations and Their Properties

This chapter is one of the most important in the book. Many structures in mathematics and computer science are formulated in terms of relations. Not only is the terminology worth learning, but the experience to be gained by working with various relations will prepare the student for the more advanced structures that he or she is sure to encounter in future work.

This section gives the basic terminology, especially the important notions of reflexivity, symmetry, antisymmetry, and transitivity. If we are given a relation as a set of ordered pairs, then reflexivity is easy to check for: we make sure that each element is related to itself. Symmetry is also fairly easy to test for: we make sure that no pair (a,b) is in the relation without its opposite (b,a) being present as well. To check for antisymmetry we make sure that no pair (a,b) with $a \neq b$ and its opposite are both in the relation. In other words, at most one of (a,b) and (b,a) is in the relation if $a \neq b$. Transitivity is much harder to verify, since there are many triples of elements to check. A common mistake to try to avoid is forgetting that a transitive relation that has pairs (a,b) and (b,a) must also include (a,a) and (b,b).

More importantly, we can be given a relation as a rule as to when elements are related. Exercises 4–7 are particularly useful in helping to understand the notions of reflexivity, symmetry, antisymmetry, and transitivity for relations given in this manner. Here you have to ask yourself the appropriate questions in order to determine whether the properties hold. Is every element related to itself? If so, the relation is reflexive. Are the roles of the variables in the definition interchangeable? If so, then the relation is symmetric. Does the definition preclude two different elements from each being related to the other? If so, then the relation is antisymmetric. Does the fact that one element is related to a second, which is in turn related to a third, mean that the first is related to the third? If so, then the relation is transitive.

In general, try to think of a relation in these two ways at the same time: as a set of ordered pairs and as a propositional function describing a relationship among objects.

1. In each case, we need to find all the pairs (a,b) with $a \in A$ and $b \in B$ such that the condition is satisfied. This is straightforward.

a) $\{(0,0),(1,1),(2,2),(3,3)\}$ **b)** $\{(1,3),(2,2),(3,1),(4,0)\}$

c) $\{(1,0),(2,0),(2,1),(3,0),(3,1),(3,2),(4,0),(4,1),(4,2),(4,3)\}$

d) Recall that $a \mid b$ means that b is a multiple of a (a is not allowed to be 0). Thus the answer is $\{(1,0),(1,1),(1,2),(1,3),(2,0),(2,2),(3,0),(3,3),(4,0)\}$.

e) We need to look for pairs whose greatest common divisor is 1—in other words, pairs that are relatively prime. Thus the answer is $\{(0,1),(1,0),(1,1),(1,2),(1,3),(2,1),(2,3),(3,1),(3,2),(4,1),(4,3)\}$.

f) There are not very many pairs of numbers (by definition only positive integers are considered) whose least common multiple is 2: only 1 and 2, and 2 and 2. Thus the answer is $\{(1,2),(2,1),(2,2)\}$.

3. a) This relation is not reflexive, since it does not include, for instance $(1,1)$. It is not symmetric, since it includes, for instance, $(2,4)$ but not $(4,2)$. It is not antisymmetric since it includes both $(2,3)$ and $(3,2)$, but $2 \neq 3$. It is transitive. To see this we have to check that *whenever* it includes (a,b) and (b,c), then it also includes (a,c). We can ignore the element 1 since it never appears. If (a,b) is in this relation, then by inspection we see that a must be either 2 or 3. But $(2,c)$ and $(3,c)$ are in the relation for all $c \neq 1$; thus (a,c) has to be in this relation whenever (a,b) and (b,c) are. This proves that the relation is transitive. Note that it is very tedious to prove transitivity for an arbitrary list of ordered pairs.

b) This relation is reflexive, since all the pairs $(1,1)$, $(2,2)$, $(3,3)$, and $(4,4)$ are in it. It is clearly symmetric, the only nontrivial case to note being that both $(1,2)$ and $(2,1)$ are in the relation. It is not antisymmetric because both $(1,2)$ and $(2,1)$ are in the relation. It is transitive; the only nontrivial cases to note are that since both $(1,2)$ and $(2,1)$ are in the relation, we need to have (and do have) both $(1,1)$ and $(2,2)$ included as well.

c) This relation clearly is not reflexive and clearly is symmetric. It is not antisymmetric since both $(2,4)$ and $(4,2)$ are in the relation. It is not transitive, since although $(2,4)$ and $(4,2)$ are in the relation, $(2,2)$ is not.

d) This relation is clearly not reflexive. It is not symmetric, since, for instance, $(1,2)$ is included but $(2,1)$ is not. It is antisymmetric, since there are no cases of (a,b) and (b,a) both being in the relation. It is not transitive, since although $(1,2)$ and $(2,3)$ are in the relation, $(1,3)$ is not.

e) This relation is clearly reflexive and symmetric. It is trivially antisymmetric since there are no pairs (a,b) in the relation with $a \neq b$. It is trivially transitive, since the only time the hypothesis $(a,b) \in R \wedge (b,c) \in R$ is met is when $a = b = c$.

f) This relation is clearly not reflexive. The presence of $(1,4)$ and absence of $(4,1)$ shows that it is not symmetric. The presence of both $(1,3)$ and $(3,1)$ shows that it is not antisymmetric. It is not transitive; both $(2,3)$ and $(3,1)$ are in the relation, but $(2,1)$ is not, for instance.

5. Recall the definitions: R is reflexive if $(a,a) \in R$ for all a; R is symmetric if $(a,b) \in R$ always implies $(b,a) \in R$; R is antisymmetric if $(a,b) \in R$ and $(b,a) \in R$ always implies $a = b$; and R is transitive if $(a,b) \in R$ and $(b,c) \in R$ always implies $(a,c) \in R$.

a) It is tautological that everyone who has visited Web page a has also visited Web page a, so R is reflexive. It is not symmetric, because there surely are Web pages a and b such that the set of people who visited a is a proper subset of the set of people who visited b (for example, the only link to page a may be on page b). Whether R is antisymmetric in truth is hard to say, but it is certainly conceivable that there are two different Web pages a and b that have had exactly the same set of visitors. In this case, $(a,b) \in R$ and $(b,a) \in R$, so R is not antisymmetric. Finally, R is transitive: if everyone who has visited a has also visited b, and everyone who has visited b has also visited c, then clearly everyone who has visited a has also visited c.

b) This relation is not reflexive, because for any page a that has links on it, $(a,a) \notin R$. The definition of R is symmetric in its very statement, so R is clearly symmetric. Also R is certainly not antisymmetric, because there surely are two different Web pages a and b out there that have no common links found on them. Finally, R is not transitive, because the two Web pages just mentioned, assuming they have links at all, give an example of the failure of the definition: $(a,b) \in R$ and $(b,a) \in R$, but $(a,a) \notin R$.

c) This relation is not reflexive, because for any page a that has no links on it, $(a,a) \notin R$. The definition of R is symmetric in its very statement, so R is clearly symmetric. Also R is certainly not antisymmetric, because there surely are two different Web pages a and b out there that have a common link found on them. Finally, R is surely not transitive. Page a might have only one link (say to this textbook), page c might have only one link different from this (say to the Erdős Number Project), and page b may have only the two links mentioned in this sentence. Then $(a,b) \in R$ and $(b,c) \in R$, but $(a,c) \notin R$.

d) This relation is probably not reflexive, because there probably exist Web pages out there with no links at all to them (for example, when they are in the process of being written and tested); for any such page a we have $(a,a) \notin R$. The definition of R is symmetric in its very statement, so R is clearly symmetric. Also R is certainly not antisymmetric, because there surely are two different Web pages a and b out there that are referenced by some third page. Finally, R is surely not transitive. Page a might have only one page that links to it, page c might also have only one page, different from this, that links to it, and page b may be cited on both of these two pages. Then there would be no page that includes links to both pages a and c, so we have $(a,b) \in R$ and $(b,c) \in R$, but $(a,c) \notin R$.

7. a) This relation is not reflexive since it is not the case that $1 \neq 1$, for instance. It is symmetric: if $x \neq y$, then of course $y \neq x$. It is not antisymmetric, since, for instance, $1 \neq 2$ and also $2 \neq 1$. It is not transitive, since $1 \neq 2$ and $2 \neq 1$, for instance, but it is not the case that $1 \neq 1$.

b) This relation is not reflexive, since $(0,0)$ is not included. It is symmetric, because the commutative property of multiplication tells us that $xy = yx$, so that one of these quantities is greater than or equal to 1 if and only if the other is. It is not antisymmetric, since, for instance, $(2,3)$ and $(3,2)$ are both included. It is transitive. To see this, note that the relation holds between x and y if and only if either x and y are both positive or x and y are both negative. So assume that (a,b) and (b,c) are both in the relation. There are two cases, nearly identical. If a is positive, then so is b, since $(a,b) \in R$; therefore so is c, since $(b,c) \in R$, and hence $(a,c) \in R$. If a is negative, then so is b, since $(a,b) \in R$; therefore so is c, since $(b,c) \in R$, and hence $(a,c) \in R$.

c) This relation is not reflexive, since $(1,1)$ is not included, for instance. It is symmetric; the equation $x = y - 1$ is equivalent to the equation $y = x + 1$, which is the same as the equation $x = y + 1$ with the roles of x and y reversed. (A more formal proof of symmetry would be by cases. If x is related to y then either $x = y + 1$ or $x = y - 1$. In the former case, $y = x - 1$, so y is related to x; in the latter case $y = x + 1$, so y is related to x.) It is not antisymmetric, since, for instance, both $(1,2)$ and $(2,1)$ are in the relation. It is not transitive, since, for instance, although both $(1,2)$ and $(2,1)$ are in the relation, $(1,1)$ is not.

d) Recall that $x \equiv y \pmod 7$ means that $x - y$ is a multiple of 7, i.e., that $x - y = 7t$ for some integer t. This relation is reflexive, since $x - x = 7 \cdot 0$ for all x. It is symmetric, since if $x \equiv y \pmod 7$, then $x - y = 7t$ for some t; therefore $y - x = 7(-t)$, so $y \equiv x \pmod 7$. It is not antisymmetric, since, for instance, we have both $2 \equiv 9$ and $9 \equiv 2 \pmod 7$. It is transitive. Suppose $x \equiv y$ and $y \equiv z \pmod 7$. This means that $x - y = 7s$ and $y - z = 7t$ for some integers s and t. The trick is to add these two equations and note that the y disappears; we get $x - z = 7s + 7t = 7(s+t)$. By definition, this means that $x \equiv z \pmod 7$, as desired.

e) Every number is a multiple of itself (namely 1 times itself), so this relation is reflexive. (There is one bit of controversy here; we assume that 0 is to be considered a multiple of 0, even though we do not consider that 0 is a divisor of 0.) It is clearly not symmetric, since, for instance, 6 is a multiple of 2, but 2 is not a multiple of 6. The relation is not antisymmetric either; we have that 2 is a multiple of -2, for instance, and -2 is a multiple of 2, but $2 \neq -2$. The relation is transitive, however. If x is a multiple of y (say $x = ty$), and y is a multiple of z (say $y = sz$), then we have $x = t(sz) = (ts)z$, so we know that x is a multiple of z.

f) This relation is reflexive, since a and a are either both negative or both nonnegative. It is clearly symmetric from its form. It is not antisymmetric, since 5 is related to 6 and 6 is related to 5, but $5 \neq 6$. Finally, it is transitive, since if a is related to b and b is related to c, then all three of them must be negative, or all three must be nonnegative.

g) This relation is not reflexive, since, for instance, $17 \neq 17^2$. It is not symmetric, since although $289 = 17^2$, it is not the case that $17 = 289^2$. To see whether it is antisymmetric, suppose that we have both (x,y) and (y,x) in the relation. Then $x = y^2$ and $y = x^2$. To solve this system of equations, plug the second into the first, to obtain $x = x^4$, which is equivalent to $x - x^4 = 0$. The left-hand side factors as $x(1-x^3) = x(1-x)(1+x+x^2)$, so the solutions for x are 0 and 1 (and a pair of irrelevant complex numbers). The corresponding solutions for y are therefore also 0 and 1. Thus the only time we have both $x = y^2$ and $y = x^2$ is when $x = y$; this means that the relation is antisymmetric. It is not transitive, since, for example, $16 = 4^2$ and $4 = 2^2$, but $16 \neq 2^2$.

h) This relation is not reflexive, since, for instance, $17 \not\geq 17^2$. It is not symmetric, since although $289 \geq 17^2$, it is not the case that $17 \geq 289^2$. To see whether it is antisymmetric, we assume that both (x,y) and (y,x)

are in the relation. Then $x \geq y^2$ and $y \geq x^2$. Since both sides of the second inequality are nonnegative, we can square both sides to get $y^2 \geq x^4$. Combining this with the first inequality, we have $x \geq x^4$, which is equivalent to $x - x^4 \geq 0$. The left-hand side factors as $x(1-x^3) = x(1-x)(1+x+x^2)$. The last factor is always positive, so we can divide the original inequality by it to obtain the equivalent inequality $x(1-x) \geq 0$. Now if $x > 1$ or $x < 0$, then the factors have different signs, so the inequality does not hold. Thus the only solutions are $x = 0$ and $x = 1$. The corresponding solutions for y are therefore also 0 and 1. Thus the only time we have both $x \geq y^2$ and $y \geq x^2$ is when $x = y$; this means that the relation is antisymmetric. It is transitive. Suppose $x \geq y^2$ and $y \geq z^2$. Again the second inequality implies that both sides are nonnegative, so we can square both sides to obtain $y^2 \geq z^4$. Combining these inequalities gives $x \geq z^4$. Now we claim that it is always the case that $z^4 \geq z^2$; if so, then we combine this fact with the last inequality to obtain $x \geq z^2$, so x is related to z. To verify the claim, note that since we are working with integers, it is always the case that $z^2 \geq |z|$ (equality for $z = 0$ and $z = 1$, strict inequality for other z). Squaring both sides gives the desired inequality.

9. The relations in parts **(a)**, **(b)**, and **(e)** all have at least one pair of the form (x,x) in them, so they are not irreflexive. The relations in parts **(c)**, **(d)**, and **(f)** do not, so they are irreflexive.

11. According to the preamble to Exercise 9, an irreflexive relation is one for which a is never related to itself; i.e., $\forall a((a,a) \notin R)$.

a) Since we saw in Exercise 5a that $\forall a((a,a) \in R)$, clearly R is not irreflexive.

b) Since there are probably pages a with no links at all, and for such pages it is true that there are no common links found on both page a and page a, this relation is probably not irreflexive.

c) This relation is not irreflexive, because for any page a that has links on it, $(a,a) \in R$.

d) This relation is not irreflexive, because for any page a that has links on it that are ever cited, $(a,a) \in R$.

13. The relation in Exercise 3a is neither reflexive nor irreflexive. It contains some of the pairs (a,a) but not all of them.

15. Of course many answers are possible. The empty relation is always irreflexive (x is never related to y). A less trivial example would be $(a,b) \in R$ if and only if a is taller than b. Since nobody is taller than him/herself, we always have $(a,a) \notin R$.

17. The relation in part **(a)** is asymmetric, since if a is taller than b, then certainly b cannot be taller than a. The relation in part **(b)** is not asymmetric, since there are many instances of a and b born on the same day (both cases in which $a = b$ and cases in which $a \neq b$), and in all such cases, it is also the case that b and a were born on the same day. The relations in part **(c)** and part **(d)** are just like that in part **(b)**, so they, too, are not asymmetric.

19. According to the preamble to Exercise 16, an asymmetric relation is one for which $(a,b) \in R$ and $(b,a) \in R$ can never hold simultaneously, even if $a = b$. Thus R is asymmetric if and only if R is antisymmetric and also irreflexive.

a) not asymmetric since $(-1,1) \in R$ and $(1,-1) \in R$

b) not asymmetric since $(-1,1) \in R$ and $(1,-1) \in R$

c) not asymmetric since $(-1,1) \in R$ and $(1,-1) \in R$

d) not asymmetric since $(0,0) \in R$

e) not asymmetric since $(2,1) \in R$ and $(1,2) \in R$

f) not asymmetric since $(0,1) \in R$ and $(1,0) \in R$

g) not asymmetric since $(1,1) \in R$

h) not asymmetric since $(2,1) \in R$ and $(1,2) \in R$

21. According to the preamble to Exercise 16, an asymmetric relation is one for which $(a,b) \in R$ and $(b,a) \in R$ can never hold simultaneously. In symbols, this is simply $\forall a \forall b \neg((a,b) \in R \land (b,a) \in R)$. Alternatively, $\forall a \forall b\,((a,b) \in R \rightarrow (b,a) \notin R)$.

23. There are mn elements of the set $A \times B$, if A is a set with m elements and B is a set with n elements. A relation from A to B is a subset of $A \times B$. Thus the question asks for the number of subsets of the set $A \times B$, which has mn elements. By the product rule, it is 2^{mn}.

25. **a)** By definition the answer is $\{\,(b,a) \mid a \text{ divides } b\,\}$, which, by changing the names of the dummy variables, can also be written $\{\,(a,b) \mid b \text{ divides } a\,\}$. (The universal set is still the set of positive integers.)

b) By definition the answer is $\{\,(a,b) \mid a \text{ does not divide } b\,\}$. (The universal set is still the set of positive integers.)

27. The inverse relation is just the graph of the inverse function. Somewhat more formally, we have $R^{-1} = \{\,(f(a),a) \mid a \in A\,\} = \{\,(b, f^{-1}(b)) \mid b \in B\,\}$, since we can index this collection just as easily by elements of B as by elements of A (using the correspondence $b = f(a)$).

29. This exercise is just a matter of the definitions of the set operations.

a) the set of pairs (a,b) where a is required to read b in a course or has read b

b) the set of pairs (a,b) where a is required to read b in a course and has read b

c) the set of pairs (a,b) where a is required to read b in a course or has read b, but not both; equivalently, the set of pairs (a,b) where a is required to read b in a course but has not done so, or has read b although not required to do so in a course

d) the set of pairs (a,b) where a is required to read b in a course but has not done so

e) the set of pairs (a,b) where a has read b although not required to do so in a course

31. To find $S \circ R$ we want to find the set of pairs (a,c) such that for some person b, a is a parent of b, and b is a sibling of c. Since brothers and sisters have the same parents, this means that a is also the parent of c. Thus $S \circ R$ is contained in the relation R. More specifically, $(a,c) \in S \circ R$ if and only if a is the parent of c, and c has a sibling (who is necessarily also a child of a). To find $R \circ S$ we want to find the set of pairs (a,c) such that for some person b, a is a sibling of b, and b is a parent of c. This is the same as the condition that a is the aunt or uncle of c (by blood, not marriage).

33. **a)** The union of two relations is the union of these sets. Thus $R_2 \cup R_4$ holds between two real numbers if R_2 holds or R_4 holds (or both, it goes without saying). Since it is always true that $a \le b$ or $b \le a$, $R_2 \cup R_4$ is all of $\mathbf{R}^2$, i.e., the relation that always holds.

b) For (a,b) to be in $R_3 \cup R_6$, we must have $a < b$ or $a \ne b$. Since this happens precisely when $a \ne b$, we see that the answer is R_6.

c) The intersection of two relations is the intersection of these sets. Thus $R_3 \cap R_6$ holds between two real numbers if R_3 holds and R_6 holds as well. Thus for (a,b) to be in $R_3 \cap R_6$, we must have $a < b$ and $a \ne b$. Since this happens precisely when $a < b$, we see that the answer is R_3.

d) For (a,b) to be in $R_4 \cap R_6$, we must have $a \le b$ and $a \ne b$. Since this happens precisely when $a < b$, we see that the answer is R_3.

e) Recall that $R_3 - R_6 = R_3 \cap \overline{R_6}$. But $\overline{R_6} = R_5$, so we are asked for $R_3 \cap R_5$. It is impossible for $a < b$ and $a = b$ to hold at the same time, so the answer is $\emptyset$, i.e., the relation that never holds.

f) Reasoning as in part (e), we want $R_6 \cap \overline{R_3} = R_6 \cap R_2$, which is clearly R_1 (since $a \ne b$ and $a \ge b$ precisely when $a > b$).

g) Recall that $R_2 \oplus R_6 = (R_2 \cap \overline{R_6}) \cup (R_6 \cap \overline{R_2})$. We see that $R_2 \cap \overline{R_6} = R_2 \cap R_5 = R_5$, and $R_6 \cap \overline{R_2} = R_6 \cap R_3 = R_3$. Thus our answer is $R_5 \cup R_3 = R_4$.

h) Recall that $R_3 \oplus R_5 = (R_3 \cap \overline{R_5}) \cup (R_5 \cap \overline{R_3})$. We see that $R_3 \cap \overline{R_5} = R_3 \cap R_6 = R_3$, and $R_5 \cap \overline{R_3} = R_5 \cap R_2 = R_5$. Thus our answer is $R_3 \cup R_5 = R_4$.

35. Recall that the composition of two relations all defined on a common set is defined as follows: $(a,c) \in S \circ R$ if and only if there is some element b such that $(a,b) \in R$ and $(b,c) \in S$. We have to apply this in each case.

a) For (a,c) to be in $R_2 \circ R_1$, we must find an element b such that $(a,b) \in R_1$ and $(b,c) \in R_2$. This means that $a > b$ and $b \geq c$. Clearly this can be done if and only if $a > c$ to begin with. But that is precisely the statement that $(a,c) \in R_1$. Therefore we have $R_2 \circ R_1 = R_1$.

b) For (a,c) to be in $R_2 \circ R_2$, we must find an element b such that $(a,b) \in R_2$ and $(b,c) \in R_2$. This means that $a \geq b$ and $b \geq c$. Clearly this can be done if and only if $a \geq c$ to begin with. But that is precisely the statement that $(a,c) \in R_2$. Therefore we have $R_2 \circ R_2 = R_2$. In particular, this shows that R_2 is transitive.

c) For (a,c) to be in $R_3 \circ R_5$, we must find an element b such that $(a,b) \in R_5$ and $(b,c) \in R_3$. This means that $a = b$ and $b < c$. Clearly this can be done if and only if $a < c$ to begin with (choose $b = a$). But that is precisely the statement that $(a,c) \in R_3$. Therefore we have $R_3 \circ R_5 = R_3$. One way to look at this is to say that R_5, the equality relation, acts as an identity for the composition operation (on the right—although it is also an identity on the left as well).

d) For (a,c) to be in $R_4 \circ R_1$, we must find an element b such that $(a,b) \in R_1$ and $(b,c) \in R_4$. This means that $a > b$ and $b \leq c$. Clearly this can always be done simply by choosing b to be small enough. Therefore we have $R_4 \circ R_1 = \mathbf{R}^2$, the relation that always holds.

e) For (a,c) to be in $R_5 \circ R_3$, we must find an element b such that $(a,b) \in R_3$ and $(b,c) \in R_5$. This means that $a < b$ and $b = c$. Clearly this can be done if and only if $a < c$ to begin with (choose $b = c$). But that is precisely the statement that $(a,c) \in R_3$. Therefore we have $R_5 \circ R_3 = R_3$. One way to look at this is to say that R_5, the equality relation, acts as an identity for the composition operation (on the left—although it is also an identity on the right as well).

f) For (a,c) to be in $R_3 \circ R_6$, we must find an element b such that $(a,b) \in R_6$ and $(b,c) \in R_3$. This means that $a \neq b$ and $b < c$. Clearly this can always be done simply by choosing b to be small enough. Therefore we have $R_3 \circ R_6 = \mathbf{R}^2$, the relation that always holds.

g) For (a,c) to be in $R_4 \circ R_6$, we must find an element b such that $(a,b) \in R_6$ and $(b,c) \in R_4$. This means that $a \neq b$ and $b \leq c$. Clearly this can always be done simply by choosing b to be small enough. Therefore we have $R_4 \circ R_6 = \mathbf{R}^2$, the relation that always holds.

h) For (a,c) to be in $R_6 \circ R_6$, we must find an element b such that $(a,b) \in R_6$ and $(b,c) \in R_6$. This means that $a \neq b$ and $b \neq c$. Clearly this can always be done simply by choosing b to be something other than a or c. Therefore we have $R_6 \circ R_6 = \mathbf{R}^2$, the relation that always holds. Note that since the answer is not R_6 itself, we know that R_6 is not transitive.

37. One earns a doctorate by, among other things, writing a thesis under an advisor, so this relation makes sense. (We ignore anomalies like someone having two advisors or someone being awarded a doctorate without having an advisor.) For (a,b) to be in R^2, we must find a c such that $(a,c) \in R$ and $(c,b) \in R$. In our context, this says that b got his/her doctorate under someone who got his/her doctorate under a. Colloquially, a is the academic grandparent of b, or b is the academic grandchild of a. Generalizing, $(a,b) \in R^n$ precisely when there is a sequence of $n+1$ people, starting with a and ending with b, such that each is the advisor of the next person in the sequence. People with doctorates like to look at these sequences (and trace their ancestry) back as far as they can. There are at least two websites where these relations are listed: one in theoretical computer science (`http://sigact.acm.org/genealogy/`, which seems to be not currently updated or maintained) and one in mathematics (`http://www.genealogy.math.ndsu.nodak.edu/`).

39. **a)** The union of two relations is the union of these sets. Thus $R_1 \cup R_2$ holds between two integers if R_1 holds or R_2 holds (or both, it goes without saying). Thus $(a,b) \in R_1 \cup R_2$ if and only if $a \equiv b \pmod 3$ or $a \equiv b \pmod 4$. There is not a good easier way to state this, other than perhaps to say that $a-b$ is a multiple of either 3 or 4, or to work modulo 12 and write $a - b \equiv 0, 3, 4, 6, 8,$ or $9 \pmod{12}$.

b) The intersection of two relations is the intersection of these sets. Thus $R_1 \cap R_2$ holds between two integers if R_1 holds and R_2 holds. Thus $(a,b) \in R_1 \cap R_2$ if and only if $a \equiv b \pmod 3$ and $a \equiv b \pmod 4$. Since this means that $a - b$ is a multiple of both 3 and 4, and that happens if and only if $a - b$ is a multiple of 12, we can state this more simply as $a \equiv b \pmod{12}$.

c) By definition $R_1 - R_2 = R_1 \cap \overline{R_2}$. Thus this relation holds between two integers if R_1 holds and R_2 does not hold. We can write this in symbols by saying that $(a,b) \in R_1 - R_2$ if and only if $a \equiv b \pmod 3$ and $a \not\equiv b \pmod 4$. We could, if we wished, state this working modulo 12: $(a,b) \in R_1 - R_2$ if and only if $a - b \equiv 3, 6,$ or $9 \pmod{12}$.

d) By definition $R_2 - R_1 = R_2 \cap \overline{R_1}$. Thus this relation holds between two integers if R_2 holds and R_1 does not hold. We can write this in symbols by saying that $(a,b) \in R_2 - R_1$ if and only if $a \equiv b \pmod 4$ and $a \not\equiv b \pmod 3$. We could, if we wished, state this working modulo 12: $(a,b) \in R_2 - R_1$ if and only if $a - b \equiv 4$ or $8 \pmod{12}$.

e) We know that $R_1 \oplus R_2 = (R_1 - R_2) \cup (R_2 - R_1)$, so we look at our solutions to part **(c)** and part **(d)**. Thus this relation holds between two integers if R_1 holds and R_2 does not hold, or vice versa. We can write this in symbols by saying that $(a,b) \in R_1 \oplus R_2$ if and only if ($a \equiv b \pmod 3$ and $a \not\equiv b \pmod 4$) or ($a \equiv b \pmod 4$ and $a \not\equiv b \pmod 3$). We could, if we wished, state this working modulo 12: $(a,b) \in R_1 \oplus R_2$ if and only if $a - b \equiv 3, 4, 6, 8$ or $9 \pmod{12}$. We could also say that $a - b$ is a multiple of 3 or 4 but not both.

41. A relation is just a subset. A subset can either contain a specified element or not; half of them do and half of them do not. Therefore 8 of the 16 relations on $\{0,1\}$ contain the pair $(0,1)$. Alternatively, a relation on $\{0,1\}$ containing the pair $(0,1)$ is just a set of the form $\{(0,1)\} \cup X$, where $X \subseteq \{(0,0),(1,0),(1,1)\}$. Since this latter set has 3 elements, it has $2^3 = 8$ subsets.

43. This is similar to Example 16 in this section.

a) A relation on a set S with n elements is a subset of $S \times S$. Since $S \times S$ has n^2 elements, we are asking for the number of subsets of a set with n^2 elements, which is 2^{n^2}. In our case $n = 4$, so the answer is $2^{16} = 65{,}536$.

b) In solving part **(a)**, we had 16 binary choices to make—whether to include a pair (x,y) in the relation or not as x and y ranged over the set $\{a,b,c,d\}$. In this part, one of those choices has been made for us: we *must* include (a,a). We are free to make the other 15 choices. So the answer is $2^{15} = 32{,}768$. See Exercise 45 for more problems of this type.

45. These are combinatorics problems, some harder than others. Let A be the set with n elements on which the relations are defined.

a) To specify a symmetric relation, we need to decide, for each unordered pair $\{a,b\}$ of distinct elements of A, whether to include the pairs (a,b) and (b,a) or leave them out; this can be done in 2 ways for each such unordered pair. Also, for each element $a \in A$, we need to decide whether to include (a,a) or not, again 2 possibilities. We can think of these two parts as one by considering an element to be an unordered pair with repetition allowed. Thus we need to make this 2-fold choice $C(n+1,2)$ times, since there are $C(n+2-1,2)$ ways to choose an unordered pair with repetition allowed. Therefore the answer is $2^{C(n+1,2)} = 2^{n(n+1)/2}$.

b) This is somewhat similar to part **(a)**. For each unordered pair $\{a,b\}$ of distinct elements of A, we have a 3-way choice—either include (a,b) only, include (b,a) only, or include neither. For each element of A we have a 2-way choice. Therefore the answer is $3^{C(n,2)}2^n = 3^{n(n-1)/2}2^n$.

c) As in part **(b)** we have a 3-way choice for $a \neq b$. There is no choice about including (a,a) in the relation—the definition prohibits it. Therefore the answer is $3^{C(n,2)} = 3^{n(n-1)/2}$.

d) For each ordered pair (a,b), with $a \neq b$ (and there are $P(n,2)$ such pairs), we can choose to include (a,b) or to leave it out. There is no choice for pairs (a,a). Therefore the answer is $2^{P(n,2)} = 2^{n(n-1)}$.

e) This is just like part **(a)**, except that there is no choice about including (a,a). For each unordered pair of distinct elements of A, we can choose to include neither or both of the corresponding ordered pairs. Therefore the answer is $2^{C(n,2)} = 2^{n(n-1)/2}$.

f) We have complete freedom with the ordered pairs (a,b) with $a \neq b$, so that part of the choice gives us $2^{P(n,2)}$ possibilities, just as in part **(d)**. For the decision as to whether to include (a,a), two of the 2^n possibilities are prohibited: we cannot include all such pairs, and we cannot leave them all out. Therefore the answer is $2^{P(n,2)}(2^n-2) = 2^{n^2-n}(2^n-2) = 2^{n^2} - 2^{n^2-n+1}$.

47. The second sentence of the proof asks us to "take an element $b \in A$ such that $(a,b) \in R$." There is no guarantee that such an element exists for the taking. This is the only mistake in the proof. If one could be guaranteed that each element in A is related to at least one element, then symmetry and transitivity would indeed imply reflexivity. Without this assumption, however, the proof and the proposition are wrong. As a simple example, take the relation $\emptyset$ on any nonempty set. This relation is vacuously symmetric and transitive, but not reflexive. Here is another counterexample: the relation $\{(1,1),(1,2),(2,1),(2,2)\}$ on the set $\{1,2,3\}$.

49. We need to show two things. First, we need to show that if a relation R is symmetric, then $R = R^{-1}$, which means we must show that $R \subseteq R^{-1}$ and $R^{-1} \subseteq R$. To do this, let $(a,b) \in R$. Since R is symmetric, this implies that $(b,a) \in R$. But since R^{-1} consists of all pairs (a,b) such that $(b,a) \in R$, this means that $(a,b) \in R^{-1}$. Thus we have shown that $R \subseteq R^{-1}$. Next let $(a,b) \in R^{-1}$. By definition this means that $(b,a) \in R$. Since R is symmetric, this implies that $(a,b) \in R$ as well. Thus we have shown that $R^{-1} \subseteq R$.

Second we need to show that $R = R^{-1}$ implies that R is symmetric. To this end we let $(a,b) \in R$ and try to show that (b,a) is also necessarily an element of R. Since $(a,b) \in R$, the definition tells us that $(b,a) \in R^{-1}$. But since we are under the hypothesis that $R = R^{-1}$, this tells us that $(b,a) \in R$, exactly as desired.

51. Suppose that R is reflexive. We must show that R^{-1} is reflexive, i.e., that $(a,a) \in R^{-1}$ for each $a \in A$. Now since R is reflexive, we know that $(a,a) \in R$ for each $a \in R$. By definition, this tells us that $(a,a) \in R^{-1}$, as desired. (Interchanging the two a's in the pair (a,a) leaves it as it was.) Conversely, if R^{-1} is reflexive, then $(a,a) \in R^{-1}$ for each $a \in A$. By definition this means that $(a,a) \in R$ (again we interchanged the two a's).

53. We prove this by induction on n. The case $n=1$ is trivial, since it is the statement $R = R$. Assume the inductive hypothesis that $R^n = R$. We must show that $R^{n+1} = R$. By definition $R^{n+1} = R^n \circ R$. Thus our task is to show that $R^n \circ R \subseteq R$ and $R \subseteq R^n \circ R$. The first uses the transitivity of R, as follows. Suppose that $(a,c) \in R^n \circ R$. This means that there is an element b such that $(a,b) \in R$ and $(b,c) \in R^n$. By the inductive hypothesis, the latter statement implies that $(b,c) \in R$. Thus by the transitivity of R, we know that $(a,c) \in R$, as desired.

Next assume that $(a,b) \in R$. We must show that $(a,b) \in R^n \circ R$. By the inductive hypothesis, $R^n = R$, and therefore R^n is reflexive by assumption. Thus $(b,b) \in R^n$. Since we have $(a,b) \in R$ and $(b,b) \in R^n$, we have by definition that (a,b) is an element of $R^n \circ R$, exactly as desired. (The first half of this proof was not really necessary, since Theorem 1 in this section already told us that $R^n \subseteq R$ for all n.)

55. We use induction on n, the result being trivially true for $n=1$. Assume that R^n is reflexive; we must show that R^{n+1} is reflexive. Let $a \in A$, where A is the set on which R is defined. By definition $R^{n+1} = R^n \circ R$. By the inductive hypothesis, R^n is reflexive, so $(a,a) \in R^n$. Also, since R is reflexive by assumption, $(a,a) \in R$. Therefore by the definition of composition, $(a,a) \in R^n \circ R$, as desired.

57. It is not necessarily true that R^2 is irreflexive when R is. We might have pairs (a,b) and (b,a) both in R, with $a \neq b$; then it would follow that $(a,a) \in R^2$, preventing R^2 from being irreflexive. As the simplest example, let $A = \{1,2\}$ and let $R = \{(1,2),(2,1)\}$. Then R is clearly irreflexive. In this case $R^2 = \{(1,1),(2,2)\}$, which is not irreflexive.

SECTION 8.2 *n*-ary Relations and Their Applications

This section is a brief introduction to relational models for data bases. The exercises are straightforward and similar to the examples. Projections are formed by omitting certain columns, and then eliminating duplicate rows. Joins are analogous to compositions of relations.

1. We simply need to find solutions of the inequality, which we can do by common sense. The set is $\{(1,2,3), (1,2,4),(1,3,4),(2,3,4)\}$.

3. The 5-tuples are just the lines of the table. Thus the relation is
$\{(\text{Nadir}, 122, 34, \text{Detroit}, 08{:}\,10), (\text{Acme}, 221, 22, \text{Denver}, 08{:}\,17), (\text{Acme}, 122, 33, \text{Anchorage}, 08{:}\,22),$
$(\text{Acme}, 323, 34, \text{Honolulu}, 08{:}\,30), (\text{Nadir}, 199, 13, \text{Detroit}, 08{:}\,47), (\text{Acme}, 222, 22, \text{Denver}, 09{:}\,10),$
$(\text{Nadir}, 322, 34, \text{Detroit}, 09{:}\,44)\}$.

5. We need to find a field that, when used along with the *Airline* field uniquely specifies a row of the table. Certainly *Flight_number* is one such field, since there is only one line of the table for each pair (*Airline*, *Flight_number*); no airline has the same flight number for two different flights. *Gate* and *Destination* do not qualify, however, since Nadir has two flights leaving from Gate 34 going to Detroit. Finally, *Departure_time* is a key by itself (no two flights leave at the same time), so it and *Airline* form a composite key as well.

7. **a)** A school would be giving itself a lot of headaches if it didn't make the student ID number different for each student, so student ID is likely to be a primary key.
b) Name could very easily not be a primary key. Two people named Jennifer Johnson might easily both be students.
c) Phone number is unlikely to be a primary key. Two roommates, or two siblings living at home, will likely have the same phone number, and they might both be students at that school.

9. **a)** Everyone has a different Social Security number, so that field will be a primary key.
b) It is unlikely that (name, street address) will be a composite key. Somewhere in the United States there could easily be two people named Jennifer Johnson both living at 123 Washington Street. In order for this to work, there must never be two people with the same name who happen to have the same street address.
c) For this to work, we must never have two people with the same name living together. Given the size of the country, one would doubt that this would work. For example, a husband and wife each named Morgan White would make this not a composite key, as would a mother and daughter living at home with the same name.

11. The selection operator picks out all the tuples that match the criteria. The 5-tuples in Table 8 that have Detroit as their destination are $(\text{Nadir}, 122, 34, \text{Detroit}, 08{:}\,10)$, $(\text{Nadir}, 199, 13, \text{Detroit}, 08{:}\,47)$, and $(\text{Nadir}, 322, 34, \text{Detroit}, 09{:}\,44)$.

13. The selection operator picks out all the tuples that match the criteria. The 5-tuples in Table 8 that have Nadir as their airline are (Nadir, 122, 34, Detroit, 08: 10), (Nadir, 199, 13, Detroit, 08: 47), and (Nadir, 322, 34, Detroit, 09: 44). The 5-tuples in Table 8 that have Denver as their destination are (Acme, 221, 22, Denver, 08: 17) and (Acme, 222, 22, Denver, 09: 10). We need the union of these two lists: (Nadir, 122, 34, Detroit, 08: 10), (Nadir, 199, 13, Detroit, 08: 47), (Nadir, 322, 34, Detroit, 09: 44), (Acme, 221, 22, Denver, 08: 17), and (Acme, 222, 22, Denver, 09: 10).

15. The subscripts on the projection mapping notation indicate which columns are to be retained. Thus if we want to delete columns 1, 2, and 4 from a 6-tuple, we need to use the projection $P_{3,5,6}$.

17. The table uses columns 1 and 4 of Table 8. We start by deleting columns 2, 3, and 5 from Table 8. At this point, rows 5, 6 and 7 are duplicates of earlier rows, so they are omitted (rather than being listed twice). Therefore the answer is as follows.

Airline	*Destination*
Nadir	Detroit
Acme	Denver
Acme	Anchorage
Acme	Honolulu

19. We need to find rows of Table 9 the last two entries of which are identical to the first two entries of rows of Table 10. We combine each such pair of rows into one row of our new table. For instance, the last two entries in the first row of Table 9 are 1092 and 1. The first two entries in the second row of Table 10 are also 1092 and 1. Therefore we combine them into the row 23, 1092, 1, 2, 2 of our new table, whose columns represent *Supplier*, *Part_number*, *Project*, *Quantity*, and *Color_code*. The new table consists of all pairs found in this way.

Supplier	*Part_number*	*Project*	*Quantity*	*Color_code*
23	1092	1	2	2
23	1101	3	1	1
23	9048	4	12	2
31	4975	3	6	2
31	3477	2	25	2
32	6984	4	10	1
32	9191	2	80	4
33	1001	1	14	8

21. Both sides of this equation pick out the subset of R consisting of those n-tuples satisfying both conditions C_1 and C_2. This follows immediately from the definition of the selection operator.

23. Both sides of this equation pick out the set of n-tuples that satisfy three conditions: they are in R, they are in S, and they satisfy condition C. This follows immediately from the definitions of intersection and the selection operator.

25. Both sides of this equation pick out the m-tuples consisting of i_1^{th}, i_2^{th}, ..., i_m^{th} components of n-tuples in either R or S (or, of course, both). This follows immediately from the definitions of union and the projection operator.

27. Note that we lose information when we delete columns. Therefore we might be taking something away when we form the second set of m-tuples that might not have been taken away if all the original information is there (forming the first set of m-tuples). A simple example would be to let $R = \{(a,b)\}$ and $S = \{(a,c)\}$, $n = 2$, $m = 1$, and $i_1 = 1$. Then $R - S = R$, so $P_1(R - S) = P_1(R) = \{(a)\}$. On the other hand, $P_1(R) = P_1(S) = \{(a)\}$, so $P_1(R) - P_1(S) = \emptyset$.

29. This is similar to Example 13.

a) Since two databases are listed in the "FROM" field, the first operation is to form the join of these two databases, specifically the join J_2 of these two databases. We then apply the selection operator with the condition "Quantity ≤ 10." This join will have eight 5-tuples in it. Finally we want just the Supplier and Project, so we are forming the projection $P_{1,3}$.

b) Four of the 5-tuples in the joined database have a quantity of no more than 10. The output, then, is the set of the four 2-tuples corresponding to these fields: $(23,1), (23,3), (31,3), (32,4)$.

31. A primary key is a domain whose value determines the values of all the other domains. For this relation, this does not happen. The third domain (the modulus) is not a primary key, because, for example, $1 \equiv 11 \pmod{10}$ and $2 \equiv 12 \pmod{10}$, so the triples $(1,11,10)$ and $(2,12,10)$ are both in the relation. Knowing that the third component of a triple is 10 does not tell us what the other two components are. Similarly, the triples $(1,11,10)$ and $(1,21,10)$ are both in the relation, so the first domain is not a key; and the triples $(1,11,10)$ and $(11,11,10)$ are both in the relation, so the second domain is not a key.

SECTION 8.3 Representing Relations

Matrices and directed graphs provide useful ways for computers and humans to represent relations and manipulate them. Become familiar with working with these representations and the operations on them (especially the matrix operation for forming composition) by working these exercises. Some of these exercises explore how properties of a relation can be found from these representations.

1. In each case we use a 3×3 matrix, putting a 1 in position (i,j) if the pair (i,j) is in the relation and a 0 in position (i,j) if the pair (i,j) is not in the relation. For instance, in part **(a)** there are 1's in the first row, since each of the pairs $(1,1)$, $(1,2)$, and $(1,3)$ are in the relation, and there are 0's elsewhere.

a) $\begin{bmatrix} 1 & 1 & 1 \\ 0 & 0 & 0 \\ 0 & 0 & 0 \end{bmatrix}$ **b)** $\begin{bmatrix} 0 & 1 & 0 \\ 1 & 1 & 0 \\ 0 & 0 & 1 \end{bmatrix}$ **c)** $\begin{bmatrix} 1 & 1 & 1 \\ 0 & 1 & 1 \\ 0 & 0 & 1 \end{bmatrix}$ **d)** $\begin{bmatrix} 0 & 0 & 1 \\ 0 & 0 & 0 \\ 1 & 0 & 0 \end{bmatrix}$

3. a) Since the $(1,1)^{\text{th}}$ entry is a 1, $(1,1)$ is in the relation. Since $(1,2)^{\text{th}}$ entry is a 0, $(1,2)$ is not in the relation. Continuing in this manner, we see that the relation contains $(1,1)$, $(1,3)$, $(2,2)$, $(3,1)$, and $(3,3)$.
b) $(1,2)$, $(2,2)$, and $(3,2)$ **c)** $(1,1)$, $(1,2)$, $(1,3)$, $(2,1)$, $(2,3)$, $(3,1)$, $(3,2)$, and $(3,3)$

5. An irreflexive relation (see the preamble to Exercise 9 in Section 8.1) is one in which no element is related to itself. In the matrix, this means that there are no 1's on the main diagonal (position m_{ii} for some i). Equivalently, the relation is irreflexive if and only if every entry on the main diagonal of the matrix is 0.

7. For reflexivity we want all 1's on the main diagonal; for irreflexivity we want all 0's on the main diagonal; for symmetry, we want the matrix to be symmetric about the main diagonal (equivalently, the matrix equals its transpose); for antisymmetry we want there never to be two 1's symmetrically placed about the main diagonal (equivalently, the meet of the matrix and its transpose has no 1's off the main diagonal); and for transitivity we want the Boolean square of the matrix (the Boolean product of the matrix and itself) to be "less than or equal to" the original matrix in the sense that there is a 1 in the original matrix at every location where there is a 1 in the Boolean square.

a) Since there are all 1's on the main diagonal, this relation is reflexive and not irreflexive. Since the matrix is symmetric, the relation is symmetric. The relation is not antisymmetric—look at positions $(1,3)$ and $(3,1)$. Finally, the Boolean square of this matrix is itself, so the relation is transitive.

b) Since there are both 0's and 1's on the main diagonal, this relation is neither reflexive nor irreflexive. Since the matrix is not symmetric, the relation is not symmetric (look at positions $(1,2)$ and $(2,1)$, for example). The relation is antisymmetric since there are never two 1's symmetrically placed with respect to the main diagonal. Finally, the Boolean square of this matrix is itself, so the relation is transitive.

c) Since there are both 0's and 1's on the main diagonal, this relation is neither reflexive nor irreflexive. Since the matrix is symmetric, the relation is symmetric. The relation is not antisymmetric—look at positions $(1,3)$ and $(3,1)$, for example. Finally, the Boolean square of this matrix is the matrix with all 1's, so the relation is not transitive (1 is related to 2, and 2 is related to 1, but 2 is not related to 2).

9. Note that the total number of entries in the matrix is $100^2 = 10{,}000$.

a) There is a 1 in the matrix for each pair of distinct positive integers not exceeding 100, namely in position (a,b) where $a > b$. Thus the answer is the number of subsets of size 2 from a set of 100 elements, i.e., $C(100,2) = 4950$.

b) There is a 1 in the matrix at each position except the 100 positions on the main diagonal. Therefore the answer is $100^2 - 100 = 9900$.

c) There is a 1 in the matrix at each entry just below the main diagonal (i.e., in positions $(2,1)$, $(3,2)$, $\ldots$, $(100,99)$. Therefore the answer is 99.

d) The entire first row of this matrix corresponds to $a = 1$. Therefore the matrix has 100 nonzero entries.

e) This relation has only the one element $(1,1)$ in it, so the matrix has just one nonzero entry.

11. Since the relation $\overline{R}$ is the relation that contains the pair (a,b) (where a and b are elements of the appropriate sets) if and only if R does not contain that pair, we can form the matrix for $\overline{R}$ simply by changing all the 1's to 0's and 0's to 1's in the matrix for R.

13. Exercise 12 tells us how to do part **(a)** (we take the transpose of the given matrix $\mathbf{M}_R$, which in this case happens to be the matrix itself). Exercise 11 tells us how to do part **(b)** (we change 1's to 0's and 0's to 1's in $\mathbf{M}_R$). For part **(c)** we take the Boolean product of $\mathbf{M}_R$ with itself.

a) $\begin{bmatrix} 0 & 1 & 1 \\ 1 & 1 & 0 \\ 1 & 0 & 1 \end{bmatrix}$ **b)** $\begin{bmatrix} 1 & 0 & 0 \\ 0 & 0 & 1 \\ 0 & 1 & 0 \end{bmatrix}$ **c)** $\begin{bmatrix} 1 & 1 & 1 \\ 1 & 1 & 1 \\ 1 & 1 & 1 \end{bmatrix}$

15. We compute the Boolean powers of $\mathbf{M}_R$; thus $\mathbf{M}_{R^2} = \mathbf{M}_R^{[2]} = \mathbf{M}_R \odot \mathbf{M}_R$, $\mathbf{M}_{R^3} = \mathbf{M}_R^{[3]} = \mathbf{M}_R \odot \mathbf{M}_R^{[2]}$, and $\mathbf{M}_{R^4} = \mathbf{M}_R^{[4]} = \mathbf{M}_R \odot \mathbf{M}_R^{[3]}$.

a) $\begin{bmatrix} 0 & 0 & 1 \\ 1 & 1 & 0 \\ 0 & 1 & 1 \end{bmatrix}$ **b)** $\begin{bmatrix} 1 & 1 & 0 \\ 0 & 1 & 1 \\ 1 & 1 & 1 \end{bmatrix}$ **c)** $\begin{bmatrix} 0 & 1 & 1 \\ 1 & 1 & 1 \\ 1 & 1 & 1 \end{bmatrix}$

17. The matrix for the complement has a 1 wherever the matrix for the relation has a 0, and vice versa. Therefore the number of nonzero entries in $\mathbf{M}_{\overline{R}}$ is $n^2 - k$, since these matrices have n rows and n columns.

19. In each case we need a vertex for each of the elements, and we put in a directed edge from x to y if there is a 1 in position (x,y) of the matrix. For simplicity we have indicated pairs of edges between the same two vertices in opposite directions by using a double arrowhead, rather than drawing two separate lines.

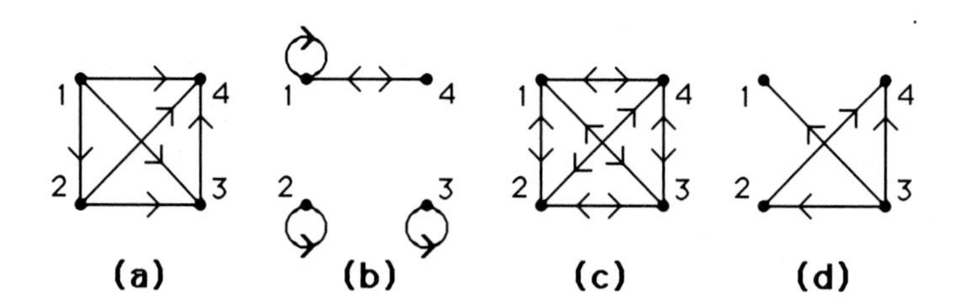

21. In each case we need a vertex for each of the elements, and we put in a directed edge from x to y if there is a 1 in position (x, y) of the matrix. For simplicity we have indicated pairs of edges between the same two vertices in opposite directions by using a double arrowhead, rather than drawing two separate lines.

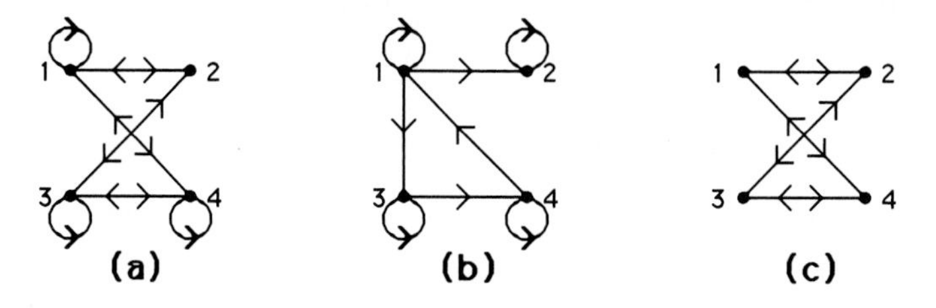

23. We list all the pairs (x, y) for which there is an edge from x to y in the directed graph: $\{(a,b),(a,c),(b,c),(c,b)\}$.

25. We list all the pairs (x, y) for which there is an edge from x to y in the directed graph: $\{(a,c),(b,a),(c,d),(d,b)\}$.

27. We list all the pairs (x, y) for which there is an edge from x to y in the directed graph: $\{(a,a),(a,b),(a,c),(b,a),(b,b),(b,c),(c,a),(c,b),(d,d)\}$.

29. An asymmetric relation is one for which it never happens that a is related to b and simultaneously b is related to a, even when $a = b$. In terms of the directed graph, this means that we must see no loops and no closed paths of length 2 (i.e., no pairs of edges between two vertices going in opposite directions).

31. Recall that the relation is reflexive if there is a loop at each vertex; irreflexive if there are no loops at all; symmetric if edges appear only in **antiparallel** pairs (edges from one vertex to a second vertex and from the second back to the first); antisymmetric if there is no pair of antiparallel edges; and transitive if all paths of length 2 (a pair of edges (x, y) and (y, z)) are accompanied by the corresponding path of length 1 (the edge (x, z)). The relation drawn in Exercise 23 is not reflexive but is irreflexive since there are no loops. It is not symmetric, since, for instance, the edge (a, b) is present but not the edge (b, a). It is not antisymmetric, since both edges (b, c) and (c, b) are present. It is not transitive, since the path $(b, c), (c, b)$ from b to b is not accompanied by the edge (b, b). The relation drawn in Exercise 24 is reflexive and not irreflexive since there is a loop at each vertex. It is not symmetric, since, for instance, the edge (b, a) is present but not the edge (a, b). It is antisymmetric, since there are no pairs of antiparallel edges. It is transitive, since the only nontrivial path of length 2 is bac, and the edge (b, c) is present. The relation drawn in Exercise 25 is not reflexive but is irreflexive since there are no loops. It is not symmetric, since, for instance, the edge (b, a) is present but not the edge (a, b). It is antisymmetric, since there are no pairs of antiparallel edges. It is not transitive, since the edges (a, c) and (c, d) are present, but not (a, d).

33. Since the inverse relation consists of all pairs (b, a) for which (a, b) is in the original relation, we just have to take the digraph for R and reverse the direction on every edge.

35. We prove this statement by induction on n. The basis step $n = 1$ is tautologically true, since $\mathbf{M}_R^{[1]} = \mathbf{M}_R$. Assume the inductive hypothesis that $\mathbf{M}_R^{[n]}$ is the matrix representing R^n. Now $\mathbf{M}_R^{[n+1]} = \mathbf{M}_R \odot \mathbf{M}_R^{[n]}$. By the inductive hypothesis and the assertion made before Example 5, that $\mathbf{M}_{S \circ R} = \mathbf{M}_R \odot \mathbf{M}_S$, the right-hand side is the matrix representing $R^n \circ R$. But $R^n \circ R = R^{n+1}$, so our proof is complete.

SECTION 8.4 Closures of Relations

This section is harder than the previous ones in this chapter. Warshall's algorithm, in particular, is fairly tricky, and Exercise 27 should be worked carefully, following Example 8. It is easy to forget to include the loops (a, a) when forming transitive closures "by hand."

1. **a)** The reflexive closure of R is R together with all the pairs (a, a). Thus in this case the closure of R is $\{(0,0),(0,1),(1,1),(1,2),(2,0),(2,2),(3,0),(3,3)\}$.

b) The symmetric closure of R is R together with all the pairs (b, a) for which (a, b) is in R. For example, since $(1, 2)$ is in R, we need to add $(2, 1)$. Thus the closure of R is $\{(0,1),(0,2),(0,3),(1,0),(1,1),(1,2),$ $(2,0),(2,1),(2,2),(3,0)\}$.

3. To form the symmetric closure we need to add all the pairs (b, a) such that (a, b) is in R. In this case, that means that we need to include pairs (b, a) such that a divides b, which is equivalent to saying that we need to include all the pairs (a, b) such that b divides a. Thus the closure is $\{\, (a, b) \mid a \text{ divides } b \text{ or } b \text{ divides } a \,\}$.

5. We form the reflexive closure by taking the given directed graph and appending loops at all vertices at which there are not already loops.

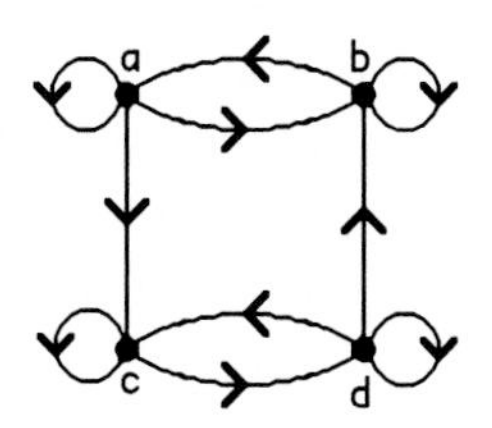

7. We form the reflexive closure by taking the given directed graph and appending loops at all vertices at which there are not already loops.

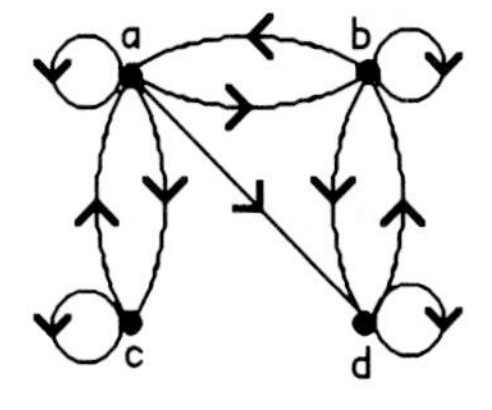

9. We form the symmetric closure by taking the given directed graph and appending an edge pointing in the opposite direction for every edge already in the directed graph (unless it is already there); in other words, we append the edge (b, a) whenever we see the edge (a, b). We have labeled the figures below (a), (b), and (c), corresponding to Exercises 5, 6, and 7, respectively.

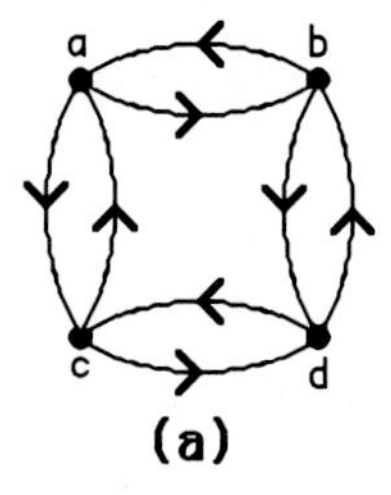

(a)

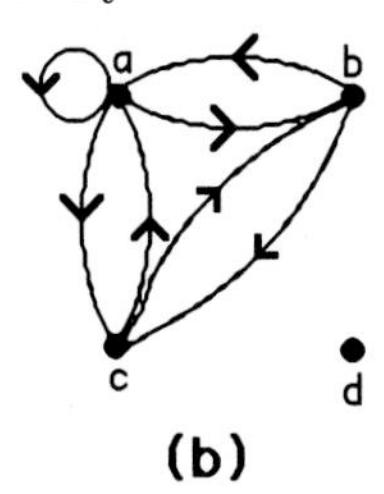

(b)

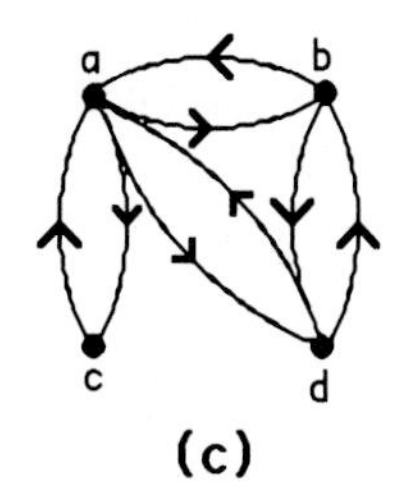

(c)

11. We are asked for the symmetric and reflexive closure of the given relation. We form it by taking the given directed graph and appending (1) a loop at each vertex at which there is not already a loop and (2) an edge pointing in the opposite direction for every edge already in the directed graph (unless it is already there). We have labeled the figures below (a), (b), and (c), corresponding to Exercises 5, 6, and 7, respectively.

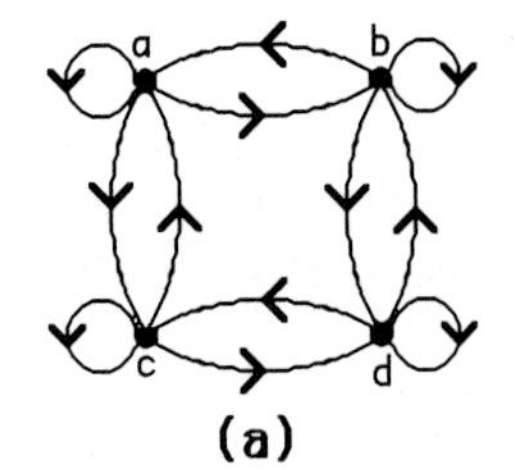

(a)

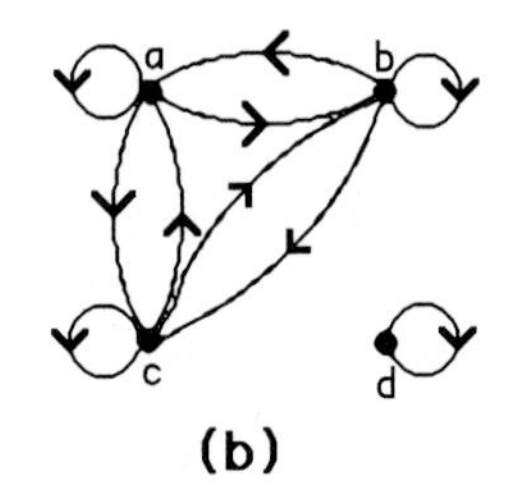

(b)

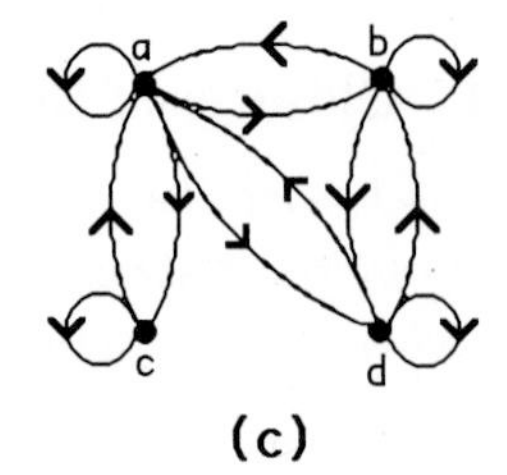

(c)

13. The symmetric closure of R is $R \cup R^{-1}$. The matrix for R^{-1} is $\mathbf{M}_R^t$, as we saw in Exercise 12 in Section 8.3. The matrix for the union of two relations is the join of the matrices for the two relations, as we saw in Section 8.3. Therefore the matrix representing the symmetric closure of R is indeed $\mathbf{M}_R \vee \mathbf{M}_R^t$.

15. If R is already irreflexive, then it is clearly its own irreflexive closure. On the other hand if R is not irreflexive, then there is no relation containing R that is irreflexive, since the loop or loops in R prevent any such relation from being irreflexive. Thus in this case R has no irreflexive closure. This exercise shows essentially that the concept of "irreflexive closure" is rather useless, since no relation has one unless it is already irreflexive (in which case it is its own "irreflexive closure").

17. A circuit of length 3 can be written as a sequence of 4 vertices, each joined to the next by an edge of the given directed graph, ending at the same vertex at which it began. There are several such circuits here, and we just have to be careful and systematically list them all. There are the circuits formed entirely by the loops: $aaaa$, $cccc$, and $eeee$. The triangles $abea$ and $adea$ also qualify. Two circuits start at b: $bccb$ and $beab$. There are two more circuits starting at c, namely $ccbc$ and $cbcc$. Similarly there are the circuits $deed$, $eede$ and $edee$, as well as the other trips around the triangle: $eabe$, $dead$, and $eade$.

19. The way to form these powers is first to form the matrix representing R, namely

$$\mathbf{M}_R = \begin{bmatrix} 0 & 0 & 1 & 0 & 0 \\ 0 & 0 & 0 & 1 & 0 \\ 1 & 0 & 0 & 0 & 1 \\ 0 & 0 & 1 & 0 & 0 \\ 1 & 1 & 0 & 1 & 0 \end{bmatrix},$$

and then take successive Boolean powers of it to get the matrices representing R^2, R^3, and so on. Finally, for part **(f)** we take the join of the matrices representing R, R^2, $\ldots$, R^5. Since the matrix is a perfectly good way to express the relation, we will not list the ordered pairs.

a) The matrix for R^2 is the Boolean product of the matrix displayed above with itself, namely

$$\mathbf{M}_{R^2} = \mathbf{M}_R^{[2]} = \begin{bmatrix} 1 & 0 & 0 & 0 & 1 \\ 0 & 0 & 1 & 0 & 0 \\ 1 & 1 & 1 & 1 & 0 \\ 1 & 0 & 0 & 0 & 1 \\ 0 & 0 & 1 & 1 & 0 \end{bmatrix}.$$

b) The matrix for R^3 is the Boolean product of the first matrix displayed above with the answer to part **(a)**, namely

$$\mathbf{M}_{R^3} = \mathbf{M}_R^{[3]} = \begin{bmatrix} 1 & 1 & 1 & 1 & 0 \\ 1 & 0 & 0 & 0 & 1 \\ 1 & 0 & 1 & 1 & 1 \\ 1 & 1 & 1 & 1 & 0 \\ 1 & 0 & 1 & 0 & 1 \end{bmatrix}.$$

c) The matrix for R^4 is the Boolean product of the first matrix displayed above with the answer to part **(b)**, namely

$$\mathbf{M}_{R^4} = \mathbf{M}_R^{[4]} = \begin{bmatrix} 1 & 0 & 1 & 1 & 1 \\ 1 & 1 & 1 & 1 & 0 \\ 1 & 1 & 1 & 1 & 1 \\ 1 & 0 & 1 & 1 & 1 \\ 1 & 1 & 1 & 1 & 1 \end{bmatrix}.$$

d) The matrix for R^5 is the Boolean product of the first matrix displayed above with the answer to part **(c)**, namely

$$\mathbf{M}_{R^5} = \mathbf{M}_R^{[5]} = \begin{bmatrix} 1 & 1 & 1 & 1 & 1 \\ 1 & 0 & 1 & 1 & 1 \\ 1 & 1 & 1 & 1 & 1 \\ 1 & 1 & 1 & 1 & 1 \\ 1 & 1 & 1 & 1 & 1 \end{bmatrix}.$$

e) The matrix for R^6 is the Boolean product of the first matrix displayed above with the answer to part **(d)**, namely

$$\mathbf{M}_{R^6} = \mathbf{M}_R^{[6]} = \begin{bmatrix} 1 & 1 & 1 & 1 & 1 \\ 1 & 1 & 1 & 1 & 1 \\ 1 & 1 & 1 & 1 & 1 \\ 1 & 1 & 1 & 1 & 1 \\ 1 & 1 & 1 & 1 & 1 \end{bmatrix}.$$

f) The matrix for R^* is the join of the first matrix displayed above and the answers to parts **(a)** through **(d)**, namely

$$\mathbf{M}_{R^*} = \mathbf{M}_R \vee \mathbf{M}_R^{[2]} \vee \mathbf{M}_R^{[3]} \vee \mathbf{M}_R^{[4]} \vee \mathbf{M}_R^{[5]} = \begin{bmatrix} 1 & 1 & 1 & 1 & 1 \\ 1 & 1 & 1 & 1 & 1 \\ 1 & 1 & 1 & 1 & 1 \\ 1 & 1 & 1 & 1 & 1 \\ 1 & 1 & 1 & 1 & 1 \end{bmatrix}.$$

21. **a)** The pair (a, b) is in R^2 if there is a person c other than a or b who is in a class with a and a class with b. Note that it is almost certain that (a, a) is in R^2, since as long as a is taking a class that has at least one other person in it, that person serves as the "c."
b) The pair (a, b) is in R^3 if there are persons c (different from a) and d (different from b and c) such that c is in a class with a, c is in a class with d, and d is in a class with b.
c) The pair (a, b) is in R^* if there is a sequence of persons, $c_0, c_1, c_2, \ldots, c_n$, with $n \geq 1$, such that $c_0 = a$, $c_n = b$, and for each i from 1 to n, $c_{i-1} \neq c_i$ and c_{i-1} is in at least one class with c_i.

23. Suppose that $(a, b) \in R^*$; then there is a path from a to b in (the digraph for) R. Given such a path, if R is symmetric, then the reverse of every edge in the path is also in R; therefore there is a path from b to a in R (following the given path backwards). This means that (b, a) is in R^* whenever (a, b) is, exactly what we needed to prove.

25. Algorithm 1 finds the transitive closure by computing the successive powers and taking their join. We exhibit our answers in matrix form as $\mathbf{M}_R \vee \mathbf{M}_R^{[2]} \vee \ldots \vee \mathbf{M}_R^{[n]} = \mathbf{M}_{R^*}$.

a) $$\begin{bmatrix} 0 & 1 & 0 & 0 \\ 1 & 0 & 1 & 0 \\ 0 & 0 & 0 & 1 \\ 1 & 0 & 0 & 0 \end{bmatrix} \vee \begin{bmatrix} 1 & 0 & 1 & 0 \\ 0 & 1 & 0 & 1 \\ 1 & 0 & 0 & 0 \\ 0 & 1 & 0 & 0 \end{bmatrix} \vee \begin{bmatrix} 0 & 1 & 0 & 1 \\ 1 & 0 & 1 & 0 \\ 0 & 1 & 0 & 0 \\ 1 & 0 & 1 & 0 \end{bmatrix} \vee \begin{bmatrix} 1 & 0 & 1 & 0 \\ 0 & 1 & 0 & 1 \\ 1 & 0 & 1 & 0 \\ 0 & 1 & 0 & 1 \end{bmatrix} = \begin{bmatrix} 1 & 1 & 1 & 1 \\ 1 & 1 & 1 & 1 \\ 1 & 1 & 1 & 1 \\ 1 & 1 & 1 & 1 \end{bmatrix}$$

b) $$\begin{bmatrix} 0 & 0 & 0 & 0 \\ 1 & 0 & 1 & 0 \\ 1 & 0 & 0 & 1 \\ 1 & 0 & 1 & 0 \end{bmatrix} \vee \begin{bmatrix} 0 & 0 & 0 & 0 \\ 1 & 0 & 0 & 1 \\ 1 & 0 & 1 & 0 \\ 1 & 0 & 0 & 1 \end{bmatrix} \vee \begin{bmatrix} 0 & 0 & 0 & 0 \\ 1 & 0 & 1 & 0 \\ 1 & 0 & 0 & 1 \\ 1 & 0 & 1 & 0 \end{bmatrix} \vee \begin{bmatrix} 0 & 0 & 0 & 0 \\ 1 & 0 & 0 & 1 \\ 1 & 0 & 1 & 0 \\ 1 & 0 & 0 & 1 \end{bmatrix} = \begin{bmatrix} 0 & 0 & 0 & 0 \\ 1 & 0 & 1 & 1 \\ 1 & 0 & 1 & 1 \\ 1 & 0 & 1 & 1 \end{bmatrix}$$

c) $$\begin{bmatrix}0&1&1&1\\0&0&1&1\\0&0&0&1\\0&0&0&0\end{bmatrix}\vee\begin{bmatrix}0&0&1&1\\0&0&0&1\\0&0&0&0\\0&0&0&0\end{bmatrix}\vee\begin{bmatrix}0&0&0&1\\0&0&0&0\\0&0&0&0\\0&0&0&0\end{bmatrix}\vee\begin{bmatrix}0&0&0&0\\0&0&0&0\\0&0&0&0\\0&0&0&0\end{bmatrix}=\begin{bmatrix}0&1&1&1\\0&0&1&1\\0&0&0&1\\0&0&0&0\end{bmatrix}$$

Note that the relation was already transitive, so its transitive closure is itself.

d) $$\begin{bmatrix}1&0&0&1\\1&0&1&0\\1&1&0&1\\0&1&0&0\end{bmatrix}\vee\begin{bmatrix}1&1&0&1\\1&1&0&1\\1&1&1&1\\1&0&1&0\end{bmatrix}\vee\begin{bmatrix}1&1&1&1\\1&1&1&1\\1&1&1&1\\1&1&0&1\end{bmatrix}\vee\begin{bmatrix}1&1&1&1\\1&1&1&1\\1&1&1&1\\1&1&1&1\end{bmatrix}=\begin{bmatrix}1&1&1&1\\1&1&1&1\\1&1&1&1\\1&1&1&1\end{bmatrix}$$

27. In Warshall's algorithm (Algorithm 2 in this section), we compute a sequence of matrices $\mathbf{W}_0$ (the matrix representing R), $\mathbf{W}_1$, $\mathbf{W}_2$, $\ldots$, $\mathbf{W}_n$, the last of which represents the transitive closure of R. Each matrix $\mathbf{W}_k$ comes from the matrix $\mathbf{W}_{k-1}$ in the following way. The $(i,j)^{\text{th}}$ entry of $\mathbf{W}_k$ is the "$\vee$" of the $(i,j)^{\text{th}}$ entry of $\mathbf{W}_{k-1}$ with the "$\wedge$" of the $(i,k)^{\text{th}}$ entry and the $(k,j)^{\text{th}}$ entry of $\mathbf{W}_{k-1}$. We will exhibit our solution by listing the matrices $\mathbf{W}_0$, $\mathbf{W}_1$, $\mathbf{W}_2$, $\mathbf{W}_3$, $\mathbf{W}_4$, in that order; $\mathbf{W}_4$ represents the answer. In each case $\mathbf{W}_0$ is the matrix of the given relation. To compute the next matrix in the solution, we need to compute it one entry at a time, using the equation just discussed (the "$\vee$" of the corresponding entry in the previous matrix with the "$\wedge$" of two entries in the old matrix), i.e., as i and j each go from 1 to 4, we need to write down the $(i,j)^{\text{th}}$ entry using this formula. Note that in computing $\mathbf{W}_k$ the kth row and the kth column are unchanged, but some of the entries in other rows and columns may change.

a) $$\begin{bmatrix}0&1&0&0\\1&0&1&0\\0&0&0&1\\1&0&0&0\end{bmatrix}\quad\begin{bmatrix}0&1&0&0\\1&1&1&0\\0&0&0&1\\1&1&0&0\end{bmatrix}\quad\begin{bmatrix}1&1&1&0\\1&1&1&0\\0&0&0&1\\1&1&1&0\end{bmatrix}\quad\begin{bmatrix}1&1&1&1\\1&1&1&1\\0&0&0&1\\1&1&1&1\end{bmatrix}\quad\begin{bmatrix}1&1&1&1\\1&1&1&1\\1&1&1&1\\1&1&1&1\end{bmatrix}$$

b) $$\begin{bmatrix}0&0&0&0\\1&0&1&0\\1&0&0&1\\1&0&1&0\end{bmatrix}\quad\begin{bmatrix}0&0&0&0\\1&0&1&0\\1&0&0&1\\1&0&1&0\end{bmatrix}\quad\begin{bmatrix}0&0&0&0\\1&0&1&0\\1&0&0&1\\1&0&1&0\end{bmatrix}\quad\begin{bmatrix}0&0&0&0\\1&0&1&1\\1&0&0&1\\1&0&1&1\end{bmatrix}\quad\begin{bmatrix}0&0&0&0\\1&0&1&1\\1&0&1&1\\1&0&1&1\end{bmatrix}$$

c) $$\begin{bmatrix}0&1&1&1\\0&0&1&1\\0&0&0&1\\0&0&0&0\end{bmatrix}\quad\begin{bmatrix}0&1&1&1\\0&0&1&1\\0&0&0&1\\0&0&0&0\end{bmatrix}\quad\begin{bmatrix}0&1&1&1\\0&0&1&1\\0&0&0&1\\0&0&0&0\end{bmatrix}\quad\begin{bmatrix}0&1&1&1\\0&0&1&1\\0&0&0&1\\0&0&0&0\end{bmatrix}\quad\begin{bmatrix}0&1&1&1\\0&0&1&1\\0&0&0&1\\0&0&0&0\end{bmatrix}$$

Note that the relation was already transitive, so each matrix in the sequence was the same.

d) $$\begin{bmatrix}1&0&0&1\\1&0&1&0\\1&1&0&1\\0&1&0&0\end{bmatrix}\quad\begin{bmatrix}1&0&0&1\\1&0&1&1\\1&1&0&1\\0&1&0&0\end{bmatrix}\quad\begin{bmatrix}1&0&0&1\\1&0&1&1\\1&1&1&1\\1&1&1&1\end{bmatrix}\quad\begin{bmatrix}1&0&0&1\\1&1&1&1\\1&1&1&1\\1&1&1&1\end{bmatrix}\quad\begin{bmatrix}1&1&1&1\\1&1&1&1\\1&1&1&1\\1&1&1&1\end{bmatrix}$$

29. a) We need to include at least the transitive closure, which we can compute by Algorithm 1 or Algorithm 2 to be (in matrix form) $\begin{bmatrix}1&1&0&1\\0&0&0&0\\0&0&1&0\\1&1&0&1\end{bmatrix}$. All we need in addition is the pair $(2,2)$ in order to make the relation reflexive. Note that the result is still transitive (the addition of a pair (a,a) cannot make a transitive relation no longer transitive), so our answer is $\begin{bmatrix}1&1&0&1\\0&1&0&0\\0&0&1&0\\1&1&0&1\end{bmatrix}$.

b) The symmetric closure of the original relation is represented by $\begin{bmatrix} 0 & 1 & 0 & 1 \\ 1 & 0 & 0 & 0 \\ 0 & 0 & 1 & 0 \\ 1 & 0 & 0 & 0 \end{bmatrix}$. We need at least the transitive closure of this relation, namely $\begin{bmatrix} 1 & 1 & 0 & 1 \\ 1 & 1 & 0 & 1 \\ 0 & 0 & 1 & 0 \\ 1 & 1 & 0 & 1 \end{bmatrix}$. Since it is also symmetric, we are done. Note that it would not have been correct to find first the transitive closure of the original matrix and then make it symmetric, since the pair $(2,2)$ would be missing. What is going on here is that the transitive closure of a symmetric relation is still symmetric, but the symmetric closure of a transitive relation might not be transitive.

c) Since the answer to part **(b)** was already reflexive, it must be the answer to this part as well.

31. Algorithm 1 has a loop executed $O(n)$ times in which the primary operation is the Boolean product computation (the join operation is fast by comparison). If we can do the product in $O(n^{2.8})$ bit operations, then the number of bit operations in the entire algorithm is $O(n \cdot n^{2.8}) = O(n^{3.8})$. Since Algorithm 2 does not use the Boolean product, a fast Boolean product algorithm is irrelevant, so Algorithm 2 still requires $O(n^3)$ bit operations.

33. There are two ways to go. One approach is to take the output of Algorithm 1 as it stands and then make sure that all the pairs (a,a) are included by forming the join with the identity matrix (specifically set $\mathbf{B} := \mathbf{B} \vee \mathbf{I}_n$). See the discussion in Exercise 29a for the justification. The other approach is to insure the reflexivity at the beginning by initializing $\mathbf{A} := \mathbf{M}_r \vee \mathbf{I}_n$; if we do this, then only paths of length strictly less than n need to be looked at, so we can change the n in the loop to $n-1$.

35. a) No relation that contains R is not reflexive, since R already contains all the pairs $(0,0)$, $(1,1)$, and $(2,2)$. Therefore there is no "nonreflexive" closure of R.

b) Suppose S were the closure of R with respect to this property. Since R does not have an odd number of elements, $S \neq R$, so S must be a proper superset of R. Clearly S cannot have more than 5 elements, for if it did, then any subset of S consisting of R and one element of $S - R$ would be a proper subset of S with the property; this would violate the requirement that S be a subset of every superset of R with the property. Thus S must have exactly 5 elements. Let T be another superset of R with 5 elements (there are $9 - 4 = 5$ such sets in all). Thus T has the property, but S is not a subset of T. This contradicts the definition. Therefore our original assumption was faulty, and the closure does not exist.

SECTION 8.5 Equivalence Relations

This section is extremely important. If you do nothing else, do Exercise 9 and understand it, for it deals with the most common instances of equivalence relations. Exercise 16 is interesting—it hints at what fractions really are (if understood properly) and perhaps helps to explain why children (and adults) usually have so much trouble with fractions: they really involve equivalence relations. Spend some time thinking about fractions in this context. (See also Writing Project 4 for this chapter.)

It is usually easier to understand equivalence relations in terms of the associated partition—it's a more concrete visual image. Thus make sure you understand exactly what Theorem 2 says. Look at Exercise 67 for the relationship between equivalence relations and closures.

1. In each case we need to check for reflexivity, symmetry, and transitivity.

a) This is an equivalence relation; it is easily seen to have all three properties. The equivalence classes all have just one element.

b) This relation is not reflexive since the pair $(1,1)$ is missing. It is also not transitive, since the pairs $(0,2)$ and $(2,3)$ are there, but not $(0,3)$.

c) This is an equivalence relation. The elements 1 and 2 are in the same equivalence class; 0 and 3 are each in their own equivalence class.

d) This relation is reflexive and symmetric, but it is not transitive. The pairs $(1,3)$ and $(3,2)$ are present, but not $(1,2)$.

e) This relation would be an equivalence relation were the pair $(2,1)$ present. As it is, its absence makes the relation neither symmetric nor transitive.

3. As in Exercise 1, we need to check for reflexivity, symmetry, and transitivity.

a) This is an equivalence relation, one of the general form that two things are considered equivalent if they have the same "something" (see Exercise 9 for a formalization of this idea). In this case the "something" is the value at 1.

b) This is not an equivalence relation because it is not transitive. Let $f(x) = 0$, $g(x) = x$, and $h(x) = 1$ for all $x \in \mathbf{Z}$. Then f is related to g since $f(0) = g(0)$, and g is related to h since $g(1) = h(1)$, but f is not related to h since they have no values in common. By inspection we see that this relation is reflexive and symmetric.

c) This relation has none of the three properties. It is not reflexive, since $f(x) - f(x) = 0 \neq 1$. It is not symmetric, since if $f(x) - g(x) = 1$, then $g(x) - f(x) = -1 \neq 1$. It is not transitive, since if $f(x) - g(x) = 1$ and $g(x) - h(x) = 1$, then $f(x) - h(x) = 2 \neq 1$.

d) This is an equivalence relation. Two functions are related here if they differ by a constant. It is clearly reflexive (the constant is 0). It is symmetric, since if $f(x) - g(x) = C$, then $g(x) - f(x) = -C$. It is transitive, since if $f(x) - g(x) = C_1$ and $g(x) - h(x) = C_2$, then $f(x) - h(x) = C_3$, where $C_3 = C_1 + C_2$ (add the first two equations).

e) This relation is not reflexive, since there are lots of functions f (for instance, $f(x) = x$) that do not have the property that $f(0) = f(1)$. It is symmetric by inspection (the roles of f and g are the same). It is not transitive. For instance, let $f(0) = g(1) = h(0) = 7$, and let $f(1) = g(0) = h(1) = 3$; fill in the remaining values arbitrarily. Then f and g are related, as are g and h, but f is not related to h since $7 \neq 3$.

5. Obviously there are many possible answers here. We can say that two buildings are equivalent if they were opened during the same year; an equivalence class consists of the set of buildings opened in a given year (as long as there was at least one building opened that year). For another example, we can define two buildings to be equivalent if they have the same number of stories; the equivalence classes are the set of 1-story buildings, the set of 2-story buildings, and so on (one class for each n for which there is at least one n-story building). In our third example, partition the set of all buildings into two classes—those in which you do have a class this semester and those in which you don't. (We assume that each of these is nonempty.) Every building in which you have a class is equivalent to every building in which you have a class (including itself), and every building in which you don't have a class is equivalent to every building in which you don't have a class (including itself).

7. Two propositions are equivalent if their truth tables are identical. This relation is reflexive, since the truth table of a proposition is identical to itself. It is symmetric, since if p and q have the same truth table, then q and p have the same truth table. There is one technical point about transitivity that should be noted. We need to assume that the truth tables, as we consider them for three propositions p, q, and r, have the same atomic variables in them. If we make this assumption (and it cannot hurt to do so, since adding information about extra variables that do not appear in a pair of propositions does not change the truth value of the propositions), then we argue in the usual way: if p and q have identical truth tables, and if q and r have identical truth tables, then p and r have that same common truth table. The proposition $\mathbf{T}$ is always true; therefore the equivalence class for this proposition consists of all propositions that are always true, no matter what truth values the atomic variables have. Recall that we call such a proposition a tautology. Therefore the equivalence class of $\mathbf{T}$ is the set of all tautologies. Similarly, the equivalence class of $\mathbf{F}$ is the set of all contradictions.

9. This is an important exercise, since very many equivalence relations are of this form. (In fact, all of them are—see Exercise 10.)

a) This relation is reflexive, since obviously $f(x) = f(x)$ for all $x \in A$. It is symmetric, since if $f(x) = f(y)$, then $f(y) = f(x)$ (this is one of the fundamental properties of equality). It is transitive, since if $f(x) = f(y)$ and $f(y) = f(z)$, then $f(x) = f(z)$ (this is another fundamental property of equality).

b) The equivalence class of x is the set of all $y \in A$ such that $f(y) = f(x)$. This is by definition just the inverse image of $f(x)$. Thus the equivalence classes are precisely the sets $f^{-1}(b)$ for every b in the range of f.

11. This follows from Exercise 9, where f is the function that takes a bit string of length 3 or more to its first 3 bits.

13. This follows from Exercise 9, where f is the function that takes a bit string of length 3 or more to the ordered pair (b_1, b_3), where b_1 is the first bit of the string and b_3 is the third bit of the string. Two bit strings agree on their first and third bits if and only if the corresponding ordered pairs for these two strings are equal ordered pairs.

15. By algebra, the given condition is the same as the condition that $f((a,b)) = f((c,d))$, where $f((x,y)) = x - y$. Therefore by Exercise 9 this is an equivalence relation. If we want a more explicit proof, we can argue as follows. For reflexivity, $((a,b),(a,b)) \in R$ because $a + b = b + a$. For symmetry, $((a,b),(c,d)) \in R$ if and only if $a + d = b + c$, which is equivalent to $c + b = d + a$, which is true if and only if $((c,d),(a,b)) \in R$. For transitivity, suppose $((a,b),(c,d)) \in R$ and $((c,d),(e,f)) \in R$. Thus we have $a + d = b + c$ and $c + e = d + f$. Adding, we obtain $a + d + c + e = b + c + d + f$. Simplifying, we have $a + e = b + f$, which tells us that $((a,b),(e,f)) \in R$.

17. a) This follows from Exercise 9, where the function f from the set of differentiable functions (from $\mathbf{R}$ to $\mathbf{R}$) to the set of functions (from $\mathbf{R}$ to $\mathbf{R}$) is the differentiation operator—i.e., f of a function g is the function g'. The best way to think about this is that any relation defined by a statement of the form "a and b are equivalent if they have the same whatever" is an equivalence relation. Here "whatever" is "derivative"; in the general situation of Exercise 9, "whatever" is "function value under f."

b) We are asking for all functions that have the same derivative that the function $f(x) = x^2$ has, i.e., all functions of x whose derivative is $2x$. In other words, we are asking for the general antiderivative of $2x$, and we know that $\int 2x = x^2 + C$, where C is any constant. Therefore the functions in the same equivalence class as $f(x) = x^2$ are all the functions of the form $g(x) = x^2 + C$ for some constant C. Indefinite integrals in calculus, then, give equivalence classes of functions as answers, not just functions.

19. This follows from Exercise 9, where the function f from the set of all URLs to the set of all Web pages is the function that assigns to each URL the Web page for that URL.

21. We need to observe whether the relation is reflexive (there is a loop at each vertex), symmetric (every edge that appears is accompanied by its antiparallel mate—an edge involving the same two vertices but pointing in the opposite direction), and transitive (paths of length 2 are accompanied by the path of length 1—i.e., edge—between the same two vertices in the same direction). We see that this relation is not transitive, since the edges (c,d) and (d,c) are missing.

23. As in Exercise 21, this relation is not transitive, since several required edges are missing (such as (a,c)).

25. This follows from Exercise 9, with f being the function from bit strings to nonnegative integers given by $f(s) =$the number of 1's in s.

27. Only parts **(a)** and **(b)** are relevant here, since the others are not equivalence relations.

a) An equivalence class is the set of all people who are the same age. (To really identify the equivalence class and the equivalence relation itself, one would need to specify exactly what one meant by "the same age." For example, we could define two people to be the same age if their official dates of birth were identical. In that case, everybody born on April 25, 1948, for example, would constitute one equivalence class.)

b) For each pair (m,f) of a man and a woman, the set of offspring of their union, if nonempty, is an equivalence class. In many cases, then, an equivalence class consists of all the children in a nuclear family with children. (In real life, of course, this is complicated by such things as divorce and remarriage.)

29. The equivalence class of 011 is the set of all bit strings that are related to 011, namely the set of all bit strings that have the same number of 1's as 011. In other words, it is the (infinite) set of all bit strings with exactly 2 1's: $\{11, 110, 101, 011, 1100, 1010, 1001, \ldots\}$.

31. a) We need the first three bits of each string in the equivalence class to agree with the first three bits of 010. Thus this equivalence class is the (infinite) set of all bit strings that begin 010, which we can list as $\{010, 0100, 0101, 01000, 01001, 01010, \ldots\}$.

b) We need the first three bits of each string in the equivalence class to agree with the first three bits of 1011. Thus this equivalence class is the (infinite) set of all bit strings that begin 101, which we can list as $\{101, 1010, 1011, 10100, 10101, 10110, \ldots\}$.

c) We need the first three bits of each string in the equivalence class to agree with the first three bits of 11111. Thus this equivalence class is the (infinite) set of all bit strings that begin 111, which we can list as $\{111, 1110, 1111, 11100, 11101, 11110, \ldots\}$.

d) This string is in the equivalence class given in part **(a)**. Therefore its equivalence class is the same.

33. This is like Example 15. Each bit string of length less than 4 is in an equivalence class by itself ($[\lambda]_{R_4} = \{\lambda\}$, $[0]_{R_4} = \{0\}$, $[1]_{R_4} = \{1\}$, $[00]_{R_4} = \{00\}$, $[01]_{R_4} = \{01\}$, $\ldots$, $[111]_{R_4} = \{111\}$). This accounts for $1+2+4+8 = 15$ equivalence classes. The remaining 16 equivalence classes are determined by the bit strings of length 4:

$$[0000]_{R_4} = \{0000, 00000, 00001, 000000, 000001, 000010, 000011, 0000000, \ldots\}$$
$$[0001]_{R_4} = \{0001, 00010, 00011, 000100, 000101, 000110, 000111, 0001000, \ldots\}$$
$$[0010]_{R_4} = \{0010, 00100, 00101, 001000, 001001, 001010, 001011, 0010000, \ldots\}$$
$$\vdots$$
$$[1111]_{R_4} = \{1111, 11110, 11111, 111100, 111101, 111110, 111111, 1111000, \ldots\}$$

35. We have by definition that $[n]_5 = \{ i \mid i \equiv n \pmod 5 \}$.
a) $[2]_5 = \{ i \mid i \equiv 2 \pmod 5 \} = \{\ldots, -8, -3, 2, 7, 12, \ldots\}$
b) $[3]_5 = \{ i \mid i \equiv 3 \pmod 5 \} = \{\ldots, -7, -2, 3, 8, 13, \ldots\}$
c) $[6]_5 = \{ i \mid i \equiv 6 \pmod 5 \} = \{\ldots, -9, -4, 1, 6, 11, \ldots\}$
d) $[-3]_5 = \{ i \mid i \equiv -3 \pmod 5 \} = \{\ldots, -8, -3, 2, 7, 12, \ldots\}$ (the same as $[2]_5$)

37. This is very similar to Example 14. There are 6 equivalence classes, namely

$$\begin{aligned}
[0]_6 &= \{\ldots, -12, -6, 0, 6, 12, \ldots\}, \\
[1]_6 &= \{\ldots, -11, -5, 1, 7, 13, \ldots\}, \\
[2]_6 &= \{\ldots, -10, -4, 2, 8, 14, \ldots\}, \\
[3]_6 &= \{\ldots, -9, -3, 3, 9, 15, \ldots\}, \\
[4]_6 &= \{\ldots, -8, -2, 4, 10, 16, \ldots\}, \\
[5]_6 &= \{\ldots, -7, -1, 5, 11, 17, \ldots\}.
\end{aligned}$$

Another way to describe this collection is to say that it is the collection of sets $\{ 6n + k \mid n \in \mathbf{Z} \}$ for $k = 0, 1, 2, 3, 4, 5$.

39. a) We observed in the solution to Exercise 15 that (a, b) is equivalent to (c, d) if $a - b = c - d$. Thus because $1 - 2 = -1$, we have $[(1, 2)] = \{ (a, b) \mid a - b = -1 \} = \{(1, 2), (3, 4), (4, 5), (5, 6), \ldots\}$.
b) Since the equivalence class of (a, b) is entirely *determined* by the integer $a - b$, which can be negative, positive, or zero, we can interpret the equivalences classes as *being* the integers. This is a standard way to *define* the integers once we have defined the whole numbers.

41. The sets in a partition must be nonempty, pairwise disjoint, and have as their union all of the underlying set.
a) This is not a partition, since the sets are not pairwise disjoint (the elements 2 and 4 each appear in two of the sets).
b) This is a partition. **c)** This is a partition.
d) This is not a partition, since none of the sets includes the element 3.

43. In each case, we need to see that the collection of subsets satisfy three conditions: they are nonempty, they are pairwise disjoint, and their union is the entire set of 256 bit strings of length 8.
a) This is a partition, since strings must begin either 1 or 0, and those that begin 0 must continue with either 0 or 1 in their second position. It is clear that the three subsets satisfy the conditions.
b) This is not a partition, since these subsets are not pairwise disjoint. The string 00000001, for example, contains both 00 and 01.
c) This is clearly a partition. Each of these four subsets contains 64 bit strings, and no two of them overlap.
d) This is not a partition, because the union of these subsets is not the entire set. For example, the string 00000010 is in none of the subsets.
e) This is a partition. Each bit string contains some number of 1's. This number can be identified in exactly one way as of the form $3k$, the form $3k + 1$, or the form $3k + 2$, where k is a nonnegative integer; it really is just looking at the equivalence classes of the number of 1's modulo 3.

45. In each case, we need to see that the collection of subsets satisfy three conditions: they are nonempty, they are pairwise disjoint, and their union is the entire set $\mathbf{Z} \times \mathbf{Z}$.
a) This is not a partition, since the subsets are not pairwise disjoint. The pair $(2, 3)$, for example, is in both of the first two subsets listed.

b) This is a partition. Every pair satisfies exactly one of the conditions listed about the parity of x and y, and clearly these subsets are nonempty.

c) This is not a partition, since the subsets are not pairwise disjoint. The pair $(2,3)$, for example, is in both of the first two subsets listed. Also, $(0,0)$ is in none of the subsets.

d) This is a partition. Every pair satisfies exactly one of the conditions listed about the divisibility of x and y by 3, and clearly these subsets are nonempty.

e) This is a partition. Every pair satisfies exactly one of the conditions listed about the positiveness of x and y, and clearly these subsets are nonempty.

f) This is not a partition, because the union of these subsets is not all of $\mathbf{Z} \times \mathbf{Z}$. In particular, $(0,0)$ is in none of the parts.

47. In each case, we need to list all the pairs we can where both coordinates are chosen from the same subset. We should proceed in an organized fashion, listing all the pairs corresponding to each part of the partition.

a) $\{(0,0),(1,1),(1,2),(2,1),(2,2),(3,3),(3,4),(3,5),(4,3),(4,4),(4,5),(5,3),(5,4),(5,5)\}$

b) $\{(0,0),(0,1),(1,0),(1,1),(2,2),(2,3),(3,2),(3,3),(4,4),(4,5),(5,4),(5,5)\}$

c) $\{(0,0), (0,1), (0,2), (1,0), (1,1), (1,2), (2,0), (2,1), (2,2), (3,3), (3,4), (3,5), (4,3), (4,4), (4,5), (5,3), (5,4), (5,5)\}$

d) $\{(0,0),(1,1),(2,2),(3,3),(4,4),(5,5)\}$

49. We need to show that every equivalence class modulo 6 is contained in an equivalence class modulo 3. We claim that in fact, for each $n \in \mathbf{Z}$, $[n]_6 \subseteq [n]_3$. To see this suppose that $m \in [n]_6$. This means that $m \equiv n \pmod 6$, i.e., that $m - n$ is a multiple of 6. Then perforce $m - n$ is a multiple of 3, so $m \equiv n \pmod 3$, which means that $m \in [n]_3$.

51. By the definition given in the preamble to Exercise 49, we need to show that every set in the first partition is a subset of some set in the second partition. Let A be a set in the first partition. So A is the set of all bit strings of length 16 that agree on their last eight bits. Pick a particular element x of A, and suppose that the last four bits of x are $abcd$. Then the set of all bit strings of length 16 whose last four bits are $abcd$ is one of the sets in the second partition, and clearly every string in A is in that set, since every string in A agrees with x on the last eight bits, and therefore perforce agrees on the last four bits.

53. We are asked to show that every equivalence class for R_{31} is a subset of some equivalence class for R_8. Let $[x]_{R_{31}}$ be an arbitrary equivalence class for R_{31}. We claim that $[x]_{R_{31}} \subseteq [x]_{R_8}$; proving this claim finishes the proof. To show that one set is a subset of another set, we choose an arbitrary element y in the first set and show that it is also an element of the second set. In this case since $y \in [x]_{R_{31}}$, we know that y is equivalent to x under R_{31}, that is, that either $y = x$ or y and x are each at least 31 characters long and agree on their first 31 characters. Because strings that are at least 31 characters long and agree on their first 31 characters perforce are at least 8 characters long and on their first 8 characters, we know that either $y = x$ or y and x are each at least 8 characters long and agree on their first 8 characters. This means that y is equivalent to x under R_8, that is, that $y \in [x]_{R_8}$.

55. We need first to make the relation symmetric, so we add the pairs (b,a), (c,a), and (e,d). Then we need to make it transitive, so we add the pairs (b,c), (c,b), (a,a), (b,b), (c,c), (d,d), and (e,e). (In other words, we formed the transitive closure of the symmetric closure of the original relation.) It happens that we have already achieved reflexivity, so we are done; if there had been some pairs (x,x) missing at this point, we would have added them as well. Thus the desired equivalence relation is the one consisting of the original 3 pairs and the 10 we have added. There are two equivalence classes, $\{a,b,c\}$ and $\{d,e\}$.

57. **a)** The equivalence class of 1 is the set of all real numbers that differ from 1 by an integer. Obviously this is the set of all integers.

b) The equivalence class of $1/2$ is the set of all real numbers that differ from $1/2$ by an integer, namely $1/2$, $3/2$, $5/2$, etc., and $-1/2$, $-3/2$, etc. These are often called **half-integers**. We could write this set as $\{\,(2n+1)/2 \mid n \in \mathbf{Z}\,\}$, among other ways.

59. This problem actually deals with a branch of mathematics called group theory; the object being studied here is related to a certain dihedral group. If this fascinates you, you might want to take a course with a title like Abstract Algebra or Modern Algebra, in which such things are studied in depth.

In order to have a way to talk about specific colorings, let us agree that a sequence of length four, each element of which is either r or b, represents a coloring of the 2×2 checkerboard, where the first letter denotes the color of the upper left square, the second letter denotes the color of the upper right square, the third letter denotes the color of the lower left square, and the fourth letter denotes the color of the lower right square. For example, the board in which every square is red except the upper right would be represented by $rbrr$. There are really only four different rotations, since after the rotation we need to end up with another checkerboard (and we can assume that the edges of the board are horizontal and vertical). If we rotate our sample coloring 90° clockwise, then we obtain the coloring $rrrb$; if we rotate it 180°, then we obtain the coloring $rrbr$; if we rotate it 270° clockwise (or 90° counterclockwise), then we obtain the coloring $brrr$; and if we rotate it 360° clockwise (or 0°—i.e., not at all), then we obtain the coloring $rbrr$ itself back. Note also that some colorings are *invariant* (i.e., unchanged) under rotations in addition to the 360° one; for example, $bbbb$ is invariant under all rotations, and $brrb$ is invariant under a 180° rotation. Similarly there are four reflections: around the center vertical axis of the board, around the center horizontal axis, around the lower-left-to-upper-right diagonal, and around the lower-right-to-upper-left diagonal. For example, applying the vertical axis reflection to $rrbb$ yields itself, while applying the lower-left-to-upper-right diagonal reflection results in $brbr$.

The definition of equivalence for this problem makes the proof rather messy, since both rotations and reflections are involved, and it is required that we reduce everything to just one or two operations. In fact, we claim that there are only eight possible motions of this square: clockwise rotations of 0°, 90°, 180°, or 270°, and reflections through the vertical, horizontal, lower-left-to-upper-right, and lower-right-to-upper-left diagonals. To verify this, we must show that the composition of every two of these operations is again an operation in our list. Below is the "group table" that shows this, where we use the symbols $r0$, $r90$, $r180$, $r270$, fv, fh, fp, and fn for these operations, respectively. (The mnemonic is that r stands for "rotation," f stands for "flip," and v, h, p, and n stand for "vertical," "horizontal", "positive-sloping," and "negative-sloping," respectively.) It is read just like a multiplication table, with the operation $\circ$ meaning "followed by." For example, if we first perform $r90$ and then perform fh, then we get the same result as if we had just performed fp (try it!).

$\circ$	$r0$	$r90$	$r180$	$r270$	fv	fh	fp	fn
$r0$	$r0$	$r90$	$r180$	$r270$	fv	fh	fp	fn
$r90$	$r90$	$r180$	$r270$	$r0$	fn	fp	fv	fh
$r180$	$r180$	$r270$	$r0$	$r90$	fh	fv	fn	fp
$r270$	$r270$	$r0$	$r90$	$r180$	fp	fn	fh	fv
fv	fv	fp	fh	fn	$r0$	$r180$	$r90$	$r270$
fh	fh	fn	fv	fp	$r180$	$r0$	$r270$	$r90$
fp	fp	fh	fn	fv	$r270$	$r90$	$r0$	$r180$
fn	fn	fv	fp	fh	$r90$	$r270$	$r180$	$r0$

So the result of this computation is that we can consider only these eight moves, and not have to worry about combinations of them—every combination of moves equals just one of these eight.

a) To show reflexivity, we note that every coloring can be obtained from itself via a 0° rotation. In technical terms, the 0° rotation is the *identity element* of our group. To show symmetry, we need to observe that rotations and reflections have inverses: If C_1 comes from C_2 via a rotation of n° clockwise, then C_2 comes from C_1 via a rotation of n° counterclockwise (or equivalently, via a rotation of $(360-n)^\circ$ clockwise); and every reflection applied twice brings us back to the position (and therefore coloring) we began with. And transitivity follows from the fact that the composition of two of these operations is again one of these operations.

b) The equivalence classes are represented by colorings that are truly distinct, in the sense of not being obtainable from each other via these operations. Let us list them. Clearly there is just one coloring using four red squares, and so just one equivalence class, $[rrrr]$. Similarly there is only one using four blues, $[bbbb]$. There is also just one equivalence class of colorings using three reds and one blue, since no matter which corner the single blue occupies in such a coloring, we can rotate to put the blue in any other corner. Thus our third and fourth equivalence classes are $[rrrb]$ and $[bbbr]$. Note that each of them contains four colorings. (For example, $[rrrb]=\{rrrb, rrbr, rbrr, brrr\}$.) This leaves only the colorings with two reds and two blues to consider. In every such coloring, either the red squares are adjacent (i.e., share a common edge), such as in $bbrr$, or they are not (e.g., $brrb$). Clearly the red squares are adjacent if and only if the blue ones are, since the only pairs of nonadjacent squares are (lower-left,upper-right) and (upper-left,lower-right). It is equally clear that there are only two colorings in which the red squares are not adjacent, namely $rbbr$ and $brrb$, and they are equivalent via a 90° rotation (among other transformations). So our fifth equivalence class is $[rbbr]=\{rbbr, brrb\}$. Finally, there is only one more equivalence class, and it contains the remaining four colorings (in which the two red squares are adjacent and the two blue squares are adjacent), namely $\{rrbb, brbr, bbrr, rbrb\}$, since each of these can be obtained from each of the others by a rotation. In summary we have partitioned the set of $2^4=16$ colorings (i.e., r-b strings of length four) into six equivalence classes, two of which have cardinality one, three of which have cardinality four, and one of which has cardinality two.

One final comment. We saw in the solution to part **(b)** that only rotations are needed to show the equivalence of every pair of equivalent colorings using just red and blue. This means that we are actually dealing with just part of the dihedral group here. If more colors had been used, then we would have needed to use the reflections as well. A complete discussion would get us into Pólya's theory of enumeration, which is studied in advanced combinatorics classes.

61. It is easier to write down a partition than it is to list the pairs in an equivalence relation, so we will answer the question using this notation. Let the set be $\{1,2,3\}$. We want to write down all possible partitions of this set. One partition is just $\{\{1,2,3\}\}$, i.e., having just one set (this corresponds to the equivalence relation in which every pair of elements are related). At the other extreme, there is the partition $\{\{1\},\{2\},\{3\}\}$, which corresponds to the equality relation (each x is related only to itself). The only other way to split up the elements of this set is into a set with two elements and a set with one element, and there are clearly three ways to do this, depending on which element we decide to put in the set by itself. Thus we get the partitions (pay attention to the punctuation!) $\{\{1,2\},\{3\}\}$, $\{\{1,3\},\{2\}\}$, and $\{\{2,3\},\{1\}\}$. If we wished to list the ordered pairs, we could; for example, the relation corresponding to $\{\{2,3\},\{1\}\}$ is $\{(2,2),(2,3),(3,2),(3,3),(1,1)\}$. We found five partitions, so the answer to the question is 5.

63. We do get an equivalence relation. The issue is whether the relation formed in this way is reflexive, transitive and symmetric. It is clearly reflexive, since we included all the pairs (a,a) at the outset. It is clearly transitive, since the last thing we did was to form the transitive closure. It is symmetric by Exercise 23 in Section 8.4.

65. We end up with the relation R that we started with. Two elements are related if they are in the same set of the partition, but the partition is made up of the equivalence classes of R, so two elements are related precisely if they are related in R.

67. We make use of Exercise 63. Given the relation R, we first form the reflexive closure R' of R by adding to R each pair (a,a) that is not already there. Next we form the symmetric closure R'' of R', by adding, for each pair $(a,b) \in R'$ the pair (b,a) if it is not already there. Finally we apply Warshall's algorithm (or Algorithm 1) from Section 8.4 to form the transitive closure of R''. This is the smallest equivalence relation containing R.

69. The exercise asks us to compute $p(n)$ for $n = 0, 1, 2, \ldots, 10$. In doing this we will use the recurrence relation, building on what we have already computed (namely $p(n-j-1)$, noting that $n-j-1 < n$), as well as using the binomial coefficients $C(n-1,j) = \dfrac{(n-1)!}{j!(n-1-j)!}$. We organize our computation in the obvious way, using the formula in Exercise 68.

$$p(0) = 1 \quad \text{(the initial condition)}$$
$$p(1) = C(0,0)p(0) = 1 \cdot 1 = 1$$
$$p(2) = C(1,0)p(1) + C(1,1)p(0) = 1 \cdot 1 + 1 \cdot 1 = 2$$
$$p(3) = C(2,0)p(2) + C(2,1)p(1) + C(2,2)p(0) = 1 \cdot 2 + 2 \cdot 1 + 1 \cdot 1 = 5$$
$$p(4) = C(3,0)p(3) + C(3,1)p(2) + C(3,2)p(1) + C(3,3)p(0) = 1 \cdot 5 + 3 \cdot 2 + 3 \cdot 1 + 1 \cdot 1 = 15$$
$$\begin{aligned} p(5) &= C(4,0)p(4) + C(4,1)p(3) + C(4,2)p(2) + C(4,3)p(1) + C(4,4)p(0) \\ &= 1 \cdot 15 + 4 \cdot 5 + 6 \cdot 2 + 4 \cdot 1 + 1 \cdot 1 = 52 \end{aligned}$$
$$\begin{aligned} p(6) &= C(5,0)p(5) + C(5,1)p(4) + C(5,2)p(3) + C(5,3)p(2) + C(5,4)p(1) + C(5,5)p(0) \\ &= 1 \cdot 52 + 5 \cdot 15 + 10 \cdot 5 + 10 \cdot 2 + 5 \cdot 1 + 1 \cdot 1 = 203 \end{aligned}$$
$$\begin{aligned} p(7) &= C(6,0)p(6) + C(6,1)p(5) + C(6,2)p(4) + C(6,3)p(3) + C(6,4)p(2) + C(6,5)p(1) + C(6,6)p(0) \\ &= 1 \cdot 203 + 6 \cdot 52 + 15 \cdot 15 + 20 \cdot 5 + 15 \cdot 2 + 6 \cdot 1 + 1 \cdot 1 = 877 \end{aligned}$$
$$\begin{aligned} p(8) &= C(7,0)p(7) + C(7,1)p(6) + C(7,2)p(5) + C(7,3)p(4) + C(7,4)p(3) + C(7,5)p(2) \\ &\quad + C(7,6)p(1) + C(7,7)p(0) \\ &= 1 \cdot 877 + 7 \cdot 203 + 21 \cdot 52 + 35 \cdot 15 + 35 \cdot 5 + 21 \cdot 2 + 7 \cdot 1 + 1 \cdot 1 = 4140 \end{aligned}$$
$$\begin{aligned} p(9) &= C(8,0)p(8) + C(8,1)p(7) + C(8,2)p(6) + C(8,3)p(5) + C(8,4)p(4) + C(8,5)p(3) \\ &\quad + C(8,6)p(2) + C(8,7)p(1) + C(8,8)p(0) \\ &= 1 \cdot 4140 + 8 \cdot 877 + 28 \cdot 203 + 56 \cdot 52 + 70 \cdot 15 + 56 \cdot 5 + 28 \cdot 2 + 8 \cdot 1 + 1 \cdot 1 = 21147 \end{aligned}$$
$$\begin{aligned} p(10) &= C(9,0)p(9) + C(9,1)p(8) + C(9,2)p(7) + C(9,3)p(6) + C(9,4)p(5) + C(9,5)p(4) \\ &\quad + C(9,6)p(3) + C(9,7)p(2) + C(9,8)p(1) + C(9,9)p(0) \\ &= 1 \cdot 21147 + 9 \cdot 4140 + 36 \cdot 877 + 84 \cdot 203 + 126 \cdot 52 \\ &\quad + 126 \cdot 15 + 84 \cdot 5 + 36 \cdot 2 + 9 \cdot 1 + 1 \cdot 1 = 115975 \end{aligned}$$

SECTION 8.6 Partial Orderings

Partial orderings (or "partial orders"—the two phrases are used interchangeably) rival equivalence relations in importance in mathematics and computer science. Again, try to concentrate on the visual image—in this case the Hasse diagram. Play around with different posets to become familiar with the different possibilities; not all posets have to look like the less than or equal relation on the integers. Exercises 32 and 33 are important, and they are not difficult if you pay careful attention to the definitions.

1. The question in each case is whether the relation is reflexive, antisymmetric, and transitive. Suppose the relation is called R.

a) Clearly this relation is reflexive because each of 0, 1, 2, and 3 is related to itself. The relation is also antisymmetric, because the only way for a to be related to b is for a to equal b. Similarly, the relation is transitive, because if a is related to b, and b is related to c, then necessarily $a = b = c$ so a is related to c (because the relation is reflexive). This is just the equality relation on $\{0, 1, 2, 3\}$; more generally, the equality relation on any set satisfies all three conditions and is therefore a partial ordering. (It is the smallest partial ordering; reflexivity insures that every partial ordering contains at least all the pairs (a, a).)

b) This is not a partial ordering, because although the relation is reflexive, it is not antisymmetric (we have $2\,R\,3$ and $3\,R\,2$, but $2 \neq 3$), and not transitive ($3\,R\,2$ and $2\,R\,0$, but 3 is not related to 0).

c) This is a partial ordering, because it is clearly reflexive; is antisymmetric (we just need to note that $(1, 2)$ is the only pair in the relation with unequal components); and is transitive (for the same reason).

d) This is a partial ordering because it is the "less than or equal to" relation on $\{1, 2, 3\}$ together with the isolated point 0.

e) This is not a partial ordering. The relation is clearly reflexive, but it is not antisymmetric ($0\,R\,1$ and $1\,R\,0$, but $0 \neq 1$) and not transitive ($2\,R\,0$ and $0\,R\,1$, but 2 is not related to 1).

3. The question in each case is whether the relation is reflexive, antisymmetric, and transitive.

a) Since nobody is taller than himself, this relation is not reflexive so (S, R) cannot be a poset.

b) To be not taller means to be exactly the same height or shorter. Two different people x and y could have the same height, in which case $x\,R\,y$ and $y\,R\,x$ but $x \neq y$, so R is not antisymmetric and this is not a poset.

c) This is a poset. The equality clause in the definition of R guarantees that R is reflexive. To check antisymmetry and transitivity it suffices to consider unequal elements (these rules hold for equal elements trivially). If a is an ancestor of b, then b cannot be an ancestor of a (for one thing, an ancestor needs to be born before any descendant), so the relation is vacuously antisymmetric. If a is an ancestor of b, and b is an ancestor of c, then by the way "ancestor" is defined, we know that a is an ancestor of b; thus R is transitive.

d) This relation is not antisymmetric. Let a and b be any two distinct friends of yours. Then $a\,R\,b$ and $b\,R\,a$, but $a \neq b$.

5. The question in each case is whether the relation is reflexive, antisymmetric, and transitive.

a) The equality relation on any set satisfies all three conditions and is therefore a partial partial ordering. (It is the smallest partial partial ordering; reflexivity insures that every partial order contains at least all the pairs (a, a).)

b) This is not a poset, since the relation is not reflexive, not antisymmetric, and not transitive (the absence of one of these properties would have been enough to give a negative answer).

c) This is a poset, as explained in Example 1.

d) This is not a poset. The relation is not reflexive, since it is not true, for instance, that $2 \not| 2$. (It also is not antisymmetric and not transitive.)

7. a) This relation is $\{(1,1), (1,2), (1,3), (2,1), (2,2), (3,3)\}$. It is not antisymmetric because $(1, 2)$ and $(2, 1)$ are both in the relation, but $1 \neq 2$. We can see this visually by the pair of 1's symmetrically placed around the main diagonal at positions $(1, 2)$ and $(2, 1)$. Therefore this matrix does not represent a partial order.

b) This matrix represents a partial order. Reflexivity is clear. The only other pairs in the relation are $(1, 2)$ and $(1, 3)$, and clearly neither can be part of a counterexample to antisymmetry or transitivity.

c) A little trial and error shows that this relation is not transitive ($(4, 1)$ and $(1, 3)$ are present, but not $(4, 3)$) and therefore not a partial order.

9. This relation is not transitive (there are arrows from a to b and from b to d, but there is no arrow from a to d), so it is not a partial order.

11. This relation is a partial order, since it has all three properties—it is reflexive (there is an arrow at each point), antisymmetric (there are no pairs of arrows going in opposite directions between two different points), and transitive (there is no missing arrow from some x to some z when there were arrows from x to y and y to z).

13. The dual of a poset is the poset with the same underlying set and with the relation defined by declaring a related to b if and only if $b \preceq a$ in the given poset.

a) The dual relation to $\leq$ is $\geq$, so the dual poset is $(\{0,1,2\}, \geq)$. Explicitly it is the set $\{(0,0),(1,0),(1,1),(2,0),(2,1),(2,2)\}$.

b) The dual relation to $\geq$ is $\leq$, so the dual poset is $(\mathbf{Z}, \leq)$.

c) The dual relation to $\supseteq$ is $\subseteq$, so the dual poset is $(P(\mathbf{Z}), \subseteq)$.

d) There is no symbol generally used for the "is a multiple of" relation, which is the dual to the "divides" relation in this part of the exercise. If we let R be the relation such that aRb if and only if $b \mid a$, then the answer can be written $(\mathbf{Z}^+, R)$.

15. We need to find elements such that the relation holds in neither direction between them. The answers we give are not the only ones possible.

a) One such pair is $\{1\}$ and $\{2\}$. These are both subsets of $\{0,1,2\}$, so they are in the poset, but neither is a subset of the other.

b) Neither 6 nor 8 divides the other, so they are incomparable.

17. We find the first coordinate (from left to right) at which the tuples differ and place first the tuple with the smaller value in that coordinate.

a) Since $1 = 1$ in the first coordinate, but $1 < 2$ in the second coordinate, $(1,1,2) < (1,2,1)$.

b) The first two coordinates agree, but $2 < 3$ in the third, so $(0,1,2,3) < (0,1,3,2)$.

c) Since $0 < 1$ in the first coordinate, $(0,1,1,1,0) < (1,0,1,0,1)$.

19. All the strings that begin with 0 precede all those that begin with 1. The 0 comes first. Next comes 0001, which begins with three 0's, then 001, which begins with two 0's. Among the strings that begin 01, the order is $01 < 010 < 0101 < 011$. Putting this all together, we have $0 < 0001 < 001 < 01 < 010 < 0101 < 011 < 11$.

21. This is a totally ordered set, so the Hasse diagram is linear.

23. We put x above y if y divides x. We draw a line between x and y, where y divides x, if there is no number z in our set, other than x or y, such that $y \mid z \wedge z \mid x$. Note that in part **(b)** the numbers other than 1 are all (relatively) prime, so the Hasse diagram is short and wide, whereas in part **(d)** the numbers all divide one another, so the Hasse diagram is tall and narrow.

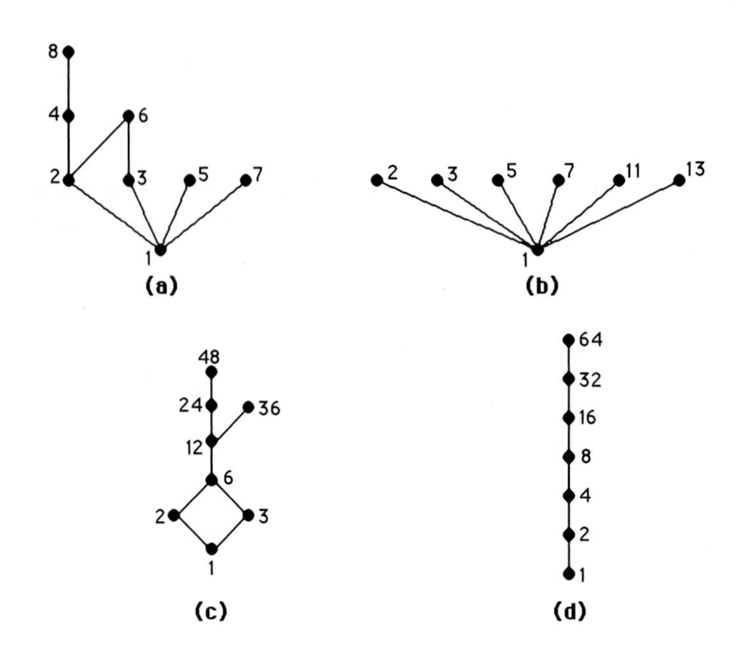

25. We need to include every pair (x,y) for which we can find a path going upward in the diagram from x to y. We also need to include all the reflexive pairs (x,x). Therefore the relation is the following set of pairs: $\{(a,a),(a,b),(a,c),(a,d),(b,b),(b,c),(b,d),(c,c),(d,d)\}$.

27. The procedure is the same as in Exercise 25: $\{(a,a),(a,d),(a,e),(a,f),(a,g),(b,b),(b,d),(b,e),(b,f),(b,g),(c,c),(c,d),(c,e),(c,f),(c,g),(d,d),(e,e),(f,f),(g,d),(g,e),(g,f),(g,g)\}$.

29. In this problem $X \preceq Y$ when $X \subseteq Y$. For (X,Y) to be in the covering relation, we need X to be a proper subset of Y but we also must have no subset strictly between X and Y. For example, $(\{a\},\{a,b,c\})$ is not in the covering relation, since $\{a\} \subset \{a,b\}$ and $\{a,b\} \subset \{a,b,c\}$. With this understanding it is easy to list the pairs in the covering relation: $(\emptyset,\{a\})$, $(\emptyset,\{b\})$, $(\emptyset,\{c\})$, $(\{a\},\{a,b\})$, $(\{a\},\{a,c\})$, $(\{b\},\{a,b\})$, $(\{b\},\{b,c\})$, $(\{c\},\{a,c\})$, $(\{c\},\{b,c\})$, $(\{a,b\},\{a,b,c\})$, $(\{a,c\},\{a,b,c\})$, and $(\{b,c\},\{a,b,c\})$.

31. Let $(S,\preceq)$ be a finite poset. We claim that this poset is just the reflexive transitive closure of its covering relation. Suppose that (a,b) is in the reflexive transitive closure of the covering relation. Then either $a=b$ or $a \prec b$ (in which cases certainly $a \preceq b$) or else there is a sequence $a \prec a_1 \prec a_2 \prec \cdots \prec a_n \prec b$, in which case again $a \preceq b$, by the transitivity of $\preceq$. Conversely, suppose that $a \preceq b$. If $a=b$, then (a,b) is certainly in the reflexive transitive closure of the covering relation. If $a \prec b$ and there is no z such that $a \prec z \prec b$, then $(a,b,)$ is in the covering relation and again therefore in its reflexive transitive closure. Otherwise, let $a \prec a_1 \prec a_2 \prec \cdots \prec a_n \prec b$ be a longest possible sequence of this form; since the poset is finite, there must be such a longest sequence. Then no intermediate elements can be inserted into this sequence (to do so would lengthen it), so each pair (a,a_1), (a_1,a_2), $\ldots$, (a_n,b) is in the covering relation, so again (a,b) is in its reflexive transitive closure. This completes the proof. Note how the finiteness of the poset was crucial here. If we let S be the set of all subsets of $\mathbf{N}$ (the set of natural numbers) under the subset relation, then we cannot recover S from its covering relation, since nothing in the covering relation allows us to relate a finite set to an infinite one; thus for example we could not recover the relationship $\{1,2\} \subset \mathbf{N}$.

33. It is helpful in this exercise to draw the Hasse diagram.

a) Maximal elements are those that do not divide any other elements of the set. In this case 24 and 45 are the only numbers that meet that requirement.

b) Minimal elements are those that are not divisible by any other elements of the set. In this case 3 and 5 are the only numbers that meet that requirement.

c) A greatest element would be one that all the other elements divide. The only two candidates (maximal elements) are 24 and 45, and since neither divides the other, we conclude that there is no greatest element.

d) A least element would be one that divides all the other elements. The only two candidates (minimal elements) are 3 and 5, and since neither divides the other, we conclude that there is no least element.

e) We want to find all elements that both 3 and 5 divide. Clearly only 15 and 45 meet this requirement.

f) The least upper bound is 15 since it divides 45 (see part **(e)**).

g) We want to find all elements that divide both 15 and 45. Clearly only 3, 5, and 15 meet this requirement.

h) The number 15 is the greatest lower bound, since both 3 and 5 divide it (see part **(g)**).

35. To help us answer the questions, we will draw the Hasse diagram, with the commas and braces eliminated in the labels, for readability.

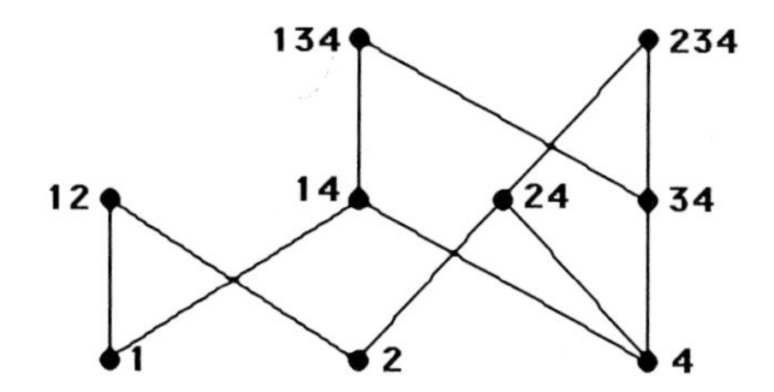

a) The maximal elements are the ones without any elements lying above them in the Hasse diagram, namely $\{1,2\}$, $\{1,3,4\}$, and $\{2,3,4\}$.

b) The minimal elements are the ones without any elements lying below them in the Hasse diagram, namely $\{1\}$, $\{2\}$, and $\{4\}$.

c) There is no greatest element, since there is more than one maximal element, none of which is greater than the others.

d) There is no least element, since there is more than one minimal element, none of which is less than the others.

e) The upper bounds are the sets containing both $\{2\}$ and $\{4\}$ as subsets, i.e., the sets containing both 2 and 4 as elements. Pictorially, these are the elements lying above both $\{2\}$ and $\{4\}$ (in the sense of there being a path in the diagram), namely $\{2,4\}$ and $\{2,3,4\}$.

f) The least upper bound is an upper bound that is less than every other upper bound. We found the upper bounds in part **(e)**, and since $\{2,4\}$ is less than (i.e., a subset of) $\{2,3,4\}$, we conclude that $\{2,4\}$ is the least upper bound.

g) To be a lower bound of both $\{1,3,4\}$ and $\{2,3,4\}$, a set must be a subset of each, and so must be a subset of their intersection, $\{3,4\}$. There are only two such subsets in our poset, namely $\{3,4\}$ and $\{4\}$. In the diagram, these are the points which lie below (in the path sense) both $\{1,3,4\}$ and $\{2,3,4\}$.

h) The greatest lower bound is a lower bound that is greater than every other lower bound. We found the lower bounds in part **(g)**, and since $\{3,4\}$ is greater than (i.e., a superset of) $\{4\}$, we conclude that $\{3,4\}$ is the greatest lower bound.

37. First we need to show that lexicographic order is reflexive, i.e., that $(a,b) \preceq (a,b)$; this is true by fiat, since we defined $\preceq$ by adding equality to $\prec$. Next we need to show antisymmetry: if $(a,b) \preceq (c,d)$ and $(a,b) \neq (c,d)$, then $(c,d) \not\preceq (a,b)$. By definition $(a,b) \prec (c,d)$ if and only if either $a \prec c$, or $a = c$ and $b \prec d$. In the first case, by the antisymmetry of the underlying relation, we know that $c \not\prec a$, and similarly in the second case we know that $d \not\prec d$. Thus there is no way that we could have $(c,d) \prec (a,b)$. Finally, for transitivity, let $(a,b) \preceq (c,d) \preceq (e,f)$. We want to show that $(a,b) \preceq (e,f)$. If one of the given inequalities is an equality, then there is nothing to prove, so we may assume that $(a,b) \prec (c,d) \prec (e,f)$. If $a \prec c$, then by the transitivity of the underlying relation, we know that $a \prec e$ and so $(a,b) \prec (e,f)$. Similarly, if $c \prec e$, then again $a \prec e$

and so $(a,b) \prec (e,f)$. The only other way for the given inequalities to hold is if $a = c = e$ and $b \prec d \prec f$. In this case the latter string of inequalities implies that $b \prec f$ and so again by definition $(a,b) \prec (e,f)$.

39. First we must show that $\preceq$ is reflexive. Since $s \preceq_1 s$ and $t \preceq_2 t$ by the reflexivity of these underlying partial orders, $(s,t) \preceq (s,t)$ by definition. For antisymmetry, assume that $(s,t) \preceq (u,v)$ and $(u,v) \preceq (s,t)$. Then by definition $s \preceq_1 u$ and $t \preceq_2 v$, and $u \preceq_1 s$ and $v \preceq_2 t$. By the antisymmetry of the underlying relations, we conclude that $s = u$ and $t = v$, whence $(s,t) = (u,v)$. Finally, for transitivity, suppose that $(s,t) \preceq (u,v) \preceq (w,x)$. This means that $s \preceq_1 u \preceq_1 w$ and $t \preceq_2 v \preceq_2 x$. The transitivity of the underlying partial orders tells us that $s \preceq_1 w$ and $t \preceq_2 x$, whence by definition $(s,t) \preceq (w,x)$.

41. **a)** We argue essentially by contradiction. Suppose that m_1 and m_2 are two maximal elements in a poset that has a greatest element g; we will show that $m_1 = m_2$. Now since g is greatest, we know that $m_1 \preceq g$, and similarly for m_2. But since each m_i is maximal, it cannot be that $m_i \prec g$; hence $m_1 = g = m_2$.

b) The proof is exactly dual to the proof in part **(a)**, so we just copy over that proof, making the appropriate changes in wording. To wit: we argue essentially by contradiction. Suppose that m_1 and m_2 are two minimal elements in a poset that has a least element l; we will show that $m_1 = m_2$. Now since l is least, we know that $l \preceq m_1$, and similarly for m_2. But since each m_i is minimal, it cannot be that $l \prec m_i$; hence $m_1 = l = m_2$.

43. In each case, we need to check whether every pair of elements has both a least upper bound and a greatest lower bound.

a) This is a lattice. If we want to find the l.u.b. or g.l.b. of two elements in the same vertical column of the Hasse diagram, then we simply take the higher or lower (respectively) element. If the elements are in different columns, then to find the g.l.b. we follow the diagonal line upward from the element on the left, and then continue upward on the right, if necessary to reach the element on the right. For example, the l.u.b. of d and c is f; and the l.u.b. of a and e is e. Finding greatest lower bounds in this poset is similar.

b) This is not a lattice. Elements b and c have f, g, and h as upper bounds, but none of them is a l.u.b.

c) This is a lattice. By considering all the pairs of elements, we can verify that every pair of them has a l.u.b. and a g.l.b. For example, b and e have g and a filling these roles, respectively.

45. As usual when trying to extend a theorem from two items to an arbitrary finite number, we will use mathematical induction. The statement we wish to prove is that if S is a subset consisting of n elements from a lattice, where n is a positive integer, then S has a least upper bound and a greatest lower bound. The two proofs are duals of each other, so we will just give the proof for least upper bound here. The basis is $n = 1$, in which case there is really nothing to prove. If $S = \{x\}$, then clearly x is the least upper bound of S. The case $n = 2$ could be singled out for special mention also, since the l.u.b. in that case is guaranteed by the definition of lattice. But there is no need to do so. Instead, we simply assume the inductive hypothesis, that every subset containing n elements has a l.u.b., and prove that every subset S containing $n+1$ elements also has a l.u.b. Pick an arbitrary element $x \in S$, and let $S' = S - \{x\}$. Since S' has only n elements, it has a l.u.b. y, by the inductive hypothesis. Since we are in a lattice, there is an element z that is the l.u.b. of x and y. We will show that in fact z is the least upper bound of S. To do this, we need to show two things: that z is an upper bound, and that every upper bound is greater than or equal to z. For the first statement, let w be an arbitrary element of S; we must show that $w \preceq z$. There are two cases. If $w = x$, then $w \preceq z$ since z is the l.u.b. of x and y. Otherwise, $w \in S'$, and so $w \preceq y$ because y is the l.u.b. of S'. But since z is the l.u.b. of x and y, we also have $y \preceq z$. By transitivity, then, $w \preceq z$. For the second statement, suppose that u is any other upper bound of S; we must show that $z \preceq u$. Since u is an upper bound of S, it is also an upper bound of x and y. But since z is the *least* upper bound of x and y, we know that $z \preceq u$.

47. The needed definitions are in Example 25.

a) No. The authority level of the first pair (1) is less than or equal to (less than, in this case) that of the second (2); but the subset of the first pair is not a subset of that of the second.

b) Yes. The authority level of the first pair (2) is less than or equal to (less than, in this case) that of the second (3); and the subset of the first pair is a subset of that of the second.

c) The classes into which information can flow are those classes whose authority level is at least as high as *Proprietary*, and whose subset is a superset of $\{$ *Cheetah, Puma* $\}$. We can list these classes: (*Proprietary*, $\{$*Cheetah, Puma*$\}$), (*Restricted*, $\{$*Cheetah, Puma*$\}$), (*Registered*, $\{$*Cheetah, Puma*$\}$), (*Proprietary*, $\{$*Cheetah, Puma, Impala*$\}$), (*Restricted*, $\{$*Cheetah, Puma, Impala*$\}$), and (*Registered*, $\{$*Cheetah, Puma, Impala*$\}$).

d) The classes from which information can flow are those classes whose authority level is at least as low as *Restricted*, and whose subset is a subset of $\{$ *Impala, Puma* $\}$, namely (*Nonproprietary*,$\{$*Impala,Puma*$\}$), (*Proprietary*,$\{$*Impala,Puma*$\}$), (*Restricted*,$\{$*Impala,Puma*$\}$), (*Nonproprietary*,$\{$*Impala*$\}$), (*Proprietary*, $\{$*Impala*$\}$), (*Restricted*,$\{$*Impala*$\}$), (*Nonproprietary*,$\{$*Puma*$\}$), (*Proprietary*,$\{$*Puma*$\}$), (*Restricted*,$\{$*Puma*$\}$), (*Nonproprietary*,$\emptyset$), (*Proprietary*,$\emptyset$), and (*Restricted*,$\emptyset$).

49. Let Π be the set of all partitions of a set S, with a relation $\preceq$ defined on Π according to the referenced preamble: a partition P_1 is a refinement of P_2 if every set in P_1 is a subset of one of the sets in P_2. We need to verify all the properties of a lattice. First we need to show that $(\Pi, \preceq)$ is a poset, that is, that $\preceq$ is reflexive, antisymmetric, and transitive. For reflexivity, we need to show that $P \preceq P$ for every partition P. This means that every set in P is a subset of one of the sets in P, and this is trivially true, since every set is a subset of itself. For antisymmetry, suppose that $P_1 \preceq P_2$ and $P_2 \preceq P_1$. We must show that $P_1 = P_2$. By the equivalent roles played here by P_1 and P_2, it is enough to show that every $T \in P_1$ (where $T \subseteq S$) is also an element of P_2. Suppose we have such a T. Then since $P_1 \preceq P_2$, there is a set $T' \in P_2$ such that $T \subseteq T'$. But then since $P_2 \preceq P_1$, there is a set $T'' \in P_1$ such that $T' \subseteq T''$. Putting these together, we have $T \subseteq T''$. But P_1 is a partition, and so the elements of P_1 are nonempty and pairwise disjoint. The only way for this to happen if one is a subset of the other is for the two subsets T and T'' to be the same. But this implies that T' (which is caught in the middle) is also equal to T. Thus $T \in P_2$, which is what we were trying to show. Finally, for transitivity, suppose that $P_1 \preceq P_2$ and $P_2 \preceq P_3$. We must show that $P_1 \preceq P_3$. To this end, we take an arbitrary element $T \in P_1$. Then there is a set $T' \in P_2$ such that $T \subseteq T'$. But then since $P_2 \preceq P_3$, there is a set $T'' \in P_3$ such that $T' \subseteq T''$. Putting these together, we have $T \subseteq T''$. This demonstrates that $P_1 \preceq P_3$.

Next we have to show that every two partitions P_1 and P_2 have a least upper bound and a greatest lower bound in Π. We will show that their greatest lower bound is their "coarsest common refinement", namely the partition P whose subsets are all the nonempty sets of the form $T_1 \cap T_2$, where $T_1 \in P_1$ and $T_2 \in P_2$. As an example, if $P_1 = \{\{1,2,3\},\{4\},\{5\}\}$ and $P_2 = \{\{1,2\},\{3,4\},\{5\}\}$, then the coarsest common refinement is $P = \{\{1,2\},\{3\},\{4\},\{5\}\}$. First, we need to check that this is a partition. It certainly is a set of nonempty subsets of S. It is pairwise disjoint, because the only way an element could be in $T_1 \cap T_2 \cap T_1' \cap T_2'$ if $T_1 \cap T_2 \neq T_1' \cap T_2'$ is for that element to be in both $T_1 \cap T_1'$ and $T_2 \cap T_2'$, which means that $T_1 = T_1'$ and $T_2 = T_2'$, a contradiction. And it covers all of S, because if $x \in S$, then $x \in T_1$ for some $T_1 \in P_1$, and $x \in T_2$ for some $T_2 \in P_2$, and so $x \in T_1 \cap T_2 \in P$. Second, we need to check that P is a refinement of both P_1 and P_2. So suppose $T \in P$. Then $T = T_1 \cap T_2$, for some $T_1 \in P_1$ and $T_2 \in P_2$. It follows that $T \subseteq T_1$ and $T \subseteq T_2$. But then T_1 and T_2 satisfy the requirements in the definition of refinement. Third, we need to check that if P' is any other common refinement of both P_1 and P_2, then P' is also a refinement of P. To this end, suppose that $T \in P'$. Then by definition of refinement, there are subsets $T_1 \in P_1$ and $T_2 \in P_2$ such that $T \subseteq T_1$ and $T \subseteq T_2$. Therefore $T \subseteq T_1 \cap T_2$. But $T_1 \cap T_2 \in P$, and our proof for greatest lower bounds is complete.

It's a little harder to state the definition of the least upper bound (which again we'll call P) of two given partitions P_1 and P_2. Essentially it is just the set of all minimal nonempty subsets of S that do not "split apart" any element of either P_1 or P_2. (In the example above, it is $\{\{1,2,3,4\},\{5\}\}$.) It will be a little easier if we define it in terms of an equivalence relation rather than a partition. Note that from this point of view, one equivalence relation is a refinement of a second equivalence relation if whenever two elements are related by the first relation, then they are related by the second. The equivalence relation determining P is the relation in which $x \in S$ is related to $y \in S$ if there is a "path" (a sequence) $x = x_0$, x_1, x_2, ..., $x_n = y$, for some $n \geq 0$, such that for each i from 1 to n, x_{i-1} and x_i are in the same element of partition P_1 or of partition P_2 (in other words, x_{i-1} and x_i are related either by the equivalence relation corresponding to P_1 or by that corresponding to P_2). It is clear that this is an equivalence relation: it is reflexive by taking $n = 0$; it is symmetric by following the path backwards; and it is transitive by composing paths. It is also clear that P_1 (and P_2 similarly) is a refinement of this partition, since if two elements of S are in the same equivalence class in P_1, then we can take $n = 1$ in our path definition to see that they are in the same equivalence class in P. Thus P is an upper bound of both P_1 and P_2. Finally, we must show that P is the *least* upper bound, that is, a refinement of every other upper bound. This is clear from our construction: we only forced two elements of S to be related (i.e., in the same class of the partition) when they *had* to be related in order to enable P_1 and P_2 to be refinements. Therefore if two elements are related by P, then they have to be related by every equivalence relation (partition) Q of which both P_1 and P_2 are refinements; so P is a refinement of Q.

51. This follows immediately from Exercise 45. To be more specific, according to Exercise 45, there is a least upper bound (respectively, a greatest lower bound) for the entire finite lattice. This element is by definition a greatest element (respectively, a least element).

53. We need to show that every nonempty subset of $\mathbf{Z}^+ \times \mathbf{Z}^+$ has a least element under lexicographic order. Given such a subset S, look at the set S_1 of positive integers that occur as first coordinates in elements of S. Let m_1 be the least element of S_1, which exists since $\mathbf{Z}^+$ is well-ordered under $\leq$. Let S' be the subset of S consisting of those pairs that have m_1 as their first coordinate. Thus S' is clearly nonempty, and by the definition of lexicographic order, every element of S' is less than every element in $S - S'$. Now let S_2 be the set of positive integers that occur as second coordinates in elements of S', and let m_2 be the least element of S_2. Then clearly the element (m_1, m_2) is the least element of S' and hence is the least element of S.

55. If x is an integer in a decreasing sequence of elements of this poset, then at most $|x|$ elements can follow x in the sequence, namely integers whose absolute values are $|x| - 1$, $|x| - 2$, ..., 1, 0. Therefore there can be no infinite decreasing sequence. This is not a totally ordered set, since 5 and -5, for example, are incomparable; from the definition given here, it is neither true that $5 \prec -5$ nor that $-5 \prec 5$, because neither one of $|5|$ or $|-5|$ is less than the other (they are equal).

57. We know from elementary arithmetic that $\mathbf{Q}$ is totally ordered by $<$, and so perforce it is a partially ordered set. To be precise, to find which of two rational numbers is larger, write them with a positive common denominator and compare numerators. To show that this set is dense, suppose $x < y$ are two rational numbers. Let z be their average, i.e., $(x + y)/2$. Since the set of rational numbers is closed under addition and division, z is also a rational number, and it is easy to show that $x < z < y$.

59. Let $(S, \preceq)$ be a partially ordered set. From the definitions of well-ordered, totally ordered, and well-founded, it is clear that what we have to show is that every nonempty subset of S contains a least element if and only if there is no infinite decreasing sequence of elements $a_1, a_2, a_3, \ldots$ in S (i.e., where $a_{i+1} \prec a_i$ for all i). One direction is clear: An infinite decreasing sequence of elements has no least element. Conversely, let A be any nonempty subset of S that has no least element. Since A is nonempty, let a_1 be any element of A. Since a_1 is not the least element of A, there is some $a_2 \in A$ smaller than a_1, i.e., $a_2 \prec a_1$. Since a_2 is not the least element of A, A must contain an element a_3 with $a_3 \prec a_2$. We continue in this manner, giving us an infinite decreasing sequence in S. Note that this proof is nonconstructive; it uses what set theorists call the Axiom of Choice.

61. We need to peel elements off the bottom of the Hasse diagram. We can begin with a, b, or c. Suppose we decide to start with a. Next we may choose any minimal element of what remains after we have removed a; only b and c meet this requirement. Suppose we choose b next. Then c, d, and e are minimal elements in what remains, so any of those can come next. We continue in this manner until we have listed and removed all the elements. One possible order, then, is $a \prec_t b \prec_t d \prec_t e \prec_t c \prec_t f \prec_t g \prec_t h \prec_t i \prec_t j \prec_t k \prec_t m \prec_t l$.

63. We follow the same reasoning as in Exercise 61. We can start with E, for instance (and this will make our answer different from the one obtained in Example 27). One such order is $E \prec C \prec A \prec B \prec F \prec D \prec G$.

65. We need to find a total order compatible with this partial order. We work from the bottom up, writing down a task (vertex in the diagram) and removing it from the diagram, so that at each stage we choose a vertex with no vertices below it. One such order is: Determine user needs $\prec$ Write functional requirements $\prec$ Set up test sites $\prec$ Develop system requirements $\prec$ Develop module A $\prec$ Develop module C $\prec$ Develop module B $\prec$ Write documentation $\prec$ Integrate modules $\prec$ α test $\prec$ β test $\prec$ Completion.

GUIDE TO REVIEW QUESTIONS FOR CHAPTER 8

1. a) See p. 521 (which refers to the definition on p. 519). **b)** See Example 6 in Section 8.1.

2. a) See p. 522. **b)** See p. 523. **c)** See p. 523. **d)** See p. 524.

3. a) $\{(1,1),(1,2),(2,1),(2,2),(2,3),(3,2),(3,3),(4,4)\}$
b) $\emptyset$ **c)** $\{(1,1),(1,2),(2,2),(2,3),(3,3),(4,4)\}$
d) $\{(1,1),(2,2),(3,3),(4,4)\}$ **e)** $\{(1,1),(2,2),(3,3),(4,4)\}$

4. a) See Example 16 in Section 8.1. **b)** See Exercise 45a in Section 8.1.
c) See Exercise 45b in Section 8.1.

5. a) See p. 531. **b)** Take the projection $P_{1,4,5}$.
c) First rearrange the order of the fields in the relations, so that the first is in the order address, telephone number, name, major, and the second is in the order name, major, student number, number of credit hours. Then form the join J_2, to get a single relation with the fields in the order address, telephone number, name, major, student number, number of credit hours. Finally, if desired, rearrange the fields to a more natural order.

6. a) See p. 538. **b)** See pp. 538–539.

7. a) See p. 541. **b)** See pp. 541–542.

8. a) See pp. 544–545. **b)** Add all the pairs (a,a).
c) Whenever a pair (a,b) is in the relation, add the pair (b,a).
d) The reflexive closure is $\{(1,1),(1,2),(2,2),(2,3),(2,4),(3,1),(3,3),(4,4)\}$, and the symmetric closure is $\{(1,2),(1,3),(2,1),(2,3),(2,4),(3,1),(3,2),(4,2)\}$.

9. a) the smallest transitive relation containing R **b)** no **c)** See Algorithms 1 and 2 in Section 8.4.
d) the relation that always holds between two elements of $\{1,2,3,4\}$ (in symbols, $\{1,2,3,4\} \times \{1,2,3,4\}$)

10. a) See p. 555.
b) One equivalence relation is the one with equivalence classes $\{a,b,d\}$ and $\{c\}$. The only other one meeting this condition is the relation that always holds, i.e., in which there is just one equivalence class, $\{a,b,c,d\}$.

11. a) See Example 3 in Section 8.5.
b) It is easy to see that this relation is
$\{(0,0),(1,1),(1,6),(6,1),(6,6),(2,2),(2,5),(5,2),(5,5),(3,3),(3,4),(4,3),(4,4)\}$.

12. a) See p. 558.
b) $[0]=\{\ldots,-5,0,5,\ldots\}$, $[1]=\{\ldots,-4,1,6,\ldots\}$, $[2]=\{\ldots,-3,2,7,\ldots\}$, $[3]=\{\ldots,-2,3,8,\ldots\}$, and $[4]=\{\ldots,-1,4,9,\ldots\}$
c) $\{0\}$, $\{1,6\}$, $\{2,5\}$, and $\{3,4\}$

13. See Theorem 2 in Section 8.5.

14. a) See p. 566. **b)** See Example 2 in Section 8.6.

15. See the definition of lexicographic ordering on p. 569.

16. a) See p. 571. **b)** Here is the Hasse diagram:

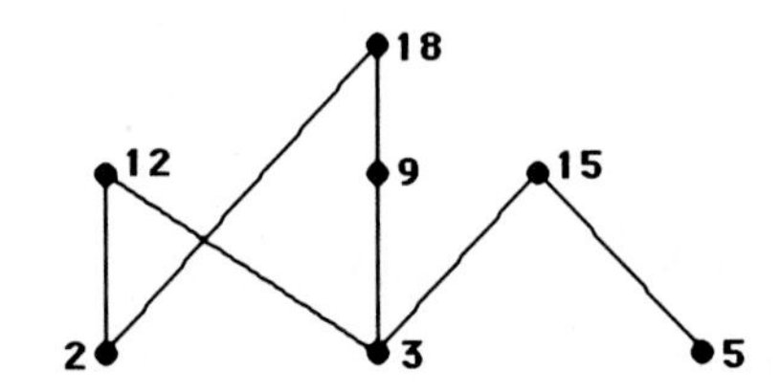

17. a) See pp. 572–573. **b)** $(\{1,2,3,4,5\},\,|\,)$ **c)** $(\{1,2,3,4,5\},\leq)$

18. a) See p. 574.
b) Extreme examples are the lattice $(\{1,2,3,4,5\},\leq)$ (every total order is a lattice) and the nonlattice $(\{1,2,3,4,5\},=)$ (no two distinct elements have an upper bound).

19. a) See Exercise 41 in Section 8.6. **b)** See Exercise 51 in Section 8.6

20. a) See p. 568. (We assume for the rest of this question that the set is finite, in which case a well-ordered set is the same as a totally ordered set, as defined also on p. 568.) **b)** See topological sorting on pp. 576–577.
c) See Example 27 in Section 8.6.

SUPPLEMENTARY EXERCISES FOR CHAPTER 8

1. a) This relation is not reflexive, since most strings have many letters in common with themselves. Whether it is irreflexive depends on whether we mean to include the empty string; the empty string is the only string s such that $(s,s)\in R_1$ (the empty string has no letters in common with itself, since it has no letters). Thus if we mean not to include the empty string in the underlying set, then the relation is irreflexive; otherwise it is not. The relation is symmetric by inspection (the roles of a and b in the sentence are symmetric). It is not antisymmetric, since there are many pairs of strings such that $(a,b)\in R_1 \wedge (b,a)\in R_1$; for instance $a=$ *bullfinch* and $b=$ *parrot*. The relation is not transitive, since, for example, although *bullfinch* and *parrot* are related, and *parrot* and *chicken* are related, *bullfinch* and *chicken* do have letters in common and so are not related.
b) This relation is very similar to the relation R_1. For no string is $(a,a)\in R_2$, so the relation is irreflexive and not reflexive. It is symmetric by inspection, and not antisymmetric (same example as above). It is also

not transitive, since, for instance, *finch* is related to *parrot*, which is related to *robin*, but *finch* is not related to *robin*, since they are the same length.

c) No string is longer than itself, so R_3 is irreflexive and not reflexive. It is not symmetric, since, for instance, *robin* is longer than *wren*, but *wren* is not longer than *robin*. It is antisymmetric: there is no way for both a to be longer than b and b to be longer than a. Finally, it is clearly transitive, since if a is longer than b which is longer than c, then a is longer than c.

3. By algebra, the given condition is the same as the condition that $f((a,b)) = f((c,d))$, where $f((x,y)) = x - y$. Therefore by Exercise 9 in Section 8.5, this is an equivalence relation.

5. Suppose that $(a,b) \in R$. We must show that $(a,b) \in R^2$. By reflexivity, we know that $(b,b) \in R$. Therefore by the definition of R^2, we combine the facts that $(a,b) \in R$ and $(b,b) \in R$ to conclude that $(a,b) \in R^2$.

7. Both of these conclusions are valid. Since each pair (a,a) is in both R_1 and R_2, we can conclude that each pair (a,a) is in $R_1 \cap R_2$ and $R_1 \cup R_2$.

9. Both of these conclusions are valid. For the first, suppose that $(a,b) \in R_1 \cap R_2$. This means that $(a,b) \in R_1$ and $(a,b) \in R_2$. By the symmetry of R_1 and R_2, we conclude that $(b,a) \in R_1$ and $(b,a) \in R_2$. Therefore (b,a) is in their intersection, as desired. For the second part, suppose that $(a,b) \in R_1 \cup R_2$. This means that $(a,b) \in R_1$ or $(a,b) \in R_2$. By the symmetry of R_1 and R_2, we conclude either that $(b,a) \in R_1$ or that $(b,a) \in R_2$. Therefore (b,a) is in their union, as desired.

11. A primary key is one for which there are no two different rows with the same value in this field. If there were two different rows with the same value after projection, then there certainly would have been two different rows with the same value before projection.

13. The key point is that $\Delta^{-1} = \Delta$, where Δ consists of all the pairs (a,a). Thus it does not matter whether we add the pairs in Δ before or after we add the reverse of every pair in the original relation.

15. a) We observed in Exercise 29b in Section 8.4 that we need to take the symmetric closure first in order to insure that the result is symmetric. The relation given in that exercise provides an example. An even simpler one is the relation $\{(0,1),(2,1)\}$; the symmetric closure of the transitive closure is $\{(0,1),(1,0),(1,2),(2,1)\}$, but the transitive closure of the symmetric closure is all of $\{0,1,2\} \times \{0,1,2\}$.

b) Suppose that (a,b) is in the symmetric closure of the transitive closure of R. We must show that (a,b) is also in the transitive closure of the symmetric closure of R. Now either (a,b) or (b,a) is in the transitive closure of R. This means that either there is a path from a to b or a path from b to a in R. In the former case, there is perforce a path from a to b in the symmetric closure of R. In the latter case, the path from b to a can be followed backwards in the symmetric closure of R, since the symmetric closure adds the reverses of all the edges in R. Therefore in either case (a,b) is in the transitive closure of the symmetric closure of R. (See also the related Exercise 23 in Section 8.4.)

17. The closure of S with respect to **P** is a relation S' which contains S as a subset and has property **P**. Since $R \subseteq S$, we conclude that $R \subseteq S'$. By definition of closure, then, the closure of R must be a subset of S', as desired.

19. We use the basic idea of Warshall's algorithm, except that $w_{ij}^{[k]}$ will be a numerical variable (taking values from 0 to ∞, inclusive) representing the length of the longest path from v_i to v_j all of whose interior vertices are labeled less than or equal to k, rather than simply a Boolean variable indicating whether such a path exists. A value of 0 for $w_{ij}^{[k]}$ will mean that there is no path from v_i to v_j all of whose interior vertices are labeled less than or equal to k. To compute $w_{ij}^{[k]}$ from the matrix $\mathbf{W}_{k-1}$, we determine, for each pair (i,j), whether there are paths from v_i to v_k and from v_k to v_j using no interior vertices labeled greater than $k-1$. If either of $w_{ik}^{[k-1]}$ or $w_{kj}^{[k-1]}$ equals 0, then such a pair of paths does not exist, so we set $w_{ij}^{[k]}$ equal to $w_{ij}^{[k-1]}$. Otherwise (if such a pair of paths does exist), then there are two possibilities. If $w_{kk}^{[k-1]} > 0$, then we now know that there are paths of arbitrary length from v_i to v_j, since we can loop around v_k as long as we please; in this case we set $w_{ij}^{[k]}$ to ∞. If $w_{kk}^{[k-1]} = 0$, then we do not yet have such looping, so we set $w_{ij}^{[k]}$ to the larger of $w_{ij}^{[k-1]}$ and $w_{ik}^{[k-1]} + w_{kj}^{[k-1]}$. (Initially we set $\mathbf{W}_0$ equal to the matrix representing the relation.)

21. There are 52 partitions in all, but that is not the question. If there are to be three equivalence classes, then the classes must have sizes $3,1,1$ or $2,2,1$. There are $C(5,3) = 10$ partitions into one set with 3 elements and the other two sets of 1 element each, since the only choice involved is choosing the 3-set. There are $C(5,2)C(3,2)/2 = 15$ ways to partition our set into sets of size 2, 2, and 1; we need to choose the 2 elements for the first set of size 2, then we need to choose the 2 elements from the 3 remaining for the second set of size 2, except that we have overcounted by a factor of 2, since we could choose these two 2-sets in either order. Therefore there are $10 + 15 = 25$ partitions into three classes.

23. There is no question that the collection defined here is a refinement of each of the given partitions, since each set $A_i \cap B_j$ is a subset of A_i and of B_j. We must show that it is actually a partition. By construction, each of the sets in this collection is nonempty. To see that their union is all of S, let $s \in S$. Since P_1 and P_2 are partitions of S, there are sets A_i and B_j such that $s \in A_i$ and $s \in B_j$. Therefore $s \in A_i \cap B_j$, which shows that s is in one of the sets in our collection. Finally, to see that these sets are pairwise disjoint, simply note that unless $i = i'$ and $j = j'$, then $(A_i \cap B_j) \cap (A_{i'} \cap B_{j'}) = (A_i \cap A_{i'}) \cap (B_j \cap B_{j'})$ is empty, since either $(A_i \cap A_{i'})$ or $(B_j \cap B_{j'})$ is empty.

25. The subset relation is a partial order on *every* collection of sets, since it is reflexive, antisymmetric, and transitive. Here the collection of sets happens to be $\mathbf{R}(S)$.

27. We need to find a total order compatible with this partial order. We work from the bottom up, writing down a task (vertex in the diagram) and removing it from the diagram, so that at each stage we choose a vertex with no vertices below it. One such order is: Find recipe $\prec$ Buy seafood $\prec$ Buy groceries $\prec$ Wash shellfish $\prec$ Cut ginger and garlic $\prec$ Clean fish $\prec$ Steam rice $\prec$ Cut fish $\prec$ Wash vegetables $\prec$ Chop water chestnuts $\prec$ Make garnishes $\prec$ Cook in wok $\prec$ Arrange on platter $\prec$ Serve.

29. Since every subset of an antichain is clearly an antichain, we will list only the maximal antichains; the actual answers will be everything we list together with all the subsets of them.

a) Here every two elements are comparable except c and d. Thus the maximal antichains are $\{c,d\}$, $\{a\}$, and $\{b\}$. (There are three more antichains which are subsets of these: $\{c\}$, $\{d\}$, and $\emptyset$.)

b) Here the maximal antichains are $\{a\}$, $\{b,c\}$, $\{c,e\}$, and $\{d,e\}$.

c) In this case there are only three maximal antichains: $\{a,b,c\}$, $\{d,e,f\}$, and $\{g\}$.

31. Let C be a maximal chain. We must show that C contains a minimal element of S. Since C can itself be viewed as a finite poset (being a subset of a poset), it contains a minimal element m. We need to show that m is also a minimal element of S. If it were not, then there would be another element $a \in S$ such that $a \prec m$. Now we claim that $C \cup \{a\}$ is a chain, which will contradict the maximality of C. We need to show that a is comparable to every element of C. We already know that a is comparable to m. Let x be any other element of C. Since m is minimal in C, it cannot be that $x \prec m$; thus since x and m have to be comparable (they are both in C), it must be that $m \prec x$. Now by transitivity we have $a \prec x$, and we are done.

33. Consider the relation R on the set of $mn+1$ people given by $(a,b) \in R$ if and only if a is a descendant of or equal to b. This makes the collection into a poset. In the terminology of Exercise 32, if there is not a subset of $n+1$ people none of whom is a descendant of any other, then $k \leq n$, since such a subset is certainly an antichain. Therefore the poset can be partitioned into $k \leq n$ chains. Now by the generalized pigeonhole principle, at least one of these chains must contain at least $m+1$ elements, and this is the desired list of descendants.

35. Recall the definition of well-founded from the preamble to Exercise 55 in Section 8.6—that there is no infinite decreasing sequence. We must show that under this hypothesis, and if $\forall x((\forall y(y \prec x \rightarrow P(y))) \rightarrow P(x))$, then $P(x)$ is true for all $x \in S$. We give a proof by contradiction. If it does not hold that $P(x)$ is true for all $x \in S$, let x_1 be an element of S such that $P(x_1)$ is not true. Then by the conditional statement given above, it must be the case that $\forall y(y \prec x_1 \rightarrow P(y))$ is not true. This means that there is some y with $y \prec x_1$ such that $P(y)$ is not true. Rename this y as x_2. So we know that $P(x_2)$ is not true. Again invoking the conditional statement, we get an $x_3 \prec x_2$ such that $P(x_3)$ is not true. And so on forever. This contradicts the well-foundedness of our poset. Therefore $P(x)$ is true for all $x \in S$.

37. We assume that R is reflexive and transitive on A, and we must show that $R \cap R^{-1}$ is reflexive, symmetric, and transitive. Reflexivity is easy: if $a \in A$, then we know that $(a,a) \in R$, so by the definition of R^{-1} as the reverses of the pairs in R, we know that $(a,a) \in R^{-1}$ as well, whence it follows that $(a,a) \in R \cap R^{-1}$. *Every* relation of the form $R \cap R^{-1}$ is symmetric, no matter what R is, since if $(a,b) \in R$, then $(b,a) \in R^{-1}$ and vice versa. For transitivity, suppose that $(a,b) \in R \cap R^{-1}$ and $(b,c) \in R \cap R^{-1}$. We must show that $(a,c) \in R \cap R^{-1}$. Since $(a,b) \in R$ and $(b,c) \in R$, and since R is transitive, $(a,c) \in R$. Similarly, since $(a,b) \in R^{-1}$ and $(b,c) \in R^{-1}$, $(b,a) \in R$ and $(c,b) \in R$. Again, since R is transitive, $(c,a) \in R$, and hence $(a,c) \in R^{-1}$. Putting these two parts together, we conclude that $(a,c) \in R \cap R^{-1}$, as desired.

39. There is not much to show in this exercise, since the definitions of greatest lower bound and least upper bound exhibit these properties by their very form.

a) The g.l.b. of x and y was defined to be the greatest element that is a lower bound of both x and y. The roles of x and y in this statement are symmetric, so it follows immediately that $x \wedge y = y \wedge x$. Similarly for least upper bound.

b) By definition, $(x \wedge y) \wedge z$ is a lower bound of x, y, and z that is greater than every other common lower bound (this is how we proceeded in Exercise 45 of Section 8.6). Since x, y, and z play interchangeable roles in this statement, grouping does not matter, so $x \wedge (y \wedge z)$ is the same element. Similarly for l.u.b.

c) The two statements are duals, so we will prove just the first one; the proof of the second can be obtained formally simply by exchanging each symbol and word for its dual. To show that $x \wedge (x \vee y) = x$, we must show that x is the greatest lower bound of x and $x \vee y$. Clearly x is a lower bound for x, and since $x \vee y$ is by definition greater than or equal to x, x is a lower bound for it as well. Therefore x is a lower bound. But every other lower bound for x has to be less than x, so x is the greatest lower bound.

d) Obviously x is a lower (upper) bound for itself and itself, and the greatest (least) such.

41. There is nothing very deep going on here—it's just a matter of applying the definitions.

a) Since 1 is the only element greater than or equal to 1, it is the only upper bound for 1 and therefore the only possible value of the least upper bound of x and 1.

b) Clearly x is a lower bound for both x and 1 (since $x \preceq 1$), and clearly no other lower bound can be greater than x, so $x \wedge 1 = x$.

c) This is the dual to part **(b)**. We formed the following proof on the word processor used to produce this solutions manual by copying the words in the solution to part **(b)** and replacing each word and symbol by its dual: Clearly x is an upper bound for both x and 0 (since $0 \preceq x$), and clearly no other upper bound can be smaller than x, so $x \vee 0 = x$.

d) This is the dual of part **(a)**: Since 0 is the only element less than or equal to 0, it is the only lower bound for 0 and therefore the only possible value of the greatest lower bound of x and 0.

43. One way to solve this problem is to play around with some small examples. Here is one counter-example that the author obtained in this way. The lattice has as its elements $\emptyset$, $\{1\}$, $\{2\}$, $\{3\}$, $\{1,2\}$, $\{2,3\}$, and $\{1,2,3\}$, with, as usual, the relation $\subseteq$. (Draw its Hasse diagram!) It is easy to check that every two elements have both a least upper bound and a greatest lower bound (note that $\emptyset$ is a lower bound for the whole lattice, and $\{1,2,3\}$ is an upper bound for the whole lattice). Take $x = \{1\}$, $y = \{2\}$, and $z = \{3\}$, and compute both sides of the equation $x \vee (y \wedge z) = (x \vee y) \wedge (x \vee z)$. Note that since we do not have the full subset lattice, least upper bounds are not just unions. The left-hand side is x, since $y \wedge z = \emptyset$. The right-hand side is the greatest lower bound of $\{1,2\}$ and $\{1,2,3\}$, which is $\{1,2\}$. Since these are different, the lattice is not distributive.

45. Yes. First, recall from Example 22 in Section 8.6 that $x \wedge y$ is the greatest common divisor (gcd) of x and y, while $x \vee y$ is their least common multiple (lcm). We can analyze this problem by looking at prime factorizations. The power to which a prime p appears in the gcd of two numbers is the minimum of the powers to which it appears in the two numbers. Similarly, the power to which p appears in the lcm is the maximum of the powers to which it appears in the two numbers. Thus if we let a, b, and c represent the powers to which p appears in x, y, and z, respectively, the first identity we need to prove is

$$\max(a, \min(b, c)) = \min(\max(a, b), \max(a, c)).$$

We consider the several cases. If a is the largest of the three numbers, then both sides equal a. If a is the smallest, then both sides equal the smaller of b and c. Otherwise, we can suppose without loss of generality (since the roles of b and c are symmetric) that $b \le a \le c$, in which case we easily compute that both sides equal a. The proof of the other statement is dual to this proof. The result now follows from the Fundamental Theorem of Arithmetic, since numbers are determined by their prime factorizations.

47. As might be expected from the name, the complement of a subset $X \subseteq S$ is its complement $S - X$. To prove this, we need to prove that $X \vee (S - X) = 1$ and $X \wedge (S - X) = 0$, which translated into our particular setting reads: $X \cup (S - X) = S$ and $X \cap (S - X) = \emptyset$. But these are trivially true.

49. Think of the rectangular grid as representing elements in a matrix. Thus we number from top to bottom and within that from left to right. For example, $(2,4)$ is the element in row 2, column 4. The partial order is that $(a,b) \preceq (c,d)$ if $a \le c$ and $b \le d$. Note that $(1,1)$ is the least element under this relation. The rules for Chomp as explained in Chapter 1 coincide with the rules stated in the preamble here. But now we can identify the point (a,b) with the natural number $p^{a-1}q^{b-1}$ for all a and b with $1 \le a \le m$ and $1 \le b \le n$. This identifies the points in the rectangular grid with the set S in this exercise, and the partial order $\preceq$ just described is the same as the divides relation, because $p^{a-1}q^{b-1} \mid p^{c-1}q^{d-1}$ if and only if the exponent on p on the left does not exceed the exponent of p on the right, and similarly for q.

WRITING PROJECTS FOR CHAPTER 8

Books and articles indicated by bracketed symbols below are listed near the end of this manual. You should also read the general comments and advice you will find there about researching and writing these essays.

1. See the same references as suggested for fuzzy logic in Writing Project 2 of Chapter 1, as well as [Zi].

2. There are numerous textbooks on databases. Try to consult one that is fairly recent, because in many areas of computer science, progress is so fast that books soon become out-dated. You will find them in the QA 76.9 area of the library's shelves. Two recommended ones are are [Da1] and [Ma1].

3. Try author or key-word search in an appropriate database (e.g., one provided by *Mathematical Reviews*, which is available on the Web as MathSciNet). Consult the oldest reference you can find that talks about these topics, and it will probably lead you to the original sources. Simultaneous discovery occurs in many branches of intellectual pursuit, not just mathematics and computer science. See Writing Project 15 in Chapter 10 for something along the same lines.

4. The abstraction and difficulty here is part of what makes fractions hard for many children (and some adults) to handle. Be careful to avoid 0 in the denominator! You should be able to figure this out without consulting other sources, and it is a good project to work on with other people.

5. See the hints for Writing Project 3.

6. Entire books have been written on security issues in computer systems ([Pf], for one), and it should not be hard to find a chapter or two on the subject in many more general books (try [De1]).

7. Textbooks on project scheduling should be a good source of information. See [Mo], for example. Scheduling is a topic in a branch of mathematics known as Operations Research. It has its own journals, conferences, subspecialties, software, etc.

8. See the suggestions for Writing Project 7.

9. We have hinted at duality in many of the exercise solutions in this *Guide*. A book on lattice theory (such as [Gr1]) will make the concept more precise.

10. As mentioned in the previous suggestion, you can find entire books on lattice theory. In fact, *Mathematical Reviews* (which is available on the Web as MathSciNet) devotes a whole category (numbered 06) to lattices and other kinds of ordered sets and ordered algebraic structures.

CHAPTER 9
Graphs

SECTION 9.1 Graphs and Graph Models

The examples and exercises give a good picture of the ways in which graphs can model various real world applications. In constructing graph models you need to determine what the vertices will represent, what the edges will represent, whether the edges will be directed or undirected, whether loops should be allowed, and whether a simple graph or multigraph is more appropriate.

1. In part **(a)** we have a simple graph, with undirected edges, no loops or multiple edges. In part **(b)** we have a multigraph, since there are multiple edges (making the figure somewhat less than ideal visually). In part **(c)** we have the same picture as in part **(b)** except that there is now a loop at one vertex; thus this is a pseudograph.

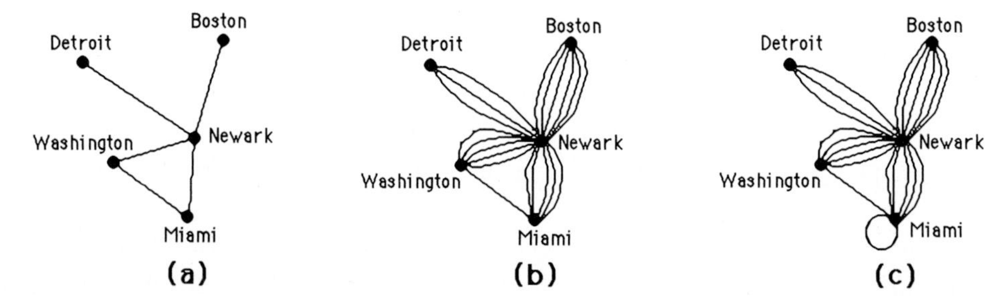

In part **(d)** we have a directed graph, the directions of the edges telling the directions of the flights; note that the **antiparallel edges** (pairs of the form (u, v) and (v, u)) are not parallel. In part **(e)** we have a directed multigraph, since there are parallel edges.

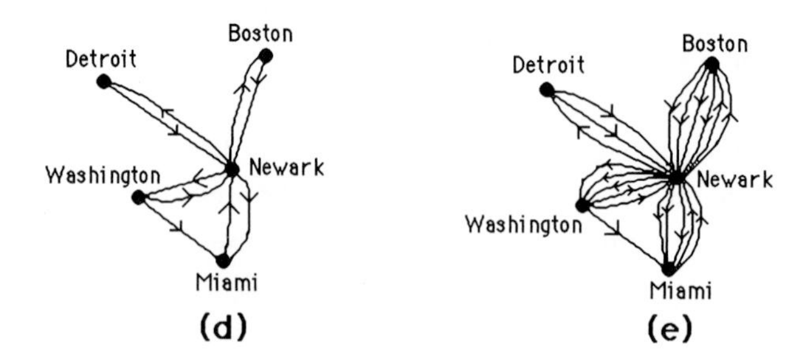

3. This is a simple graph; the edges are undirected, and there are no parallel edges or loops.

5. This is a pseudograph; the edges are undirected, but there are loops and parallel edges.

7. This is a directed graph; the edges are directed, but there are no parallel edges. (Loops and antiparallel edges—see the solution to Exercise 1d for a definition—are allowed in a directed graph.)

9. This is a directed multigraph; the edges are directed, and there is a set of parallel edges.

11. In a simple graph, edges are undirected. To show that R is symmetric we must show that if uRv, then vRu. If uRv, then there is an edge associated with $\{u, v\}$. But $\{u, v\} = \{v, u\}$, so this edge is associated with $\{v, u\}$ and therefore vRu. A simple graph does not allow loops; that is if there is an edge associated with $\{u, v\}$, then $u \neq v$. Thus uRu never holds, and so by definition R is irreflexive.

13. In each case we draw a picture of the graph in question. All are simple graphs. An edge is drawn between two vertices if the sets for the two vertices have at least one element in common. For example, in part **(a)** there is an edge between vertices A_1 and A_2 because there is at least one element common to A_1 and A_2 (in fact there are three such elements). There is no edge between A_1 and A_3 since $A_1 \cap A_3 = \emptyset$.

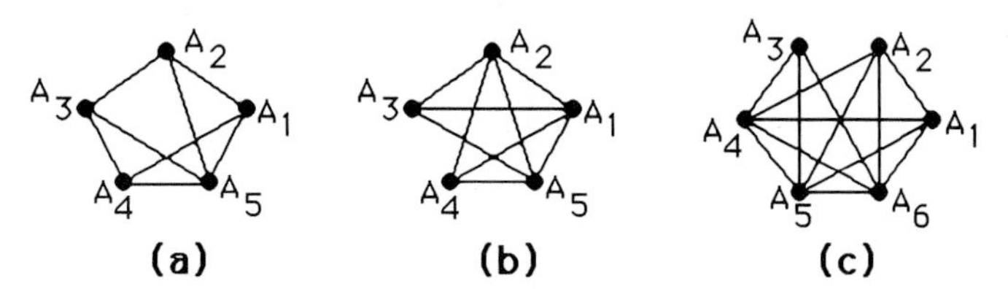

15. We draw a picture of the graph in question, which is a simple graph. Two vertices are joined by an edge if we are told that the species compete (such as robin and mockingbird) but there is no edge between pairs of species that are not given as competitors (such as robin and blue jay).

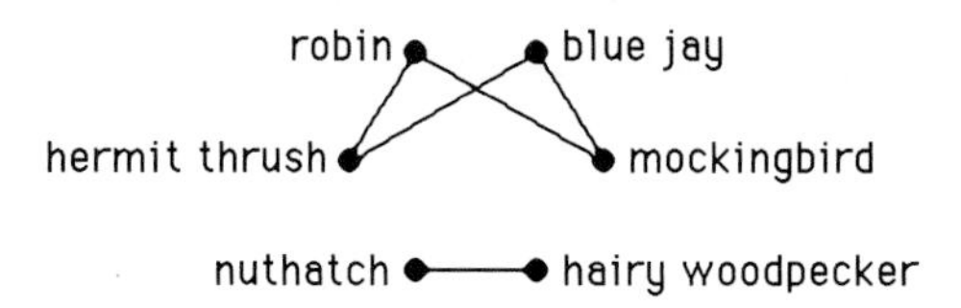

17. Here are the persons to be included, listed in order of birth year: Aristotle (384–322 B.C.E.), Euclid (325–265 B.C.E.), al-Khowarizmi (780–850), Fibonacci (1170–1250), Maurolico (1494–1575), Mersenne (1588–1648), Descartes (1596–1650), Fermat (1601–1665), Pascal (1623–1662), Goldbach (1690–1764), Vandermonde (1735–1796), Gauss (1777–1855), Lamé (1795–1870), Dirichlet (1805–1859), De Morgan (1806–1871), Lovelace (1815–1852), Boole (1815–1864), and Dodgson (1832–1898). We draw the graph by connecting two people if their date ranges overlap. Note that there is a complete subgraph (see Section 9.2) consisting of Mersenne, Descartes, Fermat, and Pascal, and a larger complete subgraph consisting of the last seven people listed. A few of the vertices are isolated (again see Section 9.2). In all our graph has 18 vertices and 31 edges. A graph like this is called an **interval graph**, since each vertex can be associated with an interval of real numbers; it is a special case of an **intersection graph**, where two vertices are adjacent if the sets associated with those vertices have a nonempty intersection (see Exercise 13).

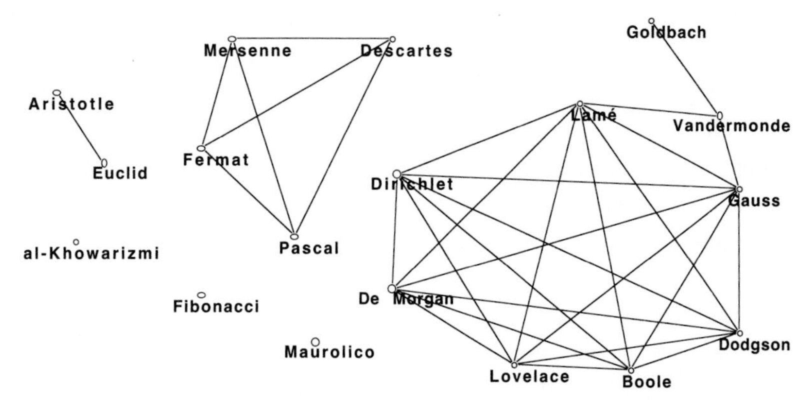

19. We draw a picture of the graph in question, which is a directed graph. We draw an edge from u to v if we are told that u can influence v. For instance the Chief Financial Officer is an isolated vertex since she is influenced by no one and influences no one.

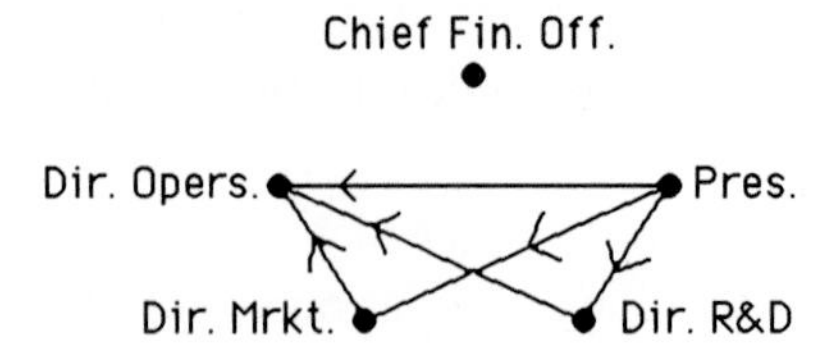

21. We draw a picture of the graph in question, which is a directed graph. We draw an edge from u to v if we are told that u beat v.

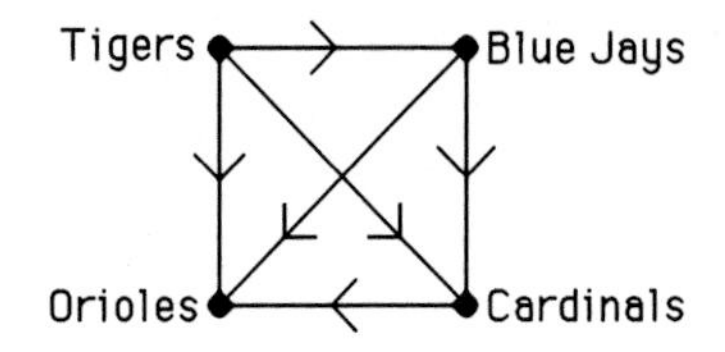

23. We could compile a list of phone numbers (the labels on the vertices) in the February call graph that were not present in January, and a list of the January numbers missing in February. For each number in each list, we could make a list of the numbers they called or were called by, using the edges in the call graphs. Then we could look for February lists that were very similar to January lists. If we found a new February number that had almost the same calling pattern as a defunct January number, then we might suspect that these numbers belonged to the same person, who had recently changed his or her number.

25. For each e-mail address (the labels on the vertices), we could make a list of the other addresses they sent messages to or received messages from. If we see two addresses that had almost the same communication pattern, then we might suspect that these addresses belonged to the same person, who had recently changed his or her e-mail address.

27. The vertices represent the people at the party. Because it is possible that a knows b's name but not vice versa, we need a directed graph. We will include an edge associated with (u, v) if and only if u knows v's name. There is no need for multiple edges (either a knows b's name or he doesn't). One could argue that we should not clutter the model with loops, because obviously everyone knows her own name. On the other hand, it certainly would not be wrong to include loops, especially if we took the instructions literally.

29. For this to be interesting, we want the graph to model all marriages, not just ones that are currently active. (In the latter case, for the Western world, there would be at most one edge incident to each vertex.) So we let the set of vertices be a set of people (for example, all the people in North America who lived at any point in the 20th century), and two vertices are joined by an edge if the two people were ever married. Since laws in the 20th century allowed only marriages between persons of the opposite sex, and ignoring complications caused by sex-change operations, we note that this graph has the property that there are two types of vertices (men and women), and every edge joins vertices of opposite types. In the next section we learn that the word used to describe a graph like this is *bipartite*.

31. We draw a picture of the directed graph in question. There is an edge from u to v if the assignment made in u can possibly influence the assignment made in v. For example, there is an edge from S_3 to S_6, since the assignment in S_3 changes the value of y, which then influences the value of z (in S_4) and hence has a bearing on S_6. We assume that the statements are to be executed in the given order, so, for example, we do not draw an edge from S_5 to S_2.

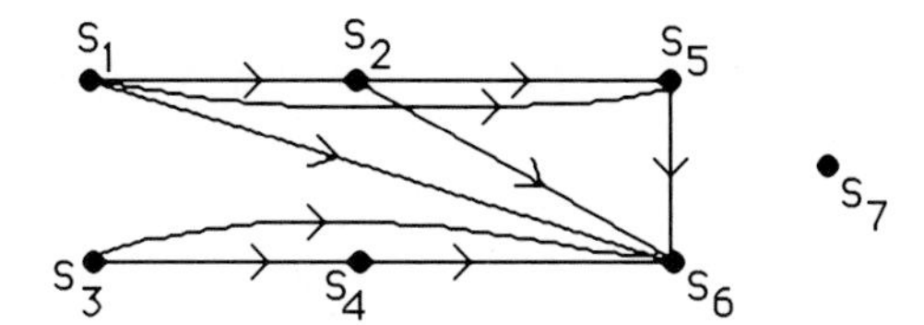

33. The vertices in the directed graph represent people in the group. We put a directed edge into our directed graph from every vertex A to every vertex $B \neq A$ (we do not need loops), and furthermore we label that edge with one of the three labels L, D, or N. Let us see how to incorporate this into the mathematical definition. Let us call such a thing a directed graph with labeled edges. It is defined to be a triple (V, E, f), where (V, E) is a directed graph (i.e., V is a set of vertices and E is a set of ordered pairs of elements of V) and f is a function from E to the set $\{L, D, N\}$. Here we are simply thinking of $f(e)$ as the attitude of the person at the tail (initial vertex—see Section 9.2) of e toward the person at the head (terminal vertex) of e.

SECTION 9.2 Graph Terminology and Special Types of Graphs

Graph theory is sometimes jokingly called the "theory of definitions," because so many terms can be—and have been—defined for graphs. A few of the most important concepts are given in this section; others appear in the rest of this chapter and the next, in the exposition and in the exercises. As usual with definitions, it is important to understand exactly what they are saying. You should construct some examples for each definition you encounter—examples both of the thing being defined and of its absence. Some students find it useful to build a dictionary as they read, including their examples along with the formal definitions.

The Handshaking Theorem (that the sum of the degrees of the vertices in a graph equals twice the number of edges), although trivial to prove, is quite handy, as Exercise 49, for example, illustrates. Be sure to look at Exercise 37, which deals with the problem of when a sequence of numbers can possibly be the degrees of the vertices of a simple graph. Some interesting subtleties arise there, as you will discover when you try to draw the graphs. Many arguments in graph theory tend to be rather ad hoc, really getting down to the nitty gritty, and Exercise 37c is a good example. Exercise 45 is really a combinatorial problem; such problems abound in graph theory, and entire books have been written on counting graphs of various types. The notion of **complementary graph**, *introduced in Exercise 53, will appear again later in this chapter, so it would be wise to look at the exercises dealing with it.*

1. There are 6 vertices here, and 6 edges. The degree of each vertex is the number of edges incident to it. Thus $\deg(a) = 2$, $\deg(b) = 4$, $\deg(c) = 1$ (and hence c is pendant), $\deg(d) = 0$ (and hence d is isolated), $\deg(e) = 2$, and $\deg(f) = 3$. Note that the sum of the degrees is $2 + 4 + 1 + 0 + 2 + 3 = 12$, which is twice the number of edges.

3. There are 9 vertices here, and 12 edges. The degree of each vertex is the number of edges incident to it. Thus $\deg(a) = 3$, $\deg(b) = 2$, $\deg(c) = 4$, $\deg(d) = 0$ (and hence d is isolated), $\deg(e) = 6$, $\deg(f) = 0$ (and hence f is isolated), $\deg(g) = 4$, $\deg(h) = 2$, and $\deg(i) = 3$. Note that the sum of the degrees is $3 + 2 + 4 + 0 + 6 + 0 + 4 + 2 + 3 = 24$, which is twice the number of edges.

5. By Theorem 2 the number of vertices of odd degree must be even. Hence there cannot be a graph with 15 vertices of odd degree 5. (We assume that the problem was meant to imply that the graph contained only these 15 vertices.)

7. This directed graph has 4 vertices and 7 edges. The in-degree of vertex a is $\deg^-(a) = 3$ since there are 3 edges with a as their terminal vertex; its out-degree is $\deg^+(a) = 1$ since only the loop has a as its initial vertex. Similarly we have $\deg^-(b) = 1$, $\deg^+(b) = 2$, $\deg^-(c) = 2$, $\deg^+(c) = 1$, $\deg^-(d) = 1$, and $\deg^+(d) = 3$. As a check we see that the sum of the in-degrees and the sum of the out-degrees are equal (both are equal to 7).

9. This directed multigraph has 5 vertices and 13 edges. The in-degree of vertex a is $\deg^-(a) = 6$ since there are 6 edges with a as their terminal vertex; its out-degree is $\deg^+(a) = 1$. Similarly we have $\deg^-(b) = 1$, $\deg^+(b) = 5$, $\deg^-(c) = 2$, $\deg^+(c) = 5$, $\deg^-(d) = 4$, $\deg^+(d) = 2$, $\deg^-(e) = 0$, and $\deg^+(e) = 0$ (vertex e is isolated). As a check we see that the sum of the in-degrees and the sum of the out-degrees are both equal to the number of edges (13).

11. To form the underlying undirected graph we simply take all the arrows off the edges. Thus, for example, the edges from e to d and from d to e become a pair of parallel edges between e and d.

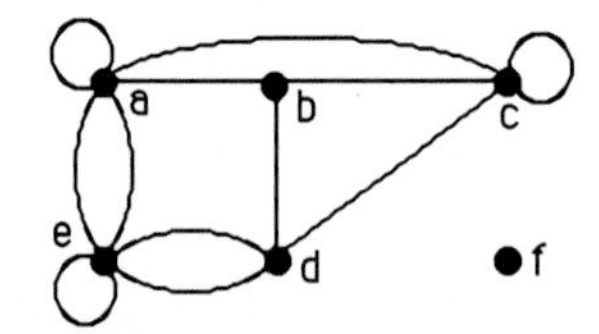

13. Since a person is joined by an edge to each of his or her collaborators, the degree of v is the number of collaborators v has. An isolated vertex (degree 0) is someone who has never collaborated. A pendant vertex (degree 1) is someone who has just one collaborator.

15. Since there is a directed edge from u to v for each call made by u to v, the in-degree of v is the number of calls v received, and the out-degree of u is the number of calls u made. The degree of a vertex in the undirected version is just the sum of these, which is therefore the number of calls the vertex was involved in.

17. Since there is a directed edge from u to v to represent the event that u beat v when they played, the in-degree of v must be the number of teams that beat v, and the out-degree of u must be the number of teams that u beat. In other words, the pair $(\deg^+(v), \deg^-(v))$ is the win-loss record of v.

19. In order to use Exercise 18, we must find a graph in which the degree of a vertex represents the number of people the given person knows. Therefore we construct the simple graph model in which V is the set of people in the group and there is an edge associated with $\{u, v\}$ if u and v know each other. In this graph the degree of vertex v is the number of people v knows. By the result of Exercise 18, there are two vertices with the same degree. Therefore there are two people who know the same number of other people in the group.

21. To show that this graph is bipartite we can exhibit the parts and note that indeed every edge joins vertices in different parts. Take $\{e\}$ to be one part and $\{a, b, c, d\}$ to be the other (in fact there is no choice in the matter). Each edge joins a vertex in one part to a vertex in the other. This graph is the complete bipartite graph $K_{1,4}$.

23. To show that a graph is not bipartite we must give a proof that there is no possible way to specify the parts. (There is another good way to characterize nonbipartite graphs, but it takes some notions not introduced until Section 9.4.) We can show that this graph is not bipartite by the pigeonhole principle. Consider the vertices b, c, and f. They form a triangle—each is joined by an edge to the other two. By the pigeonhole principle, at least two of them must be in the same part of any proposed bipartition. Therefore there would be an edge joining two vertices in the same part, a contradiction to the definition of a bipartite graph. Thus this graph is not bipartite.

An alternative way to look at this is given by Theorem 4. Because of the triangle, it is impossible to color the vertices to satisfy the condition given there.

25. As in Exercise 23, we can show that this graph is not bipartite by looking at a triangle, in this case the triangle formed by vertices b, d, and e. Each of these vertices is joined by an edge to the other two. By the pigeonhole principle, at least two of them must be in the same part of any proposed bipartition. Therefore there would be an edge joining two vertices in the same part, a contradiction to the definition of a bipartite graph. Thus this graph is not bipartite.

27. **a)** Following the lead in Example 14, we construct a bipartite graph in which the vertex set consists of two subsets—one for the employees and one for the jobs. Let $V_1 = \{\text{Zamora, Agraharam, Smith, Chou, Macintyre}\}$, and let $V_2 = \{\text{planning, publicity, sales, marketing, development, industry relations}\}$. Then the vertex set for our graph is $V = V_1 \cup V_2$. Given the list of capabilities in the exercise, we must include precisely the following edges in our graph: {Zamora, planning}, {Zamora, sales}, {Zamora, marketing}, {Zamora, industry relations}, {Agraharam, planning}, {Agraharam, development}, {Smith, publicity}, {Smith, sales}, {Smith, industry relations}, {Chou, planning}, {Chou, sales}, {Chou, industry relations}, {Macintyre, planning}, {Macintyre, publicity}, {Macintyre, sales}, {Macintyre, industry relations}.

b) Many assignments are possible. If we take it as an implicit assumption that there will be no more than one employee assigned to the same job, then we want a maximal matching for this graph. So we look for five edges in this graph that share no endpoints. A little trial and error leads us, for example, {Zamora, planning}, {Agraharam, development}, {Smith, publicity}, {Chou, sales}, {Macintyre, industry relations}. We assign the employees to the jobs given in this matching.

29. **a)** Obviously K_n has n vertices. It has $C(n,2) = n(n-1)/2$ edges, since each unordered pair of distinct vertices is an edge.

b) Obviously C_n has n vertices. Just as obviously it has n edges.

c) The wheel W_n is the same as C_n with an extra vertex and n extra edges incident to that vertex. Therefore it has $n+1$ vertices and $n+n=2n$ edges.

d) By definition $K_{m,n}$ has $m+n$ vertices. Since it has one edge for each choice of a vertex in the one part and a vertex in the other part, it has mn edges.

e) Since the vertices of Q_n are the bit strings of length n, there are 2^n vertices. Each vertex has degree n, since there are n strings that differ from any given string in exactly one bit (any one of the n different bits can be changed). Thus the sum of the degrees is $n2^n$. Since this must equal twice the number of edges (by the Handshaking Theorem), we know that there are $n2^n/2 = n2^{n-1}$ edges.

31. In each case we just record the degrees of the vertices in a list, from largest to smallest.

a) Each of the four vertices is adjacent to each of the other three vertices, so the degree sequence is $3,3,3,3$.

b) Each of the four vertices is adjacent to its two neighbors in the cycle, so the degree sequence is $2,2,2,2$.

c) Each of the four vertices on the rim of the wheel is adjacent to each of its two neighbors on the rim, as well as to the middle vertex. The middle vertex is adjacent to the four rim vertices. Therefore the degree sequence is $4,3,3,3,3$.

d) Each of the vertices in the part of size two is adjacent to each of the three vertices in the part of size three, and vice versa, so the degree sequence is $3,3,2,2,2$.

e) Each of the eight vertices in the cube is adjacent to three others (for example, 000 is adjacent to 001, 010, and 100. Therefore the degree sequence is $3,3,3,3,3,3,3,3$.

33. Each of the n vertices is adjacent to each of the other $n-1$ vertices, so the degree sequence is simply $n-1, n-1, \ldots, n-1$, with n terms in the sequence.

35. The number of edges is half the sum of the degrees (Theorem 1). Therefore this graph has $(5+2+2+2+2+1)/2 = 7$ edges. A picture of this graph is shown here (it is essentially unique).

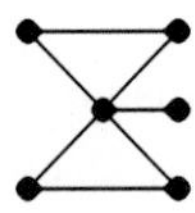

37. There is no such graph in part **(b)**, since the sum of the degrees is odd (and also because a simple graph with 5 vertices cannot have any degrees greater than 4). Similarly, the odd degree sum prohibits the existence of graphs with the degree sequences given in part **(d)** and part **(f)**. There is no such graph in part **(c)**, since the existence of two vertices of degree 4 implies that there are two vertices each joined by an edge to every other vertex. This means that the degree of each vertex has to be at least 2, and there can be no vertex of degree 1. The graphs for part **(a)** and part **(e)** are shown below; one can draw them after just a little trial and error.

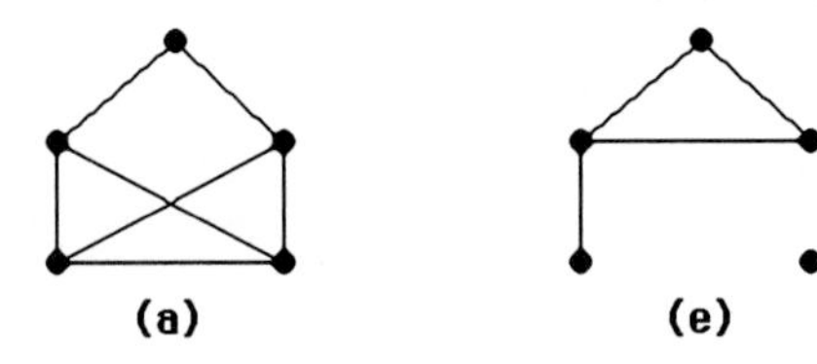

39. We need to prove two conditional statements. First, suppose that $d_1, d_2, \ldots, d_n$ is graphic. We must show that the sequence whose terms are $d_2 - 1, d_3 - 1, \ldots, d_{d_1+1} - 1, d_{d_1+2}, d_{d_1+3}, \ldots, d_n$ is graphic once it is put into nonincreasing order. Apparently what we want to do is to remove the vertex of highest degree (d_1) from a graph with the original degree sequence and reduce by 1 the degrees of the vertices to which it is adjacent, but we also want to make sure that those vertices are the ones with the highest degrees among the remaining vertices. In Exercise 38 it is proved that if the original sequence is graphic, then in fact there is a graph having this degree sequence in which the vertex of degree d_1 is adjacent to the vertices of degrees d_2, $d_3, \ldots, d_{d_1+1}$. Thus our plan works, and we have a graph whose degree sequence is as desired.

Conversely, suppose that $d_1, d_2, \ldots, d_n$ is a nonincreasing sequence such that the sequence $d_2 - 1$, $d_3 - 1, \ldots, d_{d_1+1} - 1, d_{d_1+2}, d_{d_1+3}, \ldots, d_n$ is graphic once it is put into nonincreasing order. Take a graph with this latter degree sequence, where vertex v_i has degree $d_i - 1$ for $2 \le i \le d_1 + 1$ and vertex v_i has degree d_i for $d_1 + 2 \le i \le n$. Adjoin one new vertex (call it v_1), and put in an edge from v_1 to each of the vertices $v_2, v_3, \ldots, v_{d_1+1}$. Then clearly the resulting graph has degree sequence $d_1, d_2, \ldots, d_n$.

41. Let $d_1, d_2, \ldots, d_n$ be a nonincreasing sequence of nonnegative integers with an even sum. We want to construct a pseudograph with this as its degree sequence. Even degrees can be achieved using only loops, each of which contributes 2 to the count of its endpoint; vertices of odd degrees will need a non-loop edge, but one will suffice (the rest of the count at that vertex will be made up by loops). Following the hint, we take vertices $v_1, v_2, \ldots, v_n$ and put $\lfloor d_i/2 \rfloor$ loops at vertex v_i, for $i = 1, 2, \ldots, n$. For each i, vertex v_i now has degree either d_i (if d_i is even) or $d_i - 1$ (if d_i is odd). Because the original sum was even, the number of vertices falling into the latter category is even. If there are $2k$ such vertices, pair them up arbitrarily, and put in k more edges, one joining the vertices in each pair. The resulting graph will have degree sequence $d_1, d_2, \ldots, d_n$.

43. We will count the subgraphs in terms of the number of vertices they contain. There are clearly just 3 subgraphs consisting of just one vertex. If a subgraph is to have two vertices, then there are $C(3,2) = 3$ ways to choose the vertices, and then 2 ways in each case to decide whether or not to include the edge joining them. This gives us $3 \cdot 2 = 6$ subgraphs with two vertices. If a subgraph is to have all three vertices, then there are $2^3 = 8$ ways to decide whether or not to include each of the edges. Thus our answer is $3 + 6 + 8 = 17$.

45. This graph has a lot of subgraphs. First of all, any nonempty subset of the vertex set can be the vertex set for a subgraph, and there are 15 such subsets. If the set of vertices of the subgraph does not contain vertex a, then the subgraph can of course have no edges. If it does contain vertex a, then it can contain or fail to contain each edge from a to whichever other vertices are included. A careful enumeration of all the possibilities gives the 34 graphs shown below.

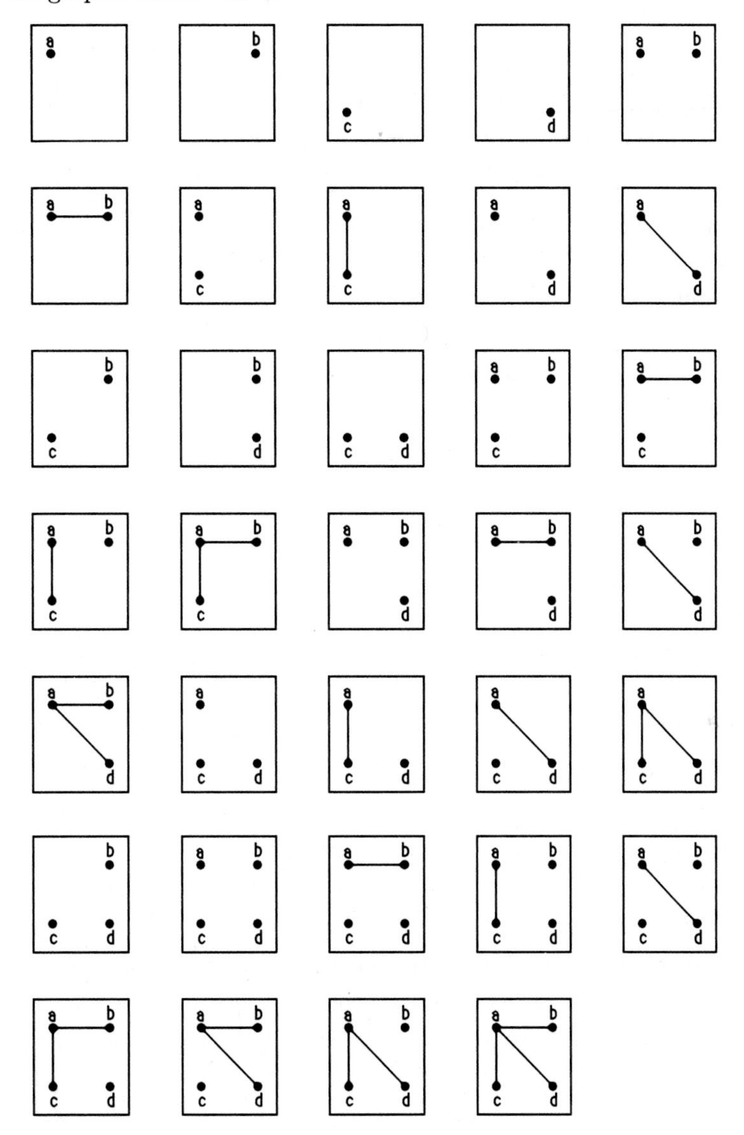

47. a) The complete graph K_n is regular for all values of $n \geq 1$, since the degree of each vertex is $n - 1$.

b) The degree of each vertex of C_n is 2 for all n for which C_n is defined, namely $n \geq 3$, so C_n is regular for all these values of n.

c) The degree of the middle vertex of the wheel W_n is n, and the degree of the vertices on the "rim" is 3. Therefore W_n is regular if and only if $n = 3$. Of course W_3 is the same as K_4.

d) The cube Q_n is regular for all values of $n \geq 0$, since the degree of each vertex in Q_n is n. (Note that Q_0 is the graph with 1 vertex.)

49. If a graph is regular of degree 4 and has n vertices, then by the Handshaking Theorem it has $4n/2 = 2n$ edges. Since we are told that there are 10 edges, we just need to solve $2n = 10$. Thus the graph has 5 vertices. The complete graph K_5 is one such graph (and the only simple one).

51. We draw the answer by superimposing the graphs (keeping the positions of the vertices the same).

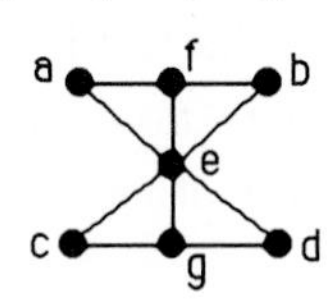

53. **a)** The complement of a complete graph is a graph with no edges.

b) Since all the edges between the parts are present in $K_{m,n}$, but none of the edges between vertices in the same part are, the complement must consist precisely of the disjoint union of a K_m and a K_n, i.e., the graph containing all the edges joining two vertices in the same part and no edges joining vertices in different parts.

c) There is really no better way to describe this graph than simply by saying it is the complement of C_n. One representation would be to take as vertex set the integers from 1 to n, inclusive, with an edge between distinct vertices i and j as long as i and j do not differ by ± 1, modulo n.

d) Again, there is really no better way to describe this graph than simply by saying it is the complement of Q_n. One representation would be to take as vertex set the bit strings of length n, with two vertices joined by an edge if the bit strings differ in more than one bit.

55. Since K_v has $C(v,2) = v(v-1)/2$ edges, and since $\overline{G}$ has all the edges of K_v that G is missing, it is clear that $\overline{G}$ has $[v(v-1)/2] - e$ edges.

57. If G has n vertices, then the degree of vertex v in $\overline{G}$ is $n-1$ minus the degree of v in G (there will be an edge in $\overline{G}$ from v to each of the $n-1$ other vertices that v is not adjacent to in G). The order of the sequence will reverse, of course, because if $d_i \geq d_j$, then $n-1-d_i \leq n-1-d_j$. Therefore the degree sequence of $\overline{G}$ will be $n-1-d_n$, $n-1-d_{n-1}$, $\ldots$, $n-1-d_2$, $n-1-d_1$.

59. Consider the graph $G \cup \overline{G}$. Its vertex set is clearly the vertex set of G; therefore it has n vertices. If u and v are any two distinct vertices of $G \cup \overline{G}$, then either the edge between u and v is in G, or else by definition it is in $\overline{G}$. Therefore by definition of union, it is in $G \cup \overline{G}$. Thus by definition $G \cup \overline{G}$ is the complete graph K_n.

61. These pictures are identical to the figures in those exercises, with one change, namely that all the arrowheads are turned around. For example, rather than there being a directed edge from a to b in #7, there is an edge from b to a. Note that the loops are unaffected by changing the direction of the arrowhead—a loop from a vertex to itself is the same, whether the drawing of it shows the direction to be clockwise or counterclockwise.

63. It is clear from the definition of converse that a directed graph $G = (V, E)$ is its own converse if and only if it satisfies the condition that $(u,v) \in E$ if and only if $(u,v) \in E$. But this is precisely the definition of symmetry for the associated relation.

65. Our picture is just like Figure 13, but with only three vertices on each side.

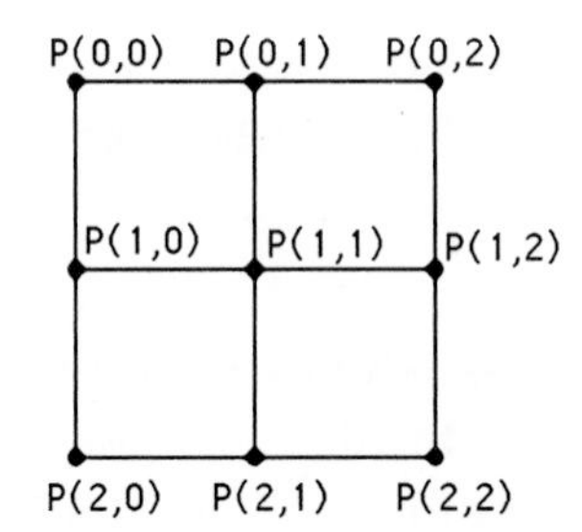

67. Suppose $P(i,j)$ and $P(k,l)$ need to communicate. Clearly by using $|i-k|$ hops we can move from $P(i,j)$ to $P(k,j)$. Then using $|j-l|$ hops we can move from $P(k,j)$ to $P(k,l)$. In all we used $|i-k|+|j-l|$ hops. But each of these absolute values is certainly less than m, since all the indices are less than m. Therefore the sum is less than $2m$, so it is $O(m)$.

SECTION 9.3 Representing Graphs and Graph Isomorphism

Human beings can get a good feeling for a small graph by looking at a picture of it drawn with points in the plane and lines or curves joining pairs of these points. If a graph is at all large (say with more than a dozen vertices or so), then the picture soon becomes too crowded to be useful. A computer has little use for nice pictures, no matter how small the vertex set. Thus people and machines need more precise—more discrete—representations of graphs. In this section we learned about some useful representations. They are for the most part exactly what any intelligent person would come up with, given the assignment to do so.

The only tricky idea in this section is the concept of graph isomorphism. It is a special case of a more general notion of isomorphism, or sameness, of mathematical objects in various settings. Isomorphism tries to capture the idea that all that really matters in a graph is the adjacency structure. If we can find a way to superimpose the graphs so that the adjacency structures match, then the graphs are, for all purposes that matter, the same. In trying to show that two graphs are isomorphic, try moving the vertices around in your mind to see whether you can make the graphs look the same. Of course there are often lots of things to help. For example, in every isomorphism, vertices that correspond must have the same degree.

A good general strategy for determining whether two graphs are isomorphic might go something like this. First check the degrees of the vertices to make sure there are the same number of each degree. See whether vertices of corresponding degrees follow the same adjacency pattern (e.g., if there is a vertex of degree 1 adjacent to a vertex of degree 4 in one of the graphs, then there must be the same pattern in the other, if the graphs are isomorphic). Then look for triangles in the graphs, and see whether they correspond. Sometimes, if the graphs have lots of edges, it is easier to see whether the complements are isomorphic (see Exercise 46). If you cannot find a good reason for the graphs not to be isomorphic (an invariant on which they differ), then try to write down a one-to-one and onto function that shows them to be isomorphic (there may be more than one such function); such a function has to have vertices of like degrees correspond, so often the function practically writes itself. Then check each edge of the first graph to make sure that it corresponds to an edge of the second graph under this correspondence.

Unfortunately, no one has yet discovered a really good algorithm for determining graph isomorphism that works on all pairs of graphs. Research in this subject has been quite active in recent years. See Writing Project 7.

1. Adjacency lists are lists of lists. The adjacency list of an undirected graph is simply a list of the vertices of the given graph, together with a list of the vertices adjacent to each. The list for this graph is as follows. Since, for instance, b is adjacent to a and d, we list a and d in the row for b.

Vertex	Adjacent vertices
a	b, c, d
b	a, d
c	a, d
d	a, b, c

3. To form the adjacency list of a directed graph, we list, for each vertex in the graph, the terminal vertex of each edge that has the given vertex as its initial vertex. The list for this directed graph is as follows. For example, since there are edges from d to each of b, c, and d, we put those vertices in the row for d.

Initial vertex	Terminal vertices
a	a, b, c, d
b	d
c	a, b
d	b, c, d

5. For Exercises 5–8 we assume that the vertices are listed in alphabetical order. The matrix contains a 1 as entry (i,j) if there is an edge from vertex i to vertex j; otherwise that entry is 0.

$$\begin{bmatrix} 0 & 1 & 1 & 1 \\ 1 & 0 & 0 & 1 \\ 1 & 0 & 0 & 1 \\ 1 & 1 & 1 & 0 \end{bmatrix}$$

7. This is similar to Exercise 5. Note that edges have direction here, so that, for example, the $(1,2)$ entry is a 1 since there is an edge from a to b, but the $(2,1)$ entry is a 0 since there is no edge from b to a. Also, the $(1,1)$ entry is a 1 since there is a loop at a, but the $(2,2)$ entry is a 0 since there is no loop at b.

$$\begin{bmatrix} 1 & 1 & 1 & 1 \\ 0 & 0 & 0 & 1 \\ 1 & 1 & 0 & 0 \\ 0 & 1 & 1 & 1 \end{bmatrix}$$

9. We can solve these problems by first drawing the graph, then labeling the vertices, and finally constructing the matrix by putting a 1 in position (i,j) whenever vertices i and j are joined by an edge. It helps to choose a nice order, since then the matrix will have nice patterns in it.

a) The order of the vertices does not matter, since they all play the same role. The matrix has 0's on the diagonal, since there are no loops in the complete graph.

$$\begin{bmatrix} 0 & 1 & 1 & 1 \\ 1 & 0 & 1 & 1 \\ 1 & 1 & 0 & 1 \\ 1 & 1 & 1 & 0 \end{bmatrix}$$

b) We put the vertex in the part by itself first.

$$\begin{bmatrix} 0 & 1 & 1 & 1 & 1 \\ 1 & 0 & 0 & 0 & 0 \\ 1 & 0 & 0 & 0 & 0 \\ 1 & 0 & 0 & 0 & 0 \\ 1 & 0 & 0 & 0 & 0 \end{bmatrix}$$

c) We put the vertices in the part of size 2 first. Notice the block structure.

$$\begin{bmatrix} 0 & 0 & 1 & 1 & 1 \\ 0 & 0 & 1 & 1 & 1 \\ 1 & 1 & 0 & 0 & 0 \\ 1 & 1 & 0 & 0 & 0 \\ 1 & 1 & 0 & 0 & 0 \end{bmatrix}$$

d) We put the vertices in the same order in the matrix as they are around the cycle.

$$\begin{bmatrix} 0 & 1 & 0 & 1 \\ 1 & 0 & 1 & 0 \\ 0 & 1 & 0 & 1 \\ 1 & 0 & 1 & 0 \end{bmatrix}$$

e) We put the center vertex first. Note that the last four columns of the last four rows represent a C_4.

$$\begin{bmatrix} 0 & 1 & 1 & 1 & 1 \\ 1 & 0 & 1 & 0 & 1 \\ 1 & 1 & 0 & 1 & 0 \\ 1 & 0 & 1 & 0 & 1 \\ 1 & 1 & 0 & 1 & 0 \end{bmatrix}$$

f) We can label the vertices by the binary numbers from 0 to 7. Thus the first row (also the first column) of this matrix corresponds to the string 000, the second to the string 001, and so on. Since Q_3 has 8 vertices, this is an 8×8 matrix.

$$\begin{bmatrix} 0 & 1 & 1 & 0 & 1 & 0 & 0 & 0 \\ 1 & 0 & 0 & 1 & 0 & 1 & 0 & 0 \\ 1 & 0 & 0 & 1 & 0 & 0 & 1 & 0 \\ 0 & 1 & 1 & 0 & 0 & 0 & 0 & 1 \\ 1 & 0 & 0 & 0 & 0 & 1 & 1 & 0 \\ 0 & 1 & 0 & 0 & 1 & 0 & 0 & 1 \\ 0 & 0 & 1 & 0 & 1 & 0 & 0 & 1 \\ 0 & 0 & 0 & 1 & 0 & 1 & 1 & 0 \end{bmatrix}$$

11. This graph has four vertices and is directed, since the matrix is not symmetric. We draw the four vertices as points in the plane, then draw a directed edge from vertex i to vertex j whenever there is a 1 in position (i,j) in the given matrix.

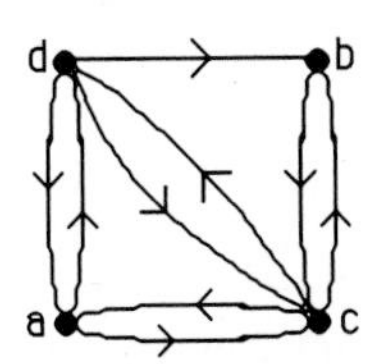

13. We use alphabetical order of the vertices for Exercises 13–15. If there are k parallel edges between vertices i and j, then we put the number k into the $(i,j)^{\text{th}}$ entry of the matrix. In this exercise, there is only one pair of parallel edges.

$$\begin{bmatrix} 0 & 0 & 1 & 0 \\ 0 & 0 & 1 & 2 \\ 1 & 1 & 0 & 1 \\ 0 & 2 & 1 & 0 \end{bmatrix}$$

15. This is similar to Exercise 13. In this graph there are loops, which are represented by entries on the diagonal. For example, the loop at c is shown by the 1 as the $(3,3)^{\text{th}}$ entry.

$$\begin{bmatrix} 1 & 0 & 2 & 1 \\ 0 & 1 & 1 & 2 \\ 2 & 1 & 1 & 0 \\ 1 & 2 & 0 & 1 \end{bmatrix}$$

17. Because of the numbers larger than 1, we need multiple edges in this graph.

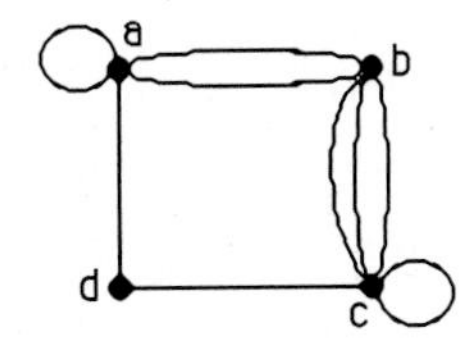

19. We use alphabetical order of the vertices. We put a 1 in position (i,j) if there is a directed edge from vertex i to vertex j; otherwise we make that entry a 0. Note that loops are represented by 1's on the diagonal.

$$\begin{bmatrix} 0 & 1 & 0 & 0 \\ 0 & 1 & 1 & 0 \\ 0 & 1 & 1 & 1 \\ 1 & 0 & 0 & 0 \end{bmatrix}$$

21. This is similar to Exercise 19, except that there are parallel directed edges. If there are k parallel edges from vertex i to vertex j, then we put the number k into the $(i,j)^{\text{th}}$ entry of the matrix. For example, since there are 2 edges from a to c, the $(1,3)^{\text{th}}$ entry of the adjacency matrix is 2; the loop at c is shown by the 1 as the $(3,3)^{\text{th}}$ entry.

$$\begin{bmatrix} 1 & 1 & 2 & 1 \\ 1 & 0 & 0 & 2 \\ 1 & 0 & 1 & 1 \\ 0 & 2 & 1 & 0 \end{bmatrix}$$

23. Since the matrix is not symmetric, we need directed edges; furthermore, it must be a directed multigraph because of the entries larger than 1. For example, the 2 in position $(3,2)$ means that there are two parallel edges from vertex c to vertex b.

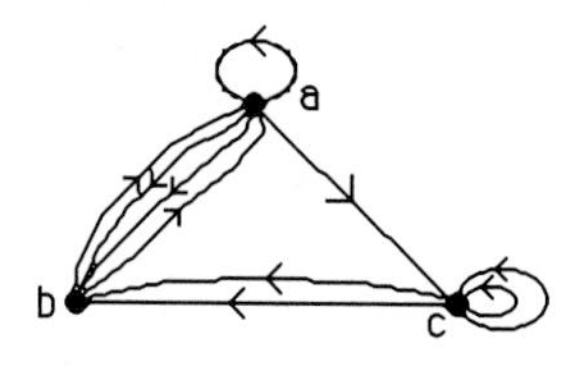

25. Since the matrix is symmetric, it has to be square, so it represents a graph of some sort. In fact, such a matrix does represent a simple graph. The fact that it is a zero-one matrix means that there are no parallel edges. The fact that there are 0's on the diagonal means that there are no loops. The fact that the matrix is symmetric means that the edges can be assumed to be undirected. Note that such a matrix also represents a directed graph in which all the edges happen to appear in antiparallel pairs (see the solution to Exercise 1d in Section 9.1 for a definition), but that is irrelevant to this question; the answer to the question asked is "yes."

27. In an incidence matrix we have one column for each edge. We use alphabetical order of the vertices. Loops are represented by columns with one 1; other edges are represented by columns with two 1's. The order in which the columns are listed is immaterial.

Exercise 13 $\begin{bmatrix} 1 & 0 & 0 & 0 & 0 \\ 0 & 1 & 1 & 1 & 0 \\ 1 & 1 & 0 & 0 & 1 \\ 0 & 0 & 1 & 1 & 1 \end{bmatrix}$ Exercise 14 $\begin{bmatrix} 1 & 1 & 1 & 1 & 0 & 0 & 0 & 0 \\ 1 & 1 & 1 & 0 & 1 & 0 & 0 & 0 \\ 0 & 0 & 0 & 0 & 1 & 1 & 1 & 1 \\ 0 & 0 & 0 & 1 & 0 & 1 & 1 & 1 \end{bmatrix}$

Exercise 15 $\begin{bmatrix} 1 & 1 & 1 & 1 & 0 & 0 & 0 & 0 & 0 & 0 \\ 0 & 0 & 0 & 0 & 1 & 1 & 1 & 1 & 0 & 0 \\ 0 & 1 & 1 & 0 & 0 & 1 & 0 & 0 & 1 & 0 \\ 0 & 0 & 0 & 1 & 0 & 0 & 1 & 1 & 0 & 1 \end{bmatrix}$

29. In an undirected graph, each edge incident to a vertex j contributes 1 in the jth column; thus the sum of the entries in that column is just the number of edges incident to j. Another way to state the answer is that the sum of the entries is the degree of j minus the number of loops at j, since each loop counts 2 toward the degree.

In a directed graph, each edge whose terminal vertex is j contributes 1 in the jth column; thus the sum of the entries in that column is just the number of edges that have j as their terminal vertex. Another way to state the answer is that the sum of the entries is the in-degree of j.

31. Since each column represents an edge, the sum of the entries in the column is either 2, if the edge has 2 incident vertices (i.e., is not a loop), or 1 if it has only 1 incident vertex (i.e., is a loop).

33. **a)** The incidence matrix for K_n has n rows and $C(n,2)$ columns. For each i and j with $1 \le i < j \le n$, there is a column with 1's in rows i and j and 0's elsewhere.

b) The matrix looks like this, with n rows and n columns.

$$\begin{bmatrix} 1 & 0 & 0 & \cdots & 0 & 1 \\ 1 & 1 & 0 & \cdots & 0 & 0 \\ 0 & 1 & 1 & \cdots & 0 & 0 \\ 0 & 0 & 1 & \cdots & 0 & 0 \\ \vdots & \vdots & \vdots & \ddots & \vdots & \vdots \\ 0 & 0 & 0 & \cdots & 1 & 0 \\ 0 & 0 & 0 & \cdots & 1 & 1 \end{bmatrix}$$

c) The matrix looks like the matrix for C_n, except with an extra row of 0's (which we have put at the end), since the vertex "in the middle" is not involved in the edges "around the outside," and n more columns for the "spokes." We show some extra space between the rim edge columns and the spoke columns; this is for human convenience only and does not have any bearing on the matrix itself.

$$\begin{bmatrix} 1 & 0 & 0 & \cdots & 0 & 1 & \quad 1 & 0 & 0 & \cdots & 0 \\ 1 & 1 & 0 & \cdots & 0 & 0 & \quad 0 & 1 & 0 & \cdots & 0 \\ 0 & 1 & 1 & \cdots & 0 & 0 & \quad 0 & 0 & 1 & \cdots & 0 \\ 0 & 0 & 1 & \cdots & 0 & 0 & \quad 0 & 0 & 0 & \cdots & 0 \\ \vdots & \vdots & \vdots & \ddots & \vdots & \vdots & \quad \vdots & \vdots & \vdots & \ddots & \vdots \\ 0 & 0 & 0 & \cdots & 1 & 0 & \quad 0 & 0 & 0 & \cdots & 0 \\ 0 & 0 & 0 & \cdots & 1 & 1 & \quad 0 & 0 & 0 & \cdots & 1 \\ 0 & 0 & 0 & \cdots & 0 & 0 & \quad 1 & 1 & 1 & \cdots & 1 \end{bmatrix}$$

d) This matrix has $m + n$ rows and mn columns, one column for each pair (i,j) with $1 \le i \le m$ and $1 \le j \le n$. We have put in some extra spacing for readability of the pattern.

$$\begin{bmatrix} 1 & 1 & \cdots & 1 & \quad 0 & 0 & \cdots & 0 & \quad \cdots & \quad 0 & 0 & \cdots & 0 \\ 0 & 0 & \cdots & 0 & \quad 1 & 1 & \cdots & 1 & \quad \cdots & \quad 0 & 0 & \cdots & 0 \\ \vdots & \vdots & \ddots & \vdots & \quad \vdots & \vdots & \ddots & \vdots & \quad \ddots & \quad \vdots & \vdots & \ddots & \vdots \\ 0 & 0 & \cdots & 0 & \quad 0 & 0 & \cdots & 0 & \quad \cdots & \quad 1 & 1 & \cdots & 1 \\ \\ 1 & 0 & \cdots & 0 & \quad 1 & 0 & \cdots & 0 & \quad \cdots & \quad 1 & 0 & \cdots & 0 \\ 0 & 1 & \cdots & 0 & \quad 0 & 1 & \cdots & 0 & \quad \cdots & \quad 0 & 1 & \cdots & 0 \\ \vdots & \vdots & \ddots & \vdots & \quad \vdots & \vdots & \ddots & \vdots & \quad \ddots & \quad \vdots & \vdots & \ddots & \vdots \\ 0 & 0 & \cdots & 1 & \quad 0 & 0 & \cdots & 1 & \quad \cdots & \quad 0 & 0 & \cdots & 1 \end{bmatrix}$$

35. These graphs are isomorphic, since each is the 5-cycle. One isomorphism is $f(u_1) = v_1$, $f(u_2) = v_3$, $f(u_3) = v_5$, $f(u_4) = v_2$, and $f(u_5) = v_4$.

37. These graphs are isomorphic, since each is the 7-cycle (this is just like Exercise 35).

39. These two graphs are isomorphic. One can see this visually—just imagine "moving" vertices u_1 and u_4 into the inside of the rectangle, thereby obtaining the picture on the right. Formally, one isomorphism is $f(u_1) = v_5$, $f(u_2) = v_2$, $f(u_3) = v_3$, $f(u_4) = v_6$, $f(u_5) = v_4$, and $f(u_6) = v_1$.

41. These graphs are not isomorphic. In the first graph the vertices of degree 3 are adjacent to a common vertex. This is not true of the second graph.

43. These are isomorphic. One isomorphism is $f(u_1) = v_1$, $f(u_2) = v_9$, $f(u_3) = v_4$, $f(u_4) = v_3$, $f(u_5) = v_2$, $f(u_6) = v_8$, $f(u_7) = v_7$, $f(u_8) = v_5$, $f(u_9) = v_{10}$, and $f(u_{10}) = v_6$.

45. We must show that being isomorphic is reflexive, symmetric, and transitive. It is reflexive since the identity function from a graph to itself provides the isomorphism (the one-to-one correspondence)—certainly the identity function preserves adjacency and nonadjacency. It is symmetric, since if f is a one-to-one correspondence that makes G_1 isomorphic to G_2, then f^{-1} is a one-to-one correspondence that makes G_2 isomorphic to G_1; that is, f^{-1} is a one-to-one and onto function from V_2 to V_1 such that c and d are adjacent in G_2 if and only if $f^{-1}(c)$ and $f^{-1}(d)$ are adjacent in G_1. It is transitive, since if f is a one-to-one correspondence that makes G_1 isomorphic to G_2, and g is a one-to-one correspondence that makes G_2 isomorphic to G_3, then $g \circ f$ is a one-to-one correspondence that makes G_1 isomorphic to G_3.

47. If a vertex is isolated, then it has no adjacent vertices. Therefore in the adjacency matrix the row and column for that vertex must contain all 0's.

49. Let V_1 and V_2 be the two parts, say of sizes m and n, respectively. We can number the vertices so that all the vertices in V_1 come before all the vertices in V_2. The adjacency matrix has $m+n$ rows and $m+n$ columns. Since there are no edges between two vertices in V_1, the first m columns of the first m rows must all be 0's. Similarly, since there are no edges between two vertices in V_2, the last n columns of the last n rows must all be 0's. This is what we were asked to prove.

51. There are two such graphs, which can be found by trial and error. (We need only look for graphs with 5 vertices and 5 edges, since a self-complementary graph with 5 vertices must have $C(5,2)/2 = 5$ edges. If nothing else, we can draw them all and find the complement of each. See the pictures for the solution of Exercise 45d in Section 9.4.) One such graph is C_5. The other consists of a triangle, together with an edge from one vertex of the triangle to the fourth vertex, and an edge from another vertex of the triangle to the fifth vertex.

53. If C_n is to be self-complementary, then C_n must have the same number of edges as its complement. We know that C_n has n edges. Its complement has the number of edges in K_n minus the number of edges in C_n, namely $C(n,2)-n = [n(n-1)/2]-n$. If we set these two quantities equal we obtain $[n(n-1)/2]-n = n$, which has $n = 5$ as its only solution. Thus C_5 is the only C_n that *might* be self-complementary—our argument just shows that it has the same number of edges as its complement, not that it is indeed isomorphic to its complement. However, it we draw C_5 and then draw its complement, then we see that the complement is again a copy of C_5. Thus $n = 5$ is the answer to the problem.

55. We need to enumerate these graphs carefully to make sure of getting them all—leaving none out and not duplicating any. Let us organize our catalog by the degrees of the vertices. Since there are only 3 edges, the largest the degree could be is 3, and the only graph with 5 vertices, 3 edges, and a vertex of degree 3 is a $K_{1,3}$ together with an isolated vertex. If all the vertices that are not isolated have degree 2, then the graph must consist of a C_3 and 2 isolated vertices. The only way for there to be two vertices of degree 2 (and therefore also 2 of degree 1) is for the graph to be three edges strung end to end, together with an isolated vertex. The only other possibility is for 2 of the edges to be adjacent and the third to be not adjacent to either of the others. All in all, then, we have the 4 possibilities shown below.

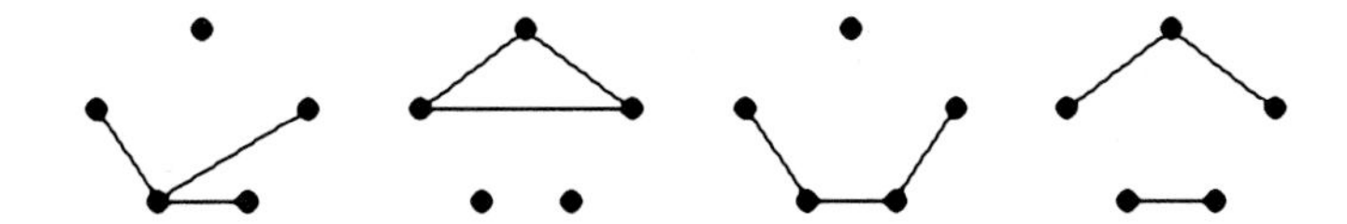

57. **a)** Both graphs consist of 2 sides of a triangle; they are clearly isomorphic.

b) The graphs are not isomorphic, since the first has 4 edges and the second has 5 edges.

c) The graphs are not isomorphic, since the first has 4 edges and the second has 3 edges.

59. There are at least two approaches we could take here. One approach is to have a correspondence not only of the vertices but also of the edges, with incidence (and nonincidence) preserved. In detail, we say that two pseudographs $G_1 = (V_1, E_1)$ and $G_2 = (V_2, E_2)$ are isomorphic if there are one-to-one and onto functions $f: V_1 \to V_2$ and $g: E_1 \to E_2$ such that for each vertex $v \in V_1$ and edge $e \in E_1$, v is incident to e if and only if $f(v)$ is incident to $g(e)$.

Another approach is simply to count the number of edges between pairs of vertices. Thus we can define $G_1 = (V_1, E_1)$ to be isomorphic to $G_2 = (V_2, E_2)$ if there is a one-to-one and onto function $f: V_1 \to V_2$ such that for every pair of (not necessarily distinct) vertices u and v in V_1, there are exactly the same number of edges in E_1 with $\{u, v\}$ as their set of endpoints as there are edges in E_2 with $\{f(u), f(v)\}$ as their set of endpoints.

61. We can tell by looking at the loop, the parallel edges, and the degrees of the vertices that if these directed graphs are to be isomorphic, then the isomorphism has to be $f(u_1) = v_3$, $f(u_2) = v_4$, $f(u_3) = v_2$, and $f(u_4) = v_1$. We then need to check that each directed edge (u_i, u_j) corresponds to a directed edge $(f(u_i), f(u_j))$. We check that indeed it does for each of the 7 edges (and there are only 7 edges in the second graph). Therefore the two graphs are isomorphic.

63. If there is to be an isomorphism, the vertices with the same in-degree would have to correspond, and the edge between them would have to point in the same direction, so we would need u_1 to correspond to v_3, and u_2 to correspond to v_1. Similarly we would need u_3 to correspond to v_4, and u_4 to correspond to v_2. If we check all 6 edges under this correspondence, then we see that adjacencies are preserved (in the same direction), so the graphs are isomorphic.

65. If f is an isomorphism from a directed graph G to a directed graph H, then f is also an isomorphism from G^c to H^c. This is clear, because (u, v) is an edge of G^c if and only if (v, u) is an edge of G if and only if $(f(v), f(u))$ is an edge of H if and only if $(f(u), f(v))$ is an edge of H^c.

67. A graph with a triangle will not be bipartite, but cycles of even length are bipartite. So we could let one graph be C_6 and the other be the union of two disjoint copies of C_3.

69. Suppose that the graph has v vertices and e edges. Then the incidence matrix is a $v \times e$ matrix, so its transpose is an $e \times v$ matrix. Therefore the product is a $v \times v$ matrix. Suppose that we denote the typical entry of this product by a_{ij}. Let t_{ik} be the typical entry of the incidence matrix; it is either a 0 or a 1. By definition

$$a_{ij} = \sum_{k=1}^{e} t_{ik} t_{jk} .$$

We can now read off the answer from this equation. If $i \neq j$, then a_{ij} is just a count of the number of edges incident to both i and j—in other words, the number of edges between i and j. On the other hand a_{ii} is equal to the number of edges incident to i.

71. Perhaps the simplest example would be to have the graphs have all degrees equaling 2. One way for this to happen is for the graph to be a cycle. But it will also happen if the graph is a disjoint union of cycles. The smallest example occurs when there are six vertices. If G_1 is the 6-cycle and G_2 is the union of two triangles, then the degree sequences are $(2,2,2,2,2,2)$ for both, but obviously the graphs are not isomorphic. If we want a connected example, then look at Exercise 41, where the degree sequence is $(3,3,2,2,1,1,1,1)$ for each graph.

SECTION 9.4 Connectivity

Some of the most important uses of graphs deal with the notion of path, as the examples and exercises in this and subsequent sections show. It is important to understand the definitions, of course. Many of the exercises here are straightforward. The reader who wants to get a better feeling for what the arguments in more advanced graph theory are like should tackle problems like Exercises 33–36.

1. a) This is a path of length 4, but it is not simple, since edge $\{b,c\}$ is used twice. It is not a circuit, since it ends at a different vertex from the one at which it began.

b) This is not a path, since there is no edge from c to a.

c) This is not a path, since there is no edge from b to a.

d) This is a path of length 5 (it has 5 edges in it). It is simple, since no edge is repeated. It is a circuit since it ends at the same vertex at which it began.

3. This graph is not connected—it has three components.

5. This graph is not connected. There is no path from the vertices in one of the triangles to the vertices in the other.

7. A connected component of an acquaintanceship graph represent a maximal set of people with the property that for any two of them, we can find a string of acquaintances that takes us from one to the other. The word "maximal" here implies that nobody else can be added to this set of people without destroying this property.

9. If a person has Erdős number n, then there is a path of length n from that person to Erdős in the collaboration graph. By definition, that means that that person is in the same component as Erdős. Conversely, if a person is in the same component as Erdős, then there is a path from that person to Erdős, and the length of a shortest such path is that person's Erdős number.

11. a) Notice that there is no path from a to any other vertex, because both edges involving a are directed toward a. Therefore the graph is not strongly connected. However, the underlying undirected graph is clearly connected, so this graph is weakly connected.

b) Notice that there is no path from c to any other vertex, because both edges involving c are directed toward c. Therefore the graph is not strongly connected. However, the underlying undirected graph is clearly connected, so this graph is weakly connected.

c) The underlying undirected graph is clearly not connected (one component has vertices b, f, and e), so this graph is neither strongly nor weakly connected.

13. The strongly connected components are the maximal sets of phone numbers for which it is possible to find directed paths between every two different numbers in the set, where the existence of a directed path from phone number x to another phone number y means that x called some number, which called another number, $\ldots$, which called y. (The number of intermediary phone numbers in this path can be any natural number.)

15. In each case we want to look for large sets of vertices all which of which have paths to all the others. For these graphs, this can be done by inspection. These will be the strongly connected components.

a) Clearly $\{a, b, f\}$ is a set of vertices with paths between all the vertices in the set. The same can be said of $\{c, d, e\}$. Every edge between a vertex in the first set and a vertex in the second set is directed from the first, to the second. Hence there are no paths from c, d, or e to a, b, or f, and therefore these vertices are not in the same strongly connected component. Therefore these two sets are the strongly connected component.

b) The circuits a, e, d, c, b, a and a, e, d, h, a show that these six vertices are all in the same component. There is no path from f to any of these vertices, and no path from g to any other vertex. Therefore f and g are not in the same strong component as any other vertex. Therefore the strongly connected components are $\{a, b, c, d, e, h\}$, $\{f\}$, and $\{g\}$.

c) It is clear that a and i are in the same strongly connected component. If we look hard, we can also find the circuit b, h, f, g, d, e, d, b, so these vertices are in the same strongly connected component. Because of edges ig and hi, we can get from either of these collections to the other. Thus $\{a, b, d, e, f, g, h, i\}$ is a strong component. We cannot travel from c to any other vertex, so c is in a component by itself.

17. One approach here is simply to invoke Theorem 2 and take successive powers of the adjacency matrix

$$\mathbf{A} = \begin{bmatrix} 0 & 1 & 1 & 1 \\ 1 & 0 & 1 & 1 \\ 1 & 1 & 0 & 1 \\ 1 & 1 & 1 & 0 \end{bmatrix}.$$

The answers are the off-diagonal elements of these powers. An alternative approach is to argue combinatorially as follows. Without loss of generality, we assume that the vertices are called $1, 2, 3, 4$, and the path is to run from 1 to 2. A path of length n is determined by choosing the $n - 1$ intermediate vertices. Each vertex in the path must differ from the one immediately preceding it.

a) A path of length 2 requires the choice of 1 intermediate vertex, which must be different from both of the ends. Vertices 3 and 4 are the only ones available. Therefore the answer is 2.

b) Let the path be denoted $1, x, y, 2$. If $x = 2$, then there are 3 choices for y. If $x = 3$, then there are 2 choices for y; similarly if $x = 4$. Therefore there are $3 + 2 + 2 = 7$ possibilities in all.

c) Let the path be denoted $1, x, y, z, 2$. If $x = 3$, then by part **(b)** there are 7 choices for y and z. Similarly if $x = 4$. If $x = 2$, then y and z can be any two distinct members of $\{1, 3, 4\}$, and there are $P(3, 2) = 6$ ways to choose them. Therefore there are $7 + 7 + 6 = 20$ possibilities in all.

d) Let the path be denoted $1, w, x, y, z, 2$. If $w = 3$, then by part **(c)** there are 20 choices for x, y, and z. Similarly if $w = 4$. If $w = 2$, then x must be different from 2, and there are 3 choices for x. For each of these there are by part **(b)** 7 choices for y and z. This gives a total of 21 possibilities in this case. Therefore the answer is $20 + 20 + 21 = 61$.

19. Graph G has a triangle (u_1, u_2, u_3). Graph H does not (in fact, it is bipartite). Therefore G and H are not isomorphic.

21. The drawing of G clearly shows it to be the cube Q_3. Can we see H as a cube as well? Yes—we can view the outer ring as the top face, and the inner ring as the bottom face. We can imagine walking around the top face of G clockwise (as viewed from above), then dropping down to the bottom face and walking around it counterclockwise, finally returning to the starting point on the top face. This is the path $u_1, u_2, u_7, u_6, u_5, u_4, u_3, u_8, u_1$. The corresponding path in H is $v_1, v_2, v_3, v_4, v_5, v_8, v_7, v_6, v_1$. We can verify that the edges not in the path do connect corresponding vertices. Therefore $G \cong H$.

23. As explained in the solution to Exercise 17, we could take powers of the adjacency matrix

$$\mathbf{A} = \begin{bmatrix} 0 & 0 & 0 & 1 & 1 & 1 \\ 0 & 0 & 0 & 1 & 1 & 1 \\ 0 & 0 & 0 & 1 & 1 & 1 \\ 1 & 1 & 1 & 0 & 0 & 0 \\ 1 & 1 & 1 & 0 & 0 & 0 \\ 1 & 1 & 1 & 0 & 0 & 0 \end{bmatrix}.$$

The answers are found in location $(1,2)$, for instance. Using the alternative approach is much easier than in Exercise 17. First of all, two nonadjacent vertices must lie in the same part, so only paths of even length can join them. Also, there are clearly 3 choices for each intermediate vertex in a path. Therefore we have the following answers:

a) $3^1 = 3$ **b)** 0 **c)** $3^3 = 27$ **d)** 0

25. There are two approaches here. We could use matrix multiplication on the adjacency matrix of this directed graph (by Theorem 2), which is

$$\mathbf{A} = \begin{bmatrix} 0 & 1 & 0 & 1 & 0 \\ 1 & 0 & 0 & 0 & 1 \\ 0 & 1 & 0 & 0 & 0 \\ 1 & 0 & 0 & 0 & 0 \\ 0 & 0 & 1 & 1 & 0 \end{bmatrix}.$$

Thus we can compute $\mathbf{A}^2$ for part **(a)**, $\mathbf{A}^3$ for part **(b)**, and so on, and look at the $(1,5)^{\text{th}}$ entry to determine the number of paths from a to e. Alternately, we can argue in an ad hoc manner, as we do below.

a) There is just 1 path of length 2, namely a, b, e.

b) There are no paths of length 3, since after 3 steps, a path starting at a must be at b, c, or d.

c) For a path of length 4 to end at e, it must be at b after 3 steps. There are only 2 such paths, a, b, a, b, e and a, d, a, b, e.

d) The only way for a path of length 5 to end at e is for the path to go around the triangle bec. Therefore only the path a, b, e, c, b, e is possible.

e) There are several possibilities for a path of length 6. Since the only way to get to e is from b, we are asking for the number of paths of length 5 from a to b. We can go around the square (a, b, e, d, a, b), or else we can jog over to either b or d and back twice—there being 4 ways to choose where to do the jogging. Therefore there are 5 paths in all.

f) As in part **(d)**, it is clear that we have to use the triangle. We can either have a, b, a, b, e, c, b, e or a, d, a, b, e, c, b, e or a, b, e, c, b, a, b, e. Thus there are 3 paths.

27. The definition given here makes it clear that u and v are related if and only if they are in the same component—in other words $f(u) = f(v)$ where $f(x)$ is the component in which x lies. Therefore by Exercise 9 in Section 8.5 this is an equivalence relation.

29. A cut vertex is one whose removal splits the graph into more components than it originally had (which is 1 in this case). Only vertex c is a cut vertex here. If it is removed, then the resulting graph will have two components. If any other vertex is removed, then the graph remains connected.

31. There are several cut vertices here: b, c, e, and i. Removing any of these vertices creates a graph with more than one component. The removal of any of the other vertices leaves a graph with just one component.

33. Without loss of generality, we can restrict our attention to the component in which the cut edge lies; other components of the graph are irrelevant to this proposition. To fix notation, let the cut edge be uv. When the cut edge is removed, the graph has two components, one of which contains v and the other of which contains u.

If v is pendant, then it is clear that the removal of v results in exactly the component containing u—a connected graph. Therefore v is not a cut vertex in this case. On the other hand, if v is not pendant, then there are other vertices in the component containing v—at least one other vertex w adjacent to v. (We are assuming that this proposition refers to a simple graph, so that there is no loop at v.) Therefore when v is removed, there are at least two components, one containing u and another containing w.

35. If every component of G is a single vertex, then clearly no vertex is a cut vertex (the removal of any of them actually decreases the number of components rather than increasing it). Therefore we may as well assume that some component of G has at least two vertices, and we can restrict our attention to that component; in other words, we can assume that G is connected. One clever way to do this problem is as follows. Define the **distance** between two vertices u and v, denoted $d(u,v)$, to be the length of a shortest path joining u and v. Now choose u and v so that $d(u,v)$ is as large as possible. We claim that neither u nor v is a cut vertex. Suppose otherwise, say that u is a cut vertex. Then v is in one component that results after u is removed, and some vertex w is in another. Since there is no path from w to v in the graph with u removed, every path from w to v must have passed through u. Therefore the distance between w and v must have been strictly greater than the distance between u and v. This is a contradiction to the choice of u and v, and our proof by contradiction is complete.

37. This problem is simply asking for the cut edges of these graphs.

a) The link joining Denver and Chicago and the link joining Boston and New York are the cut edges.

b) The following links are the cut edges: Seattle–Portland, Portland–San Francisco, Salt Lake City–Denver, New York–Boston, Boston–Bangor, Boston–Burlington.

39. A vertex basis will be a set of people who collectively can influence everyone, at least indirectly. The set consisting of Deborah is a vertex basis, since she can influence everyone except Yvonne directly, and she can influence Yvonne indirectly through Brian.

41. Since there can be no edges between vertices in different components, G will have the most edges when each of the components is a complete graph. Since K_{n_i} has $C(n_i,2)$ edges, the maximum number of edges is the sum given in the exercise.

43. Before we give a correct proof here, let us look at an incorrect proof that students often give for this exercise. It goes something like this. "Suppose that the graph is not connected. Then no vertex can be adjacent to every other vertex, only to $n-2$ other vertices. One vertex joined to $n-2$ other vertices creates a component with $n-1$ vertices in it. To get the most edges possible, we must use all the edges in this component. The number of edges in this component is thus $C(n-1,2)=(n-1)(n-2)/2$, and the other component (with only one vertex) has no edges. Thus we have shown that a disconnected graph has at most $(n-1)(n-2)/2$ edges, so every graph with more edges than that has to be connected." The fallacy here is in assuming—without justification—that the maximum number of edges is achieved when one component has $n-1$ vertices. What if, say, there were two components of roughly equal size? Might they not together contain more edges? We will see that the answer is "no," but it is important to realize that this requires proof—it is not obvious without some calculations.

Here is a correct proof, then. Suppose that the graph is not connected. Then it has a component with k vertices in it, for some k between 1 and $n-1$, inclusive. The remaining $n-k$ vertices are in one or more other components. The maximum number of edges this graph could have is then $C(k,2)+C(n-k,2)$, which, after a bit of algebra, simplifies to $k^2-nk+(n^2-n)/2$. This is a quadratic function of k. It is minimized when $k=n/2$ (the k coordinate of the vertex of the parabola that is the graph of this function) and maximized at the endpoints of the domain, namely $k=1$ and $k=n-1$. In the latter cases its value is $(n-1)(n-2)/2$.

Therefore the largest number of edges that a disconnected graph can have is $(n-1)(n-2)/2$, so every graph with more edges than this must be connected.

45. We have to enumerate carefully all the possibilities.

a) There is obviously only 1, namely K_2, the graph consisting of two vertices and the edge between them.

b) There are clearly 2 connected graphs with 3 vertices, namely K_3 and K_3 with one edge deleted, as shown.

c) There are several connected graphs with $n = 4$. If the graph has no circuits, then it must either be a path of length 3 or the "star" $K_{1,3}$. If it contains a triangle but no copy of C_4, then the other vertex must be pendant—only 1 possibility. If it contains a copy of C_4, then neither, one, or both of the other two edges may be present—3 possibilities. Therefore the answer is $2+1+3=6$. The graphs are shown below.

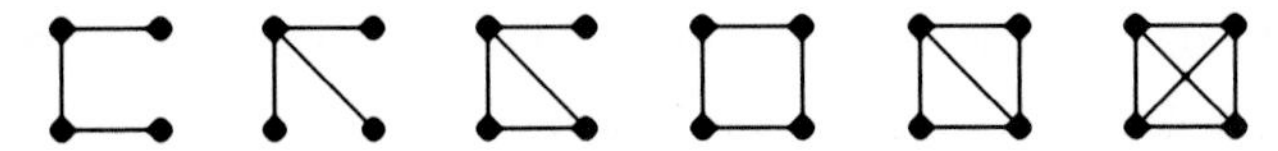

d) We need to enumerate the possibilities in some systematic way, such as by the largest cycle contained in the graph. There are 21 such graphs, as can be seen by such an enumeration, shown below. First we show those graphs with no circuits, then those with a triangle but no C_4 or C_5, then those with a C_4 but no C_5, and finally those with a C_5. In doing this problem we have to be careful not only not to leave out any graphs, but also not to list any twice.

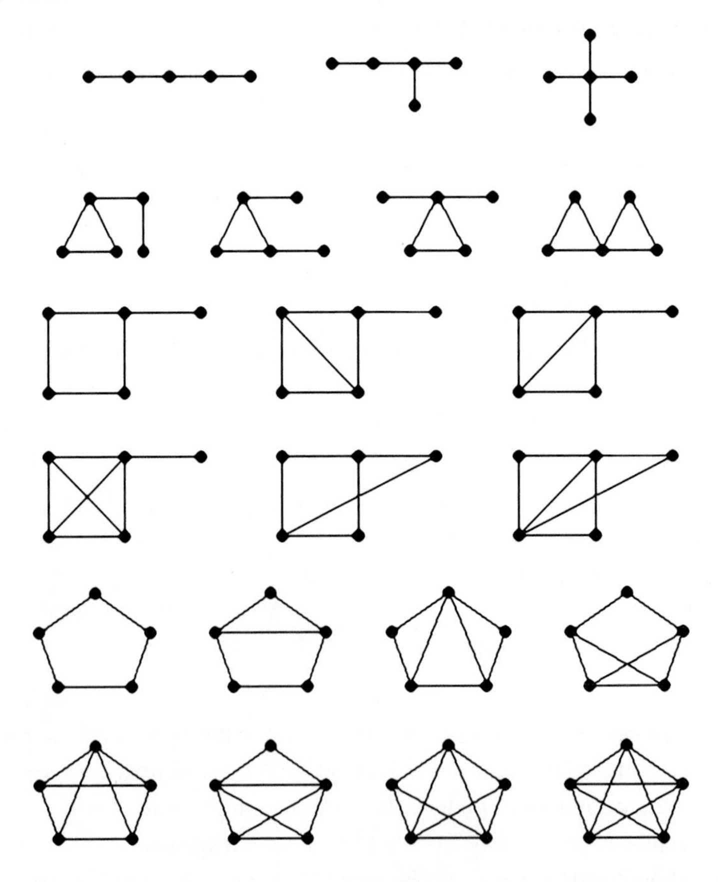

47. We need to look at successive powers of the adjacency matrix until we find one in which the $(1,6)^{\text{th}}$ entry is not 0. Since the matrix is

$$\mathbf{A} = \begin{bmatrix} 0 & 1 & 0 & 1 & 1 & 0 \\ 1 & 0 & 1 & 0 & 1 & 1 \\ 0 & 1 & 0 & 1 & 0 & 1 \\ 1 & 0 & 1 & 0 & 1 & 0 \\ 1 & 1 & 0 & 1 & 0 & 1 \\ 0 & 1 & 1 & 0 & 1 & 0 \end{bmatrix},$$

we see that the $(1,6)^{\text{th}}$ entry of $\mathbf{A}^2$ is 2. Thus there is a path of length 2 from a to f (in fact 2 of them). On the other hand there is no path of length 1 from a to f (i.e., no edge), so the length of a shortest path is 2.

49. Let the simple paths P_1 and P_2 be $u = x_0, x_1, \ldots, x_n = v$ and $u = y_0, y_1, \ldots, y_m = v$, respectively. The paths thus start out at the same vertex. Since the paths do not contain the same set of edges, they must diverge eventually. If they diverge only after one of them has ended, then the rest of the other path is a simple circuit from v to v. Otherwise we can suppose that $x_0 = y_0$, $x_1 = y_1$, $\ldots$, $x_i = y_i$, but $x_{i+1} \neq y_{i+1}$. To form our simple circuit, we follow the path y_i, y_{i+1}, y_{i+2}, and so on, until it once again first encounters a vertex on P_1 (possibly as early as y_{i+1}, no later than y_m). Once we are back on P_1, we follow it along—forwards or backwards, as necessary—to return to x_i. Since $x_i = y_i$, this certainly forms a circuit. It must be a simple circuit, since no edge among the x_k's or the y_l's can be repeated (P_1 and P_2 are simple by hypothesis) and no edge among the x_k's can equal one of the edges y_l that we used, since we abandoned P_2 for P_1 as soon as we hit P_1.

51. Let $\mathbf{A}$ be the adjacency matrix of a given graph G. Theorem 2 tells us that $\mathbf{A}^r$ counts the number of paths of length r between vertices. If an entry in $\mathbf{A}^r$ is greater than 0, then there is a path between the corresponding vertices in G. Suppose that we look at $\mathbf{A} + \mathbf{A}^2 + \mathbf{A}^3 + \cdots + \mathbf{A}^{n-1}$, where n is the number of vertices in G. If there is a path between a pair of distinct vertices in G, then there is a path of length at most $n-1$, so this sum will have a positive integer in the corresponding entry. Conversely, if there is no path, then the corresponding entry in every summand will be 0, and hence the entry in the sum will be 0. Therefore the graph is connected (i.e., there is a path between every pair of distinct vertices in G) if and only if every off-diagonal entry in this sum is strictly positive. To determine whether G is connected, therefore, we just compute this sum and check to see whether this condition holds.

53. We have to prove a statement and its converse here. One direction is fairly easy. If the graph is bipartite, say with parts A and B, then the vertices in every path must alternately lie in A and B. Therefore a path that starts in A, say, will end in B after an odd number of steps and in A after an even number of steps. Since a circuit ends at the same vertex where it starts, the length must be even. The converse is a little harder. We suppose that all circuits have even length and want to show that the graph is bipartite. We can assume that the graph is connected, because if it is not, then we can just work on one component at a time. Let v be a vertex of the graph, and let A be the set of all vertices to which there is a path of odd length starting at v, and let B be the set of all vertices to which there is a path of even length starting at v. Since the component is connected, every vertex lies in A or B. No vertex can lie in both A and B, since then following the odd-length path from v to that vertex and then back along the even-length path from that vertex to v would produce an odd circuit, contrary to the hypothesis. Thus the set of vertices has been partitioned into two sets. Now we just need to show that every edge has endpoints in different parts. If xy is an edge where $x \in A$, then the odd-length path from v to x followed by xy produces an even-length path from v to y, so $y \in B$ (and similarly if $x \in B$).

55. Suppose the couples are Bob and Carol Sanders, and Ted and Alice Henderson (these were characters in a movie from 1969). We represent the initial position by $(BCTA\bullet, \emptyset)$, indicating that all four people are on the left shore along with the boat (the dot). We want to reach the position $(\emptyset, BCTA\bullet)$. Positions will be the vertices of our graph, and legal transitions will be the edges. If Bob and Carol take the boat over, then we reach the position $(TA, BC\bullet)$. The only useful transition at that point is for someone to row back. Let's try Bob; so we have $(BTA\bullet, C)$. If Bob and Ted now row to the right shore, we reach $(A, BCT\bullet)$. Ted can take the boat back to fetch his wife, giving us $(TA\bullet, BC)$ and then $(\emptyset, BCTA\bullet)$. Notice that this path never violates the jealousy conditions imposed in this problem. The entire graph model would have many more positions, but we just need one path.

SECTION 9.5 Euler and Hamilton Paths

An Euler circuit or Euler path uses every edge exactly once. A Hamilton circuit or Hamilton path uses every vertex exactly once (not counting the circuit's return to its starting vertex). Euler and Hamilton circuits and paths have an important place in the history of graph theory, and as we see in this section they have some interesting applications. They provide a nice contrast—there are good algorithms for finding Euler paths (see also Exercises 50–53), but computer scientists believe that there is no good (efficient) algorithm for finding Hamilton paths.

Most of these exercises are straightforward. The reader should at least look at Exercises 16 and 17 to see how the concept of Euler path applies to directed graphs—these exercises are not hard if you understood the proof of Theorem 1 (given in the text before the statement of the theorem).

1. Since there are four vertices of odd degree (a, b, c, and e) and $4 > 2$, this graph has neither an Euler circuit nor an Euler path.

3. Since there are two vertices of odd degree (a and d), this graph has no Euler circuit, but it does have an Euler path starting at a and ending at d. We can find such a path by inspection, or by using the splicing idea explained in this section. One such path is $a, e, c, e, b, e, d, b, a, c, d$.

5. All the vertex degrees are even, so there is an Euler circuit. We can find such a circuit by inspection, or by using the splicing idea explained in this section. One such circuit is $a, b, c, d, c, e, d, b, e, a, e, a$.

7. All the vertex degrees are even, so there is an Euler circuit. We can find such a circuit by inspection, or by using the splicing idea explained in this section. One such circuit is $a, b, c, d, e, f, g, h, i, a, h, b, i, c, e, h, d, g, c, a$.

9. No, an Euler circuit does not exist in the graph modeling the new city either. Vertices A and B have odd degree.

11. Assuming we have just one truck to do the painting, the truck must follow an Euler path through the streets in order to do the job without traveling a street twice. Therefore this can be done precisely when there is an Euler path or circuit in the graph, which means that either zero or two vertices (intersections) have odd degree (number of streets meeting there). We are assuming, of course, that the city is connected.

13. In order for the picture to be drawn under the conditions of Exercises 13–15, the graph formed by the picture must have an Euler path or Euler circuit. Note that all of these graphs are connected. The graph in the current exercise has all vertices of even degree; therefore it has an Euler circuit and can be so traced.

15. See the comments in the solution to Exercise 13. This graph has 4 vertices of odd degree; therefore it has no Euler path or circuit and cannot be so traced.

17. If there is an Euler path, then as we follow it through the graph, each vertex except the starting and ending vertex must have equal in-degree and out-degree, since whenever we come to the vertex along some edge, we leave it along some edge. The starting vertex must have out-degree 1 greater than its in-degree, since after we have started, using one edge leading out of this vertex, the same argument applies. Similarly, the ending vertex must have in-degree 1 greater than its out-degree, since until we end, using one edge leading into this vertex, the same argument applies. Note that the Euler path itself guarantees weak connectivity; given any two vertices, there is a path from the one that occurs first along the Euler path to the other, via the Euler path.

Conversely, suppose that the graph meets the degree conditions stated here. By Exercise 16 it cannot have an Euler circuit. If we add one more edge from the vertex of deficient out-degree to the vertex of deficient in-degree, then the graph now has every vertex with its in-degree equal to its out-degree. Certainly the graph is still weakly connected. By Exercise 16 there is an Euler circuit in this new graph. If we delete the added edge, then what is left of the circuit is an Euler path from the vertex of deficient in-degree to the vertex of deficient out-degree.

19. For Exercises 18–23 we use the results of Exercises 16 and 17. By Exercise 16, we cannot hope to find an Euler circuit since vertex b has different out-degree and in-degree. By Exercise 17, we cannot hope to find an Euler path since vertex b has out-degree and in-degree differing by 2.

21. This directed graph satisfies the condition of Exercise 17 but not that of Exercise 16. Therefore there is no Euler circuit. The Euler path must go from a to e. One such path is $a, d, e, d, b, a, e, c, e, b, c, b, e$.

23. There are more than two vertices whose in-degree and out-degree differ by 1, so by Exercises 16 and 17, there is no Euler path or Euler circuit.

25. The algorithm is very similar to Algorithm 1. The input is a weakly connected directed multigraph in which either each vertex has in-degree equal to its out-degree, or else all vertices except two satisfy this condition and the remaining vertices have in-degree differing from out-degree by 1 (necessarily once in each direction). We begin by forming a path starting at the vertex whose out-degree exceeds its in-degree by 1 (in the second case) or at any vertex (in the first case). We traverse the edges (never more than once each), forming a path, until we cannot go on. Necessarily we end up either at the vertex whose in-degree exceeds its out-degree (in the first case) or at the starting vertex (in the second case). From then on we do exactly as in Algorithm 1, finding a simple circuit among the edges not yet used, starting at any vertex on the path we already have; such a vertex exists by the weak connectivity assumption. We splice this circuit into the path, and repeat the process until all edges have been used.

27. a) Clearly K_2 has an Euler path but no Euler circuit. For odd $n > 2$ there is an Euler circuit (since the degrees of all the vertices are $n-1$, which is even), whereas for even $n > 2$ there are at least 4 vertices of odd degree and hence no Euler path. Thus for no n other than 2 is there an Euler path but not an Euler circuit.
b) Since C_n has an Euler circuit for all n, there are no values of n meeting these conditions.
c) A wheel has at least 3 vertices of degree 3 (around the rim), so there can be no Euler path.
d) The same argument applies here as applied in part **(a)**. In more detail, Q_1 (which is the same as K_2) is the only cube with an Euler path but no Euler circuit, since for odd $n > 1$ there are too many vertices of odd degree, and for even $n > 1$ there is an Euler circuit.

29. Just as a graph with 2 vertices of odd degree can be drawn with one continuous motion, a graph with $2m$ vertices of odd degree can be drawn with m continuous motions. The graph in Exercise 1 has 4 vertices of odd degree, so it takes 2 continuous motions; in other words, the pencil must be lifted once. We could do this, for example, by first tracing a, c, d, e, a, b and then tracing c, b, e. The graphs in Exercises 2–7 all have Euler paths, so no lifting is necessary.

31. It is clear that a, b, c, d, e, a is a Hamilton circuit.

33. There is no Hamilton circuit because of the cut edges ($\{c, e\}$, for instance). Once a purported circuit had reached vertex e, there would be nowhere for it to go.

35. There is no Hamiltonian circuit in this graph. If there were one, then it would have to include all the edges of the graph, because it would have to enter and exit vertex a, enter and exit vertex d, and enter and exit vertex e. But then vertex c would have been visited more than once, a contradiction.

37. This graph has the Hamilton path a, b, c, f, d, e. This simple path hits each vertex once.

39. This graph has the Hamilton path f, e, d, a, b, c.

41. There are eight vertices of degree 2 in this graph. Only two of them can be the end vertices of a Hamilton path, so for each of the other six their two incident edges must be present in the path. Now if either all four of the "outside" vertices of degree 2 (a, c, g, and e) or all four of the "inside" vertices of degree 2 (i, k, l, and n) are not end vertices, then a circuit will be completed that does not include all the vertices—either the outside square or the middle square. Therefore if there is to be a Hamilton path then exactly one of the inside corner vertices must be an end vertex, and each of the other inside corner vertices must have its two incident edges in the path. Without loss of generality we can assume that vertex i is an end, and that the path begins i, o, n, m, l, q, k, j. At this point, either the path must visit vertex p, in which case it gets stuck, or else it must visit b, in which case it will never be able to reach p. Either case gives a contradiction, so there is no Hamilton path.

43. It is easy to write down a Hamiltonian path here; for example, a, d, g, h, i, f, c, e, b.

45. A Hamilton circuit in a bipartite graph must visit the vertices in the parts alternately, returning to the part in which it began. Therefore a necessary condition is certainly $m = n$. Furthermore $K_{1,1}$ does not have a Hamilton circuit, so we need $n \geq 2$ as well. On the other hand, since the complete bipartite graph has all the edges we need, these conditions are sufficient. Explicitly, if the vertices are $a_1, a_2, \ldots, a_n$ in one part and $b_1, b_2, \ldots, b_n$ in the other, with $n \geq 2$, then one Hamilton circuit is $a_1, b_1, a_2, b_2, \ldots, a_n, b_n, a_1$.

47. For Dirac's Theorem to be applicable, we need every vertex to have degree at least $n/2$, where n is the number of vertices in the graph. For Ore's Theorem, we need $\deg(x) + \deg(y) \geq n$ whenever x and y are not adjacent.

a) In this graph $n = 5$. Dirac's Theorem does not apply, since there is a vertex of degree 2, and 2 is smaller than $n/2$. Ore's Theorem also does not apply, since there are two nonadjacent vertices of degree 2, so the sum of their degrees is less than n. However, the graph does have a Hamilton circuit—just go around the pentagon. This illustrates that neither of the sufficient conditions for the existence of a Hamilton circuit given in these theorems is necessary.

b) Everything said in the solution to part **(a)** is valid here as well.

c) In this graph $n = 5$, and all the vertex degrees are either 3 or 4, both of which are at least $n/2$. Therefore Dirac's Theorem guarantees the existence of a Hamilton circuit. Ore's Theorem must apply as well, since $(n/2) + (n/2) = n$; in this case, the sum of the degrees of any pair of nonadjacent vertices (there are only two such pairs) is 6, which is greater than or equal to 5.

d) In this graph $n = 6$, and all the vertex degrees are 3, which is (at least) $n/2$. Therefore Dirac's Theorem guarantees the existence of a Hamilton circuit. Ore's Theorem must apply as well, since $(n/2) + (n/2) = n$; in this case, the sum of the degrees of any pair of nonadjacent vertices is 6.

Although not illustrated in any of the examples in this exercise, there are graphs for which Ore's Theorem applies, even though Dirac's does not. Here is one: Take K_4, and then tack on a path of length 2 between two of the vertices, say a, b, c. In all, this graph has five vertices, two with degree 3, two with degree 4, and one with degree 2. Since there is a vertex with degree less than $5/2$, Dirac's Theorem does not apply. However, the sum of the degrees of any two (nonadjacent) vertices is at least $2 + 3 = 5$, so Ore's Theorem does apply and guarantees that there is a Hamilton circuit.

49. The trick is to use a Gray code for n to build one for $n+1$. We take the Gray code for n and put a 0 in front of each term to get half of the Gray code for $n+1$; we put a 1 in front to get the second half. Then we reverse the second half so that the junction at which the two halves meet differ in only the first bit. For a formal proof we use induction on n. For $n = 1$ the code is $0, 1$ (which is not really a Hamilton circuit in Q_1). Assume the inductive hypothesis that $c_1, c_2, \ldots, c_{2^n}$ is a Gray code for n. Then $0c_1, 0c_2, \ldots, 0c_{2^n}, 1c_{2^n}, \ldots, 1c_2, 1c_1$ is a Gray code for $n+1$.

51. Turning this verbal description into pseudocode is straightforward, especially if we allow ourselves lots of words in the pseudocode. We build our *circuit* (which we think of simply as an ordered list of edges) one edge at a time, keeping track of the vertex v we are at; the subgraph containing the edges we have not yet used we will call H. We assume that the vertices of G are listed in some order, so that when we are asked to choose an edge from v meeting certain conditions, we can choose the edge to the vertex that comes first in this order among all those edges meeting the conditions. (This avoids ambiguity, which an algorithm is not supposed to have.)

```
procedure fleury(G : connected multigraph with all degrees even)
v := first vertex of G
circuit := the empty circuit
H := G
while H has edges
begin
        Let e be an edge in H with v as one of its endpoints,
                such that e is not a cut edge of H, if such an edge
                exists; otherwise let e be any edge in H with v as
                one of its endpoints.
        v := other endpoint of e
        Add e to the end of circuit
        Remove e from H
end { circuit is an Euler circuit}
```

53. If every vertex has even degree, then we can simply use Fleury's algorithm to find an Euler circuit, which is by definition also an Euler path. If there are two vertices with odd degree (and the rest with even degree), then we can add an edge between these two vertices and apply Fleury's algorithm (using this edge as the first edge to make it easier to find later), then delete the added edge.

55. A Hamilton circuit in a bipartite graph would have to look like $a_1, b_1, a_2, b_2, \ldots, a_k, b_k, a_1$, where each a_i is in one part and each b_i is in the other part, since the only edges in the graph join vertices in opposite parts. In the Hamilton circuit, no vertex is listed twice (except for the final a_1), and every vertex is listed, so the total number of vertices in the graph must be $2k$, which is not an odd number. Therefore a bipartite graph with an odd number of vertices cannot have a Hamilton circuit.

57. We draw one vertex for each of the 12 squares on the board. We then draw an edge from a vertex to each vertex that can be reached by moving 2 units horizontally and 1 unit vertically or vice versa. The result is as shown.

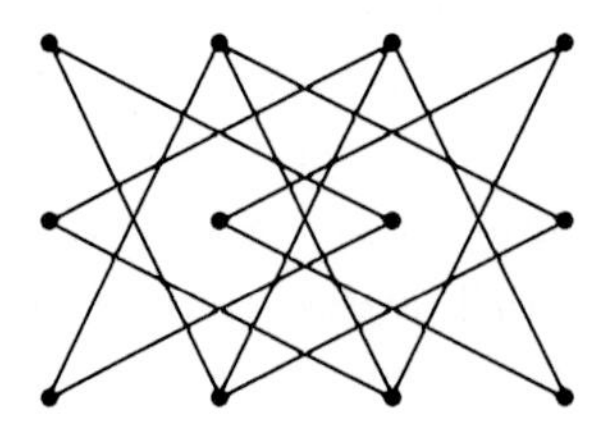

59. First let us try to find a reentrant knight's tour. Looking at the graph in the solution to Exercise 57 we see that every vertex on the left and right edge has degree 2. Therefore the 12 edges incident to these vertices would have to be in a Hamilton circuit if there were one. If we draw these 12 edges, however, we see that they form two circuits, each with six edges. Therefore there is no re-entrant knight's tour. However, we can splice these two circuits together by using an edge from a middle vertex in the top row to a middle vertex in the bottom row (and omitting two edges adjacent to this edge). The result is the knight's tour shown here.

3	6	9	12
8	11	4	1
5	2	7	10

61. We give an ad hoc argument by contradiction, using the notation shown in the following diagram. We think of the board as a graph and need to decide which edges need to be in a purported Hamilton path.

1	2	3	4
5	6	7	8
9	10	11	12
13	14	15	16

There are only two moves from each of the four corner squares. If we put in all of the edges 1-10, 1-7, 16-10, and 16-7, then a circuit is complete too soon, so at least one of these edges must be missing. Without loss of generality, then, we may assume that the endpoints of the path are 1 and either 4 or 13, and that the path contains all of the edges 1-10, 10-16, and 16-7. Now vertex (square) 3 has edges only to squares 5, 10, and 12; and square 10 already has its two incident edges. Therefore 3-5 and 3-12 must be in the Hamilton path. Similarly, edges 8-2 and 8-15 must be in the path. Now square 9 has edges only to squares 2, 7, and 15. If there were to be edges to both 2 and 15, then a circuit would be completed too soon (2-9-15-8-2). Therefore the edge 9-7 must be in the path, thereby giving square 7 its full complement of edges. But now square 14 is forced to be joined in the path to squares 5 and 12, and this completes a circuit too soon (5-14-12-3-5). Since we have reached a contradiction, we conclude that there is no Hamilton path.

63. An $m \times n$ board contains mn squares. If both m and n are odd, then it contains an odd number of squares. By Exercise 62, the corresponding graph is bipartite. Exercise 55 told us that the graph does not contain a Hamilton circuit. Therefore there is no re-entrant knight's tour (see Exercise 58b).

65. This is a proof by contradiction. We assume that G satisfies Ore's inequality that $\deg(x) + \deg(y) \geq n$ whenever x and y are nonadjacent vertices in G, but G does not have a Hamilton circuit. We will end up with a contradiction, and therefore conclude that under these conditions, G must have a Hamilton circuit.

a) Since G does not have a Hamilton circuit, we can add missing edges one at a time in such a way that we do not obtain a graph with a Hamilton circuit. We continue this process as long as possible. Clearly it cannot go on forever, because once we've formed the complete graph by adding all missing edges, there is a Hamilton circuit (recall that $n \geq 3$). Whenever the process stops, we have obtained a graph H with the desired property. (Note that H might equal G itself—in other words, we add no edges. However, H cannot be complete, as just noted.)

b) Add one more edge to H. By the construction in part **(a)**, we now have a Hamilton circuit, and clearly this circuit must use the edge we just added. The path consisting of this circuit with the added edge omitted is clearly a Hamilton path in H.

c) Clearly v_1 and v_n are not adjacent in H, since H has no Hamilton circuit. Therefore they are not adjacent in G. But the hypothesis was that the sum of the degrees of vertices not adjacent in G was at least n. This inequality can be rewritten as $n - \deg(v_n) \leq \deg(v_1)$. But $n - \deg(v_n)$ is just the number of vertices not adjacent to v_n.

d) Let's make sure we understand what this means. If, say, v_7 is adjacent to v_1, then v_6 is in S. Note that $v_1 \in S$, since v_2 is adjacent to v_1. Also, v_n is not in S, since there is no vertex following v_n in the Hamilton path. Now each one of the $\deg(v_1)$ vertices adjacent to v_1 gives rise to an element of S, so S contains $\deg(v_1)$ vertices.

e) By part **(c)** there are at most $\deg(v_1) - 1$ vertices other than v_n not adjacent to v_n, and by part **(d)** there are $\deg(v_1)$ vertices in S, none of which is v_n. So S has more vertices other than v_n than there are vertices not adjacent to v_n; in other words, at least one vertex of S is adjacent to v_n. By definition, if v_k is this vertex, then H contains edges $v_k v_n$ and $v_1 v_{k+1}$. Note that $1 < k < n-1$, since we know from part **(c)** that $v_1 v_n$ is not an edge of H.

f) We have shown that all of the edges in the circuit $v_1, v_2, \ldots, v_{k-1}, v_k, v_n, v_{n-1}, \ldots, v_{k+1}, v_1$ are in H, so H has a Hamilton circuit. That is a contradiction to the construction of H. Therefore our assumption that G did not originally have a Hamilton circuit is wrong, and our proof by contradiction is complete.

SECTION 9.6 Shortest-Path Problems

In applying Dijkstra's algorithm for finding shortest paths, it is convenient to keep track, as each vertex is labeled, of where the path comes from. You can put this information on the drawing of the graph itself, by placing a little arrow at the vertex, pointing to the vertex causing the new labeling. Remember that the labels (both the values and these arrows) may change as the algorithm proceeds (and shorter paths are found), so you need an eraser to implement the algorithm in this way. The algorithm is quite simple once you see how it goes, and these exercises are not difficult.

1. In each case we will use a directed weighted graph, since there is no reason to suppose that travel from stop A to stop B should be the same (in whatever respect) as travel from stop B to stop A.

a) We will put an edge from A to B whenever there is a train that travels from A to B without intermediate stops. The weight of that edge will be the time (in seconds, say) required for the trip, including half the stopping time at each end station. This model is not perfect. For example, the time may depend on the time of day. Also, it is not clear that allocating the waiting time at each station in this way is the best way to model the system (but we should not ignore the waiting time).

b) We assume that distance refers to the distance along the subway tracks. If so, this model is straightforward and similar to part **(a)**. We put an edge from A to B whenever there is a train that travels from A to B without intermediate stops. The weight of that edge will be the distance the train travels on that trip.

c) Under the assumption stated, we can model this problem in a manner similar to the previous parts. We put an edge from A to B whenever there is a train that travels from A to B without intermediate stops. The weight of that edge will be the fare required for that trip. Very few subway systems (if any) actually operate under this assumption.

3. We see in the solution to Exercise 5 below that a shortest path has length 16. There really is no better way to find the *length* of a shortest path than by using Dijkstra's algorithm, which for practically the same amount of work actually gives you the path.

5. We can answer these questions by applying Dijkstra's algorithm in each case, with the added feature of indicating, when a vertex is given a new label, where the new path to that vertex comes from. We will denote this by making the vertex a superscript to the distance. Then we can reconstruct the path that produces the minimum distance by tracing these superscripts backward from z to a.

We begin with the graph in Exercise 2. First a is put into S, with label 0, and vertex b is labeled 2^a, and c is labeled 3^a. Since b has the smaller label, b is put into S and d is labeled 7^b, and e is labeled 4^b. Next c is put into S, and no labels are changed. Then e is put into S and the labels of d and z become 5^e and 8^e, respectively. Next d is put into S, and the label of z is changed to 7^d. Finally, z is put into S. Now we know that a shortest path, in reverse, is z, d, e, b, a; we get this by following the superscripts, starting at z. Therefore a shortest path is a, b, e, d, z, with length 7.

We follow the same procedure for the graph in Exercise 3. A shortest path is a, c, d, e, g, z, with length 16.

For the graph in Exercise 4, we follow the same procedure. The graph is bigger, and the process takes longer, but the algorithm is the same. We find a shortest path to be $a, b, e, h, l, m, p, s, z$, having length 16.

7. We apply the variation on Dijkstra's algorithm explained in our solution to Exercise 5. In each case we start at the vertex listed first and can stop once the vertex listed last has been put into S.

a) A shortest path is a, c, d, of length 6. **b)** A shortest path is a, c, d, f, of length 11.
c) A shortest path is c, d, f, of length 8. **d)** One shortest path is b, d, e, g, z, of length 15.

9. In theory we use the variation on Dijkstra's algorithm explained in our solution to Exercise 5. In each case we start at the vertex listed first and can stop once the vertex listed last has been put into S. In practice for a network of this size with the distances having the geometric significance that they do, we solve the problem by inspection (there are usually at most two conceivable solutions, and we compute the smaller of the two).
a) The shortest trip is, not surprisingly, the direct flight from New York to Los Angeles.
b) The shortest trip is Boston to New York to San Francisco.
c) The shortest trip is Miami to Atlanta to Chicago to Denver.
d) The shortest trip is Miami to New York to Los Angeles.

11. For solution technique, see the comments for Exercise 9.
a) The shortest route is Boston to Chicago to Los Angeles.
b) The shortest route is New York to Chicago to San Francisco.
c) The shortest route is Dallas to Los Angeles to San Francisco.
d) The shortest route is Denver to Chicago to New York.

13. For solution technique, see the comments for Exercise 9.
a) The cheapest route is Boston to Chicago to Los Angeles.
b) One of the cheapest routes is New York to Chicago to San Francisco.
c) The cheapest route is Dallas to Los Angeles to San Francisco.
d) The cheapest route is Denver to Chicago to New York.

15. All we have to do is not stop once z is put into S. Thus we change the condition on the **while** statement to something like "$S \neq V$."

17. For solution technique, see the comments for Exercise 9.

a) The shortest routes are Newark to Woodbridge to Camden, and Newark to Woodbridge to Camden to Cape May. (The map is obviously not drawn to scale.)

b) The cheapest routes (in terms of tolls) are Newark to Woodbridge to Camden, and Newark to Woodbridge to Camden to Cape May.

19. One application, involving directed graphs, is in project scheduling. The vertices can represent parts of the project, and there is a directed edge from A to B if B cannot be started until A is finished. The weight on an edge is the time required to complete the initial vertex of the edge. A longest path from the start of the project to completion represents the total time required to complete the project. Another application would be in trying to find long routes through a city—something a sightseer, political canvasser, or street cleaner might want to do.

21. We can represent the distances with a 6×6 matrix, with alphabetical order. Initially it is

$$\begin{bmatrix} \infty & 4 & 2 & \infty & \infty & \infty \\ 4 & \infty & 1 & 5 & \infty & \infty \\ 2 & 1 & \infty & 8 & 10 & \infty \\ \infty & 5 & 8 & \infty & 2 & 6 \\ \infty & \infty & 10 & 2 & \infty & 3 \\ \infty & \infty & \infty & 6 & 3 & \infty \end{bmatrix}.$$

After completion of the main inner loops for $i = 1$, the matrix looks like this:

$$\begin{bmatrix} \infty & 4 & 2 & \infty & \infty & \infty \\ 4 & 8 & 1 & 5 & \infty & \infty \\ 2 & 1 & 4 & 8 & 10 & \infty \\ \infty & 5 & 8 & \infty & 2 & 6 \\ \infty & \infty & 10 & 2 & \infty & 3 \\ \infty & \infty & \infty & 6 & 3 & \infty \end{bmatrix}.$$

After completion of the main inner loops for $i = 2$, the matrix looks like this:

$$\begin{bmatrix} 8 & 4 & 2 & 9 & \infty & \infty \\ 4 & 8 & 1 & 5 & \infty & \infty \\ 2 & 1 & 2 & 6 & 10 & \infty \\ 9 & 5 & 6 & 10 & 2 & 6 \\ \infty & \infty & 10 & 2 & \infty & 3 \\ \infty & \infty & \infty & 6 & 3 & \infty \end{bmatrix}.$$

After completion of the main inner loops for $i = 3$, the matrix looks like this:

$$\begin{bmatrix} 4 & 3 & 2 & 8 & 12 & \infty \\ 3 & 2 & 1 & 5 & 11 & \infty \\ 2 & 1 & 2 & 6 & 10 & \infty \\ 8 & 5 & 6 & 10 & 2 & 6 \\ 12 & 11 & 10 & 2 & 20 & 3 \\ \infty & \infty & \infty & 6 & 3 & \infty \end{bmatrix}.$$

After completion of the main inner loops for $i = 4$, the matrix looks like this:

$$\begin{bmatrix} 4 & 3 & 2 & 8 & 10 & 14 \\ 3 & 2 & 1 & 5 & 7 & 11 \\ 2 & 1 & 2 & 6 & 8 & 12 \\ 8 & 5 & 6 & 10 & 2 & 6 \\ 10 & 7 & 8 & 2 & 4 & 3 \\ 14 & 11 & 12 & 6 & 3 & 12 \end{bmatrix}.$$

After completion of the main inner loops for $i = 5$, the matrix looks like this:

$$\begin{bmatrix} 4 & 3 & 2 & 8 & 10 & 13 \\ 3 & 2 & 1 & 5 & 7 & 10 \\ 2 & 1 & 2 & 6 & 8 & 11 \\ 8 & 5 & 6 & 4 & 2 & 5 \\ 10 & 7 & 8 & 2 & 4 & 3 \\ 13 & 10 & 11 & 5 & 3 & 6 \end{bmatrix}.$$

There is no change after the final iteration with $i = 6$. Therefore this matrix represents the distances between all pairs.

23. There are two parts to this algorithm. The first part obviously requires $O(n^2)$ operations for bookkeeping and nothing else. The second part obviously requires $O(n^3)$ operations for bookkeeping and the **if**...**then** statement. Therefore the entire procedure takes $O(n^2 + n^3) = O(n^3)$ steps.

25. The following table shows the three different Hamilton circuits and their weights:

Circuit	Weight
a-b-c-d-a	$3 + 6 + 7 + 2 = 18$
a-b-d-c-a	$3 + 4 + 7 + 5 = 19$
a-c-b-d-a	$5 + 6 + 4 + 2 = 17$

Thus we see that the circuit a-c-b-d-a (or the same circuit starting at some other point but traversing the vertices in the same or exactly opposite order) is the one with minimum total weight.

27. The following table shows the twelve different Hamilton circuits and their weights, where we abbreviate the cities with the beginning letter of their name, except that Detroit is M (for Motor City, of course!):

Circuit	Weight
S-M-N-D-L-S	$329 + 189 + 279 + 209 + 69 = 1075$
S-M-N-L-D-S	$329 + 189 + 379 + 209 + 179 = 1285$
S-M-D-N-L-S	$329 + 229 + 279 + 379 + 69 = 1285$
S-M-D-L-N-S	$329 + 229 + 209 + 379 + 359 = 1505$
S-M-L-N-D-S	$329 + 349 + 379 + 279 + 179 = 1515$
S-M-L-D-N-S	$329 + 349 + 209 + 279 + 359 = 1525$
S-N-M-D-L-S	$359 + 189 + 229 + 209 + 69 = 1055$
S-N-M-L-D-S	$359 + 189 + 349 + 209 + 179 = 1285$
S-N-D-M-L-S	$359 + 279 + 229 + 349 + 69 = 1285$
S-N-L-M-D-S	$359 + 379 + 349 + 229 + 179 = 1495$
S-D-M-N-L-S	$179 + 229 + 189 + 379 + 69 = 1045$
S-D-N-M-L-S	$179 + 279 + 189 + 349 + 69 = 1065$

As a check of our arithmetic, we can compute the total weight (price) of all the trips (it comes to 15420) and check that it is equal to 6 times the sum of the weights (which here is 2570), since each edge appears in six paths (and sure enough, $15420 = 6 \cdot 2570$). We see that the circuit S-D-M-N-L-S (or the same circuit starting at some other point but traversing the vertices in the same or exactly opposite order) is the one with minimum total weight, 1045. Note that we might have guessed this route by looking at the drawing (which is more or less to scale in terms of the actual locations of these cities).

29. If we take a triangle ABC and make one edge, say BC very weighty, then the minimum circuit will avoid that edge. So let's take the weights of AB, AC, and BC to be 1, 2, and 100, respectively. Obviously all Hamilton circuits have total weight 103, but the circuit A-B-A-C-A visits every vertex at least once and has total weight only $1+1+2+2=6$. This circuit visits vertex A an extra time in order to avoid traversing the weighty edge BC.

SECTION 9.7 Planar Graphs

As with Euler and Hamilton circuits and paths, the topic of planar graphs is a classical one in graph theory. The theory (Euler's formula, Kuratowski's Theorem, and their corollaries) is quite beautiful. It is easy to ask extremely difficult questions in this area, however—see Exercise 27, for example. In practice, there are very efficient algorithms for determining planarity that have nothing to do with Kuratowski's Theorem, but they are quite complicated and beyond the scope of this book. For the exercises here, the best way to show that a graph is planar is to draw a planar embedding; the best way to show that a graph is nonplanar is to find a subgraph homeomorphic to K_5 *or* $K_{3,3}$. *(Usually it will be* $K_{3,3}$.*)*

1. The question is whether $K_{5,2}$ is planar. It clearly is so, since we can draw it in the xy-plane by placing the five vertices in one part along the x-axis and the other two vertices on the positive and negative y-axis.

3. For convenience we label the vertices a, b, c, d, e, starting with the vertex in the lower left corner and proceeding clockwise around the outside of the figure as drawn in the exercise. This graph is just $K_{2,3}$; the picture below shows it redrawn by moving vertex c down.

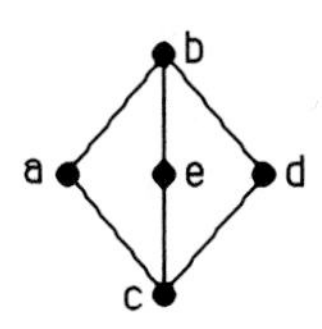

5. This is $K_{3,3}$, with parts $\{a, d, f\}$ and $\{b, c, e\}$. Therefore it is not planar.

7. This graph can be untangled if we play with it long enough. The following picture gives a planar representation of it.

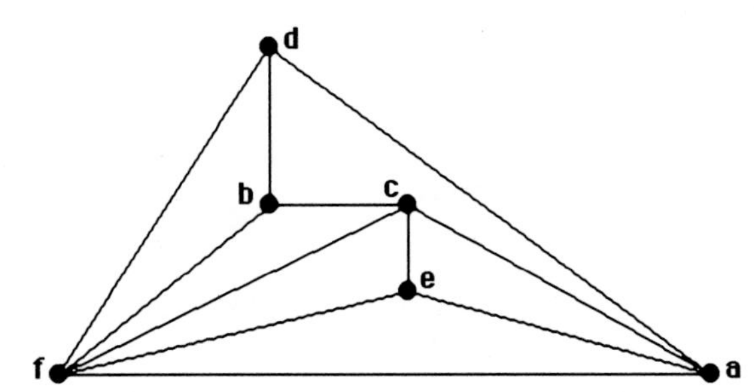

9. If one has access to software such as *The Geometer's Sketchpad*, then this problem can be solved by drawing the graph and moving the points around, trying to find a planar drawing. If we are unable to find one, then we look for a reason why—either a subgraph homeomorphic to K_5 or one homeomorphic to $K_{3,3}$ (always try the latter first). In this case we find that there is a homeomorphic copy of $K_{3,3}$, with vertices b, g, and i in one set and a, f, and h in the other; all the edges are there except for the edge bh, and it is represented by the path beh.

11. We give a proof by contradiction. Suppose that there is a planar representation of K_5, and let us call the vertices v_1, v_2, ..., v_5. There must be an edge from every vertex to every other. In particular, v_1, v_2, v_3, v_4, v_5, v_1 must form a pentagon. The pentagon separates the plane into two regions, an inside and an outside. The edge from v_1 to v_3 must be present, and without loss of generality let us assume it is drawn on the inside. Then there is no way for edges $\{v_2, v_4\}$ and $\{v_2, v_5\}$ to be in the inside, so they must be in the outside region. Now this prevents edges $\{v_1, v_4\}$ and $\{v_3, v_5\}$ from being on the outside. But they cannot both be on the inside without crossing. Therefore there is no planar representation of K_5.

13. We apply Euler's formula $r = e - v + 2$. Here we are told that $v = 6$. We are also told that each vertex has degree 4, so that the sum of the degrees is 24. Therefore by the Handshaking Theorem there are 12 edges, so $e = 12$. Solving, we find $r = 8$.

15. The proof is very similar to the proof of Corollary 1. First note that the degree of each region is at least 4. The reason for this is that there are no loops or multiple edges (which would give regions of degree 1 or 2) and no simple circuits of length 3 (which would give regions of degree 3); and the degree of the unbounded region is at least 4 since we are assuming that $v \geq 3$. Therefore we have, arguing as in the proof of Corollary 1, that $2e \geq 4r$, or simply $r \leq e/2$. Plugging this into Euler's formula, we obtain $e - v + 2 \leq e/2$, which gives $e \leq 2v - 4$ after some trivial algebra.

17. The proof is exactly the same as in Exercise 15, except that this time the degree of each region must be at least 5. Thus we get $2e \geq 5r$, which after the same algebra as before, gives the desired inequality.

19. **a)** If we remove a vertex from K_5, then we get K_4, which is clearly planar.
b) If we remove a vertex from K_6, then we get K_5, which is not planar.
c) If we remove a vertex from $K_{3,3}$, then we get $K_{3,2}$, which is clearly planar.
d) We assume the question means "Is it the case that for every v, the removal of v makes the graph planar?" Then the answer is no, since we can remove a vertex in the part of size 4 to leave $K_{3,3}$, which is not planar.

21. This graph is planar and hence cannot be homeomorphic to $K_{3,3}$.

23. The instructions are really not fair. It is hopeless to try to use Kuratowski's Theorem to prove that a graph *is* planar, since we would have to check hundreds of cases to argue that there is no subgraph homeomorphic to K_5 or $K_{3,3}$. Thus we will show that this graph is planar simply by giving a planar representation. Note that it is Q_3.

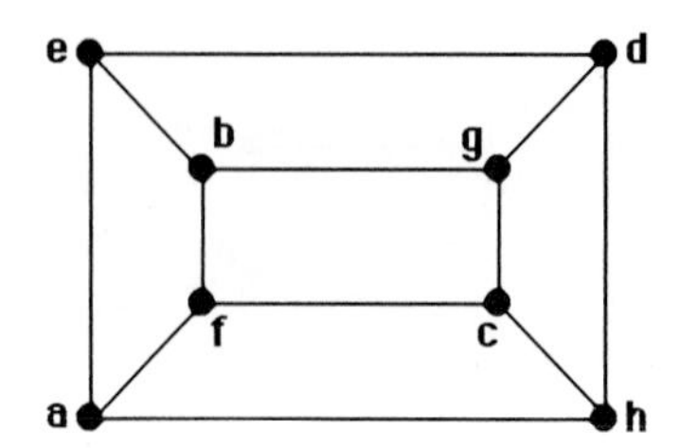

25. This graph is nonplanar, since it contains $K_{3,3}$ as a subgraph: the parts are $\{a, g, d\}$ and $\{b, c, e\}$. (Actually it contains $K_{3,4}$, and it even contains a subgraph homeomorphic to K_5.)

27. This is an extremely hard problem. We will present parts of the solution; the reader should consult a good graph theory book, such as Gary Chartrand and Linda Lesniak's *Graphs & Digraphs*, fourth edition (Chapman & Hall/CRC Press, 2005), for references and further details.

First we will state, without proof, what is known about crossing numbers for complete graphs (much is still not known about crossing numbers). If $n \le 10$, then the crossing number of K_n is given by the following product

$$\frac{1}{4}\left\lfloor\frac{n}{2}\right\rfloor\left\lfloor\frac{n-1}{2}\right\rfloor\left\lfloor\frac{n-2}{2}\right\rfloor\left\lfloor\frac{n-3}{2}\right\rfloor.$$

Thus the answers for parts **(a)**, **(b)**, and **(c)** are 1, 3, and 9, respectively. The figure below shows K_6 drawn in the plane with three crossings, which at least proves that the crossing number of K_6 is at most 3. The proof that it is not less than 3 is not easy. The embedding of K_5 with one crossing can be seen in this same picture, by ignoring the vertex at the top.

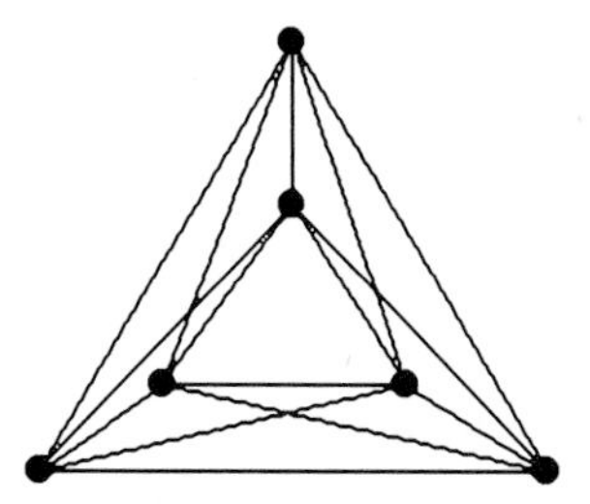

Second, for the complete bipartite graphs, what is known is that if the smaller of m and n is at most 6, then the crossing number of $K_{m,n}$ is given by the following product

$$\left\lfloor\frac{m}{2}\right\rfloor\left\lfloor\frac{m-1}{2}\right\rfloor\left\lfloor\frac{n}{2}\right\rfloor\left\lfloor\frac{n-1}{2}\right\rfloor.$$

Thus the answers for parts **(d)**, **(e)**, and **(f)** are 2, 4, and 16, respectively. The figure below shows $K_{4,4}$ drawn in the plane with four crossings, which at least proves that the crossing number of $K_{4,4}$ is at most 4. The proof that it is not less than 4 is, again, difficult. It is also easy to see from this picture that the crossing number of $K_{3,4}$ is at most 2 (by ignoring the top vertex).

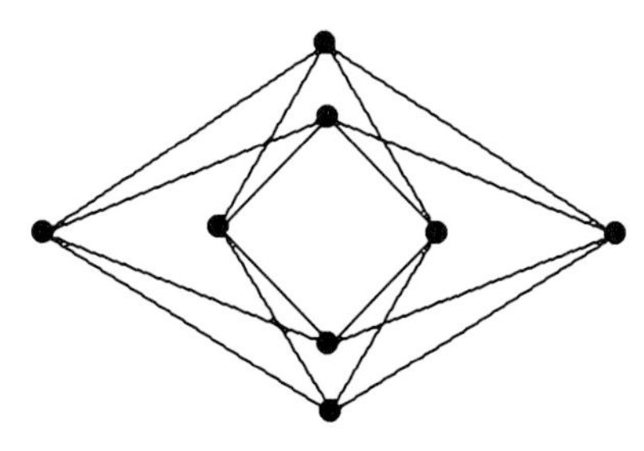

29. Let us follow the hint, and draw all the edges with straight line segments. This clearly produces a drawing of $K_{m,n}$. We will show that the number of crossings is $mn(m-2)(n-2)/16$, and that will complete the proof. (Incidentally, it is not known whether this upper bound is actually the crossing number. No one has found an embedding with fewer crossings, but only in the case in which the smaller of m and n is at most 6 has it been proved that it cannot be done. See the comments in the solution to Exercise 27.) In order to count the crossings, it is enough to count the crossings occurring in the first quadrant and multiply by 4. Let us label the points on the positive x-axis with the numbers 1 through $m/2$, and those on the y-axis with the numbers 1 through $n/2$. If we choose any two distinct numbers, say a and b with $a < b$, from 1 to $m/2$, and any two distinct numbers, say r and s with $r < s$, from 1 through $n/2$, then we get exactly one crossing in our graph, namely between the edges as and br. (There is no crossing between ar and bs.) So the number of crossings in the first quadrant is the same as the number of ways to make these choices, which is clearly $C(m/2,2)\cdot C(n/2,2)$. So the total number of crossings is 4 times this quantity, namely

$$4\cdot C(m/2,2)\cdot C(n/2,2) = 4\cdot\frac{\frac{m}{2}\left(\frac{m}{2}-1\right)}{2}\cdot\frac{\frac{n}{2}\left(\frac{n}{2}-1\right)}{2},$$

which easily simplifies to

$$\frac{mn(m-2)(n-2)}{16}.$$

31. Each of these graphs is nonplanar; the first three contain K_5, and the last three contain $K_{3,3}$. Thus if we can show how to draw each of the graphs in two planes, then we will have shown that the thickness is 2 in each case. The following picture shows that K_7 can be drawn in 2 planes, so this takes care of part **(a)**, part **(b)**, and part **(c)**.

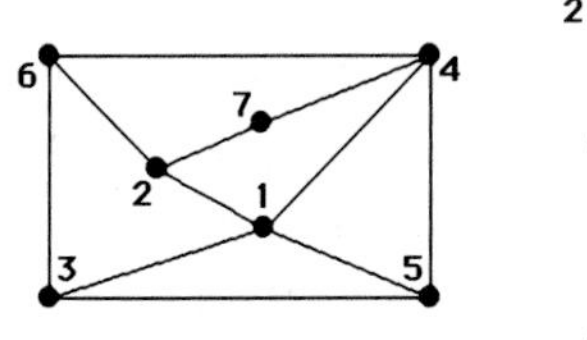

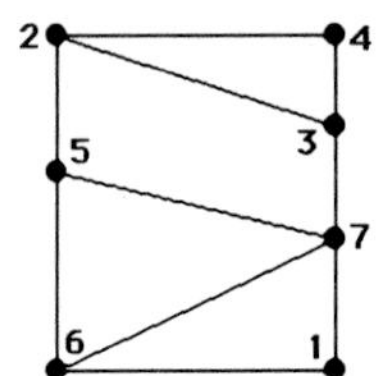

The following picture shows that $K_{5,5}$ can be drawn in 2 planes, so this takes care of part **(d)**, part **(e)**, and part **(f)**.

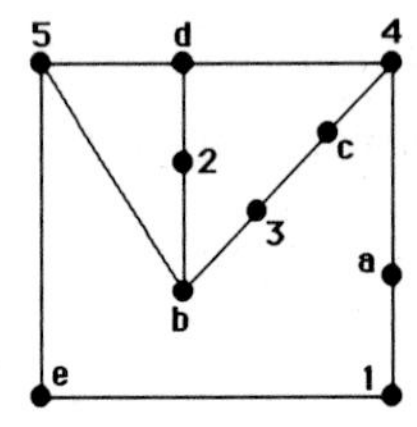

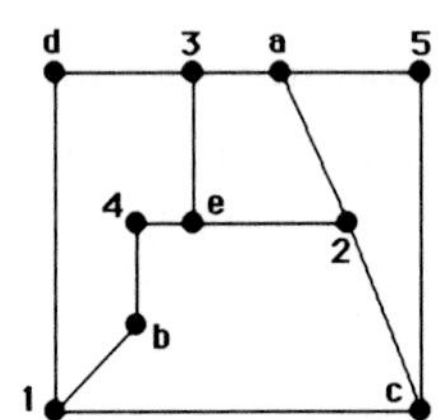

33. The formula is certainly valid for $n \leq 4$, so let us assume that $n > 4$. By Exercise 32, the thickness of K_n is at least

$$\frac{C(n,2)}{3n-6} = \frac{n(n-1)/2}{3n-6} = \frac{n(n-1)}{6(n-2)} = \frac{1}{6}\left(n+1+\frac{2}{n-2}\right)$$

rounded up. Since this quantity is never an integer, it equals one more than itself rounded down, namely

$$\frac{1}{6}\left(n+1+\frac{2}{n-2}\right)+1 = \frac{n+7}{6}+\frac{2}{6(n-2)}$$

rounded down. The last term can be ignored: it is always less than $1/6$ and therefore will not influence the rounding process (since the first term has denominator 6). Thus we have proved that the thickness of K_n is at least $\lfloor (n+7)/6 \rfloor$.

35. This follows immediately from Exercise 34, since $K_{m,n}$ has mn edges and $m+n$ vertices and, being bipartite, has no triangles.

37. We can represent the surface of a torus with a rectangle, thinking of the right-hand edge as being equal to the left-hand edge, and the top edge as being equal to the bottom edge. For example, if we travel out of the rectangle across the right-hand edge about a third of the way from the top, then we immediately reenter the rectangle across the left-hand edge about a third of the way from the top. The picture below shows $K_{3,3}$ drawn on this surface. Note that the edges that seem to leave the rectangle really reenter it from the opposite side.

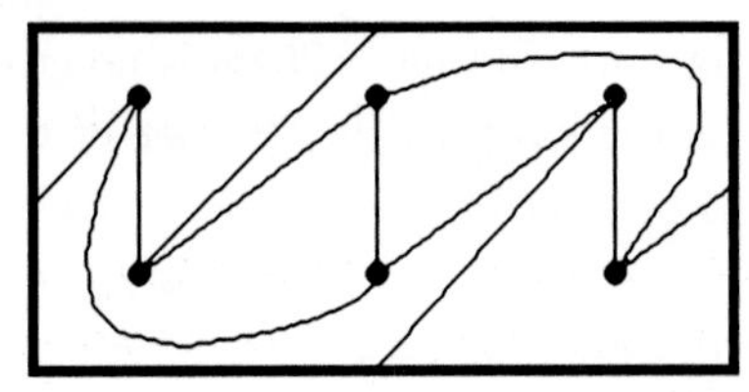

SECTION 9.8 Graph Coloring

Like the problem of finding Hamilton paths, the problem of finding colorings with the fewest possible colors probably has no good algorithm for its solution. In working these exercises, for the most part you should proceed by trial and error, using whatever insight you can gain by staring at the graph (for instance, finding large complete subgraphs). There are also some interesting exercises here on coloring the edges of graphs—see Exercises 21–22. Exercises 25–27 are worth looking at, as well: they deal with a fast algorithm for coloring a graph that is not guaranteed to produce an optimal coloring.

1. We construct the dual graph by putting a vertex inside each region (but not in the unbounded region), and drawing an edge between two vertices if the regions share a common border. The easiest way to do this is illustrated in our answer. First we draw the map, then we put a vertex inside each region and make the connections. The dual graph, then, is the graph with heavy lines.

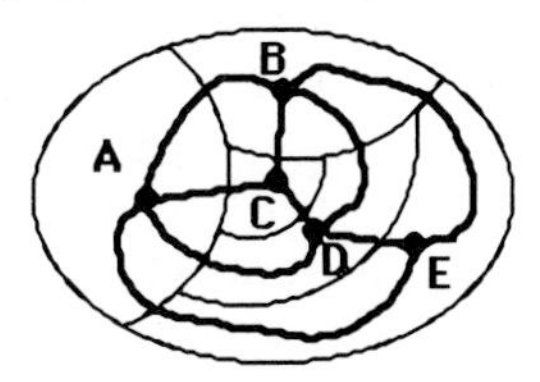

The number of colors needed to color this map is the same as the number of colors needed to color the dual graph. Since A, B, C, and D are mutually adjacent, at least four colors are needed. We can color each of the vertices (i.e., regions) A, B, C, and D a different color, and we can give E the same color as we give C.

3. We construct the dual as in Exercise 1.

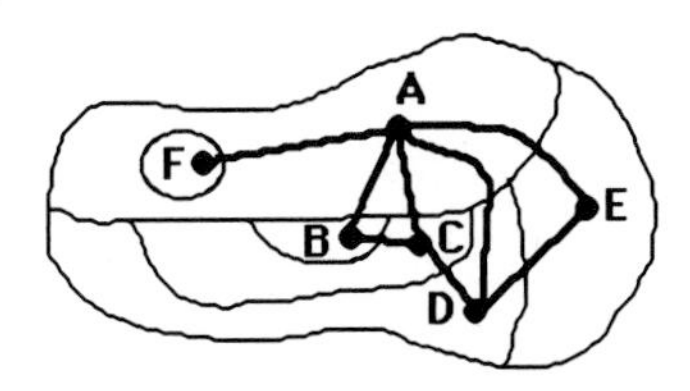

As in Exercise 1, the number of colors needed to color this map is the same as the number of colors needed to color the dual graph. Three colors are clearly necessary, because of the triangle ABC, for instance. Furthermore three colors suffice, since we can color vertex (region) A red, vertices B, D, and F blue, and vertices C and E green.

5. For Exercises 5 through 11, in order to prove that the chromatic number is k, we need to find a k-coloring and to show that (at least) k colors are needed. Here, since there is a triangle, at least 3 colors are needed. Clearly 3 colors suffice, since we can color a and d the same color.

7. Since there is a triangle, at least 3 colors are needed. Clearly 3 colors suffice, since we can color a and c the same color.

9. Since there is an edge, at least 2 colors are needed. The coloring in which b, d, and e are red and a and c blue shows that 2 colors suffice.

11. Since there is a triangle, at least 3 colors are needed. It is not hard to construct a 3-coloring. We can let a, f, h, j, and n be blue; let b, d, g, k, and m be green; and let c, e, i, l, and o be yellow.

13. If a graph has an edge (not a loop, since we are assuming that the graphs in this section are simple), then its chromatic number is at least 2. Conversely, if there are no edges, then the coloring in which every vertex receives the same color is proper. Therefore a graph has chromatic number 1 if and only if it has no edges.

15. In Example 4 we saw that the chromatic number of C_n is 2 if n is even and 3 if n is odd. Since the wheel W_n is just C_n with one more vertex, adjacent to all the vertices of the C_n along the rim of the wheel, W_n clearly needs exactly one more color than C_n (for that middle vertex). Therefore the chromatic number of W_n is 3 if n is even and 4 if n is odd.

17. Consider the graph representing this problem. The vertices are the 8 courses, and two courses are joined by an edge if there are students taking both of them. Thus there are edges between every pair of vertices except the 7 pairs listed. It is much easier to draw the complement than to draw this graph itself; it is shown below.

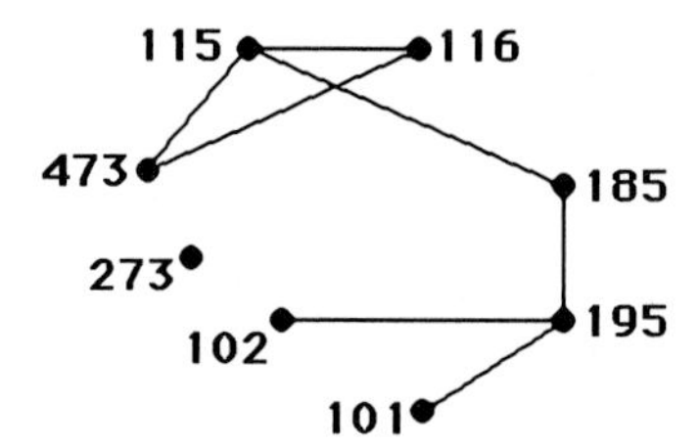

We want to find the chromatic number of the graph whose complement we have drawn; the colors will be the time periods for the exams. First note that since Math 185 and the four CS courses form a K_5 (in other words, there are no edges between any two of these in our picture), the chromatic number is at least 5. To show that it equals 5, we just need to color the other three vertices. A little trial and error shows that we can make Math 195 the same color as (i.e., have its final exam at the same time as) CS 101; and we can make Math 115 and 116 the same color as CS 473. Therefore five time slots (colors) are sufficient.

19. We model the problem with the intersection graph of these sets. Note that every pair of these intersect except for C_4 and C_5. Thus the graph is K_6 with that one edge deleted. Clearly its chromatic number is 5, since we need to color all the vertices different colors, except that C_4 and C_5 may have the same color. In other words, 5 meeting times are needed, since only committees C_4 and C_5 can meet simultaneously.

21. Note that the number of colors needed to color the edges is at least as large as the largest degree of a vertex, since the edges at each vertex must all be colored differently. Hence if we can find an edge coloring with that many colors, then we know we have found the answer. In Exercise 5 there is a vertex of degree 3, so the edge chromatic number is at least 3. On the other hand, we can color $\{a,c\}$ and $\{b,d\}$ the same color, so 3 colors suffice. In Exercise 6 the 6 edges incident to g must all get different colors. On the other hand, it is not hard to complete a proper edge coloring with only these colors (for example, color edge $\{a,f\}$ with the same color as used on $\{d,g\}$), so the answer is 6. In Exercise 7 the answer must be at least 3; it is 3 since edges that appear as parallel line segments in the picture can have the same color. In Exercise 8 clearly 4 colors are required, since the vertices have degree 4. In fact 4 colors are sufficient. Here is one proper 4-coloring (we denote edges in the obvious shorthand notation): color 1 for ac, be, and df; color 2 for ae, bd, and cf; color 3 for ab, cd, and ef; and color 4 for ad, bf, and ce. In Exercise 9 the answer must be at least 3; it is easy to construct a 3-coloring of the edges by inspection: $\{a,b\}$ and $\{c,e\}$ have the same color, $\{a,d\}$ and $\{b,c\}$ have the same color, and $\{a,e\}$ and $\{c,d\}$ have the same color. In Exercise 10 the largest degree is 6 (vertex i has degree 6); therefore at least 6 colors are required. By trial and error we come up with this coloring using 6 colors (we use the obvious shorthand notation for edges); there are many others, of course. Assign color 1 to ag, cd, and hi; color 2 to ab, cf, dg, and ei; color 3 to bh, cg, di, and ef; color 4 to ah, ci, and de; color 5 to bi, ch, and fg; and color 6 to ai, bc, and gh. Finally, in Exercise 11 it is easy to construct an edge-coloring with 4 colors; again the edge chromatic number is the maximum degree of a vertex.

Despite the appearances of these examples, it is not the case that the edge chromatic number of a graph is always equal to the maximum degree of the vertices in the graph. The simplest example in which this is not true is K_3. Clearly its edge chromatic number is 3 (since all three edges are adjacent to each other), but its maximum degree is 2. There is a theorem, however, stating that the edge chromatic number is always equal to either the maximum degree or one more than the maximum degree.

23. This problem can be modeled with the intersection graph of the sets of steps during which the variables must be stored. This graph has 7 vertices, t through z; there is an edge between two vertices if the two variables they represent must be stored during some common step. The answer to the problem is the chromatic number of this graph. Rather than considering this graph, we look at its complement (it has a lot fewer edges). Here two vertices are adjacent if the sets (of steps) do not intersect. The only edges are $\{u,w\}$, $\{u,x\}$, $\{u,y\}$, $\{u,z\}$, $\{v,x\}$, $\{x,z\}$. Note that there are no edges in the complement joining any two of $\{t,v,w,y,z\}$, so that these vertices form a K_5 in the original graph. Thus the chromatic number of the original graph is at least 5. To see that it is 5, note that vertex u can have the same color as w, and x can have the same color as z (these pairs appear as edges in the complement). Since the chromatic number is 5, we need 5 registers, with variables u and w sharing a register, and vertices x and z sharing one.

25. First we need to list the vertices in decreasing order of degree. This ordering is not unique, of course; we will pick e,a,b,c,f,h,i,d,g,j. Next we assign color 1 to e, and then to f and d, in that order. Now we assign color 2 to a, c, i, and g, in that order. Finally, we assign color 3 to b, h and j, in that order. Thus the algorithm gives a 3-coloring. Since the graph contains triangles, we know that this is the best possible, so the algorithm "worked" here (but it need not always work—see Exercise 27).

27. A simple example in which the algorithm may fail to provide a coloring with the minimum number of colors is C_6, which of course has chromatic number 2. Since all the vertices are of degree 2, we may order them v_1, v_4, v_2, v_3, v_5, v_6, where the edges are $\{v_1,v_2\}$, $\{v_2,v_3\}$, $\{v_3,v_4\}$, $\{v_4,v_5\}$, $\{v_5,v_6\}$, and $\{v_1,v_6\}$. Then v_1 gets color 1, as does v_4. Next v_2 and v_5 get color 2; and then v_3 and v_6 must get color 3.

29. We need to show that the wheel W_n when n is an odd integer greater than 1 can be colored with four colors, but that any graph obtained from it by removing one edge can be colored with three colors. Four colors are needed to color this graph, because three colors are needed for the rim (see Example 4), and the center vertex, being adjacent to all the rim vertices, will require a fourth color. To complete the proof that W_n is chromatically 4-critical, we must show that the graph obtained from W_n by deleting one edge can be colored with three colors. There are two cases. If we remove a rim edge, then we can color the rim with two colors, by starting at an endpoint of the removed edge and using the colors alternately around the portion of the rim that remains. The third color is then assigned to the center vertex. On the other hand, if we remove a spoke edge, then we can color the rim by assigning color #1 to the rim endpoint of the removed edge and colors #2 and #3 alternately to the remaining vertices on the rim, and then assign color #1 to the center.

31. We give a proof by contradiction. Suppose that G is chromatically k-critical but has a vertex v of degree $k-2$ or less. Remove from G one of the edges incident to v. By definition of "k-critical," the resulting graph can be colored with $k-1$ colors. Now restore the missing edge and use this coloring for all vertices except v. Because we had a proper coloring of the smaller graph, no two adjacent vertices have the same color. Furthermore, v has at most $k-2$ neighbors, so we can color v with an unused color to obtain a proper $(k-1)$-coloring of G. This contradicts the fact that G has chromatic number k. Therefore our assumption was wrong, and every vertex of G must have degree at least $k-1$.

33. **a)** Note that vertices d, e, and f are mutually adjacent. Therefore six different colors are needed in a 2-tuple coloring, since each of these three vertices needs a disjoint set of two colors. In fact it is easy to give a coloring with just six colors: Color a, d, and g with $\{1,2\}$; color c and e with $\{3,4\}$; and color b and f with $\{5,6\}$. Thus $\chi_2(G)=6$.

b) This one is trickier than part **(a)**. There is no coloring with just six colors, since if there were, we would be forced (without loss of generality) to color d with $\{1,2\}$; e with $\{3,4\}$; f with $\{5,6\}$; then g with $\{1,2\}$, b with $\{5,6\}$, and c with $\{3,4\}$. This gives no free colors for vertex a. Now this may make it appear that eight colors are required, but a little trial and error shows us that seven suffice: Color a with $\{2,4\}$; color b and f with $\{5,6\}$; color d with $\{1,2\}$; color c with $\{3,7\}$; color e with $\{3,4\}$; and color g with $\{1,7\}$. Thus $\chi_2(H)=7$.

c) This is similar to part **(a)**. Here nine colors are necessary and sufficient, since a, d, and g can get one set of three colors; b and f can get a second set; and c and e can get a third set. Clearly nine colors are necessary to color the triangles.

d) First we construct a coloring with 11 colors: Color a with $\{3,6,11\}$; color b and f with $\{7,8,9\}$; color d with $\{1,2,3\}$; color c with $\{4,5,10\}$; color e with $\{4,6,11\}$; and color g with $\{1,2,5\}$. To prove that $\chi_3(H)=11$, we must show that it is impossible to give a 3-tuple coloring with only ten colors. If such a coloring were possible, without loss of generality we could color d with $\{1,2,3\}$, e with $\{4,5,6\}$, f with $\{7,8,9\}$, and g with $\{1,2,10\}$. Now nine colors are needed for the three vertices a, b, and c, since they form a triangle; but colors 1 and 2 are already used in vertices adjacent to all three of them. Therefore at least $9+2=11$ colors are necessary.

35. The frequencies are the colors, the zones are the vertices, and two zones that are so close that interference would be a problem are joined by an edge in the graph. Then it is clear that a k-tuple coloring is exactly an assignment of frequencies that avoids possible interference.

37. We use induction on the number of vertices of the graph. Every graph with five or fewer vertices can be colored with five or fewer colors, since each vertex can get a different color. That takes care of the basis case(s). So we assume that all graphs with k vertices can be 5-colored and consider a graph G with $k+1$ vertices. By Corollary 2 in Section 9.7, G has a vertex v with degree at most 5. Remove v to form the graph G'. Since G' has only k vertices, we 5-color it by the inductive hypothesis. If the neighbors of v do not use all five colors, then we can 5-color G by assigning to v a color not used by any of its neighbors. The difficulty arises if v has five neighbors, and each has a different color in the 5-coloring of G'. Suppose that the neighbors of v, when considered in clockwise order around v, are a, b, c, m, and p. (This order is determined by the clockwise order of the curves representing the edges incident to v.) Suppose that the colors of the neighbors are azure, blue, chartreuse, magenta, and purple, respectively. Consider the azure-chartreuse subgraph (i.e., the vertices in G colored azure or chartreuse and all the edges between them). If a and c are not in the same component of this graph, then in the component containing a we can interchange these two colors (make the azure vertices chartreuse and vice versa), and G' will still be properly colored. That makes a chartreuse, so we can now color v azure, and G has been properly colored. If a and c are in the same component, then there is a path of vertices alternately colored azure and chartreuse joining a and c. This path together with edges av and vc divides the plane into two regions, with b in one of them and m in the other. If we now interchange blue and magenta on all the vertices in the same region as b, we will still have a proper coloring of G', but now blue is available for v. In this case, too, we have found a proper coloring of G. This completes the inductive step, and the theorem is proved.

39. We follow the hint. Because the measures of the interior angles of a pentagon total $540°$, there cannot be as many as three interior angles of measure more than $180°$ (reflex angles). If there are no reflex angles, then the pentagon is convex, and a guard placed at any vertex can see all points. If there is one reflex angle, then the pentagon must look essentially like figure (a) below, and a guard at vertex v can see all points. If there are two reflex angles, then they can be adjacent or nonadjacent (figures (b) and (c)); in either case, a guard at vertex v can see all points. (In figure (c), choose the reflex vertex closer to the bottom side.) Thus for all pentagons, one guard suffices, so $g(5)=1$.

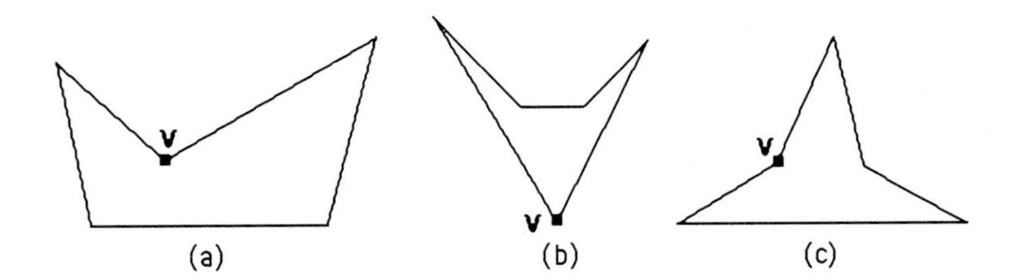

41. The figure suggested in the hint (generalized to have k prongs for any $k \geq 1$) has $3k$ vertices. Consider the set of points from which a guard can see the tip of the first prong, the set of points from which a guard can see the tip of the second prong, and so on. These are disjoint triangles (together with their interiors). Therefore a separate guard is needed for each of the k prongs, so at least k guards are needed. This shows that $g(3k) \geq k = \lfloor 3k/3 \rfloor$. To handle values of n that are not multiples of 3, let $n = 3k + i$, where $i = 1$ or 2. Then obviously $g(n) \geq g(3k) \geq k = \lfloor n/3 \rfloor$.

GUIDE TO REVIEW QUESTIONS FOR CHAPTER 9

1. **a)** See pp. 589–590 and Table 1 in Section 9.1. **b)** See Exercise 1 in Section 9.1.

2. See all the examples Section 9.1.

3. See Theorem 1 in Section 9.2.

4. See Theorem 2 in Section 9.2.

5. See Theorem 3 in Section 9.2.

6. **a)** See Example 5 in Section 9.2. **b)** See Example 13 in Section 9.2. **c)** See Example 6 in Section 9.2. **d)** See Example 7 in Section 9.2. **e)** See Example 8 in Section 9.2.

7. **a)** n, $C(n,2)$ **b)** $m+n$, mn **c)** n, n **d)** $n+1$, $2n$ **e)** 2^n, $n2^{n-1}$

8. **a)** See p. 602. **b)** K_2 and C_{2m}

c) (See also Example 12 and Exercise 60 in Section 9.2.) The following algorithm is an efficient way to determine whether a connected graph can be 2-colored (which is the same thing as saying that it is bipartite); apply it to each component of the given graph. First color any vertex red. Then color all vertices adjacent to this vertex blue. Then look at all vertices adjacent to these just-colored blue vertices. If any of them are already colored blue, then stop and declare the graph not to be bipartite; otherwise color all the uncolored ones red. Next look at all vertices adjacent to all the vertices just colored red. If any of them are already colored red, then stop and declare the graph not to be bipartite; otherwise color all the uncolored ones blue. Continue in this way until no more vertices can be colored. If we get this far, then (this component of) the graph is bipartite. (If uncolored vertices remain, they are in a different component, so we can repeat the entire process starting with any uncolored vertex.)

9. **a)** adjacency lists, adjacency matrices, incidence matrices

b) Look at Figure 1 in Section 9.3, and let G be the graph consisting of the vertices and edges shown there, together with edges $\{b,c\}$ and $\{b,e\}$. Its adjacency lists are shown on p. 612 (Table 1), once we add c and e to the list of adjacent vertices of b, and add b to the list for c and for e. Its adjacency matrix and incidence matrix are as follows (using alphabetical order):

$$\begin{bmatrix} 0 & 1 & 1 & 0 & 1 \\ 1 & 0 & 1 & 0 & 1 \\ 1 & 1 & 0 & 1 & 1 \\ 0 & 0 & 1 & 0 & 1 \\ 1 & 1 & 1 & 1 & 0 \end{bmatrix} \quad \text{and} \quad \begin{bmatrix} 1 & 1 & 1 & 0 & 0 & 0 & 0 & 0 \\ 1 & 0 & 0 & 1 & 1 & 0 & 0 & 0 \\ 0 & 1 & 0 & 1 & 0 & 1 & 1 & 0 \\ 0 & 0 & 0 & 0 & 0 & 1 & 0 & 1 \\ 0 & 0 & 1 & 0 & 1 & 0 & 1 & 1 \end{bmatrix}$$

10. **a)** See p. 615.
b) See p. 615; number of vertices, number of edges, degrees of vertices, existence of triangle (C_3 as subgraph), and existence of Hamilton circuit are all invariants.
c) See Figure 10 in Section 9.3. **d)** no

11. **a)** See p. 624. **b)** See p. 625.

12. **a)** See p. 612. **b)** See pp. 616–617. **c)** See Theorem 2 in Section 9.4.

13. **a)** See p. 633. **b)** See pp. 633–634.
c) All edges are in the same component and there are at most two vertices of odd degree (see Theorems 1 and 2 in Section 9.5).
d) All edges are in the same component and there are no vertices of odd degree (see Theorem 1 in Section 9.5).

14. **a)** See p. 638. **b)** the existence of a cut edge or a cut vertex

15. finding the shortest highway route between two cities; finding the cheapest way to lease telephone lines to join two communications centers, using intermediate switching centers

16. **a)** See pp. 649–651. **b)** See Exercise 5, third part, in Section 9.6.

17. **a)** See p. 658. **b)** K_6

18. **a)** If the planar graph has v vertices, e edges, and c components, and is embedded in the plane so as to form r regions, then $v - e + r = 1 + c$. (The theorem as stated in the text—Theorem 1 in Section 9.7—assumes that $c = 1$, but this is the more general statement.)
b) By Corollary 1 in Section 9.7, for every planar graph with at least three vertices we know that $e \leq 3v - 6$ (connectivity need not be assumed). Thus we can show that a graph is nonplanar by showing that it has too many edges, namely more than $3v - 6$ edges. For example, K_6 is nonplanar, since it has 15 edges, and $15 \not\leq 3 \cdot 6 - 6$.

19. See Theorem 2 in Section 9.7. A graph is planar if and only if it does not contain a subgraph homeomorphic to K_5 or $K_{3,3}$.

20. **a)** See p. 667. **b)** n **c)** 2 if n is even, 3 if n is odd **d)** 2

21. Every planar graph can be 4-colored (i.e., has chromatic number at most 4), but K_6, for instance, requires six colors.

22. See Examples 5–7 in Section 9.8.

SUPPLEMENTARY EXERCISES FOR CHAPTER 9

1. Every vertex has degree 50. Thus the sum of the degrees must be $50 \cdot 100 = 5000$. By the Handshaking Theorem, the graph therefore has $5000/2 = 2500$ edges.

3. Both graphs have a lot of symmetry to them, and the degrees of the vertices are the same, so we might hope that they are isomorphic. Let us try to form the correspondence f. First note that there is a 4-cycle u_1, u_5, u_2, u_6, u_1 in the first graph. Suppose that we try letting it correspond to the 4-cycle v_1, v_2, v_3, v_4, v_1 in the second graph. Thus we let $f(u_1) = v_1$, $f(u_5) = v_2$, $f(u_2) = v_3$, and $f(u_6) = v_4$. The rest of the assignments are forced: since u_7 is the other vertex adjacent to u_1, we must let $f(u_7) = v_6$, since v_6 is the other vertex adjacent to v_1 (which is $f(u_1)$). Similarly, $f(u_3) = v_7$, $f(u_8) = v_8$, and $f(u_4) = v_5$. Now we just have to check that the vertices corresponding to the vertices in the 4-cycle v_5, v_6, v_7, v_8, v_5 in the second graph form a 4-cycle in that order. Since these vertices form the 4-cycle u_4, u_7, u_3, u_8, u_4, our correspondence works.

5. These graphs are isomorphic, although the isomorphism is hard to find. One approach that can lead to the isomorphism is to draw the complement of each graph. The complements are a little simpler than the original graphs, since they have fewer edges. When we do this, it is easy to give a planar representation of each. Then by looking at the sizes of the regions we can find an isomorphism. One such correspondence is $u_1 \leftrightarrow v_5$, $u_2 \leftrightarrow v_4$, $u_3 \leftrightarrow v_7$, $u_4 \leftrightarrow v_3$, $u_5 \leftrightarrow v_8$, $u_6 \leftrightarrow v_2$, $u_7 \leftrightarrow v_6$, and $u_8 \leftrightarrow v_1$. We just need to check that all the edges are preserved by this correspondence.

7. It follows immediately from the definition that the complete m-partite graph with parts n_1, n_2, ..., n_m has $n_1 + n_2 + \cdots + n_m = \sum_{i=1}^{m} n_i$ vertices. We will organize the count of the edges by looking at which parts the edges join. Fix $1 \le i < j \le m$, and consider the edges between the ith part and the jth part. It is easy to see from the product rule that there are $n_i n_j$ edges. Therefore to get all the edges, we have to add all these products, for all possible pairs (i, j). Thus the number of edges is $\sum_{1 \le i < j \le m} n_i n_j$.

9. a) The subgraph induced by $\{a, b, c\}$ consists of those vertices and all the edges that are in the graph and join pairs of them. Thus the induced subgraph is the entire component H_1.
b) The subgraph induced by $\{a, e, g\}$ consists of those vertices and all the edges joining pairs of them in this graph. Since there are no such edges, the induced subgraph is just the graph with these three vertices and no edges.
c) The induced subgraph consists of these five vertices and the edges $\{b, c\}$, $\{f, g\}$, and $\{g, h\}$ (all the edges joining pairs of these five vertices that were in the original graph).

11. In general it is no easy task to find cliques. We need to be careful not to overlook things. We will denote a clique simply by listing the vertices in it, without punctuation. There is one K_4, namely $bcef$, and it is the largest clique. There are several K_3's not contained in this K_4, and they are all cliques: abg, adg, beg, and deg. Since every edge is contained in one of these five cliques (and there are no isolated vertices), there are no smaller cliques, so this list is complete.

13. See the comments for Exercise 11. We find the cliques by brute force and careful looking, hoping that we do not miss any. Some staring at the graph convinces us that there are no K_6's. There is one K_5, namely the clique $bcdjk$. There are two K_4's not contained in this K_5, which therefore are cliques: $abjk$, and $efgi$. All the K_3's not contained in any of the cliques listed so far are also cliques. We find abi, aij, bde, bei, bij, ghi, and hij. All the edges are in at least one of the cliques listed so far (and there are no isolated vertices), so we are done.

15. Clearly no single vertex is dominating by itself, but $\{c, d\}$ dominates, so it is a minimum dominating set (there are lots of others).

17. These graphs are quite a mess to draw, since they contain so many edges. Instead of drawing them, we describe them in set-theoretic terms. Note that we do not consider a queen on a square to control that square itself, since to do so would give us loops in the graph.
a) The vertex set consists of all pairs (i, j) with $1 \le i \le 3$ and $1 \le j \le 3$. Since a queen in any square controls all the squares in the same row and column, there are edges $\{(i, j), (i, j')\}$ and $\{(i, j), (i', j)\}$ for all i, j, i' and j' between 1 and 3 inclusive, with $i \ne i'$ and $j \ne j'$. These are not all, though, since we have to put in the diagonal controls. There are 10 such edges: $(1, 1)$, $(2, 2)$, and $(3, 3)$ are all joined to each other; there is an edge between $(1, 2)$ and $(2, 1)$, and an edge between $(2, 3)$ and $(3, 2)$; $(1, 3)$, $(2, 2)$, and $(3, 1)$ are all joined to each other; and there is an edge between $(1, 2)$ and $(2, 3)$, and an edge between $(2, 1)$ and $(3, 2)$.
b) We do the same sort of thing as in part **(a)**. The vertex set consists of all pairs (i, j) with $1 \le i \le 4$ and $1 \le j \le 4$. Since a queen in any square controls all the squares in the same row and column, there are edges

$\{(i,j),(i,j')\}$ and $\{(i,j),(i',j)\}$ for all i, j, i' and j' between 1 and 4 inclusive, with $i \neq i'$ and $j \neq j'$. Rather than listing the 28 diagonal control edges explicitly, let us use some analytic geometry. The diagonals over which the queen has control all have slope 1 or -1. Therefore there are edges between (i,j) and (i',j') if these two vertices are distinct and $i - i' = \pm(j - j')$.

19. **a)** A queen in the center controls the entire board, so the answer is 1.

b) One queen cannot control the entire board, as one can verify by considering the possible cases (there are really only three—corner, noncorner edge, or nonedge). On the other hand 2 queens will do: place one in position $(2,2)$ and the other in position $(4,4)$.

c) Three queens can control the entire board. We can place them at positions $(2,2)$, $(3,4)$, and $(5,1)$, for example. To show that two queens are not enough is tedious. One way to do this is with the help of a computer. For each pair of squares (and there are $C(25,2) = 300$ such pairs), check to see that not all the squares are under control. Such a program will show that two queens can control at most 23 of the 25 squares.

21. Suppose that G and H are isomorphic simple graphs, with f the one-to-one and onto function from the vertex set of G to the vertex set of H that gives the isomorphism. By symmetry, it is enough to show in each case that if G has the property, then so does H.

a) Assume that G is connected; then for every pair of distinct vertices in G there is a path from one to the other. We want to show that H is connected. Let u and v be two distinct vertices of H. Then there is a path in G from $f^{-1}(u)$ to $f^{-1}(v)$, which we can think of as a sequence of vertices. Applying f to each vertex in this path gives us a path from u to v in H.

b) Suppose that G has a Hamilton circuit, which we can think of as a sequence of vertices. Applying f to each vertex in this circuit gives us a Hamilton circuit in H.

c) This is the same as part **(b)**, replacing "Hamilton" by "Euler."

d) The logic of this one is slightly different from the general pattern. Suppose that we can embed G in the plane with C crossings. Then this embedding clearly gives an embedding of H in the plane with C crossings as well: we use the same picture, relabeling u by $f(u)$. Therefore the crossing number of H is no bigger than the crossing number of G. By symmetry, the crossing number of G can be no bigger than the crossing number of H, either. Therefore the crossing numbers are equal.

e) If i_1, i_2, ..., i_n are n isolated vertices in G, then $f(i_1)$, $f(i_2)$, ..., $f(i_n)$ are n isolated vertices in H.

f) If A and B are the parts for G, then $f(A) = \{\, f(v) \mid v \in A \,\}$ and $f(B) = \{\, f(v) \mid v \in B \,\}$ are the parts for H.

23. We need to consider all the possibilities carefully. First suppose that the parts are of size 1 and 3. The only connected bipartite simple graph with parts of these sizes is $K_{1,3}$. The only other possibility is that the parts are each of size 2. Then the graph could be $K_{2,2}$ or $K_{2,2}$ with one edge missing; if more than one edge is deleted, then the result will not be connected. Therefore the answer is 3.

25. **a)** It is clear from the picture that if this graph is self-converse, then the relevant isomorphism interchanges c and e and interchanges a and b. If we take the given graph, reverse all the arrows, and then apply this correspondence, then we obtain the original graph back. So the graph is self-converse.

b) It is clear from the picture that if this graph were to be self-converse, then the isomorphism must send b and c (the only vertices without loops) back onto themselves, in one order or the other. But b has its incident edges both pointing in the same direction (inwards), whereas c does not, so there is no possible isomorphism between this graph and its converse.

27. This graph is not orientable because of the cut edge $\{b,c\}$. If we orient it from b to c, then there can be no path in the resulting directed graph from c to b; if we orient it from c to b, then there can be no path in the resulting directed graph from b to c.

29. This graph is orientable. We can orient the square and each of the triangles in the clockwise direction, for instance. In other words, the edges are (a,b), (b,c), (c,d), (d,a), (c,e), (e,f), (f,c), (c,g), (g,h), and (h,c). There is now a path from every vertex to every other vertex, by traveling clockwise around the appropriate figure to vertex c, and then traveling clockwise around the other appropriate figure.

31. Suppose that $\{a,b\}$ is a cut edge of the undirected graph G. Then a and b are in separate components of G when that edge is removed; in other words, every path from a to b must go through the edge in the a to b direction, and every path from b to a must go through the path in the b to a direction. Suppose that we have an orientation of this graph. If $\{a,b\}$ is oriented as (a,b), then by what we have said, there can be no path in the resulting directed graph from b to a, so the resulting directed graph is not strongly connected. On the other hand, if $\{a,b\}$ is oriented as (b,a), then there can be no path in the resulting directed graph from a to b. Thus by definition G is not orientable. Incidentally, a kind of converse to this result is also true. The ambitious reader should try to construct a proof.

33. Let n be the number of vertices in the tournament. Since for each u different from a given vertex v there is exactly one edge with endpoints u and v (in some order), there are $n-1$ edges involving vertex v. Thus the sum of the in-degree and out-degree of v is $n-1$.

35. We make the vertices the chickens in the flock, and for distinct chickens u and v, we have the directed edge (u,v) if and only if u dominates v.

37. **a)** No matter how the vertices are labeled, we must have a_1 adjacent to a_5, so the bandwidth is $5-1=4$.
b) The best we can do is to have the vertex of degree 3 as a_2. Then the edge between a_2 and a_4 causes the bandwidth to be $4-2=2$.
c) If we make the vertices in the part with 2 vertices a_2 and a_4, then the maximum of $|i-j|$ with a_i adjacent to a_j occurs when $i=1$ and $j=4$, a maximum of 3. If we carefully consider other possibilities, then we see that there is no way to reduce this difference. Therefore the bandwidth is 3.
d) The bandwidth is 4. To see this, assume without loss of generality that a_1 is in part A. Then if a_6 is in part B, the maximum difference is 5, which is larger than 4. On the other hand, if a_6 is also in part A, then the maximum difference is 4, since either a_2 or a_5 must be in part B.
e) We can achieve a maximum difference of 4 by labeling the vertices as shown below.

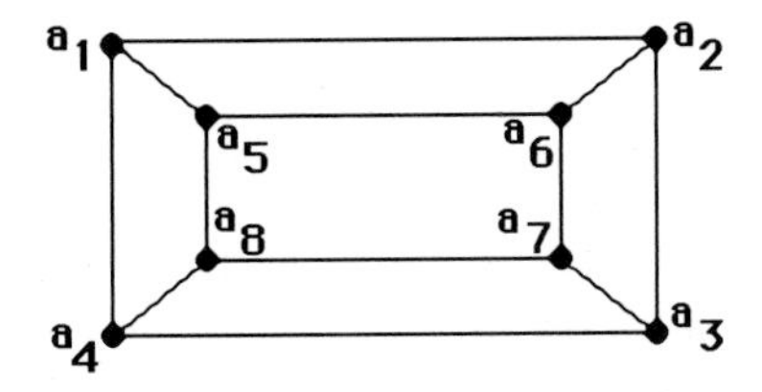

We must show that there is no way to achieve a difference of 3 or less. Now vertex a_1 is adjacent to three other vertices. If the difference is going to be 3 or less, then these have to be vertices a_2, a_3, and a_4. Now vertex a_2 is adjacent to two vertices besides these four, and therefore the index of one of them must be at least 6, giving a difference of 4.
f) The bandwidth cannot be 1, since at least one of the two vertices adjacent to vertex a_1 must have subscript at least 3, and $3-1=2$. On the other hand, if we label the vertices around the cycle as a_1, a_2, a_4, a_5, a_3 (and back to a_1), then the maximum value of $|i-j|$ with a_i and a_j adjacent is 2. Thus the bandwidth equals 2.

39. **a)** Suppose that the diameter of G is at least 4, and let u and v be two vertices whose distance apart in G is at least 4. We want to show that the diameter of $\overline{G}$ is at most 2. Let a and b be two distinct vertices of G. We need to show that the distance between a and b in $\overline{G}$ is at most 2. If $\{a,b\}$ is not an edge of G, then we are done, since then a and b are adjacent (at distance 1) in $\overline{G}$. Thus we assume that a and b are adjacent in G. Now it cannot be that $\{u,v\} = \{a,b\}$, since u and v are not adjacent in G. Without loss of generality assume that u is not in $\{a,b\}$. If u is not adjacent to a and also not adjacent to b in G, then we are done, since the path a,u,b in $\overline{G}$ shows that the distance between a and b is 2 in $\overline{G}$. We now show that the other possibility—that u is adjacent (in G) to at least one of these—leads either to the conclusion that the distance between a and b is at most 2, or to a contradiction. If u is adjacent to at least one of a and b, then v cannot be either a or b, because it would be too close to u in G. Therefore the same reasoning applies to v as applied to u, and either we are done or else we know that v is also adjacent (in G) to at least one of a and b. But this gives us a path from u to v, passing through one or both of a and b, of length less than 4, a contradiction.

b) The proof is rather similar to that in part **(a)**. Let u and v be vertices at a distance of at least 3 in G. Let a and b be arbitrary distinct vertices; we must show that the distance between a and b in $\overline{G}$ is at most 3. Assume not (we will derive a contradiction). Then certainly a and b are adjacent in G. Thus at least one of u and v is not equal to either a or b; say it is u. Now u cannot be adjacent to both a and b in $\overline{G}$, since then the distance between a and b in $\overline{G}$ would be 2. Without loss of generality assume that u is adjacent to a in G. Then v cannot be either a or b (it would be too close to u if it were), and it cannot be adjacent to a (in G), either, for the same reason. If v is not adjacent to b in G, then the path a,v,b in $\overline{G}$ makes the distance between a and b in $\overline{G}$ equal to 2. Thus we can assume that v is adjacent to b in G. It follows that u is not adjacent to b in G, since that adjacency would again make u too close to v. But now we have our contradiction, since there is the path a,v,u,b in $\overline{G}$, of length 3, from a to b.

41. There are two second shortest paths, both of length 8. One is a,b,e,z; the other is a,d,e,z,e,z.

43. Since the shortest path already went through all the vertices, the answer is that same path: a,c,b,d,e,z.

45. First we assume that G has exactly 11 vertices. Suppose that G is planar. By Corollary 1 to Euler's Theorem in Section 9.7, we know that a planar graph with 11 vertices can have at most $3 \cdot 11 - 6 = 27$ edges (if the graph is not connected, then it would have even fewer edges). Therefore G has at most 27 edges. This means that $\overline{G}$ has at least $C(11,2) - 27 = 28$ edges on its 11 vertices. By Corollary 1, again, this means that $\overline{G}$ is nonplanar, as desired. Now if in fact we were dealing with a graph G with more than 11 vertices, then let us restrict ourselves to the first 11 (in some ordering), and let H be the subgraph of G containing those 11 vertices and all the edges of G between pairs of them. Thus H is a subgraph of G, and it is easy to see that $\overline{H}$ is also a subgraph of $\overline{G}$. If G is planar, then so is H; by our argument above this means that $\overline{H}$ is not planar, so $\overline{G}$ cannot be planar.

It is actually the case that this result is still true if we replace the number 11 by the number 9. The proof, however, is very hard. The approach used here does not work, since two planes can contain the required number of edges. Indeed if we let G have 18 edges, then $\overline{G}$ will also contain $C(9,2) - 18 = 18$ edges. Corollary 1 to Euler's Theorem says that a planar graph with 9 vertices can have at most $3 \cdot 9 - 6 = 21$ edges, and $18 < 21$. The subtlety comes with trying to embed precisely the right edges for G and the right edges for $\overline{G}$, and it can be proved that this cannot be done, no matter what graph on 9 vertices G is.

47. Let n be the number of vertices in the graph, k its chromatic number, and i its independence number. Then there is a coloring of the graph with k colors. Since no two vertices in the same color class (i.e., colored the same) are adjacent, each color class is an independent set. Thus there are at most i vertices in each color class. This means there are at most ki vertices in all. In other words, $n \le ki$, as desired.

49. Think of the putting in of an edge as a success in a Bernoulli trial (see Section 6.2).

a) By Theorem 2 in Section 6.2, the probability of m successes is $C(n,m)p^m(1-p)^{n-m}$.

b) By Theorem 2 in Section 6.4, this expected value is np.

c) There is a subtle point here. Suppose $n=3$. Then there are eight different graphs on n vertices, if we view the vertices as labeled 1, 2, and 3. However, many of these graphs are isomorphic. Indeed, there are only four different graphs on three vertices, if we view them as unlabeled, and these four are not equally likely in this random generation process. So for this problem to make sense, we must be speaking of *labeled* graphs. Now we claim that each such graph has probability $1/2^{C(n,2)}$ of arising from this random generation process. Suppose we want to generate labeled graph G. As we apply the process to pairs of vertices, the random number x chosen must be less than or equal to $1/2$ when G has an edge between that pair of vertices and x must be greater than $1/2$ when G has no edge there. So the probability of the process "getting it right" is $1/2$ for each edge. Our claim follows, so all labeled graphs G are equally likely.

51. This is really an exercise in untangling the logic of what is being said. Suppose that P is monotone increasing. We must prove that not having P is monotone decreasing. That is, we must show that the property of not having P is retained whenever edges are removed from a simple graph. If this were not true, then there would be a simple graph G not having P and another simple graph G' with the same vertices but with some of the edges of G removed, which has P. But P is monotone increasing, so since G' has P, so does the graph G obtained by adding edges to G'. This contradicts the assumption that G does not have P. The converse is proved in exactly the same way.

WRITING PROJECTS FOR CHAPTER 9

Books and articles indicated by bracketed symbols below are listed near the end of this manual. You should also read the general comments and advice you will find there about researching and writing these essays.

1. The best source here is [BiLl].

2. A good source for these kinds of applications is [Ro2]. Graph theory is also being used extensively now in the human genome project and other areas of biology.

3. See the comments for Writing Project 2.

4. See [Ba3] for a recent book about research on this subject.

5. See [EaTa] and [Sk].

6. There should be some material in [Sk]. See information on the Web about a program called `nauty`, which has a good many graph algorithms.

7. The definitive work is [KöSc].

8. One of the main researchers in this area is Pavel Pevzner in the Computer Science Department at the University of California at San Diego. Look on his Web page for references.

9. Check out [Ra1].

10. Most advanced graph theory or combinatorics texts will mention this. Look at [Ro1], which also has further references. There is also a relevant chapter in [MiRo].

11. Try general graph theory references, such as [BoMu]. A specialized article on this topic is [Go2].

12. There is a whole book on this topic, namely [La3]. Up-to-date information (such as the size of the largest traveling salesman problem that has been solved) can also be found on the Web; use a search engine.

13. Try [BoMu] or, better yet, books on graph algorithms like [ChOe] or [Ev2].

14. This would be a good time to search for "book number" in *Mathematical Reviews*, which is available on the Web as MathSciNet.

15. Try [Mi] or [SaKa].

16. See [ApHa] and [WoWi]. One of the major critics of computer-based proofs is the philosopher Thomas Tymoczko. You should look at relevant articles by him and others (such as [La2], [Ty1], [Ty2], and [Sw].) Another computer-assisted proof involved projective planes of order 10.

17. One article to look at is B. Manvel's "Extremely greedy coloring algorithms" in [HaMa], which is a conference proceedings. It will be an educational experience to browse through that volume, to see what research mathematicians do. Also, books on algorithmic graph theory, such as [Mc] or those mentioned above in the suggestion for Writing Project 13, will have some material.

18. There is a relevant article in [MiRo].

19. There are entire books on random graphs and, more generally, probabilistic methods in discrete mathematics. Two books to consult are [Bo2] and the more elementary (and fun) [Pal]. There is also a paper-length introduction, [Bo3]. For something more general, try [AlSp], although it is rather advanced. This is related to the probabilistic method, discussed in Section 6.2.

CHAPTER 10
Trees

SECTION 10.1 Introduction to Trees

These exercises give the reader experience working with tree terminology, and in particular with the relationships between the height and the numbers of vertices, leaves, and internal vertices of a tree. Exercise 13 should be done to get a feeling for the structure of trees. One good way to organize your enumeration of trees (such as all nonisomorphic trees with five vertices) is to focus on a particular parameter, such as the length of a longest path in the tree. This makes it easier to include all the trees and not count any of them twice. Review the theorems in this section before working the exercises involving the relationships between the height and the numbers of vertices, leaves, and internal vertices of a tree. For a challenge that gives a good feeling for the flavor of arguments in graph theory, the reader should try Exercise 43. In many ways trees are recursive creatures, and Exercises 45 and 46 are worth looking at in this regard.

1. a) This graph is connected and has no simple circuits, so it is a tree.
b) This graph is not connected, so it is not a tree.
c) This graph is connected and has no simple circuits, so it is a tree.
d) This graph has a simple circuit, so it is not a tree.
e) This graph is connected and has no simple circuits, so it is a tree.
f) This graph has a simple circuit, so it is not a tree.

3. a) Vertex a is the root, since it is drawn at the top.
b) The internal vertices are the vertices with children, namely a, b, c, d, f, h, j, q, and t.
c) The leaves are the vertices without children, namely e, g, i, k, l, m, n, o, p, r, s, and u.
d) The children of j are the vertices adjacent to j and below j, namely q and r.
e) The parent of h is the vertex adjacent to h and above h, namely c.
f) Vertex o has only one sibling, namely p, which is the other child of o's parent, h.
g) The ancestors of m are all the vertices on the unique simple path from m back to the root, namely f, b, and a.
h) The descendants of b are all the vertices that have b as an ancestor, namely e, f, l, m, and n.

5. This is not a full m-ary tree for any m. It is an m-ary tree for all $m \geq 3$, since each vertex has at most 3 children, but since some vertices have 3 children, while others have 1 or 2, it is not full for any m.

7. We can easily determine the levels from the drawing. The root a is at level 0. The vertices in the row below a are at level 1, namely b, c, and d. The vertices below that, namely e through k (in alphabetical order), are at level 2. Similarly l through r are at level 3, s and t are at level 4, and u is at level 5.

9. We describe the answers, rather than actually drawing pictures.
a) The subtree rooted at a is the entire tree, since a is the root.
b) The subtree rooted at c consists of five vertices—the root c, children g and h of this root, and grandchildren o and p—and the four edges cg, ch, ho, and hp.
c) The subtree rooted at e is just the vertex e.

11. We find the answer by carefully enumerating these trees, i.e., drawing a full set of nonisomorphic trees. One way to organize this work so as to avoid leaving any trees out or counting the same tree (up to isomorphism) more than once is to list the trees by the length of their longest simple path (or longest simple path from the root in the case of rooted trees).

a) There is only one tree with three vertices, namely $K_{1,2}$ (which can also be thought of as the simple path of length 2).

b) With three vertices, the longest path from the root can have length 1 or 2. There is only one tree of each type, so there are exactly two nonisomorphic rooted trees with 3 vertices, as shown below.

13. We find the answer by carefully enumerating these trees, i.e., drawing a full set of nonisomorphic trees. One way to organize this work so as to avoid leaving any trees out or counting the same tree (up to isomorphism) more than once is to list the trees by the length of their longest simple path (or longest simple path from the root in the case of rooted trees).

a) If the longest simple path has length 4, then the entire tree is just this path. If the longest simple path has length 3, then the fifth vertex must be attached to one of the middle vertices of this path. If the longest simple path has length 2, then the tree is just $K_{1,4}$. Thus there are only three trees with five vertices. They can be pictured as the first, second, and fourth pictures in the top row below.

b) For rooted trees of length 5, the longest path from the root can have length 1, 2, 3 or 4. There is only one tree with longest path of length 1 (the other four vertices are at level 1), and only one with longest path of length 4. If the longest path has length 3, then the fifth vertex (after using four vertices to draw this path) can be "attached" to either the root or the vertex at level 1 or the vertex at level 2, giving us three nonisomorphic trees. If the longest path has length 2, then there are several possibilities for where the fourth and fifth vertices can be "attached." They can both be adjacent to the root; they can both be adjacent to the vertex at level 1; one can be adjacent to the root and the other to the vertex at level 1; or one can be adjacent to the root and the other to this vertex: in all there are four possibilities in this case. Thus there are a total of nine nonisomorphic rooted trees on 5 vertices, as shown below.

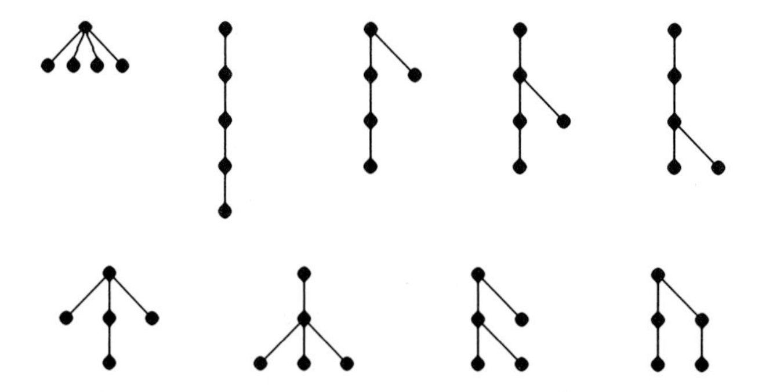

15. We will prove this statement using mathematical induction on n, the number of vertices of G. (This exercise can also be done by using Exercise 14 and Theorem 2 as the hint suggests—see the answer section of the text.) If $n = 1$, then there is only one possibility for G, it *is* a tree, it is connected, and it has $1 - 1 = 0$ edges. Thus the statement is true. Now let us assume that the statement is true for simple graphs with n vertices, and let G be a simple graph with $n + 1$ vertices.

There are two things to prove here. First let us suppose that G is a tree; we must show that G is connected and has $(n + 1) - 1 = n$ edges. Of course G is connected by definition. In order to prove that G

has the required number of edges, we need the following fact: a tree with at least one edge must contain a vertex of degree 1. (To see that this is so, let P be a simple path of greatest possible length; since the tree has no simple circuits, such a maximum length simple path exists. The ends of this path must be vertices of degree 1, since otherwise the simple path could be extended.) Let v be a vertex of degree 1 in G, and let G' be G with v and its incident edge removed. Now G' is still a tree: it has no simple circuits (since G had none) and it is still connected (the removed edge is clearly not needed to form paths between vertices different from v). Therefore by the inductive hypothesis, G', which has n vertices, has $n-1$ edges; it follows that G, which has one more edge than G', has n edges.

Conversely, suppose that G is connected and has n edges. If G is not a tree, then it must contain a simple circuit. If we remove one edge from this simple circuit, then the resulting graph (call it G') is still connected. If G' is a tree then we stop; otherwise we repeat this process. Since G had only finitely many edges to begin with, this process must eventually terminate at some tree T with $n+1$ vertices (T has all the vertices that G had). By the paragraph above, T therefore has n edges. But this contradicts the fact that we removed at least one edge of G in order to construct T. Therefore our assumption that G was not a tree is wrong, and our proof is complete.

17. Since a tree with n vertices has $n-1$ edges, the answer is 9999.

19. Each internal vertex has exactly 2 edges leading from it to its children. Therefore we can count the edges by multiplying the number of internal vertices by 2. Thus there are $2 \cdot 1000 = 2000$ edges.

21. We can model the tournament as a full binary tree. Each internal vertex represents the winner of the game played by its two children. There are 1000 leaves, one for each contestant. The root is the winner of the entire tournament. By Theorem 4(*iii*), with $m = 2$ and $l = 1000$, we see that $i = (l-1)/(m-1) = 999$. Thus exactly 999 games must be played to determine the champion.

23. Let P be a person sending out the letter. Then 10 people receive a letter with P's name at the bottom of the list (in the sixth position). Later 100 people receive a letter with P's name in the fifth position. Similarly, 1000 people receive a letter with P's name in the fourth position, and so on, until 1,000,000 people receive the letter with P's name in the first position. Therefore P should receive \$1,000,000. The model here is a full 10-ary tree.

25. No such tree exists. Suppose it did. By Theorem 4(*iii*), we know that a tree with these parameters must have $i = 83/(m-1)$ internal vertices. In order for this to be a whole number, $m-1$ must be a divisor of 83. Since 83 is prime, this means that $m = 2$ or $m = 84$. If $m = 2$, then we can have at most 15 vertices in all (the root, two at level 1, four at level 2, and eight at level 3). So m cannot be 2. If $m = 84$, then $i = 1$, which tells us that the root is the only internal vertex, and hence the height is only 1, rather than the desired 3. These contradictions tell us that no tree with 84 leaves and height 3 exists.

27. The complete binary tree of height 4 has 5 rows of vertices (levels 0 through 4), with each vertex not in the bottom row having two children. The complete 3-ary tree of height 3 has 4 rows of vertices (levels 0 through 3), with each vertex not in the bottom row having three children.

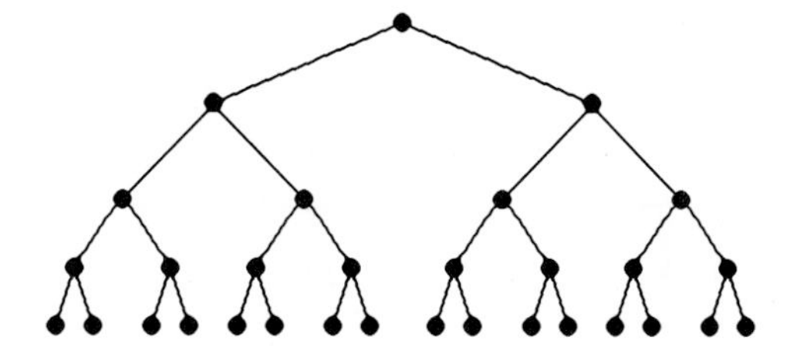

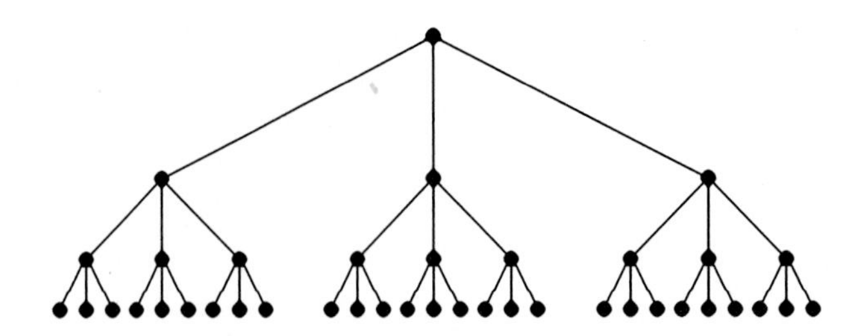

29. For both parts we use algebra on the equations $n = i + l$ (which is true by definition) and $n = mi + 1$ (which is proved in Theorem 3).

a) That $n = mi + 1$ is one of the given equations. For the second equality here, we have $l = n - i = (mi + 1) - i = (m - 1)i + 1$.

b) If we subtract the two given equations, then we obtain $0 = (1 - m)i + (l - 1)$, or $(m - 1)i = l - 1$. It follows that $i = (l - 1)/(m - 1)$. Then $n = i + l = [(l - 1)/(m - 1)] + l = (l - 1 + lm - l)/(m - 1) = (lm - 1)/(m - 1)$.

31. In each of the t trees, there is one fewer edge than there are vertices. Therefore altogether there are t fewer edges than vertices. Thus there are $n - t$ edges.

33. The number of isomers is the number of nonisomorphic trees with the given numbers of atoms. Since the hydrogen atoms play no role in determining the structure (they simply are attached to each carbon atom in sufficient number to make the degree of each carbon atom exactly 4), we need only look at the trees formed by the carbon atoms. In drawing our answers, we will show the tree of carbon atoms in heavy lines, with the hydrogen atom attachments in thinner lines.

a) There is only one tree with three vertices (up to isomorphism), the path of length 2. Thus the answer is 1. The heavy lines in this diagram of the molecule form this tree.

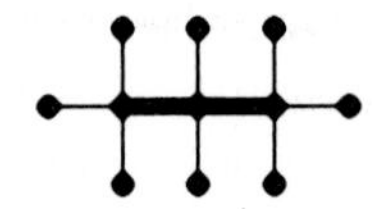

b) There are 3 nonisomorphic trees with 5 vertices: the path of length 4, the "star" $K_{1,4}$, and the tree that consists of a path of length 3 together with one more vertex attached to one of the middle vertices in the path. Thus the answer is 3. Again the heavy lines in the diagrams of the molecules form these trees.

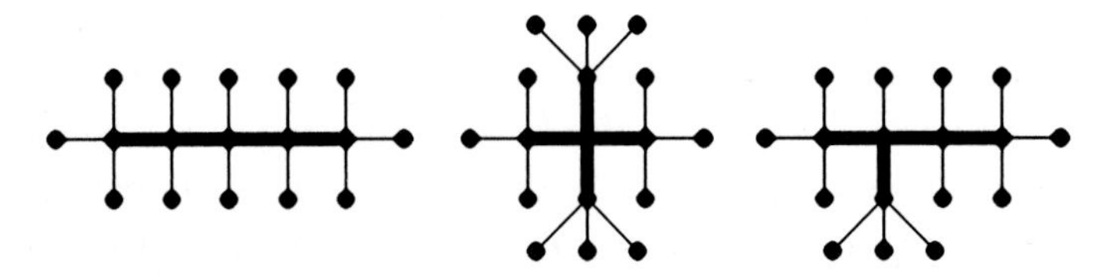

c) We need to find all the nonisomorphic trees with 6 vertices, except that we must not count the (one) tree with a vertex of degree 5 (since each carbon can only be attached to four other atoms). The complete set of trees is shown below (the heavy lines in these diagrams). Thus the answer is 5.

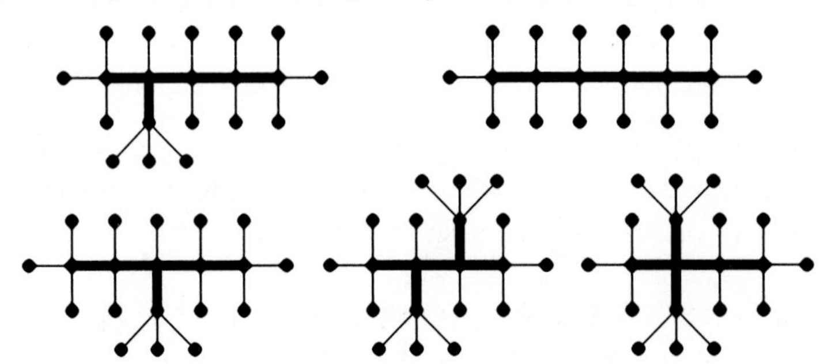

35. **a)** The parent of a vertex v is the directory in which the file or directory represented by v is contained.

b) The child of a vertex v (and v must represent a directory) is a file or directory contained in the directory that v represents.

c) If u and v are siblings, then the files or directories that u and v represent are in the same directory.

d) The ancestors of vertex v are all directories in the path from the root directory to the file or directory represented by v.

e) The descendants of a vertex v are all the files and directories either contained in v, or contained in directories contained in v, etc.

f) The level of a vertex v tells how far from the root directory is the file or directory represented by v.

g) The height of the tree is the greatest depth (i.e., level) at which a file or directory is buried in the system.

37. Suppose that $n = 2^k$, where k is a positive integer. We want to show how to add n numbers in $\log n$ steps using a tree-connected network of $n-1$ processors (recall that $\log n$ means $\log_2 n$). Let us prove this by mathematical induction on k. If $k = 1$ there is nothing to prove, since then $n = 2$ and $n - 1 = 1$, and certainly in $\log 2 = 1$ step we can add 2 numbers with 1 processor. Assume the inductive hypothesis, that we can add $n = 2^k$ numbers in $\log n$ steps using a tree-connected network of $n-1$ processors. Suppose now that we have $2n = 2^{k+1}$ numbers to add, $x_1, x_2, \ldots, x_{2n}$. The tree-connected network of $2n-1$ processors consists of the tree-connected network of $n-1$ processors together with two new processors as children of each leaf in the $(n-1)$-processor network. In one step we can use the leaves of the larger network to add $x_1 + x_2, x_3 + x_4, \ldots, x_{2n-1} + x_{2n}$. This gives us n numbers. By the inductive hypothesis we can now use the rest of the network to add these numbers using $\log n$ steps. In all, then, we used $1 + (\log n)$ steps, and, just as desired, $\log(2n) = \log 2 + \log n = 1 + \log n$. This completes the proof.

39. We need to compute the eccentricity of each vertex in order to find the center or centers. In practice, this does not involve much computation, since we can tell at a glance when the eccentricity is large. Intuitively, the center or centers are near the "middle" of the tree. The eccentricity of vertex c is 3, and it is the only vertex with eccentricity this small. Indeed, vertices a and b have eccentricities 4 and 5 (look at the paths to l); vertices d, f, g, j, and k all have eccentricities at least 4 (again look at the paths to l); and vertices e, h, i, and l also all have eccentricities at least 4 (look at the paths to k). Therefore c is the only center.

41. See the comments for the solution to Exercise 39. The eccentricity of vertices c and h are both 3. The eccentricities of the other vertices are all at least 4. Therefore c and h are the centers.

43. Certainly a tree has at least one center, since the set of eccentricities has a minimum value. First we prove that if u and v are any two distinct centers (say with minimum eccentricity e), then u and v are adjacent. Let P be the unique simple path from u to v. We will show that P is just u, v. If not, let c be any other vertex on P. Since the eccentricity of c is at least e, there is a vertex w such that the unique simple path Q from c to w has length at least e. This path Q may follow P for awhile, but once it diverges from P it cannot rejoin P without there being a simple circuit in the tree. In any case, Q cannot follow P towards both u and v, so suppose without loss of generality that it does not follow P towards u. Then the path from u to c and then on to w is simple and of length greater than e, a contradiction. Thus no such c exists, and u and v are adjacent.

Finally, to see that there can be no more than two centers, note that we have just proved that every two centers are adjacent. If there were three (or more) centers, then we would have a K_3 contained in the tree, contradicting the definition that a tree has no simple circuits.

45. We follow the recursive definition and produce the following pictures for T_3 through T_7 (of course T_1 and T_2

are both the tree with just one vertex). For example, T_3 has T_2 (a single vertex) as its left subtree and T_1 (again a single vertex) as its right subtree.

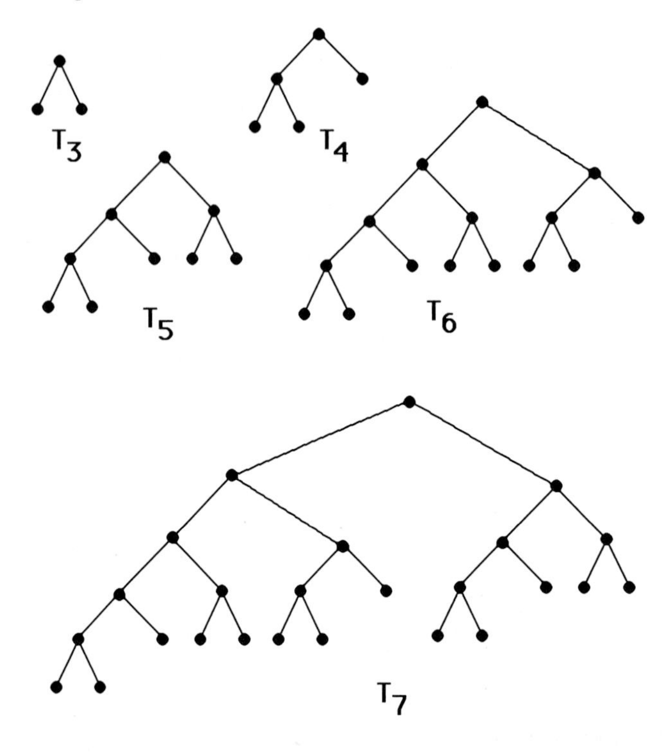

47. This "proof" shows that *there exists* a tree with n vertices having a path of length $n-1$. Note that the inductive step correctly takes the tree whose existence is guaranteed by the inductive hypothesis and correctly constructs a tree of the desired type. However, the statement was that *every* tree with n vertices has a path of length $n-1$, and this was not shown. A proof of the inductive step would need to start with an arbitrary tree with $n+1$ vertices and show that it had the required path. Of course no such proof is possible, since the statement is not true.

SECTION 10.2 Applications of Trees

Trees find many applications, especially in computer science. This section and subsequent ones deal with some of these applications. Binary search trees can be built up by adding new vertices one by one; searches in binary search trees are accomplished by moving down the tree until the desired vertex is found, branching either right or left as necessary. Huffman codes provide efficient means of encoding text in which some symbols occur more frequently than others; decoding is accomplished by moving down a binary tree. The coin-weighing problems presented here are but a few of the questions that can be asked. Try making up some of your own and answering them; it is easy to ask quite difficult questions of this type.

1. We first insert *banana* into the empty tree, giving us the tree with just a root, labeled *banana*. Next we insert *peach*, which, being greater than *banana* in alphabetical order, becomes the right child of the root. We continue in this manner, letting each new word find its place by coming down the tree, branching either right or left until it encounters a free position. The final tree is as shown.

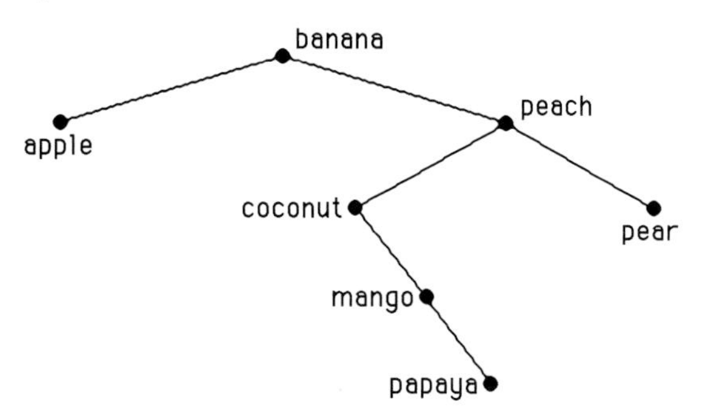

3. a) To find *pear*, we compare it with the root (*banana*), then with the right child of the root (*peach*), and finally with the right child of that vertex (*pear*). Thus 3 comparisons are needed.

b) Only 1 comparison is needed, since the item being searched for is the root.

c) We fail to locate *kumquat* by comparing it successively to *banana*, *peach*, *coconut*, and *mango*. Once we determine that *kumquat* should be in the left subtree of *mango*, and find no vertices there, we know that *kumquat* is not in the tree. Thus 4 comparisons were used.

d) This one is similar to part **(c)**, except that 5 comparisons are used. We compare *orange* successively to *banana*, *peach*, *coconut*, *mango*, and *papaya*.

5. We follow exactly the same procedure as in Exercise 1. The only unusual point is that the word "the" appears later in the sentence, after it is already in the tree. The algorithm finds that it is already in the tree, so it is not inserted again. The tree is shown below.

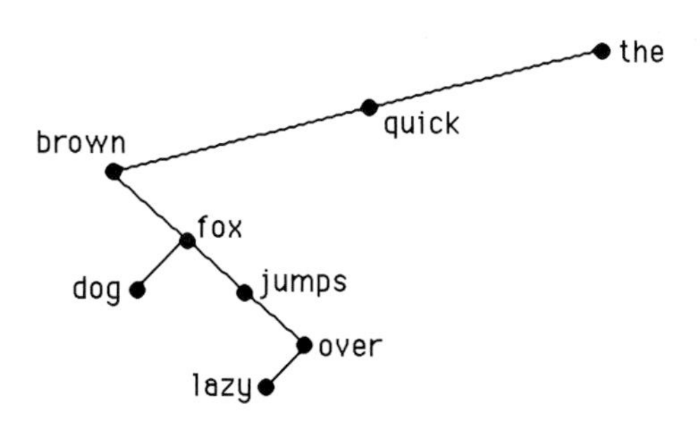

7. Since there are 4 different outcomes to the testing procedure, we need at least 2 weighings, since one weighing will give us only 3 possible outcomes (a decision tree of height 1 has only 3 leaves). Here is how to find the counterfeit coin with 2 weighings. Let us call the coins A, B, C, and D. First compare coins A and B. If they balance, then the counterfeit is among the other two. In this case, compare C with A; if they balance, then D is counterfeit; if they do not, then C is counterfeit. On the other hand if A and B do not balance, then one of them is the counterfeit. Again compare C with A. If they balance, then B is the counterfeit; if they do not, then A is counterfeit.

9. Since there are 12 different outcomes to the testing procedure, we need at least 3 weighings, since 2 weighings will only give us 9 possible outcomes (a decision tree of height 2 has only 9 leaves). Here is one way to find the counterfeit coin with 3 weighings. Divide the coins into three groups of 4 coins each, and compare two of the groups. If they balance, then the counterfeit is among the other four coins. If they do not balance, then the counterfeit is among the four coins registering lighter. In either case we have narrowed the search to 4 coins. Now by Example 3 we can, in 2 more weighings, find the counterfeit among these four coins and four of the good ones.

11. By Theorem 1 in this section, at least $\lceil \log 4! \rceil$ comparisons are needed. Since $\log_2 24 \approx 4.6$, at least five comparisons are required. We can accomplish the sorting with five comparisons as follows. (This is essentially just merge sort, as discussed in Section 4.4.) Call the elements a, b, c, and d. First compare a and b; then compare c and d. Without loss of generality, let us assume that $a < b$ and $c < d$. (If not, then relabel the elements after these comparisons.) Next we compare a and c (this is our third comparison). Whichever is smaller is the smallest element of the set. Again without loss of generality, suppose $a < c$. Now we merely need to compare b with both c and d to completely determine the ordering. This takes two more comparisons, giving us the desired five in all.

13. The first two steps are shown in the text. After 22 has been identified as the second largest element, we replace the leaf 22 by $-\infty$ in the tree and recalculate the winner in the path from the leaf where 22 used to be up to the root. The result is as shown here.

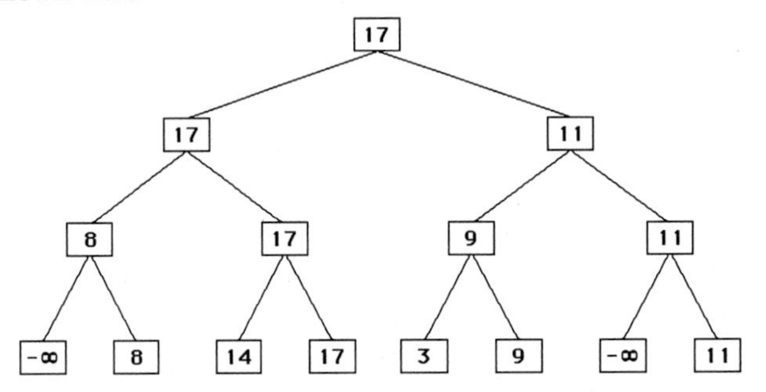

Now we see that 17 is the third largest element, so we repeat the process: replace the leaf 17 by $-\infty$ and recalculate. This gives us the following tree.

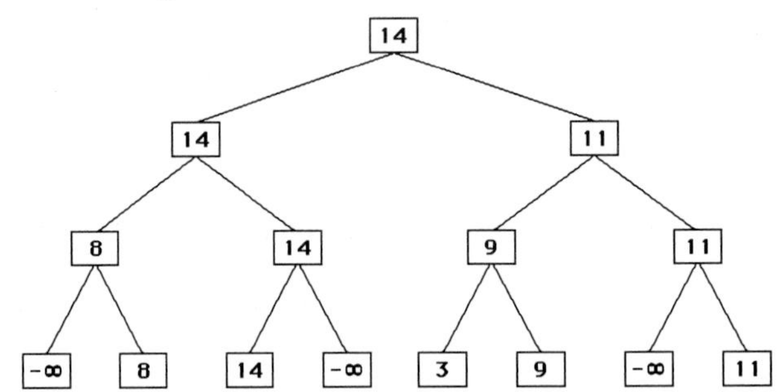

Thus we see that 14 is the fourth largest element, so we repeat the process: replace the leaf 14 by $-\infty$ and recalculate. This gives us the following tree.

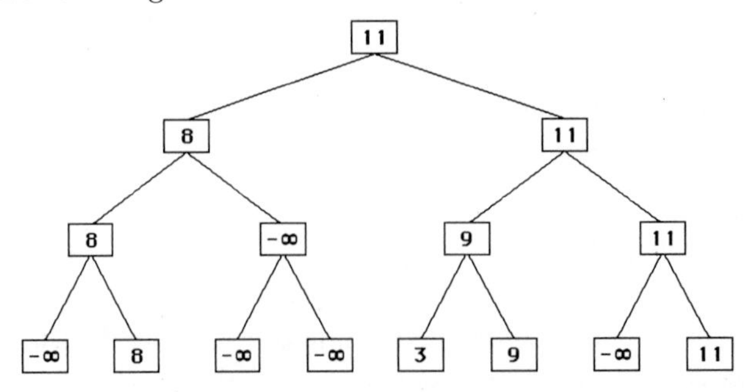

Thus we see that 11 is the fifth largest element, so we repeat the process: replace the leaf 11 by $-\infty$ and recalculate. This gives us the following tree.

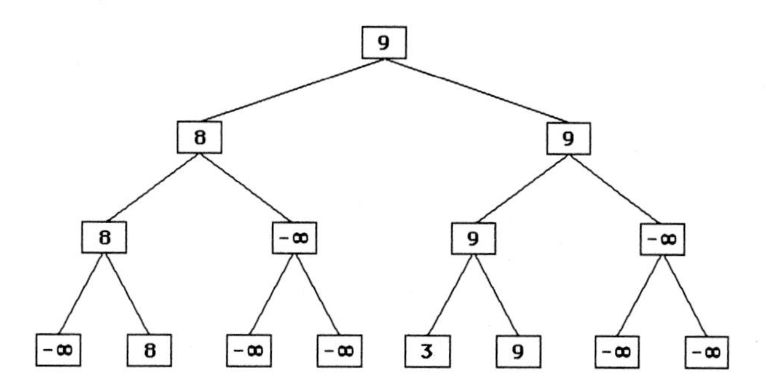

The process continues in this manner. The final tree will look like this, as we determine that 3 is the eighth largest element.

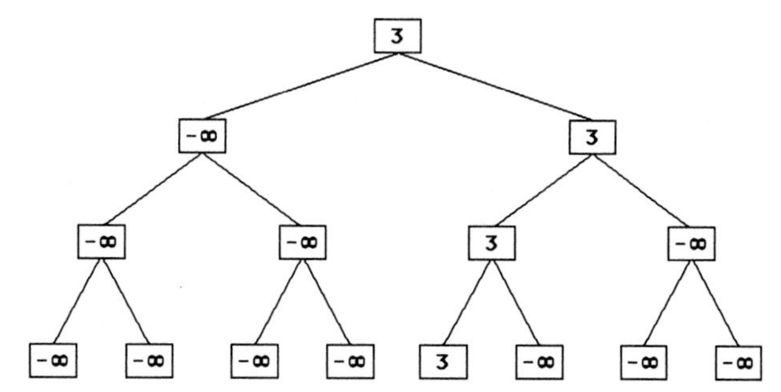

15. The heart of the matter is how to do the comparisons to work our way up the tree. We assume a data structure that allows us to access the left and right child of each vertex, as well as identify the root. We also need to keep track of which leaf is the winner of each contest so as to be able to find that leaf in one step. Each vertex of the tree, then, will have a value and a label; the value is the list element currently there, and the label is the name (i.e., location) of the leaf responsible for that value. We let $k = \lceil \log n \rceil$, which will be the height of the tree. The first part of the algorithm constructs the initial tree. The rest successively picks out the winners and recalculates the tree.

procedure *tournament sort*$(a_1, \ldots, a_n)$
$k := \lceil \log n \rceil$
build a binary tree of height k
for $i := 1$ **to** n
  set the value of the ith leaf to be a_i and its label to be itself
for $i := n+1$ **to** 2^k
  set the value of the ith leaf to be $-\infty$ and its label to be itself
for $i := k-1$ **downto** 0
  for each vertex v at level i
    set the value of v to the larger of the values of its children
    and its label to be the label of the child with the larger value
for $i := 1$ **to** n
begin
  $c_i :=$ value at the root
  let v be the label of the root
  set the value of v to be $-\infty$
    while the label at the root is still v
    begin
      $v := parent(v)$
      set the value of v to the larger of the values of its children
      and its label to be the label of the child with the larger value
    end
end $\{c_1, \ldots, c_n$ is the list in nonincreasing order$\}$

17. At each stage after the initial tree has been set up, only k comparisons are needed to recalculate the values at the vertices. (We are still assuming that $n = 2^k$ as in Exercise 16.) A leaf is replaced by $-\infty$, and, starting with that leaf's parent, one comparison is made for each vertex on the path up to the root. We can actually get by with $k-1$ comparisons, because it does not take a comparison to determine that the new leaf's sibling beats it, so no comparison is needed for the new leaf's parent.

19. **a)** This is a prefix code, since no code is the first part of another.
b) This is not a prefix code, since, for instance, the code for a is the first part of the code for t.
c) This is a prefix code, since no code is the first part of another.
d) This is a prefix code, since no code is the first part of another.

21. The code for a letter is simply the labels on the edges in the path from the root to that letter. Since the path from the root to a goes through the three edges leading left each time, all labeled 0, the code for a is 000. Similarly the codes for e, i, k, o, p and u are 001, 01, 1100, 1101, 11110 and 11111, respectively.

23. We follow Algorithm 2. Since b and c are the symbols of least weight, they are combined into a subtree, which we will call T_1 for discussion purposes, of weight $0.10 + 0.15 = 0.25$, with the larger weight symbol, c, on the left. Now the two trees of smallest weight are the single symbol a and either T_1 or the single symbol d (both have weight 0.25). We break the tie arbitrarily in favor of T_1, and so we get a tree T_2 with left subtree T_1 and right subtree a. (If we had broken the tie in the other way, our final answer would have been different, but it would have been just as correct, and the average number of bits to encode a character would be the same.) The next step is to combine e and d into a subtree T_3 of weight 0.55. And the final step is to combine T_2 and T_3. The result is as shown.

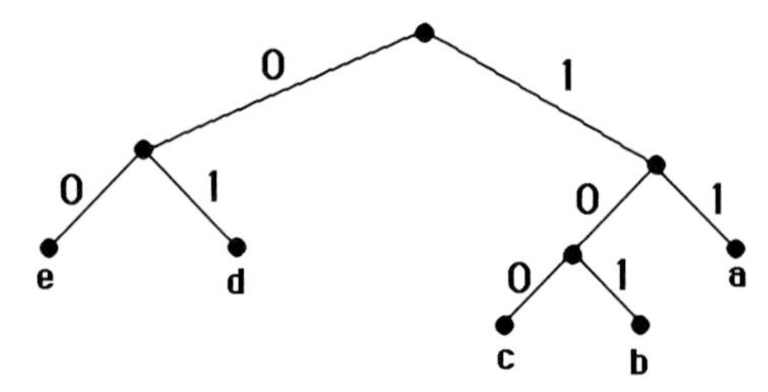

We see by looking at the tree that a is encoded by 11, b by 101, c by 100, d by 01, and e by 00. To compute the average number of bits required to encode a character, we multiply the number of bits for each letter by the weight of that latter and add. Since a takes 2 bits and has weight 0.20, it contributes 0.40 to the sum. Similarly b contributes $3 \cdot 0.10 = 0.30$. In all we get $2 \cdot 0.20 + 3 \cdot 0.10 + 3 \cdot 0.15 + 2 \cdot 0.25 + 2 \cdot 0.30 = 2.25$. Thus on the average, 2.25 bits are needed per character. Note that this is an appropriately weighted average, weighted by the frequencies with which the letters occur.

25. We proceed as in Exercise 23. The first step combines t and v (in either order—we have a choice here). At the second step we can combine this subtree either with u or with w. There are four possible answers in all, the one shown here and three more obtained from this one by swapping t and v, swapping u and w, or making both of these swaps.

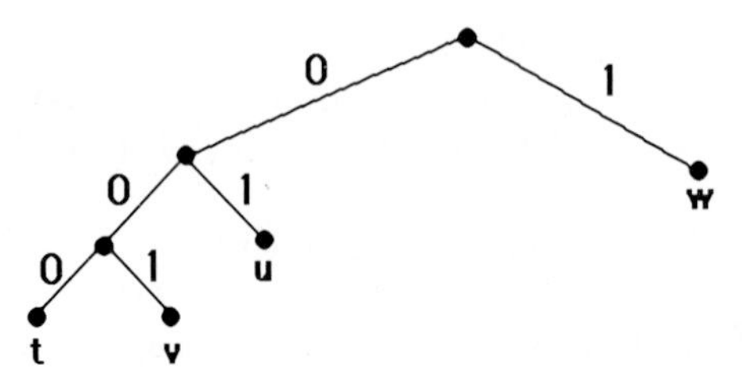

27. This is a computationally intensive exercise. We will not show the final picture here, because it is too complex. However, we show the steps below (read row by row), with the obvious notation that we are joining the tree on the left of the plus sign to the tree on the right, to produce the new tree of a certain weight. The procedure is the same as in Exercise 23. (The sum of the frequencies was only 0.9999 due to round-off.)

$\mathrm{Q}+\mathrm{Z} \to T_1(0.0014)$, $T_1+\mathrm{J} \to T_2(0.0024)$, $T_2+\mathrm{X} \to T_3(0.0039)$, $\mathrm{K}+T_3 \to T_4(0.0119)$
$T_4+\mathrm{V} \to T_5(0.0221)$, $\mathrm{P}+\mathrm{B} \to T_6(0.0301)$, $\mathrm{F}+\mathrm{G} \to T_7(0.0391)$, $T_5+\mathrm{Y} \to T_8(0.0432)$
$\mathrm{W}+\mathrm{C} \to T_9(0.0512)$, $T_6+\mathrm{M} \to T_{10}(0.0578)$, $T_7+\mathrm{U} \to T_{11}(0.0695)$, $\mathrm{D}+\mathrm{L} \to T_{12}(0.0828)$
$T_9+T_8 \to T_{13}(0.0944)$, $T_{10}+\mathrm{R} \to T_{14}(0.1150)$, $\mathrm{N}+\mathrm{S} \to T_{15}(0.1290)$, $\mathrm{I}+\mathrm{H} \to T_{16}(0.1357)$
$\mathrm{O}+T_{11} \to T_{17}(0.1476)$, $T_{12}+\mathrm{A} \to T_{18}(0.1645)$, $T_{13}+\mathrm{T} \to T_{19}(0.1849)$
$\mathrm{E}+T_{14} \to T_{20}(0.2382)$, $T_{16}+T_{15} \to T_{21}(0.2647)$, $T_{18}+T_{17} \to T_{22}(0.3121)$
$T_{20}+T_{19} \to T_{23}(0.4231)$, $T_{22}+T_{21} \to T_{24}(0.5768)$, $T_{24}+T_{23} \to T_{25}(0.9999)$

We show in the following table the resulting codes and the product of code length and frequency. Thus we see that the sum of the last column—the average number of bits required—is 4.2013. Note that this is slightly better than if we used five bits per letter (which we can do, since there are fewer than $2^5 = 32$ letters).

letter	code	length	frequency	product
A	0001	4	0.0817	0.3268
B	101001	6	0.0145	0.0870
C	11001	5	0.0248	0.1240
D	00000	5	0.0431	0.2155
E	100	3	0.1232	0.3696
F	001100	6	0.0209	0.1254
G	001101	6	0.0182	0.1092
H	0101	4	0.0668	0.2672
I	0100	4	0.0689	0.2756
J	110100101	9	0.0010	0.0090
K	1101000	7	0.0080	0.0560
L	00001	5	0.0397	0.1985
M	10101	5	0.0277	0.1385
N	0110	4	0.0662	0.2648
O	0010	4	0.0781	0.3124
P	101000	6	0.0156	0.0936
Q	1101001000	10	0.0009	0.0090
R	1011	4	0.0572	0.2288
S	0111	4	0.0628	0.2512
T	111	3	0.0905	0.2715
U	00111	5	0.0304	0.1520
V	110101	6	0.0102	0.0612
W	11000	5	0.0264	0.1320
X	11010011	8	0.0015	0.0120
Y	11011	5	0.0211	0.1055
Z	1101001001	10	0.0005	0.0050

29. Here $N = 6$ and $m = 3$, so $((N-1) \textbf{ mod } (m-1)) + 1 = 2$. Thus we start by combining the two symbols with smallest weights, **N** and **R**, into a subtree T_1 with weight 0.15. Next we combine the three items with smallest weights, namely **Z**, T_1, and **T**, into a subtree T_2 with weight 0.45. Finally we construct the tree, as shown here.

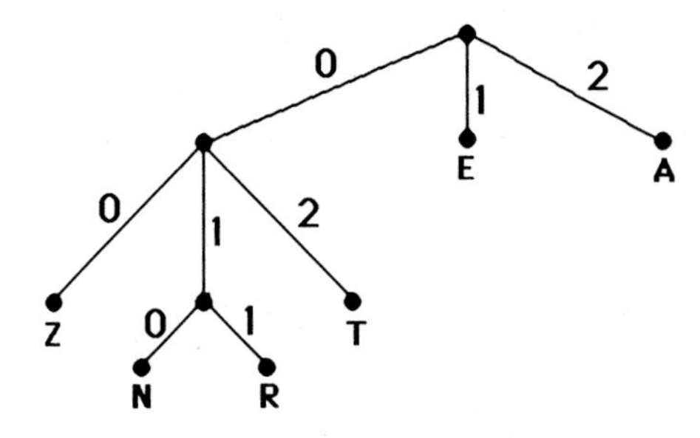

The codes, then, are as follows: A:2; E:1; N:010; R:011; T:02; Z:00.

31. We play around with this problem for small values of n, constructing the tree for the Huffman code, and we see that it will look like this (where $n = 6$).

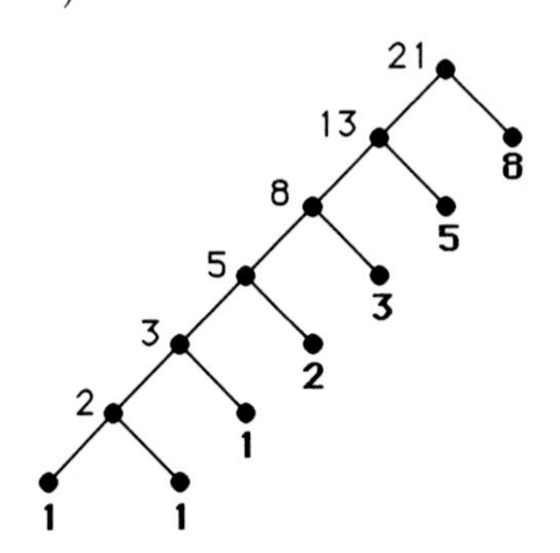

The numbers in bold font are the frequencies of the symbols, and the numbers in plain font are the weights of the subtrees. Note that at each stage the tree constructed so far is paired with the next symbol. As we see from the picture (and can be proved rigorously by induction), the maximum number of bits used to encode a symbol is n.

33. The procedure is similar to Examples 6 and 8. We draw the game tree, showing the positions within squares or circles. Since the tree is rather large, we have indicated in some places to "see text." Refer to Figure 9; the subtree rooted at these square or circle vertices is exactly the same as the corresponding subtree in Figure 9. Since the value of the root is $+1$, the first player wins the game following the optimal (minmax) strategy, by first moving to the position $2\,2$, and then leaving one pile with one stone at her second (final) move. (The figure is shown on the next page for spacing reasons.)

35. **a)** If we draw the game tree, then we see that the first player has only one winning move at her first play, namely to remove the three stones and win immediately. Thus the payoff is \$1.

b) If we draw the game tree, then we see that the first player has only one winning move at her first play, namely to remove two stones from the pile with four stones, leaving two piles of two stones each. Whatever the second player does, the first player must leave just one stone at her second move if she wants to win the game. Therefore the game will end in three moves, so the payoff to the first player is \$3.

c) If we draw the game tree we see that the first player is doomed to lose if the second player plays optimally. Since her payoff will be negative, she wants the game to end as quickly as possible. Therefore she should remove the pile of three stones, forcing her opponent to remove the pile of two stones if he wants to win the game. Therefore the game will end in two moves, so the payoff to the first player is $-\$3$.

NOTE: This is the figure for Exercise 33.

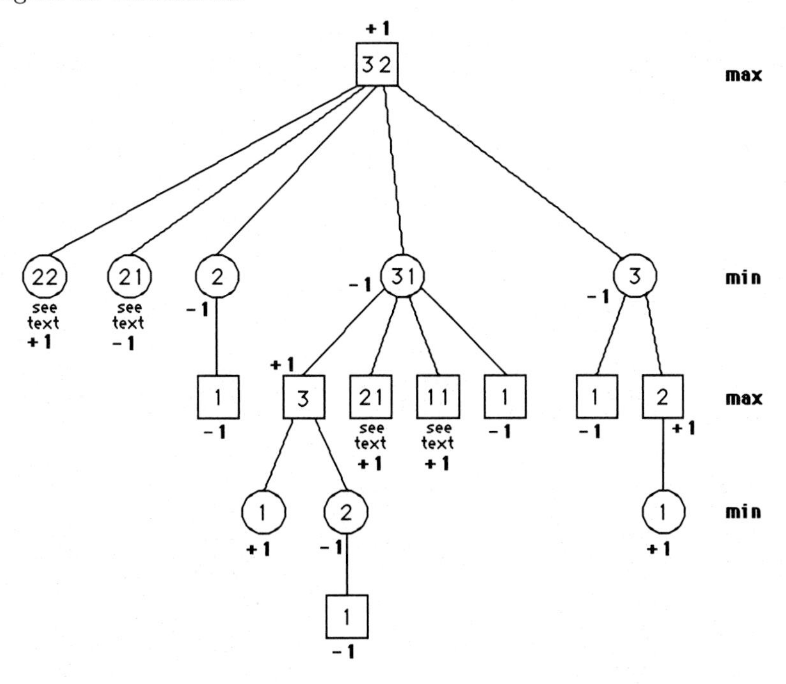

37. See the next page for the figures. Note that the second player (O) moves at the root of this subtree in each of parts **(a)** and **(b)**, and the first player (X) moves at the root of this subtree in part **(c)**. The squares and circles enclosing the positions have been suppressed for readability.

a) The value is 0, since the game must end in a draw.

b) The value is again 0, since O will choose the right branch.

c) The value is 1, since the first player will make her row of X's.

d) This is a trick question. This position cannot have occurred in a game. Note that X has three in a row, so the game was over at the point at which X made her third move. Therefore O did not make a third move, and this picture is impossible.

39. We prove this by strong induction. For the basis step, when $n = 2$ stones are in each pile, the first player is up a creek. If she takes two stones from a pile, then the second player takes one stone from the remaining pile and wins. If she takes one stone from a pile, then the second player takes two stones from the other pile and again wins. Assume the inductive hypothesis that the second player can always win if the game starts with two piles of j stones for all j between 2 and k, inclusive, where $k \geq 2$, and consider a game with two piles containing $k+1$ stones each. If the first player takes all the stones from one of the piles, then the second player takes all but one stone from the remaining pile and wins. If the first player takes all but one stone from one of the piles, then the second player takes all the stones from the other pile and again wins. Otherwise the first player takes some stones from one of the piles, leaving j stones in that pile, where $2 \leq j \leq k$, and $k+1$ stones in the other pile. The second player then takes the same number of stones from the larger pile, also leaving j stones there. At this point the game consists of two piles of j stones each, where $2 \leq j \leq k$. By the inductive hypothesis, the second player in that game, who is also the second player in our actual game, can win, and the proof by strong induction is complete.

41. In the game of checkers, each player has 12 checkers (pieces), occupying the black squares nearest her in three rows on an 8×8 checkerboard. According to the rules, a checker may move to an adjacent black square diagonally, toward the other player. Thus only the front row of checkers can move at the start, and by looking at the board, we see that one of these four checkers has only one possible move, but the others each have two choices. This gives a total of seven moves for the first player, so the root of the game tree has seven children. For each move by the first player, the second player is free to move his front row in the same manner, so each vertex at Level 1 has seven children, giving $7 \cdot 7 = 49$ grandchildren of the root of this game tree.

NOTE: These are the figures for Exercise 37.

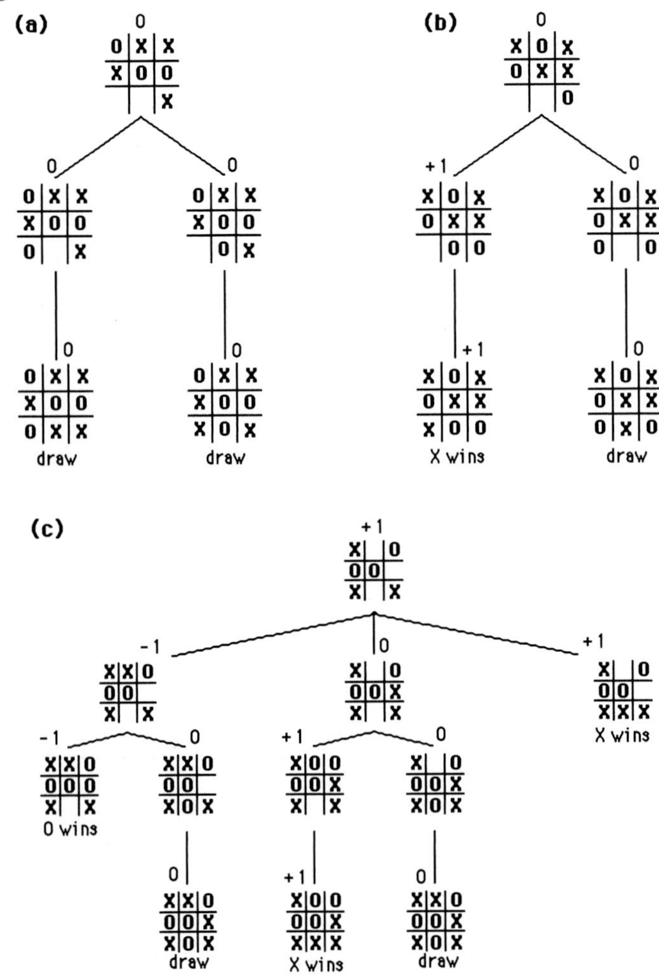

43. The game tree is too large to fit in one picture, so we show first the root, its children, and its grandchildren corresponding to a move that has only two responses (up to symmetry), and then the subtrees corresponding to the next move for O for each of the other two possible moves for X. See Figure 8 in the text. We label the second trees first, where we apply the evaluation function as described (calculation is shown), and assign the root of each subtree the minimum of the values of its children (again shown only up to symmetry). The value at the root of the tree is the maximum of the values of the three children, which is 1. Thus the first player, using this heuristic, should make her first move in the center.

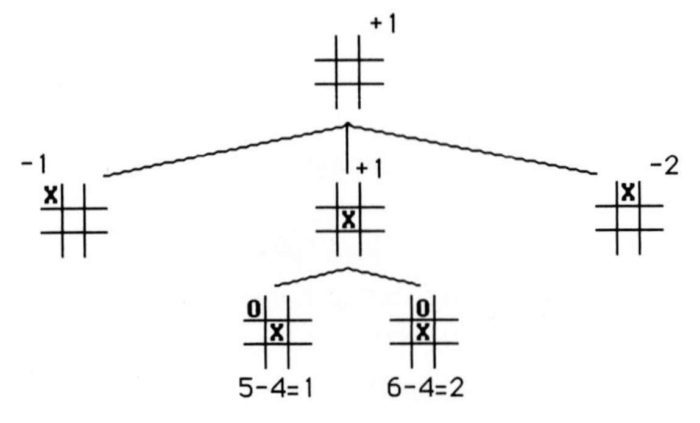

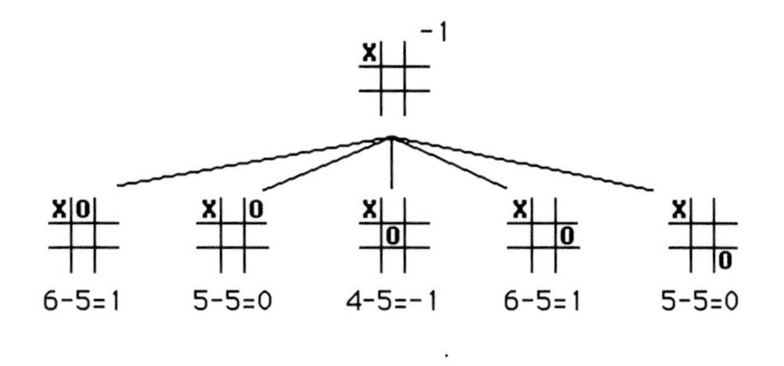

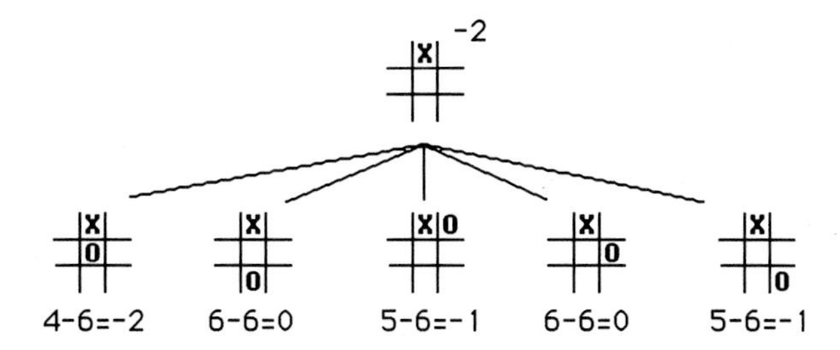

SECTION 10.3 Tree Traversal

Tree traversal is central to computer science applications. Trees are such a natural way to represent arithmetical and algebraic formulae, and so easy to manipulate, that it would be difficult to imagine how computer scientists could live without them. To see if you really understand the various orders, try Exercises 26 and 27. You need to make your mind work recursively for tree traversals: when you come to a subtree, you need to remember where to continue after processing the subtree. It is best to think of these traversals in terms of the recursive algorithms (shown as Algorithms 1, 2, and 3). A good bench-mark for testing your understanding of recursive definitions is provided in Exercises 30–34.

1. The root of the tree is labeled 0. The children of the root are labeled 1, 2, ..., from left to right. The children of a vertex labeled α are labeled $\alpha.1$, $\alpha.2$, ..., from left to right. For example, the two children of the vertex 1 here are 1.1 and 1.2. We completely label the tree in this manner, from the top down. See the figure. The lexicographic order of the labels is the preorder of the vertices: after each vertex come the subtrees rooted at its children, from left to right. Thus the order is $0 < 1 < 1.1 < 1.2 < 2 < 3$.

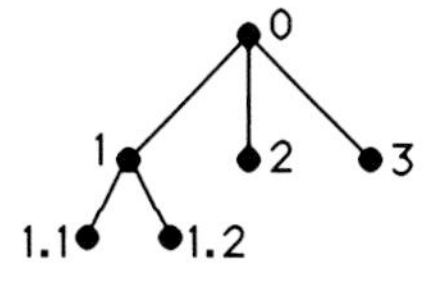

3. See the comments for the solution to Exercise 1. The order is $0 < 1 < 1.1 < 1.2 < 1.2.1 < 1.2.1.1 < 1.2.1.2 < 1.2.2 < 1.2.3 < 1.2.3.1 < 1.2.3.2 < 1.2.3.2.1 < 1.2.3.2.2 < 1.2.3.3 < 2 < 2.1$.

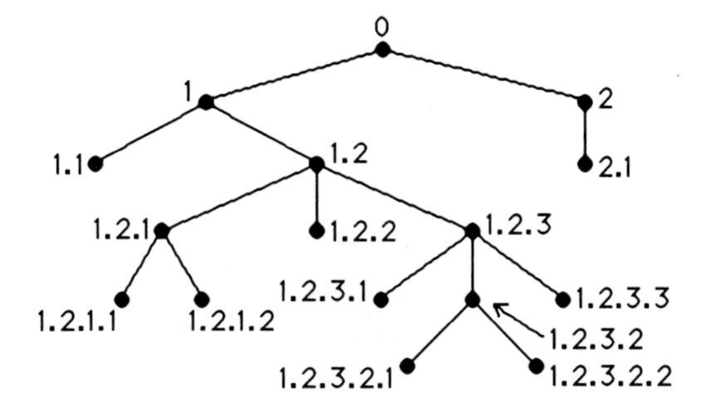

5. The given information tells us that the root has two children. We have no way to tell how many vertices are in the subtree of the root rooted at the first of these children. Therefore we have no way to tell how many vertices are in the tree.

7. In preorder, the root comes first, then the left subtree in preorder, then the right subtree in preorder. Thus the preorder is a, followed by the vertices of the left subtree (the one rooted at b) in preorder, then c. Recursively, the preorder in the subtree rooted at b is b, followed by d, followed by the vertices in the subtree rooted at e in preorder, namely e, f, g. Putting this all together, we obtain the answer a, b, d, e, f, g, c.

9. See the comments in the solution to Exercise 7 for the procedure. The only difference here is that some vertices have more than two children: after listing such a vertex, we list the vertices of its subtrees, in preorder, from left to right. The answer is $a, b, e, k, l, m, f, g, n, r, s, c, d, h, o, i, j, p, q$.

11. Inorder traversal requires that the left-most subtree be traversed first, then the root, then the remaining subtrees (if any) from left to right. Applying this principle, we see that the list must start with the left subtree in inorder. To find this, we need to start with *its* left subtree, namely d. Next comes the root of that subtree, namely b, and then the right subtree in inorder. This is i, followed by the root e, followed by the subtree rooted at j in inorder. This latter listing is m, j, n, o. We continue in this manner, ultimately obtaining: $d, b, i, e, m, j, n, o, a, f, c, g, k, h, p, l$.

13. In postorder, the root comes last, following the left subtree in postorder and the right subtree in postorder. Thus the postorder is the vertices of the left subtree (the one rooted at b) in postorder, then c, then a. Recursively, the postorder in the subtree rooted at b is d, followed by the vertices in the subtree rooted at e in postorder, namely f, g, e, followed by b. Putting this all together, we obtain the answer d, f, g, e, b, c, a.

15. This is just like Exercises 13 and 14. Note that all subtrees of a vertex are completed before listing that vertex. The answer is $k, l, m, e, f, r, s, n, g, b, c, o, h, i, p, q, j, d, a$.

17. a) For the first expression, we note that the outermost operation is the second addition. Therefore the root of the tree is this plus sign, and the left and right subtrees are the trees for the expressions being added. The first operand is the sum of x and xy, so the left subtree has a plus sign for its root and the tree for the expressions x and xy as its subtrees. We continue in this manner until we have drawn the entire tree. The second tree is done similarly. Note that the only difference between these two expressions is the placement of parentheses, and yet the expressions represent quite different operations, as can be seen from the fact that the trees are quite different.

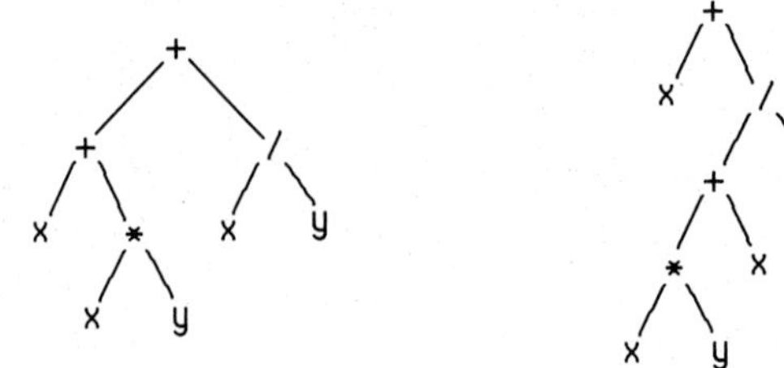

b) We can read off the answer from the picture we have just drawn simply by listing the vertices of the tree in preorder: First list the root, then the left subtree in preorder, then the right subtree in preorder. Therefore the answer is $+ + x * x\, y\, /\, x\, y$. Similarly, the second expression in prefix notation is $+ x\, /\, + * x\, y\, x\, y$.

c) We can read off the answer from the picture we have just drawn simply by listing the vertices of the tree in postorder: First list the left subtree in postorder, then the right subtree in postorder, then the root. Therefore the answer is $x\, x\, y * + x\, y\, / +$. Similarly, the second expression in postfix notation is $x\, x\, y * x + y\, / +$.

d) The infix expression is just the given expression, fully parenthesized, with an explicit symbol for multiplication. Thus the first is $((x + (x * y)) + (x/y))$, and the second is $(x + (((x * y) + x)/y))$. This corresponds to traversing the tree in inorder, putting in a left parenthesis whenever we go down to a left child and putting in a right parenthesis whenever we come up from a right child.

19. This is similar to Exercise 17, with set operations rather than arithmetic ones.

a) We construct the tree in the same way we did there, noting, for example, that the first minus is the outermost operation.

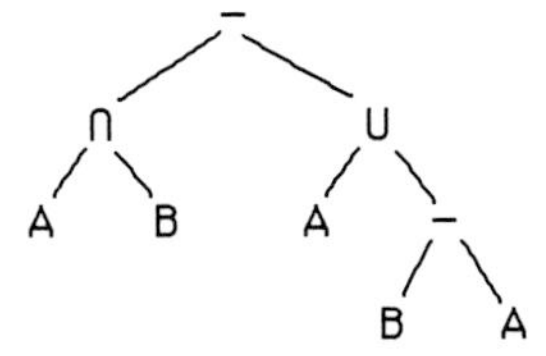

b) The prefix expression is obtained by traversing the tree in preorder: $- \cap A\,B \cup A - B\,A$.

c) The postfix expression is obtained by traversing the tree in postorder: $A\,B \cap A\,B\,A - \cup -$.

d) This is already in fully parenthesized infix notation except for needing an outer set of parentheses: $((A \cap B) - (A \cup (B - A)))$.

21. Either of the four operators can be the outermost one, so there are four cases to consider. If the first operator is the outermost one, then we need to compute the number of ways to fully parenthesize $B - A \cap B - A$. Here there are 5 possibilities: 1 in which the "$\cap$" symbol is the outermost operator and 2 with each of the "$-$" symbols as the outermost operator. If the second operator in our original expression is the outermost one, then the only choice is in the parenthesization of the second of its operands, and there are 2 possibilities. Thus there are a total 7 ways to parenthesize this expression if either of the first two operators are the outermost one. By symmetry there are another 7 if the outermost operator is one of the last two. Therefore the answer to the problem is 14.

23. We show how to do these exercises by successively replacing the first occurrence of an operator immediately followed by two operands with the result of that operation. (This is an alternative to the method suggested in the text, where the *last* occurrence of an operator, which is necessarily preceded by two operands, is acted upon first.) The final number is the value of the entire prefix expression. In part **(a)**, for example, we first replace $/\ 8\ 4$ by the result of dividing 8 by 4, namely 2, to obtain $-\ *\ 2\ 2\ 3$. Then we replace $*\ 2\ 2$ by the result of multiplying 2 and 2, namely 4, to obtain the third line of our calculation. Next we replace $-\ 4\ 3$ by its answer, 1, which is the final answer.

a)

$$- * 2 / 8\ 4\ 3$$
$$- * 2\ 2\ 3$$
$$- 4\ 3$$
$$1$$

b)

$$\uparrow - * 3\ 3 * 4\ 2\ 5$$
$$\uparrow - 9 * 4\ 2\ 5$$
$$\uparrow - 9\ 8\ 5$$
$$\uparrow 1\ 5$$
$$1$$

c)

$$+ - \uparrow 3\ 2 \uparrow 2\ 3\ /\ 6 - 4\ 2$$
$$+ - 9 \uparrow 2\ 3\ /\ 6 - 4\ 2$$
$$+ - 9\ 8\ /\ 6 - 4\ 2$$
$$+\ 1\ /\ 6 - 4\ 2$$
$$+\ 1\ /\ 6\ 2$$
$$+\ 1\ 3$$
$$4$$

d)

$$* + 3 + 3 \uparrow 3 + 3\ 3\ 3$$
$$* + 3 + 3 \uparrow 3\ 6\ 3$$
$$* + 3 + 3\ 729\ 3$$
$$* + 3\ 732\ 3$$
$$*\ 735\ 3$$
$$2205$$

25. We slowly use the clues to fill in the details of this tree, shown below. Since the preorder starts with a, we know that a is the root, and we are told that a has four children. Next, since the first child of a comes immediately after a in preorder, we know that this first child is b. We are told that b has one child, and it must be f, which comes next in the preorder. We are told that f has no children, so we are now finished with the subtree rooted at b. Therefore the second child of a must be c (the next vertex in preorder). We continue in this way until we have drawn the entire tree.

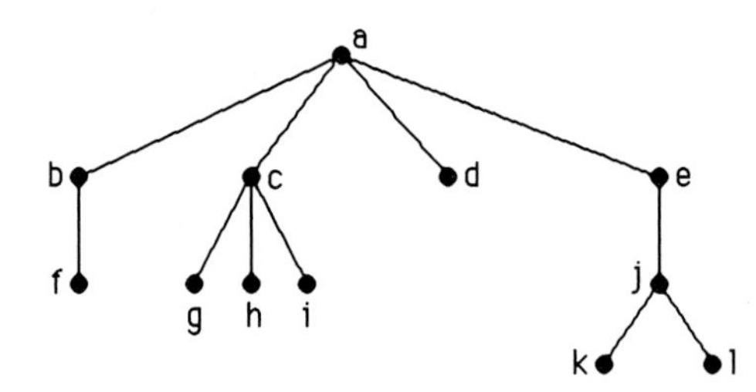

27. We prove this by induction on the length of the list. If the list has just one element, then the statement is trivially true. For the inductive step, consider the end of the list. There we find a sequence of vertices, starting with the last leaf and ending with the root of the tree, each vertex being the last child of its successor in the list. We know where this sequence starts, since we are told the number of children of each vertex: it starts at the last leaf in the list. Now remove this leaf, and decrease the child count of its parent by 1. The result is the postorder and child counts of a tree with one fewer vertex. By the inductive hypothesis we can uniquely determine this smaller tree. Then we can uniquely determine where the deleted vertex goes, since it is the last child of its parent (whom we know).

29. In each case the postorder is c, d, b, f, g, h, e, a.

31. We prove this by induction on the recursive definition, in other words, on the length of the formula, i.e., the total number of symbols and operators. The only formula of length 1 arises from the base case of the recursive definition (part (i)), and in that case we have one symbol and no operators, so the statement is true. Assume that the statement is true for formulae of length less than $n > 1$, and let F be a formula of length n. Then F arises from part (*ii*) of the definition, so F consists of $* X Y$, for some operator $*$ and some formulae X and Y. By the inductive hypothesis, the number of symbols in X exceeds the number of operators there by 1, and the same holds for Y. If we add and note that there is one more operator in F than in X and Y combined, then we see that the number of symbols in F exceeds the number of operators in F by 1, as well.

33. Any string of length n, using these six characters, is a well-formed formula as long as two conditions are met: if we read the string from left to right, the number of symbols is always at least 1 greater than the number of operators; and in all there is one more symbol than operator. We are asked to write down six such strings, with $n \geq 7$. One such set is $x\,x\,x\,x+++$, $x\,x\,x\,x\,x++++$, $x\,x+x\,x++$, $x\,x\,x\,x+x\,x++++$, $x\,x\,x+x\,x+++$, and $x\,x+x\,x\,x+++$.

SECTION 10.4 Spanning Trees

The spanning tree algorithms given here provide systematic methods for searching through graphs, and they are the foundation of many other, more complicated, algorithms. The concept of a spanning tree is quite simple and natural, of course: the problem comes with finding spanning trees efficiently. The reader should pay attention to the exercises on backtracking (Exercises 28–32) to get a feel both for the ideas behind it and for its inefficiency for large problems.

1. The graph has m edges. The spanning tree has $n-1$ edges. Therefore we need to remove $m-(n-1)$ edges.

3. We have to remove edges, one at a time. We can remove any edge that is part of a simple circuit. The answer is by no means unique. For example, we can start by removing edge $\{a,d\}$, since it is in the simple circuit $adcba$. Then we might choose to remove edge $\{a,g\}$. We can continue in this way and remove all of these edges: $\{b,e\}$, $\{b,f\}$, $\{b,g\}$, $\{d,e\}$, $\{d,g\}$, and $\{e,g\}$. At this point there are no more simple circuits, so we have a spanning tree.

5. This is similar to Exercise 3. Here is one possible set of removals: $\{a,b\}$, $\{a,d\}$, $\{a,f\}$, $\{b,c\}$, $\{b,d\}$, $\{c,d\}$, $\{d,e\}$, $\{d,g\}$, $\{d,j\}$, $\{e,g\}$, $\{e,j\}$, $\{f,g\}$, $\{h,j\}$, $\{h,k\}$, and $\{j,l\}$. As a check, note that there are 12 vertices and 26 edges. A spanning tree must have 11 edges, so we need to remove 15 of them, as we did.

7. In each case we show the original graph, with a spanning tree in heavier lines. These were obtained by trial and error. In each case except part **(c)**, our spanning tree is a simple path (but other answers are possible). In part **(c)**, of course, the graph is its own spanning tree.

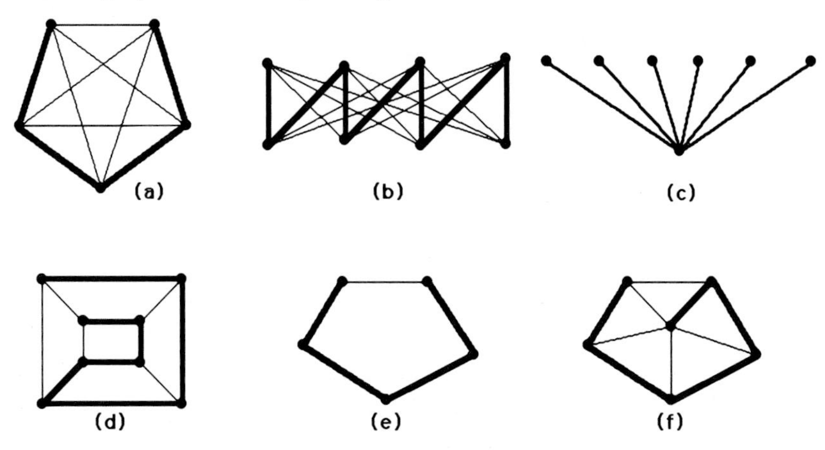

9. We can remove any one of the four edges in the square on the left, together with any one of the four edges in the square on the right. Therefore there are $4 \cdot 4 = 16$ different spanning trees, shown here.

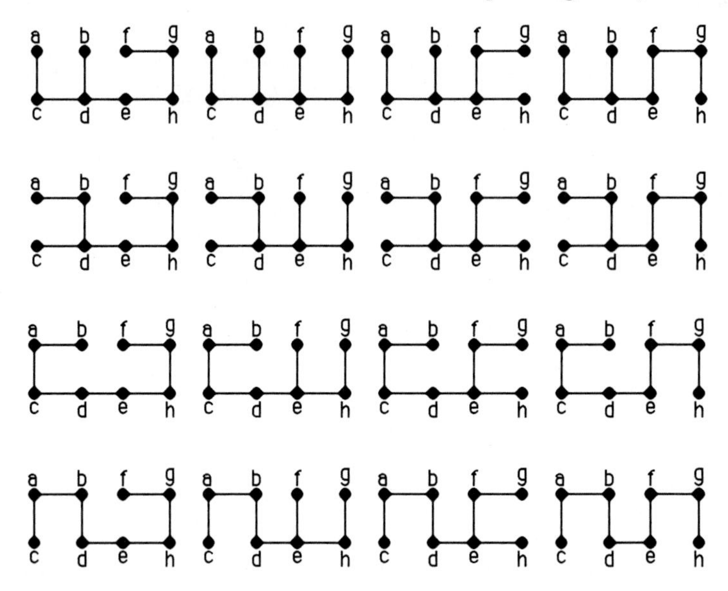

11. We approach this problem in a rather ad hoc way.

a) Every pair of edges in K_3 forms a spanning tree, so there are $C(3,2) = 3$ such trees.

b) There are 16 spanning trees; careful counting is required to see this. First, let us note that the trees can take only two shapes: the star $K_{1,3}$ and the simple path of length 3. There are 4 different spanning trees of the former shape, since any of the four vertices can be chosen as the vertex of degree 3. There are $P(4,4) = 24$ orders in which the vertices can be listed in a simple path of length 3, but since the path can be traversed in either of two directions to yield the same tree, there are only 12 trees of this shape. Therefore there are $4 + 12 = 16$ spanning trees of K_4 altogether.

c) Note that $K_{2,2} = C_4$. A tree is determined simply by deciding which of the four edges to remove. Therefore there are 4 spanning trees.

d) By the same reasoning as in part **(c)**, there are 5 spanning trees.

13. If we start at vertex a and use alphabetical order, then the depth-first search spanning tree is unique. We start at vertex a and form the path shown in heavy lines to vertex i before needing to backtrack. There are no unreached vertices from vertex h at this point, but there is an unreached vertex (j) adjacent to vertex g. Thus the tree is as shown in heavy lines.

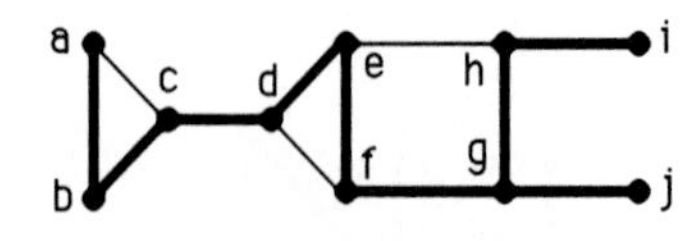

15. The procedure is the same as in Exercise 13. The spanning tree is shown in heavy lines.

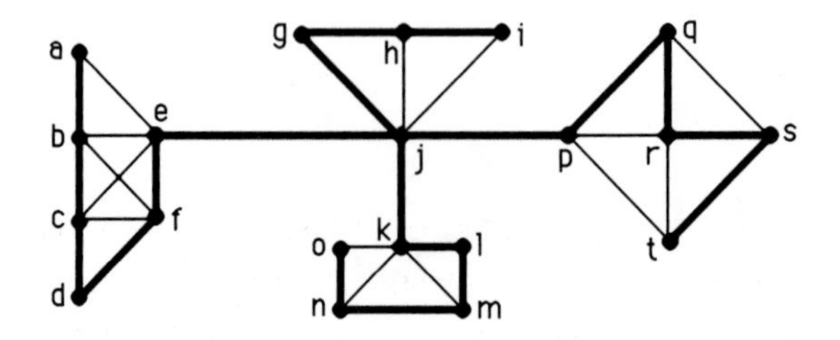

17. **a)** We start at the vertex in the middle of the wheel and visit a neighbor—one of the vertices on the rim. From there we move to an adjacent vertex on the rim, and so on all the way around until we have reached every vertex. Thus the resulting spanning tree is a path of length 6.

b) We start at any vertex, visit a neighbor, then a new neighbor, and so on until we have reached every vertex. Thus the resulting spanning tree is a path of length 5.

c) We start at a vertex in the part with four vertices, move to a vertex in the part with three vertices, then back to a vertex in the larger part, and so on. The resulting spanning tree is a path of length 6.

d) Depending on what order we choose to visit the vertices, we may or may not need to backtrack in doing this depth-first search. In most cases, the resulting tree is a path of length 7, but we could, for example, come up with the following tree.

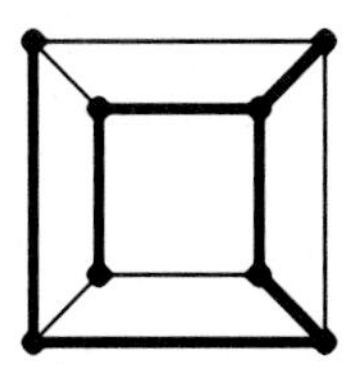

19. With breadth-first search, the initial vertex is the middle vertex, and the n spokes are added to the tree as this vertex is processed. Thus the resulting tree is $K_{1,n}$. With depth-first search, we start at the vertex in the middle of the wheel and visit a neighbor—one of the vertices on the rim. From there we move to an adjacent vertex on the rim, and so on all the way around until we have reached every vertex. Thus the resulting spanning tree is a path of length n.

21. With breadth-first search, we fan out from a vertex of degree m to all the vertices of degree n as the first step. Next a vertex of degree n is processed, and the edges from it to all the remaining vertices of degree m are added. The result is what is called a "double star": a $K_{1,n-1}$ and a $K_{1,m-1}$ with their centers joined by an edge. With depth-first search, we travel back and forth from one partite set to the other until we can go no further. If $m = n$ or $m = n - 1$, then we get a path of length $m + n - 1$. Otherwise, the path ends while some vertices in the larger partite set have not been visited, so we back up one link in the path to a vertex v and then successively visit the remaining vertices in that set from v. The result is what is called a "broom": a path with extra pendant edges coming out of one end of the path (our vertex v).

23. The question is simply asking for a spanning tree of the graph shown. There are of course many such spanning trees. One that the airline would probably not like to choose is the tree that consists of the path Bangor, Boston, New York, Detroit, Chicago, Washington, Atlanta, St. Louis, Dallas, Denver, San Diego, Los Angeles, San Francisco, Seattle. All the other 18 flights, then, would be discontinued. This set of flights would not be useful to a New York to Washington flyer, for example, nor to one who wants to fly from Chicago to Seattle. A more practical approach would be to build the tree up from nothing, using key short flights, such as the one between Detroit and Chicago. After 13 such edges had been chosen, without creating any simple circuits, we would have the desired spanning tree, and the other flights would be discontinued.

25. A connected simple graph has only one spanning tree if the graph is itself a tree (clearly its only spanning tree is itself in this case). On the other hand, if a connected simple graph is not a tree, then it has a simple circuit containing $k \geq 3$ edges, and one can find a spanning tree of the graph containing any $k - 1$ of these edges but not the other (see Exercise 23 in Section 10.5).

27. We prove this statement by induction on the length of a shortest path from v to u. If this length is 0, then $v = u$, and indeed, v is at the root of the tree. Assume that the statement is true for all vertices w for which a shortest path to v has length n, and let u be a vertex for which a shortest path has length $n + 1$. Clearly

u cannot be at a level less than $n+1$, because then a shorter path would be evident in the tree itself. On the other hand, if we let w be the penultimate vertex in a shortest path from v to u of length $n+1$, then by the inductive hypothesis, we know that w is at level n of the tree. Now when vertex w was being processed by the breadth-first search algorithm, either u was already in the tree (and therefore adjacent to a vertex at level at most n) or u was one of the vertices put into the tree adjacent to w. In either case, u is adjacent to a vertex at level at most n and therefore is at level at most $n+1$.

29. Label the squares of the $n \times n$ chessboard with coordinates (i, j), where i and j are integers from 1 to n, inclusive.

a) For the 3×3 board, we start our search by placing a queen in square $(1,1)$. The only possibility for a queen in the second column is square $(3,2)$. Now there is no place to put a queen in the third column. Therefore we backtrack and try placing the first queen in square $(2,1)$. This time there is no place to put a queen in the second column. By symmetry, we need not consider the initial choice of a queen in square $(3,1)$ (it will be just like the situation for the queen in square $(1,1)$, turned upside down). Therefore we have shown that there is no solution.

b) We start by placing a queen in square $(1,1)$. The first place a queen might then reside in the second column is square $(3,2)$, so we place a queen there. Now the only free spot in the third column is $(5,3)$, the only free spot in the fourth column is $(2,4)$, and the only free spot in the fifth column is $(4,5)$. This gives us a solution. Note that we were lucky and did not need to backtrack at all to find this solution.

c) The portion of the decision tree corresponding to placing the first queen in square $(1,1)$ is quite large here, and it leads to no solution. For example, the second queen can be in any of the squares $(3,2)$, $(4,2)$, $(5,2)$, or $(6,2)$. If the second queen is in square $(3,2)$, then the third can be in squares $(5,3)$ or $(6,3)$. After several backtracks we find that there is no solution with one queen in square $(1,1)$. Next we try square $(2,1)$ for the first queen. After a few more backtracks, we are led to the solution in which the remaining queens are in squares $(4,2)$, $(6,3)$, $(1,4)$, $(3,5)$ and $(5,6)$.

31. Assume that the graph has vertices v_1, v_2, $\ldots$, v_n. In looking for a Hamilton circuit we may as well start building a path at v_1. The general step is as follows. We extend the path if we can, to a new vertex (or to v_1 if this will complete the Hamilton circuit) adjacent to the vertex we are at. If we cannot extend the path any further, then we backtrack to the last previous vertex in the path and try other untried extensions from that vertex. The procedure for Hamilton paths is the same, except that we have to try all possible starting vertices, and we do not allow a return to the starting vertex, stopping instead when we have a path of the right length.

33. We know that every component of the graph has a spanning tree. The union of these spanning trees is clearly a spanning forest for the graph, since it contains every vertex and two vertices in the same component are joined by a path in the spanning tree for that component.

35. First we claim that the spanning forest will use $n-c$ edges in all. To see this, let n_i be the number of vertices in the ith component, for $i = 1, 2, \ldots, c$. The spanning forest uses $n_i - 1$ edges in that component. Therefore the spanning forest uses $\sum(n_i - 1) = (\sum n_i) - c = n - c$ edges in all. Thus we need to remove $m - (n - c)$ edges to form a spanning forest.

37. In effect we use the depth-first search algorithm on each component. In more detail, once that procedure wants to stop, have it search through the list of vertices in the graph to find one that is not yet in the forest. If there is such a vertex, then repeat the process starting from that vertex. Continue this until all the vertices have been included in the forest.

39. If an edge uv is not followed while we are processing vertex u during the depth-first search process, then it can only be the case that the vertex v had already been visited. There are two cases. If vertex v was visited after we started processing u, then, since we are not finished processing u yet, v must appear in the subtree rooted at u (and hence must be a descendant of u). On the other hand, if the processing of v had already begun before we started processing u, then why wasn't this edge followed at that time? It must be that we had not finished processing v, in other words, that we are still forming the subtree rooted at v, so u is a descendant of v, and hence v is an ancestor of u.

41. Certainly these two procedures produce the identical spanning trees if the graph we are working with is a tree itself, since in this case there is only one spanning tree (the whole graph). This is the only case in which that happens, however. If the original graph has any other edges, then by Exercise 39 they must be back edges and hence join a vertex to an ancestor or descendant, whereas by Exercise 40, they must connect vertices at the same level or at levels that differ by 1. Clearly these two possibilities are mutually exclusive. Therefore there can be no edges other than tree edges if the two spanning trees are to be the same.

43. Since the edges not in the spanning tree are not followed in the process, we can ignore them. Thus we can assume that the graph was a rooted tree to begin with. The basis step is trivial (there is only one vertex), so we assume the inductive hypothesis that breadth-first search applied to trees with n vertices have their vertices visited in order of their level in the tree and consider a tree T with $n+1$ vertices. The last vertex to be visited during breadth-first search of this tree, say v, is the one that was added last to the list of vertices waiting to be processed. It was added when its parent, say u, was being processed. We must show that v is at the lowest (bottom-most, i.e., numerically greatest) level of the tree. Suppose not; say vertex x, whose parent is vertex w, is at a lower level. Then w is at a lower level than u. Clearly v must be a leaf, since any child of v could not have been seen before v is seen. Consider the tree T' obtained from T by deleting v. By the inductive hypothesis, the vertices in T' must be processed in order of their level in T' (which is the same as their level in T, and the absence of v in T' has no effect on the rest of the algorithm). Therefore u must have been processed before w, and therefore v would have joined the waiting list before x did, a contradiction. Therefore v is at the bottom-most level of the tree, and the proof is complete.

45. We modify the pseudocode given in Algorithm 2 by initializing m to be 0 at the beginning of the algorithm, and adding the statements "$m := m + 1$" and "assign m to vertex v" after the statement that removes vertex v from L.

47. This is similar to Exercise 39. If a directed edge uv is not followed while we are processing its tail u during the depth-first search process, then it can only be the case that its head v had already been visited. There are three cases. If vertex v was visited after we started processing u, then, since we are not finished processing u yet, v must appear in the subtree rooted at u (and hence must be a descendant of u), so we have a forward edge. Otherwise, the processing of v must have already begun before we started processing u. If it had not yet finished (i.e., we are still forming the subtree rooted at v), then u is a descendant of v, and hence v is an ancestor of u (we have a back edge). Finally, if the processing of v had already finished, then by definition we have a cross edge.

49. There are five trees here, so there are $C(5,2) = 10$ questions. Let T be the tree in Figure 3c, and let T_1 through T_4 be the trees in Figure 4, reading from left to right. We will discuss one of the pairs at length and simply report the other answers. Let $d(T, T_1)$ denote the distance between trees T and T_1. Note that tree T_1 has edges $\{a, e\}$, $\{c, g\}$, and $\{e, f\}$ that tree T does not. Since the trees have the same number of edges, there must also be 3 edges in T that are not in T_1 (we do not need to list them). Therefore $d(T, T_1) = 6$. Similarly we have $d(T, T_2) = 4$, $d(T, T_3) = 4$, $d(T, T_4) = 2$, $d(T_1, T_2) = 4$, $d(T_1, T_3) = 4$, $d(T_1, T_4) = 6$, $d(T_2, T_3) = 4$, $d(T_2, T_4) = 2$, and $d(T_3, T_4) = 4$.

51. Let $e_1 = \{u, v\}$. The graph $T_2 \cup \{e_1\}$ contains a (unique) simple circuit C containing edge e_1. Now $T_1 - \{e_1\}$ has two components, one of which contains u and the other of which contains v. We travel along the circuit C, starting at u and not using edge e_1 first, until we first reach a vertex in the component of $T_1 - \{e_1\}$ that contains v; obviously we must reach such a vertex eventually, since we eventually reach v itself. The edge we last traversed is e_2. Clearly $T_2 \cup \{e_1\} - \{e_2\}$ is a tree, since e_2 is on C. On the other hand, $T_1 - \{e_1\} \cup \{e_2\}$ is also a tree, since e_2 reunites the two components of $T_1 - \{e_1\}$.

53. Rooted spanning trees are easy to find in all six figures, as these pictures show. There are of course many other possible correct answers. In the first five cases the tree is a path. In the last case, the root is d.

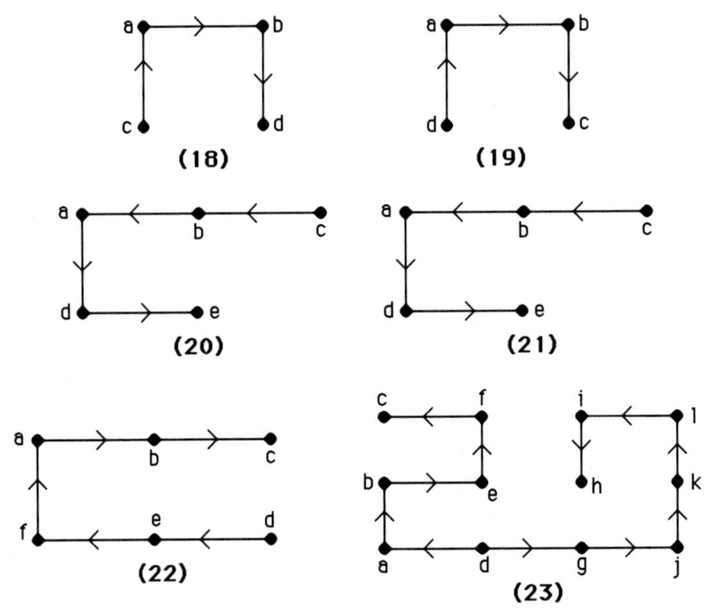

55. By Exercise 16 in Section 9.5, we know that such a directed graph has an Euler circuit. Now we traverse the Euler circuit, starting at some vertex v (which will be our root), and delete from the circuit every edge that has as its terminal vertex a vertex we have already visited on this traversal. The graph that remains is a rooted spanning tree; there is a path from the root to every other vertex, and there can be no simple circuits.

57. According to Exercise 56, a directed graph contains a circuit if and only if there are any back edges. We can detect back edges as follows. Add a marker on each vertex v to indicate what its status is: not yet seen (the initial situation), seen (i.e., put into T) but not yet finished (i.e., $visit(v)$ has not yet terminated), or finished (i.e., $visit(v)$ has terminated). A few extra lines in Algorithm 1 will accomplish this bookkeeping. Then to determine whether a directed graph has a circuit, we just have to check when looking at edge uv whether the status of v is "seen." If that ever happens, then we know there is a circuit; if not, then there is no circuit.

SECTION 10.5 Minimum Spanning Trees

The algorithms presented here are not hard, once you understand them. The two algorithms are almost identical, the only real difference being in the set of edges available for inclusion in the tree at each step. In Prim's algorithm, only those edges that are adjacent to edges already in the tree (and not completing simple circuits) may be added (so that, as a result, all the intermediate stages are trees). In Kruskal's algorithm, any edge that does not complete a simple circuit may be added (so that the intermediate stages may be forests and not trees). The reader might try to discover other methods for finding minimum spanning trees, in addition to the ones in this section.

1. We want a minimum spanning tree in this graph. We apply Kruskal's algorithm and pave the following edges: Oasis to Deep Springs, Lida to Gold Point, Lida to Goldfield, Silverpeak to Goldfield, Oasis to Dyer, Oasis to Silverpeak, Manhattan to Tonopah, Goldfield to Tonopah, Gold Point to Beatty, and Tonopah to Warm Springs. At each stage we chose the minimum weight edge whose addition did not create a simple circuit (Kruskal's algorithm).

3. We start with the minimum weight edge $\{e, f\}$. The least weight edges incident to the tree constructed so far are edges $\{c, f\}$ and $\{e, h\}$, each with weight 3, so we add one of them to the tree (we will break ties using alphabetical order, so we add $\{c, f\}$). Next we add edge $\{e, h\}$, and then edge $\{h, i\}$, which has a smaller weight but has just become eligible for addition. The edges continue to be added in the following order (note that ties are broken using alphabetical order): $\{b, c\}$, $\{b, d\}$, $\{a, d\}$, and $\{g, h\}$. The total weight of the minimum spanning tree is 22.

5. Kruskal's algorithm will have us include first the links from Atlanta to Chicago, then Atlanta to New York, then Denver to San Francisco (the cheapest links). The next cheapest link, from Chicago to New York, cannot be included, since it would form a simple circuit. Therefore we next add the link from Chicago to San Francisco, and our network is complete.

7. The edges are added in the following order (with Kruskal's algorithm, we add at each step the shortest edge that will not complete a simple circuit): $\{e, f\}$, $\{a, d\}$, $\{h, i\}$, $\{b, d\}$, $\{c, f\}$, $\{e, h\}$, $\{b, c\}$, and $\{g, h\}$. The total weight of the minimum spanning tree is 22.

9. A graph with one edge obviously cannot be the solution, and a graph with two edges cannot either, since a simple connected graph with two edges must be a tree. On the other hand, if we take a triangle (K_3) and weight all the edges equally, then clearly there are three different minimum spanning trees.

11. If we simply replace each of the occurrences of the word "minimum" with the word "maximum" in Algorithm 1, then the resulting algorithm will find a maximum spanning tree.

13. We use an analog of Kruskal's algorithm, adding at each step an edge of greatest weight that does not create a simple circuit. The answer here is unique. It uses edges $\{a, c\}$, $\{b, d\}$, $\{b, e\}$, and $\{c, e\}$.

15. There are numerous possible answers. One uses edges $\{a, d\}$, $\{b, f\}$, $\{c, g\}$, $\{d, p\}$, $\{e, f\}$, $\{f, j\}$, $\{g, k\}$, $\{h, l\}$, $\{i, j\}$, $\{i, m\}$, $\{j, k\}$, $\{j, n\}$, $\{k, l\}$, $\{k, o\}$, and $\{o, p\}$, obtained by choosing at each step an edge of greatest weight that does not create a simple circuit.

17. If we want a second "shortest" spanning tree (which may, of course, have the same weight as the "shortest" tree), then we need to use at least one edge not in some minimum spanning tree T that we have found. One way to force this is, for each edge e of T, to apply a minimum spanning tree algorithm to the graph with e deleted, and then take a tree of minimum weight among all of these. It cannot equal T, and so it must be a second "shortest" spanning tree.

19. The proof that Prim's algorithm works shows how to take any minimum spanning tree T and, if T is not identical to the tree constructed by Prim's algorithm, to find another minimum spanning tree with even more edges in common with the Prim tree than T has. The core of the proof is in the last paragraph, where we add an edge e_{k+1} to T and delete an edge e. Now if all the edges have different weights, then the result of this process is not another minimum spanning tree but an outright contradiction. We conclude that there are no minimum spanning trees with any edges not in common with the Prim tree, i.e., that the only minimum spanning tree is the Prim tree.

21. We simply apply Kruskal's algorithm, starting not from the empty tree but from the tree containing these two edges. We can add the following edges to form the desired tree: $\{c,d\}$, $\{k,l\}$, $\{b,f\}$, $\{c,g\}$, $\{a,b\}$, $\{f,j\}$, $\{a,e\}$, $\{g,h\}$, and $\{b,c\}$.

23. The algorithm is identical to Kruskal's algorithm (Algorithm 2), except that we replace the statement "$T :=$ empty graph" by the assignment to T initially of the specified set of edges, and, instead of iterating from 1 to $n-1$, we iterate from 1 to $n-1-s$, where s is the number of edges in the specified set. It is assumed that the specified set of edges forms no simple circuits.

25. **a)** First we need to find the least expensive edges incident to each vertex. These are the links from New York to Atlanta, Atlanta to Chicago, and Denver to San Francisco. The algorithm tells us to choose all of these edges. At the end of this first pass, then, we have a forest of two trees, one containing the three eastern cities, the other containing the two western cities. Next we find the least expensive edge joining these two trees, namely the link from Chicago to San Francisco, and add it to our growing forest. We now have a spanning tree, and the algorithm has finished. Note, incidentally, that this is the same spanning tree that we obtained in Example 1; by the result of Exercise 19, since the weights in this graph are all different, there was only one minimum spanning tree.
b) On the first pass, we choose all the edges that are the minimum weight edges at each vertex. This set consists of $\{a,b\}$, $\{b,f\}$, $\{c,d\}$, $\{a,e\}$, $\{c,g\}$, $\{g,h\}$, $\{i,j\}$, $\{f,j\}$, and $\{k,l\}$. At this point the forest has three components. Next we add the lowest weight edges connecting these three components, namely $\{h,l\}$ and $\{b,c\}$, to complete our tree.

27. Let $e_1, e_2, \ldots, e_{n-1}$ be the edges of the tree S chosen by Sollin's algorithm in the order chosen (arbitrarily order the edges chosen at the same stage). Let T be a minimum spanning tree that contains all the edges $e_1, e_2, \ldots, e_k$ for as large a k as possible. Thus $0 \le k \le n-1$. If $k = n-1$, then $S = T$ and we have shown that S is a minimum spanning tree. Otherwise we will construct another minimum spanning tree T' which contains edges $e_1, e_2, \ldots, e_k, e_{k+1}$, contradicting the choice of T and completing the proof.

Let S' be the forest constructed by Sollin's algorithm at the stage before e_{k+1} is added to S. Let u be the endpoint of e_{k+1} that is in a component C of S' responsible for the addition of e_{k+1} (i.e., so that e_{k+1} is the minimum weight edge incident to C). Let v be the other endpoint of e_{k+1}. We let P be the unique simple path from u to v in T. We follow P until we come to the first edge e' not in S'. Thus e' is also incident to C. Since the algorithm chose to add e_{k+1} on behalf of C, we know that $w(e_{k+1}) \le w(e')$, and that e' was added on behalf of the component of its other endpoint. Now if e' is not in $\{e_1, \ldots, e_k\}$, then we go on to the next paragraph. Otherwise, we continue following P until we come to the first edge e'' not in S'. Thus e'' is also incident to C', but was added on behalf of another component C'', and $w(e') \le w(e'')$. We continue in this way until we come to an edge $e^{(r)}$ not in $\{e_1, \ldots, e_k\}$.

Finally, let $T' = T \cup \{e_{k+1}\} - \{e^{(r)}\}$. Then by stringing together the inequalities we have obtained along the way, we know that $w(e_{k+1}) \le w(e^{(r)})$. It follows that $w(T') \le w(T)$, so T' is again a minimum spanning tree. Furthermore, T' contains $e_1, e_2, \ldots, e_k, e_{k+1}$, and we have our desired contradiction.

29. Suppose that there are r trees in the forest at some intermediate stage of Sollin's algorithm. Each new tree formed during this stage will then contain at least two of the old trees, so there are at most $r/2$ new trees. In other words, we have to reduce the number of trees by at least $r - (r/2) = r/2$. Since each edge added at this stage reduces the number of trees by exactly one, we must add at least $r/2$ edges. Finally, the number of edges added is of course an integer, so it is at least $\lceil r/2 \rceil$.

31. This follows easily from Exercises 29 and 30. After k stages of Sollin's algorithm, since the number of trees begins at n and is at least halved at each stage, there are at most $n/2^k$ trees. Thus if $n \le 2^k$, then this quantity is less than or equal to 1, so the algorithm has terminated. In other words, the algorithm terminates after at most k stages if $k \geq \log n$, which is what we wanted to prove.

GUIDE TO REVIEW QUESTIONS FOR CHAPTER 10

1. **a)** See p. 683. **b)** See p. 684.

2. No, for each ordered pair of vertices u and v, there is a unique simple path from u to v.

3. See Examples 5–8 in Section 10.1.

4. **a)** See p. 685. **b)** See p. 686. **c)** See p. 686.
d) Figure 8a in Section 10.1 is such a tree. Its root is a. The parent of each vertex is the vertex immediately above it; thus the parent of b is a, the parent of c is also a, the parent of d is b, and so on. The children of a vertex are the vertices immediately below it; thus the children of a are b and c, the children of b are d and e, the only child of h is j, e has no children, and so on. The internal vertices are the ones with children: a, b, c, d, h, i, and l. The leaves are the vertices without children: e, f, g, j, k, and m.

5. **a)** $n-1$ **b)** If c is the number of components, then $e = n - c$.

6. **a)** See p. 686. **b)** $mi+1$; $(m-1)i+1$; see Theorem 4 in Section 10.1.

7. **a)** See p. 691. **b)** See p. 691. **c)** between 1 and m^h, inclusive

8. **a)** See pp. 695–696. **b)** Repeatedly apply Algorithm 1 in Section 10.2 to insert items one by one.
c) If we insert the items in the order given, we obtain the following tree.

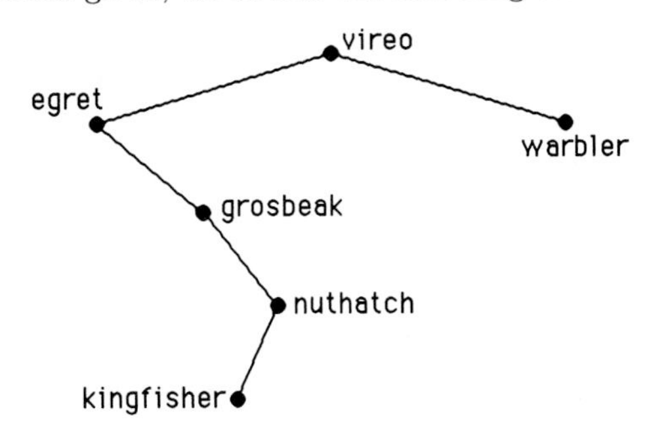

9. **a)** See pp. 700–701. **b)** See p. 701.

10. **a)** See p. 712 and p. 714. **b)** See Examples 2–4 in Section 10.3.

11. **a)** Build the expression tree. Its preorder traversal gives prefix form; its postorder traversal gives postfix form; and its inorder traversal gives infix form (if we assume that there are only binary operations, and if we put a pair of parentheses around the expression for every subtree that is not a leaf).
b) Here is the expression tree.

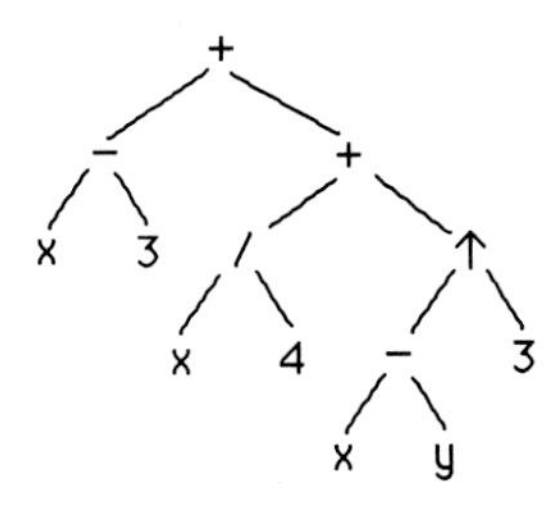

c) prefix: $+ - x\,3 + /\,x\,4 \uparrow - x\,y\,3$; postfix: $x\,3 - x\,4\,/\,x\,y - 3 \uparrow + +$

12. See Theorem 1 in Section 10.2 and the discussion preceding it.

13. a) See pp. 701–702.

b) The answer is not unique, because there is a choice of whether to make the subtree containing A and B or the leaf C the left subtree at the second step. Assuming we choose the former, the code for A is 000; that for B is 001; that for C is 01; and the code for D is 1.

14. The tree is too large to draw here. (It can be completed by referring to the solution to Exercise 33 in Section 10.2.) The children of the root are the positions 4, 31, 21, 11, and 1. Each of these has value -1, except for the position with just one pile with one stone. That is a winning position for the first player, since it is a leaf and the second player has no move. So the first player wins the game by removing the pile of four stones at her first move. Thus the value of the root of the tree is 1.

15. a) See p. 724. **b)** all connected ones **c)** road-plowing, communication network reinforcement

16. a) See pp. 726–727 (depth-first search) and p. 729 (breadth-first search).

b) See Exercises 15 and 16 in Section 10.4.

17. a) See Example 6 in Section 10.4. **b)** Apply this algorithm to W_5.

18. a) See pp. 737–738. **b)** minimum cost communications network, shortest connecting road configuration

19. a) See Algorithms 1 and 2 in Section 10.5. **b)** See Examples 2 and 3 in Section 10.5.

SUPPLEMENTARY EXERCISES FOR CHAPTER 10

1. There are of course two things to prove here. First let us assume that G is a tree. We must show that G contains no simple circuits (which is immediate by definition) and that the addition of an edge connecting two nonadjacent vertices produces a graph that has exactly one simple circuit. Clearly the addition of such an edge $e = \{u, v\}$ produces a graph with a simple circuit, namely u, e, v, P, u, where P is the unique simple path joining v to u in G. Since P is unique, moreover, this is the only simple circuit that can be formed.

To prove the converse, suppose that G satisfies the given conditions; we want to prove that G is a tree, in other words, that G is connected (since one of the conditions is already that G has no simple circuits). If G is not connected, then let u and v lie in separate components of G. Then edge $\{u, v\}$ can be added to G without the formation of any simple circuits, in contradiction to the assumed condition. Therefore G is indeed a tree.

3. Let P be a longest simple path in a given tree T. This path has length at least 1 as long as T has at least one edge, and such a longest simple path exists since T is finite. Now the vertices at the ends of P must both have degree 1 (i.e., be pendant vertices), since otherwise the simple path P could be extended to a longer simple path.

5. Since the sum of the degrees of the vertices is twice the number of edges, and since a tree with n vertices has $n-1$ edges, the answer is $2n-2$.

7. One way to prove this is simply to note that the conventional way of drawing rooted trees provides a planar embedding. Another simple proof is to observe that a tree cannot contain a subgraph homeomorphic to K_5 or $K_{3,3}$, since these must contain simple circuits. A third proof, of the fact that a tree can be embedded in the plane with straight lines for the edges, is by induction on the number of vertices. The basis step (a tree with one vertex) is trivial (there are no edges). If tree T contains $n+1$ vertices, then delete one vertex of degree 1 (which exists by Exercise 3), embed the remainder (by the inductive hypothesis), and then draw the deleted vertex and reattach it with a short straight line to the proper vertex.

9. Since the colors in one component have no effect on the colors in another component, it is enough to prove this for connected graphs, i.e., trees. In order to color a tree with 2 colors, we simply view the tree as rooted and color all the vertices at even-numbered levels with one color and all the vertices at odd-numbered levels with another color. Since every edge connects vertices at adjacent levels, this coloring is proper.

11. A B-tree of degree k and height h has the most leaves when every vertex not at level h has as many children as possible, namely k children. In this case the tree is simply the complete k-ary tree of height h, so it has k^h leaves. Thus the best upper bound for the number of leaves of a B-tree of degree k and height h is k^h. To obtain a lower bound, we want to have as few leaves as possible. This is accomplished when each vertex has as few children as possible. The root must have at least 2 children. Each other vertex not at level h must have at least $\lceil k/2 \rceil$ children. Thus there are $2\lceil k/2 \rceil^{h-1}$ leaves, so this is our best lower bound. (Of course if $h = 0$, then the tree has exactly 1 leaf.)

13. We follow the instructions, constructing each tree from the previous trees as indicated.

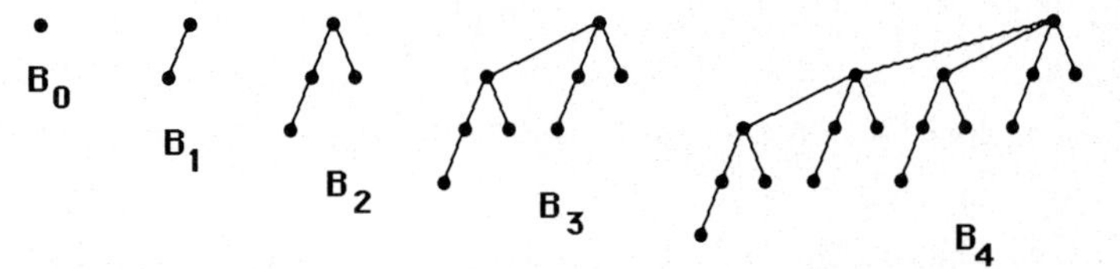

15. Since B_{k+1} is formed from two copies of B_k, one shifted down one level, the height increases by 1 as k increases by 1. Since B_0 had height 0, it follows by induction that B_k has height k.

17. Since the root of B_{k+1} is the root of B_k with one additional child (namely the root of the other B_k), the degree of the root increases by 1 as k increases by 1. Since B_0 had a root with degree 0, it follows by induction that B_k has a root with degree k.

19. We follow the recursive definition in drawing these trees. For example, we obtain S_4 by taking a copy of S_3 (on the left), adding one more child of the root (the right-most one), and putting a copy of S_3 rooted at this new child.

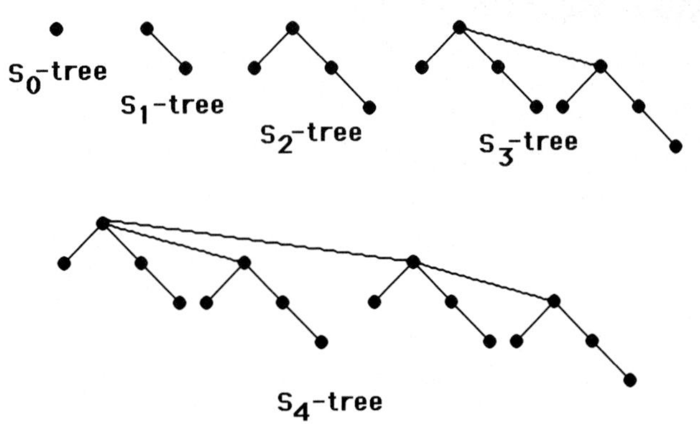

21. We prove this by induction on k. If $k = 0$ or 1, then the result is trivial. Assume the inductive hypothesis that the S_{k-1}-tree can be formed in the manner indicated, and let T be an S_k-tree. Then T consists of a copy of an S_{k-1} tree T_{k-1} with root r_{k-1}, together with another copy of an S_{k-1} tree whose root is made a child of r_{k-1}. Now by the inductive hypothesis, this latter S_{k-1}-tree can be formed from a handle v and disjoint trees T_0, T_1, ..., T_{k-2} by connecting v to r_0 and r_i to r_{i+1} for $i = 0, 1, \ldots, k-3$. Since our tree T is formed by then joining r_{k-2} to r_{k-1}, our tree is formed precisely in the manner desired, and the proof by induction is complete.

23. Essentially we just want to do a breadth-first search of the tree, starting at the root, and considering the children of a vertex in the order from left to right. Thus the algorithm is to perform breadth-first search, starting at the root of the tree. Whenever we encounter a vertex, we print it out.

25. First we determine the universal addresses of all the vertices in the tree. The root has address 0. For every leaf address, we include an address for each prefix of that address. For example, if there is a leaf with address 4.3.7.3, then we include vertices with addresses 4, 4.3, and 4.3.7. The tree structure is constructed by making all the vertices with positive integers as their addresses the children of the root, in order, and all the vertices with addresses of the form $A.i$, where A is an address and i is a positive integer, the children of the vertex whose address is A, in order by i.

27. For convenience, let us define a "very simple circuit" to be a set of edges that form a circuit all of whose vertices are distinct except that (necessarily) the last vertex is the same as the first. Thus a cactus is a graph in which every edge is in no more than one very simple circuit.

a) This is a cactus. The edge at the top is in no simple circuit. The three edges at the bottom are only in the triangle they form.

b) This is not a cactus, since the edge in the upper right-hand corner, for instance, is in more than one very simple circuit: a triangle and a pentagon.

c) This is a cactus. The edges in each of the three triangles are each in exactly one very simple circuit.

29. Adding a very simple circuit (see the solution to Exercise 27 for the definition) does not give the graph any more very simple circuits, except for the one being added. Thus the new edges are each in exactly one very simple circuit, and the old edges are still in no more than one.

31. There is clearly a spanning tree here which is a simple path (a, b, c, f, e, d, for instance); since each degree is 1 or 2, this spanning tree meets the condition imposed.

33. There is clearly a spanning tree here which is a simple path ($a, b, c, f, e, d, i, h, g$, for instance); since each degree is 1 or 2, this spanning tree meets the condition imposed.

35. We need to label these trees so that they satisfy the condition. We work by trial and error, using some common sense. For example, the labels 1 and n need to be adjacent in order to obtain the difference $n-1$.

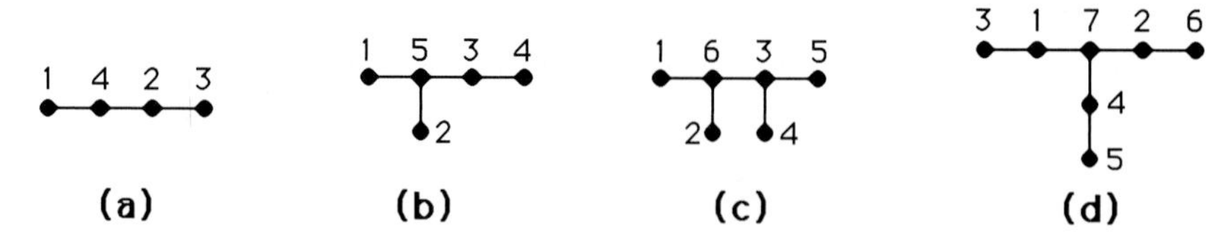

37. We count the caterpillars by drawing them all, using the length of the longest path to organize our work. In fact every tree with six vertices is a caterpillar. They are the five trees shown with heavy lines in our solution to Exercise 33c in Section 10.1, together with the star $K_{1,5}$. Thus the answer is 6.

39. a) The frequencies of the bits strings are 0.81 for 00, 0.09 for 01 and for 10, and 0.01 for 11. The resulting Huffman code uses 0 for 00, 11 for 01, 100 for 10, and 101 for 11. (The exact coding depends on how ties were broken, but all versions are equivalent.) Thus in a string of length n, the average number of bits required to send two bits of the message is $1 \cdot 0.81 + 2 \cdot 0.09 + 3 \cdot 0.09 + 3 \cdot 0.01 = 1.29$, so the average number of bits required to encode the string is $1.29 \cdot (n/2) = 0.645n$.

b) The frequencies of the bits strings are 0.729 for 000, 0.081 for 001, 010, and 100, 0.009 for 011, 101, and 110, and 0.001 for 111. The resulting Huffman code uses 0 for 000, 100 for 001, 101 for 010, 110 for

100, 11100 for 011, 11101 for 101, 11110 for 110, and 11111 for 111. (The exact coding depends on how ties were broken, but all versions are equivalent.) Thus in a string of length n, the average number of bits required to send three bits of the message is $1 \cdot 0.729 + 3 \cdot 3 \cdot 0.081 + 5 \cdot 3 \cdot 0.009 + 5 \cdot 0.001 = 1.598$, so the average number of bits required to encode the string is $1.598 \cdot (n/3) \approx 0.533n$.

41. Let T be a minimum spanning tree. If T contains e, then we are done. If not, then adding e to T creates a simple circuit, and then deleting another edge e' of this simple circuit gives us another spanning tree T'. Since $w(e) \leq w(e')$, the weight of T' is no larger than the weight of T. Therefore T' is a minimum spanning tree containing e.

43. The proof uses the same idea as in the solution to Exercise 18 in Section 10.5. Suppose that edge e is the edge of least weight incident to vertex v, and suppose that T is a spanning tree that does not include e. Add e to T, and delete from the simple circuit formed thereby the other edge of the circuit that contains v. The result will be a spanning tree of strictly smaller weight (since the deleted edge has weight greater than the weight of e). This is a contradiction, so T must include e.

WRITING PROJECTS FOR CHAPTER 10

Books and articles indicated by bracketed symbols below are listed near the end of this manual. You should also read the general comments and advice you will find there about researching and writing these essays.

1. You can probably find something useful in [BiLl]. Also, take a look at [WhCl].

2. Consult books on data structures or algorithms, such as [Kr] or [Ma2].

3. There is a relevant article in [De2], pp. 290–295.

4. See hints for Writing Project 2.

5. A Web search will turn up some good expositions.

6. An elementary exposition can be found in [Gr2]. See also algorithm or data structures texts, such as [Sm].

7. A Web search will turn up some good sites, including one by Dr. A. N. Walker.

8. See books on parallel processing, such as [Le2], to learn what meshes of trees are and how they are applied.

9. Books are beginning to appear in this new field; see [Ma5], for instance.

10. Good books on algorithms will contain material for this. See [CoLe], for example.

11. Good books on algorithms will contain material for this. See [CoLe], for example.

12. Each search company wants to keep the details of its techniques secret, but you should be able to find some general information on the Web. See also [Sz].

13. This problem is much more difficult than the plain minimum spanning tree problem. You may have to go to the research literature here (search the *Mathematical Reviews*, database available on the Web as MathSciNet, for keywords “spanning tree” and “constrained” or “degree”).

14. Any good data structures or algorithms book would have loads of material on this topic, including those mentioned in Writing Project 2 (or others given in the reference list). See also volume 3 of the classic [Kn].

15. See the article [GrHe].

16. See books on random graph theory ([Bo2] or [Pa1]), or the article [Ti].

CHAPTER 11
Boolean Algebra

SECTION 11.1 Boolean Functions

The first 28 of these exercises are extremely straightforward and should pose no difficulty. The next four relate to the interconnection between duality and De Morgan's laws; they are a bit subtle. Boolean functions can be proved equal by tables of values (as we illustrate in Exercise 13), by using known identities (as we illustrate in Exercise 11), or by taking duals of equal Boolean functions (Exercise 28 justifies this). To show that two Boolean functions are not equal, we need to find a counterexample, i.e., values of the variables that give the two functions different values. Exercises 35–43 deal with abstract Boolean algebras, and the proofs are of the formal, "symbol-pushing" variety.

1. a) $1 \cdot \overline{0} = 1 \cdot 1 = 1$ **b)** $1 + \overline{1} = 1 + 0 = 1$ **c)** $\overline{0} \cdot 0 = 1 \cdot 0 = 0$ **d)** $\overline{(1+0)} = \overline{1} = 0$

3. a) We compute $(1 \cdot 1) + (\overline{0 \cdot 1} + 0) = 1 + (\overline{0} + 0) = 1 + (1 + 0) = 1 + 1 = 1$.
b) Following the instructions, we have $(\mathbf{T} \wedge \mathbf{T}) \vee (\neg(\mathbf{F} \wedge \mathbf{T}) \vee \mathbf{F}) \equiv \mathbf{T}$.

5. In each case, we compute the various components of the final expression and put them together as indicated. For part **(a)** we have

x	y	z	$\overline{x}$	$\overline{x}y$
1	1	1	0	0
1	1	0	0	0
1	0	1	0	0
1	0	0	0	0
0	1	1	1	1
0	1	0	1	1
0	0	1	1	0
0	0	0	1	0

For part **(b)** we have

x	y	z	yz	$x + yz$
1	1	1	1	1
1	1	0	0	1
1	0	1	0	1
1	0	0	0	1
0	1	1	1	1
0	1	0	0	0
0	0	1	0	0
0	0	0	0	0

For part **(c)** we have

x	y	z	$\overline{y}$	$x\overline{y}$	xyz	$\overline{xyz}$	$x\overline{y}+\overline{xyz}$
1	1	1	0	0	1	0	0
1	1	0	0	0	0	1	1
1	0	1	1	1	0	1	1
1	0	0	1	1	0	1	1
0	1	1	0	0	0	1	1
0	1	0	0	0	0	1	1
0	0	1	1	0	0	1	1
0	0	0	1	0	0	1	1

For part **(d)** we have

x	y	z	$\overline{y}$	$\overline{z}$	yz	$\overline{y}\,\overline{z}$	$yz+\overline{y}\,\overline{z}$	$x(yz+\overline{y}\,\overline{z})$
1	1	1	0	0	1	0	1	1
1	1	0	0	1	0	0	0	0
1	0	1	1	0	0	0	0	0
1	0	0	1	1	0	1	1	1
0	1	1	0	0	1	0	1	0
0	1	0	0	1	0	0	0	0
0	0	1	1	0	0	0	0	0
0	0	0	1	1	0	1	1	0

7. In each case, we note from our solution to Exercise 5 which vertices need to be blackened in the cube, as in Figure 1.

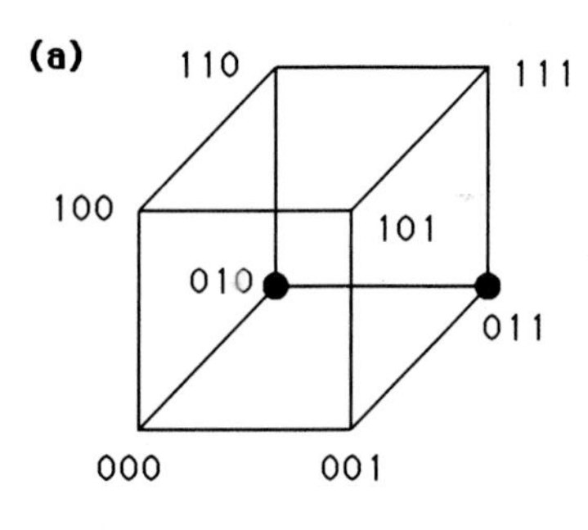

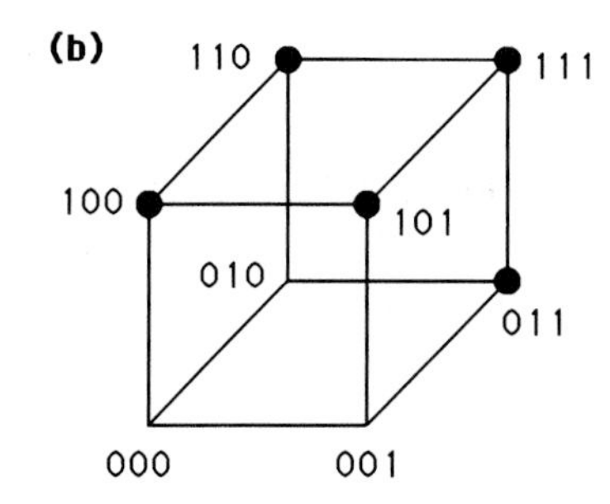

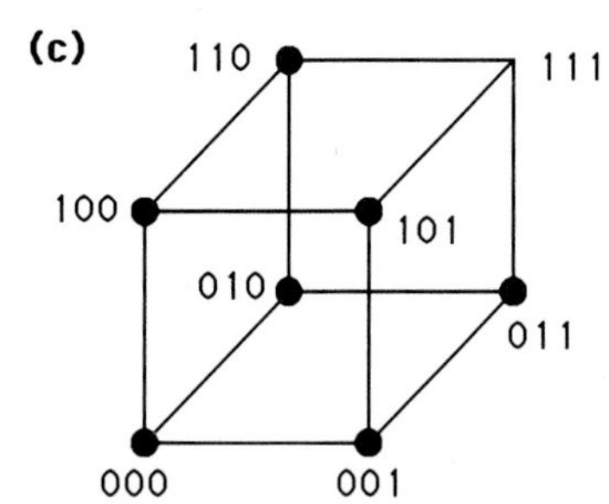

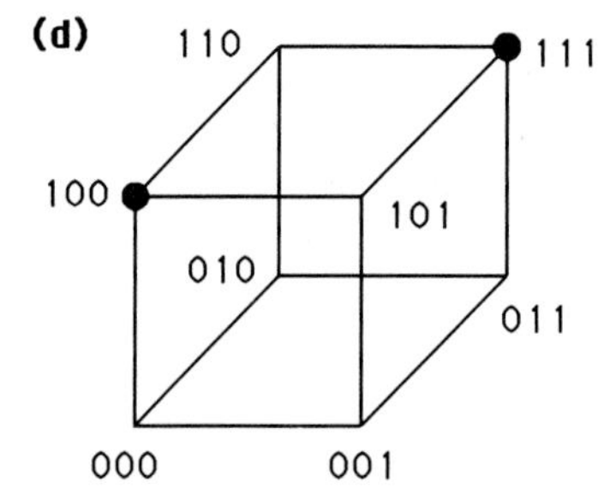

9. By looking at the definitions, we see that this equation is satisfied if and only if $x = y$, i.e., $x = y = 0$ or $x = y = 1$.

11. First we "factor" out an x by using the identity and distributive laws: $x + xy = x \cdot 1 + x \cdot y = x \cdot (1 + y)$. Then we use the commutative law, the domination law, and finally the identity law again to write this as $x \cdot (y + 1) = x \cdot 1 = x$.

13. Probably the simplest way to do this is by use of a table, as in Example 8. We list all the possible values for the triple (x, y, z) (there being eight such), and for each compute both sides of this equation. For example, for $x = y = z = 0$ we have $x\overline{y} + y\overline{z} + \overline{x}z = 0 \cdot 1 + 0 \cdot 1 + 1 \cdot 0 = 0 + 0 + 0 = 0$, and similarly $\overline{x}y + \overline{y}z + x\overline{z} = 0$. We do this for all eight lines of the table to conclude that the two functions are equal.

x	y	z	$x\overline{y}+y\overline{z}+\overline{x}z$	$\overline{x}y+\overline{y}z+x\overline{z}$
1	1	1	0	0
1	1	0	1	1
1	0	1	1	1
1	0	0	1	1
0	1	1	1	1
0	1	0	1	1
0	0	1	1	1
0	0	0	0	0

15. The idempotent laws state that $x \cdot x = x$ and $x + x = x$. There are only four things to check: $0 \cdot 0 = 0$, $0 + 0 = 0$, $1 \cdot 1 = 1$, and $1 + 1 = 1$, all of which are part of the definitions. The relevant tables, exhibiting these calculations, have only two rows.

17. The domination laws state that $x + 1 = 1$ and $x \cdot 0 = 0$. There are only four things to check: $0 + 1 = 1$, $0 \cdot 0 = 0$, $1 + 1 = 1$, and $1 \cdot 0 = 0$, all of which are part of the definitions. The relevant tables, exhibiting these calculations, have only two rows.

19. We can verify each associative law by constructing the relevant table, which will have eight rows, since there are eight combinations of values for x, y, and z in the equations $x+(y+z) = (x+y)+z$ and $x(yz) = (xy)z$. Rather than write down these tables, let us observe that in the first case, both sides are equal to 1 unless $x = y = z = 0$ (in which case both sides equal 0), and, dually, in the second case, both sides are equal to 0 unless $x = y = z = 1$ (in which case both sides equal 1).

21. We construct the relevant tables (as in Exercise 13) and compute the quantities shown. Since the fourth and seventh columns are equal, we conclude that $\overline{(xy)} = \overline{x} + \overline{y}$; since the ninth and tenth columns are equal, we conclude that $\overline{(x+y)} = \overline{x}\,\overline{y}$.

x	y	xy	$\overline{(xy)}$	$\overline{x}$	$\overline{y}$	$\overline{x}+\overline{y}$	$x+y$	$\overline{(x+y)}$	$\overline{x}\,\overline{y}$
1	1	1	0	0	0	0	1	0	0
1	0	0	1	0	1	1	1	0	0
0	1	0	1	1	0	1	1	0	0
0	0	0	1	1	1	1	0	1	1

23. The zero property states that $x \cdot \overline{x} = 0$. There are only two things to check: $0 \cdot \overline{0} = 0 \cdot 1 = 0$ and $1 \cdot \overline{1} = 1 \cdot 0 = 0$. The relevant table, exhibiting this calculation, has only two rows.

25. We could prove these by constructing tables, as in Exercise 21. Instead we will argue directly.

a) The left-hand side is equal to 1 if $x \neq y$. In this case the right-hand side is necessarily $1 \cdot \overline{0} = 1$, as well. On the other hand if $x = y = 1$, then the left-hand side is by definition equal to 0, and the right-hand side equals $1 \cdot 0 = 0$; similarly if $x = y = 0$, then the left-hand side is by definition equal to 0, and the right-hand side equals $0 \cdot 1 = 0$.

b) The left-hand side is equal to 1 if $x \neq y$. In this case the right-hand side is necessarily $1 + 0 = 1$ or $0 + 1 = 1$, as well. On the other hand if $x = y$, then the left-hand side is by definition equal to 0, and the right-hand side equals $0 + 0 = 0$.

27. a) We can prove this by constructing the appropriate table, as in Exercise 21. What we will find is that each side equals 1 if and only if an odd number of the variables are equal to 1. Thus the two functions are equal.

b) This is not an identity. If we let $x = y = z = 1$, then the left-hand side is $1 + 0 = 1$, while the right-hand side is $1 \oplus 1 = 0$.

c) This is not an identity. If we let $x = y = 1$ and $z = 0$, then the left-hand side is $1 \oplus 1 = 0$, while the right-hand side is $0 + 1 = 1$.

29. Let B be a Boolean expression representing F, and let D be the dual of B. We want to show that for every set of values assigned to the variables $x_1, x_2, \ldots, x_n$, the value D equals the opposite of the value of B with the opposites of these values assigned to $x_1, x_2, \ldots, x_n$. The trick is to look at $\overline{B}$. Then by repeatedly applying De Morgan's laws from the outside in, we see that $\overline{B}$ is the same as the expression obtained by replacing each occurrence of x_i in D by $\overline{x}_i$. Thus for any values of $x_1, x_2, \ldots, x_n$, the value of D is the same as the value of $\overline{B}$ for the corresponding value of $\overline{x}_1, \overline{x}_2, \ldots, \overline{x}_n$. This tells us that $\overline{B}$ represents the function whose values are exactly those of the function represented by D when the opposites of each of the values of the variables x_i are used, and that is exactly what we wanted to prove.

31. Because of the stated condition, we are free to specify $F(1, y, z)$ for all pairs (y, z), but then all the values of $F(0, y, z)$ are thereby determined. There are 4 such pairs (y, z) (each one can be either 0 or 1), and for each such pair we have 2 choices as to the value of $F(1, y, z)$. Therefore the answer is $2^4 = 16$.

33. We need to replace each 0 by **F**, 1 by **T**, + by $\vee$, $\cdot$ (or Boolean product implied by juxtaposition) by $\wedge$, and $\overline{\ }$ by $\neg$. We also replace x by p and y by q so that the variables look like they represent propositions, and we replace the equals sign by the logical equivalence symbol. Thus for the first De Morgan law in Table 5, $\overline{xy} = \overline{x} + \overline{y}$ becomes $\neg(p \wedge q) \equiv \neg p \vee \neg q$, which is the first De Morgan law in Table 6 of Section 1.2. Dually, $\overline{x + y} = \overline{x}\,\overline{y}$ becomes $\neg(p \vee q) \equiv \neg p \wedge \neg q$ for the other De Morgan law.

35. We need to play around with the symbols until the desired results fall out. To prove that $x \vee x = x$, let us compute $x \vee (x \wedge \overline{x})$ in two ways. On the one hand,

$$x \vee (x \wedge \overline{x}) = x \vee 0 = x$$

by the complement law and the identity law. On the other hand, by using the distributive law followed by the complement and identity laws we have

$$\begin{aligned} x \vee (x \wedge \overline{x}) &= (x \vee x) \wedge (x \vee \overline{x}) \\ &= (x \vee x) \wedge 1 = x \vee x. \end{aligned}$$

By transitivity of equality, $x = x \vee x$. The other property is the dual of this one, and its proof can be obtained by formally replacing every $\vee$ with $\wedge$ and replacing every 0 with 1, and vice versa. Thus our proof, shortened to one line, becomes

$$x = x \wedge 1 = x \wedge (x \vee \overline{x}) = (x \wedge x) \vee (x \wedge \overline{x}) = (x \wedge x) \vee 0 = x \vee x.$$

37. By Exercise 36 we know that the complement of an element is that unique element that obeys the complement laws. Therefore to show that $\overline{0} = 1$ we just need to prove that $0 \vee 1 = 1$ and $0 \wedge 1 = 0$. But these follow immediately from the identity laws (and, for the first, the commutative law). The other half follows in the same manner.

39. Before we prove this, we will find the following lemma useful: $x \wedge 0 = 0$ and $x \vee 1 = 1$ for all x. To prove the first half of the lemma, we invoke the results of Exercises 35 and 36 and compute as follows:

$$\begin{aligned} x \wedge 0 &= x \wedge (x \wedge \overline{x}) \\ &= (x \wedge x) \wedge \overline{x} \\ &= x \wedge \overline{x} = 0. \end{aligned}$$

The second half is similar, using the duality replacements mentioned in the solution to Exercise 35.

Now for the exercise at hand, by Exercise 36 it is enough to show that the claimed complements behave correctly. That is, we must show that $(x \vee y) \vee (\overline{x} \wedge \overline{y}) = 1$ and $(x \vee y) \wedge (\overline{x} \wedge \overline{y}) = 0$. For the second of these we compute as follows, using the lemma at the end and using the various defining properties freely (in particular, we use distributivity, associativity and commutativity a lot).

$$\begin{aligned}(x \vee y) \wedge (\overline{x} \wedge \overline{y}) &= \overline{y} \wedge [\overline{x} \wedge (x \vee y)] \\ &= \overline{y} \wedge [(\overline{x} \wedge x) \vee (\overline{x} \wedge y)] \\ &= \overline{y} \wedge [0 \vee (\overline{x} \wedge y)] \\ &= \overline{y} \wedge \overline{x} \wedge y = \overline{x} \wedge (y \wedge \overline{y}) \\ &= \overline{x} \wedge 0 = 0\end{aligned}$$

The first statement is proved in a similar way:

$$\begin{aligned}(x \vee y) \vee (\overline{x} \wedge \overline{y}) &= y \vee [x \vee (\overline{x} \wedge \overline{y})] \\ &= y \vee [(x \vee \overline{x}) \wedge (x \vee \overline{y})] \\ &= y \vee [1 \wedge (x \vee \overline{y})] \\ &= y \vee x \vee \overline{y} = (y \vee \overline{y}) \vee x \\ &= 1 \vee x = 1\end{aligned}$$

And of course the other half of this problem is proved in a manner completely dual to this.

41. Using the hypothesis, we compute as follows. $x = x \vee 0 = x \vee (x \vee y) = (x \vee x) \vee y = x \vee y = 0$. Similarly for y, by the commutative law. Note that we used the result of Exercise 35. The other statement is proved in the dual manner: $x = x \wedge 1 = x \wedge (x \wedge y) = (x \wedge x) \wedge y = x \wedge y = 1$ and similarly for y.

43. Most of this work was done in the Supplementary Exercises for Chapter 8 (or else is immediate from the definition). We need to verify the five laws (each one consisting of a dual pair, of course). The identity laws are Exercise 41. The complement laws are part of the definition of "complemented". The associative and commutative laws are Exercise 39. Finally, the distributive laws are again part of the definition.

SECTION 11.2 Representing Boolean Functions

The first six exercises are straightforward practice dealing with sum-of-products expansions. These are obtained by writing down one product term for each combination of values of the variables that makes the function have the value 1, and taking the sum of these terms. The dual to the sum-of-products expansion is discussed in Exercises 7–11, and these should be looked at. Since the concept of complete sets of operators plays an important role in the logical circuit design in sections to follow, Exercises 12–20 are also important.

1. **a)** We want $\overline{x}$, $\overline{y}$, and z all to have the value 1; therefore we take the product $\overline{x}\,\overline{y}\,z$. The other parts are similar, so we present only the answers.

b) $\overline{x}\,y\,\overline{z}$ **c)** $\overline{x}\,y\,z$ **d)** $\overline{x}\,\overline{y}\,\overline{z}$

3. **a)** We want the function to have the value 1 whenever at least one of the variables has the value 1. There are seven minterms that achieve this, so the sum has seven summands: $x\,y\,z+x\,y\,\overline{z}+x\,\overline{y}\,z+\overline{x}\,y\,z+x\,\overline{y}\,\overline{z}+\overline{x}\,y\,\overline{z}+\overline{x}\,\overline{y}\,z$.
b) Here is another way to think about this problem (rather than just making a table and reading off the minterms that make the value equal to 1). If we expand the expression by the distributive law (and use the commutative law), we get $xy+yz$. Now invoking the identity laws, the law that $s+\overline{s}=1$, and the distributive and commutative laws again, we write this as $xy1 + 1yz = xy(z+\overline{z}) + (x+\overline{x})yz = xyz + xy\overline{z} + xyz + \overline{x}yz$. Finally, we use the idempotent law to collapse the first and third term, to obtain our answer: $xyz+xy\overline{z}+\overline{x}yz$.

c) We can use either the straightforward approach or the idea used in part **(b)**. The answer is $xyz + xy\overline{z} + x\overline{y}z + x\overline{y}\,\overline{z}$.

d) The method discussed in part **(b)** works well here, to obtain the answer $x\overline{y}z + x\overline{y}\,\overline{z}$.

5. We need to list all minterms that have an odd number of the variables without bars (and hence an odd number with bars). There are $C(4,1)+C(4,3) = 8$ terms. The answer is $wxy\overline{z} + wx\overline{y}z + w\overline{x}yz + \overline{w}xyz + w\overline{x}\,\overline{y}\,\overline{z} + \overline{w}x\overline{y}\,\overline{z} + \overline{w}\,\overline{x}y\overline{z} + \overline{w}\,\overline{x}\,\overline{y}z$.

7. This exercise is dual to Exercise 1.

a) Note that $\overline{x}$ will have the value 0 if and only if $x = 1$. Similarly, $\overline{y}$ will have the value 0 if and only if $y = 1$. Therefore the expression $\overline{x} + \overline{y} + z$ will have the value 0 precisely in the desired case. The remaining parts are similar, so we list only the answers.

b) $x + y + z$ **c)** $x + \overline{y} + z$

9. By the definition of "+," the sum $y_1 + \cdots + y_n$ has the value 0 if and only if each $y_i = 0$. This happens precisely when $x_i = 0$ for those cases in which $y_i = x_i$ and $x_i = 1$ in those cases in which $y_i = \overline{x}_i$.

11. **a)** This function is already written in its product-of-sums form (with one factor).

b) This function has the value 0 in case $y = 0$ or both x and z equal 0. Therefore we need maxterms $x + y + z$, $x + y + \overline{z}$, $\overline{x} + y + z$, and $\overline{x} + y + \overline{z}$ (to take care of $y = 0$), and also $x + \overline{y} + z$. Therefore the answer is the product of these five maxterms.

c) This function has the value 0 in case $x = 0$. Therefore we need four maxterms, and the answer is $(x + y + z)(x + y + \overline{z})(x + \overline{y} + z)(x + \overline{y} + \overline{z})$.

d) Let us indicate another way to solve problems like this. In Exercise 3d we found the sum-of-products expansion of this function. Suppose that we take take sum-of-products expansion of the function that is the opposite of this one. It will have all the minterms other than the ones in the answer to Exercise 3d, so it will be $xyz + xy\overline{z} + \overline{x}yz + \overline{x}y\overline{z} + \overline{x}\,\overline{y}z + \overline{x}\,\overline{y}\,\overline{z}$. If we now take the complement of this (put a big bar over it), then we will have an expression for the function we want. Then we push the complementations inside, using De Morgan's laws and the fact that $\overline{\overline{s}} = s$. This will give us the desired product-of-sums expansion. Formally, all we do is put parentheses around the minterms, erase all the plus signs, put plus signs between all the variables (where there used to be implied products), and change every complemented variable to its uncomplemented version and vice versa. The answer is thus $(\overline{x} + \overline{y} + \overline{z})(\overline{x} + \overline{y} + z)(x + \overline{y} + \overline{z})(x + \overline{y} + z)(x + y + \overline{z})(x + y + z)$.

13. To do this exercise we need to use De Morgan's law to replace st by $\overline{(\overline{s} + \overline{t})}$. Thus we just do this formally in the expressions in Exercise 12, and we obtain the answers. It is also good to simplify double complements, of course.

a) This is already in the desired form, having no products.

b) $x + \overline{y}(\overline{x} + z) = x + \overline{\left(\overline{\overline{y}} + \overline{(\overline{x} + z)}\right)} = x + \overline{\left(y + \overline{(\overline{x} + z)}\right)}$

c) This is already in the desired form, having no products.

d) $\overline{x}(x + \overline{y} + \overline{z}) = \overline{\left(\overline{\overline{x}} + \overline{(x + \overline{y} + \overline{z})}\right)} = \overline{\left(x + \overline{(x + \overline{y} + \overline{z})}\right)}$

15. **a)** We use the definition of $\downarrow$. If $x = 1$, then $x \downarrow x = 0$; and if $x = 0$, then $x \downarrow x = 1$. These are precisely the corresponding values of $\overline{x}$.

b) We can construct a table to look at all four cases, as follows. Since the fifth and sixth columns are equal, the expressions are equivalent.

x	y	$x \downarrow x$	$y \downarrow y$	$(x \downarrow x) \downarrow (y \downarrow y)$	xy
1	1	0	0	1	1
1	0	0	1	0	0
0	1	1	0	0	0
0	0	1	1	0	0

c) We can construct a table to look at all four cases, as follows. Since the fourth and fifth columns are equal, the expressions are equivalent.

x	y	$x \downarrow y$	$(x \downarrow y) \downarrow (x \downarrow y)$	$x + y$
1	1	0	1	1
1	0	0	1	1
0	1	0	1	1
0	0	1	0	0

17. a) Since $x + y + z = (x + y) + z$, we first write this as $((x + y) \mid (x + y)) \mid (z \mid z)$, using the identity in Exercise 14c. Then we rewrite $x + y$ as $(x \mid x) \mid (y \mid y)$, using the same identity. This gives us

$$(((x \mid x) \mid (y \mid y)) \mid ((x \mid x) \mid (y \mid y))) \mid (z \mid z).$$

b) First we write this as $((x + z) \mid y) \mid ((x + z) \mid y)$, using the identity in Exercise 14b. Then we rewrite $x + z$ as $(x \mid x) \mid (z \mid z)$, using the identity in Exercise 14c. This gives us

$$(((x \mid x) \mid (z \mid z)) \mid y) \mid (((x \mid x) \mid (z \mid z)) \mid y).$$

c) There are no operators, so nothing needs to be done; the expression as given is the answer.

d) First we write this as $(x \mid \overline{y}) \mid (x \mid \overline{y})$, using the identity in Exercise 14b. Then we rewrite $\overline{y}$ as $y \mid y$, using the identity in Exercise 14a. This gives us $(x \mid (y \mid y)) \mid (x \mid (y \mid y))$.

19. We claim that it is impossible to write a Boolean expression for $\overline{x}$ involving x, $+$, and $\cdot$. The reason is that if $x = 1$, then any combination of these two operators applied to x (and the results of previous calculations) can yield only the value 1. But $\overline{1} = 0$. (This problem is somewhat harder if we allow the use of the constants 0 and 1. The argument given here is then no longer valid, since $x \cdot 0 = 0$. Nevertheless it is possible to show that even with these constants allowed, the set $\{+, \cdot\}$ is not functionally complete—i.e., it is impossible to write down a Boolean expression using only the operators $+$ and $\cdot$ (together with x, 0, and 1) that is equivalent to $\overline{x}$. What one can do is to prove by induction on the length of the expression that any such Boolean expression that has the value 1 when $x = 0$ must also have the value 1 when $x = 1$.)

SECTION 11.3 Logic Gates

This section is a brief introduction to circuit design using AND and OR gates and inverters. In real life, circuits will have thousands of these components, but you will get some of the flavor in these exercises. Notice particularly Exercise 9, which shows how circuits already constructed can be further combined to give more complex, useful circuits. If we want to get by with fewer types of gates (but more gates), then we can use NOR or NAND gates, as illustrated in Exercises 15–18. Exercise 20 introduces the notion of the **depth** *of a circuit.*

1. The output of the OR gate at the top is $x+y$. This and $\overline{y}$ are the inputs to the final AND gate. Therefore the output of the circuit is $(x+y)\overline{y}$.

3. The idea is the same as in the previous two exercises. The final output is an OR with two inputs. The first of these inputs is the result of inverting xy, and the second is $\overline{z}+x$. Therefore the answer is $\overline{(xy)}+(\overline{z}+x)$.

5. The outputs from the three OR gates are $x+y+z$, $\overline{x}+y+z$, and $\overline{x}+\overline{y}+\overline{z}$. Therefore the output from the final OR gate is $(x+y+z)+(\overline{x}+y+z)+(\overline{x}+\overline{y}+\overline{z})$. It is not hard to see that this is always 1, since if $x=1$ then the output of the top initial OR gate is 1, whereas if $x=0$ then the output of the middle initial OR gate is 1.

7. Let v, w, x, y, and z be the votes of the five individuals, with a 1 representing a yes vote and a 0 representing a no vote. Then the majority will be a yes vote (represented by an output of 1) if and only if there are at least three yes votes. Thus we make an AND gate for each of the $C(5,3)=10$ triples of voters, and combine the outputs from these 10 gates with an OR. We turn the picture on its side for convenience.

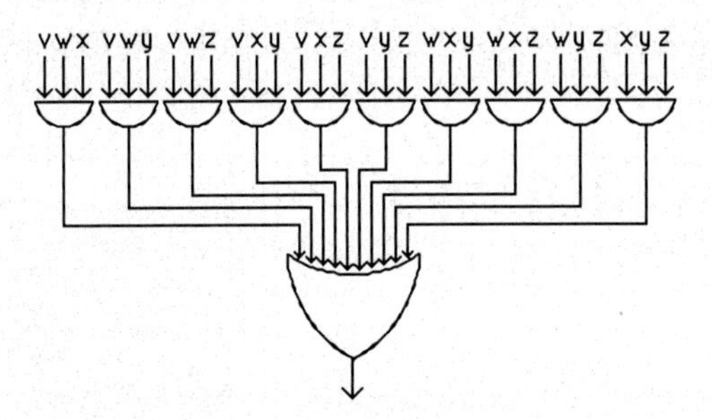

9. The circuit is identical to Figure 10, expanded by two more units to accommodate the two additional bits. To get the computation started, x_0 and y_0 are the inputs to the half adder. Thereafter, the carry bit from each column is input, together with the next pair (x_i, y_i) to a full adder to find the output and carry for the next column. The final carry bit (c_4) is the final answer bit (s_5).

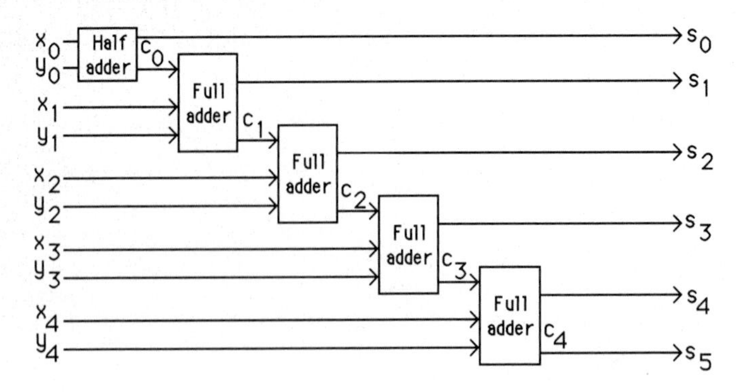

11. We will construct the full subtractor directly. Suppose the input bits are x, y, and b, where we are computing $x-y$ with borrow b. The output is a bit, z, and a borrow from the next column, b'. Then by looking at the eight possibilities for the bits x, y, and b, we see that $z = x\,y\,b + x\,\overline{y}\,\overline{b} + \overline{x}\,y\,\overline{b} + \overline{x}\,\overline{y}\,b$; and that $b' = x\,y\,b + \overline{x}\,y\,b + \overline{x}\,y\,\overline{b} + \overline{x}\,\overline{y}\,b$. Therefore a full subtractor can be formed by using AND gates, OR gates, and inverters to represent these expressions. We obtain the circuits shown below.

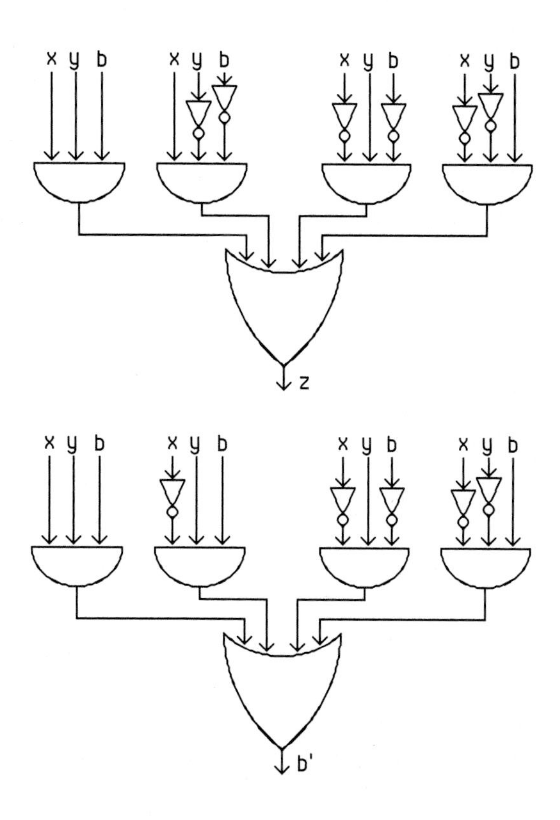

13. The first number is larger than the second if $x_1 > y_1$ (which means $x_1 = 1$ and $y_1 = 0$), or if $x_1 = y_1$ (which means either that $x_1 = 1$ and $y_1 = 1$, or that $x_1 = 0$ and $y_1 = 0$) and also $x_0 > y_0$ (which means $x_0 = 1$ and $y_0 = 0$). We translate this sentence into a circuit in the obvious manner, obtaining the picture shown.

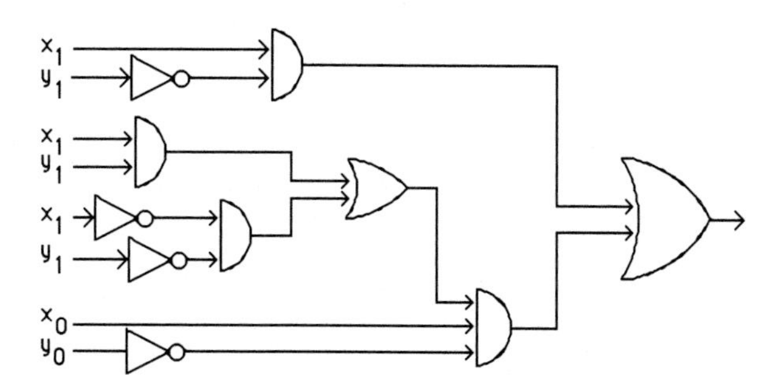

15. Note that in this exercise the usual operation symbol $|$ is used for the *NAND* operation.

a) By Exercise 14a in Section 11.2, $\overline{x} = x \mid x$. Therefore the gate for $\overline{x}$ is as shown below.

b) By Exercise 14c in Section 11.2, $x + y = (x \mid x) \mid (y \mid y)$. Therefore the gate for $x + y$ is as shown below.

c) By Exercise 14b in Section 11.2, $xy = (x \mid y) \mid (x \mid y)$. Therefore the gate for xy is as shown below.

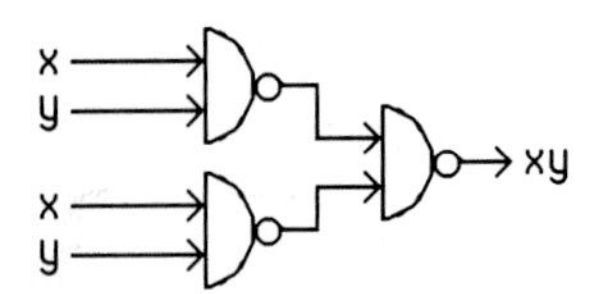

d) First we note that $x \oplus y = x\overline{y} + \overline{x}\,y = \overline{\overline{((\overline{x\overline{y}})\,(\overline{\overline{x}\,y}))}} = (\overline{x\overline{y}}) \mid (\overline{\overline{x}\,y}) = (x \mid \overline{y}) \mid (\overline{x} \mid y)$. We constructed the gate for inverting in part **(a)**. Therefore the gate for xy is as shown below.

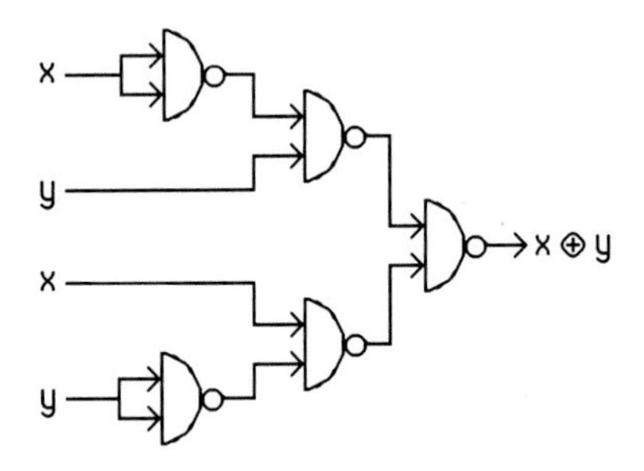

17. We know that the sum bit in the half adder is $s = x\overline{y} + \overline{x}\,y$. The answer to Exercise 15d shows precisely this gate constructed from *NAND* gates, so it gives us this part of the answer. Also, the carry bit in the half adder is $c = xy$. The answer to Exercise 15c shows precisely this gate constructed from *NAND* gates, so it gives us this part of the answer.

19. We can set this up so that the value of x_i "gets through" to a final *OR* gate if and only if $(c_1c_0)_2 = i$. For example, we want x_2 to get through if and only if $c_1 = 1$ and $c_0 = 0$, since the Base 2 numeral for 2 is 10. We can do this by combining the x_i input with either c_0 or its inversion and either c_1 or its inversion, using an *AND* gate with three inputs. Thus the output of each of these gates is either 0 (if $(c_1c_0)_2 \neq i$) or is the value of x_i (if $(c_1c_0)_2 = i$). So if we combine the result of these four outputs, using an *OR* gate, we will get the desired result (since at most one of the four outputs can possibly be nonzero). Here is the circuit.

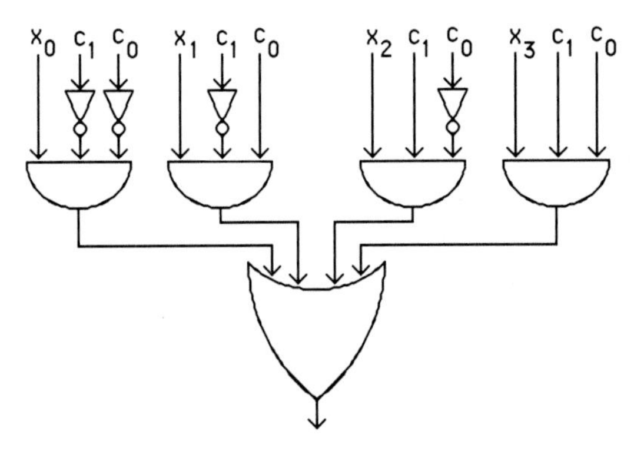

SECTION 11.4 Minimization of Circuits

The two methods presented here for minimizing circuits are really the same, just looked at from two different points of view—one geometric (K-maps) and one algebraic (the Quine–McCluskey method). In each case the idea is to get larger blocks, which represent simpler terms, to cover several minterms. The calculations can get very messy, but if you follow the examples and organize your work carefully, you should not have trouble with them. The hard part in these algorithms from a theoretical point of view is finding the set of products that cover all the minterms. In these small examples, this tends not to be a problem.

1. a) The K-map we draw here has the variable x down the side and variable y across the top.

	y	$\bar{y}$
x		
$\bar{x}$	1	

b) The upper left-hand corner cell (whose minterm is $x\,y$) and the lower right-hand corner cell (whose minterm is $\overline{x}\,\overline{y}$) are adjacent to this cell.

3. The 2×2 square is used in each case. We put a 1 in those cells whose minterms are listed.

	y	$\bar{y}$
x		1
$\bar{x}$		

(a)

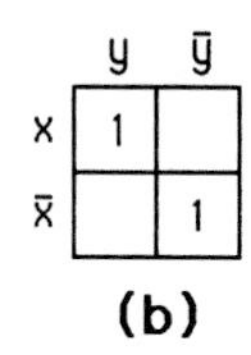

(b)

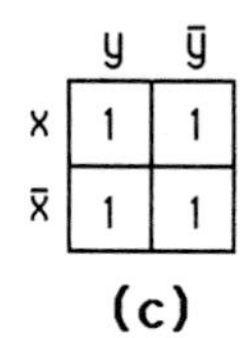

(c)

5. a) We can draw a K-map for three variables in the manner shown here, with the x variable down the side and the y and z variables across the top, in the order shown. We have placed a 1 in the requested position.

	yz	$y\bar{z}$	$\bar{y}\bar{z}$	$\bar{y}z$
x				
$\bar{x}$		1		

b) There are three cells adjacent to every cell (since there are three variables). The minterms of the adjacent cells can be read off the picture: $\overline{x}\,y\,z$, $x\,y\,\overline{z}$, and $\overline{x}\,\overline{y}\,\overline{z}$.

7. The 2×4 square is used in each case. We put a 1 in those cells whose minterms are listed.

	yz	$y\bar{z}$	$\bar{y}\bar{z}$	$\bar{y}z$
x			1	
$\bar{x}$				

(a)

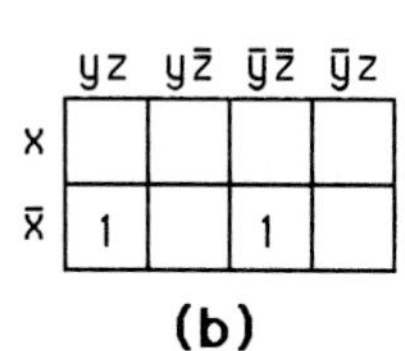

(b)

	yz	$y\bar{z}$	$\bar{y}\bar{z}$	$\bar{y}z$
x	1	1		
$\bar{x}$		1		1

(c)

9. In the figure below we have drawn the K-map. For example, since one of the terms was $x\overline{z}$, we put a 1 in each cell whose address contained x and $\overline{z}$. Note that this meant two cells, one for y and one for $\overline{y}$. Each cell with a 1 in it is an implicant, as are the pairs of cells that form blocks, namely xy, $x\overline{z}$, and $y\overline{z}$. Since each cell by itself is contained in a block with two cells, none of them is prime. Each of the mentioned blocks with two cells is prime, since none is contained in a larger block. Furthermore, each of these blocks is essential, since each contains a cell that no other prime implicant contains: xy contains xyz, $x\overline{z}$ contains $x\,\overline{y}\,\overline{z}$, and $y\overline{z}$ contains $\overline{x}\,y\,\overline{z}$.

	yz	$y\bar{z}$	$\bar{y}\bar{z}$	$\bar{y}z$
x	1	1	1	
$\bar{x}$		1		

11. The figure below shows the 4-cube Q_4, labeled as requested. Compare with Figure 1 in Section 11.1. A complemented Boolean variable corresponds to 0, and an uncomplemented Boolean variable corresponds to 1. For example, the lower right corner of the sub-3-cube on the left corresponds to 0001.

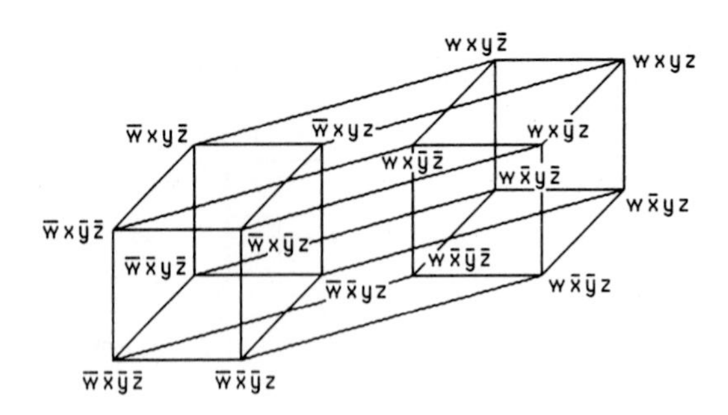

The 3-cube on the right corresponds to w, since all of its vertices are labeled w. Similarly, the 3-cube given by the top surface of the whole figure represents x; the 3-cube given by the back surface of the whole figure represents y; and the 3-cube given by the right surfaces of both the left and the right 3-cube represents z. We show the last of these with heavy lines below. The "opposite 3-face" in each case represents the complemented literal.

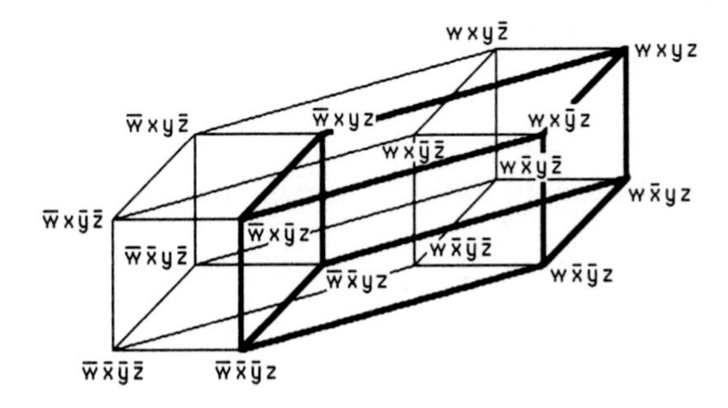

The 2-cube (i.e., square) that represents wz, in the same way, is the set of vertices that have w and z as part of their labels, rather than $\overline{w}$ and/or $\overline{z}$. This is the right face of the 3-cube on the right. Similarly, the 2-cube that represents $\overline{x}\,y$ is bottom rear, and the 2-cube that represents $\overline{y}\,\overline{z}$ is front left.

13. **a)** We can draw a K-map for four variables in the manner shown here, with the w and x variables down the side and the y and z variables across the top, in the order shown. We have placed a 1 in the requested position.

	yz	$y\overline{z}$	$\overline{y}\overline{z}$	$\overline{y}z$
wx				
$w\overline{x}$				
$\overline{w}\overline{x}$				
$\overline{w}x$		1		

b) There are four cells adjacent to every cell (since there are four variables). The minterms of the adjacent cells can be read off the picture: $\overline{w}\,x\,y\,z$, $\overline{w}\,\overline{x}\,y\,\overline{z}$, $\overline{w}\,x\,\overline{y}\,\overline{z}$, and (recalling that the top row is adjacent to the bottom row) $w\,x\,y\,\overline{z}$.

15. A K-map for five variables needs 32 cells in all. We arrange them as shown below, following the discussion in Example 6. Reread that example to understand what rows and columns have to be considered adjacent. (We use rectangles rather than squares to save vertical space.)

a) We want to put a 1 in all cells the correspond to x_1, x_2, x_3, and x_4 all being uncomplemented. Looking at the diagram, we see that we need 1's in precisely the two cells shown.

	$x_3x_4x_5$	$x_3x_4\bar{x}_5$	$x_3\bar{x}_4\bar{x}_5$	$x_3\bar{x}_4x_5$	$\bar{x}_3\bar{x}_4x_5$	$\bar{x}_3\bar{x}_4\bar{x}_5$	$\bar{x}_3x_4\bar{x}_5$	$\bar{x}_3x_4x_5$
x_1x_2	1	1						
$x_1\bar{x}_2$								
$\bar{x}_1\bar{x}_2$								
$\bar{x}_1x_2$								

b) We want to put a 1 in all cells the correspond to x_1 being complemented and x_3 and x_5 both being uncomplemented. Looking at the diagram, we see that we need 1's in precisely the four cells shown. Note that these are all adjacent, even though they don't look adjacent, since the first and fourth columns are considered adjacent.

	$x_3x_4x_5$	$x_3x_4\bar{x}_5$	$x_3\bar{x}_4\bar{x}_5$	$x_3\bar{x}_4x_5$	$\bar{x}_3\bar{x}_4x_5$	$\bar{x}_3\bar{x}_4\bar{x}_5$	$\bar{x}_3x_4\bar{x}_5$	$\bar{x}_3x_4x_5$
x_1x_2								
$x_1\bar{x}_2$								
$\bar{x}_1\bar{x}_2$	1			1				
$\bar{x}_1x_2$	1			1				

c) We want to put a 1 in all cells the correspond to x_2 and x_4 both being uncomplemented. Looking at the diagram, we see that we need 1's in precisely the eight cells shown.

	$x_3x_4x_5$	$x_3x_4\bar{x}_5$	$x_3\bar{x}_4\bar{x}_5$	$x_3\bar{x}_4x_5$	$\bar{x}_3\bar{x}_4x_5$	$\bar{x}_3\bar{x}_4\bar{x}_5$	$\bar{x}_3x_4\bar{x}_5$	$\bar{x}_3x_4x_5$
x_1x_2	1	1					1	1
$x_1\bar{x}_2$								
$\bar{x}_1\bar{x}_2$								
$\bar{x}_1x_2$	1	1					1	1

d) We want to put a 1 in all cells the correspond to x_3 and x_4 both being complemented. Looking at the diagram, we see that we need 1's in precisely the eight cells shown.

	$x_3x_4x_5$	$x_3x_4\bar{x}_5$	$x_3\bar{x}_4\bar{x}_5$	$x_3\bar{x}_4x_5$	$\bar{x}_3\bar{x}_4x_5$	$\bar{x}_3\bar{x}_4\bar{x}_5$	$\bar{x}_3x_4\bar{x}_5$	$\bar{x}_3x_4x_5$
x_1x_2					1	1		
$x_1\bar{x}_2$					1	1		
$\bar{x}_1\bar{x}_2$					1	1		
$\bar{x}_1x_2$					1	1		

e) We want to put a 1 in all cells the correspond to x_3 being uncomplemented. Looking at the diagram, we see that we need 1's in precisely the 16 cells shown.

	$x_3x_4x_5$	$x_3x_4\bar{x}_5$	$x_3\bar{x}_4\bar{x}_5$	$x_3\bar{x}_4x_5$	$\bar{x}_3\bar{x}_4x_5$	$\bar{x}_3\bar{x}_4\bar{x}_5$	$\bar{x}_3x_4\bar{x}_5$	$\bar{x}_3x_4x_5$
x_1x_2	1	1	1	1				
$x_1\bar{x}_2$	1	1	1	1				
$\bar{x}_1\bar{x}_2$	1	1	1	1				
$\bar{x}_1x_2$	1	1	1	1				

f) We want to put a 1 in all cells the correspond to x_5 being complemented. Looking at the diagram, we see that we need 1's in precisely the 16 cells shown.

	$x_3x_4x_5$	$x_3x_4\bar{x}_5$	$x_3\bar{x}_4\bar{x}_5$	$x_3\bar{x}_4x_5$	$\bar{x}_3\bar{x}_4x_5$	$\bar{x}_3\bar{x}_4\bar{x}_5$	$\bar{x}_3x_4\bar{x}_5$	$\bar{x}_3x_4x_5$
x_1x_2		1	1			1	1	
$x_1\bar{x}_2$		1	1			1	1	
$\bar{x}_1\bar{x}_2$		1	1			1	1	
$\bar{x}_1x_2$		1	1			1	1	

17. **a)** There are clearly 2^n cells in a K-map for n variables, since in specifying a minterm each variable can appear either complemented or uncomplemented. Thus the answer is $2^6 = 64$.

b) There are n cells adjacent to each cell in the K-map for n variables, because each of the variables can be changed (from complemented to uncomplemented or vice versa) to produce an adjacent cell. Thus the answer is 6.

19. Clearly we need to consider the first and fourth rows adjacent. The columns are more complicated, since each column needs to be adjacent to three others (those that differ from it in one bit). Thus we must declare the following columns adjacent (using the obvious notation): 1–4, 1–12, 1–16, 2–11, 2–15, 3–6, 3–10, 4–9, 5–8, 5–16, 6–15, 7–10, 7–14, 8–13, 9–12, 11–14, and 13–16. The complexity of this makes it virtually impossible for a human to use this visual aid.

21. We could ignore the stated condition, but then our circuit would be more complex than need be. If we use the condition, then there are really only three inputs; let us call them x, y, and z, where z represents Marcus's vote and x and y represent the votes of the unnamed people. Since Smith and Jones always vote against Marcus, a majority will occur if either Marcus assents with both of the other two (i.e., the minterm xyz), or else if Marcus votes no but at least one of the other two vote yes (which we can represent by $(x+y)\bar{z}$). Thus we design our circuit to be the *OR* of these two expressions.

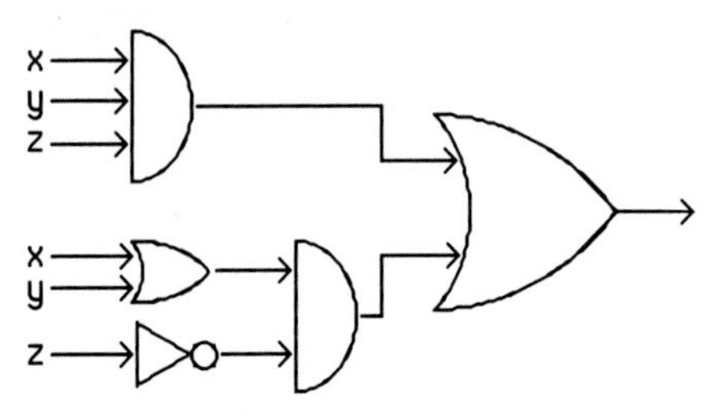

23. We organize our work as in the text.

a)

			Step 1	
	Term	String	Term	String
1	$\bar{x}yz$	011	$(1,2)\,\bar{x}z$	0–1
2	$\bar{x}\bar{y}z$	001		

In this case we have one product that covers both of the minterms, so this product ($\bar{x}z$) is our answer.

b)

			Step 1		Step 2	
	Term	String	Term	String	Term	String
1	xyz	111	$(1,2)\,xy$	11–	$(1,2,3,4)\,y$	–1–
2	$xy\bar{z}$	110	$(1,3)\,yz$	–11		
3	$\bar{x}yz$	011	$(2,4)\,y\bar{z}$	–10		
4	$\bar{x}y\bar{z}$	010	$(3,4)\,\bar{x}y$	01–		

Again we have one product that covers all the minterms, so that is our answer (y).

c)

	Term	String	Step 1 Term	Step 1 String
1	$x\,y\,\overline{z}$	110	$(1,4)\;x\,\overline{z}$	1–0
2	$x\,\overline{y}\,z$	101	$(2,4)\;x\,\overline{y}$	10–
3	$\overline{x}\,y\,z$	011	$(2,5)\;\overline{y}\,z$	–01
4	$x\,\overline{y}\,\overline{z}$	100	$(3,5)\;\overline{x}\,z$	0–1
5	$\overline{x}\,\overline{y}\,z$	001		

(Note that we reordered the minterms so that the number of 1's decreased as we went down the list.) No further combinations are possible at this point, so there are four products that must be used to cover our five minterms, each product covering two minterms. Clearly, then, two of these are not enough, but it is easy to find three whose sum covers all the minterms. One possible answer is to choose the first, third, and fourth of the products, namely $x\,\overline{z}+\overline{y}\,z+\overline{x}\,z$.

d)

	Term	String	Step 1 Term	Step 1 String
1	$x\,y\,z$	111	$(1,2)\;x\,z$	1–1
2	$x\,\overline{y}\,z$	101	$(1,3)\;y\,z$	–11
3	$\overline{x}\,y\,z$	011	$(2,4)\;x\,\overline{y}$	10–
4	$x\,\overline{y}\,\overline{z}$	100	$(3,5)\;\overline{x}\,y$	01–
5	$\overline{x}\,y\,\overline{z}$	010	$(4,6)\;\overline{y}\,\overline{z}$	–00
6	$\overline{x}\,\overline{y}\,\overline{z}$	000	$(5,6)\;\overline{x}\,\overline{z}$	0–0

(Note that we reordered the minterms so that the number of 1's decreased as we went down the list.) No further combinations are possible at this point, so there are six products that must be used to cover our six minterms, each product covering two minterms. Clearly, then, two of these are not enough, but it is not hard to find three whose sum covers all the minterms. One possible answer is to choose the first, fourth, and fifth of the products, namely $x\,z+\overline{x}\,y+\overline{y}\,\overline{z}$.

25. We follow the procedure and notation given in the text.

a)

	Term	String	Step 1 Term	Step 1 String
1	$w\,x\,y\,z$	1111	$(1,2)\;w\,x\,z$	11–1
2	$w\,x\,\overline{y}\,z$	1101	$(2,3)\;w\,x\,\overline{y}$	110–
3	$w\,x\,\overline{y}\,\overline{z}$	1100	$(2,5)\;w\,\overline{y}\,z$	1–01
4	$w\,\overline{x}\,y\,\overline{z}$	1010		
5	$w\,\overline{x}\,\overline{y}\,z$	1001		

The three products in the last column as well as minterm #4 are possible products in the desired expansion, since they are not contained in any other product. We make a table of which products cover which of the original minterms.

	1	2	3	4	5
$w\,x\,z$	X	X			
$w\,x\,\overline{y}$		X	X		
$w\,\overline{y}\,z$		X			X
$w\,\overline{x}\,y\,\overline{z}$				X	

Since only the first of our products covers minterm #1, it must be included. Similarly, the other three products must be included since they are the only ones that cover minterms #3, #4, and #5. If we do include them all, then of course all the minterms are covered. Therefore our answer is $w\,x\,z+w\,x\,\overline{y}+w\,\overline{y}\,z+w\,\overline{x}\,y\,\overline{z}$.

b)

	Term	String	Step 1 Term	String
1	$w\,x\,y\,\overline{z}$	1110	$(2,4)\;x\,\overline{y}\,z$	−101
2	$w\,x\,\overline{y}\,z$	1101	$(4,6)\;\overline{w}\,\overline{y}\,z$	0−01
3	$w\,\overline{x}\,y\,z$	1011		
4	$\overline{w}\,x\,\overline{y}\,z$	0101		
5	$\overline{w}\,\overline{x}\,y\,\overline{z}$	0010		
6	$\overline{w}\,\overline{x}\,\overline{y}\,z$	0001		

Since minterms #1, #3, and #5 are not contained in any others, these, along with the two products in the last column, are the products that we look at to cover the original minterms. It is not hard to see that all five are needed to cover all the original minterms. Therefore the answer is $x\,\overline{y}\,z+\overline{w}\,\overline{y}\,z+w\,x\,y\,\overline{z}+w\,\overline{x}\,y\,z+\overline{w}\,\overline{x}\,y\,\overline{z}$.

c)

	Term	String	Step 1 Term	String	Step 2 Term	String
1	$w\,x\,y\,z$	1111	$(1,2)\;w\,x\,y$	111− ⇐	$(3,4,5,8)\;\overline{y}\,z$	−−01
2	$w\,x\,y\,\overline{z}$	1110	$(1,3)\;w\,x\,z$	11−1 ⇐		
3	$w\,x\,\overline{y}\,z$	1101	$(3,4)\;w\,\overline{y}\,z$	1−01		
4	$w\,\overline{x}\,\overline{y}\,z$	1001	$(3,5)\;x\,\overline{y}\,z$	−101		
5	$\overline{w}\,x\,\overline{y}\,z$	0101	$(4,6)\;w\,\overline{x}\,\overline{y}$	100− ⇐		
6	$w\,\overline{x}\,\overline{y}\,\overline{z}$	1000	$(4,8)\;\overline{x}\,\overline{y}\,z$	−001		
7	$\overline{w}\,\overline{x}\,y\,\overline{z}$	0010	$(5,8)\;\overline{w}\,\overline{y}\,z$	0−01		
8	$\overline{w}\,\overline{x}\,\overline{y}\,z$	0001				

The product in the last column, as well as the products in Step 1 that are marked with an arrow, as well as minterm #7, are possible products in the desired expansion, since they are not contained in any other product. We make a table of which products cover which of the original minterms.

	1	2	3	4	5	6	7	8
$\overline{y}\,z$			X	X	X			X
$w\,x\,y$	X	X						
$w\,x\,z$	X		X					
$w\,\overline{x}\,\overline{y}$				X		X		
$\overline{w}\,\overline{x}\,y\,\overline{z}$							X	

In order to cover minterms #5, #2, #6, and #7, we need the first, second, fourth, and fifth of the products in this table. If we include these four, then all the minterms are covered, and we do not need the third one. Therefore our answer is $\overline{y}\,z+w\,x\,y+w\,\overline{x}\,\overline{y}+\overline{w}\,\overline{x}\,y\,\overline{z}$.

d)

	Term	String	Step 1 Term	String	Step 2 Term	String
1	$w\,x\,y\,z$	1111	$(1,2)\;w\,x\,y$	111−	$(1,2,4,6)\;w\,y$	1−1−
2	$w\,x\,y\,\overline{z}$	1110	$(1,3)\;w\,x\,z$	11−1 ⇐	$(1,4,5,7)\;y\,z$	−−11
3	$w\,x\,\overline{y}\,z$	1101	$(1,4)\;w\,y\,z$	1−11	$(4,6,7,8)\;\overline{x}\,y$	−01−
4	$w\,\overline{x}\,y\,z$	1011	$(1,5)\;x\,y\,z$	−111		
5	$\overline{w}\,x\,y\,z$	0111	$(2,6)\;w\,y\,\overline{z}$	1−10		
6	$w\,\overline{x}\,y\,\overline{z}$	1010	$(4,6)\;w\,\overline{x}\,y$	101−		
7	$\overline{w}\,\overline{x}\,y\,z$	0011	$(4,7)\;\overline{x}\,y\,z$	−011		
8	$\overline{w}\,\overline{x}\,y\,\overline{z}$	0010	$(5,7)\;\overline{w}\,y\,z$	0−11		
9	$\overline{w}\,\overline{x}\,\overline{y}\,z$	0001	$(6,8)\;\overline{x}\,y\,\overline{z}$	−010		
			$(7,8)\;\overline{w}\,\overline{x}\,y$	001−		
			$(7,9)\;\overline{w}\,\overline{x}\,z$	00−1 ⇐		

The products in the last column, as well as the products in Step 1 that are marked with an arrow, are possible products in the desired expansion, since they are not contained in any other product. We make a table of which products cover which of the original minterms.

	1	2	3	4	5	6	7	8	9
$w\,y$	X	X		X		X			
$y\,z$	X			X	X		X		
$\overline{x}\,y$				X		X	X	X	
$w\,x\,z$	X		X						
$\overline{w}\,\overline{x}\,z$							X		X

In order to cover minterms #2, #5, #8, #3, and #9, we need the first, second, third, fourth, and fifth of the products in this table, respectively—i.e., all of them. Therefore our answer is $w\,y + y\,z + \overline{x}\,y + w\,x\,z + \overline{w}\,\overline{x}\,z$.

27. Using the method of Exercise 26, we draw the following picture, putting a 0 in each cell that represents a maxterm in our product-of-sums expansion.

	$y+z$	$y+\bar{z}$	$\bar{y}+\bar{z}$	$\bar{y}+z$
x	0	0	0	0
$\bar{x}$	0			

We then combine them into larger blocks, as shown, obtaining two large blocks: the entire first row, which represents the factor x, and the entire first column, which represents the factor $(y+z)$. Since neither of these blocks alone covers all the maxterms, we need to use both. Therefore the simplified product is $x(y+z)$.

29. We need to review Example 8 and note that the various positions in the 4×4 square correspond to the various decimal digits. For example, the digit 6 corresponds to the box labeled $\overline{w}\,x\,y\,\overline{z}$. There are "don't care" positions as well, for 4-digit binary numerals that exceed 9. (We do not care what output results for such inputs.) Using the techniques of this section, we obtain the following maximal blocks.

	yz	$y\bar{z}$	$\bar{y}\bar{z}$	$\bar{y}z$
wx	d	d	d	d
$w\bar{x}$	d	d		1
$\bar{w}\bar{x}$	1		1	
$\bar{w}x$		1		

Each one corresponds to a minterm, and our minimal sum-of-products expansion is therefore $w\,z + \overline{x}\,y\,z + x\,y\,\overline{z} + \overline{w}\,\overline{x}\,\overline{y}\,\overline{z}$. Since we were asked for a circuit, we turn the products into *AND* gates and the sum into one big *OR* gate.

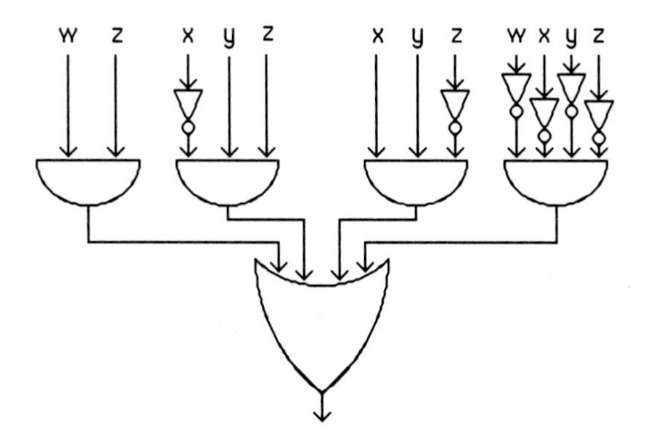

31. We can cover all the 1's with two large blocks here, one consisting of the middle four cells, and the other consisting of the four corners (it is a block because of the wrap-around nature of the figure). In doing so, we happened to have covered all the *d*'s as well, but that is irrelevant. The point is that we obtained the largest possible blocks in this manner, and if we had chosen not to cover the *d*'s, then the blocks would have been smaller. It is clear from this covering that the minimal sum-of-products expansion is just $\overline{x}\,\overline{z} + x\,z$.

33. A formal proof here proceeds by induction on n. If $n = 1$, then we are looking at the 1-cube, which is a line segment, labeled 0 at one end and 1 at the other end. The only possible value of k is also 1, and if the literal is x_1, then the subcube we have is the 0-dimensional subcube consisting of the endpoint labeled 1, and if the literal is $\overline{x}_1$, then the subcube we have is the 0-dimensional subcube consisting of the endpoint labeled 0. Now assume that the statement is true for n; we must show that it is true for $n+1$. If the literal x_{n+1} (or its complement) is not part of the product, then by the inductive hypothesis, the product when viewed in the setting of n variables corresponds to an $(n-k)$-dimensional subcube of the n-dimensional cube, and the Cartesian product of that subcube with the line segment $[0,1]$ gives us a subcube one dimension higher in our given $(n+1)$-dimensional cube, namely having dimension $(n+1)-k$, as desired. On the other hand, if the literal x_{n+1} (or its complement) is part of the product, then the product of the remaining $k-1$ literals corresponds to a subcube of dimension $n-(k-1) = (n+1)-k$ in the n-dimensional cube, and that slice, at either the 1-end or the 0-end in the last variable, is the desired subcube.

GUIDE TO REVIEW QUESTIONS FOR CHAPTER 11

1. See p. 750.

2. 16 (see Table 3 in Section 11.1)

3. See p. 751.

4. a) See p. 754. **b)** See p. 754.

5. See pp. 757–758.

6. a) See p. 759. **b)** This set is not functionally complete; see Exercise 19 in Section 11.2.
c) yes, such as $\{\downarrow\}$ (see Exercise 16 in Section 11.2)

7. See Example 3 in Section 11.3.

8. See Figure 8 in Section 11.3.

9. yes—*NAND* gates or *NOR* gates

10. a) See pp. 768–770. **b)** The following figure shows that the simplification is $x\,\overline{y} + y\,z + \overline{y}\,\overline{z}$.

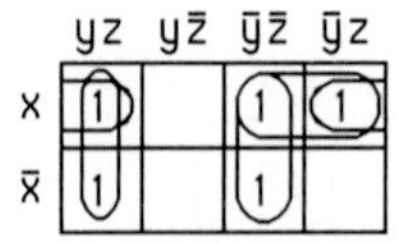

	yz	$y\bar{z}$	$\bar{y}\bar{z}$	$\bar{y}z$
x	1		1	1
$\bar{x}$	1		1	

11. a) See pp. 770–772. **b)** The following figure shows that the simplification is $w\,x + y\,z + w\,z + \overline{w}\,\overline{x}\,y$.

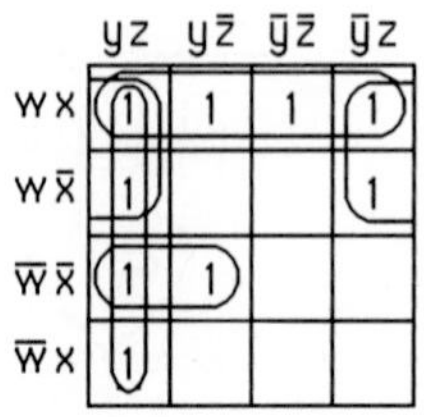

	yz	$y\bar{z}$	$\bar{y}\bar{z}$	$\bar{y}z$
wx	1	1	1	1
$w\bar{x}$	1			1
$\bar{w}\bar{x}$	1	1		
$\bar{w}x$	1			

12. **a)** See p. 774.

b) The following figure shows that the simplification is $w + xy$. The resulting circuit needs just one *AND* gate, one *OR* gate, and no inverters.

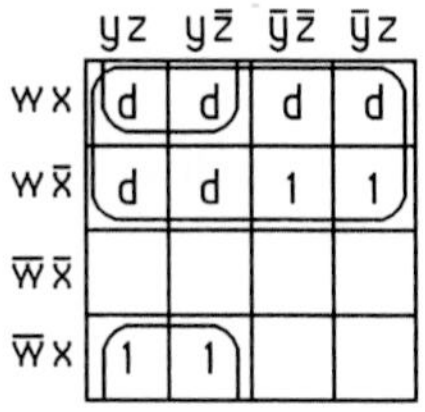

	yz	$y\bar{z}$	$\bar{y}\bar{z}$	$\bar{y}z$
wx	d	d	d	d
$w\bar{x}$	d	d	1	1
$\bar{w}\bar{x}$				
$\bar{w}x$	1	1		

13. **a)** See pp. 775–779.

b)

			Step 1		Step 2	
	Term	String	Term	String	Term	String
1	$xy\bar{z}$	110	$(1,2)\,y\bar{z}$	-10	$(1,2,3,5)\,\bar{z}$	$--0$
2	$\bar{x}y\bar{z}$	010	$(1,3)\,x\bar{z}$	$1-0$		
3	$x\bar{y}\bar{z}$	100	$(2,5)\,\bar{x}\bar{z}$	$0-0$		
4	$\bar{x}\bar{y}z$	001	$(3,5)\,\bar{y}\bar{z}$	-00		
5	$\bar{x}\bar{y}\bar{z}$	000	$(4,5)\,\bar{x}\bar{y}$	$00-$		

The product $\bar{z}$ in the last column covers all the minterms except #4, and the fifth product in Step 1 ($\bar{x}\bar{y}$) covers it. Thus the answer is $\bar{z} + \bar{x}\bar{y}$.

SUPPLEMENTARY EXERCISES FOR CHAPTER 11

1. **a)** Suppose that the equation holds. If $x = 1$, then the left-hand side is 1; therefore the right-hand side must be 1, as well, and that forces y and z to be 1. Similarly, if $x = 0$, then equality holds if and only if $y = z = 0$. Therefore the only solutions are $(1,1,1)$ and $(0,0,0)$.

b) If $x = 1$, then the right-hand side equals 1, so in order for there to be equality we need $y + z = 1$, whence $y = 1$ or $z = 1$. Thus $(1,1,0)$, $(1,0,1)$, and $(1,1,1)$ are all solutions. If $x = 0$, then the left-hand side is 0, so we need $yz = 0$ in order for equality to hold, whence $y = 0$ or $z = 0$. Thus $(0,0,0)$, $(0,0,1)$, and $(0,1,0)$ are also solutions.

c) There are no solutions here. If any of the variables equals 1, then the left-hand side is 0 and the right-hand side is 1; if all the variables are 0, then the left-hand side is 1 and the right-hand side is 0.

3. In each case we form $\overline{F(\bar{x}_1, \ldots, \bar{x}_n)}$ and simplify. If we get back to what we originally started with, then the function is self-dual; if what we obtain is not equivalent to what we began with, then the function is not self-dual. The simplification is done using the identities for Boolean algebra, especially De Morgan's laws.

a) $\bar{\bar{x}} = x$, so the function is self-dual.

b) $\overline{(\bar{x}\,\bar{\bar{y}} + \bar{\bar{x}}\,\bar{y})} = (\bar{x}\,\bar{y} + x\,y)$, which is the complement of what we originally had. Thus this is as far from being self-dual as it can possibly be.

c) $\overline{(\bar{x} + \bar{y})} = \bar{\bar{x}}\,\bar{\bar{y}} = x\,y$, which is certainly not equivalent to $x + y$. Therefore this is not self-dual.

d) We first simplify the expression, using the distributive law and the fact that $x + \bar{x} = 1$ to rewrite our function as $F(x,y) = y$. Now, as in part **(a)**, we see that it is indeed self-dual.

5. The reasoning here is essentially the same as in Exercise 31 in Section 11.1. To specify all the values of a self-dual function, we are free to specify the values of $F(1, x_2, x_3, \ldots, x_n)$, and we can do this in $2^{2^{n-1}}$ ways, since there are 2^{n-1} different elements at which we can choose to make the function value either 0 or 1. Once we have specified these values, the values of $F(0, x_2, x_3, \ldots, x_n)$ are all determined by the definition of self-duality, so no further choices are possible. Therefore the answer is $2^{2^{n-1}}$.

7. a) At every point in the domain, it is certainly the case that if $F(x_1,\ldots,x_n)=1$, then $(F+G)(x_1,\ldots,x_n)=F(x_1,\ldots,x_n)+G(x_1,\ldots,x_n)=1+G(x_1,\ldots,x_n)=1$, no matter what value G has at that point. Thus by definition $F\le F+G$.

b) This is dual to the first part. At every point in the domain, it is certainly the case that if $F(x_1,\ldots,x_n)=0$, then $(FG)(x_1,\ldots,x_n)=F(x_1,\ldots,x_n)G(x_1,\ldots,x_n)=0\cdot G(x_1,\ldots,x_n)=0$, no matter what value G has at that point. The contrapositive of this statement is that if $(FG)(x_1,\ldots,x_n)=1$, then $F(x_1,\ldots,x_n)=1$. Thus by definition $FG\le F$.

9. We need to show that this relation is reflexive, antisymmetric, and transitive. That $F\le F$ (reflexivity) is simply the tautology "if $F(x_1,\ldots,x_n)=1$, then $F(x_1,\ldots,x_n)=1$." For antisymmetry, suppose that $F\le G$ and $G\le F$. Then the definition of the relation says that $F(x_1,\ldots,x_n)=1$ if and only if $G(x_1,\ldots,x_n)=1$, which is the definition of equality between functions, so $F=G$. Finally, for transitivity, suppose that $F\le G$ and $G\le H$. We want to show that $F\le H$. So suppose that $F(x_1,\ldots,x_n)=1$. Then by the first inequality $G(x_1,\ldots,x_n)=1$, whence by the second inequality $H(x_1,\ldots,x_n)=1$, as desired.

11. None of these are identities. The counterexample $x=1,\ y=z=0$ works for all three.

a) We have $1\mid(0\mid 0)=1\mid 1=0$, whereas $(1\mid 0)\mid 0=1\mid 0=1$.

b) We have $1\downarrow(0\downarrow 0)=1\downarrow 1=0$, whereas $(1\downarrow 0)\downarrow(1\downarrow 0)=0\downarrow 0=1$.

c) We have $1\downarrow(0\mid 0)=1\downarrow 1=0$, whereas $(1\downarrow 0)\mid(1\downarrow 0)=0\mid 0=1$.

13. This is clear from the definitions. The given operation applied to x and y is defined to be 1 if and only if $x=y$, while *XOR* applied to x and y is defined (preamble to Exercise 24 in Section 11.1) to be 1 if and only if $x\ne y$.

15. We show this to be true with a table. Since the fifth and seventh columns are equal, the equation is an identity.

x	y	z	$x\odot y$	$(x\odot y)\odot z$	$y\odot z$	$x\odot(y\odot z)$
1	1	1	1	1	1	1
1	1	0	1	0	0	0
1	0	1	0	0	0	0
1	0	0	0	1	1	1
0	1	1	0	0	1	0
0	1	0	0	1	0	1
0	0	1	1	1	0	1
0	0	0	1	0	1	0

17. In each case we can actually list all the functions.

a) The only function values we can get are x, y, 0, 1, $\overline{x}$, and $\overline{y}$, since applying complementation twice does not give us anything new. Therefore the answer is 6.

b) Since $s\cdot s=s$, $s\cdot 1=s$, and $s\cdot 0=0$ for all s, the only functions we can get are x, y, 0, 1, and xy. Therefore the answer is 5.

c) By duality the answer here has to be the same as the answer to part **(b)**, namely 5.

d) We can get the 6 distinct functions x, y, 0, 1, xy and $x+y$. Any further applications of these operations, however, returns us to one of these functions. For example, $xy+x=x$.

19. The sum bit is the exclusive *OR* of the inputs, and the carry bit is their product. Therefore we need only two gates to form the half adder if we allow an *XOR* gate and an *AND* gate.

21. We need to figure out which combinations of values for x_1, x_2, and x_3 cause the inequality $-x_1+x_2+2x_3 \geq 1/2$ to be satisfied. Clearly this will be true if $x_3 = 1$. If $x_3 = 0$, then it will be true if and only if $x_2 = 1$ and $x_1 = 0$. Thus a Boolean expression for this function is $x_3 + \overline{x}_1 x_2$.

23. We prove this by contradiction. Suppose that a, b, and T are such that $ax+by \geq T$ if and only if $x \oplus y = 1$, i.e., if and only if either $x = 1$ and $y = 0$, or else $x = 0$ and $y = 1$. Thus for the first case we need $a \geq T$, and for the second we need $b \geq T$. Since we need $ax+by < T$ for $x = y = 0$, we know that $T > 0$. Hence in particular b is positive. Therefore we have $a+b > a \geq T$, which contradicts the fact that $1 \oplus 1 = 0$ (requiring $a+b \leq T$).

WRITING PROJECTS FOR CHAPTER 11

Books and articles indicated by bracketed symbols below are listed near the end of this manual. You should also read the general comments and advice you will find there about researching and writing these essays.

1. Martin Gardner is probably the best writer of mathematical material for the general public. For this project, look at [Ga1].

2. Try a good book on circuits, such as [Ch].

3. Try a good book on circuits, such as [Ch].

4. Try a good book on circuits, such as [Ch].

5. Try a good book on circuits, such as [Ch].

6. Try a good book on circuits, such as [Ch].

7. See Section 9.5 of [HiPe1].

8. Try a good book on switching theory, such as [Ko1].

9. Try a good book on switching theory, such as [Ko1]. Also see [HiPe1].

10. Try a good book on switching theory, such as [Ko1].

11. The classic paper on this algorithm is [RuSa], but good books on circuits should mention it (the second author of that paper has written some—check his Web page).

12. Try a good book on switching theory, such as [Ko1].

CHAPTER 12 Modeling Computation

SECTION 12.1 Languages and Grammars

There is no magical way to come up with the grammars to generate a language described in English. In particular, Exercises 15 and 16 are challenging and very worthwhile. Exercise 21 shows how grammars can be combined. In constructing grammars, we observe the rule that every production must contain at least one nonterminal symbol on the left. This allows us to know when a derivation is completed—namely, when the string we have generated contains no nonterminal symbols.

1. The following sequences of lines show that each is a valid sentence.

a) **sentence**
noun phrase intransitive verb phrase
article adjective noun intransitive verb phrase
article adjective noun intransitive verb
the **adjective noun intransitive verb**
the happy **noun intransitive verb**
the happy hare **intransitive verb**
the happy hare runs

b) **sentence**
noun phrase intransitive verb phrase
article adjective noun intransitive verb phrase
article adjective noun intransitive verb adverb
the **adjective noun intransitive verb adverb**
the sleepy **noun intransitive verb adverb**
the sleepy tortoise **intransitive verb adverb**
the sleepy tortoise runs **adverb**
the sleepy tortoise runs quickly

c) **sentence**
noun phrase transitive verb phrase noun phrase
article noun transitive verb phrase noun phrase
article noun transitive verb noun phrase
article noun transitive verb article noun
the **noun transitive verb article noun**
the tortoise **transitive verb article noun**
the tortoise passes **article noun**
the tortoise passes the **noun**
the tortoise passes the hare

d) **sentence**
noun phrase transitive verb phrase noun phrase
article adjective noun transitive verb phrase noun phrase
article adjective noun transitive verb noun phrase

article adjective noun transitive verb article adjective noun
the **adjective noun transitive verb article adjective noun**
the sleepy **noun transitive verb article adjective noun**
the sleepy hare **transitive verb article adjective noun**
the sleepy hare passes **article adjective noun**
the sleepy hare passes the **adjective noun**
the sleepy hare passes the happy **noun**
the sleepy hare passes the happy tortoise

3. Since *runs* is only an **intransitive verb**, it can only occur in a sentence of the form **noun phrase intransitive verb phrase**. Such a sentence cannot have anything except an **adverb** after the **intransitive verb**, and *the sleepy tortoise* cannot be an **adverb**.

5. a) It suffices to give a derivation of this string. We write the derivation in the obvious way. $S \Rightarrow 1A \Rightarrow 10B \Rightarrow 101A \Rightarrow 1010B \Rightarrow 10101$.

b) This follows from our solution to part **(c)**, because 10110 has two 1's in a row and is not of the form discussed there.

c) Notice that the only production with A on the left is $A \rightarrow 0B$. Furthermore, the only productions with B on the left are $B \rightarrow 1A$ and $B \rightarrow 1$. Combining these, we see that we can eliminate B and replace these three rules by $A \rightarrow 01A$ and $A \rightarrow 01$. This tells us that every string in the language generated by G must end with some number of repetitions of 01 (at least one). Furthermore, because of the rules $S \rightarrow 0A$ and $S \rightarrow 1A$, the string must start with either a 0 or a 1 preceding the repetitions of 01. Therefore the strings in this language consist of a 0 or a 1 followed by one or more repetitions of 01. We can write this as $\{0(01)^n \mid n \geq 0\} \cup \{1(01)^n \mid n \geq 0\}$

7. We write the derivation in the obvious way. $S \Rightarrow 0S1 \Rightarrow 00S11 \Rightarrow 000S111 \Rightarrow 000111$. We used the rule $S \rightarrow 0S1$ in the first three steps and $S \rightarrow \lambda$ in the last step.

9. a) Using G_1, we can add 0's on the left or 1's on the right of S. Thus we have $S \Rightarrow 0S \Rightarrow 00S \Rightarrow 00S1 \Rightarrow 00S11 \Rightarrow 00S111 \Rightarrow 00S1111 \Rightarrow 001111$.

b) In this grammar we must add all the 0's first to S, then change to an A and add the 1's, again on the left. Thus we have $S \Rightarrow 0S \Rightarrow 00S \Rightarrow 001A \Rightarrow 0011A \Rightarrow 00111A \Rightarrow 001111$.

11. First we apply the first rule twice and the rule $S \rightarrow \lambda$ to get $00ABAB$. We can then apply the rule $BA \rightarrow AB$, to get $00AABB$. Now we can apply the rules $0A \rightarrow 01$ and $1A \rightarrow 11$ to get $0011BB$; and then the rules $1B \rightarrow 12$ and $2B \rightarrow 22$ to end up with 001122, as desired.

13. In each case we will list only the productions, because V and T will be obvious from the context, and S speaks for itself.

a) For this finite set of strings, we can simply have $S \rightarrow 0$, $S \rightarrow 1$, and $S \rightarrow 11$.

b) We assume that "only 1's" includes the case of no 1's. Thus we can take simply $S \rightarrow 1S$ and $S \rightarrow \lambda$.

c) The middle can be anything we like, and we will let A represent the middle. Then our productions are $S \rightarrow 0A1$, $A \rightarrow 1A$, $A \rightarrow 0A$, and $A \rightarrow \lambda$.

d) We will let A represent the pairs of 1's. Then our productions are $S \rightarrow 0A$, $A \rightarrow 11A$, and $A \rightarrow \lambda$.

15. a) We need to add the 0's two at a time. Thus we can take the rules $S \rightarrow S00$ and $S \rightarrow \lambda$.

b) We can use the same first rule as in part **(a)**, namely $S \rightarrow S00$, to increase the number of 0's. Since the string must begin 10, we simply adjoin to this the rule $S \rightarrow 10$.

c) We need to add 0's and 1's two at a time. Furthermore, we need to allow for 0's and 1's to change their order. Since we cannot have a rule $01 \to 10$ (there being no nonterminal symbol on the left), we make up nonterminal analogs of 0 and 1, calling them A and B, respectively. Thus our rules are as follows: $S \to AAS$, $S \to BBS$, $AB \to BA$, $BA \to AB$, $S \to \lambda$, $A \to 0$, and $B \to 1$. (There are also totally different ways to approach this problem, which are just as effective.)

d) This one is fairly simple: $S \to 0000000000A$, $A \to 0A$, $A \to \lambda$. This assures at least 10 0's and allows for any number of additional 0's.

e) We need to invoke the trick used in part **(c)** to allow 0's and 1's to change their order. Furthermore, since we need at least one extra 0, we use $S \to A$ as our vanishing condition, rather than $S \to \lambda$. Our solution, then, is $S \to AS$, $S \to ABS$, $S \to A$, $AB \to BA$, $BA \to AB$, $A \to 0$, and $B \to 1$.

f) This is identical to part **(e)**, except that the vanishing condition is $S \to \lambda$, rather than $S \to A$, and there is no rule $S \to AS$.

g) We just put together two copies of a solution to part **(e)**, one in which there are more 0's than 1's, and one in which there are more 1's than 0's. The rules are as follows: $S \to ABS$, $S \to T$, $S \to U$, $T \to AT$, $T \to A$, $U \to BU$, $U \to B$, $AB \to BA$, $BA \to AB$, $A \to 0$, and $B \to 1$.

17. In each case we will list only the productions, because V and T will be obvious from the context, and S speaks for itself.

a) It suffices to have $S \to 0S$ and $S \to \lambda$.

b) We let A represent the string of 1's. Thus we take $S \to A0$, $A \to 1A$, and $A \to \lambda$. Notice that $A \to A1$ works just as well $A \to 1A$ here, so either one is fine.

c) It suffices to have $S \to 000S$ and $S \to \lambda$.

19. a) This is a type 2 grammar, because the left-hand side of each production has a single nonterminal symbol. It is not a type 3 grammar, because the right-hand side of the productions are not of the required type.

b) This meets the definition of a type 3 grammar.

c) This is only a type 0 grammar; it does not fit the definition of type 1 because the right side of the second production does not maintain the context set by the left side.

d) This is a type 2 grammar, because the left-hand side of each production has a single nonterminal symbol. It is not a type 3 grammar, because the right-hand side of the productions are not of the required type.

e) This meets the definition of a type 2 grammar. It is not of type 3, because of the production $A \to B$.

f) This is only a type 0 grammar; it does not fit the definition of type 1 because the right side of the second production does not maintain the context set by the left side.

g) This meets the definition of a type 3 grammar. Note, however, that it does not meet the definition of a type 1 grammar because of $S \to \lambda$.

h) This is only a type 0 grammar; it does not fit the definition of type 1 because the right side of the third production does not maintain the context set by the left side.

i) This is a type 2 grammar because each left-hand side is a single nonterminal. It is not type 3 because of the production $B \to \lambda$.

j) This is a type 2 grammar because each left-hand side is a single nonterminal. It is not type 3; each of the productions violates the conditions imposed for a type 3 grammar.

21. Let us assume that the nonterminal symbols of G_1 and G_2 are disjoint. (If they are not, we can give those in G_2, say, new names so that they will be; obviously this does not change the language that G_2 generates.) Call the start symbols S_1 and S_2. In each case we will define G by taking all the symbols and rules for G_1 and G_2, a new symbol S, which will be the start symbol for G, and the rules listed below.

a) Since we want strings that either G_1 or G_2 generate, we add the rules $S \to S_1$ and $S \to S_2$.

b) Since we want strings that consist of a string that G_1 generates followed by a string that G_2 generates, we add the rule $S \to S_1S_2$.

c) This time we add the rules $S \to S_1S$ and $S \to \lambda$. This clearly gives us all strings that consist of the concatenation of any number of strings that G_1 generates.

23. We simply translate the derivations we gave in the solution to Exercise 1 to tree form, obtaining the following pictures.

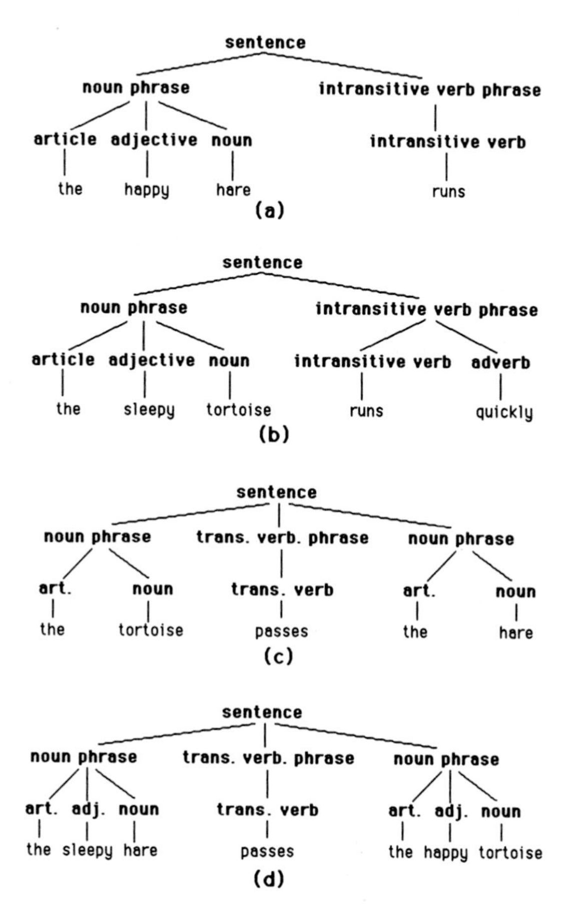

25. We can assume that the derivation starts $S \Rightarrow AB \Rightarrow CaB \Rightarrow cbaB$, or $S \Rightarrow AB \Rightarrow CaB \Rightarrow baB$. This shows that neither the string in part **(b)** nor the string in part **(d)** is in the language, since they do not begin cba or ba. In order to derive the string in part **(a)**, we need to turn B into ba, and this is easy, using the rule $B \to Ba$ and then the rule $B \to b$. Finally, for part **(c)**, we again simply apply these two rules to change B into ba.

27. This is straightforward. The $-$ is the sign and the 109 is an integer, so the tree starts as shown. Then we decompose the integer 109 into the digit 1 and the integer 09, then in turn to the digit 0 and the integer (digit) 9.

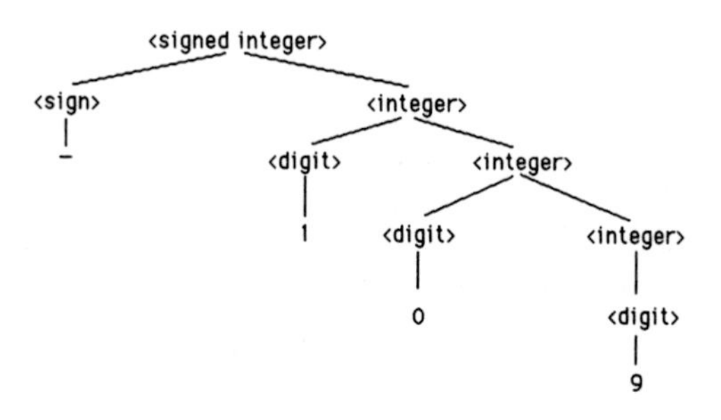

29. a) Note that a string such as "34." is not allowed by this definition, but a string such as -02.780 is. This is pretty straightforward using the following rules. As can be seen, we are using $\langle integer\rangle$ to stand for a nonnegative integer.

$S \to \langle sign\rangle\langle integer\rangle$
$S \to \langle sign\rangle\langle integer\rangle.\langle positive\ integer\rangle$
$\langle sign\rangle \to +$
$\langle sign\rangle \to -$
$\langle integer\rangle \to \langle integer\rangle\langle digit\rangle$
$\langle integer\rangle \to \langle digit\rangle$
$\langle positive\ integer\rangle \to \langle integer\rangle\langle nonzero\ digit\rangle\langle integer\rangle$
$\langle positive\ integer\rangle \to \langle integer\rangle\langle nonzero\ digit\rangle$
$\langle positive\ integer\rangle \to \langle nonzero\ digit\rangle\langle integer\rangle$
$\langle positive\ integer\rangle \to \langle nonzero\ digit\rangle$
$\langle digit\rangle \to \langle nonzero\ digit\rangle$
$\langle digit\rangle \to 0$
$\langle nonzero\ digit\rangle \to 1$
$\langle nonzero\ digit\rangle \to 2$
$\langle nonzero\ digit\rangle \to 3$
$\langle nonzero\ digit\rangle \to 4$
$\langle nonzero\ digit\rangle \to 5$
$\langle nonzero\ digit\rangle \to 6$
$\langle nonzero\ digit\rangle \to 7$
$\langle nonzero\ digit\rangle \to 8$
$\langle nonzero\ digit\rangle \to 9$

b) We combine rows of the previous answer with the same left-hand side, and we change the notation to produce the answer to this part.

$\langle signed\ decimal\ number\rangle ::= \langle sign\rangle\langle integer\rangle \mid \langle sign\rangle\langle integer\rangle.\langle positive\ integer\rangle$
$\langle sign\rangle ::= + \mid -$
$\langle integer\rangle ::= \langle integer\rangle\langle digit\rangle \mid \langle digit\rangle$
$\langle positive\ integer\rangle ::= \langle integer\rangle\langle nonzero\ digit\rangle\langle integer\rangle \mid \langle integer\rangle\langle nonzero\ digit\rangle$
$\qquad \mid \langle nonzero\ digit\rangle\langle integer\rangle \mid \langle nonzero\ digit\rangle$
$\langle digit\rangle ::= \langle nonzero\ digit\rangle \mid 0$
$\langle nonzero\ digit\rangle ::= 1 \mid 2 \mid 3 \mid 4 \mid 5 \mid 6 \mid 7 \mid 8 \mid 9$

c) We easily produce the following tree.

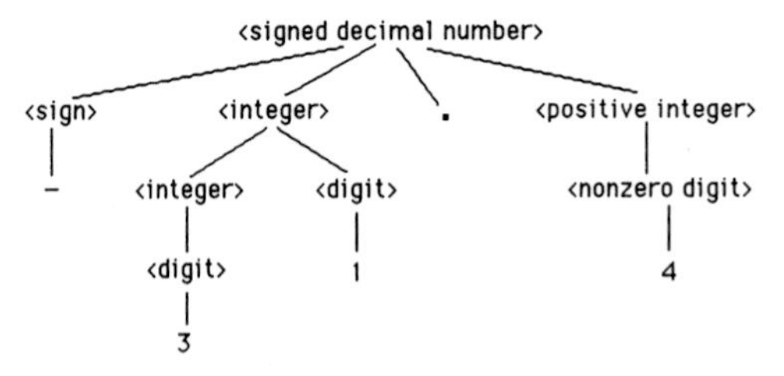

31. a) We can think of appending letters to the end at each stage:

$\langle identifier\rangle ::= \langle lcletter\rangle \mid \langle identifier\rangle\langle lcletter\rangle$
$\langle lcletter\rangle ::= a \mid b \mid c \mid \cdots \mid z$

b) We need to be more explicit here than in part **(a)** about how many letters are used:

$\langle identifier\rangle ::= \langle lcletter\rangle\langle lcletter\rangle\langle lcletter\rangle \mid \langle lcletter\rangle\langle lcletter\rangle\langle lcletter\rangle\langle lcletter\rangle \mid$
$\qquad \langle lcletter\rangle\langle lcletter\rangle\langle lcletter\rangle\langle lcletter\rangle\langle lcletter\rangle \mid$

⟨*lcletter*⟩⟨*lcletter*⟩⟨*lcletter*⟩⟨*lcletter*⟩⟨*lcletter*⟩⟨*lcletter*⟩

⟨*lcletter*⟩ ::= *a* | *b* | *c* | ··· | *z*

c) This is similar to the part **(b)**, allowing for two types of letters:

⟨*identifier*⟩ ::= ⟨*ucletter*⟩ | ⟨*ucletter*⟩⟨*letter*⟩ | ⟨*ucletter*⟩⟨*letter*⟩⟨*letter*⟩ |
⟨*ucletter*⟩⟨*letter*⟩⟨*letter*⟩⟨*letter*⟩ | ⟨*ucletter*⟩⟨*letter*⟩⟨*letter*⟩⟨*letter*⟩⟨*letter*⟩ |
⟨*ucletter*⟩⟨*letter*⟩⟨*letter*⟩⟨*letter*⟩⟨*letter*⟩⟨*letter*⟩

⟨*letter*⟩ ::= ⟨*lcletter*⟩ | ⟨*ucletter*⟩

⟨*lcletter*⟩ ::= *a* | *b* | *c* | ··· | *z*

⟨*ucletter*⟩ ::= *A* | *B* | *C* | ··· | *Z*

d) This is again similar to previous parts. We need to invent a name for "digit or underscore."

⟨*identifier*⟩ ::= ⟨*lcletter*⟩⟨*digitorus*⟩⟨*alphanumeric*⟩⟨*alphanumeric*⟩⟨*alphanumeric*⟩ |
⟨*lcletter*⟩⟨*digitorus*⟩⟨*alphanumeric*⟩⟨*alphanumeric*⟩⟨*alphanumeric*⟩⟨*alphanumeric*⟩

⟨*digitorus*⟩ ::= ⟨*digit*⟩ | _

⟨*alphanumeric*⟩ ::= ⟨*letter*⟩ | ⟨*digit*⟩

⟨*letter*⟩ ::= ⟨*lcletter*⟩ | ⟨*ucletter*⟩

⟨*lcletter*⟩ ::= *a* | *b* | *c* | ··· | *z*

⟨*ucletter*⟩ ::= *A* | *B* | *C* | ··· | *Z*

⟨*digit*⟩ ::= 0 | 1 | 2 | ··· | 9

33. We create a name for "letter or underscore" and then define an identifier to consist of one of those, followed by any number of other allowed symbols. Note that an underscore by itself is a valid identifier, and there is no prohibition on consecutive underscores.

⟨*identifier*⟩ ::= ⟨*letterorus*⟩ | ⟨*identifier*⟩⟨*symbol*⟩

⟨*symbol*⟩ ::= ⟨*letterorus*⟩ | ⟨*digit*⟩

⟨*letterorus*⟩ ::= ⟨*letter*⟩ | _

⟨*letter*⟩ ::= ⟨*lcletter*⟩ | ⟨*ucletter*⟩

⟨*lcletter*⟩ ::= *a* | *b* | *c* | ··· | *z*

⟨*ucletter*⟩ ::= *A* | *B* | *C* | ··· | *Z*

⟨*digit*⟩ ::= 0 | 1 | 2 | ··· | 9

35. We assume that leading 0's are not allowed in the whole number part, since the problem explicitly mentioned them only for the decimal part. Our rules have to allow the optional sign using the question mark, the integer part consisting of one or more digits, not beginning with a 0 unless 0 is the entire whole number part, and then either the decimal part or not. Note that the decimal part has a decimal point followed by zero or more digits.

numeral ::= *sign*? *nonzerodigit digit*∗ *decimal*? | *sign*? 0 *decimal*?

decimal ::= .*digit*∗

digit ::= 0 | *nonzerodigit*

sign ::= + | −

nonzerodigit ::= 1 | 2 | ··· | 9

37. We can simplify the answer given in Exercise 33 using the asterisk for repeating optional elements.

identifier ::= *letterorus symbol*∗

symbol ::= *letterorus* | *digit*

letterorus ::= *letter* | _

letter ::= *lcletter* | *ucletter*

lcletter ::= *a* | *b* | *c* | ··· | *z*

ucletter ::= *A* | *B* | *C* | ··· | *Z*

digit ::= 0 | 1 | 2 | ··· | 9

39. **a)** This string is generated by the grammar. The substring $bc*$ is a term, since it consists of the factor b followed by the factor c followed by the mulOperator $*$. Thus the entire expression consists of two terms followed by an addOperator. We can show the steps in the following sequence:

$\langle \mathit{expression} \rangle$
$\langle \mathit{term} \rangle\langle \mathit{term} \rangle\langle \mathit{addOperator} \rangle$
$\langle \mathit{factor} \rangle\langle \mathit{factor} \rangle\langle \mathit{factor} \rangle\langle \mathit{mulOperator} \rangle\langle \mathit{addOperator} \rangle$
$\langle \mathit{identifier} \rangle\langle \mathit{identifier} \rangle\langle \mathit{identifier} \rangle\langle \mathit{mulOperator} \rangle\langle \mathit{addOperator} \rangle$
$a\, b\, c * +$

b) This string is not generated by the grammar. The second plus sign needs two terms preceding it, and $xy+$ can only be deconstructed to be one term.

c) This string is generated by the grammar. The substring $xy-$ is a factor, since it is an expression, namely the term x followed by the term y followed by the addOperator $-$. Thus the entire expression consists of two factors followed by a mulOperator. We can show the steps in the following sequence:

$\langle \mathit{expression} \rangle$
$\langle \mathit{term} \rangle$
$\langle \mathit{factor} \rangle\langle \mathit{factor} \rangle\langle \mathit{mulOperator} \rangle$
$\langle \mathit{expression} \rangle\langle \mathit{factor} \rangle\langle \mathit{mulOperator} \rangle$
$\langle \mathit{term} \rangle\langle \mathit{term} \rangle\langle \mathit{addOperator} \rangle\langle \mathit{factor} \rangle\langle \mathit{mulOperator} \rangle$
$\langle \mathit{factor} \rangle\langle \mathit{factor} \rangle\langle \mathit{addOperator} \rangle\langle \mathit{factor} \rangle\langle \mathit{mulOperator} \rangle$
$\langle \mathit{identifier} \rangle\langle \mathit{identifier} \rangle\langle \mathit{addOperator} \rangle\langle \mathit{identifier} \rangle\langle \mathit{mulOperator} \rangle$
$x\, y - z *$

d) This is similar to part **(c)**. The entire expression consists of two factors followed by a mulOperator; the first of these factors is just w, and the second is the term $x\, y\, z - *$. That term, in turn, deconstructs as in previous parts. We can show the steps in the following sequence:

$\langle \mathit{expression} \rangle$
$\langle \mathit{term} \rangle$
$\langle \mathit{factor} \rangle\langle \mathit{factor} \rangle\langle \mathit{mulOperator} \rangle$
$\langle \mathit{factor} \rangle\langle \mathit{expression} \rangle\langle \mathit{mulOperator} \rangle$
$\langle \mathit{factor} \rangle\langle \mathit{term} \rangle\langle \mathit{mulOperator} \rangle$
$\langle \mathit{factor} \rangle\langle \mathit{factor} \rangle\langle \mathit{factor} \rangle\langle \mathit{mulOperator} \rangle\langle \mathit{mulOperator} \rangle$
$\langle \mathit{factor} \rangle\langle \mathit{factor} \rangle\langle \mathit{expression} \rangle\langle \mathit{mulOperator} \rangle\langle \mathit{mulOperator} \rangle$
$\langle \mathit{factor} \rangle\langle \mathit{factor} \rangle\langle \mathit{term} \rangle\langle \mathit{term} \rangle\langle \mathit{addOperator} \rangle\langle \mathit{mulOperator} \rangle\langle \mathit{mulOperator} \rangle$
$\langle \mathit{factor} \rangle\langle \mathit{factor} \rangle\langle \mathit{factor} \rangle\langle \mathit{factor} \rangle\langle \mathit{addOperator} \rangle\langle \mathit{mulOperator} \rangle\langle \mathit{mulOperator} \rangle$
$\langle \mathit{identifier} \rangle\langle \mathit{identifier} \rangle\langle \mathit{identifier} \rangle\langle \mathit{identifier} \rangle\langle \mathit{addOperator} \rangle\langle \mathit{mulOperator} \rangle\langle \mathit{mulOperator} \rangle$
$w\, x\, y\, z - * /$

e) This string is generated as follows (similar to previous parts of this exercise):

$\langle \mathit{expression} \rangle$
$\langle \mathit{term} \rangle$
$\langle \mathit{factor} \rangle\langle \mathit{factor} \rangle\langle \mathit{mulOperator} \rangle$
$\langle \mathit{factor} \rangle\langle \mathit{expression} \rangle\langle \mathit{mulOperator} \rangle$
$\langle \mathit{factor} \rangle\langle \mathit{term} \rangle\langle \mathit{term} \rangle\langle \mathit{addOperator} \rangle\langle \mathit{mulOperator} \rangle$
$\langle \mathit{factor} \rangle\langle \mathit{factor} \rangle\langle \mathit{factor} \rangle\langle \mathit{addOperator} \rangle\langle \mathit{mulOperator} \rangle$
$\langle \mathit{identifier} \rangle\langle \mathit{identifier} \rangle\langle \mathit{identifier} \rangle\langle \mathit{addOperator} \rangle\langle \mathit{mulOperator} \rangle$
$a\, d\, e - *$

41. The answers will depend on the grammar given as the solution to Exercise 40. We assume here that the answer to that exercise is very similar to the preamble to Exercise 39. The only difference is that the operators are placed between their operands, rather than behind them, and parentheses are required in expressions used as factors.

a) This string is not generated by the grammar, because the addition operator can only be applied to two terms, and terms that are themselves expressions must be surrounded by parentheses.

b) This string is generated by the grammar. The substrings a/b and c/d are terms, so they can be combined to form the expression. We show the steps in the following sequence:

$\langle \mathit{expression} \rangle$
$\langle \mathit{term} \rangle \langle \mathit{addOperator} \rangle \langle \mathit{term} \rangle$
$\langle \mathit{factor} \rangle \langle \mathit{mulOperator} \rangle \langle \mathit{factor} \rangle \langle \mathit{addOperator} \rangle \langle \mathit{factor} \rangle \langle \mathit{mulOperator} \rangle \langle \mathit{factor} \rangle$
$\langle \mathit{identifier} \rangle \langle \mathit{mulOperator} \rangle \langle \mathit{identifier} \rangle \langle \mathit{addOperator} \rangle \langle \mathit{identifier} \rangle \langle \mathit{mulOperator} \rangle \langle \mathit{identifier} \rangle$
$a/b + c/d$

c) This string is generated by the grammar. The substring $(n + p)$ is a factor, since it is an expression surrounded by parentheses. We show the steps in the following sequence:

$\langle \mathit{expression} \rangle$
$\langle \mathit{term} \rangle$
$\langle \mathit{factor} \rangle \langle \mathit{mulOperator} \rangle \langle \mathit{factor} \rangle$
$\langle \mathit{factor} \rangle \langle \mathit{mulOperator} \rangle (\langle \mathit{expression} \rangle)$
$\langle \mathit{factor} \rangle \langle \mathit{mulOperator} \rangle (\langle \mathit{term} \rangle \langle \mathit{addOperator} \rangle \langle \mathit{term} \rangle)$
$\langle \mathit{factor} \rangle \langle \mathit{mulOperator} \rangle (\langle \mathit{factor} \rangle \langle \mathit{addOperator} \rangle \langle \mathit{factor} \rangle)$
$\langle \mathit{identifier} \rangle \langle \mathit{mulOperator} \rangle (\langle \mathit{identifier} \rangle \langle \mathit{addOperator} \rangle \langle \mathit{identifier} \rangle)$
$m * (n + p)$

d) There are several reasons that this string is not generated, among them the fact that it is impossible for an expression to start with an operator in this grammar.

e) This is very similar to part **(c)**:

$\langle \mathit{expression} \rangle$
$\langle \mathit{term} \rangle$
$\langle \mathit{factor} \rangle \langle \mathit{mulOperator} \rangle \langle \mathit{factor} \rangle$
$(\langle \mathit{expression} \rangle) \langle \mathit{mulOperator} \rangle (\langle \mathit{expression} \rangle)$
$(\langle \mathit{term} \rangle \langle \mathit{addOperator} \rangle \langle \mathit{term} \rangle) \langle \mathit{mulOperator} \rangle (\langle \mathit{term} \rangle \langle \mathit{addOperator} \rangle \langle \mathit{term} \rangle)$
$(\langle \mathit{factor} \rangle \langle \mathit{addOperator} \rangle \langle \mathit{factor} \rangle) \langle \mathit{mulOperator} \rangle (\langle \mathit{factor} \rangle \langle \mathit{addOperator} \rangle \langle \mathit{factor} \rangle)$
$(\langle \mathit{identifier} \rangle \langle \mathit{addOperator} \rangle \langle \mathit{identifier} \rangle) \langle \mathit{mulOperator} \rangle (\langle \mathit{identifier} \rangle \langle \mathit{addOperator} \rangle \langle \mathit{identifier} \rangle)$
$(m + n) * (p - q)$

SECTION 12.2 Finite-State Machines with Output

Finding finite-state machines to do specific tasks is in essence computer programming. There is no set method for doing this. You have to think about the problem for awhile, ask yourself what it might be useful for the states to represent, and then very carefully proceed to construct the machine. Expect to have several false starts. "Bugs" in your machines are also very common. There are of course many machines that will accomplish the same task. The reader should look at Exercises 20–25 to see that it is also possible to build finite-state machines with the output associated with the states, rather than the transitions.

1. We draw the state diagrams by making a node for each state and a labeled arrow for each transition. In part **(a)**, for example, since under input 1 from state s_2 we are told that we move to state s_1 and output a 0, we draw an arrow from s_2 to s_1 and label it $1,0$. It is assumed that s_0 is always the start state.

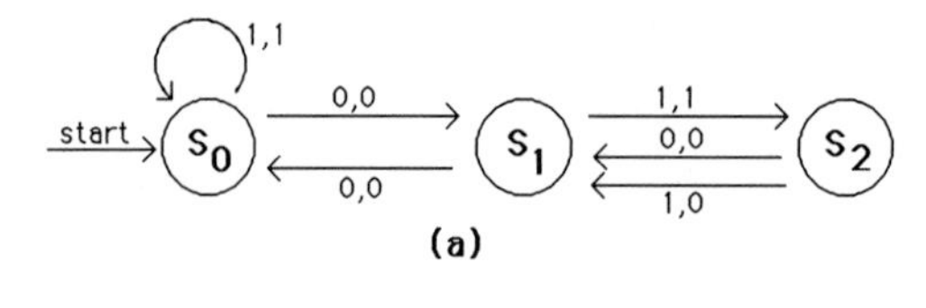

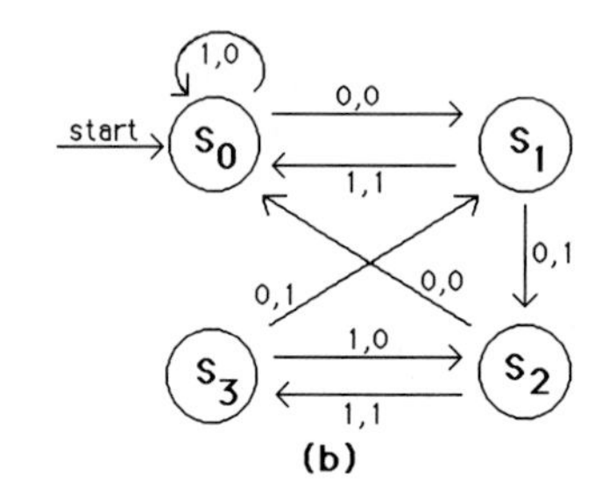

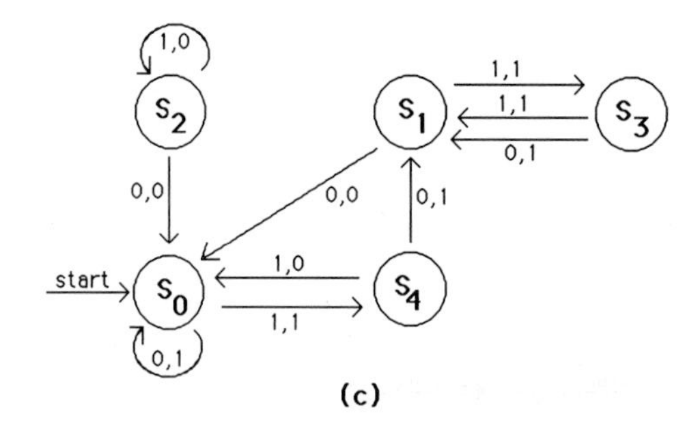

3. **a)** The machine starts in state s_0. Since the first input symbol is 0, the machine moves to state s_1 and gives 0 as output. The next input symbol is 1, so the machine moves to state s_2 and gives 1 as output. The next input is 1, so the machine moves to state s_1 and gives 0 as output. The fourth input is 1, so the machine moves to state s_2 and gives 1 as output. The fifth input is 0, so the machine moves to state s_1 and gives 0 as output. Thus the output is 01010.

b) The machine starts in state s_0. Since the first input symbol is 0, the machine moves to state s_1 and gives 0 as output. The next input symbol is 1, so the machine moves to state s_0 and gives 1 as output. The next input is 1, so the machine stays in state s_0 and gives 0 as output. The fourth input is 1, so the machine again stays in state s_0 and gives 0 as output. The fifth input is 0, so the machine moves to state s_1 and gives 0 as output. Thus the output is 01000.

c) The machine starts in state s_0. Since the first input symbol is 0, the machine stays in state s_0 and gives 1 as output. The next input symbol is 1, so the machine moves to state s_4 and gives 1 as output. The next input is 1, so the machine moves to state s_0 and gives 0 as output. The fourth input is 1, so the machine moves to state s_4 and gives 1 as output. The fifth input is 0, so the machine moves to state s_1 and gives 1 as output. Thus the output is 11011.

5. **a)** The machine starts in state s_0. Since the first input symbol is 0, the machine moves to state s_1 and gives 1 as output. (This is what the arrow from s_0 to s_1 with label $0,1$ means.) The next input symbol is 1. Because of the edge from s_1 to s_0, the machine moves to state s_0 and gives 1 as output. The next input is 1. Because of the loop at s_0, the machine stays in state s_0 and gives output 0. The same thing happens on the fourth input symbol. Therefore the output is 1100 (and the machine ends up in state s_0).

b) This is similar to part **(a)**. The first two symbols of input cause the machine to output two 0's and remain in state s_0. The third symbol causes an output of 1 as the machine moves into state s_1. The fourth input takes us back to state s_0 with output 1. The next four symbols of input cause the machine to give output 0110 as it goes to states s_0, s_1, s_0, and s_0, respectively. Therefore the output is 00110110.

c) This is similar to the other parts. The machine alternates between states s_0 and s_1, outputting 1 for each input. Thus the output is 11111111111.

7. We model this machine as follows. There are four possible inputs, which we denote by 5, 10, 25, and b, standing for a nickel, a dime, a quarter, and a button labeled by a kind of soda pop, respectively. (Actually the model is a bit more complicated, since there are three kinds of pop, but we will ignore that; to incorporate the kind of pop into the model, we would simply have three inputs in place of just b.) The output can either be an amount of money in cents—0, 5, 10, 15, 20, or 25—or can be a can of soda pop, which we denote c. There will be eight states. Intuitively, state s_i will represent the state in which the machine is indebted to the customer by $5i$ cents. Thus s_0, the start state, will represent that the machine owes the customer nothing; state s_1 will represent that the machine has accepted 5 cents from the customer, and so on. State s_7 will mean that the machine owes the customer 35 cents, which will be paid with a can of soda pop, at which time the machine will return to state s_0, owing nothing. The following picture is the state diagram of this machine, simplified even further in that we have eliminated quarters entirely for sake of readability. For example, the transition from state s_6 (30 cents credit) on input of a dime is to state s_7 (35 cents credit) with the return of 5 cents in change. We have also used a to stand for any monetary input: if you deposit any amount when the machine already has your 35 cents, then you get that same amount back. Thus the transition a, a really stands for three transitions: $5, 5$ and $10, 10$ and $25, 25$.

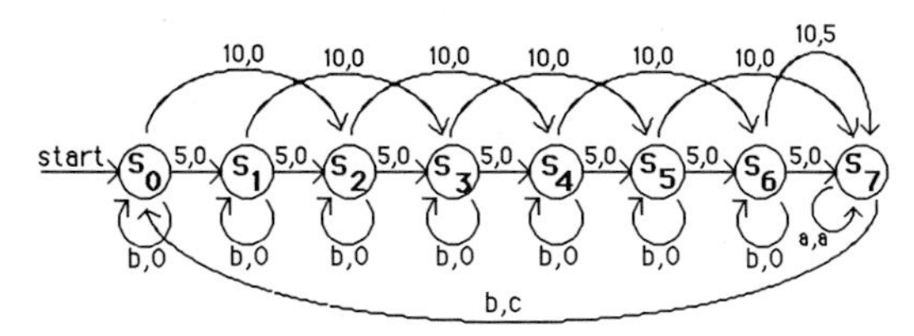

9. We draw the diagram for this machine. Intuitively, we need four states, corresponding to the four possibilities for what the last two bits have been. In our picture, state s_1 corresponds to the last two bits having been 00; state s_2 corresponds to the last two bits having been 01; state s_3 corresponds to the last two bits having been 10; state s_4 corresponds to the last two bits having been 11. We also need a state s_0 to get started, to account for the delay. Let us see why some of the transitions are what they are. If you are in state s_3, then the last two bits have been 10. If you now receive an input 0, then the last two bits will be 00, so we need to move to state s_1. Furthermore, since the bit received two pulses ago was a 1 (we know this from the fact that we are in state s_3), we need to output a 1. Also, since we are told to output 00 at the beginning, it is right to have transitions from s_0 as shown.

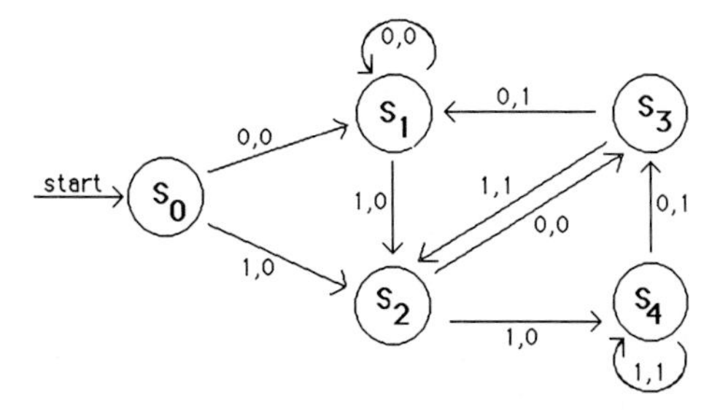

If we look at this machine, we observe that states s_0 and s_1 are equivalent, i.e., they cause exactly the same transitions and outputs. Therefore a simpler answer would be a machine like this one, but without state s_0, where state s_1 is the start state.

11. This machine is really only part of a machine; we are not told what happens after a successful log-on. Also, the machine is really much more complicated than we are indicating here, because we really need a separate state for each user. We assume that there is only one user. We also assume that an invalid user ID is rejected immediately, without a request for a password. (The alternate assumption is also reasonable, that the machine requests a password whether or not the ID is valid. In that case we obtain a different machine, of course.) We need only two states. The initial state waits for the valid user ID. We let i be the valid user ID, and we let j be any other input. If the input is valid, then we enter state s_1, outputting the message e: "enter your password." If the input is not valid, then we remain in state s_0, outputting the message t: "invalid ID; try again." From state s_1 there are only two relevant inputs: the valid password p and any other input q. If the input is valid, then we output the message w: "welcome" and proceed. If the input is invalid, then we output the message a: "invalid password; enter user ID again" and return to state s_0 to await another attempt at logging-on.

13. This exercise is similar to Exercise 7. We let state s_i for $i = 0, 1, 2, 3, 4$ represent the fact that $5i$ cents has been deposited. When at least 25 cents has been deposited, we return to state s_0 and open the gate. Nickels (input 5), dimes (input 10) and quarters (input 25) are available. We let o and c be the outputs: the gate is opened (for a limited time, of course), or remains closed. After the gate is opened, we return to state s_0. (We assume that the gate closes after the car has passed.)

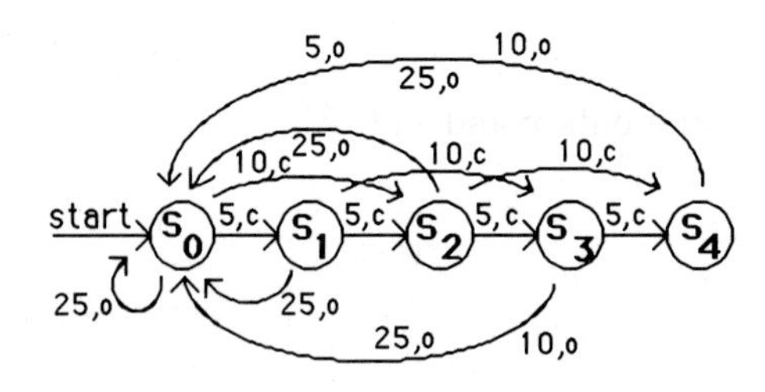

15. The picture for this machine would be too complex to draw. Instead, we will describe the machine verbally, and even then we won't give every last gory detail. We assume that possible inputs are the ten digits. We will let s_0 be the start state and let s_1 be the state representing a successful call (so we will not list any outputs from s_1). From s_0, inputs of 2, 3, 4, 5, 6, 7, or 8 send the machine back to s_0 with output of an error message for the user. From s_0 an input of 0 sends the machine to state s_1, with the output being that the 0 is sent to the network. From s_0 an input of 9 sends the machine to state s_2 with no output; from there an input of 1 sends the machine to state s_3 with no output; from there an input of 1 sends the machine to state s_1 with the output being that the 911 is sent to the network. All other inputs while in states s_2 or s_3 send the machine back to s_0 with output of an error message for the user. From s_0 an input of 1 sends the machine to state s_4 with no output; from s_4 an input of 2 sends the machine to state s_5 with no output; and this path continues in a similar manner to the 911 path, looking next for 1, then 2, then any seven digits, at which point the machine goes to state s_1 with the output being that the ten-digit input is sent to the network. Any "incorrect" input while in states s_5 or s_6 (that is, anything except a 1 while in s_5 or a 2 while in s_6) sends the machine back to s_0 with output of an error message for the user. Similarly, from s_4 an input of 8 followed by appropriate successors drives us eventually to s_1, but inappropriate outputs drive us back to s_0 with an error message. Also, inputs while in state s_4 other than 2 or 8 send the machine back s_0 with output of an error message for the user.

17. We interpret this problem as asking that a 1 be output if the conditions are met, and a 0 be output otherwise. For this machine, we need to keep track of what the last two inputs have been, and we need four states to "store" this information. Let the states s_3, s_4, s_5, and s_6 be the states corresponding to the last two inputs having been 00, 10, 01, and 11, respectively. We also need some states to get started—to get us into one of these four states. There are only two cases in which the output is 1: if we are in states s_3 or s_5 (so that the last two inputs have been 00 or 01) and we receive a 1 as input. The transitions in our machine are the obvious ones. For example, if we are in state s_5, having just read 01, and receive a 0 as input, then the last two symbols read are now 10, so we move to state s_4.

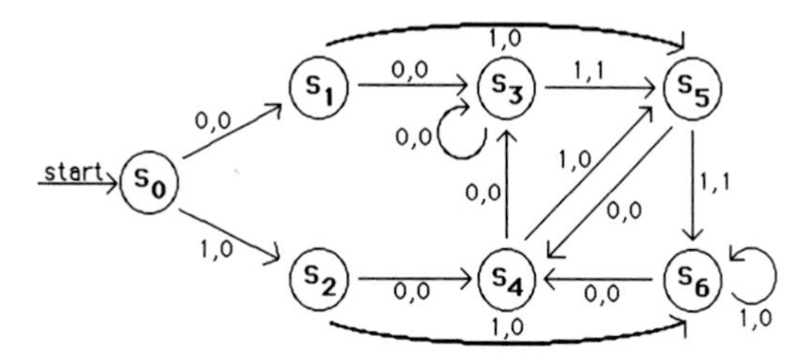

As in Exercise 9, we can actually get by with a smaller machine. Note that here states s_1 and s_4 are equivalent, as are states s_2 and s_6. Thus we can merge each of these pairs into one state, producing a machine with only five states. At that point, furthermore, state s_0 is equivalent to the merged s_2 and s_6, so we can omit state s_0 and make this other state the start state. The reader is urged to draw the diagram for this simpler machine.

19. We need some notation to make our picture readable. The alphabet has 26 symbols in it. If α is a letter, then by $\overline{\alpha}$ we mean any letter other than α. Thus an arrow labeled $\overline{\alpha}$ really stands for 25 arrows. The output is to be 1 when we have just finished reading the word *computer*. Thus we need eight states, to stand for the various stages of having read part of that word. The picture below gives the details, except that we have omitted all the outputs except on inputs r and $\overline{r}$; all the omitted ones are intended to be 0. The reader might contemplate why this problem would have been harder if the word in question were something like *baboon*.

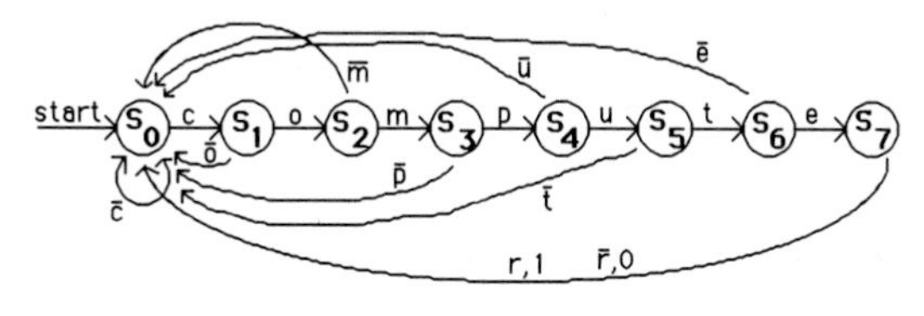

21. We construct the state table by having one row for each state. The arrows tell us what the values of the transition function are. For example, since there is an arrow from s_0 to s_1 labeled 0, the transition from s_0 on input 0 is to s_1. Similarly, the transition from s_0 on input 1 is to s_2. The output function values are shown next to each state. Thus the output for state s_0 is 1, the output for state s_1 is 1, and the output for state s_2 is 0. The table is therefore as shown here.

State	Input 0	Input 1	Output
s_0	s_1	s_2	1
s_1	s_1	s_0	1
s_2	s_1	s_2	0

23. a) The input drives the machine successively to states s_1, s_0, s_1, and s_0. The output is the output of the start state, followed by the outputs of these four states, namely 11111.

b) The input drives the machine to state s_2, where it remains because of the loop. The output is the output of the start state, followed by the output at state s_2 six times, namely 1000000.

c) The states visited after the start state are, in order, s_2, s_2, s_2, s_1, s_0, s_2, s_2, s_1, s_0, s_2, and s_2. Therefore the output is 100011001100.

25. We can use a machine with just two states, one to indicate that there is an even number of 1's in the input string, the other to indicate that there is an odd number of 1's in the string. Since the empty string has an even number of 1's, we make s_0 (the start state) the state for an even number of 1's. The output for this state will be 1, as directed. The output from state s_1 will be 0 to indicate an odd number of 1's. The input 1 will drive the machine from one state to the other, while the input 0 will keep the machine in its current state. The diagram below gives the desired machine.

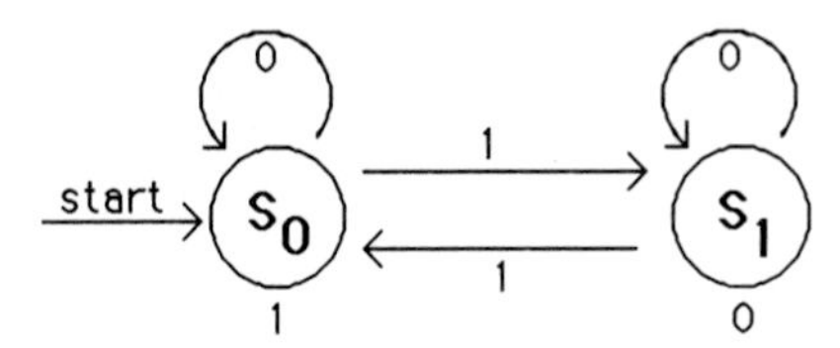

SECTION 12.3 Finite-State Machines with No Output

As in the previous section, many of these exercises are really exercises in programming. There is no magical way to become a good programmer, but experience helps. The converse problem is also hard—finding a good verbal description of the set recognized by a given finite-state automaton.

1. a) This is the set of all strings ab, where $a \in A$ and $b \in B$. Thus it contains precisely 000, 001, 1100, and 1101.

b) This is the set of all strings ba, where $a \in A$ and $b \in B$. Thus it contains precisely 000, 0011, 010, and 0111.

c) This is the set of all strings a_1a_2, where $a_1 \in A$ and $a_2 \in A$. Thus it contains precisely 00, 011, 110, and 1111.

d) This is the set of all strings $b_1b_2b_3$, where each $b_i \in B$. Thus it contains precisely 000000, 000001, 000100, 000101, 010000, 010001, 010100 and 010101.

3. Two possibilities are of course to let A be this entire set and let $B = \{\lambda\}$, and to let B be this entire set and let $A = \{\lambda\}$. Let us find more. With a little experimentation we see that $A = \{\lambda, 10\}$ and $B = \{10, 111, 1000\}$ also works, and it can be argued that there are no other solutions in which λ appears in either set. Finally, there is the solution $A = \{1, 101\}$ and $B = \{0, 11, 000\}$. It can be argued that there are no more. (Here is how the first of these arguments goes. If $\lambda \in A$, then necessarily $\lambda \notin B$. Hence the shortest string in B has length at least 2, from which it follows that $10 \in B$. Now since the only other string in AB that ends with 10 is 1010, the only possible other string in A is 10. This leads to the third solution mentioned above. On the other hand, if $\lambda \in B$, then $\lambda \notin A$, so it must be that the shortest string in A is 10. This forces 111 to be in A, and now there can be no other strings in B. The second argument is similar.)

5. a) One way to write this answer is $\{\,(10)^n \mid n = 0, 1, 2, \ldots\}$. It is the concatenation of zero or more copies of the string 10.

b) This is like part **(a)**. This set consists of all copies of zero or more concatenations of the string 111. In other words, it is the set of all strings of 1's of length a multiple of 3. In symbols, it is $\{\,(111)^n \mid n = 0, 1, 2, \ldots\} = \{\,1^{3n} \mid n = 0, 1, 2, \ldots\}$.

c) A little thought will show that this consists of all bit strings in which every 1 is immediately preceded by a 0. No other restrictions are imposed, since $0 \in A$.

d) Because the 0 appears only in 101, the strings formed here have the property that there are at least two 1's between every pair of 0's in the string, and the string begins and ends with a 1. All strings satisfying this property are in A^*.

7. This follows directly from the definition. Every string w in A^* consists of the concatenation of one or more strings from A. Since $A \subseteq B$, all of these strings are also in B, so w is the concatenation of one or more strings from B, i.e., is in B^*.

9. a) This set contains all bit strings, so of course the answer is yes.

b) This set contains all strings consisting of any number of 1's, followed by any number of 0's, followed by any number of 1's. Since 11101 is such a string, the answer is yes.

c) Any string belonging to this set must start 110, and 11101 does not, so the answer is no.

d) All the strings in this set must in particular have even length. The given string has odd length, so the answer is no.

e) The answer is yes. Just take one copy of each of the strings 111 and 0, together with the required string 1.

f) The answer is yes again. Just take 11 from the first set and 101 from the second.

11. In each case we will list the states in the order that they are visited, starting with the initial state. All we need to do then is to note whether the place we end up is a final state (s_0 or s_3) or a nonfinal state. (It is interesting to note that there are no transitions to s_3, so this state can never be reached.)

a) We encounter $s_0s_1s_2s_0$, so this string is accepted.

b) We encounter $s_0s_0s_0s_1s_2$, so this string is not accepted.

c) We encounter $s_0s_1s_0s_1s_0s_1s_2s_0$, so this string is accepted.

d) We encounter $s_0s_0s_1s_2s_0s_1s_2s_0s_1s_2$, so this string is not accepted.

13. a) The set in question is the set of all strings of zero or more 0's. Since the machine in Figure 1 has s_0 as a final state, and since there is a transition from s_0 to itself on input 0, every string of zero or more 0's will leave the machine in state s_0 and will therefore be accepted. Therefore the answer is yes.

b) Since this set is a subset of the set in part **(a)**, the answer must be yes.

c) One string in this set is the string 1. Since an input of 1 drives the machine to the nonfinal state s_1, not every string in this set is accepted. Therefore the answer is no.

d) One string in this set is the string 01. Since an input of 01 drives the machine to the nonfinal state s_1, not every string in this set is accepted. Therefore the answer is no.

e) The answer here is no for exactly the same reason as in part **(d)**.

f) The answer here is no for exactly the same reason as in part **(c)**.

15. We use structural induction on the input string y. The basis step is $y = \lambda$, and for the inductive step we write $y = wa$, where $w \in I^*$ and $a \in I$. For the basis step, we have $xy = x$, so we must show that $f(s,x) = f(f(s,x),\lambda)$. But part (*i*) of the definition of the extended transition function says that this is true. We then assume the inductive hypothesis that the equation holds for shorter strings and try to prove that $f(s,xwa) = f(f(s,x),wa)$. By part (*ii*) of the definition, the left-hand side of this equation equals $f(f(s,xw),a)$. By the inductive hypothesis (because w is shorter than y), $f(s,xw) = f(f(s,x),w)$, so $f(f(s,xw),a) = f(f(f(s,x),w),a)$. On the other hand, the right-hand side of our desired equality is, by part (*ii*) of the definition, equal to $f(f(f(s,x),w),a)$. We have shown that the two sides are equal, and our proof is complete.

17. The only final state is s_2, so we need to determine which strings drive the machine to state s_2. Clearly the strings 0, 10, and 11 do so, as well as any of these strings followed by anything else. Thus we can write the answer as $\{0,10,11\}\{0,1\}^*$.

19. A string is accepted if and only if it drives this machine to state s_1. Thus the string must consist of zero or more 0's, followed by a 1, followed by zero or more 1's. In short, the answer is $\{\,0^m1^n \mid m \geq 0 \land n \geq 1\,\}$.

21. Because s_0 is final, the empty string is accepted. The strings that drive the machine to final state s_3 are precisely $\{0\}\{1\}^*\{0\}$. There are three ways to get to final state s_4, and once we get there, we stay there. The path through s_2 tells us that strings in $\{10,11\}\{0,1\}^*$ are accepted. The path $s_0s_1s_3s_4$ tells us that strings in $\{0\}\{1\}^*\{01\}\{0,1\}^*$ are accepted. And the path $s_0s_1s_3s_5s_4$ tells us that strings in $\{0\}\{1\}^*\{00\}\{0\}^*\{1\}\{0,1\}^*$ are accepted. Thus the language recognized by this machine is $\{\lambda\}\cup\{0\}\{1\}^*\{0\}\cup\{10,11\}\{0,1\}^*\cup\{0\}\{1\}^*\{01\}\{0,1\}^*\cup\{0\}\{1\}^*\{00\}\{0\}^*\{1\}\{0,1\}^*$.

23. We want to accept only the strings that begin 01. Let s_2 be the only final state, and put transitions from s_2 to itself on either input. We want to reach s_2 after encountering 01, so put a transition from the start state s_0 to s_1 on input 0, and a transition from s_1 to s_2 on input 1. Finally make a "graveyard" state s_3, and have the other transitions from s_0 and s_1 (as well as both transitions from s_3) lead to s_3.

25. We can have a sequence of three states to record the appearance of 101. State s_1 will signify that we have just seen a 1; state s_2 will signify that we have just seen a 1 followed by a 0; state s_3 will be the only final state and will signify that we have seen the string 101. Put transitions from s_3 to itself on either input (it doesn't matter what follows the appearance of 101). Put a transition from the start state s_0 to itself on input 0, because we are still waiting for a 1. Put a transition from s_0 to s_1 on input 1 (because we have just seen a 1). From s_1 on input 0 we want to go to state s_2, but on input 1 we stay at s_1 because we have still just seen a 1. Finally, from s_2, put a transition on input 1 to the final state s_3 (success!), but on input 0 we have to start over looking for 101, so this transition must be back to s_0.

27. We can let state s_i, for $i = 0, 1, 2, 3$ represent that exactly i 0's have been seen, and state s_4 will represent that four or more 0's have been seen. Only s_3 will be final. For $i = 0, 1, 2, 3$, we transition from s_i to itself on input 1 and to s_{i+1} on input 0. Both transitions from s_4 are to itself.

29. We can let state s_i, for $i = 0, 1, 2, 3$ represent that i consecutive 1's have been seen. Only s_3 will be final. For $i = 0, 1, 2$, we transition from s_i to s_{i+1} on input 1 but back to s_0 on input 0. Both transitions from s_3 are to itself.

31. This is a little tricky. We want states at the start that prevent us from accepting a string if it does not start with 11. Once we have seen the first two 1's, we can accept the string if we do not encounter a 0 (after all, the strings 11 and 111 do satisfy the condition). We can also accept the string if it has anything whatsoever in the middle, as long as it ends 11. The machine shown below accomplishes all this. Note that s_3 is a graveyard state, and state s_4 is where we "start over" looking for the final 11.

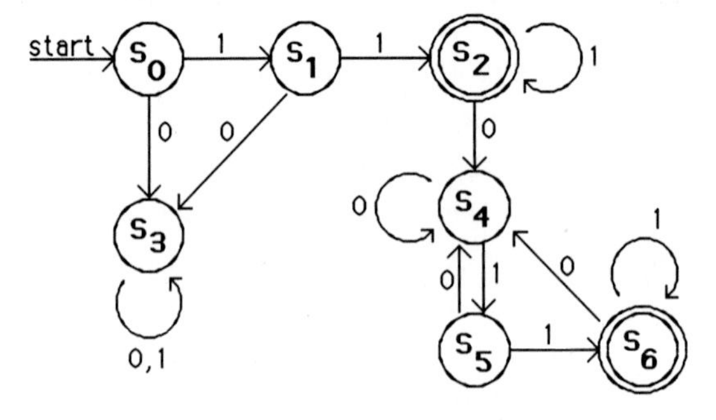

33. We need just two states, s_0 to represent having seen an even number of 0's (this will be the start state, because to begin we have seen no 0's), and s_1 to represent having seen an odd number of 0's (this will be the only final state). The transitions are from each state to itself on input 1, and from each state to the other on input 0.

35. This is similar to Exercise 33, except that we need to look for the initial 0. Note that s_3 is the graveyard.

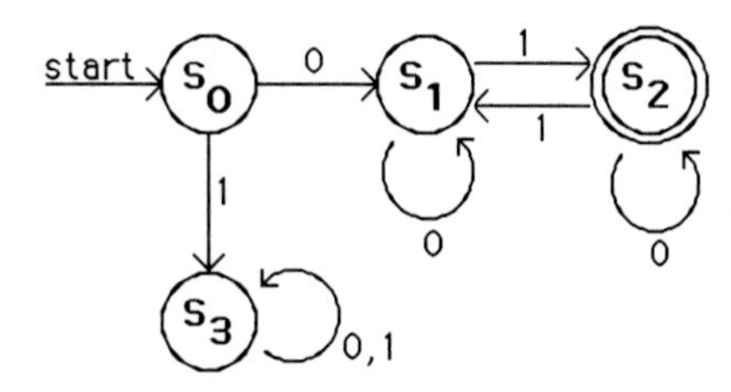

37. We prove this by contradiction. Suppose that such a machine exists, with start state s_0 and other state s_1. Because the empty string is not in the language but some strings are accepted, we must have s_1 as the only final state, with at least one transition from s_0 to s_1. Because the string 0 is not in the language, any transition from s_0 on input 0 must be to itself, so there must be a transition from s_0 to s_1 on input 1. But this cannot happen, because the string 1 is not in the language. Having obtained a contradiction, we conclude that no such finite-state automaton exists.

39. We want the new machine to accept exactly those strings that the original machine rejects, and vice versa. So we simply change each final state to a nonfinal state and change each nonfinal state to a final state.

41. We use exactly the same machine as in Exercise 25, but make s_0, s_1, and s_2 the final states and make s_3 nonfinal.

43. First some general comments on Exercises 43–49: In general it is quite hard to describe succinctly languages recognized by machines. An ad hoc approach is usually best. In this exercise there is only one final state, s_2, and only three ways to get there, namely on input 0, 01, or 11. Therefore the language recognized by this machine is $\{0, 01, 11\}$.

45. Clearly the empty string is accepted. There are essentially two ways to get to the final state s_2. We can go through state s_1, and every string of the form 0^n1^m, where n and m are positive integers, will take us through state s_1 on to s_2. We can also bypass state s_1, and every string of the form 01^m for $m \geq 0$ will take us directly to s_2. Thus our answer is $\{\lambda\} \cup \{\, 0^n1^m \mid n, m \geq 1\,\} \cup \{\, 01^m \mid m \geq 0\,\}$. Note that this can also be written as $\{\lambda, 0\} \cup \{\, 0^n1^m \mid n, m \geq 1\,\}$.

47. First it is easy to see that all strings of the form 10^n for $n \geq 0$ can drive the machine to the final state s_1. Next we see that all strings of the form 10^n10^m for $n, m \geq 0$ can drive the machine to state s_3. No other strings can drive the machine to a final state. Therefore the answer is $\{\, 10^n \mid n \geq 0\,\} \cup \{\, 10^n10^m \mid n, m \geq 0\,\}$.

49. We notice first that state s_2 is a final state, that once we get there, we can stay there, and that any string that starts with a 0 can lead us there. Therefore all strings that start with a 0 are in the language. If the string starts with a 1, then we must go first to state s_1. If we ever leave state s_1, then the string will not be accepted, because there are no paths out of s_1 that lead to a final state. Therefore the only other strings that are in the language are the empty string (because s_0 is final) and those strings that can drive the machine to state s_1, namely strings consisting of all 1's (we've already included those of the form 01^*). Therefore the language accepted by this machine is the union of the set of all strings that start with a 0 and the set of all strings that have no 0's.

51. One way to do Exercises 50–54 is to construct a machine following the proof of Theorem 1. Rather than do that, we construct the machines in an ad hoc way, using the answers obtained in Exercises 43–47. Since λ, 0, and 1 are accepted by the nondeterministic automaton in Exercise 44, we make states s_0, s_1, and s_2 in the following diagram final. States s_3 and s_4 provide for the acceptance of strings of the form 1^n0 for all $n \geq 1$. State s_5, the graveyard state, assures that no other strings are accepted.

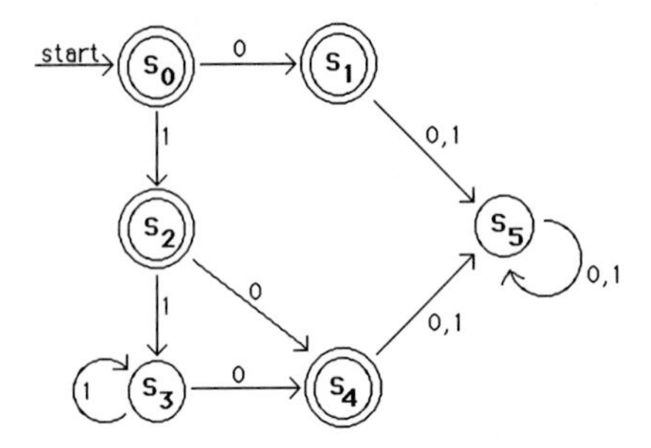

53. This machine is practically deterministic already, since there are no cases of ambiguous transitions (a given input allowing transition to more than one state). All that keeps this machine from being deterministic is that there are no transitions from certain states on certain inputs. Therefore to make this machine deterministic, we just need to add a "graveyard" state, s_3, with transitions from s_0 on input 0 and from s_1 on input 1 to this graveyard state, and transitions from s_3 to itself on input 0 or 1. The graveyard state is not final, of course.

55. a) We want to accept only the string 0. Let s_1 be the only final state, where we reach s_1 on input 0 from the start state s_0. Make a "graveyard" state s_2, and have all other transitions (there are five of them in all) lead there.

b) This uses the same idea as in part **(a)**, but we need a few more states. The graveyard state is s_4. See the picture for details.

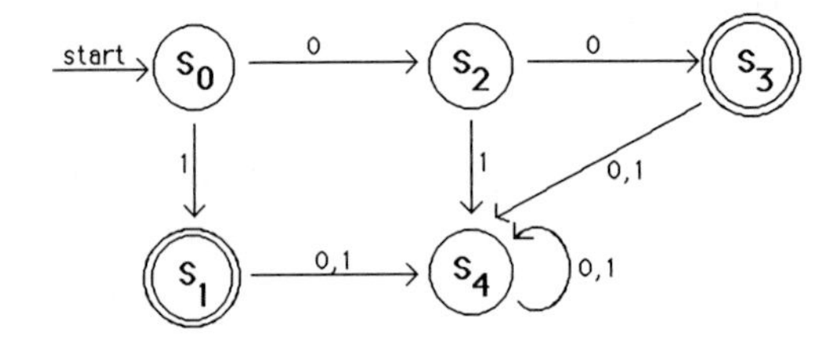

c) In the picture of our machine, we show a transition to the graveyard state whenever we encounter a 0. The only final state is s_2, which we reach after 11 and remain at as long as the input consists just of 1's.

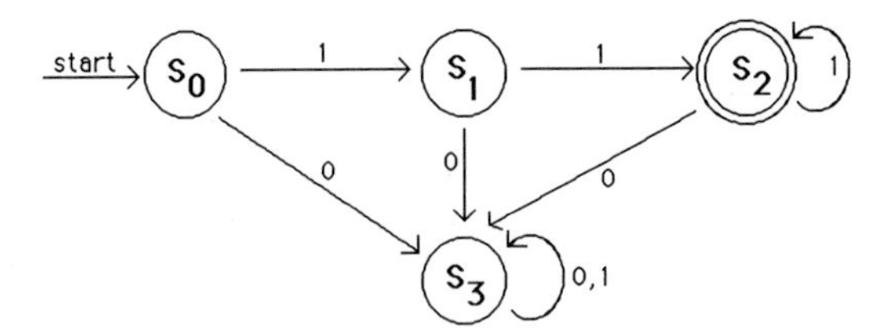

57. Intuitively, the reason that a finite-state automaton cannot recognize the set of bit strings containing an equal number of 0's and 1's is that there is not enough "memory" in the machine to keep track of how many extra 0's or 1's the machine has read so far. Of course, this intuition does not constitute a proof—maybe we are just not being clever enough to see how a machine could do this with a finite number of states. Instead, we must give a proof of this assertion. See Exercises 22–25 of Section 12.4 for a development of what are called "pumping lemmas" to handle various problems like this. (See also Example 6 in Section 12.4.)

The natural way to prove a negative statement such as this is by contradiction. So let us suppose that we do have a finite-state automaton M that accepts precisely the set of bit strings containing an equal number of 0's and 1's. We will derive a contradiction by showing that the machine must accept some illegal strings. The idea behind the proof is that since there are only finitely many states, the machine must repeat some states as it computes. In this way, it can get into arbitrarily long loops, and this will lead us to a contradiction. To be specific, suppose that M has n states. Consider the string $0^{n+1}1^{n+1}$. As the machine processes this string, it must encounter the same state more than once as it reads the first $n+1$ 0's (by the pigeonhole principle). Say that it hits state s twice. Then some positive number, say k, of 0's in the input drives M from state s back to state s. But then the machine will end up at exactly the same place after reading $0^{n+1+k}1^{n+1}$ as it will after reading $0^{n+1}1^{n+1}$, since those extra k 0's simply drove it in a loop. Therefore since M accepts $0^{n+1}1^{n+1}$, it also accepts $0^{n+1+k}1^{n+1}$. But this is a contradiction, since this latter string does not have the same number of 0's as 1's.

59. We know from Exercise 58d that the equivalence classes of R_k are a refinement of the equivalence classes of R_{k-1} for each positive integer k. The equivalence classes are finite sets, and finite sets cannot be refined indefinitely (the most refined they can be is for each equivalence class to contain just one state). Therefore this sequence of refinements must stabilize and remain unchanged from some point onward. It remains to show that as soon as we have $R_n = R_{n+1}$, then $R_n = R_m$ for all $m > n$, from which it follows that $R_n = R_*$, and so the equivalence classes for these two relations will be the same. By induction, it suffices to show that if $R_n = R_{n+1}$, then $R_{n+1} = R_{n+2}$. By way of contradiction, suppose that $R_{n+1} \neq R_{n+2}$. This means that there are states s and t that are $(n+1)$-equivalent but not $(n+2)$-equivalent. Thus there is a string x of length $n+2$ such that, say, $f(s,x)$ is final but $f(t,x)$ is nonfinal. Write $x = aw$, where $a \in I$. Then $f(s,a)$ and $f(t,a)$ are not $(n+1)$-equivalent, because w drives the first to a final state and the second to a nonfinal state. But $f(s,a)$ and $f(t,a)$ are n-equivalent, because s and t are $(n+1)$-equivalent. This contradicts the fact that $R_n = R_{n+1}$, and our proof is complete.

61. **a)** By the way the machine $\overline{M}$ was constructed, a string will drive M from the start state to a final state if and only if that string drives $\overline{M}$ from the start state to a final state.

b) For a proof of this theorem, see a source such as *Introduction to Automata Theory, Languages, and Computation* (2nd Edition) by John E. Hopcroft, Rajeev Motwani, and Jeffrey D. Ullman (Addison Wesley, 2000).

SECTION 12.4 Language Recognition

Finding good verbal descriptions of the set of strings generated by a regular expression is not easy; neither is finding a good regular expression for a given verbal description. What Kleene's Theorem says is that these problems of "programming" in regular expressions are really the same as the programming problems for machines discussed in the previous section. The **pumping lemma**, *discussed in Exercise 22 and the three exercises that follow it, is an important technique for proving that certain sets are not regular.*

1. **a)** This regular expression generates all strings consisting of zero or more 1's, followed by a lone 0.

b) This regular expression generates all strings consisting of zero or more 1's, followed by one or more 0's.

c) This set has only two elements, 111 and 001.

d) This set contains all strings in which the 0's come in pairs.

e) This set consists of all strings in which every 1 is preceded by at least one 0, with the proviso that the string ends in a 1 if it is not the empty string.

f) This gives us all strings of length at least 3 that end 00.

3. In each case we try to view 0101 as fitting the regular expression description.

a) The strings described by this regular expression have at most three "blocks" of different digits—a 0, then some 1's, then some 0's. Thus we cannot get the string 0101, which has four blocks.

b) The 1's that might come between the first and second 0 in any string described by this regular expression must come in pairs (because of the $(11)^*$). Therefore we cannot get 0101. Alternatively, note that every string described by this regular expression must have odd length.

c) We can get this string as $0(10)^1 1^1$.

d) We can get this string as $0^1 10(1)$, where the final 1 is one of the choices in $(0 \cup 1)$.

e) We can get this string as $(01)^2(11)^0$.

f) We cannot get this string, because every string with any 1's at all described by this regular expression must end with 10 or 11.

g) We cannot get this string, because every string described by this regular expression must end with 11.

h) We can get this string as $01(01)1^0$, where the second 01 is one of the choices in $(01 \cup 0)$.

5. a) We just need to take a union: $\mathbf{0} \cup \mathbf{11} \cup \mathbf{010}$.

b) More simply put, this is the set of strings of five or more 0's, so the regular expression is $\mathbf{000000}^*$.

c) We can use $(\mathbf{0} \cup \mathbf{1})$ to represent any symbol and $(\mathbf{00} \cup \mathbf{01} \cup \mathbf{10} \cup \mathbf{11})$ to represent any string of even length. We need one symbol followed by any string of even length, so we can take $(\mathbf{0} \cup \mathbf{1})(\mathbf{00} \cup \mathbf{01} \cup \mathbf{10} \cup \mathbf{11})^*$.

d) The one 1 can be preceded and/or followed by any number of 0's, so we have $\mathbf{0}^*\mathbf{10}^*$.

e) This one is a little harder. In order to prevent 000 from appearing, we must have every group of one or two 0's followed by a 1 (if we note that the entire string ends with a 1 as well). Thus we can break our string down into groups of 1, 01, or 001, and we get $(\mathbf{1} \cup \mathbf{01} \cup \mathbf{001})^*$ as our regular expression.

7. a) We can translate "one or more 0's" into $\mathbf{00}^*$. Therefore the answer is $\mathbf{00}^*\mathbf{1}$.

b) We can translate "two or more symbols" into $(\mathbf{0} \cup \mathbf{1})(\mathbf{0} \cup \mathbf{1})(\mathbf{0} \cup \mathbf{1})^*$. Therefore the answer is $(\mathbf{0} \cup \mathbf{1})(\mathbf{0} \cup \mathbf{1})(\mathbf{0} \cup \mathbf{1})^*\mathbf{0000}^*$.

c) A little thought tells us that we want all strings in which all the 0's come before all the 1's or all the 1's come before all the 0's. Thus the answer is $\mathbf{0}^*\mathbf{1}^* \cup \mathbf{1}^*\mathbf{0}^*$.

d) The string of 1's can be represented by $\mathbf{11}(\mathbf{111})^*$; the string of 0's, by $(\mathbf{00})^*$. Thus the answer is $\mathbf{11}(\mathbf{111})^*(\mathbf{00})^*$.

9. a) The simplest solution here is to have just the start state s_0, nonfinal, with no transitions.

b) The simplest solution here is to have just the start state s_0, final, with no transitions.

c) The simplest solution here is to have just two states—the nonfinal start state s_0 (since we do not want to accept the empty string) and a final state s_1—and just the one transition from s_0 to s_1 on input a.

11. We can prove this by induction on the length of a regular expression for A. If this expression has length 1, then it is either $\emptyset$ or λ or x (where x is some symbol in the alphabet). In each case A is its own reversal, so there is nothing to prove. There are three inductive steps. If the regular expression for A is $\mathbf{BC}$, then $A = BC$, where B is the set generated by $\mathbf{B}$ and C is the set generated by $\mathbf{C}$. By the inductive hypothesis, we know that there are regular expressions $\mathbf{B}'$ and $\mathbf{C}'$ that generate B^R and C^R, respectively. Now $A^R = (BC)^R = (C^R)(B^R)$. Therefore a regular expression for A^R is $\mathbf{C}'\mathbf{B}'$. The case of union is handled similarly. Let the regular expression for A be $\mathbf{B} \cup \mathbf{C}$, with B, C, $\mathbf{B}'$, and $\mathbf{C}'$ as before. Then a regular expression for A^R is $\mathbf{B}' \cup \mathbf{C}'$, since clearly $(B \cup C)^R = (B^R) \cup (C^R)$. Finally, if the regular expression for A is $\mathbf{B}^*$, then, with the same notation as before, it is easy to see that $(\mathbf{B}')^*$ is a regular expression for A^R.

13. a) We can build machines to recognize $\mathbf{0}^*$ and $\mathbf{1}^*$ as shown in the second row of Figure 3. Next we need to put these together to make a machine that recognizes $\mathbf{0}^*\mathbf{1}^*$. We place the first machine on the left and the second machine on the right. We make each final state in the first machine nonfinal (except for the start state, since $\lambda \in \mathbf{0}^*\mathbf{1}^*$), but leave the final states in the second machine final. Next we copy each transition to a state that was formerly final in the first machine into a transition (on the same input) to the start state of the second machine. Lastly, since $\lambda \in \mathbf{0}^*$, we add the transition from the start state to the state to which there is a transition from the start state of the machine for $\mathbf{1}^*$. The result is as shown. (In all parts of this exercise we have not put names on the states in our state diagrams.)

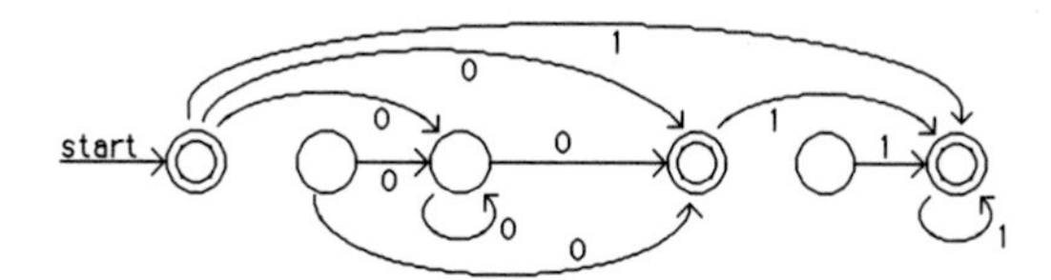

b) This machine is quite messy. The upper portion is for $\mathbf{0}$, and the lower portion is for $\mathbf{11}$. They are combined to give a machine for $\mathbf{0} \cup \mathbf{11}$. Finally, to incorporate the Kleene star, we added a new start state (on the far left), and adjusted the transitions according to the procedure shown in Figure 2.

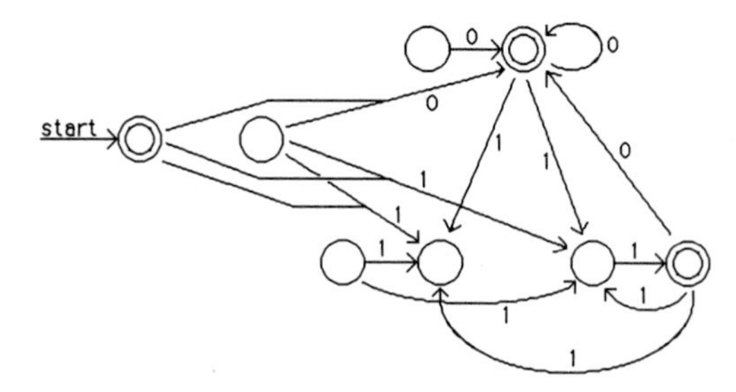

c) This is similar to the other parts. We grouped the expression as $\mathbf{01}^* \cup (\mathbf{00}^*)\mathbf{1}$. The answer is as shown.

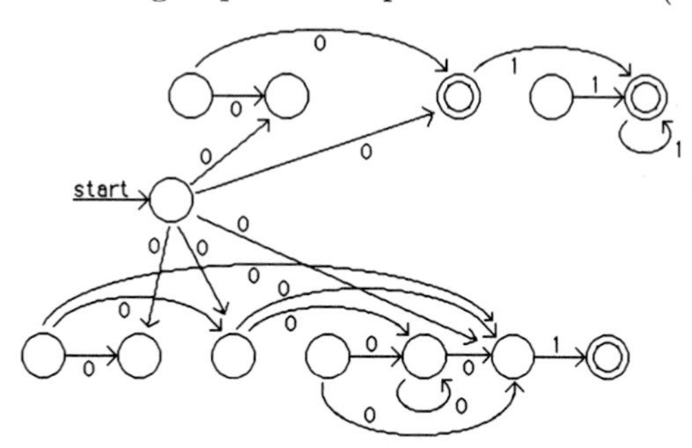

15. We choose as the nonterminal symbols corresponding to states s_0, s_1, and s_2 the symbols S, A, and B, respectively. Thus S is our start symbol. The terminal symbols are of course 0 and 1. We construct the rules for our grammars by following the procedure described in the proof of the second half of Theorem 2: putting in rules of the form $X \rightarrow aY$ for each transition from the state corresponding to X to the state corresponding to Y, on input a, and putting in a rule of the form $X \rightarrow a$ for a transition from the state corresponding to X to the final state, on input a. Specifically, since there is a transition from s_0 to s_1 on input 0, we include the rule $S \rightarrow 0A$. Similarly, the other transitions give us the rules $S \rightarrow 1B$, $A \rightarrow 0B$, $A \rightarrow 1B$, $B \rightarrow 0B$, and $B \rightarrow 1B$. Also, the transition to the final state from S on input 0 gives rise to the rule $S \rightarrow 0$. Thus our grammar contains these seven rules.

17. This is similar to Exercise 15—see the discussion there for the approach. We let C correspond to state s_3. The set of rules contains $S \rightarrow 0C$, $S \rightarrow 1A$, $A \rightarrow 1A$, $A \rightarrow 0C$, $B \rightarrow 0B$, $B \rightarrow 1B$, $C \rightarrow 0C$, and $C \rightarrow 1B$ (for the transitions from a state to another state on a given input), as well as $S \rightarrow 1$, $A \rightarrow 1$, $B \rightarrow 0$, $B \rightarrow 1$, and $C \rightarrow 1$ (for the transitions to the final states that can end the computation).

19. This is clear, since the operation of the machine is exactly mimicked by the grammar. If the current string in the derivation in the grammar is $v_1v_2\ldots v_kA_s$, then the machine has seen input $v_1v_2\ldots v_k$ and is currently in state s. If the current string in the derivation in the grammar is $v_1v_2\ldots v_k$, then the machine has seen input $v_1v_2\ldots v_k$ and is currently in some final state. Hence the machine accepts precisely those strings that the grammar generates. (The empty string does not fit this discussion, but it is handled separately—and correctly—since we take $S \to \lambda$ as a production if and only if we are supposed to.)

21. First suppose that the language recognized by M is infinite. Then the length of the words recognized by M must be unbounded, since there are only a finite number of symbols. Thus $l(x)$ is greater than the finite number $|S|$ for some word $x \in L(M)$.

Conversely, let x be such a word, and let s_0, s_{i_1}, s_{i_2}, ..., s_{i_n} be the sequence of states that the machine goes through on input x, where $n = l(x)$ and s_{i_n} is a final state. By the pigeonhole principle, some state occurs twice in this sequence, i.e., there is a loop from this state back to itself during the computation. Let y be the substring of x that causes the loop, so that $x = uyv$. Then for every nonnegative integer k, the string uy^kv is accepted by the machine M (i.e., is in $L(M)$), since the computation is the same as the computation on input x, except that the loop is traversed k times. Thus $L(M)$ is infinite.

23. We apply the pumping lemma in a proof by contradiction. Suppose that this set were regular. Clearly it contains arbitrarily long strings. Thus the pumping lemma tells us that for some strings u, $v \neq \lambda$, and w, the string uv^iw is in our set for every i. Now if v contains both 0's and 1's, then uv^2w cannot be in the set, since it would have a 0 following a 1, which no string in our set has. On the other hand, if v contains only 0's (or only 1's), then for large enough i, it is clear that uv^iw has more than (or less than) twice as many 0's as 1's, again contradicting the definition of our set. Thus the set cannot be regular.

25. We will give a proof by contradiction, using the pumping lemma. Following the hint, let x be the palindrome 0^N10^N, for some fixed $N > |S|$, where S is the set of states in a machine that recognizes palindromes. By the lemma, we can write $x = uvw$, with $l(uv) \leq |S|$ and $l(v) \geq 1$, so that for all i, uv^iw is a palindrome. Now since $0^N10^N = uvw$ and $l(uv) \leq |S| < N$, it must be the case that v is a string consisting solely of 0's, with the 1 lying in w. Then uv^2w cannot be a palindrome, since it has more 0's before its sole 1 than it has 0's following the 1.

27. It helps to think of L/x in words—it is the set of "ends" of strings in L that start with the string x; in other words, it is the set of strings obtained from strings in L by stripping away an initial piece x. To show that 11 and 10 are distinguishable, we need to find a string z such that $11z \in L$ and $10z \notin L$ or vice versa. A little thought and trial and error shows us that $z = 1$ works: $111 \notin L$ but $101 \in L$. To see that 1 and 11 are indistinguishable, note that the only way for $1z$ to be in L is for z to end with 01, and that is also the only way for $11z$ to be in L.

29. By Exercise 28, if two strings are distinguishable, then they drive the machine from the start state to different states. Therefore, if x_1, x_2, ..., x_n are all distinguishable, the states $f(s_0, x_1)$, $f(s_0, x_2)$, ..., $f(s_0, x_n)$ are all different, so the the machine has at least n states.

31. We claim that any two distinct strings of the same length are distinguishable with respect to the language P of all palindromes. Indeed, if x and y are distinct strings of length n, let $z = x^R$ (the reverse of string x). Then $xz \in P$ but $yz \notin P$. Note that there are 2^n different strings of length n. By Exercise 29, this tells us that any deterministic finite-state automaton for recognizing palindromes must have at least 2^n states. Because n is arbitrary (we want our machine to recognize *all* palindromes), this tells us that no finite-state machine can recognize P.

SECTION 12.5 Turing Machines

In this final section of the textbook, we have studied a machine that has all the computing capabilities possible (if one believes the Church–Turing thesis). Most of these exercises are really programming assignments, and the programming language you are stuck with is not a nice, high-level, structured language like Java or C, nor even a nice assembly language, but something much messier and less efficient. One point of the exercises is to convince you that even in this horrible setting you can, with enough time and patience, instruct the computer—the Turing machine—to do whatever you wish computationally. Keep in mind that in many senses, a Turing machine is just as powerful as any computer running programs written in any language. One reason for talking about Turing machines at all, rather than just using high-level languages, is that their simplicity makes it feasible to prove some very interesting things about them (and therefore about computers in general). For example, one can prove that computers cannot solve the halting problem (see also Section 3.1), and one can prove that a large class of problems have efficient algorithmic solutions if and only if certain very specific problems, such as a decision version of the traveling salesman problem, do (the NP-complete problems—see also Section 3.3). This is part of what makes Turing machines so important in theoretical computer science, and time spent becoming acquainted with them will not go unrewarded as you progress in this field.

1. We will indicate the configuration of the Turing machine using a notation such as $0[s_2]1B1$. This string of symbols means that the tape is blank except for a portion which reads $01B1$ from left to right; that the machine is currently in state s_2; and that the tape head is reading the left 1 (the currently scanned symbol will always be the one following the bracketed state information).

a) The initial configuration is $[s_0]0011$. Because of the five-tuple $(s_0, 0, s_1, 1, R)$ and the fact that the machine is in state s_0 and the tape head is looking at a 0, the machine changes the 0 to a 1 (i.e., writes a 1 in that square), moves to the right, and enters state s_1. Therefore the configuration at the end of one step of the computation is $1[s_1]011$. Next the transition given by the five-tuple $(s_1, 0, s_2, 1, L)$ occurs, and we reach the configuration $[s_2]1111$. There are no five-tuples starting with s_2, so the machine halts at this point. The nonblank portion of the tape contains 1111.

b) The initial configuration is $[s_0]101$. Because of the five-tuple $(s_0, 1, s_1, 0, R)$ and the fact that the machine is in state s_0 and the tape head is looking at a 1, the machine changes the 1 to a 0, moves to the right, and enters state s_1. Therefore the configuration at the end of one step of the computation is $0[s_1]01$. At this time transition $(s_1, 0, s_2, 1, L)$ kicks in, resulting in configuration $[s_2]011$, and the machine halts, with 011 on its tape.

c) We seem to have the idea from the first two parts, so let us just list the configurations here, using the notation "$\rightarrow$" to show the progression from one to the next. $[s_0]11B01 \rightarrow 0[s_1]1B01 \rightarrow 00[s_1]B01 \rightarrow 0[s_2]0001$. Therefore the final output is 00001.

d) $[s_0]B \rightarrow 0[s_1]B \rightarrow [s_2]00$. So the final tape reads 00.

3. Note that all motion is from left to right.

a) The machine starts in state s_0 and sees the first 1. Therefore using the second five-tuple, it replaces the 1 by a 0, moves to the right, and enters state s_1. Now it sees the second 1, so, using the fifth five-tuple, it replaces the 1 by a 1 (i.e., leaves it unchanged), moves to the right, and enters state s_0. The third five-tuple now tells it to leave the blank it sees alone, move to the right, and enter state s_2, which is a final (accepting) state (because it is not the first state in any five-tuple). Since there are no five-tuples telling the machine what to do in state s_2, it halts. Note that 01 is on the tape, and the input was accepted.

b) When in state s_0 the machine skips over 0's, ignoring them, until it comes to a 1. When (and if) this happens, the machine changes this 1 to a 0 and enters state s_1. Note also that if the machine hits a blank (B) while in state s_0 or s_1, then it enters the final (accepting) state s_2. Next note that s_1 plays a role similar to that played by s_0, causing the machine to skip over 0's, but causing it to go back into state s_0 if and when it encounters a 1. In state s_1, however, the machine does not change the 1 it sees to a 0. Thus the machine will alternate between states s_0 and s_1 as it encounters 1's in the input string, changing half of these 1's to 0's. To summarize, if the machine is given a bit string as input, it scans it from left to right, changing every other occurrence of a 1, if any, starting with the first, to a 0, and otherwise leaving the string unchanged; it halts (and accepts) when it comes to the end of the string.

5. **a)** The machine starts in state s_0 and sees the first 1. Therefore using the first five-tuple, it replaces the 1 by a 0, moves to the right, and enters state s_1. Now it sees the second 1, so, using the second five-tuple, it replaces the 1 by a 1 (i.e., leaves it unchanged), moves to the right, and stays in state s_1. Since there are no five-tuples telling the machine what to do in state s_1 when reading a blank, it halts. Note that 01 is on the tape, and the input was not accepted, because s_1 is not a final state; in fact, there are no final states (states that begin no 5-tuples).

b) This is essentially the same as part **(a)**. The first 1 (if any) is changed to a 0 and the others are left alone. The input is not accepted.

7. The machine needs to search for the first 0 and when (and if) it finds it, replace it with a 1. So let's have the machine stay in its initial state (s_0) as long as it reads 1's, constantly moving to the right. If it ever reads a 0 it will enter state s_1 while changing the 0 to a 1. No further action is required. Thus we can get by with just the following two five-tuples: $(s_0, 0, s_1, 1, R)$ and $(s_0, 1, s_0, 1, R)$. Note that if the input string consists of just 1's, then the machine eventually sees the terminating blank and halts.

9. The machine should scan the tape, leaving it alone until it has encountered the first 1. At that point, it needs to enter a phase in which it changes all the 1's to 0's, until it reaches the end of the input. So we'll have tuples $(s_0, 0, s_0, 0, R)$ and $(s_0, 1, s_1, 1, R)$ to complete the first phase, and then have tuples $(s_1, 0, s_1, 0, R)$ and $(s_1, 1, s_1, 0, R)$ to complete the second phase. When the machine encounters the end of the input (a blank on the tape) it halts, since there are no transitions given with a blank as the scanned symbol.

11. We can have the machine scan the input tape until it reaches the first blank, "remembering" what the last symbol was that it read. Let us use state s_0 to represent that last symbol's being a 1, and s_1 to represent its being a 0. It doesn't matter what gets written, so we'll just leave the tape unchanged as we move from left to right. Thus our first few five-tuples are $(s_0, 0, s_1, 0, R)$, $(s_0, 1, s_0, 1, R)$, $(s_1, 0, s_1, 0, R)$, $(s_1, 1, s_0, 1, R)$. Now suppose the machine encounters the end of the input, namely the blank at the end of the input string. If it is in state s_0, then the last symbol read was not a 0, so we want to not accept the string. If it is in state s_1, then the last symbol read was a 0, so we want to accept the string. Recall that the convention presented in this section was that acceptance is indicated by halting in a final state, i.e., one with no transitions out of it. So let's add the five-tuple (s_1, B, s_2, B, R) for accepting when we should. To make sure we don't accept when we shouldn't, we need do nothing else, because the machine will halt in the nonfinal state s_0 in this case.

An alternative approach to this problem is to have the machine scan to the right until it reaches the end of the tape, then back up, "look" at the last symbol, and take the appropriate action.

13. This is very similar to Exercise 11. We want the machine to "remember" whether it has seen an even number of 1's or not. We'll let s_0 be the state representing that an even number of 1's have been seen (which is of course true at the start of the computation), and let s_1 be the state representing that an odd number of 1's have been seen. So we put in the following tuples: $(s_0, 0, s_0, 0, R)$, $(s_0, 1, s_1, 1, R)$, $(s_1, 0, s_1, 0, R)$, and $(s_1, 1, s_0, 1, R)$. When the machine encounters the terminating blank, we want it to accept if it is in state s_0, so we add the tuple (s_0, B, s_2, B, R). Thus the machine will halt in final state s_2 if the input string has an even number of 0's, and it will halt in nonfinal state s_1 otherwise.

15. You need to play with this machine to get a feel for what is going on. After doing so, you will understand that it operates as follows. If the input string is blank or starts with a 1, then the machine halts in state s_0, which is not final, and therefore every such string is not accepted (which is a good thing, since it is not in the set to be recognized). Otherwise the initial 0 is changed to an M, and the machine skips past all the intervening 0's and 1's until it either comes to the end of the input string or else comes to an M (which, as we will see, has been written over the right-most remaining bit). At this point it backs up (moves left) one square and is in state s_2. Since the acceptable strings must have a 1 at the right for each 0 at the left, there had better be a 1 here if the string is acceptable. Therefore the only transition out of state s_2 occurs when this square contains a 1. If it does, then the machine replaces it with an M, and makes its way back to the left. (If this square does not contain a 1, then the machine halts in the nonfinal state s_2, as appropriate.) On its way back, it stays in state s_3 as long as it sees 1's, then stays in s_4 as long as it sees 0's. Eventually either it encounters a 1 while in state s_4, at which point it (appropriately) halts without accepting (since the string had a 0 to the right of a 1); or else it reaches the right-most M that had been written over a 0 near the beginning of the string. If it is in state s_3 when this happens, then there are no more 0's in the string, so it had better be the case (if we want to accept this string) that there are no more 1's either; this is accomplished by the transitions (s_3, M, s_5, M, R) and (s_5, M, s_6, M, R), and s_6 is a final state. Otherwise, the machine halts in nonfinal state s_5. If it is in state s_4 when this M is encountered, then we need to start all over again, except that now the string will have had its left-most remaining 0 and its right-most remaining 1 replaced by M's. So the machine moves (staying in state s_4) to the left-most remaining 0 and goes back into state s_0 to repeat the process.

17. This will be similar to the machine in Example 3, in that we will change the digits one at a time to a new symbol M. We can't work from the outside in as we did there, however, so we'll replace all three digits from left to right. Furthermore, we'll put a new symbol, E, at the left end of the input in order to tell more easily when we have arrived back at the starting point. Here is our plan for the states and the transitions that will accomplish our goal. State s_9 is our (accepting) final state. States s_0 and s_1 will write an E to the left of the initial input and return to the first input square, entering state s_2. (If, however, the tape is blank, then the machine will accept immediately, and if the first symbol is not a 0, then it will reject immediately.) The five-tuples are (s_0, B, s_9, B, L), $(s_0, 0, s_1, 0, L)$, and (s_1, B, s_2, E, R). State s_2 will skip past any M's until it finds the first 0, change it to an M, and enter state s_3. The transitions are (s_2, M, s_2, M, R) and $(s_2, 0, s_3, M, R)$. Similarly, state s_3 will skip past any remaining 0's and any M's until it finds the first 1, change it to an M, and enter state s_4. The transitions are $(s_3, 0, s_3, 0, R)$, (s_3, M, s_3, M, R) and $(s_3, 1, s_4, M, R)$. State s_4 will do the same for the first 2 (skipping past remaining 1's and M's, and ending in state s_5), with transitions $(s_4, 1, s_4, 1, R)$, (s_4, M, s_4, M, R) and $(s_4, 2, s_5, M, R)$. State s_5 then will skip over any remaining 2's and (if there is any chance of accepting this string) encounter the terminating blank. The transitions are $(s_5, 2, s_5, 2, R)$ and (s_5, B, s_6, B, L). Note that once this blank has been seen, we back up to the last symbol before it and enter state s_6. There are now two possibilities. If the scanned square is an M, then we should accept if and only if the entire string consists of M's at this point. We will enter state s_8 to check this, with the transition (s_6, M, s_8, M, L). Otherwise, there will be a 2 here, and we want to go back to the start of the string to begin the cycle all over; we'll use state s_7 to accomplish this, so we put in the five-tuple $(s_6, 2, s_7, 2, L)$. In this latter case, the machine should skip over everything until it sees the marker E that we put at the left end of the input, then move back to the initial input square, and start over in state s_2. The transitions $(s_7, 0, s_7, 0, L)$, $(s_7, 1, s_7, 1, L)$, $(s_7, 2, s_7, 2, L)$, (s_7, M, s_7, M, L), and (s_7, E, s_2, E, R) accomplish this. But if we entered state s_8, then we need to make sure that there is nothing but M's all the way back to the starting point; we add the five-tuples (s_8, M, s_8, M, L) and (s_8, E, s_9, E, L), and we're finished.

19. Recall that functions are computed in a funny way using unary notation. The string representing n is a string of $n + 1$ 1's. Thus we want our machine to erase three of these 1's (or all but one of them, if there are fewer than four), and then halt. One way to accomplish this is as follows. If $n \geq 3$, then the five-tuples $(s_0, 1, s_1, B, R)$, $(s_1, 1, s_2, B, R)$, $(s_2, 1, s_3, B, R)$, and $(s_3, 1, s_4, 1, R)$ will do the trick (s_4 is just a halting state). To account for the possibilities that $n < 3$, we add transitions $(s_1, B, s_4, 1, R)$, $(s_2, B, s_4, 1, R)$, and $(s_3, B, s_4, 1, R)$. In each of these three cases, we needed to restore one 1 before halting (since the "answer" was to be 0).

21. The machine here first needs to "decide" whether $n \geq 5$. If it finds that $n \geq 5$, then it needs to leave exactly four 1's on the tape (according to our rules for representing numbers in unary); otherwise it needs to leave exactly one 1. We'll use states s_0 through s_6 for this task, with the following five-tuples, which erase the tape as they move from left to right through the input: $(s_0, 1, s_1, B, R)$, $(s_1, 1, s_2, B, R)$, (s_1, B, s_6, B, R), $(s_2, 1, s_3, B, R)$, (s_2, B, s_6, B, R), $(s_3, 1, s_4, B, R)$, (s_3, B, s_6, B, R), $(s_4, 1, s_5, B, R)$, (s_4, B, s_6, B, R). At this point, the machine is either in state s_5 (and $n \geq 5$), or in state s_6 with a blank tape (and $n < 5$). To finish in the latter case, we just write a 1 and halt: $(s_6, B, s_{10}, 1, R)$. For the former case, we erase the rest of the tape, write four 1's, and halt: $(s_5, 1, s_5, B, R)$, $(s_5, B, s_7, 1, R)$, $(s_7, B, s_8, 1, R)$, $(s_8, B, s_9, 1, R)$, and $(s_9, B, s_{10}, 1, R)$.

23. We start with a string of $n+1$ 1's, and we want to end up with a string of $3n+1$ 1's. Our idea will be to replace the last 1 with a 0, then for each 1 to the left of the 0, write a pair of new 1's to the right of the 0. To keep track of which 1's we have processed so far, we will change each left-side 1 to a 0 as we process it. At the end, we will change all the 0's back to 1's. Basically our states will mean the following ("first" means "first encountered"): s_0, scan right for last 1; s_1, change the last 1 to 0; s_2, scan left to first 1; s_3, scan right for end of input (having replaced the 1 where we started with a 0); s_3 and s_4, write the two more 1's; s_5, scan left to first 0; s_6, replace the remaining 0's with 1's; s_7, halt.

The needed five-tuples are as follows: $(s_0, 1, s_0, 1, R)$, (s_0, B, s_1, B, L), $(s_1, 1, s_2, 0, L)$, $(s_2, 0, s_2, 0, L)$, $(s_2, 1, s_3, 0, R)$, (s_2, B, s_6, B, R), $(s_3, 0, s_3, 0, R)$, $(s_3, 1, s_3, 1, R)$, $(s_3, B, s_4, 1, R)$, $(s_4, B, s_5, 1, L)$, $(s_5, 1, s_5, 1, L)$, $(s_5, 0, s_2, 0, L)$, $(s_6, 0, s_6, 1, R)$, $(s_6, 1, s_7, 1, R)$, (s_6, B, s_7, B, R).

25. The idea here is to match off the 1's in the two inputs (changing the 1's to 0's from the left, say, to keep track), until one of them is exhausted. At that point, we need to erase the larger input entirely (as well as the asterisk) and change the 0's back to 1's. Here is how we'll do it. In state s_0 we skip over any 0's until we come to either a 1 or the $*$. If it's the $*$, then we know that the second input (n_2) is at least as large as the first (n_1), so we enter a clean-up state s_5, which erases the asterisk and all the 0's and 1's to its right. The five-tuples for this much are $(s_0, 0, s_0, 0, R)$, $(s_0, *, s_5, B, R)$, $(s_5, 1, s_5, B, R)$, and $(s_5, 0, s_5, B, R)$. Once this erasing is finished, we need to go over to the part of the tape where the first input was and change all the 0's back to 1's; the following transitions accomplish this: (s_5, B, s_6, B, L), (s_6, B, s_6, B, L), $(s_6, 0, s_7, 1, L)$, and $(s_7, 0, s_7, 1, L)$. Eventually the machine halts in state s_7 when the blank preceding the original input is encountered.

The other possibility is that the machine encounters a 1 while in state s_0. We want to change this 1 to a 0, skip over any remaining 1's as well as the asterisk, skip over any 0's to the right of the asterisk (these represent parts of n_2 that have already been matched off against equal parts of n_1), and then either find a 1 in n_2 (which we change to a 0) or else come to the blank at the end of the input. Here are the transitions: $(s_0, 1, s_1, 0, R)$, $(s_1, 1, s_1, 1, R)$, $(s_1, *, s_2, *, R)$, $(s_2, 0, s_2, 0, R)$, $(s_2, 1, s_3, 0, L)$, (s_2, B, s_4, B, L). At this point we are either in state s_3, ready to go back for the next iteration, or in state s_4 ready for some cleanup. In the former case, we want to skip back over the nonblank symbols until we reach the start of the string, so we add five-tuples $(s_3, *, s_3, *, L)$, $(s_3, 0, s_3, 0, L)$, $(s_3, 1, s_3, 1, L)$, and (s_3, B, s_0, B, R). In the latter case, we know that the first string is longer than the second. Therefore we want to change the 0's in the second input string back to 1's and then erase the asterisk and remnants of the first input string. Here are the transitions: $(s_4, 0, s_4, 1, L)$, $(s_4, *, s_8, B, L)$, $(s_8, 0, s_8, B, L)$, $(s_8, 1, s_8, B, L)$.

27. The discussion in the preamble tells how to take the machines from Exercises 22 and 18 and create a new machine. The only catch is that the tape head needs to be back at the leftmost 1. Suppose that s_m, where m is the largest index, is the state in which the Turing machine for Exercise 22 halts after completing its work, and suppose that we have designed that machine so that when the machine halts the tape head is reading the leftmost 1 of the answer. Then we renumber each state in the machine for Exercise 18 by adding m to each subscript, and take the union of the two sets of five-tuples.

29. If the answer is yes/no, then the problem is a decision problem.

a) No, the answer here is a number, not yes or no.

b) Yes, the answer is either yes or no.

c) Yes, the answer is either yes or no.

d) Yes, the answer is either yes or no.

31. This is a fairly hard problem, which can be solved by patiently trying various combinations. The following five-tuples will do the trick: $(s_0, B, s_1, 1, L)$, $(s_0, 1, s_1, 1, R)$, $(s_1, B, s_0, 1, R)$.

GUIDE TO REVIEW QUESTIONS FOR CHAPTER 12

1. a) See p. 787. **b)** See p. 787.

2. a) See p. 788. **b)** $\{0^{3n}1 \mid n \geq 0\}$

c) The vocabulary is $\{S, 0, 1\}$; the terminals are $T = \{0, 1\}$; the start symbol is S; and the productions are $S \to S1$ and $S \to 0$.

3. a) See p. 789 (see also the top of p. 790). **b)** a grammar that contains a production like $AB \to C$

c) See p. 789. **d)** a grammar that contains a production like $Sa \to Sbc$

e) See p. 789. **f)** a grammar that contains a production like $S \to SS$

4. a) See p. 789. **b)** See p. 789.

c) See Example 8 in Section 12.1.

5. a) See p. 792. **b)** See Example 14 in Section 12.1.

6. a) See p. 798 (machines with output) and p. 805 (machines without output). See also p. 801 for comments on other types of finite-state machines.

b) Have three states and only one input symbol, Q (quarter). The start state s_0 has a transition to state s_1 on input Q and outputs nothing; state s_1 has a transition to state s_2 on input Q and outputs nothing; state s_2 has a transition back to state s_1 on input Q and outputs a drink.

7. $1^* \cup 1^*00$

8. Have four states, with only s_2 final. From the start state s_0, go to a graveyard state s_1 on input 0, and go to state s_2 on input 1. From both states s_2 and s_3, go to s_2 on input 1 and to s_3 on input 0.

9. a) See p. 804.

b) the set of all strings in which all the maximal blocks of consecutive 1's (if any) have an even number of 1's

10. a) See p. 805. **b)** See p. 806.

11. a) See p. 811. **b)** See Theorem 1 in Section 12.3.

12. a) See p. 818. **b)** See p. 818.

13. See Theorem 1 in Section 12.4.

14. See the proof of Theorem 2 in Section 12.4.

15. See Example 6 in Section 12.4.

16. See p. 828.

17. See p. 830.

18. See p. 831.

19. See p. 834. The halting problem is unsolvable; see page 835.

SUPPLEMENTARY EXERCISES FOR CHAPTER 12

1. **a)** We simply need to add two 0's on the left and three 1's on the right at the same time. Thus the rules can be $S \to 00S111$ and $S \to \lambda$.

b) We need to add two 0's for every 1 and also allow the symbols to change places at will. Following the trick in our solution to Exercise 15c in Section 12.1, we let A and B be nonterminal symbols representing 0 and 1, respectively. Our rules are $S \to AABS$, $AB \to BA$, $BA \to AB$, $A \to 0$, $B \to 1$, and $S \to \lambda$.

c) Our trick here is first to generate a string that looks like $Ew(w^R)$, with A in the place of 0, and B in the place of 1, in the second half. The rules $S \to ET$, $T \to 0TA$, $T \to 1TB$, and $T \to \lambda$ will accomplish this much. Then we force the A's and B's to march to the left, across all the 0's and 1's, until they bump into the left-hand wall (E), at which point they turn into their terminal counterparts. Finally, the wall disappears. The rules for doing this are $0A \to A0$, $1A \to A1$, $0B \to B0$, $1B \to B1$, $EA \to E0$, $EB \to E1$, and $E \to \lambda$.

3. For part **(a)** note that (()) can come from (B), which in turn can come from (A), which can come from B, and we can start $S \Rightarrow A \Rightarrow B$. Thus the tree can be as shown in the first picture. For part **(b)** we need to use the rule $A \to AB$ early in the derivation, with the A turning into (), and the B turning into (()). The ideas in part **(c)** are similar.

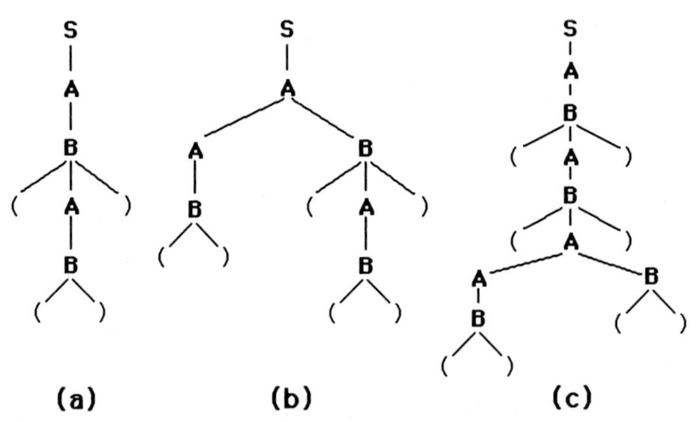

5. The idea is that the rules enable us to add 0's to either the right or the left. Thus we can get three 0's in many ways, depending on which side we add the 0's on.

7. It is not true that $|AB|$ is always equal to $|A| \cdot |B|$, since a string in AB may be formed in more than one way. After a little experimentation, we might come up with the following example to show that $|AB|$ need not equal $|BA|$ and that $|AB|$ need not equal $|A| \cdot |B|$. Let $A = \{0, 00\}$, and let $B = \{01, 1\}$. Then $AB = \{01, 001, 0001\}$ (there are only 3 elements, not $2 \cdot 2 = 4$, since 001 can be formed in two ways), whereas $BA = \{010, 0100, 10, 100\}$ has 4 elements.

9. This is clearly not necessarily true. For example, we could take $A = V^*$ and $B = V$. Then A^* is again V^*, so it is true that $A^* \subseteq B^*$, but of course $A \not\subseteq B$ (for one thing, A is infinite and B is finite).

11. In each case we apply the definition to rewrite $h(\mathbf{E})$ in terms of h applied to the subexpressions of $\mathbf{E}$.

a) $h(\mathbf{0^*1}) = \max(h(\mathbf{0^*}), h(\mathbf{1})) = \max(h(\mathbf{0}) + 1, 0) = \max(0 + 1, 0) = 1$

b) $h(\mathbf{0^*1^*}) = \max(h(\mathbf{0^*}), h(\mathbf{1^*})) = \max(h(\mathbf{0}) + 1, h(\mathbf{1}) + 1) = \max(0 + 1, 0 + 1) = 1$

c) $h((\mathbf{0^*01})^*) = h(\mathbf{0^*01}) + 1 = 1 + 1 = 2$

d) This is similar to part **(c)**; the answer is 3.

e) There are three "factors," and by the definition we need to find the maximum value that h takes on them. It is easy to compute that these values are 1, 2, and 2, respectively, so the answer is 2.

f) A calculation similar to that in part **(c)** shows that the answer is 4.

13. We need to have states to represent the number of 1's read in so far. Thus s_i, for $i = 0, 1, 2, 3$, will "mean" that we have seen exactly i 1's so far, and s_4 will signify that we have seen at least four 1's. We draw only the finite-state automaton; the machine with output is exactly the same, except that instead of a state being designated final, there is an output for each transition; all the outputs are 0 except for the outputs to our final state, and all of the outputs to this final state are 1.

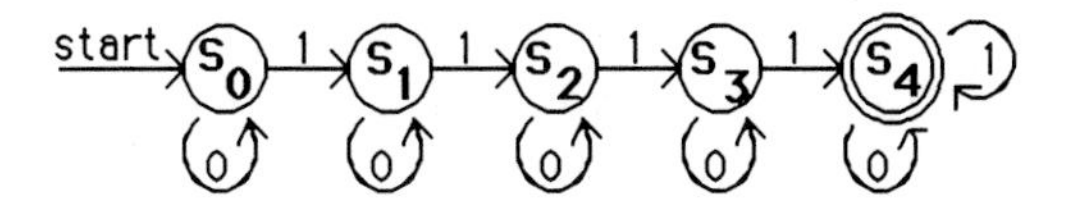

15. This is similar to Exercise 13, except that we need to return to the starting state whenever we encounter a 0, rather than merely remaining in the same state. As in Exercise 13, we draw only the automaton, since the machine with output is practically the same.

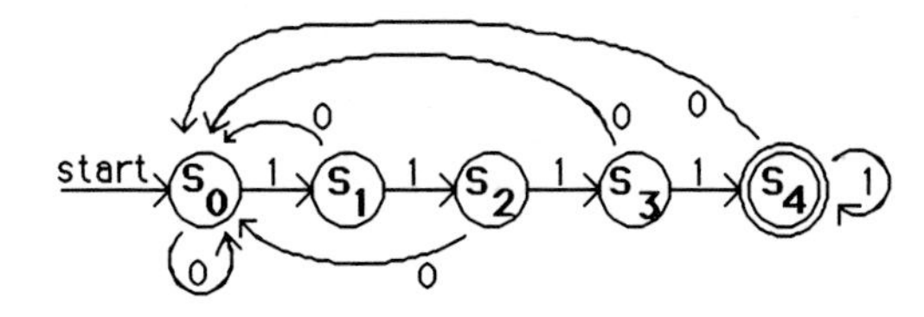

17. a) To specify a machine, we need to pick a start state (this can be done in n ways), and for each pair (state, input) (and there are nk such pairs), we need to choose a state and an output. By the product rule, therefore, the answer is $n \cdot n^{nk} \cdot m^{nk}$. (We are answering the question as it was asked. A much harder question is to determine how many "really" different machines there are, since two machines that really do the same thing and just have different names on the states should perhaps be considered the same. We will not pursue this question.)

b) This is just like part **(a)**, except that we do not need to choose an output for each transition, only an output for each state. Thus the term m^{nk} needs to be replaced by m^n, and the answer is $n \cdot n^{nk} \cdot m^n$.

19. This machine has no final states. Therefore no strings are accepted. Any deterministic machine with no final states will be equivalent to this one. We show one such machine below.

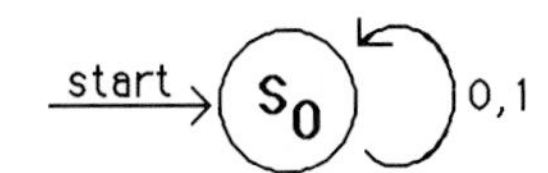

21. The answers are not unique, of course. There are two ways to approach this exercise. We could simply apply the algorithm inherent in the proof of Kleene's Theorem, but that would lead to machines much more complicated than they need to be. Alternately, we just try to be clever and make the machines "do what the expressions say." This is essentially computer programming, and it takes experience to be able to do it well. In part **(a)**, for example, we want to accept every string of 0's, so we make the start state a final state, with returns to this state on input 0; and we want to accept every string that has this beginning and then consists of any number of copies of 10—which is precisely what the rest of our machine does. These pictures can either be viewed as nondeterministic machines, or else for all the missing transitions we assume a transition to a new state (the graveyard), which is not final and which has transitions to itself on both inputs. Also, as usual, having two labels on an edge is an abbreviation for two edges, one with each label.

a) This one is pretty simple. State s_0 represents the condition that only 0's have been read so far; it is final. After we have read in as many 0's as desired, we still want to accept the string if we read in any number of copies of 10. This is accomplished with the other two states.

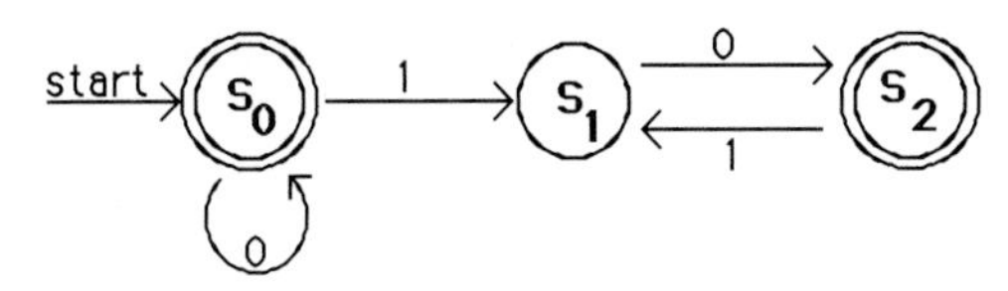

b) In this machine, we keep returning to s_0 as long as we are reading 01 or 111, corresponding to the first factor in our regular expression. Then we move to s_4 for the term $\mathbf{10}^*$, and finally to s_5 for the factor $(\mathbf{0} \cup \mathbf{1})$.

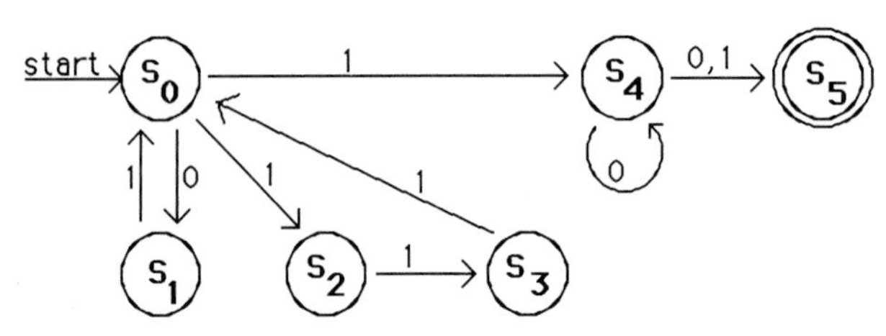

c) Note that the inner star in this regular expression is irrelevant; we get the same set whether it is there or not. Our machine returns us to s_0 after we have read either 001 or 11, so we accept every string consisting of any number of copies of these strings.

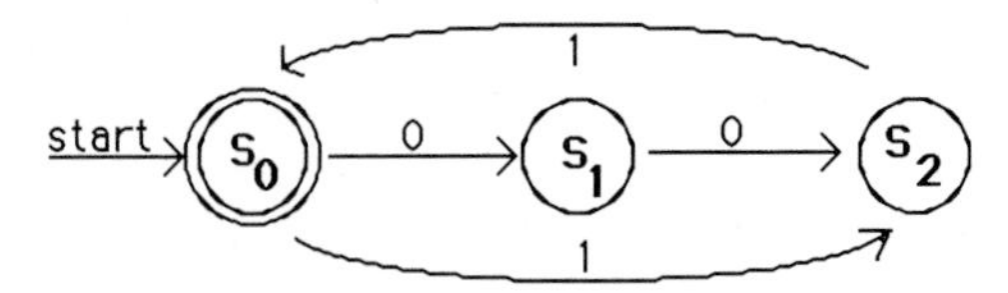

23. We invoke the power of Kleene's Theorem here. If A is a regular set, then there is a deterministic finite automaton that accepts A. If we take the same machine but make all the final states nonfinal and all the nonfinal states final, then the result will accept precisely $\overline{A}$. Therefore $\overline{A}$ is regular.

25. See the comments for Exercise 21. Here the problem is even harder, since we are given just verbal descriptions of the sets. Thus there *is* no general algorithm we can invoke. We just have to be clever programmers. See the comments on the solution to Exercise 21 for how to interpret missing transitions.

a) The top part of our machine (as drawn) takes us to a graveyard if there are more than three consecutive 0's at the beginning. The rest assures that there are at least two consecutive 1's.

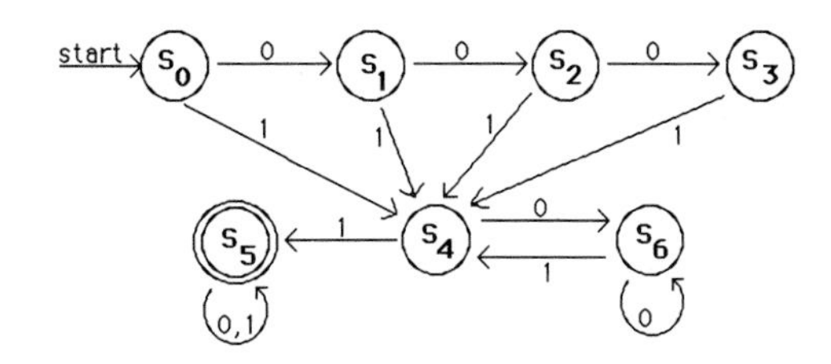

b) This one is rather complicated. The states represent what has been seen recently in the input. For example, states s_2 and s_6 represent the condition in which the last two symbols have been 10. Thus if we encounter a 1 from either of these states, we move to a graveyard. (Note that we could have combined states s_3 and s_7 into one, or, under our conventions, we could have omitted them altogether; the answers to these exercises are by no means unique.) States s_0, s_2 and s_5 all represent conditions in which an even number of symbols have been read in, whereas s_1, s_4 and s_6 represent conditions in which an odd number of symbols have been read.

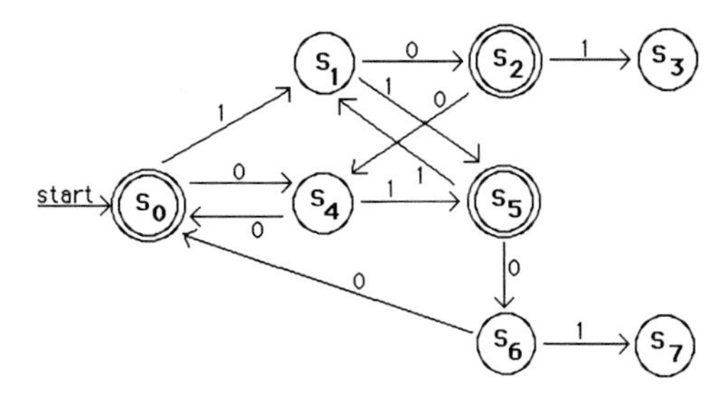

c) This one is really not as bad as it looks. The first row in our machine (as drawn) represents conditions before any 0's have been read; the second row after one 0, and the third row after two or more 0's. The horizontal direction takes care of looking for the blocks of 1's.

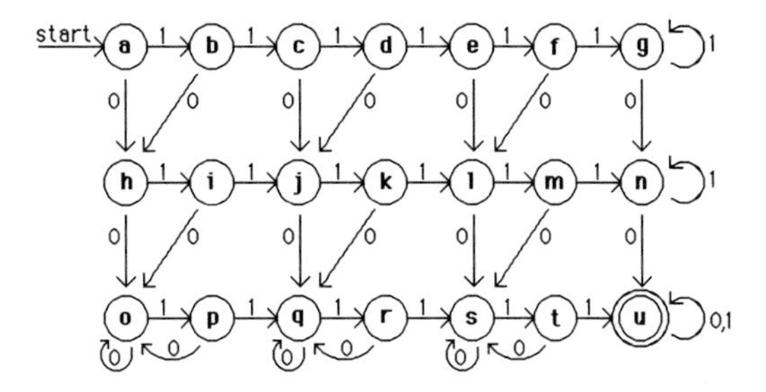

27. Suppose that $\{\, 1^p \mid p \text{ is prime} \,\}$ is regular. Then by the pumping lemma, we can find a prime p such that $1^p = uvw$, with $l(v) \geq 1$, so that uv^iw is a string of a prime number of 1's for all i. If we let a be the number of 1's in uw and $b > 0$ the number of 1's in v, then this means that $a + bi$ is prime for all i. In other words, the gap between two consecutive primes (once we are looking at numbers greater than a) is at most b. This contradicts reality, however, since for every n, all the numbers from $n! + 2$ through $n! + n$ are not prime—in other words the gaps between primes can be arbitrarily large.

29. The idea here is to match off the 1's in the two inputs (changing the 1's to 0's from the left, say, to keep track), until one of them is exhausted. At that point, we need to erase the smaller input entirely (as well as the asterisk) and change the 0's back to 1's. Here is how we'll do it. In state s_0 we skip over any 0's until we come to either a 1 or the $*$. If it's the $*$, then we know that the second input (n_2) is at least as large as the first (n_1), so we enter a clean-up state s_5, which erases the asterisk and all the 0's and 1's to its left. The five-tuples for this much are $(s_0, 0, s_0, 0, R)$, $(s_0, *, s_5, B, L)$, and $(s_5, 0, s_5, B, L)$. Once this erasing is finished, we need to go over to the part of the tape where the second input was and change all the 0's back to 1's; the following transitions accomplish this: (s_5, B, s_6, B, R), (s_6, B, s_6, B, R), $(s_6, 0, s_7, 1, R)$, $(s_7, 0, s_7, 1, R)$, and $(s_7, 1, s_7, 1, R)$. Eventually the machine halts in state s_7 when the blank following the original input is encountered.

The other possibility is that the machine encounters a 1 while in state s_0. We want to change this 1 to a 0, skip over any remaining 1's as well as the asterisk, skip over any 0's to the right of the asterisk (these represent parts of n_2 that have already been matched off against equal parts of n_1), and then either find a 1 in n_2 (which we change to a 0) or else come to the blank at the end of the input. Here are the transitions: $(s_0, 1, s_1, 0, R)$, $(s_1, 1, s_1, 1, R)$, $(s_1, *, s_2, *, R)$, $(s_2, 0, s_2, 0, R)$, $(s_2, 1, s_3, 0, L)$, (s_2, B, s_4, B, L). At this point we are either in state s_3, ready to go back for the next iteration, or in state s_4 ready for some cleanup. In the former case, we want to skip back over the nonblank symbols until we reach the start of the string, so we add five-tuples $(s_3, *, s_3, *, L)$, $(s_3, 0, s_3, 0, L)$, $(s_3, 1, s_3, 1, L)$, and (s_3, B, s_0, B, R). In the latter case, we know that the first string is longer than the second. Therefore we want to erase remnants of the second input string and the asterisk, and change the 0's in the first input string back to 1's. Here are the transitions: $(s_4, 0, s_4, B, L)$, $(s_4, *, s_8, B, L)$, $(s_8, 0, s_8, 1, L)$, $(s_8, 1, s_8, 1, L)$. The machine halts in state s_8 when the blank preceding the original input is encountered.

WRITING PROJECTS FOR CHAPTER 12

Books and articles indicated by bracketed symbols below are listed near the end of this manual. You should also read the general comments and advice you will find there about researching and writing these essays.

1. See the chapter on generative grammars in [De2]. Lindenmeyer systems are a special kind of generative grammar.

2. Most textbooks on programming languages should discuss this, as well as books on the specific languages mentioned. The call number for programming languages is QA 76.7. As usual, you can find websites with this information using a search engine; for example, search for the three keywords *backus*, *naur*, and *java*.

3. See [BeKa], for example.

4. One book on network protocols that discusses finite state machines is [Ho3].

5. Mehryar Mohri of the Department of Computer Science at the Courant Institute of Mathematical Sciences is an expert in this area. See his Web page.

6. There are several textbooks on automata theory and finite-state machines of all kinds that cover topics such as this. Try [Br1], [Co], [DeDe], [HoUl], or [LePa], for example. These are also good sources for supplementing the material in this chapter. This subject in general (including Turing machines, computability, and computational complexity), is usually called "the theory of computation," and, again, there are numerous books with essentially this title; see [Si] for a fairly recent and readable one that takes you from the beginning to a fairly advanced level.

7. Again, there are entire books on this subject (see [PrDu], for instance). The Game of Life was invented by the British mathematician John H. Conway, and was the subject of several articles in Martin Gardner's *Scientific American* column during the 1970s. Three of them are collected in [Ga2], which also mentions other books and articles. See also [BeCo], which covers lots of solitaire and two-person games, as well as Life.

8. See standard references on automata theory, such as those mentioned in Writing Project 6.

9. See standard references on automata theory, such as those mentioned in Writing Project 6.

10. Turing's first article is [Tu2], and it is actually quite readable. Keep in mind as you read it that real computers had not yet been invented.

11. See standard references on automata theory, such as those mentioned in Writing Project 6.

12. This actually opens up the door to most of the important modern-day research in theoretical computer science. For an elementary, nontechnical account, see [Gr2]. See the references given in Writing Project 4 for more detail. The big question is whether deterministic Turing machines can compute functions as efficiently as nondeterministic ones. You should definitely consult [GaJo], the classic work in this area.

13. See the references given for Writing Project 12.

14. See standard references on automata theory, such as those mentioned in Writing Project 6.

15. Searching under "lambda calculus" or "recursive function theory" in your library should turn up a place to start.

16. See standard references on automata theory, such as those mentioned in Writing Project 6.

17. See standard references on automata theory, such as those mentioned in Writing Project 6.

APPENDIXES

APPENDIX 1 Axioms for the Real Numbers and the Positive Integers

These facts are all things you know to be true. The point here is to prove them rigorously using the axioms and the theorems stated or proved in this section. Throughout this section we will omit parentheses that clarify order of operations (writing $a \cdot b + c \cdot d$ rather than $(a \cdot b) + (c \cdot d)$, for instance).

1. This proof is similar to the proof of Theorem 1, that the additive identity for real numbers is unique. In fact, we can just mimic that proof, changing addition to multiplication and 0 to 1 throughout. So suppose that $1'$ is also a multiplicative identity for the real numbers. This means that $1' \cdot x = x \cdot 1' = x$ whenever x is a real number. In particular, letting $x = 1$, we have $1 \cdot 1' = 1$. By the multiplicative identity law, we also have $1 \cdot 1' = 1'$. It follows that $1' = 1$, because both equal $1 \cdot 1'$. This shows that 1 is the unique multiplicative identity for the real numbers.

3. To show that a number equals $-(x \cdot y)$, the additive inverse of $x \cdot y$, it suffices to show that this number plus $x \cdot y$ equals 0, because Theorem 2 guarantees that additive inverses are unique. For the first part, then, we have

$$\begin{aligned}(-x) \cdot y + x \cdot y &= (-x + x) \cdot y \quad \text{(by the distributive law)}\\ &= 0 \cdot y \quad \text{(by the inverse law)}\\ &= y \cdot 0 \quad \text{(by the commutative law)}\\ &= 0 \quad \text{(by Theorem 5).}\end{aligned}$$

The second part is essentially identical.

5. This is similar to Exercise 3. We will show that $(-x) \cdot (-y)$ has the property that when we add it to $-(x \cdot y)$ we get 0. This will guarantee that it is the unique additive inverse of $-(x \cdot y)$; but because $x \cdot y$ is the additive inverse of $-(x \cdot y)$, this means that it must equal $x \cdot y$, which is what we are asked to show.

$$\begin{aligned}(-x) \cdot (-y) + (-(x \cdot y)) &= (-x) \cdot (-y) + (-x) \cdot y \quad \text{(by Exercise 3)}\\ &= (-x) \cdot ((-y) + y) \quad \text{(by the distributive law)}\\ &= (-x) \cdot 0 \quad \text{(by the inverse law)}\\ &= 0 \quad \text{(by Theorem 5)}\end{aligned}$$

7. By definition, $-(-x)$ is the additive inverse of $-x$. But $-x$ is the additive inverse of x, so x is the additive inverse of $-x$. By Theorem 2 additive inverses are unique, so we must have $-(-x) = x$.

9. This is similar to Exercise 3. We will show that $-x - y$ has the property that when we add it to $x + y$ we get 0. This will guarantee that it is the unique additive inverse of $x + y$ and so must equal $-(x + y)$, which is what we are asked to show.

$$\begin{aligned}(-x - y) + (x + y) &= ((-x) + (-y)) + (x + y) \quad \text{(by definition of subtraction)}\\ &= ((-y) + (-x)) + (x + y) \quad \text{(by the commutative law)}\\ &= (-y) + ((-x) + (x + y)) \quad \text{(by the associative law)}\\ &= (-y) + ((-x + x) + y) \quad \text{(by the associative law)}\\ &= (-y) + (0 + y) \quad \text{(by the inverse law)}\\ &= (-y) + y \quad \text{(by the identity law)}\\ &= 0 \quad \text{(by the inverse law)}\end{aligned}$$

11. By definition of division and uniqueness of multiplicative inverses (Theorem 4) it suffices to prove that $((w/x)+(y/z))\cdot(x\cdot z)=w\cdot z+x\cdot y$. But this follows after several steps, using the definition that division is the same as multiplication by the inverse, as well as the distributive law and the associative, commutative, identity, and inverse laws for multiplication: $((w\cdot(1/x))+(y\cdot(1/z)))\cdot(x\cdot z)=(w\cdot(1/x))\cdot(x\cdot z)+(y\cdot(1/z))\cdot(x\cdot z)=(w\cdot(1/x))\cdot(x\cdot z)+(y\cdot(1/z))\cdot(z\cdot x)=w\cdot((1/x)\cdot(x\cdot z))+y\cdot((1/z)\cdot(z\cdot x))=w\cdot(((1/x)\cdot x)\cdot z)+y\cdot(((1/z)\cdot z)\cdot x)=w\cdot(1\cdot z)+y\cdot(1\cdot x)=w\cdot z+y\cdot x=w\cdot z+x\cdot y$.

13. We must show that if $x>0$ and $y>0$, then $x\cdot y>0$. By the multiplicative compatibility law, we have $x\cdot y>0\cdot y$, but by the commutative law and Theorem 5, $0\cdot y=0$.

15. First let's prove a lemma, that if $z<0$, then $-z>0$. This follows immediately by adding $-z$ to both sides of the hypothesis. Now given $x>y$ and $-z>0$, we have $x\cdot(-z)>y\cdot(-z)$ by the multiplicative compatibility law. But by Exercise 3 this is equivalent to $-(x\cdot z)>-(y\cdot z)$. By the additive compatibility law we can add $x\cdot z$ and $y\cdot z$ to both sides and obtain another valid inequality. Then we apply the various laws in the obvious ways to simplify and obtain $x\cdot z<y\cdot z$.

17. The additive compatibility law tells us that $w+y<x+y$ and (together with the commutative law) that $x+y<x+z$. By the transitivity law, this gives the desired conclusion.

19. A simple proof using Theorem 8 is given in the answer key in the textbook. For a proof from first principles, we can use the completeness property and a proof by contradiction. Suppose that $n\cdot x\le 1$ for all positive integers n. Then the set $S=\{n\cdot x \mid n \text{ is a positive integer}\}$ is bounded above and therefore has a least upper bound; call it b. Now for all positive integers n we have $(n+1)\cdot x\le b$, so $n\cdot x+x\le b$. This tells us that $n\cdot x\le b-x$, which says that $b-x$ is also an upper bound for S, contradicting the definition of least upper bound. Therefore it must be true that $n\cdot x>1$ for some positive integer n.

21. The proof practically writes itself if we just use the definitions. We want to show that if $(w,x)\sim(w',x')$ and $(y,z)\sim(y',z')$, then $(w+y,x+z)\sim(w'+y',x'+z')$ and that $(w\cdot y+x\cdot z,x\cdot y+w\cdot z)\sim(w'\cdot y'+x'\cdot z',x'\cdot y'+w'\cdot z')$. Thus we are given that $w+x'=x+w'$ and that $y+z'=z+y'$, and we want to show that $w+y+x'+z'=x+z+w'+y'$ and that $w\cdot y+x\cdot z+x'\cdot y'+w'\cdot z'=x\cdot y+w\cdot z+w'\cdot y'+x'\cdot z'$ (note that grouping does not matter, because of the associative law). For the first of the desired conclusions, add the two given equations. For the second, rewrite the given equations as $w-x=w'-x'$ and $y-z=y'-z'$, multiply them, and do the algebra.

APPENDIX 2 Exponential and Logarithmic Functions

This material should all be familiar from high school algebra. When working with exponents it is important to remember the rules $b^x b^y = b^{x+y}$ *and* $(b^x)^y = b^{xy}$. *When working with logarithms it is important to remember that* $\log_b(xy) = \log_b x + \log_b y$ *and* $\log_b x^y = y \log_b x$.

1. a) We use the facts that $2^1 = 2$ and $2^{x+y} = 2^x \cdot 2^y$. Thus we have $2 \cdot 2^2 = 2^1 \cdot 2^2 = 2^{1+2} = 2^3$.

b) We use the second part of Theorem 1 to write $(2^2)^3 = 2^{(2 \cdot 3)} = 2^6$.

c) There is no way to use a rule of exponents to simplify this. However, since $2^2 = 4$, we can write this as 2^4.

3. a) We use Theorem 3. Thus $\log_2 x = (\log_4 x)/(\log_4 2) = (\log_4 x)/(1/2) = 2\log_4 x = 2y$.

b) This is just like part **(a)**. We have $\log_8 x = (\log_4 x)/(\log_4 8) = (\log_4 x)/(3/2) = (2\log_4 x)/3 = 2y/3$.

c) This is just like part **(a)**. We have $\log_{16} x = (\log_4 x)/(\log_4 16) = (\log_4 x)/2 = y/2$.

5. We can draw these graphs by plotting points, following the general shape shown in Figure 1. For part **(a)** the graph is similar to the graph of $f(x) = 2^x$. For part **(b)** the graph is similar to the graph of $f(x) = (1/2)^x$. Finally for part **(c)** the function is constant, since $1^x = 1$ for all x.

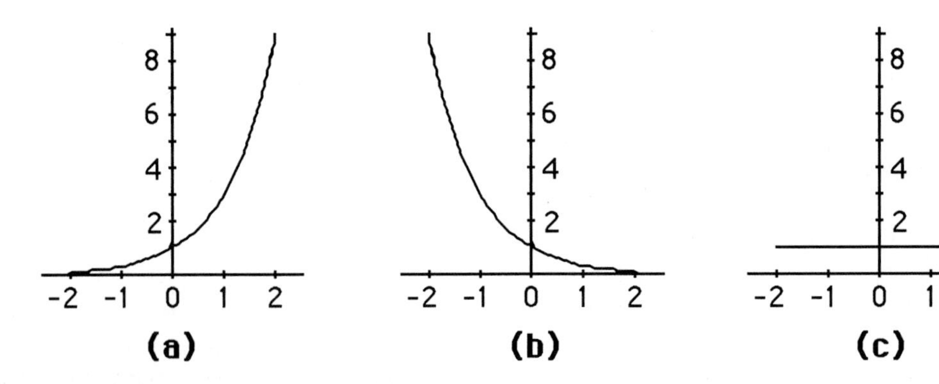

APPENDIX 3 Pseudocode

Obviously the text cannot give an entire course in programming in just a few pages. One nice thing about pseudocode is that it is readable even when the reader has no specialized knowledge of the conventions being followed.

1. In the first case, the value of a is first set to the value of b, and then the value of b is changed. Thus the final value of a is the same as the original value of b. In the second case, the value of b is changed first (to the original value of c), then this new value is assigned to a. Thus the final value of a is the same as the original value of c.

3. According to the definition, the way this loop is executed is that first i is assigned the *initial value*. Then as long as i is less than or equal to *final value*, the statement in the loop is executed, and the value of i is incremented by 1. Thus the code is equivalent to the following code.

```
i := initial value
while i ≤ final value
begin
        statement
        i := i + 1
end
```

A Guide to Proof-Writing

by Ron Morash, University of Michigan–Dearborn

At the end of Section 1.7, the text states, "We have not given a procedure that can be used for proving theorems in mathematics. It is a deep theorem of mathematical logic that there is no such procedure." This is true, but does not mean that proof-writing is purely an art, so that only those with exceptional talent and insight can possibly write proofs. Most proofs that students are asked to write in elementary courses fall into one of several categories, each calling for a systematic approach that can be demonstrated, imitated, and eventually mastered. We present some of these categories and techniques for working within them, organized as follows. This material supplements that found in the text and is intended to help get you started creating your own proofs. Also, studying the material in this Guide will help you understand better the proofs you read.

1. Deducing conclusions having the form "For every x, if $P(x)$, then $Q(x)$."

Many defining properties in mathematics have the form $\forall x[P(x) \rightarrow Q(x)]$, representing the idea "All P's are Q's." (Cf. Examples 23, 26, and 27 in Section 1.3 of the text.) Some definitions involving this form are:

(i) A set A is a *subset* of a set B: In symbols, $A \subseteq B$ if and only if $\forall x[(x \in A) \rightarrow (x \in B)]$. This is read in words, "A is a subset of B if and only if, for every x, if $x \in A$, then $x \in B$." Less formally, A is a subset of B if and only if every element of A is also an element of B. (Cf. Definition 4 in Section 2.1 of the text.)

(ii) A function f is *one-to-one*: f is one-to-one if and only if, for every x_1 and x_2 in the domain of f, if $f(x_1) = f(x_2)$, then $x_1 = x_2$. (Cf. Definition 5 in Section 2.3 of the text.)

(iii) A relation R on a set A is *symmetric*: R is symmetric if and only if, for every $x, y \in A$, if $(x,y) \in R$, then $(y,x) \in R$. (Cf. Definition 4 in Section 8.1 of the text.)

Many mathematical propositions that students are asked to prove have as their conclusion a statement involving a definition of the form just described. Some examples are:

(a) Prove that for all sets A and B, $A \subseteq A \cup B$.

(b) Prove that for all sets X, Y, and Z, if $X \subseteq Y$, then $X \cap Z \subseteq Y \cap Z$.

(c) Prove that for every function f whose domain and codomain are subsets of the set of real numbers, if f is strictly increasing (cf. Definition 6 in Section 2.3), then f is one-to-one.

(d) Prove that for all relations R_1 and R_2 on a set A, if R_1 and R_2 are symmetric, then the relation $R_1 \cap R_2$ is symmetric.

Note that the *desired conclusion* in each of the propositions (a)–(d) is a statement involving one of the definitions (i)–(iii). Furthermore, propositions (b), (c), and (d) each have a *hypothesis*, a statement we are allowed to assume true and whose assumed truth, presumably, will play a role in deriving the conclusion. We begin our study of proof-writing methods by considering the very broad category known as *direct proof*.

1.1. Direct proof

An argument in which we prove a proposition in its originally-stated form is called a *direct proof*. Some forms of direct proof are discussed in Sections 1.5 and 1.6 of the text. In the sections of this Guide that follow, we present various techniques for creating direct proofs. Attempting to write a direct proof of a proposition is usually our first line of attack. Direct proof contrasts with *indirect proof*, in which we prove a proposition by proving a different, but logically equivalent, form of the original proposition. We will introduce indirect proofs later, but will focus, in Examples 1–11, on various approaches to direct proofs.

1.1.1. Propositions having no hypothesis

Direct proofs of propositions like (a), having no hypothesis, tend to be simpler in their structure than the proofs that are required for propositions (b)–(d). Examples 1 and 2 demonstrate proofs for this simpler case.

Example 1. Prove Proposition (a): For all sets A and B, $A \subseteq A \cup B$.

Solution. The proof proceeds as follows: Let A and B be arbitrary sets. To prove $A \subseteq A \cup B$, let x be an arbitrarily chosen element of A. [Note: We are assuming that $x \in A$.] We must prove that $x \in A \cup B$. By the definition of "union," this means we must prove that either $x \in A$ or $x \in B$. Since we know $x \in A$, by our assumption, the desired conclusion $x \in A$ or $x \in B$ follows immediately. ■

Let's dissect the proof in Example 1 and analyze what we did. Our starting point "... assume $x \in A$ ..." is an application of one of the most widely-used approaches to proof-writing, known as the *choose method*. The basic approach to deriving a conclusion of the form $\forall x[P(x) \rightarrow Q(x)]$, is to begin by choosing an arbitrary object (giving it a specific name such as "x") for which it is assumed that $P(x)$ is true. Our goal is to deduce that $Q(x)$ must consequently also be true. In Example 1, $P(x)$ is the assumption "$x \in A$" and $Q(x)$ is the desired conclusion "$x \in A \cup B$." We make the following additional observations:

• The object x is a fixed, but arbitrarily chosen, element of the universe of discourse of $P(x)$ and $Q(x)$. We do not assign any specific value to x; rather we give the name "x" to a generic object [that is assumed to satisfy the propositional function $P(x)$] and use that name to keep track of the object as we proceed through the steps of the proof. The power of this approach is that any conclusion we draw about "this x" applies to every object a for which the assumption $P(a)$ is true. This is valid by the rule of universal generalization; see Table 2 of Section 1.5.

• In the first part of a proof of a conclusion of the form $\forall x[P(x) \rightarrow Q(x)]$, called the "setting-up" of the proof, we choose x, assume $P(x)$ is true, and then write out what it would mean for $Q(x)$ to be true (in our example, "... to prove $x \in A \cup B$, we must prove that either $x \in A$ or $x \in B$"). Learning the process of setting up a proof in this category provides a fairly standardized, predictable, and almost mechanical beginning of a prospective direct proof. Furthermore, once we have written out these details, the remainder of the proof—the path from the assumption $P(x)$ to the desired conclusion $Q(x)$—is sometimes obvious.

• In the proof in Example 1, the path from what we assumed (i.e., $x \in A$) to the conclusion (i.e., $x \in A \cup B$) was obvious. In our proof, we stated that the conclusion "follows immediately." But was there something more than just "common sense" to justify that conclusion? Yes! The rule of inference $p \rightarrow p \vee q$ ("Law of Addition") is the underlying logical tool that justifies this step. This rule and other rules of inference are stated in the text, in Table 1 of Section 1.5. In the sample proofs that follow, we will make explicit reference to the rules of inference used (in an increasingly less obvious way as the proofs become more complex), even though it is common practice to apply these rules only implicitly, that is, without specific mention. To become proficient at writing proofs, you need to know how to use these rules of inference and when to use them.

Example 2. Prove that for all sets X and Y, $X \cap (Y \cup \overline{X}) \subseteq Y$.

Discussion. Let X and Y be arbitrary sets. To prove $X \cap (Y \cup \overline{X}) \subseteq Y$, let a be an arbitrarily chosen element of $X \cap (Y \cup \overline{X})$. [We could also say "Assume $a \in X \cap (Y \cup \overline{X})$."] We must prove $a \in Y$. [This concludes the setting-up of the proof. Now we must figure out how to get from the assumption to the conclusion. To do that, we begin by analyzing what our assumption means.]

Since $a \in X \cap (Y \cup \overline{X})$, we know that $a \in X$ and $a \in Y \cup \overline{X}$. The latter, in turn, tells us that either $a \in Y$ or $a \in \overline{X}$, that is, either $a \in Y$ or $a \notin X$. [Note that the preceding sentence makes the first mention of the desired conclusion $a \in Y$.] Now, can we infer the conclusion $a \in Y$ from the known "either $a \in Y$ or $a \notin X$"? This would require a rule of inference "$(p \vee q) \rightarrow p$" (the converse of the Law of Addition). This is not a valid inference since $(p \vee q) \rightarrow p$ is not a tautology, so this approach does not work. Note, however, that not only do we know "either $a \in Y$ or $a \notin X$" from our assumption, but we also know that $a \in X$. Thus, what we know from our assumption has the form $(p \vee q) \wedge \neg q$. [Note: p is "$a \in Y$" and q is "$a \notin X$," so $\neg q$ is "$a \in X$."] Does Table 1 in Section 1.5 give us a conclusion that follows from this premise? It does! The Law of Disjunctive Syllogism, $[(p \vee q) \wedge \neg q] \rightarrow p$, enables us to draw the conclusion p, that is, $a \in Y$, the desired conclusion. [Note: As stated in Table 1, the roles of p and q are reversed from what we have here, but that is of no consequence.] ∎

Before moving on, we rewrite the preceding proof, leaving out explanatory comments. What remains provides a representative view of what a typical proof looks like:

"Let X and Y be arbitrary sets. To prove $X \cap (Y \cup \overline{X}) \subseteq Y$, assume $a \in X \cap (Y \cup \overline{X})$. We must prove $a \in Y$. By our assumption, we know $a \in X$ and $a \in Y \cup \overline{X}$; therefore $a \in X$ and either $a \in Y$ or $a \in \overline{X}$. Thus we know that either $a \in Y$ or $a \notin X$; but we also know that $a \in X$, so $a \notin X$ is false. Hence we conclude $a \in Y$, as desired."

At this point, you may wish to try some relevant exercises in the text, such as Exercises 16(a,c) and 18(a,b,c) in Section 2.2. You should find the principles from Examples 1 and 2 helpful in attempting these exercises.

1.1.2. Propositions having one or more hypotheses

As we work through the steps of a prospective proof, the tools at our disposal in moving toward a desired conclusion are

- the assumption(s) we are entitled to make at the outset in setting up the proof,
- assumed axioms and previously-proved theorems (if any), and
- rules of inference from logic, such as $p \rightarrow (p \vee q)$, used in Example 1, and $[(p \vee q) \wedge \neg q] \rightarrow p$, used in Example 2. (See Table 1 in Section 1.5 for additional rules of this type.)

In addition to these, most propositions we are asked to prove contain

- one or more *hypotheses*, statements whose truth is to be assumed in the proof and which, we expect, will be used as part of the argument leading to the conclusion.

Example 3 provides our first instance of a proposition in which a hypothesis is to be assumed in deriving a conclusion of the form $\forall x[P(x) \rightarrow Q(x)]$.

Example 3. Prove Proposition (b): For all sets X, Y, and Z, if $X \subseteq Y$, then $X \cap Z \subseteq Y \cap Z$.

Proof. Let X, Y, and Z be sets such that $X \subseteq Y$. To prove $X \cap Z \subseteq Y \cap Z$, assume $b \in X \cap Z$. To prove $b \in Y \cap Z$, we must prove $b \in Y$ and $b \in Z$. [This marks the end of "setting up the proof." Now we must return to our assumption and the hypothesis, and begin to analyze what they mean and what information we can draw from them.] By our assumption, we know that $b \in X$ and $b \in Z$, so, in particular, $b \in Z$, one of our two desired conclusions. Furthermore, since $b \in X$ (part of our assumption) and since $X \subseteq Y$ [here we are, for the first time, bringing in the hypothesis], we may conclude that $b \in Y$, our other desired conclusion. ∎

The following feature of the proof in Example 3 is very important. In setting up the argument at the outset, we applied the choose method to the desired conclusion, not the hypothesis. Thus our initial statement was "... assume $b \in X \cap Z$." A common mistake by beginning students is to begin with "... assume $b \in X$ ...," erroneously focusing at the start of the proof on the hypothesis rather than on the desired conclusion. Note that we did not employ the hypothesis until the very end of the proof!

In the last sentence of the proof in Example 3, we concluded $b \in Y$ from knowing $b \in X$ and $X \subseteq Y$. Let us consider why this conclusion is justified. The truth of $X \subseteq Y$ means that the proposition $\forall x[(x \in X) \rightarrow (x \in Y)]$ is true. Thus, in particular, the proposition $(b \in X) \rightarrow (b \in Y)$ is true, where b is the specific object we are working with in the proof. Since $b \in X$ is true and the "if...then" statement $(b \in X) \rightarrow (b \in Y)$ is true, the truth of $b \in Y$ follows from the rule of inference modus ponens (cf. Table 1 in Section 1.5 of the text). Note, once again, that a rule of inference has played an important, though implicit, role in a proof!

The principles discussed thus far apply to every proof of a proposition whose conclusion has the logical form $\forall x[P(x) \to Q(x)]$, and not just to proofs that one set is a subset of another. Examples 4 and 5 illustrate this.

Example 4. Prove that every nonconstant linear function $f(x) = Mx + B$, $M \neq 0$, is one-to-one.

Proof. Let M be a nonzero real number. Let x_1 and x_2 be real numbers and assume that $f(x_1) = f(x_2)$. We must prove that $x_1 = x_2$. Since $f(x_1) = Mx_1 + B$ and $f(x_2) = Mx_2 + B$, we have $Mx_1 + B = Mx_2 + B$. By a rule of elementary algebra, if $Mx_1 + B = Mx_2 + B$, then $Mx_1 = Mx_2$. Since $Mx_1 = Mx_2$ and $M \neq 0$, by hypothesis, we conclude by another rule of elementary algebra that $x_1 = x_2$, as desired. ∎

Example 5. Prove Proposition (d): For all relations R_1 and R_2 on a set A, if R_1 and R_2 are symmetric, then the relation $R_1 \cap R_2$ is symmetric.

Proof. Let A be an arbitrary set and let R_1 and R_2 be symmetric relations on A. To prove that the relation $R_1 \cap R_2$ is symmetric, let x and y be arbitrary elements of A and assume that $(x, y) \in R_1 \cap R_2$. We must prove $(y, x) \in R_1 \cap R_2$, that is, $(y, x) \in R_1$ and $(y, x) \in R_2$. [End of set-up!] Now since $(x, y) \in R_1 \cap R_2$ (by assumption), we know that $(x, y) \in R_1$ and $(x, y) \in R_2$. Since $(x, y) \in R_1$ and R_1 is symmetric (by hypothesis), $(y, x) \in R_1$. This is one of our desired conclusions. Since $(x, y) \in R_2$ and R_2 is symmetric (by hypothesis), $(y, x) \in R_2$, the second of our two desired conclusions. With this, the proposition is proved. ∎

At this point, you may want to practice applying the principles from Examples 3–5 in the following exercises:

- Prove that for all sets A and B, if $A \cap B = A$, then $A \subseteq B$.

- Prove that for all sets A, B, and C, if $A \subseteq B$ and $B \subseteq C$, then $A \subseteq C$. (This is Exercise 15 in Section 2.1 of the text.)

- Prove that the function $g(x) = x^3$ is one-to-one.

1.1.3. Disproving false propositions having conclusions of the form $\forall x[P(x) \to Q(x)]$

Sometimes we are faced with a proposition that we must either "prove or disprove." We are not told in advance whether the proposition is true. If it is false, then it will of course be impossible to write a correct proof of the proposition. Time spent trying to do so may provide insight, but cannot ultimately lead to a valid proof. Example 6 illustrates how we should approach this type of problem.

Example 6. Prove or disprove the converse of Proposition (b): For all sets X, Y, and Z, if $X \cap Z \subseteq Y \cap Z$, then $X \subseteq Y$. (Cf. Example 3.)

Discussion. Suppose we try first to approach this proposition in the manner of previous examples. Our setting up of a "proof" would read as follows: "Let X, Y, and Z be arbitrary sets such that $X \cap Z \subseteq Y \cap Z$. To prove $X \subseteq Y$, let $w \in X$; we must prove $w \in Y$." At this point, we must return to the hypothesis $X \cap Z \subseteq Y \cap Z$ and ask whether, in combination with the assumption $w \in X$, it leads to the conclusion $w \in Y$. If we could get w to lie in $X \cap Z$, then we could invoke the hypothesis to conclude $w \in Y \cap Z$, which would imply $w \in Y$, the desired conclusion. However, we know from our assumption only that $w \in X$; in order to conclude $w \in X \cap Z$, we would need to know that $w \in Z$, which we do not!

With this, our attempt to write a direct proof breaks down, leaving us with two possibilities. Either there is another route to a proof, or perhaps, the proposition we are trying to prove is false. If we do not know whether a general proposition is true or false, and our initial attempts at a proof fail, we should do some experimenting to see whether we can find a *counterexample*, i.e., a specific example that contradicts the truth of the proposition. Before we can do that, we must formulate precisely the negation of the proposition. Logically the negation of a proposition "for every x, if $P(x)$, then $Q(x)$" is "there exists x such that $P(x)$ but not $Q(x)$." In symbols, $\neg\forall x[P(x) \to Q(x)]$ is equivalent to $\exists x[P(x) \wedge \neg Q(x)]$. In this example, the negation is "there exist sets X, Y, and Z such that $X \cap Z \subseteq Y \cap Z$, but X is not a subset of Y." Can we find specific sets X, Y, and Z satisfying this statement? Consider the sets $X = \{4, 7, 8, 11\}$, $Y = \{2, 7, 8, 9, 11\}$, and $Z = \{1, 7, 8, 9, 10, 11, 12\}$. Note that $X \cap Z = \{7, 8, 11\}$ and $Y \cap Z = \{7, 8, 9, 11\}$, so $X \cap Z \subseteq Y \cap Z$. However X is clearly not a subset of Y. Hence we have a counterexample; the proposition in question is false! ∎

Note that a single counterexample to a general proposition is sufficient to prove that proposition false. This is a far cry from what is required to prove a general proposition true, when the domain (universe of discourse) is infinite. Since we can never exhaust all the possible examples, no number of specific cases that affirm a proposition are sufficient to establish its truth in general. We must write a general proof in order to do that—the process of writing such proofs is the major topic you are now studying, and one to which we will return shortly.

In the last sentence of the Discussion of Example 6, we stated that "X is clearly not a subset of Y." How do we justify this statement formally? Recall that "$X \subseteq Y$" is defined by "$\forall w[(w \in X) \rightarrow (w \in Y)]$." Hence the proposition "$\exists w[(w \in X) \wedge (w \notin Y)]$" corresponds to "$X$ is not a subset of Y." For the sets $X = \{4, 7, 8, 11\}$ and $Y = \{2, 7, 8, 9, 11\}$, we note that, choosing $w = 4$, we have $4 \in X$, but $4 \notin Y$. This particular example is all that is needed to conclude that X is not a subset of Y. (Note incidentally that the choice of w used to prove that X is not a subset of Y, $w = 4$, has the property that $w \notin Z$. This is not surprising since, in our initial attempt to prove the false proposition in Example 6, the obstacle we could not overcome was that our arbitrarily chosen w did not need to lie in Z.)

The following exercises are germane to the issues raised by Example 6 and the paragraphs following it.

- Prove or disprove: For all sets X, Y, and Z, $X \cup (Y \cap Z) \subseteq (X \cup Y) \cap Z$.
- Prove or disprove: For all sets X, Y, and Z, if $X \subseteq Z$, then $X \cup (Y \cap Z) \subseteq (X \cup Y) \cap Z$.

1.1.4. The tactic of *division into cases*

As propositions we are asked to prove become more complex, we must expand our arsenal of tools that are effective for proceeding toward the desired conclusion of a proposition, once we have finished setting up the argument. The applications of the choose method that occur in Examples 7 and 8, which follow, demonstrate a new such tool, known as *division into cases*, which is useful in some proofs.

Example 7. Prove that for all sets A and B, $(A \cap B) \cup (A \cap \overline{B}) \subseteq A$.

Proof. Let A and B be arbitrary sets. To prove $(A \cap B) \cup (A \cap \overline{B}) \subseteq A$, assume $x \in (A \cap B) \cup (A \cap \overline{B})$. We must prove $x \in A$. By our assumption, we know that either $x \in A \cap B$ or $x \in A \cap \overline{B}$, that is, either $x \in A$ and $x \in B$, or else $x \in A$ and $x \in \overline{B}$. [Note: We don't know which of these two is the case, but we do know that at least one of them must be true.] Hence, at this point, we divide the argument into two exhaustive cases:

Case I Suppose that $x \in A$ and $x \in B$. Then, in particular, $x \in A$ [by the rule of inference $(p \wedge q) \rightarrow p$], so the desired conclusion obtains in this case.

Case II Suppose that $x \in A$ and $x \in \overline{B}$. Then, again, $x \in A$, so the desired conclusion is again verified.

Under either of the only two possible cases, we have $x \in A$, the desired conclusion. ∎

Example 8. Prove that for all sets A and B, $A \subseteq (A \cap B) \cup (A \cap \overline{B})$.

Proof. Let A and B be arbitrary sets. To prove $A \subseteq (A \cap B) \cup (A \cap \overline{B})$, assume $x \in A$. We must prove $x \in (A \cap B) \cup (A \cap \overline{B})$. To do this, we must prove that either $x \in A \cap B$ or $x \in A \cap \overline{B}$, that is, either $x \in A$ and $x \in B$, or else $x \in A$ and $x \in \overline{B}$. [Recall that our assumption is that $x \in A$. This assumption involves only the set A. The problem we must solve is how to bring the relationship between x and the set B into the discussion.] We note that, necessarily, either $x \in B$ or $x \notin B$, by the tautology $p \vee \neg p$. Having noted this, we consider two cases:

Case I Suppose that $x \in B$. Then, since $x \in A$ [by our assumption], we have $x \in A$ and $x \in B$, one of the two alternatives in our desired conclusion.

Case II Suppose that $x \notin B$. Equivalently, $x \in \overline{B}$. Then, since $x \in A$, we have $x \in A$ and $x \in \overline{B}$, the other of the two alternatives in our desired conclusion. ∎

Compare the proofs in Examples 7 and 8, both involving division into cases. The second proof illustrates somewhat more creativity than does the first. It requires a slightly more active role on our part, involving the tautology $p \vee \neg p$. The truth of this tautology is "common sense." Once the idea of bringing this division into cases into the argument is suggested, virtually anyone would agree that it is correct and, furthermore, is an effective step at this stage. The difficult part for you, as a beginning student just learning to write proofs, is thinking of this idea on your own.

Many complex proofs require that some creative idea be brought in from outside the basic structure (i.e., the setting up) of the argument. This is the aspect of proof-writing that is not mechanical. It is learned from experience and by an active interest in the "why" in mathematics. It is fostered by developing the habit of having firmly in mind all the statements, pertaining to the problem at hand, that we know to be true, and by being willing to try to apply these statements until we find one that works.

Here are some exercises that involve the choose method and division into cases:

- Prove that for all sets X, Y, and Z, if $X \subseteq Z$ and $Y \subseteq Z$, then $X \cup Y \subseteq Z$.
- Prove that for all sets A, B, and C, if $A \subseteq B$, then $A \cup C \subseteq B \cup C$.
- Prove that for all sets X, Y, and Z, if $X \cap Z \subseteq Y \cap Z$ and $X \cap \overline{Z} \subseteq Y \cap \overline{Z}$, then $X \subseteq Y$.

1.1.5. Proving equality of sets

Three approaches to proving equality of sets are discussed in Examples 14–18 in Section 2.2 of the text. Of these, the first, known as *mutual inclusion* (introduced in Example 10), is the most generally applicable. In addition, it expands naturally on the earlier material in this Guide, so we give that approach additional emphasis here.

A formal version of the definition of equality of sets given in the text (cf. Definition 3 in Section 2.1) is

$$A = B \quad \text{if and only if} \quad \forall x[(x \in A) \leftrightarrow (x \in B)].$$

The latter proposition is equivalent to

$$\forall x\{[(x \in A) \rightarrow (x \in B)] \wedge [(x \in B) \rightarrow (x \in A)]\},$$

which, in turn, is equivalent to

$$\{\forall x[(x \in A) \rightarrow (x \in B)]\} \ \wedge \ \{\forall x[(x \in B) \rightarrow (x \in A)]\},$$

which is the definition of

$$A \subseteq B \quad \text{and} \quad B \subseteq A.$$

We may prove that two sets are equal by proving that each is a subset of the other. We illustrate this approach in Examples 9 and 10.

Example 9. Prove that for all sets A and B, $(A \cap B) \cup (A \cap \overline{B}) = A$.

Proof. We may prove the desired equality by proving mutual inclusion, i.e., that each of the two sets is a subset of the other. The inclusion $(A \cap B) \cup (A \cap \overline{B}) \subseteq A$, however, is precisely what we proved already in Example 7. The other inclusion $A \subseteq (A \cap B) \cup (A \cap \overline{B})$ was proved in Example 8. Having written these two proofs, we have established the desired equality. ■

In Example 10, we encounter a conclusion of equality that is preceded by a hypothesis.

Example 10. Prove that for all sets A and B, if $A \subseteq B$, then $A \cup B = B$.

Proof. Let A and B be arbitrary sets such that $A \subseteq B$. We may prove $A \cup B = B$ by proving $B \subseteq A \cup B$ and $A \cup B \subseteq B$.

1. $B \subseteq A \cup B$ is essentially Proposition (a), proved earlier in Example 1. [Note that the hypothesis of the theorem is not required to establish this inclusion. Like Example 6 in Section 1.6 of the text, the proof in this direction is *trivial*.]

2. To prove $A \cup B \subseteq B$, given the hypothesis $A \subseteq B$, we proceed by the choose method. Assume $x \in A \cup B$. We must prove that$x \in B$. By our assumption, we know that either $x \in A$ or $x \in B$. Since we do not know which of these two statements is true, we divide the argument into cases:

Case I Suppose that $x \in A$. Then, since $A \subseteq B$, by hypothesis, we have $x \in B$, as desired. [Recall the middle paragraph of the discussion between Examples 3 and 4, on the role of modus ponens.]

Case II Suppose that $x \in B$. Since this is the desired conclusion, the result is trivially true in this case.

We conclude $x \in B$, as desired. ∎

In some circumstances, it is possible to prove set equality using a single chain of valid equations, thus avoiding the sometimes cumbersome mutual-inclusion approach. One such circumstance is in proving any of the set identities in Table 1 in Section 2.2 of the text. Each of these results is a set-theory version of a corresponding equivalence of propositions in logic, covered in Section 1.2. Each can be proved using the approach of Example 12 in Section 2.2 of the text.

Another approach to proving an equality of sets is to use other equalities proved previously. Suppose the following identities of set theory have already been proved:

$$\text{For all sets } A,\ B, \text{ and } C,\ A \cap (B \cup C) = (A \cap B) \cup (A \cap C). \qquad [1]$$

$$\text{For every set } B,\ B \cup \overline{B} = U \quad (U \text{ represents the universal set.}) \qquad [2]$$

$$\text{For every set } C,\ C \cap U = C. \qquad [3]$$

On these bases, we can give a proof of the proposition proved in Example 9 that does not use mutual inclusion.

Example 11. Prove that for all sets A and B, $(A \cap B) \cup (A \cap \overline{B}) = A$.

Proof. Let A and B be arbitrary sets. Then we have

$$\begin{aligned} (A \cap B) \cup (A \cap \overline{B}) &= A \cap (B \cup \overline{B}) && \text{(by [1])} \\ &= A \cap U && \text{(by [2])} \\ &= A, && \text{(by [3])} \end{aligned}$$

as desired. ∎

Note that, in applying the result [1], we used the special case $C = \overline{B}$ of the identity [1] (which is the property of *distributivity* of intersection over union—see Table 1 in Section 2.2 of the text). The technique of using a special case of a known result is called *specialization*. Whenever you find yourself saying "in particular," in making an inference from a known general fact in the course of an argument, you are using the specialization tactic. Like division into cases, specialization is a sometimes-useful technique for proceeding beyond the initial setting up of a proof toward the desired conclusion. Formally, it is justified by the rule of universal instantiation, shown in Table 2 of Section 1.5.

1.2. Indirect proof

Sometimes it is convenient, or even necessary, to prove a form of a proposition that is different from the original, but logically equivalent to it. Whenever we write a proof in such a form, we are writing an indirect proof . Three common forms of indirect proof are based on three logical equivalences of pairs of propositions:

- $(\neg q) \to (\neg p)$ is equivalent to $p \to q$ [4]
- $(\neg p) \to (q \wedge \neg q)$ is equivalent to p [5]
- $(p \wedge \neg q) \to r$ is equivalent to $p \to (q \vee r)$. [6]

The equivalence [4] is the basis for the form of an indirect proof known as *proof by contraposition*. The equivalence [5] justifies the form of an indirect proof called *proof by contradiction*. The equivalence [6] underlies a standard approach to deriving a conclusion involving alternatives, i.e., having the form "either q or r."

1.2.1. Proof by contraposition

Sometimes it is difficult to see how to prove a proposition of the form $\forall x[P(x) \to Q(x)]$ by starting with the assumption that $P(x)$ is true. (Recall Section 1.1.2 of this Guide.) What can you do if you cannot see how to deduce the conclusion $Q(x)$ from the assumption $P(x)$ and any additional given hypotheses (if there are any)? Sometimes, in such cases, assuming the negation $\neg Q(x)$ of the conclusion provides a better match with known facts or the other given hypotheses, and the two together lead readily to the negation $\neg P(x)$ of the original assumption. An argument in this form is an instance of proof by contraposition. See Example 3 in Section 1.6 of the text for an example of such a proof. Our Example 12 provides another illustration of the method.

Example 12. Prove Proposition (c): For every function f whose domain and codomain are subsets of the set of real numbers, if f is strictly increasing, then f is one-to-one.

Discussion. Let f be any function that is strictly increasing. To show that f is one-to-one using the original form of the definition, we would let x_1 and x_2 be real numbers in the domain of f and assume $f(x_1) = f(x_2)$. We would then have to prove $x_1 = x_2$. We have completed the setting up of a direct proof, but have no way of using the hypothesis that f is strictly increasing [i.e., "... if $x_1 < x_2$, then $f(x_1) < f(x_2)$"]. The assumption $f(x_1) = f(x_2)$ simply does not "match up" with the "if part" of the hypothesis in a way that permits us to proceed anywhere from that hypothesis.

However, suppose we decide instead to derive the contrapositive of the definition of one-to-one. Under this approach, we will begin by assuming that $x_1 \neq x_2$. Our goal will then be to prove that $f(x_1) \neq f(x_2)$. We proceed from here as follows. Since $x_1 \neq x_2$ (by assumption), then it must be that either $x_1 < x_2$ or $x_2 < x_1$. We consider two cases:

Case I If $x_1 < x_2$, then since f is strictly increasing, we may conclude $f(x_1) < f(x_2)$, so, in particular, $f(x_1) \neq f(x_2)$, as desired.

Case II If $x_2 < x_1$, then since f is strictly increasing, we may conclude $f(x_2) < f(x_1)$, so $f(x_1) \neq f(x_2)$, again as desired. ∎

Another situation in which a proof by contraposition is sometimes called for is one in which the conclusion of the proposition has a much simpler logical form than does the hypothesis. We illustrate this in the following example.

Example 13. Suppose that a is a real number satisfying the property $\forall M{>}0\,(|a| < M)$. Then $a = 0$.

Discussion. Note the simple form of the conclusion. Rather than involving a definition having an "if... then" form, as in most of our earlier examples, it is simply the flat statement that $a = 0$. If we try to begin a direct proof by focusing on the desired conclusion, then there is really no place to begin, no basis for making the kind of assumption that is needed to get the proof "off the ground."

So instead we proceed by contraposition. Our approach will be to assume $a \neq 0$ and try to deduce the negation of the hypothesis. This negation may be formulated (cf. Table 2 in Section 1.3 of the text)

$$\exists M{>}0\,(|a| \geq M). \qquad [7]$$

We need only produce a positive M whose value does not exceed that of the absolute value of the nonzero a. We take $M = |a|$, noting that this value of M clearly satisfies the condition [7]. ∎

Here is a third circumstance in which a proof by contraposition is appropriate. Suppose a proposition of the form "if p and q, then r" is known to be true, and we are asked to prove that p and the negation of r together imply the negation of q. We may always proceed, using contraposition, by assuming that the negation of q is false, that is, that q is true. Then since p is true, by hypothesis, we have that p and q are both true, so, by the known proposition, we may conclude that r is true, contradicting the fact that $\neg r$ is one of the hypotheses. You will have an opportunity to apply this approach in the third and fourth of the exercises that follow.

- Prove that for all sets A and B, if $A \subseteq B$, then $\overline{B} \subseteq \overline{A}$.

- Prove that if a linear function $f(x) = Mx + B$ is one-to-one, then $M \neq 0$.

- Suppose it is known that "every sum or difference of two integers is an integer." Use this result to prove that for all real numbers x and y, if x is an integer and $x + y$ is an integer, then y is an integer. Prove also that the sum of an integer and a noninteger must be a noninteger.

- Prove that for all sets A and B and for every object x, if $x \in A$ and $x \notin A \cap B$, then $x \notin B$.

1.2.2. Proof by contradiction

The idea behind the equivalence [5] is that we may prove a conclusion p by showing that the denial of p leads to a contradiction. Actually, proof by contraposition is a form of proof by contradiction. For if, in a proof that p implies q, we assume the truth of p (as we are entitled to do) and then use the negation of q to derive $\neg p$, then we have obtained the contradiction $p \wedge \neg p$. Another circumstance in which proof by contradiction is the standard approach is any proof of a theorem in set theory in which the conclusion asserts that some set equals the empty set.

Example 14. Prove that for all sets A and B, if $A \subseteq B$, then $A \cap \overline{B} = \emptyset$.

Discussion. A direct approach would be to establish the equality $A \cap \overline{B} = \emptyset$ using mutual inclusion. Indeed the containment in one direction, $\emptyset \subseteq A \cap \overline{B}$, is true automatically, based on the principle that the empty set is a subset of every set (this result is vacuously true—recall Example 5 in Section 1.6 of the text). However, for the containment in the other direction, $A \cap \overline{B} \subseteq \emptyset$, the approach "assume $x \in A \cap \overline{B}$ ... we must prove $x \in \emptyset$" is doomed to failure, since the conclusion "$x \in \emptyset$" can never be reached.

Since we are unable to write a direct proof, we proceed by contradiction. Let A and B be arbitrary sets such that $A \subseteq B$. Assume that $A \cap \overline{B} \neq \emptyset$. Then there exists some object that lies in $A \cap \overline{B}$; let us call it c. Since $c \in A \cap \overline{B}$, we know that $c \in A$ and $c \in \overline{B}$. Since $c \in A$ and $A \subseteq B$, by hypothesis, we have $c \in B$. Thus we have $c \in B$ and $c \in \overline{B}$, so $c \in B$ and $c \notin B$. This is a contradiction of the form $p \wedge \neg p$, so our proof is complete . ∎

A classic example of a proof by contradiction is provided in the text in Example 10 of Section 1.6, which shows that $\sqrt{2}$ is irrational. Here are some exercises:

- Prove that for every set A, $A \cap \overline{A} = \emptyset$.
- Prove that for all sets A and B, if $(B \cap \overline{A}) \cup (\overline{B} \cap A) = B$, then $A = \emptyset$.
- Prove that for all sets A and B, $(\overline{A} \cup \overline{B}) \cap (\overline{A} \cup B) \cap (A \cup \overline{B}) \cap (A \cup B) = \emptyset$.

1.2.3. Deriving conclusions of the form "q or r"

The equivalence [6] becomes relevant to the writing of proofs when we must derive a conclusion involving alternatives, q or r. For a proposition of this type, there may be no circumstance under which we can be sure which of the alternatives is true, only that at least one of them must be true under every circumstance in which the hypothesis is true. Because of this, we are unable to determine whether to set up a direct proof based on the conclusion q or on the conclusion r. (Indeed, usually no such proof is possible.) Fortunately there is an indirect approach, based on the equivalence [6], that enables us to get around this difficulty. Rather than attempting to prove directly that q or r follows from p, we may replace this problem by the problem of showing that r follows from p and $\neg q$ (or else that q follows from p and $\neg r$—either approach will do the job).

A classic example of a proof in this category is the following theorem from elementary algebra: "For all real numbers x and y, if $xy = 0$, then $x = 0$ or $y = 0$." Clearly we should set up this proof by letting x and y be real numbers such that $xy = 0$. But at this point there is no evident way of proceeding toward the conclusion that one or the other of x and y (we know not which) equals 0. The escape is to make the additional assumption that $x \neq 0$, with the goal of proving that therefore y must equal 0. Since $x \neq 0$, its reciprocal $1/x$ must exist, and we may write the chain of equations $y = 1 \cdot y = [(1/x)(x)](y) = (1/x)(xy) = (1/x)(0) = 0$, so $y = 0$, as desired. [Note that this chain of equations also uses the facts that multiplication of real numbers is associative and that the product of every real number with 0 equals 0.]

A problem in set theory in which this approach is sometimes useful is proving that one set is a subset of the union of two other sets. This is demonstrated in Example 15.

Example 15. Prove that for all sets A and B, $A \subseteq B \cup (A \cap \overline{B})$.

Proof. Let A and B be arbitrary sets. To prove $A \subseteq B \cup (A \cap \overline{B})$, assume that $x \in A$. We must prove that $x \in B \cup (A \cap \overline{B})$, that is, either $x \in B$ or $x \in A \cap \overline{B}$. Since our desired conclusion is now seen to have the form "either q or r," we take the approach suggested by the equivalence [6], and assume that $x \notin B$. Our goal now becomes to prove that, on the basis of this additional assumption, it must be true that $x \in A \cap \overline{B}$, that is, $x \in A$ and $x \in \overline{B}$. We already know $x \in A$, by our initial assumption in the proof. As for $x \in \overline{B}$, that follows immediately from our additional assumption $x \notin B$. ∎

If a desired conclusion has more than two alternatives, the strategy suggested by [6] is generalized as follows: Assume the negation of all but one of the alternative conclusions and, on that basis, try to prove that the remaining one must be true. We illustrate this in Example 16.

Example 16. Prove that for all sets A and B, if $A \times B = B \times A$, then either $A = \emptyset$ or $B = \emptyset$ or $A = B$.

Sketch of Proof. [Note first that the notation $A \times B$ refers to the *cartesian product* of sets A and B, defined as the set of all ordered pairs (a, b), where $a \in A$ and $b \in B$. This definition is the basis of the content of Chapter 6 of the text, on relations.] Let A and B be any sets such that $A \times B = B \times A$. Assume further that $A \neq \emptyset$ and $B \neq \emptyset$. With these two additional assumptions, our goal then becomes to prove that the third alternative $A = B$ must be true. The remainder of this proof is left as an exercise. ■

The following exercises provide the opportunity to use the strategy suggested by the equivalence [6].

- Complete the proof in Example 16.
- Prove that if A, B, and C are any sets such that $A \times B = A \times C$, then either $A = \emptyset$ or $B = C$.
- Prove that if A and B are any sets such that $A \times B = \emptyset$, then either $A = \emptyset$ or $B = \emptyset$.
- Prove that for all sets X, Y, and Z, $(X \cup Y) \cap Z \subseteq X \cup (Y \cap Z)$.

2. Remarks on additional methods of proof

Not all propositions we may wish to prove have conclusions involving the form $\forall x[P(x) \to Q(x)]$. Nonetheless, beginning students who are able to write correctly the proofs called for in the exercises in Section 1 of this Guide are well prepared to deal with the new issues that arise in writing other types of proofs. One reason for this is that many of the tactics (e.g., division into cases, specialization) and strategies (e.g., the choose method, indirect proof), highlighted in Section 1, have application beyond proving propositions whose conclusion is of the form $\forall x[P(x) \to Q(x)]$. In this section, we discuss briefly two additional types of propositions.

2.1. Deducing conclusions having the form "For every x, there exists y such that $P(x, y)$."

Many defining properties in mathematics have one of the forms $\exists x\, P(x)$ or $\forall x\, \exists y\, P(x, y)$. Elementary definitions of these types include:

(i) Let m and n be integers. We say that m *divides* n, denoted $m \mid n$, if and only if there exists an integer p such that $n = mp$. (Cf. Definition 1 in Section 3.4 of the text.)

(ii) A function f is *onto*: f is onto if and only if, for every y in the codomain of f, there exists x in the domain of f such that $f(x) = y$. (Cf. Definition 7 in Section 2.3 of the text.)

(iii) A real number x is said to be *rational* if and only if there exist integers p and q, with $q \neq 0$, such that $x = p/q$.

Many important mathematical propositions whose proofs should be within the capabilities of students working through this Guide have as their conclusion a statement involving one of the preceding definitions. Some examples are:

(a) Prove that if m, n, and p are integers such that m divides n and m divides p, then m divides $n + p$.

(b) Prove that if functions f and g, having the real numbers as their domain and codomain, are both onto, then their composition $f \circ g$ is also onto.

(c) Prove that if x and y are rational, then xy and $x + y$ are rational.

The new issue involved in proving propositions like (a)–(c) is *existence*. At a key point of each of these proofs, we must "produce," or define, an appropriate object of the type whose existence is asserted in the desired conclusion. In doing this, it is important to realize that, for a conclusion of the form $\forall x\, \exists y\, P(x, y)$, the y whose existence is to be proved usually depends on the given x; we should expect it to be defined in terms of x or else in terms of some other object that is defined in terms of x. This principle is demonstrated in Examples 17 and 18, which follow.

Example 17. Prove the first part of Proposition (c): If x and y are rational, then xy is rational.

Discussion. Assume that real numbers x and y are rational. To prove that their product xy is rational, we must show that $xy = p/q$, where p and q are integers with $q \neq 0$. Our job in this proof is to produce, literally to build, the integers p and q whose quotient p/q equals xy. As in most proofs, once the argument is set up, we must next assess what we have available to work with. In the case of a proof of existence, this includes asking whether what we have to work with provides any "building blocks." We have at our disposal only the hypotheses that x and y are rational. This means we can state that there exist integers p_1 and q_1, with $q_1 \neq 0$, such that $x = p_1/q_1$; and there exist integers p_2 and q_2, with $q_2 \neq 0$, such that $y = p_2/q_2$. We note that, therefore, $xy = (p_1/q_1)(p_2/q_2)$, which, by rules of algebra, equals $(p_1p_2)/(q_1q_2)$. Noting that p_1p_2 and q_1q_2 are necessarily integers and that $q_1q_2 \neq 0$ (Why?), we declare that $p = p_1p_2$ and $q = q_1q_2$ are the required integers. ∎

Example 18. Prove Proposition (b): If functions f and g, having the real numbers as their domain and codomain, are both onto, then their composition $f \circ g$ is also onto.

Proof. Assume that the functions f and g are onto. To prove that their composition $f \circ g$ is onto, let z be an arbitrary real number. We must prove that there exists $x \in \mathbf{R}$ such that $(f \circ g)(x) = z$. Now since f is onto, we know that, corresponding to the given z, there exists a real number y such that $z = f(y)$. Next, since g is onto, then corresponding to this y, there must exist a real number w such that $y = g(w)$. Note therefore that $z = f(y) = f(g(w)) = (f \circ g)(w)$. Hence our choice of the desired real number x becomes evident, namely choose $x = w$. ∎

Theorem 1 in Section 3.4 of the text contains several propositions related to Definition (i), including a proof of Proposition (a). You should study that proof, noting its similarities to the proofs in Examples 17 and 18, and then attempt the following exercises.

- Prove parts (*ii*) and (*ii*) of Theorem 1 in Section 3.4 of the text.
- Prove that for every integer n, there exists an integer m such that $m \mid n$.
- Prove that for every positive integer m, there exists a positive integer n such that $m \mid n$.
- Prove the second part of Proposition (c): If x and y are rational, then $x + y$ is rational.
- Prove that if f and g are functions having the real numbers as their domain and codomain, and if $f \circ g$ is onto, then f is onto.

2.2. Proof by mathematical induction

We use mathematical induction to prove a proposition whose conclusion has the form $\forall n\, P(n)$, where n is a positive integer (or, sometimes, a nonnegative integer). Thus proof by induction is an appropriate approach when the universe of discourse for a predicate quantified by "for every" is the set of all positive integers. If you review earlier sections of this Guide, you will note that this has not usually been the case in most of the examples and exercises covered, so induction would not have been an appropriate approach at those earlier stages.

Note that the *inductive step* in every proof by mathematical induction involves a proposition of the form $\forall n[P(n) \rightarrow Q(n)]$, where $Q(n)$ is $P(n+1)$. Thus the basic approach to be taken in the second part of a proof by induction is the same approach that was emphasized throughout Section 1 of this Guide, namely the choose method. We start by letting n be an arbitrary positive integer for which it is assumed that $P(n)$ is true. We must prove, on the basis of that assumption and whatever else is available (e.g., hypotheses), that $P(n+1)$ is also true.

For more on mathematical induction, see Sections 4.1 and 4.2 of the text.

General Advice on the Writing Projects

If your instructor assigns one or more of the writing projects, you are fortunate. Written communication skills are of utmost importance in today's world of information. In doing the research for such essays, you will become familiar with the literature in many areas of mathematics and computer science, and you will hone your library and information gathering skills. In this section we offer some helpful advice and provide a list of information resources—including books, articles, and Internet resources—to get you started. At the end of the solutions section of each chapter in this *Guide*, we give specific suggestions of where you might look when working on the various writing projects. We do not guarantee that you will find exactly what you are looking for in the references we suggest, but at least our pointers will start you in the right direction. Tracking down the information is half the challenge!

Here are several ideas and points to bear in mind as you do the research for the writing projects:

- There is a meta-source for information today: the "information superhighway." If you don't already have access to the Internet, ask the appropriate people for access privileges and some guidance on how to use the Internet. Then explore its nooks and crannies, such as e-mail, Usenet (try the `sci.math` newsgroup, accessible from `http://news-reader.org/sci.math/`), the World Wide Web, and so on. You will find lots of sources of information, and you will get to communicate with other people who have the kinds of information you want, or know where to get it. Ask around—people tend to be very friendly and helpful in this community. It's a fascinating social dynamic!

- The *first* place to search for material on any of the writing projects is probably the World Wide Web. In fact, the existence of the World Wide Web and search engines makes it embarrassingly easy to find sources of information on any topic one desires. To use a search engine, you type in one or more key words or phrases (such as "graph theory" inside quotation marks), and the search engine looks over the billions of websites around the world to find those that mention these words or phrases; then you can visit those sites. The whole process takes only a few seconds. One particularly good search engine is called Google. It shows you the "best" hits (those sites that contain your words or phrases most prominently, and/or are sites that many people link to). The URL is `http://www.google.com`.

- You should definitely check the website for this textbook: `http://www.mhhe.com/rosen`. There you will find many useful links that can get you started on researching the writing projects. Notice the various Web icons throughout the textbook, which indicates relevant material on this website.

- Most libraries have on-line search facilities that allow you to look for key words in titles of books in their collection. For example, to find resources on fuzzy sets or fuzzy logic, you could search on the word *fuzzy*. You can also search for authors or titles, of course. Ask a librarian for assistance if necessary. Also, catalogs to many university libraries (and the Library of Congress) are available on the Internet; see `http://www.libdex.com`.

- The following library research technique should come in handy. If the source you are looking at does not deal in enough detail with the topic you are investigating, then consult the references given in that source. Continue this process backwards as deeply and broadly as necessary. Of course this is particularly easy to do on the Web.

- There is a comprehensive set of brief reviews of essentially every mathematical research paper and book written since 1940 (and being kept up-to-date at the rate of over 60,000 items per year), in a journal called *Mathematical Reviews*, published out of Ann Arbor, Michigan, by the American Mathematical Society. It comes in various forms—in paper volumes, on CD-ROM, and on the Internet. Ask your librarian for the forms you can have access to. In the best formats, you can search for key words, as well as authors or titles. The hypertext version on the World Wide Web (called MathSciNet) lets you quickly follow leads from one review to another. Your institution must be a subscriber in order for the computers on your campus to have access.

- We shouldn't need to mention obvious things, like using the index and table of contents of any book you consult. When looking up items in an index, don't forget to try possible variations of what you are looking for (e.g., you may find one of the entries "induction, mathematical" or "mathematical induction" but not the other).

- An excellent source for many of these writing projects is [MiRo] (see the bibliography that follows), which is published as a companion to this textbook. It has articles covering various parts of pure and applied discrete mathematics, at levels varying from elementary to intermediate. It is worth browsing through this book, even if you do not find anything in it relevant to a project you are working on. It will give you a feeling for the breadth of the subject you are studying.

- Popular accounts of mathematical topics often make their way into *The New York Times*. This premier of American newspapers has a detailed index, which is available in most libraries that carry the newspaper; it can also be accessed from the Web, at `http://www.nytimes.com/ref/membercenter/nytarchive.html`. One prolific writer of mathematical articles is Gina Kolata. The *Times* also produces on a regular basis a special edition of mathematics-related articles; ask your mathematics department whether they have a recent issue. The *Times* and most other major newspapers are also available on the Web, usually for free.

- Many of the essays assigned in this textbook deal with the history of mathematical topics. Most books on the general history of mathematics are filed under the call letters QA 21. See [Bo4] and [Ev3] for two good sources. There are also wonderful extensive collections of essays about mathematics, both historical and expository. A classic is the four-volume treatise [Ne]. A more recent one of high quality is [DaHe]. Perhaps the best resource for the history of mathematics is the MacTutor History of Mathematics archive on the Web; its URL is `http://www-history.mcs.st-and.ac.uk`. It has biographies of hundreds of mathematicians, as well as references, articles, links, and an unbelievable amount of information.

- The Mathematical Association of America (MAA) has a website with lots of interesting articles (updated monthly) and special sections for students. Its URL is `http://www.maa.org`.

- Some of these projects go into depth on various topics in discrete mathematics. There are several good, more advanced textbooks on combinatorics and graph theory, such as [Bo1], [BoMu], [Br2], [ChLe], [Ro1], and [Tu1]. The library classifications here are QA 164 and QA 166, where you will also find specialized books, research monographs, and conference proceedings. In addition, there are dozens of other discrete mathematics textbooks at a level comparable with or slightly more or less advanced than your textbook. An excellent one is [Gr2]. It has comprehensive discussions of most discrete mathematical topics and a wide variety of interesting problems, including some challenging and open-ended ones. It also has a bibliography of 335 books and articles, and a detailed index that will lead you to the right source for further reading. Another, slightly different, more advanced book to take a look at (if nothing else, for its style!) is [GrKn].

- There is an intimate relationship between discrete mathematics and computer science. Computer science books of all sorts, whether dealing with hardware and circuit issues, programming, data structures, algorithms, complexity, theoretical foundations, operating systems, compilers, artificial intelligence, or other topics, may well be relevant to many of these projects. QA 76 is where many such books are housed in the library, although specialized topics will have their own call numbers (e.g., Q 335 for artificial intelligence or the high TK 7800's for circuit design). Our list that follows includes several textbooks on data structures and algorithms. Another lively source is [De2], a collection of essays on various aspects of computer science and related mathematics, each with references for further reading. You will find those essays relevant to a large number of the writing projects, and you should definitely try to have a look at this collection.

Here are several points to bear in mind about writing essays (whether in mathematics or in other subjects):

- All the rules and advice you have learned over the years about good writing apply to technical writing as well as to other forms of prose. It is often more difficult to express mathematical ideas clearly and precisely, so do not expect these writing projects to be easy.

- Know your reader! Keep in mind for whom you are writing, and pitch the level to that audience. Think about how much you will assume your reader knows and how much you will need to fill in. (When in doubt, do not assume the reader knows much.)

- Organize, organize, organize! Essays need to have an introduction, a body, and a conclusion. If the work is going to be long, it probably makes sense to have labeled sections covering the different points. Make an outline of what you plan to say, and think a lot about how to order it, both before you start writing and throughout the process.

- Use a word processor if you have access to one. This makes it much easier to revise and edit your work numerous times, until it is just the way you want it. Make sure to take advantage of special features like spelling, grammar and usage checkers. Pay some attention to the format (fonts, spacing, layout, etc.); most word processors let you design a very pleasing document. Print your essay on a laser-quality printer if you can. If your essay will contain much mathematical symbolism, try to use a mathematical word processor or typesetter. The best of these is TEX, which it would definitely be worth your while to learn to use (although it is not easy). To give you an idea of how nice TEX can look, note that this solutions manual was produced using TEX.
- Finally, be careful to give credit to the sources you use. Plagiarism has become a major problem, and if you copy material from the Internet or other sources and present it as your own, you are stealing another person's property. The consequences can include suspension from your school. When in doubt, ask you instructor about proper procedures for citations.

List of References for the Writing Projects

[AlSp] Noga Alon and Joel H. Spencer, *The Probabilistic Method*, with an Appendix on Open Problems by Paul Erdős (Wiley, 1982; second edition, 2000)

[ApHa] K. Appel and W. Haken, "The solution of the four-color-map problem," *Scientific American* **237,4** (1977) 108–121

[Ba1] Roland C. Backhouse, *Program Construction and Verification* (Prentice-Hall, 1986)

[Ba2] Paul Bachmann, *Analytische Zahlentheorie* (Leipzig, 1900)

[Ba3] Albert László Barabási, *Linked: The New Science of Networks* (Perseus, 2002)

[BaGo] Hans Bandemer and Siegfried Gottwald, *Fuzzy Sets, Fuzzy Logic, Fuzzy Methods with Applications* (Wiley, 1995)

[BeKa] Kenneth R. Beesley and Lauri Karttunen, *Finite State Morphology* (Center for the Study of Language and Information, 2003)

[BeCo] Elwyn R. Berlekamp, John H. Conway, and Richard K. Guy, *Winning Ways for Your Mathematical Plays*, in two volumes (Academic Press, 1982)

[Be] Albrecht Beutelspacher, *Cryptology* (Mathematical Association of America, 1994)

[BiLl] Norman L. Biggs, E. Keith Lloyd, and Robin J. Wilson, *Graph Theory 1736–1936* (Clarendon Press, 1976)

[Bo1] Kenneth P. Bogart, *Introductory Combinatorics*, second edition (Harcourt Brace, 1990)

[BoDo] Kenneth P. Bogart and Peter G. Doyle, "Nonsexist solution of the menage problem," *The American Mathematical Monthly* **93** (1986) 514–519

[Bo2] Bela Bollobás, *Random Graphs* (Academic Press, 1985)

[Bo3] Bela Bollobás, "Random graphs," in Bela Bollobás, ed., *Probabilistic Combinatorics and Its Applications*, Proceedings of Symposia in Applied Mathematics **44** (American Mathematical Society, 1991) 1–20

[BoMu] John A. Bondy and U. S. R. Murty, *Graph Theory with Application* (American Elsevier, 1976)

[Bo4] Carl B. Boyer, *A History of Mathematics*, second edition (Wiley, 1991)

[BrBr] Gilles Brassard and Paul Bratley, *Algorithmics: Theory and Practice* (Prentice-Hall, 1988)

[Br1] J. Glenn Brookshear, *Theory of Computation: Formal Languages, Automata, and Complexity* (Prentice-Hall, 1988)

[Br2] Richard A. Brualdi, *Introductory Combinatorics*, second edition (North Holland, 1992)

[Ca1] Lewis Carroll, *Lewis Carroll's Symbolic Logic* (Crown, 1977)

[Ca2] Lewis Carroll, *Mathematical Recreations of Lewis Carroll* (Dover, 1958)

[Ca3] Lewis Carroll, *Symbolic Logic and the Game of Logic* (Dover, 1958)

[ChLe] Gary Chartrand and Linda Lesniak, *Graphs & Digraphs*, fourth edition (Chapman & Hall/CRC, 2005)

[ChOe] Gary Chartrand and Ortrud R. Oellermann, *Applied and Algorithmic Graph Theory* (McGraw-Hill, 1993)

[Ch] Paul M. Chirlian, *Analysis and Design of Integrated Electronic Circuits*, second edition (Harper & Row, 1987)

[Co] Daniel I. A. Cohen, *Introduction to Computer Theory* (Wiley, 1986)

[CoLe] Thomas H. Cormen, Charles E. Leiserson, and Ronald L Rivest, *Introduction to Algorithms*, second edition (MIT Press and McGraw-Hill, 2001)

[Da1] C. J. Date, *An Introduction to Database Systems* (Addison-Wesley, 1990)

[Da2] Karl David, "Rencontres reencountered," *The College Mathematics Journal* **19** (1988) 139–148

[DaHe] Philip J. Davis and Reuben Hersh, *The Mathematical Experience* (Birkhäuser, 1981)

[De1] Dorothy Elizabeth Robling Denning, *Cryptography and Data Security* (Addison-Wesley, 1982)

[DeDe] Peter J. Denning, Jack B. Dennis, and Joseph E. Qualitz, *Machines, Languages, and Computation* (Prentice-Hall, 1978)

[De2] A. K. Dewdney, *The New Turing Omnibus: 66 Excursions in Computer Science* (Freeman, 1993)

[Di] Edsger W. Dijkstra, *A Discipline of Programming* (Prentice-Hall, 1976)

[DuPr] Didier Dubois, Henri Prade, and Ronald R. Yager, eds., *Fuzzy Information Engineering: A Guided Tour of Applications* (Wiley, 1997)

[Du] David A. Duffy, *Principles of Automated Theorem Proving* (Wiley, 1991)

[EaTa] P. Eades and R. Tamassia, *Algorithms for Drawing Graphs: An Annotated Bibliography*, Technical Report CS-89-09 (Department of Computer Science, Brown University, Providence, RI, 1989)

[En] Herbert Enderton, *A Mathematical Introduction to Logic*, second edition (Harcourt/Academic Press, 2000)

[Ev1] Shimon Even, *Algorithmic Combinatorics* (Macmillan, 1973)

[Ev2] Shimon Even, *Graph Algorithms* (Computer Science Press, 1979)

[Ev3] Howard Eves, *An Introduction to the History of Mathematics* (Saunders, 1990)

[Fr] Roger L. Freeman, ed., *Reference Manual for Telecommunications Engineering* (Wiley, 1994)

[Ga1] Martin Gardner, *Logic Machines and Diagrams* (McGraw-Hill, 1958)

[Ga2] Martin Gardner, *Wheels, Life and Other Mathematical Amusements* (Freeman, 1983)

[GaJo] Michael R. Garey and David S. Johnson, *Computers and Intractability: A Guide to the Theory of NP-Completeness* (Freeman, 1979)

[Gi] Alan M. Gibbons, *Efficient Parallel Algorithms* (Cambridge University Press, 1988)

[Go1] Solomon W. Golomb, *Polyominoes: Puzzles, Patterns, Problems, and Packings*, second edition (Princeton University Press, 1994)

[Go2] Ronald J. Gould, "Updating the Hamiltonian problem—a survey," *Journal of Graph Theory* **15** (1991) 121–151

[GrHe] Ronald L. Graham and Pavol Hell, "On the history of the minimum spanning tree problem," *Annals of the History of Computing* **7** (1985) 43–57

[GrKn] Ronald L. Graham, Donald E. Knuth, and Oren Patashnik, *Concrete Mathematics: A Foundation for Computer Science*, second edition (Addison-Wesley, 1994)

[GrRo] Ronald L. Graham, Bruce L. Rothschild, and Joel H. Spencer, *Ramsey Theory*, second edition (Wiley, 1990)

[Gr1] George Gratzer, *Lattice Theory: First Concepts and Distributive Lattices* (Freeman, 1971)

[Gr2] Jerrold W. Grossman, *Discrete Mathematics: An Introduction to Concepts, Methods, and Applications* (Macmillan, 1990)

[GrZe] Jerrold W. Grossman and R. Suzanne Zeitman, "An inherently iterative computation of Ackermann's function," *Theoretical Computer Science* **57** (1988) 327–330

[GrSh] Branko Grünbaum and G. C. Shephard, *Tilings and Patterns* (Freeman, 1987)

[Gu] Richard K. Guy, *Unsolved Problems in Number Theory*, second edition (Springer-Verlag, 1994)

[HaRi] H. Halberstam and H.-E. Richert, *Sieve Methods* (Academic Press, 1974)

[Ha1] Paul R. Halmos, *Naive Set Theory* (Springer-Verlag, 1974)

[HaMa] Frank Harary and John S. Maybee, eds., *Graphs and Applications: Proceedings of the First Colorado Symposium on Graph Theory* (Wiley, 1985)

[Ha2] David Harel, *Algorithmics: The Spirit of Computing* (Addison-Wesley, 1987)

[He] Michael Henle, *A Combinatorial Introduction to Topology* (Freeman, 1979)

[HiPe1] Frederick J. Hill and Gerald R. Peterson, *Computer Aided Logical Design with Emphasis on VLSI*, fourth edition (Wiley, 1993)

[HiPe2] Peter Hilton and Jean Pedersen, "Catalan numbers, their generalization, and their uses," *The Mathematical Intelligencer* **13,2** (1991) 64–75

[Ho1] C. A. R. Hoare, "An axiomatic basis for computer programming," *Communications of the Association for Computing Machinery* **12** (1969) 576–580, 583

[Ho2] John Hogger, *Essentials of Logic Programming* (Oxford University Press, 1990)

[Ho3] Gerard J. Holzmann, *Design and Validation of Computer Protocols* (Prentice-Hall, 1990)

[HoUl] John E. Hopcroft and Jeffrey D. Ullman, *Introduction to Automata Theory, Languages, and Computation* (Addison-Wesley, 1979)

[Ka] Abraham Kandel, *Fuzzy Mathematical Techniques with Applications* (Addison-Wesley, 1986)

[Kn] Donald E. Knuth, *The Art of Computer Programming*, in three volumes, some in second edition (Addison-Wesley, 1968–73)

[KöSc] Johannes Köbler, Uwe Schöning, and Jacobo Torán, *The Graph Isomorphism Problem: Its Structural Complexity* (Birkhäuser, 1993)

[Ko1] Zvi Kohavi, *Switching and Finite Automata Theory*, second edition (McGraw-Hill, 1978)

[Ko2] Israel Koren, *Computer Arithmetic Algorithms* (Prentice-Hall, 1993)

[Ko3] Bart Kosko, *Fuzzy Thinking: The New Science of Fuzzy Logic* (Hyperion, 1993)

[Kr] Robert L. Kruse, *Data Structures and Program Design*, second edition (Prentice-Hall, 1987)

[La1] Jeffrey Lagarias, "Pseudorandom number generators," in Carl Pomerance, ed., *Cryptology and Computational Number Theory, Proceedings of Symposia in Applied Mathematics* **42** (American Mathematical Society, 1990) 115–143

[La2] Clement W. H. Lam, "How reliable is a computer-based proof?" *Mathematical Intelligencer* **12,1** (1990) 8–12

[La3] Eugene L. Lawler et al., eds., *The Traveling Salesman Problem: A Guided Tour of Combinatorial Optimization* (Wiley, 1985)

[Le1] D. H. Lehmer, "The machine tools of combinatorics," in E. F. Beckenbach, ed., *Applied Combinatorial Mathematics* (Wiley, 1964)

[Le2] F. Thomson Leighton, *Introduction to Parallel Algorithms and Architectures: Arrays, Trees, Hypercubes* (Kaufman, 1992)

[Le3] Arjen K. Lenstra, "Primality testing," in Carl Pomerance, ed., *Cryptology and Computational Number Theory, Proceedings of Symposia in Applied Mathematics* **42** (American Mathematical Society, 1990) 13–26

[LePa] Harry R. Lewis and Christos H. Papadimitriou, *Elements of the Theory of Computation* (Prentice-Hall, 1981)

[Li] Mario Livio, *The Golden Ratio: The Story of Phi, the World's Most Astonishing Number* (Broadway, 2002)

[Ma1] David Maier, *The Theory of Relational Databases* (Computer Science Press, 1983)

[Ma2] Udi Manber, *Introduction to Algorithms* (Addison-Wesley, 1989)

[Ma3] Eli Maor, *e: The Story of a Number* (Princeton University Press, 1994)

[Ma4] Eli Maor, *To Infinity and Beyond* (Birkhäuser, 1987)

[Ma5] Thomas Maufer, *Deploying IP Multicast in the Enterprise* (Prentice-Hall, 1998)

[Mc] James A. McHugh, *Algorithmic Graph Theory* (Prentice-Hall, 1990)

[McCh] L. E. McMahon, L. L. Cherry, and R. Morris, "Statistical text processing," *Bell System Technical Journal* **57** (1978) 2137–2154

[McFr] Daniel McNeill and Paul Freiberger, *Fuzzy Logic* (Simon and Schuster, 1993)

[MeOo] Alfred J. Menezes, Paul C. van Oorschot, and Scott A. Vanstone, *Handbook of Applied Cryptography* (CRC Press, 1996)

[Me1] Susan Merritt, "An Inverted Taxonomy of Sorting Algorithms," *Communications of the ACM* **20** (1985) 96–99

[**Me2**] W. Meyer, "Huffman codes and data compression," *UMAP Journal* **5** (1984) 278–296

[**MiRo**] John G. Michaels and Kenneth H. Rosen, *Applications of Discrete Mathematics* (McGraw-Hill, 1991)

[**Mi**] J. Mitchem, "On the history and solution of the four-color map problem," *Two-Year College Mathematics Journal* **12** (1981) 108–112

[**Mo**] Joseph J. Moder, *Project Management with CPM, PERT, and Precedence Diagramming* (Van Nostrand Reinhold, 1983)

[**Ne**] James R. Newman, *The World of Mathematics*, in four volumes (Simon and Schuster, 1956)

[**Ni**] Ivan Niven, *Irrational Numbers* (Wiley, 1956)

[**O'R**] Joseph O'Rourke, *Computational Geometry in C*, second edition (Cambridge University Press, 1998)

[**Pa1**] Edgar M. Palmer, *Graphical Evolution* (Wiley, 1985)

[**Pa2**] Cheng Dong Pan and Cheng Biao Pan, *Goldbach Conjecture* (Science Press, 1992)

[**PeWi**] Marko Petkovsek, Herbert S. Wilf, and Doron Zeilberger, $A = B$ (A. K. Peters, 1996)

[**Pf**] Charles P. Pfleeger, *Security in Computing* (Prentice-Hall, 1989)

[**Po**] Carl Pomerance, "Factoring," in Carl Pomerance, ed., *Cryptology and Computational Number Theory, Proceedings of Symposia in Applied Mathematics* **42** (American Mathematical Society, 1990) 27–47

[**PrDu**] Kendall Preston, Jr., and Michael J. B. Duff, *Modern Cellular Automata: Theory and Applications* (Plenum Press, 1984)

[**Ra1**] Anthony Ralston, "De Bruijn sequences—a model example of the intersection of discrete mathematics and computer science," *Mathematics Magazine* **55** (1982) 131–143

[**Ra2**] K. Ramachandra, "Many famous conjectures on primes; meagre but precious progress of a deep nature," *The Mathematics Student* **67** (1998) 187–199

[**Re**] John H. Reif, "Successes and Challenges," *Science* **296** (19 April 2002) 478–479

[**ReNi**] Edward M. Reingold, Jurg Nievergelt, and Narsingh Deo, *Combinatorial Algorithms: Theory and Practice* (Prentice-Hall, 1977)

[**Ri1**] Paulo Ribenboim, *The Book of Prime Number Records* (Springer-Verlag, 1989)

[**Ri2**] John Riordan, *Combinatorial Identities* (Wiley, 1968)

[**Ro1**] Fred S. Roberts, *Applied Combinatorics* (Prentice-Hall, 1984)

[**Ro2**] Fred S. Roberts, *Discrete Mathematical Models, with Applications to Social, Biological, and Environmental Problems* (Prentice-Hall, 1976)

[**Ro3**] Kenneth H. Rosen, *Number Theory and Its Applications*, fifth edition (Addison-Wesley, 2004)

[**Ru**] Rudy Rucker, *Infinity and the Mind* (Birkhäuser, 1982)

[**RuSa**] R. Rudell and A. Sangiovanni-Vincentelli, "Espresso-MV: Algorithms for multiple-valued logic minimization," *Proc. Custom International Circuit Conf., Portland* (1985) 230–234

[**SaKa**] Thomas L. Saaty and Paul C. Kainen, *The Four-Color Problem: Assaults and Conquest* (McGraw-Hill, 1977)

[**Sa1**] Patrick Saint-Dizier, *An Introduction to Programming in Prolog* (Springer-Verlag, 1990)

[Sa2] András Sárközy, "Unsolved problems in number theory," *Periodica Mathematica Hungarica* **42** (2001) 17–35

[Sc] E. D. Schell,"Samuel Pepys, Isaac Newton, and Probability," *The American Statistician* **14** (1960) 27–30

[Si] Michael Sipser, *An Introduction to the Theory of Computation* (PWS, 1997)

[Sk] Steven Skiena, *Implementing Discrete Mathematics: Combinatorics and Graph Theory with Mathematica* (Addison-Wesley, 1990)

[Sl] N. J. A. Sloane and Simon Plouffe, *Encyclopedia of Integer Sequences* (Academic Press, 1995)

[Sm] Jeffrey D. Smith, *Design and Analysis of Algorithms* (PWS-Kent, 1989)

[St] Thomas A. Standish, *Data Structure Techniques* (Addison-Wesley, 1980)

[St2] Douglas R. Stinson, *Cryptography: Theory and Practice* (CRC Press, 1995)

[Sw] Edward R. Swart, "The philosophical implications of the four-color problem," *The American Mathematical Monthly* **87** (1980) 697–707

[Sz] Bohdan O. Szuprowicz, *Search Engine Technologies for the World Wide Web and Intranets* (Computer Technology Research Corp., 1997)

[Ti] G. Tinhofer, "Generating graphs uniformly at random," in G. Tinhofer et al., eds., *Computational Graph Theory*, Computing Supplementum (Springer-Verlag, 1990) 235–255

[Tu1] Alan Tucker, *Applied Combinatorics*, third edition (Wiley, 1995)

[Tu2] Alan M. Turing, "On computable numbers with an application to the Entscheidungsproblem," *Proceedings of the London Mathematical Society* **2** (1936) 230–265

[Ty1] Thomas Tymoczko, "Computers, proofs and mathematics: a philosophical investigation of the four-color problem," *Mathematics Magazine* **53** (1980) 131–138

[Ty2] Thomas Tymoczko, "The four-color problem and its philosophical significance," *Journal of Philosophy* **76** (1979) 57–83

[WhCl] J. R. C. White and M. J. Clugston, "The enumeration of isomers—with special reference to the stereoisomers of Decane," *Journal of Chemical Education* **70** (1993) 874–876

[Wi1] Raymond Wilder, *The Foundations of Mathematics*, second edition (Wiley, 1965)

[Wi2] Herbert Wilf, *generatingfunctionology* (Academic Press, 1990)

[WoWi] D. R. Woodall and Robin J. Wilson, "The Appel–Haken proof of the four-color problem," in Lowell W. Beineke and Robin J. Wilson, eds., *Selected Topics in Graph Theory* (Academic Press, 1978) 83–101

[Zi] H.-J. Zimmermann, *Fuzzy Set Theory and Its Applications* (Kluwer, 1991)

Sample Tests

This section of the *Student Solutions Guide* contains sample tests based on the text. A test is included for each of the 12 chapters. Solutions are provided, of course.

You can use these tests to prepare for a midterm or final examination in your course, perhaps simulating an actual test environment. Also, you can use them for general review or to see whether you already know certain material from previous study.

It is understood that general directions for a test in a mathematics course such as this include showing your work and justifying your answers. Remember that mathematics involves communication as well as computation.

Each sample test has been printed on its individual page.

Sample Test for Chapter 1

1. Consider the proposition "Alice will win the game only if she plays by the rules."

(a) Restate this proposition in English in three different equivalent ways.

(b) State the converse of this proposition.

(c) State the contrapositive of this proposition.

(d) Suppose that Alice plays by the rules but loses. Determine with justification whether the original proposition is true or false.

2. Determine with justification whether $p \to (q \vee r)$ and $(p \wedge \neg q) \to r$ are logically equivalent.

3. If the propositional function $E(p,t)$ is "Person p does task t correctly," write a proposition in symbols using quantifiers that expresses the idea "Nobody's perfect."

4. Consider the propositions $\forall x \exists y P(x,y)$ and $\exists y \forall x P(x,y)$.

(a) Write out the first of these completely in English.

(b) Repeat part (a) for the second of the propositions.

(c) Give an example that shows that the two propositions need not be logically equivalent; explain how your example shows this.

5. Prove or disprove each statement.

(a) If x is an irrational number, then $3x + 2$ is an irrational number.

(b) If x is an irrational number, then $3x^2 + 2$ is an irrational number.

6. Prove or disprove each of the following propositions.

(a) If n is a multiple of 4 and k is a multiple of 3, then nk is a multiple of 12.

(b) If n is a multiple of 4 and k is a multiple of 3, then $n + k$ is a multiple of 7.

7. Use a proof by contraposition or a proof by contradiction to show that if $3n + 5$ is a multiple of 7, then n is not a multiple of 7.

Solutions for Chapter 1 Sample Test

1. (a) If Alice wins the game, she plays by the rules. Playing by the rules is necessary for Alice to win the game. Winning the game is a sufficient condition for having played by the rules.

   (b) If Alice plays by the rules, then she will win the game.

   (c) If Alice doesn't play by the rules, then she won't win the game.

   (d) It is true, since the hypothesis is false and the conclusion is true.

2. The only case in which the first proposition is false is the case in which the hypothesis is true and the conclusion is false. This means that p is true and q and r are both false. The only case in which the second proposition is false is again the case in which the hypothesis is true and the conclusion is false. Here this means that p and $\neg q$ must both be true and r must be false—which is the same as saying that p is true and q and r are both false. Since these two propositions are true in exactly the same cases, they are logically equivalent.

3. We want to express the idea that everybody makes mistakes—in other words, that everybody does some task incorrectly. More precisely, for every person, there is some task that that person does not do correctly. Therefore the answer is $\forall p \exists t \neg E(p,t)$. Another way to express this is that it is not the case that there exists a perfect person, i.e., a person who does every task correctly. In symbols, $\neg \exists p \forall t E(p,t)$.

4. (a) For every x there exists a y such that $P(x,y)$.

   (b) There exists a y such that for every x, $P(x,y)$.

   (c) Let $P(x,y)$ be $y > x$, where the context is real numbers. Then for every x there is certainly a y larger than x, for example, $y = x+1$. The y depends on x. Thus the first proposition is true. However, there does not exist a constant y that is larger than every x, so the second proposition is false.

5. (a) We will use a proof by contraposition. If $3x+2$ were a rational number r, then we would have $x = (r-2)/3$. Since the rational numbers are closed under the operations of subtraction and division (by a nonzero number), this means that x is rational. Thus we have proved the contrapositive of the desired statement, and our proof by contraposition is complete.

   (b) This is not true. If $x = \sqrt{2}$, then $3x^2 + 2 = 8$, which is a rational number. This counterexample disproves the statement.

6. (a) This is true. The hypotheses tell us that $n = 4s$ and $k = 3t$ for some integers s and t. Therefore $nk = 12st$, so nk is a multiple of 12.

   (b) This statement is false. For a counterexample, let $n = 4$ and $k = 6$. Then n is divisible by 4 and k is divisible by 3, but $n + k = 10$ is not divisible by 7.

7. Assume the given hypothesis that $3n + 5$ is a multiple of 7, and suppose to the contrary of what we wish to show that n *is* a multiple of 7. Then $n = 7k$ for some integer k. This means that $3n+5 = 21k+5 = 7(3k)+5$, which tells us that $3n + 5$ leaves a remainder of 5 when divided by 7 and hence is not divisible by 7. This contradicts the hypothesis that $3n+5$ is a multiple of 7. Therefore our supposition was wrong, and we conclude that n is not a multiple of 7.

Sample Test for Chapter 2

1. Prove or disprove that $A - (B \cup C) = (A - B) \cap (A - C)$, where A, B, and C are sets.

2. Suppose that $A = P(\{a,b,c,d,e\})$ and $B = P(\{c,d,e,f,g\})$. (Recall that $P(X)$ is the power set of X.) Compute $|A \cup B|$. [*Hint:* Determine first what $A \cap B$ is. Then use the fact that $|P(X)| = 2^n$ if $|X| = n$.]

3. Let $A = \{1,2,3,4,5,6\}$ and $B = \{a,b,c\}$.

(a) Give an example of a function from A to B that is neither one-to-one nor onto, and explain why it meets these conditions.

(b) State the domain, codomain, and range of the function you gave in part (a).

4. Compute $\sum_{n=0}^{6} \left\lfloor \frac{n}{2} \right\rfloor$.

5. Let S be the set of countries of the world, and let C be the set of cities of the world.

(a) Give an example of a function f from C to S.

(b) Give an example of a function g from S to C.

(c) One of $f \circ g$(Venice) and $g \circ f$(Venice) makes sense and one does not. Compute the value of the one that does, and explain why the other one does not.

6. (a) Is the set of all finite bit strings countable or uncountable? Prove that your answer is correct.

(b) Is the set of all infinite bit strings countable or uncountable? Prove that your answer is correct.

Solutions for Chapter 2 Sample Test

1. $A-(B\cup C)=A\cap\overline{B\cup C}=A\cap(\overline{B}\cap\overline{C})=A\cap A\cap\overline{B}\cap\overline{C}=(A\cap\overline{B})\cap(A\cap\overline{C})=(A-B)\cap(A-C)$, using, respectively, the definition of difference, De Morgan's law, idempotence, the commutativity and associativity of intersection, and again the definition of difference. Both sides of this equality consist of those elements in A that fail to be in either B or C.

2. Following the hint, we note that $A\cap B=P(\{c,d,e\})$, since a set is a subset of $\{a,b,c,d,e\}$ and also a subset of $\{c,d,e,f,g\}$ if and only if its elements are chosen from among $\{c,d,e\}$. Therefore $|A\cap B|=2^3$. On the other hand, $|A|=2^5$ and $|B|=2^5$. Therefore $|A\cup B|=|A|+|B|-|A\cap B|=2^5+2^5-2^3=32+32-8=56$.

3. **(a)** If we send every element of A to c, for example (i.e., define the function by $f(x)=c$ for all $x\in A$), then this function is not one-to-one since $f(1)=f(2)$, and not onto since for no $x\in A$ does $f(x)=b$.

(b) The domain of this function is A; its codomain is B, and its range is $\{c\}$.

4. We need to plug in the values of n from 0 to 6 and add: $\lfloor 0/2\rfloor+\lfloor 1/2\rfloor+\lfloor 2/2\rfloor+\lfloor 3/2\rfloor+\lfloor 4/2\rfloor+\lfloor 5/2\rfloor+\lfloor 6/2\rfloor=0+0+1+1+2+2+3=9$.

5. **(a)** The obvious choice is to let $f(x)=$ the country in which city x lies. Thus we would have $f(\text{Perth})=$ Australia, and $f(\text{Vienna})=$ Austria, for example. One alternative (not very interesting) correct answer is to let $f(x)=$ France for all cities x.

(b) One obvious choice is to let $g(x)=$ the capital of country x. Thus we would have $g(\text{Argentina})=$ Buenos Aires, and $g(\text{Russia})=$ Moscow, for example. One alternative (not very interesting) correct answer is to let $g(x)=$ Detroit for all countries x.

(c) The expression $f\circ g(\text{Venice})$ makes no sense because Venice is not a country and so not an element of S, which is the domain of g. However $g\circ f(\text{Venice})=g(f(\text{Venice}))=g(\text{Italy})=$ Rome.

6. **(a)** We can list all finite bit strings in order of increasing length, with only a finite number of each length: λ, 0, 1, 00, 01, 10, 11, 000, 001, ... This provides a one-to-one correspondence between the set of positive integers and the set of all finite bit strings, which by definition means that the set of all finite bit strings is countable: $1\leftrightarrow\lambda$, $2\leftrightarrow 0$, $3\leftrightarrow 1$, $4\leftrightarrow 00$, $5\leftrightarrow 01$, $6\leftrightarrow 10$, $7\leftrightarrow 11$, $8\leftrightarrow 000$, $9\leftrightarrow 001$, ...

(b) This set is uncountable. The proof is by contradiction, using Cantor's diagonal argument. Assume that the set of all infinite bit strings is countable, that is, that there exists a one-to-one correspondence f from the set of positive integers to the set of all infinite bit strings—in other words, a list of all infinite bit strings. For example, it might happen that $f(1)=010101\ldots$, $f(2)=110110110\ldots$, and so on. Form a new bit string w whose i^{th} bit is a 0 if the ith bit of $f(i)$ is 1, and whose i^{th} bit is a 1 if the ith bit of $f(i)$ is 0. In our example, w starts out 10.... Then w differs from every element in the list, since it differs from $f(i)$ in the ith bit. In other words, our list is incomplete, and f is not a one-to-one correspondence because it is not onto. Having obtained a contradiction, we conclude that the set of all infinite bit strings is uncountable.

Sample Test for Chapter 3

1. **(a)** Write an algorithm to find the sum $a_1a_1 + a_1a_2 + a_1a_3 + \cdots + a_1a_n + a_2a_1 + a_2a_2 + a_2a_3 + \cdots + a_2a_n + \cdots + a_na_1 + a_na_2 + a_na_3 + \cdots + a_na_n$.

(b) Analyze the time complexity of the algorithm you developed in part (a).

2. Compute the following.

(a) $\gcd(742, 1908)$

(b) the prime factorization of 3080

(c) an inverse of 4 modulo 13

(d) $-50 \textbf{ mod } 7$

3. Give an example of two 2×2 matrices $\mathbf{A}$ and $\mathbf{B}$ such that $\mathbf{AB} \neq \mathbf{BA}$. (Make sure to justify your answer.)

4. Explain how the binary search algorithm would look for 7 in the list $(1, 2, 3, 4, 8, 10)$.

5. Show that if $a \equiv b \pmod{m}$ and $c \equiv d \pmod{m}$ then $a + c \equiv b + d \pmod{m}$.

6. **(a)** Express 425 in base 6.

(b) Express the binary numeral $(101001000)_2$ in base 10.

7. Suppose that $\mathbf{A}$, $\mathbf{B}$, and $\mathbf{C}$ are 3×7, 7×2, and 2×5 matrices of numbers, respectively. Is it more efficient to compute the product $\mathbf{ABC}$ as $(\mathbf{AB})\mathbf{C}$ or as $\mathbf{A}(\mathbf{BC})$? Justify your answer by computing the number of multiplications of numbers needed each way.

8. Suppose that a message is encrypted using the shift cipher function $f(p) = (p + 10) \textbf{ mod } 26$, where the letters A through Z are represented by the integers 0 through 25. The message "ZKCC DRSC DOCD" is received. Determine what the original message was.

9. Let $f(n) = 10n^2 + 5n$ and $g(n) = n^3 + 2$.

(a) State the definition of $f(n)$ being $O(g(n))$.

(b) Determine with justification whether $f(n)$ is $O(g(n))$ and also whether $g(n)$ is $O(f(n))$.

Solutions for Chapter 3 Sample Test

1. **(a)** We can write the algorithm in pseudocode as follows. It is straightforward: we just have a loop within a loop to compute each product $a_i a_j$, and accumulate the sum in the variable *sum*.

```
procedure SumOfProducts(a_1, a_2, ..., a_n : integers)
sum := 0
for i := 1 to n
    for j := 1 to n
        sum := sum + a_i a_j
{we are done: sum holds the desired sum}
```

(b) The outer loop is executed n times, and for each pass through the outer loop the inner loop is executed n times. Therefore the assignment statement is executed n^2 times. Since essentially no other work is done by this algorithm, the number of steps is $O(n^2)$. [It is interesting to note that the sum can be obtained in $O(n)$ steps by adding all the a_i's together and then squaring the sum.]

2. **(a)** We use the Euclidean algorithm: $\gcd(742, 1908) = \gcd(742, 424) = \gcd(424, 318) = \gcd(318, 106) = \gcd(106, 0) = 106$.

(b) This number is clearly even, so we divide by 2. The quotient is even again, so we divide by 2 again, and repeat this process once more, obtaining $3080 = 2^3 \cdot 385$. Now clearly 385 is divisible by 5, so we have $3080 = 2^3 \cdot 5 \cdot 77 = 2^3 \cdot 5 \cdot 7 \cdot 11$, and this last product contains only primes.

(c) In order to find this, we need to perform the Euclidean algorithm to find $\gcd(13, 4)$ and keep track of the results, in order to write 1 in the form $4s + 13t$. Only one step is needed, since we see that $13 = 3 \cdot 4 + 1$, so we have $1 = 13 + (-3) \cdot 4$. From this we immediately see that -3 is an inverse of 4 modulo 13, since the last equation shows that $1 - (-3) \cdot 4$ is divisible by 13, i.e., that $1 \equiv (-3) \cdot 4 \pmod{13}$. If we want a positive answer, we can add 13, obtaining 10. Note that $4 \cdot 10 = 40 \equiv 1 \pmod{13}$.

(d) We just divide -50 by 7, obtaining -8 (note that we need to have a positive remainder, so that -7 is not enough), with a remainder of 6. In other words, $-50 = (-8) \cdot 7 + 6$. Then $-50 \textbf{ mod } 7$ is the remainder, namely 6.

3. If we choose two matrices more or less at random, then we are very likely to find the desired pair, since it almost never happens that the product is the same when the order of multiplication is reversed. Suppose

$$\mathbf{A} = \begin{bmatrix} 0 & 3 \\ 1 & 2 \end{bmatrix} \quad \text{and} \quad \mathbf{B} = \begin{bmatrix} 2 & 1 \\ 1 & 3 \end{bmatrix}.$$

Then

$$\mathbf{AB} = \begin{bmatrix} 3 & 9 \\ 4 & 7 \end{bmatrix} \quad \text{and} \quad \mathbf{BA} = \begin{bmatrix} 1 & 8 \\ 3 & 9 \end{bmatrix},$$

clearly not equal.

4. The active portion of the list starts as the entire list, $(1, 2, 3, 4, 8, 10)$. The sought-after item, 7, is compared to the middle term in the list (actually the term just below the middle, since there are an even number of items), namely 3. Since $7 > 3$, we restrict the search to the portion of the list above 3, namely $(4, 8, 10)$, and repeat the process. This time we compare 7 with 8, and since $7 \not> 8$, we restrict the list to $(4, 8)$. The next comparison shows that $7 > 4$, so the active portion of the list becomes (8). At this point there is only one number left, so we check to see whether this number equals 7. Since it does not, we conclude that 7 is not in the original list.

5. To say that $a \equiv b \pmod{m}$ is to say that $a - b = sm$ for some integer s (i.e., $a - b$ is a multiple of m). Similarly, to say that $c \equiv d \pmod{m}$ is to say that $c - d = tm$ for some integer t (i.e., $c - d$ is also a multiple of m). Adding these two equations and doing some trivial algebra, we have $(a + c) - (b + d) = (s + t)m$, which tells us immediately from the definition that $a + c \equiv b + d \pmod{m}$, as desired.

6. **(a)** We note that the relevant powers of 6 are $6^0 = 1$, $6 = 6$, $6^2 = 36$, and $6^3 = 216$. Thus we see that 425 contains one 6^3, leaving $425 - 216 = 209$, that 209 contains five 6^2's, leaving $209 - 5 \cdot 36 = 29$. Similarly, $29 = 4 \cdot 6 + 5$, so this number contains four 6's and 5 left over. Therefore $425 = (1545)_6$. An alternative way to do this is to divide repeatedly by 6 and note the remainders. These are the digits of the base 6 expansion of 425, from right to left.

(b) We see by eye-straining counting that the leftmost digit in this numeral represents 2^8. Proceeding from there, we see that the number represented by this expansion is then $2^8 + 2^6 + 2^3 = 256 + 64 + 8 = 328$.

7. To multiply $\mathbf{A}$ by $\mathbf{B}$, we will need $3 \cdot 7 \cdot 2 = 42$ multiplications. The result is a 3×2 matrix. To multiply it by $\mathbf{C}$ will require $3 \cdot 2 \cdot 5 = 30$ multiplications. This gives a total of $42 + 30 = 72$ steps. On the other hand, if we multiply $\mathbf{B}$ by $\mathbf{C}$ first, we use $7 \cdot 2 \cdot 5 = 70$ multiplications just for that, and then another $3 \cdot 7 \cdot 5 = 105$ to multiply $\mathbf{A}$ by the 7×5 matrix $\mathbf{BC}$. This method uses a total of $70 + 105 = 175$ steps. Therefore the first method is faster (more than twice as fast).

8. To decrypt, we need the inverse function for p, which is clearly $g(q) = (q - 10) \ \mathbf{mod}\ 26$. The encrypted message, translated into numbers, is 25-10-2-2 3-17-18-2 3-14-2-3 (remembering that A$\leftrightarrow$ 0, B$\leftrightarrow$ 1, and so on). Subtracting 10 (and adding 26 if the difference is negative, in order to compute the answer **mod** 26), we obtain the numerical version of the plaintext: 15-0-18-18 19-7-8-18 19-4-18-19. This is quickly seen to read PASS THIS TEST.

9. **(a)** There exist constants C and k such $|10n^2 + 5n| \le C|n^3 + 2|$ for all $n > k$.

(b) To see that $f(n)$ is $O(g(n))$, we observe that $|10n^2 + 5n| = 10n^2 + 5n \le 10n^3 + 5n^3 = 15n^3 \le 15|n^3 + 2|$ for all $n \ge 1$. Thus we have witnesses $C = 15$ and $k = 1$. On the other hand, $n^3 + 2$ is not $O(10n^2 + 5n)$. To see this, note that the ratio of these two functions, $(n^3 + 2)/(10n^2 + 5n)$, is greater than $n^3/(15n^2)$ for all $n \ge 1$. This latter quantity equals $n/15$, and clearly it cannot be bounded above by a constant. Hence there is no constant C such that $|n^3 + 2| \le C|10n^2 + 5n|$ for all sufficiently large n.

Sample Test for Chapter 4

1. **(a)** State the principle of mathematical induction completely.

(b) Suppose that a sequence is defined recursively by $A_0 = 2$ and $A_n = (A_{n-1} + 1)/2$ for all $n \geq 1$. Show that this sequence is decreasing (i.e., $A_n < A_{n-1}$ for all n).

2. **(a)** State the principle of strong induction completely.

(b) For each integer n greater than 1, let $s(n)$ be the sum of the primes in the prime factorization of n (with repetitions). So, for example, $s(12) = 7$, because the prime factorization of 12 is $2 \cdot 2 \cdot 3$, and $2+2+3 = 7$. Prove that $n \geq s(n)$ for all $n \geq 2$. [*Hint:* First show that if x and y are integers greater than 1, then $x \cdot y \geq x + y$; do this by starting with the inequality $(x-1)(y-1) \geq 1$. Second, how is $s(x \cdot y)$ related to $s(x) + s(y)$?]

3. Give a recursive definition of the set of all postages that can be obtained using only 25-cent stamps and 44-cent stamps.

4. A certain function of two nonnegative integer variables, $C(a, b)$, defined whenever $a \geq b$, has the property that $C(a, b) = C(a-1, b) + C(a-1, b-1)$ for all nonnegative integers a and b, with $a > b > 0$, with the further conditions that $C(a, a) = C(a, 0) = 1$ for all a.

(a) Write in pseudocode a recursive algorithm for computing this function.

(b) Compute $C(4, 2)$ using this definition.

5. Verify that the program segment

if n is odd **then** $y := (x + y)/2$
else $x := (x + y)/2$

is correct with respect to the initial assertion "n is an integer and $0 \leq y - x \leq \epsilon$," and the final assertion "n is an integer and $0 \leq y - x \leq \epsilon/2$."

Solutions for Chapter 4 Sample Test

1. **(a)** If $S(n)$ is a predicate defined for all positive integers n, and if $S(1)$ is true, and if for every positive integer k, $S(k)$ implies $S(k+1)$, then $S(n)$ is true for every n.

(b) Let $S(n)$ be the statement that $A_n < A_{n-1}$. We will show by mathematical induction that $S(n)$ holds for all n. By the definition of this sequence, we see that $A_0 = 2$ and $A_1 = (A_0 + 1)/2 = 1.5$, so $S(1)$ is true; this takes care of the basis step. For the inductive step, assume the inductive hypothesis that $S(k)$ is true, i.e., $A_k < A_{k-1}$. We want to show that $S(k+1)$ is true, i.e., $A_{k+1} < A_k$. Adding 1 to both sides of the given inequality and dividing by 2, we obtain another valid inequality, $(A_k + 1)/2 < (A_{k-1} + 1)/2$. But by the definition of this sequence, this says precisely that $A_{k+1} < A_k$, as desired.

2. **(a)** If $S(n)$ is a predicate defined for all positive integers n, and if $S(1)$ is true, and if for every positive integer k, $\forall j{\le}k\, S(j)$ implies $S(k+1)$, then $S(n)$ is true for every n.

(b) Following the hint, suppose that x and y are integers greater than 1. Then $x - 1 \ge 1$ and $y - 1 \ge 1$, so $1 = 1 \cdot 1 \le (x-1)(y-1) = x \cdot y - x - y + 1$, from which it follows that $x \cdot y \ge x + y$. We will prove that $n \ge s(n)$ for all $n \ge 2$ by strong induction. The basis step is trivial, because $s(2) = 2$. Suppose that $s(j) \ge j$ whenever $2 \le j \le k$. We must show that $s(k+1) \ge k+1$. If $k+1$ is prime, then it is its own prime factorization, so $s(k+1) = k+1$. Otherwise, $k+1$ can be factored, say as $x \cdot y$, where $2 \le x \le k$ and $2 \le y \le k$. The prime factorization of $k+1$ is the product of the prime factorizations of x and y, so clearly $s(k+1) = s(x) + s(y)$. Both x and y are in the range of the inductive hypothesis, so we know that $x \ge s(x)$ and $y \ge s(y)$. Putting this all together, we have $k+1 = x \cdot y \ge x + y \ge s(x) + s(y) = s(k+1)$, and our proof by strong induction is complete.

3. By using no stamps, we can get 0 cents postage. This our basis case. The recursive part of the definition is that if we can obtain p cents postage, then we can also obtain $p + 25$ cents postage and $p + 44$ cents postage.

4. **(a)** We just incorporate the definition into the pseudocode.

procedure $C(a, b :$ nonnegative integers with $a \ge b)$
if $b = 0$ or $b = a$ **then** $C(a,b) := 1$
else $C(a,b) := C(a-1,b) + C(a-1,b-1)$

(b) We apply the recursive definition to compute that $C(4,2) = C(3,2) + C(3,1) = [C(2,2) + C(2,1)] + [C(2,1) + C(2,0)] = 1 + 2C(2,1) + 1 = 2 + 2[C(1,1) + C(1,0)] = 2 + 2(1+1) = 6$.

5. Since n is not changed, the final assertion that n is an integer is clearly true if the initial assertion that n is an integer is true. Suppose that $y - x$ had some value before the program segment was executed. If n is odd, then after the segment is executed the new value of $y - x$ is $[(x+y)/2] - x$ which equals $(y-x)/2$. If n is not odd, then after the segment is executed the new value of $y - x$ is $y - [(x+y)/2]$ which again equals $(y-x)/2$. Thus in either case the value of $y - x$ was cut in half. Therefore if $y - x$ was between 0 and ϵ before the program segment was executed (our initial assertion), then $y - x$ is between 0 and $\epsilon/2$ after the program segment is executed (our final assertion); this statement is what we were asked to prove.

Sample Test for Chapter 5

1. Determine the number of bit strings of length 10 that either start with six consecutive 0's or end with six consecutive 0's. (Recall that "or" is used in the inclusive sense in mathematics, unless specifically stated otherwise.)

2. There are more than 298,000,000 people in the United States. Prove that at least six of them were born during the exact same minute of the same hour of the same day of the same year.

3. Determine the number of 5-card hands from a standard deck of cards that contain two cards from each of two suits and an ace from a third suit.

4. How many ways are there to distribute 25 identical Christmas tree ornaments among 5 (distinct) children, if each child must get at least 3 ornaments and no child may get 13 or more?

5. How many ways are there to arrange 4 oranges, 5 apples, 6 pears, and 7 peaches in a row?

6. Give a combinatorial proof that $\sum_{k=0}^{n} 2^k C(n,k) = 3^n$ by computing in two different ways the number of ways to donate some (possibly all or none) of your collection of n distinct baseball cards to 2 different museums. [Hint: each card can go to one of 3 places; on the other hand, you can donate k cards in all.]

7. Find the coefficient of x^{10} in $(3x+2)^{15}$.

Solutions for Chapter 5 Sample Test

1. If a 10-bit string is to start 000000, then we are free to choose the last four bits, so there are $2^4 = 16$ such strings. Similarly, there are 16 strings that end with six consecutive 0's. Furthermore, there is one string (0000000000) that meets both of these conditions, so it is counted twice. By the inclusion–exclusion principle, the answer is therefore $16 + 16 - 1 = 31$.

2. We can assume that essentially all of the 298,000,000 people are less than 100 years old. Thus there are at most $100 \cdot 366 \cdot 24 \cdot 60 = 52{,}704{,}000$ possible minutes for these people to have been born during. By the generalized pigeonhole principle, there are therefore at least $\lceil 298000000/52704000 \rceil = \lceil 5.6\ldots \rceil = 6$ people with the same minute of birth.

3. We need to make several choices in order to describe such a hand. First, there are $C(4,2) = 6$ ways to choose the two suits. There are $C(13,2) = 78$ ways to choose the two cards from each of these suits, so there are 78^2 ways to specify these four cards in all, once the suits have been selected. Finally, there are 2 ways to decide which of the other two suits the ace is to come from. Therefore by the multiplication principle the answer is $6 \cdot 78^2 \cdot 2 = 73{,}008$.

4. After the required 15 ornaments are distributed (3 to each of the 5 children), 10 ornaments remain. There are $C(5 + 10 - 1, 10) = C(14, 10) = C(14, 4) = 1001$ ways to distribute these 10 ornaments among the 5 children. However, 5 of these distributions are forbidden—giving all 10 to one child—since this would give that child a total of 13 ornaments, in violation of the condition stated in the problem. Therefore the answer is $1001 - 5 = 996$.

5. This problem asks for the number of permutations of 22 objects, consisting of 4 objects of one type, 5 of a second type, 6 of a third type, and 7 of a fourth type. By Theorem 3 in Section 5.5, the answer is $22!/(4!5!6!7!) \approx 1.1 \times 10^{11}$.

6. For the right-hand side we just note that each card can end up in one of 3 places—Museum #1, Museum #2, or still with me. Therefore there are 3^n ways to distribute the cards. On the other hand, for each k from 0 to n, I can decide to donate k cards. There are $C(n,k)$ ways to choose the cards to donate, and then for each card chosen there are 2 ways to decide which museum it will go to. Therefore there are $C(n,k)2^k$ ways to donate k cards. Summing over all possible values of k (certainly different values of k lead to mutually exclusive outcomes), we obtain the left-hand side of the identity. Therefore the two sides are equal.

7. By the Binomial Theorem, the term containing x^{10} is the term $\binom{15}{5}(3x)^{10}2^5$. Therefore the coefficient of x^{10} is $\binom{15}{5}3^{10}2^5 = 3003 \cdot 59049 \cdot 32 = 5{,}674{,}372{,}704$.

Sample Test for Chapter 6

1. What is the probability that a positive integer less than 100 picked at random has at least five prime factors? (The factors need not be distinct; for example, 24 has four prime factors, since $24 = 2 \cdot 2 \cdot 2 \cdot 3$.)

2. Suppose that a fair die is rolled and a card is drawn from a standard deck of cards. What is the probability that the number shown on the die is the same as the number shown on the card (if any—note that only the numbers 2 through 10 are shown on cards, since aces, kings, queens, and jacks are represented by letters).

3. What is the probability of getting exactly four heads when a fair coin in flipped eight times?

4. A fair red die and a fair blue die are rolled. What is the expected value of the sum of the number on the red die plus twice the number on the blue die?

5. Suppose that the probability that a thumbtack lands point up is 0.4. What is the expected number of times we must toss the tack until it lands point up? medskip

6. In Sun and Fun Retirement Community, 60% of the residents play golf, 45% of the golfers smoke, and 30% of the non-golfers smoke. If a random non-smoker is selected, what is the probability that he or she plays golf? Report the answer to the nearest whole percent.

Solutions for Chapter 6 Sample Test

1. It is easy to see by considering all possible cases that the only numbers meeting this condition are $2^6 = 64$, $2^5 = 32$, $2^4 \cdot 3 = 48$, $2^4 \cdot 5 = 80$, and $2^3 \cdot 3^2 = 72$. Since there are 5 outcomes in this event, out of 99 outcomes in all, the desired probability is $5/99$.

2. For this event to occur, the die must show something other than a 1, and the probability of this happening is $5/6$. Once a 2 through 6 is rolled, the probability that a card of that rank is drawn is $4/52 = 1/13$, since there are four cards of each rank. (Note that the draw of the card is independent of the roll of the die.) Therefore the answer is $(5/6)(1/13) = 5/78 \approx 6.4\%$.

3. The number of heads follows a binomial distribution. Therefore, by Theorem 2 in Section 6.2, the answer is $C(8,4)\left(\frac{1}{2}\right)^4\left(\frac{1}{2}\right)^4 = 70/256 \approx 27\%$.

4. Let X_r and X_b be the random variables for the numbers shown on the dice. We are asked for $E(X_r + X_b + X_b)$. Since these dice are fair, we know that $E(X_r) = E(X_b) = (1+2+3+4+5+6)/6 = 3.5$. By linearity of expectation (note that independence is not required), we have $E(X_r + X_b + X_b) = 3.5 + 3.5 + 3.5 = 10.5$.

5. The number of tosses required has a geometric distribution with $p = 0.4$. Therefore, by Theorem 4 in Section 6.4, the expected value is $1/0.4 = 2.5$. medskip

6. Let G be the event that a randomly chosen resident plays golf. We know that $p(G) = 0.60$ and therefore $p(\overline{G}) = 0.40$. Let S be the event that a randomly chosen resident smokes. We are told that $p(S \mid G) = 0.45$ and $p(S \mid \overline{G}) = 0.30$. From these last facts, we know that $p(\overline{S} \mid G) = 0.55$ and $p(\overline{S} \mid \overline{G}) = 0.70$. We are asked for $p(G \mid \overline{S})$. We use Bayes' Theorem:

$$p(G \mid \overline{S}) = \frac{p(\overline{S} \mid G)p(G)}{p(\overline{S} \mid G)p(G) + p(\overline{S} \mid \overline{G})p(\overline{G})} = \frac{(0.55)(0.60)}{(0.55)(0.60) + (0.70)(0.40)} \approx 54\%$$

Sample Test for Chapter 7

1. Let s_n be the number of ways to write the positive integer n as the sum of an ordered list of 2's and 3's. For example, $s_8 = 4$, since $8 = 2+2+2+2 = 2+3+3 = 3+2+3 = 3+3+2$.

(a) Write down a recurrence relation for $\{s_n\}$.

(b) Write down initial conditions that, together with the recurrence relation, determine the entire sequence $\{s_n\}$.

(c) Compute s_{11}.

2. Solve the recurrence relations $a_n = 4a_{n-1} + 12a_{n-2}$ with initial conditions $a_0 = 3$ and $a_1 = 1$.

3. How many positive integers less than or equal to 10,000 are divisible by 2, 5, or 7?

4. Let D_n denote the number of derangements of n objects.

(a) Define what a derangement is.

(b) State an explicit formula for calculating D_n, and use it to compute D_5.

(c) Five married couples are having a party. The names of the men are written on slips of paper and put into a hat, and each woman draws a name at random (the slips are not replaced), so that each woman gets one name. Each woman dances the first dance with the person whose name she drew. How likely is it that no one dances with his or her spouse?

5. Suppose that f satisfies the divide-and-conquer relation $f(n) = 3f(n/2)+2n$ and the initial condition $f(1) = 0$.

(a) Compute $f(32)$.

(b) What can be said about the asymptotic behavior of f if we know that f is an increasing function? Give a good estimate of the form "$f(n)$ is $O(g(n))$."

6. How many ways are there to distribute seven distinct Christmas tree ornaments to four distinct children if each child must get at least one ornament?

7. What is the generating function for the number of ways to make change using pennies, nickels, and dimes, if no more than two nickels and no more than five pennies may be used? Give your answer in closed form.

Solutions for Chapter 7 Sample Test

1. (a) A sequence that sums to n can begin with a 2 and continue with a sequence that sums to $n-2$, or it can begin with a 3 and continue with a sequence that sums to $n-3$. Therefore $s_n = s_{n-2} + s_{n-3}$. This is valid if $n \geq 4$.

(b) We need initial conditions for s_1, s_2, and s_3. It is clear that $s_1 = 0$, since there is no way to write 1 as the sum of 2's and 3's. It is equally clear that $s_2 = 1$ and $s_3 = 1$.

(c) We start with the initial conditions and successively compute using the recurrence relation: $s_4 = s_2 + s_1 = 1 + 0 = 1$, $s_5 = s_3 + s_2 = 1 + 1 = 2$, $s_6 = s_4 + s_3 = 1 + 1 = 2$, $s_7 = s_5 + s_4 = 2 + 1 = 3$, $s_8 = s_6 + s_5 = 2 + 2 = 4$, $s_9 = s_7 + s_6 = 3 + 2 = 5$, $s_{11} = s_9 + s_8 = 5 + 4 = 9$.

2. We write down the characteristic equation $r^2 - 4r - 12 = 0$ and find by factoring that the roots are 6 and -2. Therefore the general solution is $a_n = A \cdot 6^n + B \cdot (-2)^n$. The initial conditions tell us that $3 = A + B$ and $1 = 6A - 2B$, whence we find $A = 7/8$ and $B = 17/8$. Therefore the solution is $a_n = (7/8) \cdot 6^n + (17/8) \cdot (-2)^n$. As a check we can compute $a_2 = 4 \cdot 1 + 12 \cdot 3 = 40$ from the recurrence relation and $a_2 = (7/8) \cdot 36 + (17/8) \cdot 4 = 40$ using the formula.

3. Using the principle of inclusion–exclusion, we need to compute as follows:

$$\left\lfloor \frac{10000}{2} \right\rfloor + \left\lfloor \frac{10000}{5} \right\rfloor + \left\lfloor \frac{10000}{7} \right\rfloor - \left\lfloor \frac{10000}{2 \cdot 5} \right\rfloor - \left\lfloor \frac{10000}{2 \cdot 7} \right\rfloor - \left\lfloor \frac{10000}{5 \cdot 7} \right\rfloor + \left\lfloor \frac{10000}{2 \cdot 5 \cdot 7} \right\rfloor$$
$$= 5000 + 2000 + 1428 - 1000 - 714 - 285 + 142 = 6571$$

4. (a) A derangement of the objects $i_1, i_2, \ldots, i_n$ is a permutation of these objects, so that for no k is i_k the kth object. In other words, the first object does not end up first, the second object does not end up second, $\ldots$, and the nth object does not end up last.

(b) By the formula developed in this chapter, we have

$$D_n = n! \left[1 - \frac{1}{1!} + \frac{1}{2!} - \cdots + (-1)^n \frac{1}{n!} \right].$$

For $n = 5$ we compute

$$D_5 = 120 \left[1 - \frac{1}{1} + \frac{1}{2} - \frac{1}{6} + \frac{1}{24} - \frac{1}{120} \right] = 44.$$

(c) No one dances with his or her spouse if and only if the permutations of the husbands determined by the wives' selection is a derangement. Therefore the desired probability is $D_5/5! = 44/120 \approx 0.367$.

5. (a) We compute successively $f(2) = 3f(1) + 2 \cdot 2 = 3 \cdot 0 + 4 = 4$, $f(4) = 3f(2) + 2 \cdot 4 = 3 \cdot 4 + 8 = 20$, $f(8) = 3f(4) + 2 \cdot 8 = 3 \cdot 20 + 16 = 76$, $f(16) = 3f(8) + 2 \cdot 16 = 3 \cdot 76 + 32 = 260$, $f(32) = 3f(16) + 2 \cdot 32 = 3 \cdot 260 + 64 = 844$.

(b) The Master Theorem in Section 7.3 applies, with $a = 3$, $b = 2$, $c = 2$, and $d = 1$. Since $a > b^d$ $(3 > 2^1)$, we have $f(n)$ is $O(n^{\log_b a}) = O(n^{\log 3}) \approx O(n^{1.6})$.

6. We are asked for the number of onto functions from a 7-set to a 4-set. By Theorem 1 in Section 7.6, the answer is $4^7 - C(4,1)3^7 + C(4,2)2^7 - C(4,3)1^7 = 8400$.

7. The generating function for choosing pennies is $1 + x + x^2 + \cdots + x^5$, since anywhere from 0 to 5 pennies may be used. The generating function for choosing nickels is $1 + x^5 + x^{10}$, since anywhere from 0 to 2 nickels may be used and a nickel is worth 5 cents. The generating function for choosing dimes is $1 + x^{10} + x^{20} + x^{30} + \cdots$, since any number of dimes may be used. A choice consists of making each of these choices, so the generating function we seek is the product of these functions, which we can put into closed form by using the identities given in Table 1 of Section 7.4:

$$(1 + x + x^2 + \cdots + x^5)(1 + x^5 + x^{10})(1 + x^{10} + x^{20} + x^{30} + \cdots) = \frac{1 - x^6}{1 - x} \cdot \frac{1 - x^{15}}{1 - x^5} \cdot \frac{1}{1 - x^{10}}.$$

The coefficient of x^n in the power series for this function gives the number of ways to make change of n cents. (A computer algebra package will report that this series starts $1 + x + x^2 + x^3 + x^4 + 2x^5 + x^6 + x^7 + x^8 + x^9 + 3x^{10}$, so, for example, there are 3 ways to give change of 10 cents under the rules laid out here.)

Sample Test for Chapter 8

1. Define the relation R on the set of real numbers by xRy if and only if $x^3 = -y^3$.

(a) Determine whether R is reflexive, symmetric, antisymmetric, and/or transitive.

(b) State a simple rule for when two numbers are related by $R \circ R$.

2. Explain how to tell whether a relation is symmetric under each of the following representations.

(a) matrix **(b)** digraph

3. Find the following closures of the relation $<$ on the integers, stating each as succinctly as possible.

(a) reflexive **(b)** symmetric **(c)** transitive symmetric

4. Define a relation on the set of real numbers by setting xRy if and only if $x^2 - 5x + 6 = y^2 - 5y + 6$.

(a) Show that R is an equivalence relation.

(b) Write down the equivalence class $[2]$.

(c) Write down a "formula" (explicit description) for $[r]$ that holds for an arbitrary real number r.

5. (a) Write down the definition of what it means for a collection π to be a partition of a set A.

(b) Explain how a partition π on a set A determines an equivalence relation R on A (give an explicit definition of R).

(c) The set $\pi = \{\{1,3,5\},\{2,4\},\{6\}\}$ is a partition of $\{1,2,3,4,5,6\}$. Write down the equivalence relation determined by this partition, and then find the partition determined by this equivalence relation.

6. Consider the partial order on the set of all nonempty subsets of $\{a,b,c\}$.

(a) Identify all the maximal elements, greatest elements, minimal elements, and least elements of this poset.

(b) Find a total order compatible with this partial order.

(c) Give a simple way to determine the least upper bound of two elements of this poset (which will work for every poset consisting of a collection of sets with the $\subseteq$ relation).

7. (a) State what additional properties a poset must have in order for it to be a lattice.

(b) Give an example of a poset which is not a lattice but in which every pair of elements has an upper bound and a lower bound.

Solutions for Chapter 8 Sample Test

1. (a) Since $1^3 \neq -1^3$, R is not reflexive. Suppose that $x^3 = -y^3$. Then multiplying both sides by -1 gives us $y^3 = -x^3$; this says that $xRy \to yRx$, so R is symmetric. The relation is not antisymmetric, since $1R(-1)$ and $(-1)R1$, but $1 \neq -1$. Finally, R is not transitive, since $1R(-1)$ and $(-1)R1$, but 1 is not related to 1.

(b) By taking the cube root of both sides of $x^3 = -y^3$, we obtain the equivalent equation $x = -y$ (since cubing is a one-to-one function). We claim that $R \circ R$ is the identity relation. Indeed, every number is related to itself by $R \circ R$, since $xR(-x)$ and $(-x)Rx$. Conversely, if $x(R \circ R)z$, then for some y we have $x = -y$ and $y = -z$; it follows that $x = -(-z) = z$.

2. (a) Look at each entry not on the main diagonal, and compare it to the entry in the corresponding location after reflecting across the main diagonal. In order for the relation to be symmetric, these entries must be the same (both 0 or both 1) in each case.

(b) All nonloop edges must appear in antiparallel pairs; in other words, whenever there is an edge from one vertex to another vertex, an edge in the opposite direction (from the second vertex back to the first) must also be present.

3. (a) We need to arrange for each integer to be related to itself as well as to the integers to which it is already related. Therefore the answer is that x is related to y if and only if $x \leq y$. In brief, the reflexive closure of $<$ is $\leq$.

(b) We need to arrange for x to be related to y when y is related to x, i.e., $y < x$ (in addition to those cases in which x is already related to y). Thus x is related to y in the symmetric closure if and only if $x < y$ or $y < x$, or, more succinctly, $x \neq y$.

(c) The transitive symmetric closure must contain at least the pairs in the symmetric closure, which we just found to be $\{ (x, y) \mid x \neq y \}$. In addition, for every x, we have $x \neq x + 1$ and $x + 1 \neq x$, so x will have to be related to x in the transitive closure. But this means that every pair of numbers are related, so the transitive symmetric closure is all of $\mathbf{Z} \times \mathbf{Z}$, i.e., the relation that always holds.

4. (a) By Exercise 9 in Section 8.5, every relation defined by "xRy if and only if $f(x) = f(y)$" is an equivalence relation. That is the situation here, with $f(x) = x^2 - 5x + 6$. Therefore this is an equivalence relation.

(b) We need to find the set of all numbers x such that $x^2 - 5x + 6 = 2^2 - 5 \cdot 2 + 6 = 0$. Solving this quadratic equation for x gives $x = 2$ and $x = 3$. Therefore $[2] = \{2, 3\}$.

(c) This is like the previous part. We need to solve $x^2 - 5x + 6 = r^2 - 5r + 6$, which reduces to $x^2 - r^2 - 5x + 5r = 0$, or, after factoring, $(x + r)(x - r) - 5(x - r) = 0$, or more simply $(x - r)(x + r - 5) = 0$. This equation clearly has two solutions, namely $x = r$ and $x = 5 - r$. Therefore $[r] = \{r, 5 - r\}$. For each r other than $2\frac{1}{2}$, this equivalence class has two elements.

5. (a) π is a set of nonempty, pairwise disjoint subsets of A whose union is A.

(b) For $x, y \in A$, xRy if and only if x and y are in the same set in the partition, i.e., $\exists B \in \pi(x \in B \wedge y \in B)$.

(c) The equivalence relation determined by π is $R = \{(1,1), (1,3), (1,5), (3,1), (3,3), (3,5), (5,1), (5,3), (5,5), (2,2), (2,4), (4,2), (4,4), (6,6)\}$, and the partition determined by R (its set of equivalence classes) is then π.

6. (a) The only maximal element, which is also a greatest element, is the set itself, $\{a, b, c\}$. There is no least element, but there are three minimal elements, namely the sets $\{a\}$, $\{b\}$, and $\{c\}$.

(b) The answer is not unique; here is one such order: $\{b\} \prec \{a\} \prec \{a, b\} \prec \{c\} \prec \{a, c\} \prec \{b, c\} \prec \{a, b, c\}$.

(c) The least upper bound of two sets is their union.

7. (a) Every pair of elements must have a least upper bound and a greatest lower bound.

(b) We need to find a situation where upper bounds and lower bounds exist, but in which *least* upper bounds or *greatest* lower bounds do not. One example is to take the following sets, under the $\subseteq$ relation: $\emptyset$, $\{1\}$, $\{2\}$, $\{1, 2, 3\}$, $\{1, 2, 4\}$, $\{1, 2, 3, 4\}$. Then $\emptyset$ is a lower bound for every two of these, and $\{1, 2, 3, 4\}$ is an upper bound. However, $\{1\}$ and $\{2\}$ have no least upper bound, since $\{1, 2, 3\}$ and $\{1, 2, 4\}$ are both upper bounds, but neither is less than the other.

Sample Test for Chapter 9

1. The *intersection graph* of a collection of sets is the graph whose vertices are the sets in the collection, in which an edge joins two sets if the two sets are not disjoint. Consider the intersection graph of the set of all two-element subsets of $\{1,2,3,4\}$. You will find that this graph has 6 vertices and 12 edges.
   (a) Draw a picture of this graph.
   (b) Determine the degree of each vertex, and verify that the Handshaking Theorem is satisfied.
   (c) Write down the adjacency matrix of this graph.
   (d) Determine whether this graph is bipartite.
   (e) What is wrong with the following "proof" that this graph is planar? *Since* $v = 6$ *and* $e = 12$, *the inequality for planar graphs* $e \le 3v - 6$ *is satisfied; therefore the graph is planar.* Determine whether this is in fact a planar graph.
   (f) Find the chromatic number of this graph.
   (g) Determine whether this graph has a Hamilton circuit and whether it has an Euler circuit.

2. There are two notions of connectivity in directed graphs.
   (a) Define what it means for a directed graph to be strongly connected, and define what it means for a directed graph to be weakly connected.
   (b) Give an example of a directed graph that is weakly connected but not strongly connected, and give an example of a directed graph that is not weakly connected.

3. Find all nonisomorphic graphs with 6 vertices and 4 edges.

4. Show how to apply Dijkstra's algorithm to find the shortest path from a to z in the following weighted graph.

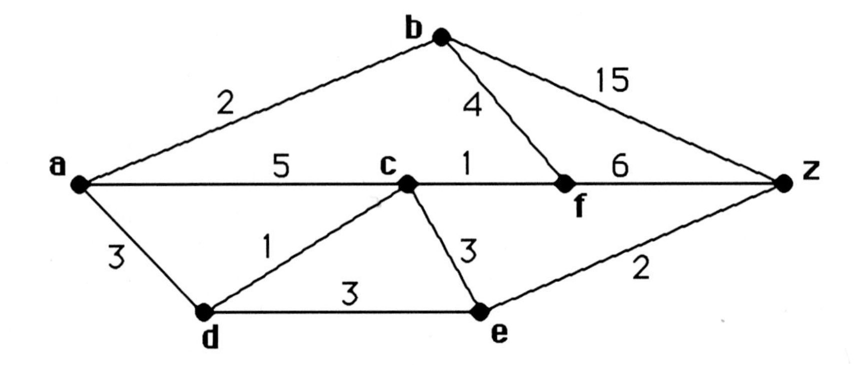

Solutions for Chapter 9 Sample Test

1. (a) There is more than one way to draw this picture. We have labeled each vertex with the two-element subset of $\{1,2,3,4\}$ that it represents (omitting the comma and braces for simplicity).

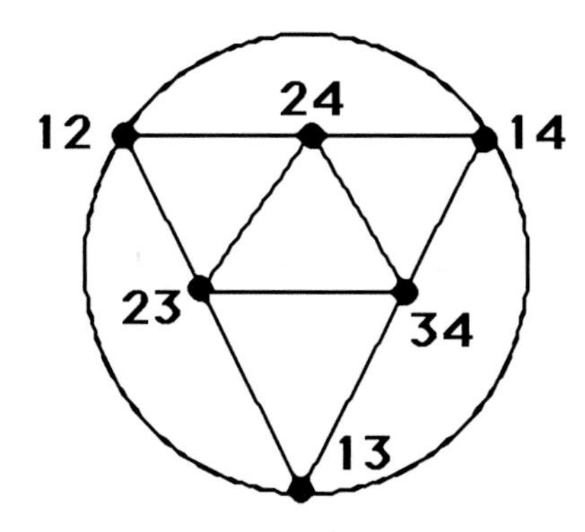

(b) We observe that each of the 6 vertices has degree 4, so that there should be $6 \cdot 4/2 = 12$ edges, as there are.

(c) The adjacency matrix is as follows, where we are using the following order of the vertices: 12, 13, 14, 23, 24, 34.

$$\begin{bmatrix} 0 & 1 & 1 & 1 & 1 & 0 \\ 1 & 0 & 1 & 1 & 0 & 1 \\ 1 & 1 & 0 & 0 & 1 & 1 \\ 1 & 1 & 0 & 0 & 1 & 1 \\ 1 & 0 & 1 & 1 & 0 & 1 \\ 0 & 1 & 1 & 1 & 1 & 0 \end{bmatrix}$$

(d) A bipartite graph can contain no triangle (K_3). This graph contains many triangles, such as 12–13–23. Therefore it is not bipartite.

(e) The theorem being applied here gives a necessary condition for a graph to be planar, not a sufficient condition. There are many graphs that satisfy this condition and yet are not planar ($K_{3,3}$, for instance). In fact, though, the graph is planar, as we see in our drawing above.

(f) Clearly at least 3 colors are required to properly color this graph, since it has triangles. On the other hand, if we put 12 and 34 into one color class, 13 and 24 into a second, and 14 and 23 into a third, then we have colored the graph with 3 colors. Therefore the chromatic number is 3.

(g) This graph must have an Euler circuit because it is connected and the degree of every vertex is even. We can also easily find a Hamilton circuit in this graph by trial and error: 12-13-34-23-24-14-12.

2. (a) A directed graph is strongly connected if given any two vertices u and v there is a path from u to v. (Note that this really means that there are paths both ways, since we can interchange the roles of u and v.) A directed graph is weakly connected if the underlying undirected graph is connected—in other words, given any two vertices u and v, there is a path from u to v in the underlying directed graph (i.e., if the directions on the edges are ignored).

(b) The digraph consisting of just two vertices u and v and the directed edge from u to v (i.e., edge (u,v)) is weakly connected but not strongly connected (because there is no path from v to u). The digraph consisting of these two vertices and no edges is not even weakly connected, because there is no path from u to v in the underlying undirected graph (since it has no edges).

3. There are 9 nonisomorphic graphs with 6 vertices and 4 edges, as shown here.

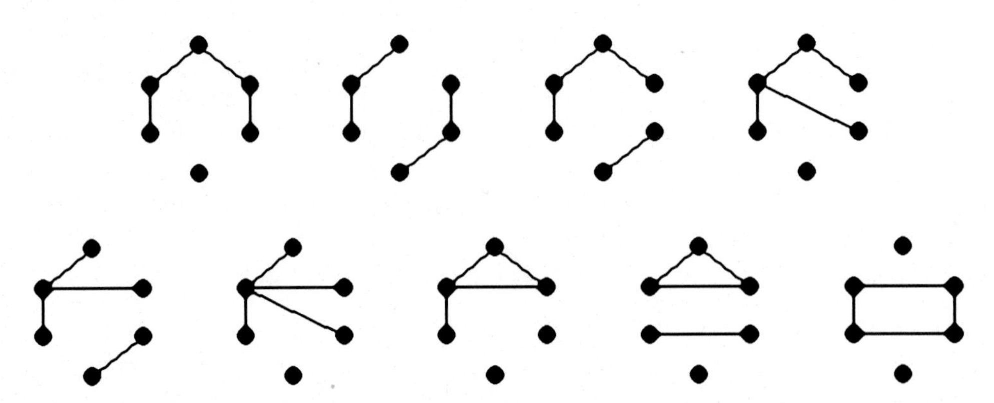

4. The following table shows the operation of the algorithm. Each line shows the labels on the vertices as they change from iteration to iteration. The number is the distance from vertex a (the source) and the letter is the vertex from which the currently known best path reaches this vertex. An asterisk indicates that the vertex was selected during that iteration (the vertex not yet selected having the smallest label).

a :	$0*$						
b :	∞	$2a*$					
c :	∞	$5a$	$5a$	$4d*$			
d :	∞	$3a$	$3a*$				
e :	∞	∞	∞	$6d$	$6d$	$6d*$	
f :	∞	∞	$6b$	$6b$	$5c*$		
z :	∞	∞	$17b$	$17b$	$17b$	$11f$	$8e*$

We now read off the path in reverse, from z back to e, back to d, back to a. Its length is given by the label on z, namely 8.

Sample Test for Chapter 10

1. **(a)** State the definition of a tree.

(b) State the definition of a rooted tree.

(c) Give an example of two trees that are isomorphic as trees but not isomorphic as rooted trees.

2. Suppose that a full 4-ary tree has 27 internal vertices.

(a) How many leaves does it have?

(b) What is the smallest height it could possibly have?

(c) What is the largest height it could possibly have?

3. Insert the words of the following sentence into a binary search tree in the order they are encountered, using alphabetical order to define the order relation in the tree. *The only thing we have to fear is fear itself.* (Note that the second occurrence of the word *fear* is merely found by the insertion algorithm and thereby skipped.)

4. Consider the expression $4 * 5 - 4 * (4 + 1) * 2$. Recall the usual rules for precedence of operations in absence of full parenthesization.

(a) Draw the expression tree for this expression.

(b) Write this expression in prefix form, and show how to evaluate it directly in that form.

(c) Write this expression in postfix form.

(d) Write this as a fully parenthesized infix expression.

5. Consider the complete bipartite graph whose vertex sets are $\{2, 3, 5, 8\}$ and $\{1, 4, 7, 9, 10\}$. In the following parts, use vertex number to break ties in selecting vertices at each point in an algorithm.

(a) Find the breadth-first search spanning tree of this graph, starting at vertex 1.

(b) Find the depth-first search spanning tree of this graph, starting at vertex 1.

(c) Consider this a weighted graph by assigning the weight $|u - v|$ to edge uv. (For example, edge $\{5, 9\}$ has weight 4, but edge $\{5, 8\}$ is not present.) Find a minimum spanning tree of this weighted graph, using Kruskal's algorithm.

Solutions for Chapter 10 Sample Test

1. (a) A tree is a connected undirected graph with no simple circuits.

   (b) A rooted tree is a tree in which one vertex has been specified as the root. It can be thought of as a directed graph, in which all edges are considered to be directed away from the root.

   (c) Let T_1 and T_2 both consist of vertices a, b, and c, with edges ab and bc. As trees they are clearly isomorphic (in fact they are equal). Make T_1 a rooted tree by specifying a as the root, and make T_2 a rooted tree by specifying b as the root. As rooted trees these are not isomorphic, since in T_1 the root has degree 1, whereas in T_2 the root has degree 2.

2. (a) Since each internal vertex has 4 children, the total number of children must be $4 \cdot 27 = 108$. Therefore there are a total of 109 vertices, since only the root is not a child. Thus it has $109 - 27 = 82$ leaves.

   (b) Recall that the height of an m-ary tree is at least $\lceil \log_m l \rceil$, where l is the number of leaves. Therefore the height of this tree is at least $\lceil \log_4 82 \rceil = 4$.

   (c) The tree's maximum height would be achieved if only one vertex at each level had children. Therefore the 27 internal vertices would occur at levels 0, 1, ..., 26, and the height would be 27.

3. The tree will end up looking like this:

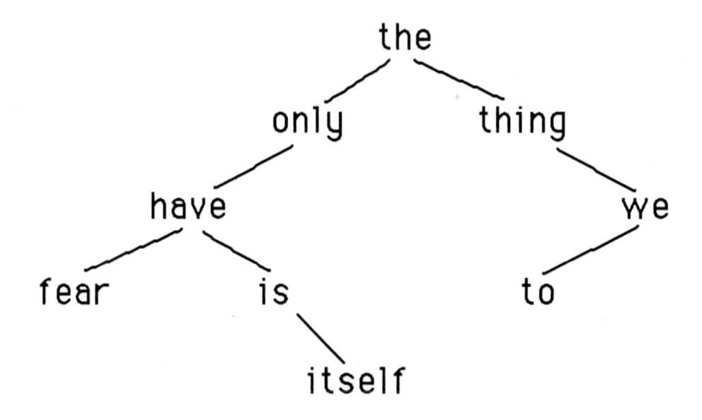

4. (a) The outermost operation is the subtraction, and within the second term, the first multiplication is done before the second. We are led to the following tree.

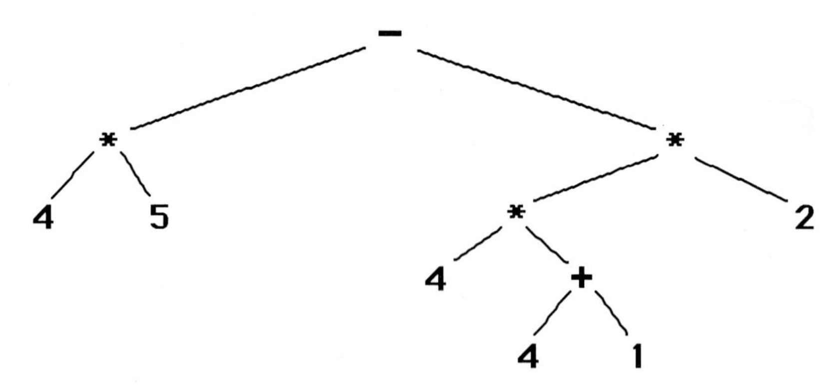

   (b) We traverse the tree in preorder, obtaining $- * 4\, 5 * * 4 + 4\, 1\, 2$. To evaluate this, we can look for the first occurrence of an operator immediately followed by two operands (numbers) and replace these three items by the result of the operation applied to the operands (from left to right). If we do this on the left-most occurrence at each stage, then we have successively $-\,20 * * 4 + 4\,1\,2 = -\,20 * * 4\,5\,2 = -\,20 * 20\,2 = -\,20\,40 = -20$.

   (c) We traverse the tree in postorder, obtaining $4\,5 * 4\,4\,1 + * 2 * -$.

   (d) $((4 * 5) - ((4 * (4 + 1)) * 2))$

5. (a) We include the following edges on the first pass: 1–2, 1–3, 1–5, and 1–8. Then we "fan out" from vertex 2, adding the edges 2–4, 2–7, 2–9 and 2–10, completing the tree.

(b) The search starts by adding edge 1–2 and proceeding from vertex 2. It then moves to the first unused vertex adjacent to vertex 2, namely vertex 4, adding edge 2–4 to the tree and continuing from vertex 4. Continuing in this way, we add the following edges: 4–3, 3–7, 7–5, 5–9, 9–8, and 8–10.

(c) Several edges of weight 1 are added: 1–2, 3–4, 4–5, 7–8, and 8–9. Then some edges of weight 2 that do not form a simple circuit are added: 1–3, 5–7, and 8–10. Since we have formed a tree (having added $9-1=8$ edges to the 9 vertices already present), we are done. The total weight is $1+1+1+1+1+2+2+2=11$.

Sample Test for Chapter 11

1. Use Boolean identities to simplify $x + \overline{x + y}$.

2. Find a sum-of-products expansion for the Boolean expression in problem 1.

3. Draw a combinatorial circuit for the Boolean expression obtained in problem 2.

4. Consider the Boolean expression $x\,y\,z + x\,y\,\overline{z} + x\,\overline{y}\,z + x\,\overline{y}\,\overline{z} + \overline{x}\,y\,z + \overline{x}\,\overline{y}\,z$.

(a) Draw a K-map for this expression, circle the blocks, and find a minimal expansion by finding a minimal set of blocks covering all the 1's in your diagram.

(b) Use the Quine–McCluskey method to find a minimal expansion.

5. One of the following equations always holds in a Boolean algebra, but the other need not always hold. Prove the valid identity, and give an example of a Boolean algebra in which the invalid one fails.

(a) $x \wedge (y \vee z) = (x \wedge y) \vee z$ **(b)** $\overline{(x \wedge \overline{y})} = \overline{x} \vee y$

Solutions for Chapter 11 Sample Test

1. First we apply De Morgan's law to write this as $x + (\overline{x}\,\overline{y})$. Then we can apply the distributive law (addition over multiplication) to expand this into $(x+\overline{x})(x+\overline{y})$. Next we use the identity that $x+\overline{x} = 1$ to rewrite this as $1 \cdot (x+\overline{y})$, and finally use the identity law to obtain $x+\overline{y}$.

2. We just saw that this expression equals $x+\overline{y}$. In order to rewrite this in sum-of-products form, we can multiply the first term by 1, written as $y+\overline{y}$, and we can multiply the second term by 1, written as $x+\overline{x}$. This gives us $x(y+\overline{y}) + (x+\overline{x})\overline{y}$. Then expanding by the distributive law we have $x\,y + x\,\overline{y} + x\,\overline{y} + \overline{x}\,\overline{y}$, which upon deleting the duplication by the idempotent law gives us the desired answer: $x\,y + x\,\overline{y} + \overline{x}\,\overline{y}$.

3. The appropriate inputs (some complemented) are fed to the three *AND* gates, and the outputs from these three gates provide the input to the *OR* gate.

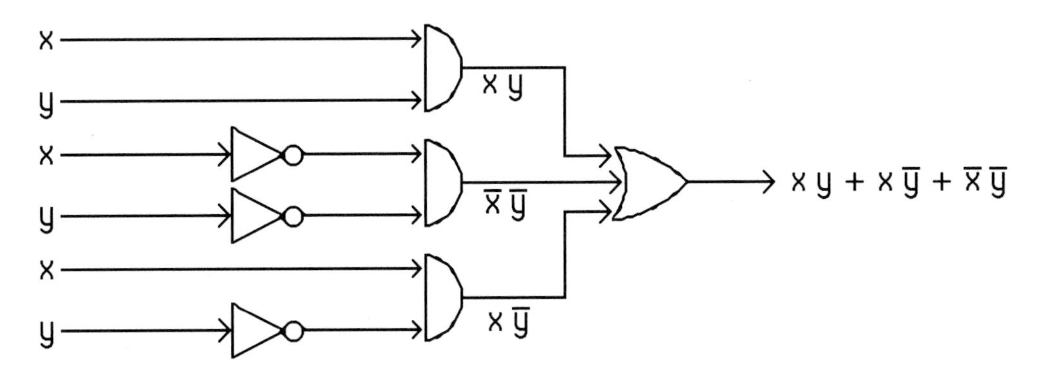

4. (a) The picture below shows that two blocks of four squares—the top row and the 2×2 square consisting of the first and last columns—cover all the terms. Therefore the solution is the sum of the Boolean expressions for these two blocks, namely $x+z$.

	yz	$y\overline{z}$	$\overline{y}\,\overline{z}$	$\overline{y}z$
x	1	1	1	1
$\overline{x}$	1			1

(b) Here is the Quine–McCluskey calculation. We see that the two terms in the last column cover all the minterms, so the answer is $x+z$.

			Step 1		Step 2	
	Term	String	Term	String	Term	String
1	$x\,y\,z$	111	$(1,2)\,x\,y$	11–	$(1,2,3,5)\,x$	1––
2	$x\,y\,\overline{z}$	110	$(1,3)\,x\,z$	1–1	$(1,3,4,6)\,z$	––1
3	$x\,\overline{y}\,z$	101	$(1,4)\,y\,z$	–11		
4	$\overline{x}\,y\,z$	011	$(2,5)\,x\,\overline{z}$	1–0		
5	$x\,\overline{y}\,\overline{z}$	100	$(3,5)\,x\,\overline{y}$	10–		
6	$\overline{x}\,\overline{y}\,z$	001	$(3,6)\,\overline{y}\,z$	–01		
			$(4,6)\,\overline{x}\,z$	0–1		

5. (a) This is invalid. As an example, take $x = 0$ and $z = 1$ (y can be anything). Then the left-hand side is $0 \wedge (y \vee z) = 0$, while the right-hand side is $(x \wedge y) \vee 1 = 1$.

(b) This follows from De Morgan's law and the double complement law: $\overline{(x \wedge \overline{y})} = \overline{x} \vee \overline{\overline{y}} = \overline{x} \vee y$.

Sample Test for Chapter 12

1. Consider the following grammar. The terminal symbols are a and b; the nonterminal symbols are S, A, B, and C; the start symbol is S; and the productions are $S \to ABC$; $A \to Aa$; $A \to a$; $B \to BB$; $B \to b$; $C \to \lambda$; $C \to aaC$.

   (a) Describe the language generated by this grammar.

   (b) Which of the terms type 0, type 1, type 2, and type 3 apply to this grammar?

   (c) Find a deterministic finite-state machine that recognizes the language generated by this grammar.

   (d) Describe this language using a regular expression.

2. Write down the rules for a grammar, using Backus–Naur form, that will generate all positive integer multiples of 5. Leading 0's are not permitted. Assume that the rule

$$\langle nonzero\ digit \rangle ::= 1 \mid 2 \mid 3 \mid 4 \mid 5 \mid 6 \mid 7 \mid 8 \mid 9$$

   has already been included.

3. Describe in two or three sentences how nondeterministic finite-state automata are really no more powerful than deterministic ones. Be specific about the construction involved.

4. State precisely the relationship among regular grammars, regular sets, and finite-state machines.

5. State whether the following statement is true or false, and explain in a short paragraph, being as specific as possible. *Every set of strings can be recognized by some finite-state automaton.*

6. Construct a Turing machine that computes the function $f(n_1, n_2) = n_2 - n_1$ whenever $n_2 \geq n_1 \geq 0$, using the notational convention for representing nonnegative integers and inputs to functions adopted in this chapter. It does not matter how the machine behaves if $n_1 > n_2$.

Solutions for Chapter 12 Sample Test

1. (a) From the first production we see that the strings generated by this language seem to have three blocks to them, spawned by A, B, and C, respectively. It is clear that A leads to any string of one or more a's. Similarly, B leads to any string of one or more b's. Finally, C is seen to lead to a string consisting of an even number (possibly 0) of a's. Thus we can describe this language as $\{ a^m b^n a^{2k} \mid m > 0,\ n > 0,\ k \geq 0 \}$.

(b) This is certainly a type 0 grammar, since type 0 grammars have no restrictions. It is also a type 1 grammar, since no production shortens unless it has λ as its right-hand side. In fact, it is a context-free (type 2) grammar, since the left-hand sides are all single nonterminal symbols. It is not a type 3 (regular) grammar, since, for instance, the first production is not of the required form.

(c) The following nondeterministic machine does the trick, since it leads to an accepting state, s_2 or s_4, precisely when there have been some a's followed by some b's, followed by an even number of a's. To make this machine deterministic, simply add a graveyard state, s_5, with transitions to it on all inputs not shown here (for example, from state s_3 on input b).

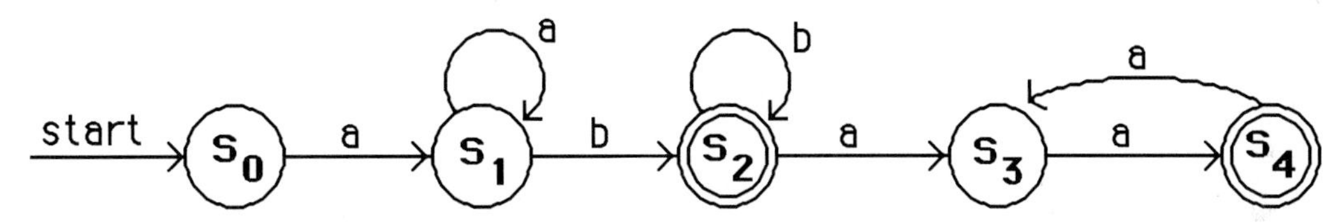

(d) The following regular expression is easily seen to work: $\mathbf{aa^*bb^*(aa)^*}$.

2. The key is to realize that a positive integer multiple of 5 is either 5 itself or else consists of a positive integer followed by either a 0 or a 5. We can define a digit by $\langle \textit{digit} \rangle$ ::= 0 | $\langle \textit{nonzero digit} \rangle$, positive integer recursively by $\langle \textit{positive integer} \rangle$ ::= $\langle \textit{nonzero digit} \rangle$ | $\langle \textit{positive integer} \rangle \langle \textit{digit} \rangle$, and finally multiple of five by $\langle \textit{multiple of five} \rangle$ ::= 5 | $\langle \textit{positive integer} \rangle 5$ | $\langle \textit{positive integer} \rangle 0$.

3. A nondeterministic finite-state automaton can be converted into a deterministic finite-state automaton that recognizes exactly the same language. This latter machine has as its states all the sets of states of the former machine, with a transition from A to B when B is the set of all states that can be reached from states in A on the given input, and has as its final states all the sets of states that include final states of the nondeterministic machine.

4. Kleene's Theorem tells us that regular sets (i.e., sets that can be formed using concatenations, unions, and Kleene closures starting from singletons and the empty set) are precisely the sets recognized by finite-state machines. Another theorem of this chapter tells us that these are precisely the sets of strings that can be generated by regular grammars. Explicit constructions allow one to go back and forth among a regular expression describing the language, a regular grammar generating the language, and a finite state machine accepting the language.

5. The statement is false. For one thing, the orders of infinity of these two collections are different: there are an uncountable number of sets of strings and only a countable number of finite-state machines. More to the point, though, the pumping lemma and special cases of it allow one to show that infinite languages accepted by finite-state machines have to have a certain kind of periodicity to them, since the computations in these machines will need to cycle around and around in the same loop for arbitrarily many repetitions. In particular, the set of strings of the form $0^n 1^n$ does not have this property, because there is no way for the cycling to preserve the same number of 0's as 1's and at the same time have all the 0's precede all the 1's.

6. The idea is that we need to erase $n_1 + 1$ 1's from both inputs to perform this subtraction, and then change the $*$ to a 1 in order to take into account the fact that n is represented by $n + 1$ 1's. For example, if the input is 111∗1111111, then we want $6 - 2 = 4$, so we need the output to be 11111. As in Example 3 in Section 12.5, we will wander between the ends of the string, erasing 1's from each end, until the first input (n_1) is exhausted. Here are some five-tuples that will do the job: $(s_0, 1, s_1, B, R)$, $(s_0, *, s_4, 1, R)$, $(s_1, 1, s_1, 1, R)$, $(s_1, *, s_1, *, R)$, (s_1, B, s_2, B, L), $(s_2, 1, s_3, B, L)$, $(s_3, 1, s_3, 1, L)$, $(s_3, *, s_3, *, L)$, (s_3, B, s_0, B, R).

Common Mistakes in Discrete Mathematics

In this section of the *Guide* we list many common mistakes that people studying discrete mathematics sometimes make. The list is organized chapter by chapter, based on when they first occur, but sometimes mistakes made early in the course perpetuate in later chapters. Also, some of these mistakes are remnants of misconceptions from high school mathematics (such as the impulse to assume that every operation distributes over every other operation).

In most cases we describe the mistake, give a concrete example, and then offer advice about how to avoid it. Note that additional advice about common mistakes in given, implicitly or explicitly, in the solutions to the odd-numbered exercises, which constitute the bulk of this *Guide*.

Solving Problems in Discrete Mathematics

Before getting to the common mistakes, we offer some general advice about solving problems in mathematics, which are particularly relevant to working in discrete mathematics. The problem-solving process should consist of the following steps. (This four-step approach is usually attributed to the mathematician George Pólya (1887–1985).)

1. Read and understand the problem at hand. Play around with it to get a feeling for what is going on and what is being asked.
2. Apply one or more problem-solving strategies to attack the problem. Do not give up when one particular tactic doesn't work. This phase of the process can take a long time.
3. Carefully write up the solution when you have solved the problem. Make sure to communicate your ideas clearly.
4. Look back on what you have done. Make sure that your answer is correct (think of creative ways to test it). Also consider other ways of solving the problem, and think about how you might generalize your results.

Your bag of problem-solving strategies may include drawing a picture or diagram; looking at special cases or simpler instances of your problem; looking at related problems; searching for patterns; making tables of what you know (and in general, being organized); giving names to what you don't know and writing equations about it (or, more generally, applying the mathematical tools you have learned in algebra and other courses); working backwards and setting subgoals ("I could solve this problem if I could do such-and-such"); using trial and error (usually called "guess and check" in educational circles); using indirect reasoning and looking for counterexamples ("if this weren't true, then what would have to happen?"); jumping out of the system and trying something totally different; or just going away from the problem and coming back to it later. You will find it useful to read over this description of the problem-solving process repeatedly during your study of discrete mathematics.

Here is an example of applying the problem-solving process to a problem in discrete mathematics. Suppose we want to find the number of squares (of all sizes) on a checkerboard. We understand this to mean not only the obvious 64 little squares, but also the 2×2 squares (of which there will be many), the 3×3 squares, and so on, up to the entire 8×8 board itself. We should draw a picture to see what's going on here. Let's begin by playing with a smaller version of this problem, using a board of size 2×2. Then obviously there are 4 small squares and the entire board, for a total of $4 + 1 = 5$. Let's try a 3×3 board. Here there are 9 little squares, and it is easy to see that there are 4 squares of size 2×2 (nuzzled in the upper left, the upper right, the lower left, or the lower right), as well as the entire board; so the answer here is $9 + 4 + 1 = 14$. A pattern seems to be emerging—we seem to be adding perfect squares to get the answer. Maybe for the 4×4 board there will be $16 + 9 + 4 + 1 = 30$ squares in all. We draw the 4×4 picture and verify that this is correct. In fact, we can see exactly what is going on, the way the upper left corner of a $k \times k$ square can be in any of the first $5 - k$ rows and first $5 - k$ columns of the board, for $k = 1, 2, 3, 4$, and so there are $(5 - k)^2$ squares of size k in the 4×4 board, exactly as our sum indicated. We have now solved the problem and can write up a solution explaining why the answer is $8^2 + 7^2 + 6^2 + 5^2 + 4^2 + 3^2 + 2^2 + 1^2 = 204$. In the looking back stage (step 4) we would certainly want to notice that for an $n \times n$ board, there are $\sum_{i=1}^{n} i^2$ squares. To continue our investigation, we might want to explore such further questions as allowing rectangles rather than squares, looking at rectangular checkerboards rather than square ones (or counting triangles in boards made up of triangles), or moving to 3-dimensional space and counting cubes in a large cube. Notice how we followed the process outlined above and used many of the strategies listed.

List of Common Mistakes

If students or instructors have items to add to the lists below, please let the author know (visit the companion website for *Discrete Mathematics and Its Applications* at `http://www.mhhe.com/rosen`).

Chapter 1

- *Incorrectly translating English statements into symbolic form.* There are many errors of this type. For example, there are difficulties with the use of the word "or" in English; be sure to differentiate between inclusive and exclusive versions (see pages 4–5 of the text). A conditional statement is quite different from a conjunction, but some speakers fail to distinguish them; to say that B will happen *if* A happens is quite different from saying that A and/or B *will* happen. Perhaps the most common mistake is confusing $p \to q$ with $q \to p$. To say, for example, that I will go to the movie *if* I finish my homework means something quite different from asserting that I will go to the movie *only if* I finish my homework.
- *Incorrectly negating compound statements without using De Morgan's laws—in effect saying, for example, that* $\neg(p \vee q)$ *is logically equivalent to* $\neg p \vee \neg q$, *or that* $\neg(p \wedge q)$ *is logically equivalent to* $\neg p \wedge \neg q$. For example, if it is not true that John is over 18 years old or lives away from home, then it is true that he is not over 18 years old *and* (not *or*) he does not live away from home. The correct statements are that $\neg(p \vee q)$ is logically equivalent to $\neg p \wedge \neg q$, and that $\neg(p \wedge q)$ is logically equivalent to $\neg p \vee \neg q$. This mistake is a general instance of assuming that every operation distributes over every other operation, here that negation distributes over disjunction (or conjunction).
- *Misinterpreting the meaning of the word "any" in a mathematical statement.* This word is ambiguous in many situations, and so should usually be avoided in mathematical writing. If you are not sure whether the writer meant "every" or "some" when the word "any" was used, get the statement clarified. As a corollary, of course, you should avoid using this word yourself. Here is an example: What would one mean if she defined a *purple* set of integers to be one "in which any integer in the set has at least three distinct prime divisors"? Does "any" mean "every" here (in which case the set $\{30, 40\}$ is not purple), or does "any" mean "some" here (in which case the set $\{30, 40\}$ is purple)?
- *Incorrectly writing the symbolic form of an existential statement as* $\exists x(A(x) \to B(x))$ *instead of* $\exists x(A(x) \wedge B(x))$. For example, the symbolic form of "There exists an even number that is prime" is $\exists x(E(x) \wedge P(x))$, not $\exists x(E(x) \to P(x))$, where we are letting $E(x)$ mean "x is even" and $P(x)$ mean "x is prime." As a rule of thumb, existential quantifiers are usually followed by conjunctions.
- *Incorrectly writing the symbolic form of a universal statement as* $\forall x(A(x) \wedge B(x))$ *instead of* $\forall x(A(x) \to B(x))$. For example, the symbolic form of "Every odd number is prime" is $\forall x(O(x) \to P(x))$, not $\forall x(O(x) \wedge P(x))$, where we are letting $O(x)$ mean "x is odd" and $P(x)$ mean "x is prime." As a rule of thumb, universal quantifiers are usually followed by conditional statements.
- *Incorrectly putting predicates inside predicates, such as* $P(O(x))$. For example, if $P(x)$ means "x is prime," and $O(x)$ means "x is odd," then it would never make sense to write $P(O(x))$ in trying to express a statement such as "x is an odd prime" or to write $\forall x\, P(O(x))$ to say "all odd numbers are prime." The notation $P(O(x))$ would mean that *the assertion* that x is odd is a prime number, and clearly an assertion isn't any kind of number at all. Functional notation has a wonderful internal beauty and consistency to it—the thing inside the parentheses has to be what the thing outside the parentheses applies to.
- *Failure to change the quantifier when negating a quantified proposition, especially in English.* For example, the negation of the statement that some cats like liver is *not* the statement that some cats do not like liver; it is that no cats like liver, or that all cats dislike liver.
- *Overusing the term "by definition" in justifying statements in a proof.* For example, Franklin Roosevelt was not the President of the United States at the start of the country's entry into World War II in December, 1941, "by definition"; he was the President because he had been inaugurated as such early in 1941 and had not died or left office.
- *Not going back to carefully check the definitions in justifying statements in a proof.* For example, if one is trying to prove something about odd integers, then it is important to correctly use the meaning of that notion (that an odd integer is one that can be written as $2k + 1$ for some integer k) at one or more places in the proof.
- *Incorrectly starting a proof by assuming what is to be proved.* A common occurrence of this in an earlier course is trying to prove trigonometric identities by starting with the identity and using algebra to reach $A = A$; this is not valid. Similarly, if we are trying to prove a set identity in Chapter 2, such as $A \subseteq A \cup B$, it would be

invalid to start with the statement $A \subseteq A \cup B$.

- *Invalidly assuming that a few (or even a large number of) examples of a universally quantified proposition imply that the proposition is true.* There is an example from number theory of an intriguing proposition about a positive integer n that is true for every $n \le 4{,}000{,}000$ with the sole exception of $n = 1969$. A proof of the universally quantified proposition $\forall x\, P(x)$ consists of showing that the property $P(x)$ holds no matter what x is chosen from the domain (universe of discourse).
- *Incorrectly assuming that an arbitrary object has a particular property when all you know is that there exists an object with the property.* For example, suppose we are trying to prove the assertion that x^2 always leaves a remainder of 1 when divided by 8. It would be invalid to start our proof by assuming that $x = 2n+1$ for some integer n; even though *some* integers have this property of being odd, it is not true that all of them do, so we would be proving the assertion to be true only some of the time, rather than always.

Chapter 2

- *Incorrectly forming complements of sets without using De Morgan's laws—in effect saying, for example, that* $\overline{A \cap B} = \overline{A} \cap \overline{B}$. The correct statements are $\overline{A \cap B} = \overline{A} \cup \overline{B}$ and $\overline{A \cup B} = \overline{A} \cap \overline{B}$. This is another general instance of assuming that every operation distributes over every other operation, here that complementation distributes over intersection (or union). Students sometimes make similar errors in algebra, such as asserting, incorrectly, that $\sqrt{a^2+b^2} = a+b$ or $\sin(\alpha+\beta) = \sin\alpha + \sin\beta$.
- *Using incorrect notation regarding elements and subsets of power sets and confusing the notions of "element" and "subset" when dealing with a power set.* If A is a subset of S, then A is an element of the power set of S. For example, if $S = \{p,q,r\}$, then $\{p,r\} \subseteq S$, so $\{p,r\} \in P(S)$. On the other hand, $\{p,r\} \not\subseteq P(S)$, and $\{p,r\} \notin S$. Also, note that $\emptyset \notin S$, $\emptyset \in P(S)$, and $\{\emptyset\} \notin P(S)$.
- *Incorrectly writing $\{\emptyset\}$ to represent the empty set.* One reason this cannot be the empty set is that it has one element in it. Correct notation for the empty set is either $\{\ \}$ or $\emptyset$. It is a set with no elements in it.
- *Incorrectly omitting parentheses in expressions involving intersection, union, and difference of sets.* In absence of a default order of operations, an expression such as $A \cap B \cup C$ is ambiguous, since it might mean either $(A \cap B) \cup C$ or $A \cap (B \cup C)$, and these are not the same sets. It is important to put in parentheses so that the reader knows what you mean.
- *Failing to regard a function as an object at a higher order of abstraction than an element, an ordered pair, or a set.* The notation $f : A \to B$ means that f is a process—a rule that must apply to every element of A and in each case yield a result in B. The function is not A, it is not B, it is not an element of A or B, nor is it the action on just one element of A. For example, if f is the function from $\{1,2,3\}$ to the natural numbers that has the rule $f(x) = x^2$, then f is the entire process that takes 1 to 1, 2 to 4, and 3 to 9; it is not, for example, just the number 9 or just the pair $(3,9)$.
- *Confusing the idea that a function must be well defined with the concept of one-to-oneness.* A function in general need not have the property that distinct elements of the domain be sent to distinct elements of the range. What a function must satisfy is the requirement that two different elements of the range cannot both be the image of the same element of the domain. Thus $f(x) = x^2$ *is* a function from the set of real numbers to the set of nonnegative real numbers (it's onto but not one-to-one), but $f(x) = \pm\sqrt{x}$ is not a function from the nonnegative real numbers to the real numbers.
- *Incorrectly calculating values of floor and ceiling functions, especially for negative values.* The floor function always rounds down, and the ceiling function always rounds up. Thus, for example, $\lfloor -3.2 \rfloor = -4$, and $\lceil -3.2 \rceil = -3$.
- *Incorrectly including diagonal portions in sketches of floor and ceiling functions and their variations.* These graphs almost always consist entirely of horizontal segments.
- *Misusing function notation, especially the dummy variable in a defining equation.* The input goes inside the parentheses following the function name, and the entire expression then represents the output. For example, if $f(x) = x^2$, then it is correct to write $f(7) = 49$, not things like $f = 49$ or $f = 7$ or $7 = 49$. The equals sign in mathematics should be used to mean that two things are equal, and not as a shorthand for "and then we have ...."
- *Confusing function notation with multiplication.* Although writing two mathematical expressions next to each other, especially with parentheses, often implies multiplication, this definitely does not apply when the first object is a function. For example, if a function is defined by $f(x) = x + 10$, then $f(7)$ means "eff *of* seven"

(which is 17 in this case) and not "eff *times* seven" (which makes no sense). Math has to make sense!

- *Thinking that all functions are linear.* The distributive law for numbers says that $t(x+y) = t(x)+t(y)$ if here t, x, and y are numbers and the juxtaposition means multiplication. For example, $8(2+10) = 8 \cdot 2 + 8 \cdot 10$; both sides equal 96. However, if t is the squaring function and again x and y are numbers, then it is not true that $t(x+y) = t(x)+t(y)$; if we again take $x = 2$ and $y = 10$, then the left-hand side is $(2+10)^2 = 12^2 = 144$, whereas the right-hand side is $2^2 + 10^2 = 4 + 100 = 104$.

Chapter 3

- *Forgetting to increment counters inside loops of procedures when constructing algorithms.* If the counter is not incremented, the loop will probably run forever.
- *Thinking of a big-O estimate as if it were a big-Theta estimate, i.e., thinking that is provides both a lower bound and an upper bound on the size of the function.* Big-O estimates are only upper bounds. It is not correct to say that such-and-such an algorithm is inefficient because it runs in time $O(n^7)$, since, for instance, a linear algorithm (one with running time proportional to n) also satisfies this estimate. What one would want to say in such a case (if it is true) is that the algorithm runs in time $\Theta(n^7)$.
- *Not understanding that the constant in a big-O estimate can be very large.* For example, if an algorithm has running time $O(n^2)$, then it might take $10^{55}n^2$ steps and so be impractical even for small values of n.
- *Being misled by exponents when comparing dissimilar functions.* For example, it might at first glance look as if $(\log n)^{100}$ is growing faster than $n^{1/2}$, but in fact the latter is growing faster, since $\log n$ grows so slowly compared to n. Thus $(\log n)^{100}$ is $O(n^{1/2})$, but $n^{1/2}$ is not $O((\log n)^{100})$.
- *Adding when one should be multiplying, or vice versa, when analyzing running times of algorithms.* If one loop is nested inside another, then the running times are multiplied; if one loop follows another, the times are added.
- *Incorrectly making a* **mod** *m negative.* For example, -16 **mod** 5 is 4, not -1.
- *Confusing a/b and $a \mid b$.* The slanted slash is an operation, and the result of the operation is a *number*. For example, $6/3$ is the number 2. The vertical bar is the *verb* of a sentence. For example, $3 \mid 6$ is asserting that 3 is a divisor of 6; it is not speaking of the result of actually carrying out the division.
- *Writing $a \mid b$ when one means $b \mid a$.* It is true that $3 \mid 6$, but it is not true that $6 \mid 3$.
- *Incorrectly considering 1 to be a prime number.* It's as much a matter of convention as anything else, but the number 1 is, by definition, neither prime nor composite.
- *Forgetting that all positive integers are divisors of 0, and therefore that $\gcd(a, 0) = a$ for all positive integers a.* For example, $\gcd(6, 0) = 6$. Of course 0 is not a divisor of any nonzero number, and division by 0 is undefined.
- *When carrying out the Euclidean algorithm, incorrectly using the last quotient as the final answer (the gcd), rather than the last divisor.* For example, if the last step is to divide 4 into 8, giving a quotient of 2 and remainder of 0, then the gcd is the final divisor, 4, not the final quotient, 2.
- *Incorrectly assuming that everything that works for equality works for congruence.* For example, it is not true that just because r and s are congruent modulo m, $a^r \equiv a^s \pmod{m}$. Try $m = 3$, $a = 2$, $r = 1$, and $s = 4$. Also, note that even though $8 \equiv 14 \pmod{6}$, it would be wrong to divide both sides by 2 and claim that $4 \equiv 7 \pmod{6}$.
- *Incorrectly assuming that multiplication of matrices means entry-by-entry multiplication.* Addition of matrices is done term by term, but multiplication is more complex.

Chapter 4

- *Erroneously believing that all infinite sets are countable.* Although it is possible to give an infinite list of certain infinite sets (such as the set of rational numbers, using an order like $0, \frac{1}{1}, -\frac{1}{1}, \frac{1}{2}, -\frac{1}{2}, \frac{2}{1}, -\frac{2}{1}, \frac{1}{3}, -\frac{1}{3}, \frac{3}{1}, -\frac{3}{1}, \frac{1}{4}, -\frac{1}{4}, \frac{2}{3}, -\frac{2}{3}, \frac{3}{2}, -\frac{3}{2}, \frac{4}{1}, -\frac{4}{1}, \frac{1}{5}, \ldots$, it is not possible to do this for certain larger sets, such as the set of real numbers. There is no list of all the real numbers, as was proved in Section 2.4.
- *Incorrectly applying to infinite sets intuition that is valid for finite sets.* For example, if a *finite set* A can be put into one-to-one correspondence with a proper subset of B, then clearly $|A| < |B|$. This is not true if A is infinite, however: Let A be the even natural numbers and let B be the natural numbers—then A can be put into one-to-one correspondence with (in fact *is*) a proper subset of B, but A can also be put into one-to-one correspondence with all of B (pairing $2n$ with n for $n = 0, 1, 2, \ldots$), so in fact $|A| = |B|$.
- *Forgetting to do the basis step in a proof by mathematical induction.* The inductive step goes through fine, for

example, if we try to prove that $n = n + 1$ for all positive integers n, but this proposition is obviously not true. The catch is that the basis step (when $n = 1$) fails, since $1 \neq 1 + 1$.

- *Failing to do more than one case in the basis step in a proof by mathematical induction in certain situations, such as when the inductive step needs two or more previous conditions.* For example, when proving statements about the Fibonacci sequence, it usually is necessary to check the first two basis cases (say $n = 1$ and $n = 2$), since the inductive step relies on the equation $f_n = f_{n-1} + f_{n-2}$.
- *Confusing a summation with the propositional function $P(n)$ in an induction proof.* For example, in trying to prove $1 + 2 + 3 + \cdots + n = n(n+1)/2$ by induction, $P(n)$ is this entire equation, not its left-hand side.
- *Not being organized when attempting to write a recursive definition.* Good advice here is to think about how you want to build up the items under discussion, step by step. The inductive rules of the definition need to be formulated to permit each such step, and base cases are needed to get the process off the ground. Common mistakes include not including enough base cases (for example, the recursive definition of regular expressions in Chapter 12 requires three base cases), having conflicting cases (for example, having one clause to handle n divisible by 2 and another clause to handle n divisible by 3, and thereby not having a unique definition for those n divisible by 6), or having a function value at n depend on a function value at an input larger than n (e.g., trying to set $f(n) = f(3n + 1) - 2$).
- *In a proof by mathematical induction, writing $P(k+1)$ incorrectly.* Once $P(n)$ is properly formulated, writing $P(k+1)$ can usually be done more or less mechanically by plugging $k + 1$ in for n. For example, if $P(n)$ is the statement $2+4+6+\cdots+2n = n(n+1)$, then $P(k+1)$ is $2+4+6+\cdots+2k+2(k+1) = (k+1)(k+2)$.
- *In a proof by mathematical induction, making errors in basic algebra, especially in simplifying expressions.* For example, you might have to use the fact that $2^n + 2^n = 2^{n+1}$, or to simplify $(n+1)^3 + 5(n+1)^2$, which is best done by factoring, not by first expanding each term. Carefully check your algebraic manipulations when you have trouble with the inductive step of such a proof.
- *In a recursive algorithm, failing to let the computer do the recursing.* The hardest thing to overcome in thinking about recursive algorithms is the reluctance to believe that the smaller case will be handled correctly. For example, suppose that you want to write a recursive algorithm to compute a^n, using the facts that $a^{2k} = (a^k)^2$ and $a^{2k+1} = (a^k)^2 \cdot a$. The recursive call will handle the calculation of a^k, and so you don't need to worry about how the computer will repeatedly recurse, all the way down to the base case, in order to do that. As long as your recursive step and basis case (here, $a^0 = 1$) are correct, your algorithm is correct.

Chapter 5

- *Drawing an incorrect diagram when solving counting problems.* Diagrams are very useful in all of mathematics, and drawing a diagram is almost always a good way to start solving a problem; thus *failing to draw a diagram* could also be considered a common mistake. For example, you should draw a row of six blanks (not five) as a template for constructing words of length 6 whose symbols are chosen from a set of five elements. Tree diagrams are also sometimes quite helpful.
- *Not determining whether or not order matters in solving a counting problem.* For example, if we are asked for the number of ways to write 7 as the sum of positive integers, then we need to know whether $3 + 2 + 2$ and $2 + 3 + 2$ are to be considered the same way or distinct ways. Read the problem very carefully to understand what is being counted. Resolve any ambiguities ahead of time by explicitly stating any assumptions that seem to be missing from the problem formulation.
- *Not determining whether or not repetitions are allowed in solving a counting problem.* For example, if we are asked for the number of ways to choose five donuts from a shop selling eight varieties of donuts, we need to know whether we are allowed to choose more than one donut of the same variety. Read the problem very carefully to understand what is being counted. Resolve any ambiguities ahead of time by explicitly stating any assumptions that seem to be missing from the problem formulation.
- *Counting each item in the set under discussion more than once, not recognizing that an adjustment needs to be made for double counting.* For example, if we count handshakes person by person, then we need to recognize that each shake has been counted twice, once for each of its participants.
- *Counting each item in the set under discussion more than once, not recognizing that the inclusion–exclusion principle is needed.* For example, if we are told that there are 26 computer science majors and 34 mathematics majors at a certain university, then there may not be $26 + 34 = 60$ people majoring in either computer science or mathematics, since these two numbers might both include the double majors. To correctly compute the

total number of people majoring in these subjects, we would need to subtract the number of double majors from this sum.

- *Using the pigeonhole or generalized pigeonhole principle incorrectly.* It helps to explicitly identify the pigeons and the holes. For example, to find the minimum number of cards that must be chosen in order to guarantee that at least six of the same suit are picked, the holes are the suits, and the cards are the pigeons (the answer is 21).

Chapter 6

- *When trying to calculate the probability that one of two events will occur, incorrectly taking the sum of the individual probabilities.* For example, the probability that a 3 will show up if a fair die is rolled twice is not $\frac{1}{6}+\frac{1}{6}$, the sum of the probability that the 3 occurs on the first roll and the probability that the 3 occurs on the second roll. Instead, we should calculate this as 1 minus the probability that the 3 fails to appear on either roll (which is $\frac{5}{6}\cdot\frac{5}{6}$, since the rolls are independent). Thus the correct answer is $1-\frac{25}{36}=\frac{11}{36}$, rather than $\frac{2}{6}$.
- *Assuming that all events in a probability calculation are disjoint.* Doing so can lead to absurd conclusions, such as a probability greater than 1. (This is really a generalization of the previously listed mistake.) For example, to calculate the probability that someone else in your graduating class of 400 students shares your birthday (assuming that all birthdays are equally likely and ignoring February 29), you cannot argue that since each of them has a probability of 1/365 of sharing your birthday, the probability is 399/365 (probabilities can never exceed 1, and in any case, this event is not a certainty).
- *Assuming that all events in a probability calculation are independent.* For example, to calculate the probability that we get two hearts when drawing two cards from a deck of cards, without replacing the first card before drawing the second, we cannot simply note that the probability of drawing a heart is $\frac{13}{52}$ on each draw (that much is true) and therefore conclude that the answer is $\frac{13}{52}\cdot\frac{13}{52}=\frac{1}{16}$. Instead, we must determine that for the second draw, the probability of drawing a heart, given that we drew a heart on the first draw, is $\frac{12}{51}$, and therefore that the probability of drawing a heart both times is $\frac{13}{52}\cdot\frac{12}{51}=\frac{1}{17}$.
- *Getting misled by the subtle assumptions inherent in probability problems.* The most famous example here is the Monty Hall Three Door Problem (see Example 10 in Section 6.1 of the text). Unless one is very careful about the assumptions one makes about the game host's protocol, one cannot calculate the probability that switching doors will change your chances of winning. For example, if the host (who knows where the prize lies) were to offer you a switch if and only if you had chosen the correct door, then obviously it would be wrong for you to switch when he makes the offer. A national debate about this problem raged for many months when it was popularized in a magazine article.
- *Letting intuition interfere with reason in working with probability.* For example, it might seem counter-intuitive that among a group of 23 people, the odds favor two of them having the same birthday, but the calculation shows this to be true.
- *Forgetting that units for variance are squares of units for the underlying random variable.* For example, if the heights of adults have a mean of 67 inches and a variance of 9, this is 9 square inches, not 9 inches. One should take the square root and restate this as "the standard deviation is 3 inches," which is a more meaningful way to measure the spread of the distribution of heights.
- *Confusing $p(A \mid B)$ with $p(B \mid A)$.* For example, the probability that a person who tests positive for a disease actually has the disease is usually much smaller than the probability that a person who has the disease tests positive for it. One can often use Bayes' Theorem to compute the former probability, given the latter (and additional information about the prevalence of the disease and the false positive rate for the test).

Chapter 7

- *Failing to note the need for the inclusion–exclusion principle.* To believe that $|A\cup B|=|A|+|B|$ is always true is related to the wishful thinking that every operation distributes (or otherwise behaves in some simple, agreeable way) with respect to every other operation. This equality holds only when A and B are disjoint.
- *Confusing the signs of the terms when applying the inclusion–exclusion principle.* Note that the signs alternate as we take larger and larger unions.
- *Not including all the terms when applying the inclusion–exclusion principle.* If there are n sets involved, then there are nearly 2^n different terms in the equation altogether.

- *Giving up too easily when trying to write down a recurrence relation to model a problem situation.* Ask yourself how one can obtain an instance of the problem of size n from instances of sizes $n-1$ (or sometimes also smaller instances). Make sure to consider all the possibilities, and make sure to include enough initial conditions. For example, if a_n is the number of ways to climb n stairs if we are allowed to take them either one at a time or three at a time, then clearly $a_1 = 1$, $a_2 = 1$, and $a_3 = 2$, and then $a_n = a_{n-1} + a_{n-3}$ for $n \geq 4$, since the first step could be a single step or a triple step.
- *Misapplying the algorithm for solving linear homogeneous recurrence relations with constant coefficients when there are repeated roots of the characteristic equation.* One needs to multiply by powers of n in this case. For example, if the characteristic equation is $r^2 - 6r + 9 = (r-3)^2 = 0$, then the general solution is $a_n = c_1 \cdot 3^n + c_2 n \cdot 3^n$.
- *Finding a bogus particular solution of a linear nonhomogeneous recurrence relation with constant coefficients.* It is always advisable to check the solutions you obtain. For example, if you had computed that $a_n = 2^n$ was a particular solution to $a_n = 2a_{n-1} + 2^n$, then plugging this in would show you that you must have made an error, since it is not true that $2^n = 2 \cdot 2^{n-1} + 2^n$.
- *Forgetting to use the inclusion–exclusion principle when counting solutions to an equation in nonnegative integers.* For example, to count the number of solutions to $x + y + z = 58$ where $0 \leq x < 8$, $0 \leq y < 10$, and $0 \leq z < 15$, one needs to count the number of solutions when the upper bound restrictions are lifted, then subtract the number of solutions in which each such restriction is violated, then add back the number of solutions in which two such restrictions are violated simultaneously, and finally subtract the number of solutions in which all three restrictions are violated.
- *Forgetting to worry about the first few terms of a power series.* When solving a recurrence relation by using generating functions, the recurrence relation usually kicks in only for $k \geq 1$ or 2; thus the first term or two must be handled explicitly.
- *Failing to change the variable in a power series when necessary.* For example, if a power series has x^{k-1} and you need it to be x^k, you can replace k by $k+1$ throughout the summation (including the limits) and simplify algebraically: $\sum_{k=1}^{\infty} k\, x^{k-1} = \sum_{k+1=1}^{\infty} (k+1)\, x^{(k+1)-1} = \sum_{k=0}^{\infty} (k+1)\, x^k$.
- *Setting up the wrong model when solving counting problems with generating functions.* You need to carefully work out what each factor of the generating function needs to be, worrying about how much repetition is allowed and whether order matters. See, for instance, Example 12 in Section 7.4 of the text, where the proper generating function depends on whether or not we take order into account.
- *Making algebraic errors in working with generating functions.* When expanding a generating function to find the coefficient of x^n, one must of course use the distributive law. The algebraic manipulations can get messy, as the number of terms can grow rapidly. One solution to this problem is to use a computer algebra package such as *Maple* to do the algebra. For example, to multiply out $(1+x+x^2)(1+x^2+x^4+x^6)$, you end up with 12 terms, which then simplify to $1 + x + 2x^2 + x^3 + 2x^4 + x^5 + 2x^6 + x^7 + x^8$.
- *Not knowing how to use partial fraction decomposition when dealing with generating functions, or making errors in the procedure, such as forgetting to include terms of the form $(x-a)^k$ for all k such that $1 \leq k \leq n$ when the factor $(x-a)^n$ appears in the denominator of the fraction to be expanded.* This subject is traditionally taught in calculus courses, even though it has little to do with calculus (other than the fact that it is used as a technique of integration). Therefore those students who have not yet studied enough calculus (partial fractions are usually covered in the second semester), or who have taken a course in which this topic is not covered, may need to find a source of instruction for this useful tool (or rely on a computer algebra package such as *Maple* to perform the task). Any traditional calculus text will probably have a section from which this material can be learned or reviewed.

Chapter 8

- *Failing to draw a picture when dealing with relations.* The digraph of a relation on a set gives an excellent way to visualize what is going on. This common mistake can be generalized: Failing to draw a picture when dealing with any mathematical object. See the list of general problem-solving strategies given in the introduction to this section of the *Guide*.
- *Forgetting to think about pairs (a,b) and (b,a) when checking for transitivity of a relation or forming the transitive closure.* In this case, one needs to have (or add) the loops (a,a) and (b,b) as well.
- *Failing to recognize that symmetry or transitivity often hold vacuously.* For example, the relation $\{(1,2),(1,3)\}$

is transitive, because it is vacuously true that whenever (a, b) and (b, c) are in the relation, so is (a, c)—i.e., there are no pairs making the hypothesis of this conditional statement true.

- *Forgetting that order of operations matters when forming closures.* The symmetric, transitive closure is not the same as the transitive, symmetric closure, for example.
- *Incorrectly assuming that every relation has desired properties, such as symmetry or transitivity.* It is certainly not true that every relation is a total order or an equivalence relation. Many partial orders are not total, so elements can be incomparable (for example, it is true neither that $\{1,2\} \subseteq \{1,3\}$ nor that $\{1,3\} \subseteq \{1,2\}$). If I know you and you know Mary, that doesn't mean that I necessarily know Mary. If $f(x)$ is $O(g(x))$, that doesn't mean that $g(x)$ is $O(f(x))$.
- *Confusing equivalence relations and partitions.* These two concepts are closely related—two elements in the underlying set are related if and only if they are in the same set of the partition. An equivalence relation is a set of *ordered pairs* of elements of the underlying set (satisfying certain conditions), and a partition is a set of *subsets* of the underlying set (again satisfying certain conditions).
- *Making the understanding of equivalence relations harder than necessary.* In most cases you can think of an equivalence relation as a relation of the form "two elements are related if and only if they have the same something-or-other." For example, the equivalence relation on the set of integers given by "a is related to b if and only if $a \equiv b \pmod 7$" can be thought of as "a and b are related if and only if they have the same remainder when divided by 7." See the important Exercise 9 in Section 8.5.
- *Invalidly applying to infinite partially ordered sets intuition that is valid for finite posets.* Hasse diagrams can be misleading for infinite posets (or they may not exist at all), so be wary about thinking of all posets in finite diagrammatic terms. For example, in the poset consisting of the positive real numbers under the $\leq$ relation, there is no immediate successor to any element, so there would be no edges in the Hasse diagram if we were to try to draw one.
- *Forgetting to eliminate all the implied edges when drawing a Hasse diagram.* If there is an edge from a to b and one from b to c, where a lies above b and b lies above c, then you must not show an edge between a and c.
- *Forgetting that certain pairs in a relation are implied in a Hasse diagram.* If there is an edge from a to b and one from b to c, where a lies above b and b lies above c, then c is related to a even though there is no edge between a and c.
- *Incorrectly supposing that all partial orders have least or greatest elements, or that least upper bounds or greatest lower bounds always exist.* For example, in the poset $(\{a,b,c,d,e\}, \{(a,a),(b,b),(c,c),(d,d),(e,e),(a,b),(c,b),(c,d),(e,d)\})$ there is no least element, no greatest element, and no least upper bound for $\{a, e\}$. Similarly, in the poset consisting of the integers under $\leq$ together with two extra elements x and y that are defined to be less than all the integers and unrelated to each other, there is no least upper bound for $\{x, y\}$, even though every integer is an upper bound.

Chapter 9

- *Getting confused by some of the terminology in graph theory, such as the distinction between path and simple path.* Here is one place where memorization is required. Making your own glossary on file cards or in a computer file may be helpful.
- *Overcounting the edges in a graph by forgetting to divide by 2 when adding the degrees of the vertices.* Each edge is counted twice, once for each end.
- *Being unsure about what kind of graph model to use.* If the relationship between objects in the situation you are trying to model is symmetric, then an undirected graph is probably appropriate; otherwise a directed graph usually works best. For example, highways joining major cities can be traversed in either direction, so an undirected graph is appropriate for this model. The predator-prey relation among species of animals is definitely not symmetric, so a digraph seems right here.
- *Incorrectly thinking that if two graphs share many of the same attributes (invariants like number of vertices, number of edges, etc.) then they must be isomorphic.* See Section 9.3 for some counterexamples.
- *Incorrectly interchanging the definitions of Hamilton and Euler paths and circuits.* Remember that there is a simple test for Euler paths and circuits, but no one knows of any simple tests for Hamilton paths and circuits.
- *Ignoring the fact that having an Euler [Hamilton] circuit implies the existence of an Euler [Hamilton] path.* Look carefully at the definitions.

- *Confusing theorems in graph theory with their converses.* For example, if in a connected simple graph with $n \geq 3$ vertices each vertex has degree at least $n/2$, then the graph has a Hamilton circuit; but the converse or inverse of this statement is not true (there are plenty of graphs having Hamilton circuits in which the vertex degrees are small). Here is another example: If a connected simple graph is planar, then it must satisfy $e \leq 3v - 6$, where e is the number of edges and v is the number of vertices. Therefore (by the contrapositive), we know that if a graph has too many edges ($e > 3v - 6$), then it cannot be planar. What we *cannot* conclude is the converse—it is not a theorem that if $e \leq 3v - 6$, then the graph has to be planar.
- *Using the nonexistent word "vertice" instead of the correct word "vertex" to talk about just one of the dots in a graph.* Similarly, there is no such thing as a "matrice."
- *Mistakenly thinking that a graph is nonplanar just because it is drawn one way with two edges crossing.* If it is possible to redraw the graph without edges crossing, then the graph is planar. For example, K_4 is planar, even though drawing it as the vertices of a square with straight line segments representing the edges causes a crossing in the middle of the picture. (Redraw it as the vertices of a triangle with one more vertex in the interior.)
- *Invalidly concluding that once one has found a coloring of a graph with n colors, its chromatic number has to be n.* In fact, all we know in that case is that its chromatic number is at most n. It may be possible to find another coloring with fewer than n colors. For example, one could color C_4 with four colors (a different color for each vertex), but its chromatic number is in fact 2.
- *Mistakenly believing that greedy algorithms always produce the optimal solution to a problem.* It often happens that the simple-minded greedy approach *does* find the best solution (e.g., in looking for minimum spanning trees in Section 10.5), but often the greedy approach fails (e.g., in finding a coloring of a graph using as few colors as possible).
- *Failing to recognize that writing down a procedure doesn't guarantee that it does what you want it to do.* For example, one cannot write down a greedy algorithm to color a graph and then claim without justification that this procedure finds the coloring with the fewest possible colors. In fact, it won't, as fairly simple counterexamples can show.

Chapter 10

- *Incorrectly setting up a decision tree for a problem such as identifying counterfeit coins by weighing them, and thereby drawing the wrong conclusions.* Each possible situation must correspond to a path from the root of the tree to a leaf.
- *Not realizing what type of tree is needed for a particular mathematical model.* Issues to consider are whether there should be a root (a starting point for some process), whether the children of a vertex are ordered, and whether each child should be classified as a right child or a left child.
- *Incorrectly omitting parentheses in expressions written in infix notation.* In absence of a default order of operations, an expression such as $A \cap B \cup C$ is ambiguous, since it might mean either $(A \cap B) \cup C$ or $A \cap (B \cup C)$, and these are not the same sets. With prefix or postfix notation, no such ambiguities arise.
- *Forgetting that when doing an inorder traversal of an ordered rooted tree that is not binary, the root of each subtree comes after the first subtree but before all the other subtrees.* For a binary tree, inorder traversal is rather obvious—left, root, right. When the tree isn't binary, we can still define inorder traversals, but the definition isn't as natural.
- *When applying Prim's algorithm for finding minimum spanning trees, forgetting that edges become eligible for inclusion in the tree gradually (as opposed to Kruskal's algorithm, in which they are all eligible from the start).* If there is a low-cost edge that does not currently have any endpoint in the tree constructed so far, then it cannot yet be added to the tree. When one of its endpoints finally becomes part of the tree, it suddenly becomes eligible and can then be added to the tree if it is the lowest cost edge currently eligible. It is easy to overlook such edges when performing the algorithm.

Chapter 11

- *Being off by one level of abstraction when thinking about Boolean functions.* A Boolean function with n variables can be represented by a table with 2^n rows; therefore there are 2^{2^n} different Boolean functions with n variables.

- *Putting inverters in the wrong place when building combinational circuits.* If we want to invert the value of the output of a gate, the inverter needs to go after the gate.
- *Forgetting to apply De Morgan's laws correctly when evaluating the output of a combinational circuit.* The output of a circuit is a certain Boolean expression of the input variables. When simplifying this expression, it is important to remember than $\overline{xy} = \overline{x} + \overline{y}$ and $\overline{x+y} = \overline{x}\,\overline{y}$.
- *Not finding the largest possible blocks when looking for minimum Boolean expressions using K-maps.* Since there is no known efficient algorithm for solving this problem in general (with more than just a few variables), it should not be surprising that this procedure seems to involve some ugly "guessing" to it.
- *Not finding the best cover when looking for minimum Boolean expressions using the Quine–McCluskey method.* Since there is no known efficient algorithm for solving this problem in general (with more than just a few variables), it should not be surprising that this procedure seems to involve some ugly "guessing" to it. It might be very hard to make sure that a covering we have found with, say, five minterms is really the best possible—that there isn't another covering with four minterms.

Chapter 12

- *Incorrectly constructing grammars to generate a desired language.* There is no algorithm for doing this (this statement is a theorem in the theory of computation, similar to Turing's theorem on the unsolvability of the halting problem). Constructing grammars is like writing computer programs, and all the advice given in a programming course (such as thinking from the top down in a structured way) applies.
- *Incorrectly constructing finite-state machines (including Turing machines) to perform a desired task.* There is no algorithm for doing this (this statement is a theorem in the theory of computation, similar to Turing's theorem on the unsolvability of the halting problem). Constructing machines is like writing computer programs, and all the advice given in a programming course (such as thinking from the top down in a structured way) applies.
- *Not including all the strings that are accepted by a given finite-state automaton.* Sometimes students will follow some paths that the machine can take to reach an accepting state and forget to consider others. This will lead to a claim that the language recognized by this automaton is a proper subset of what it really is. Make sure to "play computer" and follow all the branches.
- *Forgetting to have one arrow leaving each state for each input symbol when constructing deterministic finite-state automata.* You usually want to have a "graveyard" state to which the machine goes when it is clear that the input is not acceptable. There needs to be a loop from the graveyard state to itself for each alphabet symbol.
- *Failing to realize that a nondeterministic finite-state automaton can accept a string even when some computation paths on a certain input drive the machine to a nonaccepting state.* As long as at least one path leads to an accepting state, the input string is accepted.
- *Failing to keep track of all the possible states in which a nondeterministic finite-state automaton can enter at each step, when constructing the corresponding deterministic automaton.* Make good use of all your fingers in analyzing what can happen!
- *Failing to check that a machine or a grammar or a regular expression presented as the solution of some problem actually works.* This is similar to debugging a computer program. Many test cases should be tried, so that you can be confident that your machine or grammar or expression really works.
- *When constructing Turing machines, forgetting to include all the cases.* If the machine can ever reach a certain state and be viewing a particular input symbol, then a transition is needed to handle that case. Using top-down programming methodology is advisable to make sure your machines do what you want them to do.
- *Getting so bogged down in the details of constructing Turing machines that you lose sight of the main points of the theory.* The main point is given in the Church–Turing thesis: that every conceivable computation can be performed by any reasonable computational model, be it a Turing machine, your favorite high-level programming language, or yourself working with pencil and paper. And on that note of keeping the "big picture" in mind, we'll bring this list of common mistakes to a close.

Crib Sheet for Chapter 1

Logical and: $p \wedge q$ is true when both p and q are true, is false when at least one of p and q is false.

Logical or (inclusive): $p \vee q$ is true when at least one of p and q is true, is false when both p and q are false.

Exclusive or: $p \oplus q$ is true when exactly one of p and q is true, is false otherwise.

Conditional statement: $p \rightarrow q \equiv$ "If p, then q" $\equiv$ "p only if q" $\equiv$ "p is a sufficient condition for q" $\equiv$ "q is a necessary condition for p." $p \rightarrow q$ is false when p is true and q is false, is true otherwise. $\neg(p \rightarrow q) \equiv p \wedge (\neg q)$. $p \rightarrow q$ is equivalent to **contrapositive** $\neg q \rightarrow \neg p$, but not to **converse** $q \rightarrow p$ or **inverse** $\neg p \rightarrow \neg q$.

Biconditional statement: $p \leftrightarrow q$, means $(p \rightarrow q) \wedge (q \rightarrow p)$, usually read "if and only if" and sometimes written "iff" in English.

De Morgan's laws: $\neg(p \vee q) \equiv (\neg p) \wedge (\neg q)$; $\neg(p \wedge q) \equiv (\neg p) \vee (\neg q)$.

Quantifiers: $\forall x(P(x) \rightarrow Q(x)) \equiv$ "for all x, if $P(x)$ then $Q(x)$"; $\exists x(P(x) \wedge Q(x)) \equiv$ "there exists an x such that $P(x)$ and $Q(x)$." Here $P(x)$ and $Q(x)$ are **propositional functions**, and there is always a **domain** or **universe of discourse**, either implicit or explicitly stated, over which the variable ranges.

Negations of quantified propositions: $\neg\forall x P(x) \equiv \exists x \neg P(x)$; $\neg\exists x P(x) \equiv \forall x \neg P(x)$.

Theorem: a proposition that can be proved; **lemma:** a simple theorem used to prove other theorems; **proof:** a demonstration that a proposition is true; **corollary:** a proposition that can be proved as a consequence of a theorem that has just been proved.

A **valid** argument—one using correct rules of inference based on tautologies—will always give correct conclusions if the hypotheses used are correct. Invalid arguments, relying on **fallacies**, such as affirming the conclusion, denying the hypothesis, begging the question, or circular reasoning, can lead to false conclusions.

Some rules of inference: $[p \wedge (p \rightarrow q)] \rightarrow q$ (**modus ponens**); $[\neg q \wedge (p \rightarrow q)] \rightarrow \neg p$ (**modus tollens**); $[(p \rightarrow q) \wedge (q \rightarrow r)] \rightarrow (p \rightarrow r)$ (**hypothetical syllogism**); $[(p \vee q) \wedge (\neg p)] \rightarrow q$ (**disjunctive syllogism**); $\{P(a) \wedge \forall x\,[P(x) \rightarrow Q(x)]\} \rightarrow Q(a)$ (**universal modus ponens**); $\{\neg Q(a) \wedge \forall x\,[P(x) \rightarrow Q(x)]\} \rightarrow \neg P(a)$ (**universal modus tollens**); $(\forall x\, P(x)) \rightarrow P(c)$ (**universal instantiation**); ($P(c)$ for an arbitrary c) $\rightarrow \forall x\, P(x)$ (**universal generalization**); $(\exists x\, P(x)) \rightarrow$ ($P(c)$ for some element c) (**existential instantiation**); ($P(c)$ for some element c) $\rightarrow \exists x\, P(x)$ (**existential generalization**).

Trivial proof: a proof of $p \rightarrow q$ that just shows that q is true without using the hypothesis p.

Vacuous proof: a proof of $p \rightarrow q$ that just shows that the hypothesis p is false.

Direct proof: a proof of $p \rightarrow q$ that shows that the assumption of the hypothesis p implies the conclusion q.

Proof by contraposition: a proof of $p \rightarrow q$ that shows that the assumption of the negation of the conclusion q implies the negation of the hypothesis p (i.e., proof of contrapositive).

Proof by contradiction: a proof of p that shows that the assumption of the negation of p leads to a contradiction.

Proof by cases: a proof of $(p_1 \vee p_2 \vee \cdots \vee p_n) \rightarrow q$ that shows that each conditional statement $p_i \rightarrow q$ is true.

Statements of the form $p \leftrightarrow q$ require that both $p \rightarrow q$ and $q \rightarrow p$ be proved. It is sometimes necessary to give two separate proofs (usually a direct proof or a proof by contraposition); other times a string of equivalences can be constructed starting with p and ending with q: $p \leftrightarrow p_1 \leftrightarrow p_2 \leftrightarrow \cdots \leftrightarrow p_n \leftrightarrow q$.

To give a **constructive proof** of $\exists x P(x)$ is to show how to find an element x that makes $P(x)$ true. **Nonconstructive existence proofs** are also possible, often using proof by contradiction.

One can **disprove** a universally quantified proposition $\forall x P(x)$ simply by giving a **counterexample**, i.e., an object x such that $P(x)$ is false. One cannot *prove* it with an example, however.

Fermat's Last Theorem: There are no positive integer solutions of $x^n + y^n = z^n$ if $n > 2$.

An integer is **even** if it can be written as $2k$ for some integer k; an integer is **odd** if it can be written as $2k + 1$ for some integer k; every number is even or odd but not both. A number is **rational** if it can be written as p/q, with p an integer and q a nonzero integer.

Crib Sheet for Chapter 2

Empty set: the set with no elements, $\{\ \}$, is denoted $\emptyset$; this is <u>not</u> the same as $\{\emptyset\}$, which has one element.

Subset: $A \subseteq B \equiv \forall x(x \in A \rightarrow x \in B)$; **proper subset:** $A \subset B \equiv (A \subseteq B) \wedge (A \neq B)$ (i.e., B has at least one element not in A).

Equality of sets: $A = B \equiv (A \subseteq B \wedge B \subseteq A) \equiv \forall x(x \in A \leftrightarrow x \in B)$.

Power set: $P(A) = \{ B \mid B \subseteq A \}$ (the set of all subsets of A). A set with n elements has 2^n subsets.

Cardinality: $|S|$ = number of elements in S.

Some specific sets: $\mathbf{R}$ is the set of real numbers, all of which can be represented by finite or infinite decimals; $\mathbf{N} = \{0, 1, 2, 3, \ldots\}$ (natural numbers); $\mathbf{Z} = \{\ldots, -2, -1, 0, 1, 2, \ldots\}$ (integers); $\mathbf{Z}^+ = \{1, 2, \ldots\}$ (positive integers); $\mathbf{Q} = \{ p/q \mid p, q \in \mathbf{Z} \wedge q \neq 0 \}$ (rational numbers); $\mathbf{Q}^+ = \{ p/q \mid p, q \in \mathbf{Z}^+ \}$ (positive rational numbers).

Set operations: $A \times B = \{ (a, b) \mid a \in A \wedge b \in B \}$ (**Cartesian product**); $\overline{A}$ = the set of elements in the universe not in A (**complement**); $A \cap B = \{ x \mid x \in A \wedge x \in B \}$ (**intersection**); $A \cup B = \{ x \mid x \in A \vee x \in B \}$ (**union**); $A - B = A \cap \overline{B}$ (**difference**); $A \oplus B = (A - B) \cup (B - A)$ (**symmetric difference**).

Inclusion–exclusion (simple case): $|A \cup B| = |A| + |B| - |A \cap B|$.

De Morgan's laws for sets: $\overline{A \cap B} = \overline{A} \cup \overline{B}$; $\overline{A \cup B} = \overline{A} \cap \overline{B}$.

A **function** f from A (the **domain**) to B (the **codomain**) is an assignment of a unique element of B to each element of A. Write $f : A \rightarrow B$. Write $f(a) = b$ if b is assigned to a. **Range** of f is $\{ f(a) \mid a \in A \}$; f is **onto (surjective)** $\equiv$ range$(f) = B$; f is **one-to-one (injective)** $\equiv \forall a_1 \forall a_2 [f(a_1) = f(a_2) \rightarrow a_1 = a_2]$.

If f is one-to-one and onto (**bijective**), then the **inverse** function $f^{-1} : B \rightarrow A$ is defined by $f^{-1}(y) = x \equiv f(x) = y$.

If $f : B \rightarrow C$ and $g : A \rightarrow B$, then the **composition** $f \circ g$ is the function from A to C defined by $f \circ g(x) = f(g(x))$.

Rounding functions: $\lfloor x \rfloor$ = the largest integer less than or equal to x (**floor function**); $\lceil x \rceil$ = the smallest integer greater than or equal to x (**ceiling function**).

Summation notation: $\sum_{i=1}^{n} a_i = a_1 + a_2 + \cdots + a_n$.

Sum of first n positive integers: $\sum_{j=1}^{n} j = 1 + 2 + \cdots + n = \dfrac{n(n+1)}{2}$.

Sum of squares of first n positive integers: $\sum_{j=1}^{n} j^2 = 1^2 + 2^2 + \cdots + n^2 = \dfrac{n(n+1)(2n+1)}{6}$.

Sum of geometric progression: $\sum_{j=0}^{n} ar^j = a + ar + ar^2 + \cdots + ar^n = \dfrac{ar^{n+1} - a}{r - 1}$ if $r \neq 1$.

A set is **countable** if it is finite or there is a bijection from the natural numbers to the set—in other words, if the elements of the set can be listed $a_1, a_2, \ldots$. The empty set, the integers, and the rational numbers are countable; the real numbers and the power set of the natural numbers are uncountable.

Crib Sheet for Chapter 3

Big-O notation: "$f(x)$ is $O(g(x))$" means $\exists C\,\exists k\,\forall x(x > k \rightarrow |f(x)| \leq C|g(x)|)$. Big-$O$ of a sum is largest (fastest growing) of the functions in the sum; big-O of a product is the product of the big-O's of the factors. If f is $O(g)$, then g is $\Omega(f)$ ("big-Omega"). If f is both big-O and big-Omega of g, then f is $\Theta(g)$ ("big-Theta").

Algorithm: definite procedure for solving a problem using a finite number of steps; can be expressed in **pseudocode**; **properties:** having input, having output, definiteness, finiteness, effectiveness, generality. Some greedy algorithms work; some don't.

The **halting problem** is unsolvable: There is no algorithm to test whether a given computer program with a given input will ever halt.

Binary search has **time complexity** $O(\log n)$, whereas **linear search** has (**worst case** and **average case**) time complexity $O(n)$; both have **space complexity** $O(1)$ (not counting input). Bubble sort and insertion sort have $O(n^2)$ worst case time complexity.

Divisibility: $a \,|\, b \equiv a \neq 0 \wedge \exists c(ac = b)$ (a is a **divisor** or **factor** of b; b is a **multiple** of a).

Integer $n > 1$ is **prime** iff its only factors are 1 and itself $(2, 3, 5, 7, \ldots)$; otherwise it is **composite** $(4, 6, 8, 9, \ldots)$. There are infinitely many primes, but it is not known whether there are infinitely many twin primes (primes that differ by2) or whether every even positive integer greater than 2 is the sum of two primes (**Goldbach's Conjecture**).

Naive **test for primeness** (and method for finding **prime factorization**): To find prime factorization of n, successively divide it by all primes less than $\sqrt{n}$ $(2, 3, 5, \ldots)$; if none is found, then n is prime. If a prime factor p is found, then continue the process to find the prime factorization of the remaining factor, namely n/p; this time the trial divisions can start with p. Continue until a prime factor remains. The **Prime Number Theorem** states that there are approximately $n/\ln n$ primes less than or equal to n.

Fundamental Theorem of Arithmetic: Every integer greater than 1 can be written as a product of one or more primes, and the product is unique except for the order of the factors.

Division "algorithm": $\forall a\,\forall d\!>\!0\,\exists q\,\exists r(a = dq + r \wedge 0 \leq r < d)$; q is the quotient and r is the remainder; we write a **mod** d for the remainder. Example: $-18 = 5 \cdot (-4) + 2$, so -18 **mod** $5 = 2$.

Congruent modulo m: $a \equiv b \pmod m$ iff $m \,|\, a - b$ iff a **mod** $m = b$ **mod** m.

Euclidean algorithm for greatest common divisor: $\gcd(x, y) = \gcd(y, x \textbf{ mod } y)$ if $y \neq 0$; $\gcd(x, 0) = x$.

Two integers are **relatively prime** if their greatest common divisor (gcd) is 1. The integers $a_1, a_2, \ldots, a_n$ are **pairwise relatively prime** iff $\gcd(a_i, a_j) = 1$ whenever $1 \leq i < j \leq n$.

Chinese Remainder Theorem: If $m_1, m_2, \ldots, m_n$ are pairwise relatively prime, then the system $\forall i(x \equiv a_i \pmod{m_i})$ has unique solution modulo $m_1 m_2 \cdots m_n$. Application: handling very large integers on a computer.

Fermat's Little Theorem: $a^{p-1} \equiv 1 \pmod p$ if p is prime and does not divide a. The converse is not true; for example $2^{340} \equiv 1 \pmod{341}$, so 341 $(=11 \cdot 31)$ is a **pseudoprime**.

If a and b are positive integers, then there exist integers s and t such that $as + bt = \gcd(a, b)$ (**linear combination**). This theorem allows one to compute the **multiplicative inverse** $\overline{a}$ of a modulo b (i.e., $\overline{a}a \equiv 1 \pmod b$) as long as a and b are relatively prime, which enables one to solve **linear congruences** $ax \equiv c \pmod b$.

If p is prime and $p \,|\, a_1 a_2 \cdots a_n$, then $p \,|\, a_i$ for some i.

A common **hashing function:** $h(k) = k$ **mod** m, where k is the key.

Pseudorandom numbers can be generated by the **linear congruential method:** $x_{n+1} = (ax_n + c)$ **mod** m, where x_0 is arbitrarily chosen **seed**. Then $\{x_n/m\}$ will be rather randomly distributed numbers between 0 and 1.

Shift cipher: $f(p) = (p + k)$ **mod** 26 [A $\leftrightarrow$ 0, B $\leftrightarrow$ 1, ...]. Julius Caesar used $k = 3$.

RSA public key encryption system: An integer M representing the plaintext is translated into an integer C representing the ciphertext using the function $C = M^e$ **mod** b, where n is a public number that is the product of two large (maybe 100-digit or so) primes, and e is a public number relatively prime to $(p-1)(q-1)$; the primes p and q are kept secret. Decryption is accomplished via $M = C^d$ **mod** n, where d is an inverse of e modulo $(p-1)(q-1)$. It is infeasible to compute d without knowing p and q, which are infeasible to compute from n.

Addition of two **binary numerals** each of n bits $((a_{n-1}a_{n-2}\ldots a_2a_1a_0)_2)$ requires $O(n)$ bit operations. **Multiplication** requires $O(n^2)$ bit operations if done naively, $O(n^{1.585})$ steps by more sophisticated algorithms.

Base b representations: $(a_{n-1}a_{n-2}\ldots a_2a_1a_0)_b = a_{n-1}b^{n-1} + \cdots + a_2b^2 + a_1b + a_0$. To convert from base 10 to base b, continually divide by b and record remainders as $a_0, a_1, a_2, \ldots$ ($b = 16$ is **hexadecimal**, using A through F for digits 10 through 15). Convert from binary to hexadecimal by grouping bits by fours, from the right.

Matrix multiplication: The $(i, j)^{\text{th}}$ entry of **AB** is $\sum_{t=1}^{k} a_{it}b_{tj}$ for $1 \leq i \leq m$ and $1 \leq j \leq n$, where **A** is an $m \times k$ matrix and **B** is a $k \times n$ matrix. **Identity matrix** $\mathbf{I}_n$ with 1's on main diagonal and 0's elsewhere is the multiplicative identity.

Matrix addition (+), Boolean meet ($\wedge$) and join ($\vee$) are done entry-wise; Boolean matrix product ($\odot$) is like matrix multiplication using Boolean operations.

Transpose: $\mathbf{A}^t$ is the matrix whose $(i, j)^{\text{th}}$ entry is a_{ji} (the $(j, i)^{\text{th}}$ entry of **A**); **A** is **symmetric** if $\mathbf{A}^t = \mathbf{A}$.

Crib Sheet for Chapter 4

The well-ordering property: Every nonempty set of nonnegative integers has a least element.

Principle of mathematical induction: Let $P(n)$ be a propositional function in which the domain (universe of discourse) is the set of positive integers. Then if one can show that $P(1)$ is true (**basis step** or **base case**) and that for every positive integer k the conditional statement $P(k) \to P(k+1)$ is true (**inductive step**), then one has proved $\forall n P(n)$. The hypothesis $P(k)$ in a proof of the inductive step is called the **inductive hypothesis**. More generally, the induction can start at any integer, and there can be several base cases.

Strong induction: Let $P(n)$ be a propositional function in which the domain (universe of discourse) is the set of positive integers. Then if one can show that $P(1)$ is true (**basis step** or **base case**) and that for every positive integer k the conditional statement $[P(1) \land P(2) \land \cdots \land P(k)] \to P(k+1)$ is true (**inductive step**), then one has proved $\forall n P(n)$. The hypothesis $\forall j \leq k\, P(j)$ in a proof of the inductive step is called the (strong) **inductive hypothesis**. Again, the induction can start at any integer, and there can be several base cases.

Sometimes **inductive loading** is needed, where we must prove by mathematical induction or strong induction something stronger than the desired statement so as to have a powerful enough inductive hypothesis (this concept was introduced in the exercises).

Inductive or **recursive** definition of a function f with the set of nonnegative integers as its domain: specification of $f(0)$, together with, for each $n > 0$, a rule for finding $f(n)$ from values of $f(k)$ for $k < n$. Example: $0! = 1$ and $(n+1)! = (n+1) \cdot n!$ (**factorial function**).

Inductive or **recursive** definition of a set S: a rule specifying one or more particular elements of S, together with a rule for obtaining more elements of S from those already in it. It is understood that S consists precisely of those elements that can be obtained by applying these two rules.

Structural induction can be used to prove facts about recursively defined objects.

Fibonacci numbers: $f_0, f_1, f_2, \ldots$: $f_0 = 0$, $f_1 = 1$, and $f_n = f_{n-1} + f_{n-2}$ for all $n \geq 2$.

Lamé's Theorem: The number of divisions used by the Euclidean algorithm to find $\gcd(a, b)$ is $O(\log b)$.

An algorithm is **recursive** if it solves a problem by reducing it to an instance of the same problem with smaller input. It is **iterative** if it is based on the repeated use of operations in a loop.

There is an efficient recursive algorithm for computing b^n **mod** m, based on computing $b^{\lfloor n/2 \rfloor}$ **mod** m.

Merge sort is an efficient recursive algorithm for sorting a list: break the list into two parts, recursively sort each half, and merge them together in order. It has $O(n \log n)$ time complexity in all cases.

A program segment S is **partially correct** with respect to **initial assertion** p and **final assertion** q, written $p\{S\}q$, if whenever p is true for the input values of S and S terminates, q is true for the output values of S.

A **loop invariant** for **while** *condition* S is an assertion p that remains true each time S is executed in the loop; i.e., $(p \land \mathit{condition})\{S\}p$. If p is true before the program segment is executed, then p and $\neg\mathit{condition}$ are true after it terminates (if it does). In symbols, $p\{\textbf{while } \mathit{condition}\ S\}(\neg\mathit{condition} \land p)$.

Crib Sheet for Chapter 5

Sum rule: Given t mutually exclusive tasks, if task i can be done in n_i ways, then the number of ways to do exactly one of the tasks is $n_1 + n_2 + \cdots + n_t$.

Size of union of disjoint sets: $|A_1 \cup A_2 \cup \cdots \cup A_n| = |A_1| + |A_2| + \cdots + |A_n|$.

Two-set case of inclusion–exclusion: $|A \cup B| = |A| + |B| - |A \cap B|$.

Product rule: If a task consists of successively performing t tasks, and if task i can be done in n_i ways (after previous tasks have been completed), then the number of ways to do the task is $n_1 \cdot n_2 \cdots n_t$.

A set with n elements has 2^n subsets (equivalently, there are 2^n bit strings of length n).

Tree diagrams can be used to organize counting problems.

Pigeonhole principle: If more than k objects are placed in k boxes, then some box will have more than 1 object. **Generalized** version: If N objects are placed in k boxes, then some box will have at least $\lceil N/k \rceil$ objects.

Ramsey number $R(m, n)$ is the smallest number of people there must be at a party so that there exist either m mutual friends or n mutual enemies (assuming each pair of people are either friends or enemies). $R(3, 3) = 6$.

***r*-permutation** of set with n objects: *ordered* arrangement of r of the objects from the set (no repetitions allowed); there are $P(n, r) = n!/(n-r)!$ such permutations.

***r*-combination** of set with n objects: *unordered* selection (i.e., subset) of r of the objects from the set (no repetitions allowed); there are $C(n, r) = n!/[r!(n-r)!]$ such combinations. Alternative notation is $\binom{n}{r}$, called **binomial coefficient**.

Pascal's Identity: $C(n, k-1) + C(n, k) = C(n+1, k)$ if $n \geq k \geq 1$; allows construction of **Pascal's triangle** of binomial coefficients, using $C(n, 0) = C(n, n) = 1$ along the sides.

Combinatorial identities often have combinatorial proofs: $C(n, r) = C(n, n-r)$; $(a+b)^n = \sum_{k=0}^{n} C(n, k) a^{n-k} b^k$ (**Binomial Theorem**), with corollary $\sum_{k=0}^{n} C(n, k) = 2^n$; $C(m+n, r) = \sum_{k=0}^{r} C(m, r-k) C(n, k)$ (**Vandermonde's Identity**).

Number of r-permutations of an n-set **with repetitions allowed** is n^r; number of r-combinations of an n-set **with repetitions allowed** is $C(n+r-1, r)$ (equals number of solutions in nonnegative integers to $x_1 + x_2 + \cdots + x_n = r$).

Number of n-permutations of an n-set with n_i indistinguishable objects of type t for $1 \leq t \leq k$ is $n!/(n_1! n_2! \cdots n_k!)$ (equals number of ways to distribute n distinguishable objects into k distinguishable boxes so that box t gets n_t objects).

For distributing distinguishable object into distinguishable boxes, use product rule (or the formula $n!/(n_1! n_2! \cdots n_k!)$ if the number in each box is specified). For distributing indistinguishable object into distinguishable boxes, use formula for the number of combinations with repetitions allowed. For distributing distinguishable object into indistinguishable boxes, there is no good closed formula; **Stirling numbers of the second kind** are involved. Distributing indistinguishable object into indistinguishable boxes involves partitions of positive integers, and there is no good closed formula.

There are good algorithms for finding the lexicographically "next" permutation or combination and thereby for generating all permutations or combinations.

Crib Sheet for Chapter 6

If all outcomes are equally likely in a **sample space** S with n outcomes, then the **probability of an event** E is $p(E) = |E|/n$; more generally, if $p(s_i)$ is probability of ith outcome s_i, then $p(E) = \sum_{s_i \in E} p(s_i)$.

Probability distributions satisfy these conditions: $0 \leq p(s) \leq 1$ for each $s \in S$, and $\sum_{s \in S} p(s) = 1$.

For **complementary** event, $p(\overline{E}) = 1 - p(E)$; for **union** of two events (either one or both happen), $p(E \cup F) = p(E) + p(F) - p(E \cap F)$; for **independent events**, $p(E \cap F) = p(E)p(F)$.

The **conditional probability** of E given F (probability that E will happen after it is known that F happened) is $p(E|F) = p(E \cap F)/p(F)$.

Bernoulli trials: If only two outcomes are **success** and **failure**, with $p(\text{success}) = p$ and $p(\text{failure}) = q = 1-p$, then the **binomial distribution** applies, with probability of exactly k success in n trials being $b(k; n, p) = C(n, k)p^k q^{n-k}$.

Bayes' Theorem: If E and F are events such that $p(E) \neq 0$ and $p(F) \neq 0$, then

$$p(F \mid E) = \frac{p(E \mid F)p(F)}{p(E \mid F)p(F) + p(E \mid \overline{F})p(\overline{F})}.$$

A **random variable** assigns a number to each outcome. **Expected value** (expectation) of random variable X is $E(X) = \sum_{i=1}^{n} p(s_i)X(s_i)$; alternatively, $E(X) = \sum_{r \in X(S)} p(X = r)r$ Expected number of successes for n Bernoulli trials is pn.

Variance of random variable X is $V(X) = \sum_{i=1}^{n} p(s_i)(X(s_i) - E(X))^2$; variance can also be computed as $V(X) = E(X^2) - E(X)^2$; square root of variance is **standard deviation** $\sigma(X)$; variance of number of successes for n Bernoulli trials is npq.

X_1 and X_2 are **independent** if $p(X_1 = r_1 \text{ and } X_2 = r_2) = p(X_1 = r_1) \cdot p(X_2 = r_2)$. In this case $E(XY) = E(X)E(Y)$.

Expectation is linear even when the variables are not independent. This means that the expectation of a sum is sum of expectations, and $E(aX + b) = aE(X) + b$.

Variance of a sum is the sum of the variances ($V(X_1 + X_2) = V(X_1) + V(X_2)$) *when the variables are independent.*

The random variable X that gives the number of flips needed before a coin lands tails, when the probability of tails is p and the flips are independent, has the **geometric distribution**: $p(X = k) = (1 - p)^{k-1}p$ for $k = 1, 2, \ldots$; $E(X) = 1/p$.

Chebyshev's inequality: $p(|X(s) - E(X)| \geq r) \leq V(X)/r^2$.

A **probabilistic algorithm** is an algorithm that might give the incorrect answer but only with small probability. For example, there are good probabilistic tests as to whether or not a natural number is prime.

The **probabilistic method** is a proof technique that shows the existence of an object with a given property by showing that there is a nonzero probability of choosing such an object if choices are made at random. For example, there is a probabilistic proof that the Ramsey number $R(k, k)$ is at least $2^{k/2}$.

Crib Sheet for Chapter 7

A **recurrence relation** for a sequence a_0, a_1, a_2, ..., is a formula expressing each a_n in terms of previous terms (for all $n > n_0$); **initial conditions** specify a_0 through a_{n_0}; a **solution** to such a system is an explicit formula for a_n in terms of n that satisfies the recurrence relation and initial conditions.

The **Fibonacci numbers** are defined recursively by $f_0 = 0$, $f_1 = 1$, and $f_n = f_{n-1} + f_{n-2}$ for $n \geq 2$; continues $f_2 = 1$, $f_3 = 2$, $f_4 = 3$, $f_5 = 5$, $f_6 = 8$, $f_7 = 13$, $f_8 = 21$, ...; explicit formula is that f_n equals nearest integer to $((1+\sqrt{5})/2)^n/\sqrt{5}$.

A recurrence relation is **linear** of degree k if it is of the form $a_n = c_1 a_{n-1} + c_2 a_{n-2} + \cdots + c_k a_{n-k} + f(n)$; if all the c_i's are constants, then it has **constant coefficients**; if $f(n)$ is identically 0, then it is **homogeneous**. Such a recurrence relation and k initial conditions completely determine the sequence.

Recurrence relations of degree 1 can often be solved by **iteration**. Given a_n expressed in terms of a_{n-1}, rewrite a_{n-1} in this equation using the same recurrence relation with $n-1$ in place of n. This expresses a_n in terms of a_{n-2}. Then rewrite a_{n-2} in terms of a_{n-3}, again using the recurrence relation. Continue in this way, noting the pattern that evolves, until finally you have a_n written explicitly in terms of a_1 (or a_0), probably as a series. This gives the explicit solution (preferably with the series summed in closed form).

To **solve** linear homogeneous recurrence relation with constant coefficients: (1) write down the **characteristic equation** $r^k - c_1 r^{k-1} - c_2 r^{k-2} - \cdots - c_k = 0$ and find all its roots, with multiplicities; (2) each distinct root (**characteristic root**) r gives rise to a solution $a_n = r^n$; if a root is repeated, occurring s times, then there are solutions $a_n = n^i r^n$ for $i = 0, 1, \ldots, s-1$; (3) take arbitrary linear combination of all solutions so obtained, with coefficients α_1, α_2, ..., α_k; (4) plug in the k initial conditions to solve for the α's.

To **solve** linear nonhomogeneous recurrence relation with constant coefficients: (1) solve the **associated homogeneous recurrence relation** (with the $f(n)$ term omitted) to obtain a general solution $a_n^{(h)}$ with some yet-to-be-calculated constants α_i; (2) obtain a **particular solution** $a_n^{(p)}$ of the nonhomogeneous recurrence relation using the method of undetermined coefficients (the form to use depends on $f(n)$ and on the solution of the associated homogeneous recurrence relation); (3) write down the general solution: $a_n = a_n^{(h)} + a_n^{(p)}$; (4) plug in the k initial conditions to solve for the α's.

Divide-and-conquer relation: $f(n) = af(n/b) + g(n)$. If f is an increasing function satisfying this relation whenever n is a power of b, where $g(n) = cn^d$, then $f(n) = O(n^d)$ if $a < b^d$, $f(n) = O(n^d \log n)$ if $a = b^d$, and $f(n) = O(n^{\log_b a})$ if $a > b^d$ (Master Theorem).

Matrix multiplication can be done in $O(n^{\log 7}) \approx O(n^{2.8})$ steps, rather than the naive $O(n^3)$, using a divide-and-conquer algorithm.

Generating functions are expressions of the form $f(x) = \sum_{k=0}^{\infty} a_k x^k$, associated with an infinite sequence $\{a_k\}$. They can be used to solve recurrence relations, prove combinatorial identities, and solve counting problems. To model a combinatorial situation (such as counting how many ways there are to distribute cookies), let a_k be the quantity of interest (the answer when there are k cookies); choices are modeled by adding (corresponding to "or" situations—the person can get 1, 2, or 3 cookies) or multiplying (corresponding to "and" situations—Tom, Dick, and Jane must each receive cookies) polynomials in x. The combinatorics is then replaced by the algebra of multiplying out the polynomials or obtaining a **closed form** expression. The most important generating function in applications is $1/(1-ax)^n = \sum_{k=0}^{\infty} C(n+k-1, k) a_k x^k$. **Partial fractions** must sometimes be used when solving recurrence relations using generating functions.

Inclusion–exclusion principle: (for $n = 3$) $|A \cup B \cup C| = |A| + |B| + |C| - |A \cap B| - |A \cap C| - |B \cap C| + |A \cap B \cap C|$; (general case) $|A_1 \cup A_2 \cup \cdots \cup A_n| = \sum_i |A_i| - \sum_{i<j} |A_i \cap A_j| + \sum_{i<j<k} |A_i \cap A_j \cap A_k| - \cdots + (-1)^{n+1} |A_1 \cap A_2 \cap \cdots \cap A_n|$. Can be applied to counting number of objects among N objects with none of a collection of properties: $N(P_1' P_2' \cdots P_n') = N - \sum_i N(P_i) + \sum_{i<j} N(P_i P_j) - \sum_{i<j<k} N(P_i P_j P_k) + \cdots + (-1)^n N(P_1 P_2 \cdots P_n)$. Specific application: counting the number of primes up to n found by using **sieve of Eratosthenes**.

Number of onto functions from an m-set to an n-set is $n^m - C(n,1)(n-1)^m + C(n,2)(n-2)^m - \cdots + (-1)^{n-1} C(n, n-1) 1^m$.

A **derangement** is a permutation leaving no object in its original position; there are $D_n = n![1 - 1/1! + 1/2! - 1/3! + \cdots + (-1)^n/n!]$ derangements of n objects; as $n \to \infty$, $D_n/n!$ quickly approaches $1/e \approx 0.368$.

Crib Sheet for Chapter 8

(Binary) relation R from A to B: subset of $A \times B$. Write aRb for $(a, b) \in R$; relation **on** A is relation from A to A; graph of a function from A to B is a relation such that $\forall a \in A \ \exists$ exactly one pair (a, b) in the relation.

Relation R on a set A is **reflexive** if aRa for all $a \in A$; **irreflexive** if aRa for no $a \in A$; **symmetric** if aRb implies bRa for all $a, b \in A$; **asymmetric** if aRb implies that b is not related to a, for all $a, b \in A$; **antisymmetric** if $aRb \wedge bRa$ implies $a = b$ for all $a, b \in A$; **transitive** if $aRb \wedge bRc$ implies aRc for all $a, b, c \in A$.

Inverse relation to R is given by $bR^{-1}a$ if and only if aRb.

If R is a relation from A to B, and S is a relation from B to C, then the **composite** is the relation $S \circ R$ from A to C in which a is related to c if and only if there exists a $b \in B$ such that aRb and bSc. $R^n = R \circ R \circ \cdots \circ R$.

n-ary relation on domains A_1, A_2, ..., A_n is a subset of $A_1 \times A_2 \times \cdots \times A_n$; data bases using the relational data model are just sets of n-ary relations in which each n-tuple is called a **record** made up of **fields** (n is the **degree**); a domain is a **primary key** if the corresponding field uniquely determines the record.

The **projection** $P_{i_1, i_2, \ldots, i_m}$ maps an n-tuple to the m-tuple formed by deleting all fields not in the list $i_1, i_2, \ldots, i_m$.

The **join** J_p takes a relation R of degree m and a relation S of degree n and produces a relation of degree $m + n - p$ by finding all tuples $(a_1, a_2, \ldots, a_{m+n-p})$ such that $(a_1, a_2, \ldots, a_m) \in R \wedge (a_{m-p+1}, a_{m-p+2}, \ldots, a_{m-p+n}) \in S$.

A relation R on $A = \{a_1, a_2, \ldots, a_n\}$ can be represented by an $n \times n$ matrix $\mathbf{M}_R$ whose $(i, j)^{\text{th}}$ entry is 1 if $a_i R a_j$ and is 0 otherwise. Reflexivity, symmetry, antisymmetry, transitivity can easily be read off the matrix. Boolean products of the matrices give the matrix for the composite ($\mathbf{M}_{S \circ R} = \mathbf{M}_R \odot \mathbf{M}_S$).

Digraph for R on A: a vertex for each element of A and an arrow (**arc, edge**) from a to b whenever aRb (**loop** at a when aRa). Reflexivity, symmetry, antisymmetry, transitivity can easily be read off the digraph.

Closure of relation R on set A with respect to property P: smallest relation on A containing R and having property P; **reflexive closure**—add all pairs (a, a) if not already in R ($R \cup \Delta_A$, where $\Delta_A = \{ (a, a) \mid a \in A \}$); **symmetric closure**—add pair (b, a) whenever (a, b) is in R, if (b, a) is not already in R ($R \cup R^{-1}$).

Transitive closure of R equals R^* (**connectivity relation** for R), defined as $\bigcup_{n=1}^{\infty} R^n$; can be computed efficiently at the matrix level using **Warshall's algorithm**, and visually at the digraph level by considering **paths**.

A relation R on set A is an **equivalence relation** if R is reflexive, symmetric, and transitive; usually equivalence relations can be recognized by their definition's being of the form "two elements are related if and only if they have the same [something]." For each $a \in A$ the set of elements in A related to (i.e., **equivalent to**) a is the **equivalence class** of a, denoted $[a]$. (Canonical example: congruence modulo m.) The set of equivalence classes partitions A into pairwise disjoint nonempty sets; conversely, every partition of A induces an equivalence relation by declaring two elements to be related if they are in the same set of the partition.

A relation R on set A is a **partial order** if R is reflexive, antisymmetric, and transitive; **total** or **linear** order if in addition every pair of elements are **comparable** (either aRb or bRa). (A, R) is called a **partially ordered set** or **poset**, and R is denoted $\preceq$ ($\prec$ means $\preceq$ but not equal); canonical example is $\subseteq$ on $P(S)$.

Partial orders on A_i induce **lexicographic order** on $A_1 \times A_2 \times \cdots \times A_n$ given by $(a_1, a_2, \ldots, a_n) \prec (b_1, b_2, \ldots, b_n)$ if for some $i \geq 0$, $a_1 = b_1$, ..., $a_i = b_i$, and $a_{i+1} \prec b_{i+1}$. A partial order on A induces **lexicographic order** on strings A^* given by $a_1 a_2 \ldots a_n \prec b_1 b_2 \ldots b_m$ if $(a_1, a_2, \ldots, a_t) \prec (b_1, b_2, \ldots, b_t)$ where $t = \min(m, n)$ or if $(a_1, a_2, \ldots, a_n) = (b_1, b_2, \ldots, b_n)$ and $n < m$.

Hasse diagram represents poset on finite set by placing x above y whenever $y \prec x$, and also in this case drawing line from x to y if there is no z such that $y \prec z \prec x$; then $a \prec b$ if and only if there is an upward path from a to b in the Hasse diagram.

m in a poset is **maximal** if there is no x with $m \prec x$ (**minimal** in the dual situation); m is a **greatest element** if $x \preceq m$ for all x (dually for **least element**). An **upper bound** for a subset B is an element u such that $b \preceq u$ for all $b \in B$ (dually for **lower bound**), and u is a **least** upper bound if it is an upper bound such that $u \preceq v$ for every upper bound v (dually for **greatest** lower bound). A poset is a **lattice** if l.u.b.'s and g.l.b.'s always exist.

Every finite poset has a minimal element that can be found by moving down the Hasse diagram. Iterating this observation gives an algorithm for obtaining a total order **compatible** with the partial order (**topological sorting**)—keep peeling off a minimal element.

Crib Sheet for Chapter 9

A **simple graph** $G = (V, E)$ consists of nonempty set of **vertices** (singular: **vertex**) and set of unordered pairs of distinct vertices called **edges**. A **multigraph** allows more than one edge joining same pair of vertices (**multiple** or **parallel edges**)—E is just a set with an endpoint function f taking each edge e to its two *distinct* endpoints. A **pseudograph** is like a multigraph but endpoints $f(e)$ need not be distinct, allowing for **loops**. A **directed graph** (**digraph**) is just like a simple graph except that edges are directed (each e is an *ordered pair*, and loops are allowed). **Directed multigraph** is just like multigraph (parallel edges allowed) except that edges are directed (loops allowed). Given directed graph, we can ignore order and look at **underlying undirected graph**.

Graphs can be used to model relationships of many kinds, such as acquaintances, food webs, telephone calls, road systems, the Internet, tournaments, organizational structure.

Vertices joined by an edge are **adjacent** and the edge is **incident** to them. **Degree of a vertex** ($\deg(v)$) is the number of incident edges, with loops counted double; **isolated vertex** has degree 0, **pendant vertex** has degree 1; **regular graph** has all degrees equal. In digraph $\deg^-(v)$ is number of edges leading into v (**in-degree**; v is **terminal** vertex) and $\deg^+(v)$ is number of edges leading out of v (**out-degree**; v is **initial** vertex).

Handshaking Theorem: Undirected: $2e = \sum_{v \in V} \deg(v)$; corollary—number of vertices of odd degree is even. Directed: $e = \sum_{v \in V} \deg^-(v) = \sum_{v \in V} \deg^+(v)$.

In **bipartite graph** vertex set can be partitioned into two nonempty sets with no edges joining vertices in same set.

The **complete** graph K_n has n vertices and an edge joining every pair ($n(n-1)/2$ edges in all); **complete bipartite** graph $K_{m,n}$ has $m+n$ vertices in parts of sizes m and n and an edge joining every pair of vertices in different parts (mn edges in all). The **cycle** C_n has n vertices and n edges, joined in a circle; the **wheel** W_n is C_n with one more vertex joined to these n vertices. The **cube** Q_n has all n-bit binary strings for vertices, with an edge between every pair of vertices differing in only one bit position.

(V, E) is **subgraph** of (W, F) if $V \subseteq W$ and $E \subseteq F$; **union** of two simple graph is formed by taking union of corresponding vertex sets and corresponding edge sets.

Graph can be represented by **adjacency matrix** (m in position (i, j) denotes m parallel edges from i to j), **adjacency lists**, **incidence matrix** (1 is position (i, j) says that vertex i is incident to edge j).

Two graphs are **isomorphic** if there is a bijection between their vertex sets that preserves all adjacencies and all nonadjacencies; to show that two graphs are *not* isomorphic, find an **invariant** on which they differ (e.g., degree sequence or existence of cycles).

A **path** of length n from u to v is a sequence of n edges leading successively from u to v; is a **circuit** if $n > 0$ and $u = v$, is **simple** if no edge occurs more than once. A graph is **connected** if every pair of vertices is joined by a path; digraph is **strongly connected** if every pair of vertices is joined by a path in each direction, **weakly connected** if underlying undirected graph is connected. **Components** are maximal connected subgraphs. Removal of **cut edge** (**bridge**) or **cut vertex** (**articulation point**) creates more components.

$(i, j)^{\text{th}}$ entry of $\mathbf{A}^r$, where $\mathbf{A}$ is adjacency matrix, counts numbers of paths of length r from i to j.

An **Euler circuit** [**path**] is simple circuit [path] containing all edges. Connected graph has an Euler circuit [path] if and only if the vertex degrees are all even [the vertex degrees are all even except for at most two vertices]. Splicing algorithm or Fleury's algorithm finds them efficiently.

A **Hamilton path** is path containing all vertices exactly once; Hamilton path together with edge back to starting vertex is **Hamilton circuit**. No good necessary and sufficient conditions for existence of these, or algorithms for finding them, are known. Q_n has Hamilton circuit for all $n \geq 2$ (**Gray code**).

Weighted graphs have **lengths** assigned to edges; one can find shortest path from u to v (minimum sum of weights of edges in the path) using **Dijkstra's algorithm**.

A **planar graph** is a graph having a **planar representation** (drawing in plane without edges crossing); **Kuratowski's Theorem:** graph is planar if and only if it has no subgraph **homeomorphic** to (formed by performing **elementary subdivisions** on edges of) K_5 or $K_{3,3}$.

Euler's formula: Given planar representation with v vertices, e edges, c components, r regions, $v - e + r = c + 1$; corollary: in planar graph with at least 3 vertices, $e \leq 3v - 6$.

A graph is **colored** by assigning colors to vertices with adjacent vertices getting distinct colors; minimum number of colors required is **chromatic number**. Every planar graph can be colored with four colors.

Crib Sheet for Chapter 10

A **tree** is a connected undirected graph with no simple circuits; characterized by having unique simple path between every pair of vertices, and by being connected and satisfying $e = v - 1$. A **forest** is an undirected graph with no simple circuits—each component is a tree, and $e = v -$ number of components. Every tree has at least two vertices of degree 1.

A **rooted tree** is a tree with one vertex specified as root; can be viewed as directed graph away from root. If uv is a directed edge, then u is **parent** and v is **child**; **ancestor, descendant, sibling** defined genealogically. Vertices without children are **leaves**; others are **internal**. Draw trees with root at the top, so that vertices occur at **levels**, with root at level 0; **height** is maximum level number. A tree is **balanced** if all leaves occur only at bottom or next-to-bottom level, **complete** if only at bottom level. The **subtree rooted at** a is the tree involving a and all its descendants.

An **m-ary tree** is a rooted tree in which every vertex has at most m children (**binary** tree when $m = 2$); a *full* m-ary tree has exactly m children at each internal vertex. Full m-ary tree with i internal vertices and l leaves has $n = i + l = mi + 1$ vertices. An m-ary tree with height h satisfies $l \leq m^h$, so $h \geq \lceil \log_m l \rceil$ (equality in latter inequality if tree is balanced).

An **ordered rooted tree** has an order among the children of each vertex, drawn left-to-right; in ordered binary tree, each child is a **right child** or **left child**, and subtree rooted at right [left] child is called **right [left] subtree**.

A **binary search tree (BST)** is binary tree with a key at each vertex so that at each vertex, all keys in left subtree are less and all keys in right subtree are greater than key at the vertex; $O(\log n)$ algorithm for insertion and search in BST.

Decision trees provide lower bounds on number of questions an algorithm needs to ask to accomplish its task for all inputs (e.g., coin-weighing, searching).

Binary trees can be used to encode **prefix codes**, binary codes for symbols so that no code word is the beginning of another code word. **Huffman codes** are efficient prefix codes for data compression.

Game trees can be used to find optimal strategies for two-person games, using the **minmax** principle. Value of a leaf is payoff to first player. Value of an internal vertex at an even level (square) is maximum of values of its children; value of an internal vertex at an odd level (circle) is minimum of values of its children.

Universal address system: root is labeled 0; children of root are labeled 1, 2, ...; children of vertex labeled x are labeled $x.1$, $x.2$, ... Addresses are ordered using preorder traversal.

Preorder visits root, then subtrees (recursively) in preorder; **postorder** visits subtrees (recursively) in postorder, then root; **inorder** visits first subtree (recursively) in inorder, then root, then remaining subtrees (recursively) in inorder.

The **expression tree** for a calculation has constants at leaves, operations at internal vertices (evaluated by applying operation to values of its children).

Prefix (Polish) notation corresponds to preorder traversal of expression tree (operator precedes operands); **postfix (reverse Polish) notation** corresponds to postorder traversal of expression tree (operands precede operator); both of these allow unambiguous expressions without using parentheses. **Infix** form is normal notation but requires full parenthesization (corresponds to inorder traversal of expression tree).

Naive sorting routines like **bubble sort** require $O(n^2)$ steps in worst case; the best that can be done with comparison-based sorting is $O(n \log n)$ (a huge improvement), using something like **merge sort**.

A **spanning tree** is a tree containing all vertices of a connected given graph; can be found by **depth-first** search (recursively search the unvisited neighbors) or **breadth-first** search (fan out). Depth-first search can be modified to implement **backtracking** algorithms for exhaustive consideration of all cases of a problem (like graph coloring). **Minimum spanning trees** can be found in weighted graphs using greedy **Prim** or **Kruskal** algorithms (choose least costly edge at each stage that doesn't get you into trouble).

Crib Sheet for Chapter 11

Boolean operations: sum, product, and complementation defined on $\{0,1\}$ by $0+0=0$, $0+1=1+0=1+1=1$, $1\cdot 1=1$, $0\cdot 1=1\cdot 0=0\cdot 0=0$ (also write product using concatenation, without the dot), $\overline{0}=1$, $\overline{1}=0$; also *XOR* defined by $1\oplus 1=0\oplus 0=0$ and $1\oplus 0=0\oplus 1=1$, *NAND* defined by $1\mid 1=0$ and $1\mid 0=0\mid 1=0\mid 0=1$, *NOR* defined by $0\downarrow 0=1$ and $1\downarrow 0=0\downarrow 1=1\downarrow 1=0$. A **Boolean algebra** is an abstraction of this, with operations $\vee$, $\wedge$, and $\overline{}$, which also applies to other situations (e.g., sets).

Boolean operations obey same identities (commutative, associative, idempotent, distributive, De Morgan, etc.) with $\cup$ replaced by $+$, $\cap$ replaced by $\cdot$, U replaced by 1, and $\emptyset$ replaced by 0.

Boolean variables are variables taking on only values 0 and 1; **Boolean expression** is expression made up from Boolean constants (0 and 1), variables and operations combined in usual ways with parentheses where desired to override the natural precedence that products are evaluated before sums.

Dual of Boolean expression: interchange 0 and 1, $+$ and $\cdot$; dual of an identity is an identity.

Boolean functions are functions from n-tuples of variables to $\{0,1\}$. They can be represented using Boolean expressions, and in particular, in **disjunctive normal form** as **sums of products** or in **conjunctive normal form** as **products of sums**. In sum of products form, each product is a **minterm** $y_1y_2\cdots y_n$, where each y_i is a **literal**, either x_i or $\overline{x}_i$. Two expressions are called **equivalent** if they compute the same function.

Set of operators is **functionally complete** if every Boolean function can be represented using them. Examples are $\{+,\cdot,\overline{}\}$, $\{+,\overline{}\}$, $\{\cdot,\overline{}\}$, $\{\mid\}$, and $\{\downarrow\}$.

The standard Boolean operations can be represented by **gates**:

x → x̄ (inverter); x, y → xy (AND); x, y → x+y (OR); x, y → x|y (NAND); x, y → x↓y (NOR)

They can be combined to make **combinational circuits** to represent any Boolean function, such as **(full) adders** (take two bits and a carry and produce a sum bit and a carry) or **half-adders** (same, without the carry as input).

Minimization of circuits: given Boolean function in sum of products form, find an expression as simple as possible to represent it (i.e., use as few literals and operations as possible, meaning using few gates to produce the circuit). Geometric method (**Karnaugh maps** or **K-maps**) and tabular method (**Quine-McCluskey procedure**) organize this task efficiently for small n. Typical K-maps with blocks circled:

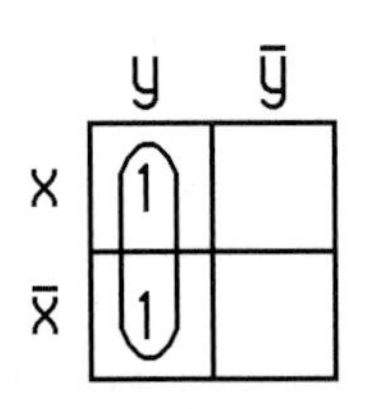

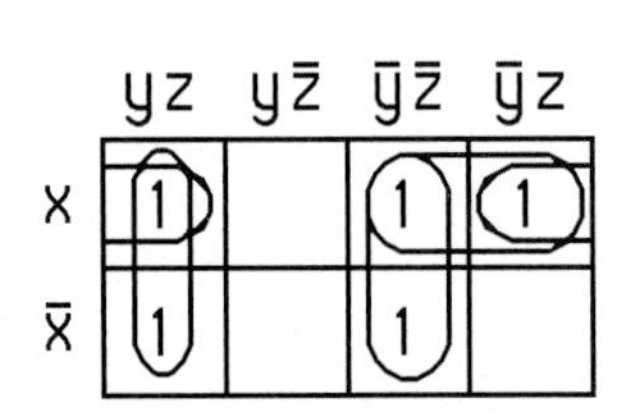

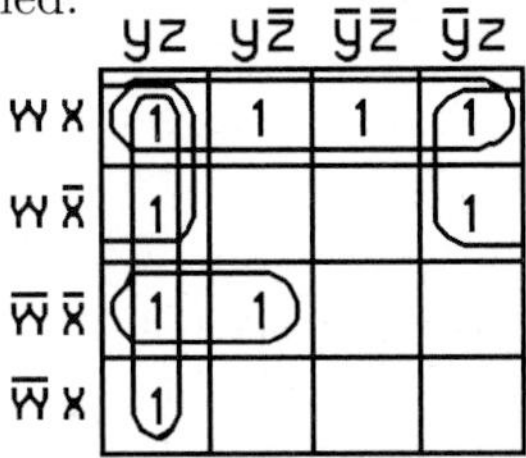

Typical Quine–McCluskey calculation:

	Term	String	Step 1 Term	Step 1 String	Step 2 Term	Step 2 String
1	$wxyz$	1111	$(1,2)\,wxy$	111− ⇐	$(3,4,5,8)\,\overline{y}z$	−−01
2	$wxy\overline{z}$	1110	$(1,3)\,wxz$	11−1 ⇐		
3	$wx\overline{y}z$	1101	$(3,4)\,w\overline{y}z$	1−01		
4	$w\overline{x}\,\overline{y}z$	1001	$(3,5)\,x\overline{y}z$	−101		
5	$\overline{w}x\overline{y}z$	0101	$(4,6)\,w\overline{x}\,\overline{y}$	100− ⇐		
6	$w\overline{x}\,\overline{y}\,\overline{z}$	1000	$(4,8)\,\overline{x}\,\overline{y}z$	−001		
7	$\overline{w}\,\overline{x}y\overline{z}$	0010 ⇐	$(5,8)\,\overline{w}\,\overline{y}z$	0−01		
8	$\overline{w}\,\overline{x}\,\overline{y}z$	0001				

	1	2	3	4	5	6	7	8
$\overline{y}z$			X	X	X			X
wxy	X	X						
wxz	X		X					
$w\overline{x}\,\overline{y}$				X		X		
$\overline{w}\,\overline{x}y\overline{z}$							X	

$\overline{y}z+wxy+w\overline{x}\,\overline{y}+\overline{w}\,\overline{x}y\overline{z}$ covers all minterms

Crib Sheet for Chapter 12

A **vocabulary** or **alphabet** V is just a finite nonempty set of symbols; strings of symbols from V, including the empty string λ, are **words**; the set of *all* words is denoted V^*; a **language** is any subset of V^*.

If L_1 and L_2 are languages, then so are the **union** $L_1 \cup L_2$, **intersection** $L_1 \cap L_2$, **concatenation** $L_1L_2 = \{ uv \mid u \in L_1 \wedge v \in L_2 \}$ ($L^2 = LL$, etc.), **Kleene closure** $L^* = \{\lambda\} \cup L \cup L^2 \cup L^3 \cdots$, **complement** $\overline{L} = V^* - L$.

Phrase-structure grammar: $G = (V, T, S, P)$, where V is a vocabulary, $T \subseteq V$ is the set of **terminal** symbols (the **nonterminal symbols** are $N = V - T$), $S \in V$ is the **start symbol**, P is a set of **productions**, which are rules of the form $w_1 \to w_2$, where $w_1, w_2 \in V^*$. (The convention is to use capital letters for the nonterminals.)

Derivations: If a string u can be transformed to a string v by applying some production (i.e., replacing a substring w_1 in u by w_2 where $w_1 \to w_2$ is a production), then write $u \Rightarrow v$ and say v is **directly derivable** from u; if $u_1 \Rightarrow u_2 \Rightarrow \cdots \Rightarrow u_n$, then write $u_1 \stackrel{*}{\Rightarrow} u_n$ and say u_n is **derivable** from u_1.

The **language generated by** G is the set of strings *of terminal symbols* derivable from S (the start symbol).

Types of grammars: type 0—no restrictions; **type 1 (context-sensitive)**—productions are all of the form $lAr \to lwr$, where A is a nonterminal symbol, l and r are strings of zero or more terminal or nonterminal symbols, and w is a nonempty string of terminal or nonterminal symbols, or of the form $S \to \lambda$ as long as S does not appear on the right-hand side of any other production; **type 2 (context-free)**—left side of each production must be single nonterminal symbol; **type 3 (regular)**—only allowed productions are $A \to bB$, $A \to b$, and $S \to \lambda$, where A, B, and S are nonterminals, b is terminal, and S is start symbol; each type is included in previous type. Productions in type 2 grammars can also be represented in **Backus–Naur form**; nonterminals have angled brackets surrounding them; a vertical bar means "or"; arrows are replaced by ::=; e.g., $\langle A \rangle ::= \langle A \rangle c \langle A \rangle \langle B \rangle \mid 3 \mid \langle B \rangle \langle C \rangle$.

Derivation (or **parse**) **tree** shows transformation of start symbol into string of terminals (the leaves, read from left to right), invoking a production at each internal vertex.

Palindrome: string that reads the same forward as backward, i.e., $w = w^R$ (string whose first half is the reverse of its last half, either of the form uu^R or uxu^R for x a symbol). The set of palindromes is context-free but not regular.

Finite state machine with output on the transitions (Mealy machines): $M = (S, I, O, f, g, s_0)$, where S is set of **states**, I and O are **input and output alphabets**, $s_0 \in S$ is **start state**, $f : S \times I \to S$ and $g : S \times I \to O$ are **transition function** and **output function**; can be represented in **state table**, or by **state diagram** with i, o labeling edge (s, t) if $f(s, i) = t$ and $g(s, i) = o$. Machine "moves" from state to state as it reads an input string, producing an output string of the same length; can be used for language recognition (output symbol 1 if input string read so far is in language, 0 if not).

Finite state machine with output on the states (Moore machines): same as Mealy machine except $g : S \to O$ assigns an output to every state rather than to every transition; output is one symbol longer than input.

Finite state machine with no output (deterministic automaton): $M = (S, I, f, s_0, F)$—same as Mealy machine except that there is no output function but rather a set of **final states** $F \subseteq S$. A string w is **recognized** or **accepted** by M if M ends up in a final state on input w; the **language recognized** or **accepted** by M, written $L(M)$, is the set of all strings accepted by M. (Plural of "automaton" is "automata.")

Nondeterministic automaton: same as deterministic one except that the transition function f sends a state and input symbol to a *set of states*; think of the machine as choosing which state to go into next from among the possibilities provided by f. A string is accepted if some sequence of choices leads to a final state at the end of the input; $L(M)$ defined as before. For automata of either type, we say that M_1 and M_2 are equivalent if $L(M_1) = L(M_2)$. **Theorem:** Given a nondeterministic finite automaton, there is an equivalent deterministic one.

Regular expressions over a set I are built up from symbols for the elements of I, a symbol for the empty set, and a symbol for the empty string by the operations of concatenation, union, and Kleene closure. The expressions represent the corresponding sets of strings. **Regular sets (languages)** are sets represented by regular expressions. Regular languages are **closed under** intersection, union, concatenation, Kleene closure, complement; **context-free languages are closed under** union, concatenation, Kleene closure.

Theorem: A set is regular if and only if it is generated by some regular grammar if and only if it is accepted by some finite automaton.

Pumping Lemma: If z is a string in $L(M)$ of length longer than the number of states in M, then we can write $z = uvw$, with $v \neq \lambda$ so that $uv^iw \in L(M)$ for all i. This allows us to prove, for example, that $\{ 0^n1^n \mid n = 1, 2, 3, \ldots \}$ is not regular.

A **Turing machine** is specified by a set of 5-tuples (s, x, s', x', d): if in state s scanning symbol x on tape, it writes x' on tape, enters state s', and moves tape head R or L according to d. TM's can recognize all computable (Type 0) languages, can compute all computable functions (Church–Turing thesis).

Notes

Notes

Notes

Notes

Notes